International Series in Operations Research & Management Science

Volume 353

The book series **International Series in Operations Research and Management Science** encompasses the various areas of operations research and management science. Both theoretical and applied books are included. It describes current advances anywhere in the world that are at the cutting edge of the field. The series is aimed especially at researchers, advanced graduate students, and sophisticated practitioners.

The series features three types of books:

- Advanced expository books that extend and unify our understanding of particular areas.
- Research monographs that make substantial contributions to knowledge.
- Handbooks that define the new state of the art in particular areas. Each handbook will be edited by a leading authority in the area who will organize a team of experts on various aspects of the topic to write individual chapters. A handbook may emphasize expository surveys or completely new advances (either research or applications) or a combination of both.

The series emphasizes the following four areas:

Mathematical Programming: Including linear programming, integer programming, nonlinear programming, interior point methods, game theory, network optimization models, combinatorics, equilibrium programming, complementarity theory, multiobjective optimization, dynamic programming, stochastic programming, complexity theory, etc.

Applied Probability: Including queuing theory, simulation, renewal theory, Brownian motion and diffusion processes, decision analysis, Markov decision processes, reliability theory, forecasting, other stochastic processes motivated by applications, etc.

Production and Operations Management: Including inventory theory, production scheduling, capacity planning, facility location, supply chain management, distribution systems, materials requirements planning, just-in-time systems, flexible manufacturing systems, design of production lines, logistical planning, strategic issues, etc.

Applications of Operations Research and Management Science: Including telecommunications, health care, capital budgeting and finance, economics, marketing, public policy, military operations research, humanitarian relief and disaster mitigation, service operations, transportation systems, etc.

This book series is indexed in Scopus.

Faiz Hamid
Editor

Optimization Essentials

Theory, Tools, and Applications

Springer

Editor
Faiz Hamid
Department of Industrial and Management Engineering
Indian Institute of Technology Kanpur
Kanpur, Uttar Pradesh, India

ISSN 0884-8289 ISSN 2214-7934 (electronic)
International Series in Operations Research & Management Science
ISBN 978-981-99-5490-2 ISBN 978-981-99-5491-9 (eBook)
https://doi.org/10.1007/978-981-99-5491-9

This Springer imprint is published by the registered company Springer Nature Singapore Pte Ltd.
The registered company address is: 152 Beach Road, #21-01/04 Gateway East, Singapore 189721, Singapore

To my parents…

Foreword

The birth of modern operations research (OR) can be traced back to radar detection activities in the UK during World War II, although several important concepts such as queueing theory and the economic order quantity were already known. After the war, OR witnessed a significant growth and was applied to a wide variety of sectors, with manufacturing, natural resources, agriculture, telecommunications, and transportation attracting the most attention.

Many practical problems in these areas can be modeled as large systems which must be optimized, and mathematical programming is the most common optimization tool. The advent of the simplex algorithm for linear programs in the late 1940s lies at the heart of mathematical programming and has spawned several subfields such as integer programming, nonlinear programming, and stochastic programming which sustain most application areas. Classical mathematical programming methodologies borrow extensively from concepts belonging to fundamental mathematics, such as calculus, linear algebra, and probability and statistics, and have greatly benefitted from advances in discrete optimization. Branch-and-bound, branch-and-cut, column generation, and several other decomposition techniques are now routinely employed for the solution of most problems.

Several key developments have prompted the growth of OR in the past 40–50 years. One is the advent and generalized use of micro-computing which has considerably facilitated the creation of new OR methodologies, as well as their dissemination and adoption by the scientific community and end-users. Another key development is the advent of sophisticated and highly efficient solvers, such as CPLEX and Gurobi, as well as open-source software which enables the resolution of large-scale and hard problems by non-experts. In the past few decades, we have also witnessed the creation of powerful metaheuristics based on diverse paradigms such as local search, genetic algorithms, and their hybridization. It is evident that concepts borrowed from machine learning and artificial intelligence will play an increasingly important role in modeling and problem-solving in the near future.

The book *Optimization Essentials: Theory, Tools, and Applications* edited by Faiz Hamid contains 15 chapters devoted to theory and 10 to applications, written by reputed authors. The topics cover several of the aforementioned methodological and application areas, but they do so through a novel lens, introducing original and thought-provoking perspectives. The methodological chapters include a survey and a tutorial, as well as a description of several improvements to existing mathematical programming techniques, and novel applications of these techniques to known problems and new variants. The applications chapters cover a wide variety of both established and emerging domains. A common theme underlying several of the chapters is how to make classical solution methodologies more efficient through the inclusion of diverse algorithmic enhancements.

Graduate students, researchers, professors, and practitioners should all benefit from this well-crafted, creative, and highly interesting book which constitutes a welcome addition to the ever-growing field of operations research.

August 2023

Gilbert Laporte
Professor Emeritus HEC Montréal
Canada

Professor, University of Bath
United Kingdom

Professor II, Molde University College
Norway

Preface

The field of optimization has evolved continuously since its inception. The success of Operations Research (OR) during World War II spurred its applications to various organizations in business, industry, and government. Now the art and science of optimization lie at the heart of countless endeavors. The results have been spectacular. At the same time, complexity of algorithms and breadth of applications have grown by leaps and bounds. The role of optimization techniques stands as a pivotal force that shapes outcomes across a multitude of domains. The advent of the computer revolution has been a critical factor, which provided great impetus to the growth of the field enabling even personal computers to solve real-world OR problems. This era has presented us with an opportunity to solve a variety of problems that could scarcely have been envisioned in the past. This book is an endeavor to cover state-of-the-art optimization techniques in depth and breadth.

Why Another Book?

There are leading high-quality textbooks available for teaching optimization theory to graduate students and practitioners. Moreover, there are several excellent reference books on specialized topics. Therefore, the natural question is: Why another book?

From my experience of several years of teaching optimization courses in the Department of Industrial and Management Engineering at the Indian Institute of Technology Kanpur and as an applied researcher, I have learnt three facts. First, the field of optimization is evolving rapidly. However, only a handful of leading textbooks or specialized reference books cover the recent developments in a simplified manner. There are not many books that offer the reader an opportunity to learn state-of-the-art optimization techniques in a reasonably simple manner from the point of view of a practitioner, rather than taking a dry theoretical approach. On the other hand, the illustrations or case studies presented in traditional textbooks are too toyish (simplified versions of bigger problems) to inspire any confidence in the reader when confronted with problems arising in the real world. Some of these techniques may work well for small instances; however, they fail to solve larger real-life instances. Students following these books do not have clarity on how they will benefit their research; thus, the learning experience is passive, at best. They cannot connect the tools and techniques learnt from these books with the skills required to solve real-life problems. This book is an attempt to bridge this gap by reviewing a collection of mathematical tools from Operations Research and Computer Science (with an emphasis on optimization techniques) and demonstrating, using real-life cases, how they can be applied to address challenging problems.

There is an adequate number of textbooks on special topics, but those are too focused on a particular topic. If an instructor or a student is exploring different research areas in the optimization field, i.e., looking for one book that offers a gentle introduction to various state-of-the-art optimization techniques and modeling approaches with detailed applications, such a reference material is hard to find. Although this book is designed to provide broad coverage of the optimization techniques and their applications, a single-volume book cannot possibly cover all techniques exhaustively. Nor is it my intent to cover everything in one book.

The third motivation behind this book is to impart learning in a student-friendly language without neglecting rigor. A reader may refer to journal articles for state-of-the-art techniques and applications; nevertheless, he/she may be bogged down soon by the theoretical results and theorem-proof style of the articles. From our teaching experience, we know that even the brightest student in the class may find it challenging to assimilate analytically rich concepts found in research papers that are not explained well in student-friendly language. A scholar engaged in applied research or a practitioner may prefer a simple introduction to an advanced method, a straightforward explanation of theorems, and a more in-depth discussion on practical applications.

Hence, in my humble opinion, this book is an attempt to bridge multiple gaps. The book is intended to be a teaching aid for post-graduate students enrolled in Operations Research and Mathematical Optimization courses. It will also serve as a valuable reference for researchers who wish to explore various optimization techniques as part of their research methodology. The book covers various state-of-the-art methods (both exact and heuristics) and modeling approaches in sync with the current research trends that are not typically discussed in graduate-level textbooks. Additionally, the chapters offer valuable insights for future research endeavors. A significant emphasis has been placed on optimization tools, such as the use of commercial optimization solvers to model and solve large-sized problems. Most of the chapters include supplementary material to implement the models and algorithms. Applications covered in the book span the realms of resource planning, telecommunications, scheduling, logistics, education, environmental conservation, and many others. The topics covered reflect the diversity and depth of the optimization landscape. The volume brings together a rich tapestry of contributions from experts in diverse fields, spanning mathematics, computer science, engineering, management, and beyond. The learning from the book should enable professionals to apply optimization theory and algorithms to their fields of interest. Given these aspects, my belief is that this volume's scope and content will be appealing to a broad audience.

Plan and Organization of the Book

The book systematically examines recent developments and exciting challenges and provides the following in a unified and carefully developed presentation: (a) significant methodological advances for solving existing optimization problems and (b) novel challenging problems that have emerged in the real-life optimization domain.

Most of the chapters in this book begin with an introduction to the underlying optimization technique. The chapter then explores a real-life case study to which the technique is applied. The objective of each chapter is to demonstrate how the underlying technique can be applied to solve a challenging problem. Each chapter offers details on how to formulate a research problem as a formal optimization model, reformulate or transform it (if required) to improve computational tractability, and apply necessary customizations to the optimization technique specific to the underlying problem to obtain an optimal or near-optimal solution.

The chapters in the book are organized into two sections: the first section being devoted to method-oriented chapters and the second to application-oriented ones. Although each chapter covers both theory and application, chapters in the first group are focused on "teaching" exact methods with a special emphasis on large-scale integer programming problems. On the other hand, the focus of the chapters in the latter group is largely on specific real-life optimization problems, with discussion on one or more solution techniques to solve the problem. Applications of several metaheuristics have been demonstrated in this section. Thus, the reader will traverse a spectrum that ranges from algorithmic intricacies to the tangible impact of optimization on real-world problems.

Possible Use of the Book

The book is relatively advanced at places. It assumes a decent understanding of matrix algebra, linear and integer programming, nonlinear programming, computational complexity, and graph theory. The book aspires to cater to a diverse audience, including postgraduate students of mathematics, engineering, computer science, operations research, and management departments, as well as practicing professionals in the domain of optimization. For the novice, the chapters provide an initial foundation in optimization, unraveling its terminology, principles, and methodologies. Seasoned practitioners will find in these pages a source of inspiration, as they encounter innovative applications and novel perspectives that reinvigorate their approach to problem-solving. Although the book includes only selected topics and will not serve as a textbook for an optimization theory course, it can be used as an easy-to-read reference for a semester-long "Operations Research" or "Optimization" course. The extensive coverage of this volume provides an instructor with much flexibility in choosing 10–12 topics for such a one-semester course. In addition, the book can be used as a complementary reference along with standard textbooks.

Kanpur, India Faiz Hamid

Acknowledgement

A project of this magnitude could not be accomplished without the support and encouragement of several individuals. The compilation of this volume would not have been possible without the enthusiasm and expertise of the contributing authors. Their dedication to their respective fields of research and willingness to share their insights and knowledge have enriched this book immensely. It is worth noting that the majority of the chapters were written during the challenging period of the pandemic, with some authors even contending with multiple bouts of COVID-19. Thus, I wholeheartedly commend the authors for their unwavering dedication to this endeavor. I also express my gratefulness to the reviewers who diligently assessed the chapters, ensuring their scholarly rigor and relevance. Their thoughtful comments have significantly improved the quality of the chapters and enhanced the overall coherence of the volume.

I would also like to extend my gratitude to Deep Mukherjee from IIT Kanpur and Sachin Jayaswal from IIM Ahmedabad for their guidance and assistance throughout the editorial process. They provided invaluable insights and helped me shape the content into a coherent and comprehensive collection. I am also grateful to the numerous students I have instructed over the years at IIT Kanpur. Their feedback and the challenges they have encountered while pursuing their research endeavors or professional pursuits have served as the driving force behind the creation of this book. I acknowledge my institute/university for providing the necessary autonomy, flexibility, and cooperation which are prerequisites for passionate teaching and thinking. I am sure I will stay fortunate on this front, like umpteen others!

Working with the Springer team was amazing! Special thanks to the publishing editor, Nupoor Singh, for showing confidence in the book proposal, exhibiting remarkable patience and cooperation throughout the book's development, and providing assistance whenever it was needed.

Finally, on a more personal front, I wish to express my gratitude to my family. I will be forever indebted to my parents who have always been there when I needed them and instilled in me the love for mathematics, science, and engineering. I am profoundly thankful to my wife for her support, encouragement, and understanding throughout this project.

Despite our best efforts, it is possible that the book could contain certain areas open for improvement, primarily arising from typographical errors and similar mistakes. I would greatly appreciate it if readers could kindly share their comments, recommendations, and criticisms via my e-mail address (fhamid@iitk.ac.in). A list of errata will be made available if required.

Thank you all for being part of this journey.

August 2023

Faiz Hamid
Indian Institute of Technology Kanpur
India

List of Reviewers of Chapters

1. Ajinkya Tanksale, Clarkson University, United States
2. Amit Upadhyay, Indian Institute of Technology Roorkee, India
3. Arnab Bhattacharya, Indian Institute of Technology Kanpur, India
4. Atanu Chaudhuri, Durham University Business School, United Kingdom
5. Divya Padmanabhan, Indian Institute of Technology Goa, India
6. Garima Mittal, Indian Institute of Management Lucknow, India
7. Goutam Sen, Indian Institute of Technology Kharagpur, India
8. Ishwar Murthy, Indian Institute of Management Bangalore, India
9. Martin Schmidt, Trier University, Germany
10. Raouia Taktak, University of Sfax, Tunisia
11. Reshma Chandrasekharan, Indian Institute of Management Bangalore, India
12. Silke Jütte, IU International University of Applied Sciences, Germany
13. Surya Prakash Singh, Indian Institute of Technology Delhi, India
14. Tarun Rambha, Indian Institute of Science Bangalore, India
15. Walid Benameur, Telecom SudParis, France

Contents

About the Editor

Faiz Hamid is an Associate Professor in the Department of Industrial and Management Engineering at the Indian Institute of Technology (IIT) Kanpur, where he joined as an Assistant Professor in 2014. After completing his PhD in 2012 at the Indian Institute of Management Lucknow with a specialization in Operations Research, he worked on optimization problems for 4G wireless networks as a postdoctoral research fellow at Telecom SudParis, France. Prior to joining IIT Kanpur, he served as a Functional Architect at JDA Software (now Blue Yonder), Hyderabad, for a project on revenue management for passenger railways.

Faiz's research expertise lies primarily in integer programming, combinatorial optimization, data science, and their applications to transportation planning, railways, communications network design, and sports. His research has been published in several international journals of repute. At IIT Kanpur, Faiz has been actively involved with teaching activities at various levels, from undergraduates to Masters and PhD students. He teaches in the broad areas of operations research, management science, and data science, with a strong emphasis on modeling and applying theoretical concepts to real-world problems.

Chapter 1
Introduction

Faiz Hamid

1.1 Book Overview

There are numerous textbooks on optimization techniques. Several of them excel at explaining theories at a beginner's level using small illustrative examples. However, they often fail to connect them to real-world problems and the associated challenges. At the same time, there are several advanced-level books targeted at graduate students that focus heavily on abstract mathematical theories with little emphasis on their connections with practice. This book falls somewhere in between the above two, establishing a connection between the theoretical underpinnings of optimization techniques and their pragmatic applications in complex real-world scenarios. Therefore, it is neither a book that overly emphasizes abstract mathematical theory nor a textbook focusing on solving small mathematical programming problems manually. Rather, the book imparts an understanding of how theory and practice converge to refine standard formulations and/or fine-tune state-of the-art solvers when confronted with complex real-world problems.

The authors bring extensive experience in applying mathematical programming to real-world problems, sharing their knowledge and expertise throughout the chapters. Unlike journal articles, the emphasis has been on writing the chapters in a style that is accessible to students without neglecting rigor. The book is ideally suited for post-graduate students in mathematics, engineering, computer science, operations research, and management. The extensive coverage of this volume provides an instructor with much flexibility in choosing 10–12 topics for a one-semester course designed for such students. Additionally, the book can serve as a complementary reference to standard textbooks on operations research. Practicing professionals in the domain of optimization will find this book useful for upskilling themselves.

F. Hamid (✉)
Department of Industrial and Management Engineering, Indian Institute of Technology Kanpur, Kanpur 208016, Uttar Pradesh, India
e-mail: fhamid@iitk.ac.in

F. Hamid (ed.), *Optimization Essentials*, International Series in Operations Research & Management Science 353, https://doi.org/10.1007/978-981-99-5491-9_1

1.2 Book Organization

The chapters in the book are arranged to provide readers with a coherent progression from foundational theories to advanced applications, offering a comprehensive understanding of optimization techniques and their real-world applications. The chapters are organized in two sections:

- Section 1, consisting of 15 chapters, is devoted to optimization techniques that can provide optimal solutions. A special emphasis is placed on large-scale integer programming problems. Each chapter considers one technique in depth and illustrates it with a real-life application.
- Section 2, comprising 10 chapters, is application oriented in nature. Although these chapters cover both theory and application, they are primarily concerned with solving a specific real-life problem applying multiple or hybrid optimization techniques. The majority of the chapters in this section are based on heuristics; however, there are a few exact methods as well.

1.3 Structure of Each Chapter

Most of the chapters begin with an introduction to the underlying optimization technique, followed by an application of the technique to a real-world case study. This structure is followed by the majority of chapters to ensure uniformity in the presentation of the material, while allowing authors to tailor the content to the specific nuances of their chosen optimization technique. The chapters offer details on how to formulate a research problem into a formal optimization model, reformulate or transform it (if required) to improve computational tractability, and apply necessary customizations to the optimization technique specific to the underlying problem to obtain an optimal or near-optimal solution.

The introductory overview of the optimization technique in each chapter is by no means comprehensive. It is not intended to substitute formal courses on the topic. Each technique is substantial enough to merit its own extensive exploration. The introduction provides a balanced outline of the technique and guides students toward areas they should explore more thoroughly in formal coursework or textbooks. Nevertheless, each chapter is comprehensive in demonstrating how to apply the technique to an application problem.

1.4 Key Characteristics of the Book

1. The book delves into both theoretical and practical aspects of several state-of-the-art optimization techniques that are not covered in widely used graduate-level textbooks.

2. It demonstrates several tricks and step-by-step instructions for developing optimization models and reformulations, particularly in the context of large-scale optimization problems.
3. It presents the application of metaheuristics to solve complex real-world optimization problems.
4. Even though some topics are advanced, they are explained in a student-friendly language.
5. Applications covered in the book span the realms of resource planning, telecommunications, scheduling, logistics, education, environmental conservation, and many others.
6. The chapters provide software implementation guidance and practical exercises to enhance hands-on experience.

1.5 Book Outline

Here we briefly describe the contribution of each chapter.

Part I consists of Chaps. 2–16. These chapters delve deep into general modeling and solution techniques for linear and integer programming problems. In order to demonstrate their applicability, real-life problems are considered as well.

Chapter 2 describes the author's personal experience of learning linear programming (LP) from the perspective of convex programming. The chapter discusses two different approaches to LP, namely, the simplex method and the interior point methods based on the log-barrier technique. The approach to simplex method is based on the work of Manfred Padberg; unlike traditional discourse of LP, this approach is tableau free and nonlinear programming flavored. The author presents an extensive discussion on the KKT conditions for LPs, providing several insights throughout the chapter that are not commonly available in textbooks.

Chapters 3–7 discuss different formulation approaches for a given problem, analyzing their strengths and weaknesses in developing solution strategies, especially those based on valid inequalities. Chapter 3 reviews basic techniques for modeling scheduling optimization problems as Mixed-Integer Linear Programs (MILP). The authors demonstrate how various modeling strategies used for commonly occurring scheduling problems can be directly applied, modified, or combined to formulate novel scheduling problems. These problems arise in radar pulse interleaving, identifying schedules that are robust to changes in data, and scheduling path flows in a dynamic flow optimization problem.

In Chap. 4, the authors discuss how to identify weak MILP formulations based on the contrast between the physical systems modeled by the MILP and the LP relaxation that initiates the branch-and-bound (B&B) algorithm. Although MILP solvers can generate cuts and decompose problems to solve them efficiently, they are generic in nature and work by identifying common structures and characteristics in MIP models from diverse sources. Some problems have uncommon characteristics that elude the implemented cuts. In such situations, leveraging problem-specific knowledge can assist practitioners in developing more customized cuts that strengthen the formulation and significantly improve the solver's performance. The authors illustrate the methodology using a number of practical business applications. Several concepts are explained using a visual approach.

Chapter 5 briefly introduces the concept of valid inequalities and reviews the integer-rounding and Chvátal–Gomory methods for generating valid inequalities. The author considers a real-life problem of designing a backbone telecommunications network using optical fibers. A method for generating valid inequalities from a problem of reduced size, obtained from shrinking the underlying grapgh, is described. The inequalities thus obtained remain valid for the original problem. The shrinking-based method is applied to sub-structures (p-partitions) of size $p = 2$, 3, and 4 nodes to derive valid inequalities. The author nicely illustrates the Chvátal–Gomory method to derive several valid inequalities for the network design problem. A separation heuristic is proposed to find the p-partitions that are likely to result in violated valid inequalities.

Chapter 6 discusses different modeling and solution strategies, along with their advantages for the same problem. The authors consider real-life resource management problems involving the assignment of resources to execute a given set of ordered tasks while adhering to policies and restrictions governing the deployment of these resources. The problem is modeled as several alternate network-based integer programs. The models are compared in terms of their size and tightness of LP relaxations. The authors discuss the advantages of developing extended formulations and embedding constraints within the network representation. The solution procedure includes problem reduction, valid inequalities, coefficient lifting, and optimization-based heuristics.

In Chap. 7, the authors study a generalization of the distance-based critical node detection problem, which considers pairwise distances between nodes for identifying critical nodes of a network. The problem is modeled with fewer entities than the models in the literature. A new version of the problem is introduced where edges, rather than nodes, are to be deleted. This version is useful in application contexts where deleting nodes is impractical. The authors illustrate several practical aspects of the problem on social and transportation networks.

Chapters 8–11 discuss solution techniques for solving large-scale MILPs. Chapter 8 begins with a mathematical foundation of the basic Benders decomposition method and illustrates it using a real-life humanitarian operation problem within a stochastic programming framework. The chapter concludes by discussing a variety of implementation-related challenges using the CPLEX solver and a few theoretical acceleration strategies.

Chapter 9 discusses another popular large-scale optimization technique known as delayed column generation. This chapter delves into the mathematical nuances of delayed column generation and branch-and-price computational procedures. The technique is used to solve the aircraft routing and scheduling problems appearing in the airline industry. The following chapter describes the use of column generation and Lagrangian relaxation to solve the crew scheduling and crew rescheduling problems for railways. Lagrangian relaxation is applied to compute dual bounds for a minimization problem. Column generation is used to deal with the large number of columns. A detailed illustration of the subgradient optimization algorithm is also presented.

Chapter 11 describes Dantzig–Wolfe reformulation and column generation using a shipping industry problem as an illustration. The problem is to assign chemicals to the compartments of a cargo ship that transports chemicals between ports. Several cargo-cargo and cargo-compartment compatibility rules, and tanker balancing safety restrictions make this assignment problem difficult. An MILP model for the assignment problem is presented along with insights into its structure and complexity. A Dantzig–Wolfe decomposition based reformulation is derived to take advantage of multiple special structures in the problem. The reformulation is solved using a customized column generation algorithm.

Chapter 12 discusses an algorithm based on the subgradient method, known as the Volume Algorithm (VA), which produces fast approximate solutions for large-scale LPs. The primary benefit of the algorithm is that it produces approximate primal and dual solutions. The authors present three applications of the algorithm. First, to train neural networks. The results are improved when the widely used stochastic gradient descent method is replaced with the VA. The algorithm is then used to solve a bus routing problem for a city. The LP relaxation of this problem is too large to be solved using conventional LP algorithms. The VA provides tight lower bounds, and the approximate primal solution serves as the starting point for several heuristics that generate the final routing. Finally, the algorithm is used to solve the p-median problem for restructuring semistructured data.

Chapters 13–16 deal with nonlinear programming problems. Chapter 13 discusses techniques to solve mixed-integer nonlinear programming problems. Two approximation algorithms are presented: piece-wise linearization and linearization based on the McCormick envelope. Next, two exact solution methods are discussed: outer approximation and generalized Benders decomposition. The authors demonstrate how non-convex problems can be convexified by transforming non-convex functions. The application of these methods is presented in a real-world case study for the location-allocation of shared wastewater treatment plants.

Chapter 14 examines several mixed-integer nonlinear formulations of the unit commitment problem in electrical power production from the perspective of size and tightness trade-offs. The authors discuss Lagrangian relaxation based algorithms for single-unit versions of the problem. A number of guidelines are provided to choose the best approach depending on the application.

Chapter 15 discusses an interesting category of problems, known as bilevel optimization, which contains an inner optimization problem as a constraint in an outer optimization problem. The chapter discusses classical solution approaches based on reformulating a bilevel problem into a single-level problem. In addition, alternative single-level formulations for the application problems considered in this chapter are discussed. The chapter presents several real-life examples of bilevel optimization problems arising in various contexts, such as the product line selection problem and the shortest path interdiction problem.

Chapter 16 considers robust optimization problems and describes modeling and solution approaches for binary programs involving chance constraints and ambiguously defined probability distributions. The theory is explained using an illustrative example of a stochastic bin packing problem where the bin capacity may be exceeded with the desired probability. A probability cut-based algorithm is also presented, and computational results demonstrate its effectiveness.

Part II includes Chaps. 17–26. In each chapter of this section, the emphasis shifts to the application of optimization techniques to specific problems. The chapters discuss the modeling and solution approach for the given problem.

Chapter 17 covers the problem of production planning and scheduling on parallel machines with sequence-dependent setup times. The chapter first reviews the literature on this problem, along with their different solution approaches. Next, a real-life case of production planning and scheduling from the aluminum industry is presented. The authors discuss the reasons behind model's low computational time.

Chapter 18 explores a popular class of optimization problems in the literature of operations research, namely the facility location problem. A real-life problem of school consolidation is considered where one has to decide which small schools to close and merge with large schools, obeying certain constraints. An important aspect to consider is that schools should not become inaccessible after the merging process. The problem is modeled as a multi-objective integer linear program and solved using lexicographic ordering in two stages, along with the ε-constraint method.

Chapters 19 and 20 focus on optimization problems arising in railways. Chapter 19 begins with a discussion of the fundamentals of railway operations management, and the optimization problems involved at various levels of planning. The chapter then explores large-scale timetabling, the rolling stock circulation problem for long-distance passenger operations, and station-level planning. The chapter describes the application of simulation and optimization tools to solve these problems. Chapter 20 studies the rail yard optimization problem. Rail yards serve as the nerve center for freight railroads where railcars are switched and new trains are formed. The challenge lies in deciding the optimal placement of incoming trains within the terminal, determining the locations for constructing blocks within the yard, and establishing the sequence for assembling outbound trains. Exact methods cannot solve large-scale versions of these problems. The authors propose metaheuristic-based algorithms, including a very large-scale neighborhood search for optimization. The algorithms have been validated for an extremely large rail yard in North America. The results are analyzed extensively using several key performance indicators.

Chapter 21 considers the problem of tool indexing occurring in automated machining centers. The objective of the problem is to minimize the non-machining time by positioning tools optimally so that the tool changer can quickly pick the required tool. Solving large instances of the problem is computationally difficult, and the chapter presents local search and tabu search algorithms to obtain good-quality solutions to the problem. The authors propose two innovative bookkeeping techniques to reduce the computation time by two orders of magnitude.

Chapters 22–24 examine a few variants of the widely popular Vehicle Routing Problem (VRP). Chapter 22 considers the multi-depot VRP for a real-life distributor in Turkey with three depots and 502 customers. The problem is solved heuristically by the cluster-first, route-second approach. The first subproblem assigns customers to each depot using the capacitated p-median problem for clustering. Next, a single-depot VRP is utilized to solve the routing problem for each depot and its cluster of customers. For this, a guided local search metaheuristic is implemented and Google-OR-Tool is utilized as a solver.

Chapter 23 discusses a variant of VRP where the vehicles must collect blood samples (perishable products) from donation centers and transport them to processing centers within the stipulated time limit. Two metaheuristic methods, namely the hybrid flower pollination algorithm and simulated annealing, are applied to determine the optimal route of vehicles to maximize the amount of collected blood. Computational results demonstrate the efficacy of the proposed algorithms. Chapter 24 considers a multi-objective VRP with simultaneous pickup and delivery. The objectives are minimization of traveled distance and fuel consumption; thus, the problem is environmentally concerned. The problem is inspired by an industry partner. A novel constructive heuristic integrating Non-Dominated Sorting Genetic Algorithm II (NSGA-II), along with Technique for Order Preference by Similarity to Ideal Solution (TOPSIS) is applied to solve the problem. TOPSIS helps generate a better initial population as opposed to randomly generated solutions.

Chapter 25 presents a real-life problem that defies typical mathematical programming formulations. These types of problems demand the incorporation of external procedure-based declarations to assess system performance or establish relationships among various subcomponents of the problem. These problems are frequently solved using optimization metaheuristics such as genetic algorithms, which are structurally quite versatile and typically require only system performance information for optimization. The authors illustrate the approach using a real-life problem of urban mobility where the objective is to minimize the travel time of commuters using a combination of one-way and two-way roads.

Chapter 26 presents the problem of revenue-sharing contracts between a risk-neutral supplier and a behavioral retailer in a dyadic supply chain. To model the retailer's behavior and derive the optimal behavioral retailer option, the prospect theory framework is utilized, involving nonlinear optimization in the stochastic demand setting. The chapter discusses managerial implications and future research directions.

Let us now commence this enlightening expedition where theory and practice facilitate the exploration of novel frontiers. As we proceed through the diverse chapters that follow, we aim to inspire innovative thinking and foster a deeper appreciation of the connection between optimization theory and real-world practice. Enjoy the reading!

Part I
Methods

Chapter 2
The Many Guises of Linear Programming

Joydeep Dutta

Abstract This is a survey on linear programming. However, this is not a survey of a single approach to linear programming, but rather of two distinct approaches: the simplex method and interior point methods based on the log-barrier technique. These two topics are in fact extremely vast, and a survey such as this can only provide a glimpse of the huge edifice known as linear programming. Our approach here is unconventional since instead of viewing linear programming as a stand-alone subject, we examine it through the lens of convex programming. This allows us additional insights that are not part of the traditional discourse of linear programming. The approach to the simplex method given in this survey is credited to Manfred Padberg. The freshness of his approach is that it is tableau-free and has a non-linear programming flavor of using a rank-one update of a basic feasible matrix if it is not optimal. We also provide a very simple proof of the strong duality theorem for linear programming due to J. E. Martinez-Legaz. In this chapter we have an extensive discussion on the KKT conditions for linear programming problems with detailed proofs. The convex programming approach will be readily apparent in the discussion of the interior point methods. The chapter also contains the author's personal experience of learning linear programming, beginning with an aversion to the tableau approach and progressing to a profound appreciation for the Padberg and interior point methods.

2.1 Linear Programming: A Personal Story

This may be a very unusual way of starting a technical paper. Nevertheless, I chose to do it as I strongly believe that research level exposition must be essentially a form of storytelling. My introduction to linear programming came about on a sultry August afternoon in 1989 on the 3rd floor of the celebrated St. Xavier's College, Calcutta (now Kolkata). You must be finding it strange that when it was introduced, we were

---

J. Dutta (✉)
Department of Economics Sciences, Indian Institute of Technology Kanpur, Kanpur, India
e-mail: jdutta@iitk.ac.in

F. Hamid (ed.), *Optimization Essentials*, International Series in Operations Research & Management Science 353, https://doi.org/10.1007/978-981-99-5491-9_2

never told that this was a *constrained optimization problem*. Though I had "learned" the Lagrange multiplier method in a mechanical way in my calculus course, I had not heard the term constrained optimization. My mathematical thinking capability was very slow, and before I could understand anything about the problem or why it was interesting, I was bombarded with words like *basic feasible solution*, *corner solutions*, and *simplex method*. In those days, the simplex method was the first formal algorithm that I was encountering without even understanding why it was needed. The most dreaded thing was the *simplex tableau*, which was a mechanical way of solving a linear programming problem. It was filled with strange calculations with a strange operation called pivoting, and finding the leaving and entering variables, and so on. The dread of the tableau developed a strange dislike for the simplex method in my mind, and I still dread the simplex tableau. One might dismiss this view of the simplex method, but in my fear of the tableau, I stand in the exalted company of Manfred Padberg, a great name in optimization and also a student of George B. Dantzig, the inventor of the simplex method. Let me take this opportunity to quote from Padberg's book titled: **Linear Optimization and Extensions** [17].

> The equation format is useful to teach the elements of linear programming at an undergraduate level and communicate linear programming better than its much publicized predecessor, the *tableau format*, which is obtained from the equation format by simply dropping the explicit form of the equations and arranging the resulting quantities in a rectangular "tableau". As an aside, when in the academic year 1963/64 the author was introduced to the simplex algorithm via the tableau format, he was completely "lost" when one of his academic teachers (a non-mathematician) was pivoting around numbers in a huge blackboard in some classroom at the University of Munster in front of a class of about 100 students. Just imagine the students' average distance from the blackboard!

Nowadays, there are a large number of software available for solving linear programming problems. But most of them really do not use the simplex tableau to design the software.

My interest in linear programming as a subject of optimization developed when I came across the celebrated book by Stephen. J. Wright, titled: **Primal-Dual Interior Point Method** [24]. I started enjoying the ideas there since it was a nonlinear programming approach to linear programming. The key idea was to obtain the solution to the primal-dual pair of linear programming problem by solving the Karush-Kuhn-Tucker (KKT) conditions using the celebrated Newton's method. This approach, which gives rise to modern interior point methods, is based on the fact that linear programming problems form a subclass of convex programming problems, and for a convex programming problem, any solution of the KKT system is a minimizer of the problem. In fact, as we will see below, the dual solutions are the Lagrange multipliers associated with the KKT conditions. Actually, my appreciation of the simplex method came much after I was fascinated by the interior point approach. This was due to Padberg's approach to developing the simplex method without the need for any tableau. His approach is very simple for any student of optimization to understand. We take a feasible point and check if it is optimal. If not, go to another feasible point so that the objective function value improves. With this naive view, Padberg [17] builds up the simplex method in a very simple and beautiful way. Ever since I came across this approach, I not only have a much better appreciation of the simplex method but also feel much liberated by the thought that I never have to speak about the tableau again.

It is important to tell at the outset who my target audience is. I would like to say that this exposition is not for the uninitiated. At least some exposure to basic optimization would indeed be extremely beneficial. It is mainly for those using linear programming in their scientific research as a tool, for example, in economics, operations research, finance, and management. I am simply attempting to outline the various avenues through which one can explore the fascinating subject of linear programming. Let me conclude this section by listing the sections that will appear in this chapter.

Section 2: A Primer on Convexity
Section 3: Linear Programming: Basic Structure and Results
Section 4: Simplex Method without Tableau
Section 5: KKT Conditions for Linear Programming
Section 6: Interior-Point Methods

The notations that we will use in this chapter are fairly standard. For example, given a set $A \subset \mathbb{R}^n$ we denote by $int A$ and $\overline{A}$, the interior and closure of the set A respectively. By $\langle x, y \rangle$, the inner-product or the dot product of two vectors in $\mathbb{R}^n$. Let $\|x\|$ denote the norm of a vector in $\mathbb{R}^n$, and for $x \in \mathbb{R}^n_+$, the symbol $x \geq 0$ is used to denote $x_i \geq 0$ for all $i = 1, \ldots, n$.

2.2 A Primer on Convexity

Convexity is a key tool for the whole of optimization, not just linear programming. It might be more wise now to call the subject "linear optimization" so as to eliminate any confusion that one might have with the term computer programming. However, I will hang on to the traditional name "linear programming" since this was suggested by the great economist T. C. Koopmans to George Dantzig. The study of convexity has two fundamental objects: convex sets and convex functions, which we will briefly review here.

Definition 2.2.1 The set $C \subset \mathbb{R}^n$ is said to be a convex set if for any $x, y \in C$ and $\lambda \in [0, 1]$ we have $\lambda y + (1 - \lambda)x \in C$.

Geometrically, this means that for any two points in a convex set, the line segment joining those two points is also inside the set. It is clear that when $\lambda = 0$ we are at x, and as λ is moved from zero to one, we actually walk along the line segment from x toward y and reach y when $\lambda = 1$ without leaving the set C. The reader should look around himself and see that there are many artificial objects that are convex sets, while natural objects aren't. Observe that the human body is not a convex subset of $\mathbb{R}^3$. For our purposes, the most important categories of convex sets that we will need to study are polyhedral convex sets, which are formed by the intersection of simpler

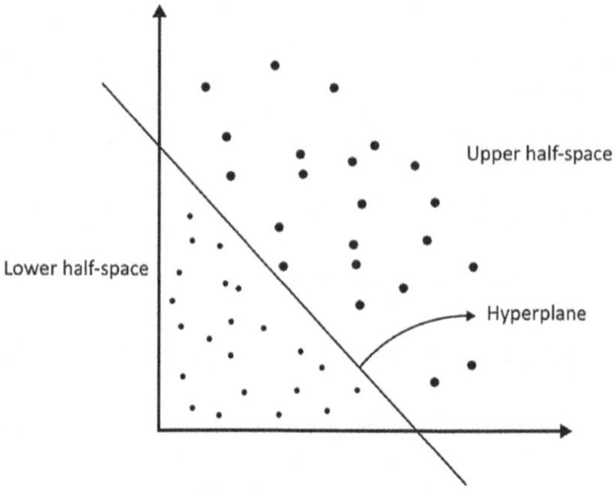

Fig. 2.1 Hyperplane and half-spaces

convex sets called *half spaces*, and half spaces arise when we construct possibly the most simple non-trivial convex set, the *hyperplane*. A hyperplane is a convex set which we denote by H, and this is given by

$$H = \{x \in \mathbb{R}^n : \langle a, x \rangle = b\},$$

where $a \in \mathbb{R}^n$ and $b \in \mathbb{R}$. In other words, the set H is the solution set of the linear equation $\langle a, x \rangle = b$. The term linear equation means an equation involving a linear function. A function $f : \mathbb{R}^n \to \mathbb{R}$ is said to be a linear function if

(a) $f(x + y) = f(x) + f(y)$, for all $x, y \in \mathbb{R}^n$, and
(b) $f(\lambda x) = \lambda f(x)$ for any $\lambda \in \mathbb{R}$.

It is a simple exercise from linear algebra to show that any linear function $f : \mathbb{R}^n \to \mathbb{R}$ can be expressed as an inner product. This means that there exists a vector $a \in \mathbb{R}^n$ such that $f(x) = \langle a, x \rangle$.

Let us go back to the hyperplane. If $n = 2$, then a hyperplane represents a straight line on the plane, and if $n = 3$, it represents a plane in three dimensions. Each hyper plane divides the space $\mathbb{R}^n$ into two half-spaces, namely the upper and lower half-spaces (Fig. 2.1), which we denote as H^+ and H^- and are given as

$$H^+ = \{x \in \mathbb{R}^n : \langle a, x \rangle \geq b\}$$

and

$$H^- = \{x \in \mathbb{R}^n : \langle a, x \rangle \leq b\}.$$

It goes without saying that H^+ and H^- are convex sets. In linear programming, a special type of convex set, called **polyhedral set**, plays a crucial role. Linear programming, after all, deals with the minimization or maximization of linear function over a polyhedral set. A polyhedral set is formed by the intersection of a finite number

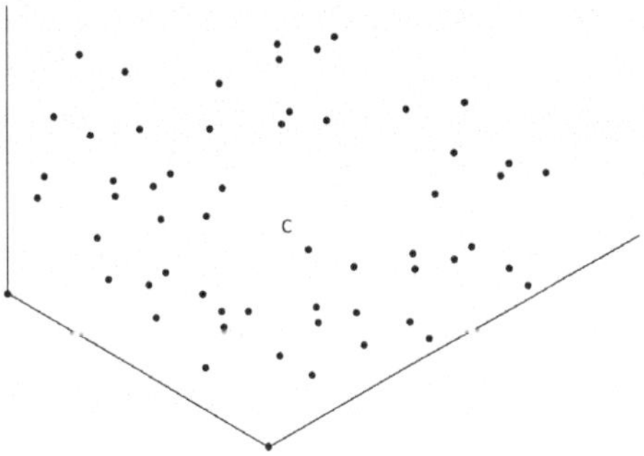

Fig. 2.2 Polyhedral set

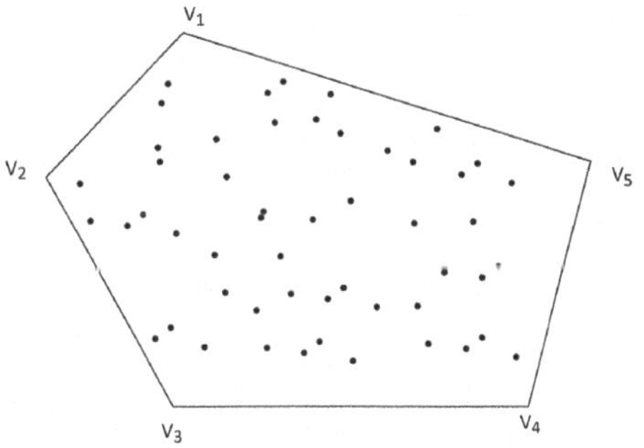

Fig. 2.3 A polytope with five vertices

of half spaces (Fig. 2.2), and in general, a polyhedral set P can be expressed as

$$P = \{x \in \mathbb{R}^n : \langle a_i, x \rangle \geq b_i, i = 1, \ldots, m\}.$$

In fact, the representation P can also be given by less than equal to type inequality format or even in an equality format. However, both formats can always be rewritten in the way we have represented P. A bounded polyhedral set is called a **polytope** (Fig. 2.3). We shall discuss polyhedral sets in more details in the next section. We shall present the definition of a convex function which plays such a crucial role in optimization.

Definition 2.2.2 A function $f : C \to \mathbb{R}$, where C is a convex subset of $\mathbb{R}^n$ is called a convex function if for any $x, y \in C$ and $\lambda \in [0, 1]$ we have

$$f(\lambda y + (1 - \lambda)x) \leq \lambda f(y) + (1 - \lambda) f(x).$$

One of the key features of a convex function is that whenever we seek to minimize it over a convex set, every local minimizer is global. From this very property stems the importance of convexity in optimization. A linear function is, of course, a convex function, and there are abundant examples of convex functions in the literature; hence, we do not separately provide them here. However, we would like to mention two good sources for examples of convex functions. It will be instructive to look at Borwein and Lewis [4] and Boyd and Vendenberghe [5]. Economists, however, would prefer to look at concave functions for which every local maximizer is a global maximizer. A function $f : C \to \mathbb{R}$, where C is a convex set in $\mathbb{R}^n$ is called concave if for any $x, y \in C$ and $\lambda \in [0, 1]$,

$$f(\lambda y + (1 - \lambda)x) \geq \lambda f(x) + (1 - \lambda) f(x).$$

Thus, f is concave if and only if $-f$ is convex. In many areas of investigation in engineering, business and economics we seek to minimize a convex function over a convex set. To be a bit more erudite, we are looking at the optimization problem (CP)

$$\min f(x) \quad \text{for all} \quad x \in C,$$

where $f : \mathbb{R}^n \to \mathbb{R}$ is a convex function and C is a convex subset of $\mathbb{R}^n$. Here, the function f is called the *objective function*, and C is the *feasible set*. Often, the convex set C can be described using convex inequalities, i.e.

$$C = \{x \in X : g_i(x) \leq 0\},$$

where each g_i is a real-valued convex function of $\mathbb{R}^n$ and X is a convex subset of $\mathbb{R}^n$. However, the problem (CP) can be recast as a problem of minimizing a linear function over a convex set. Such a problem, which we denote as (LCP), is given as

$$\min t, \quad \text{subject to} \quad f(x) \leq t, x \in C.$$

Observe that here we have two decision variables x and t. The feasible set of (LCP) is given as

$$F = \{(x, t) \in \mathbb{R}^n \times \mathbb{R} : f(x) \leq t, x \in C\}.$$

Note that the problem (LCP) is in $\mathbb{R}^{n+1}$ while (CP) is in $\mathbb{R}^n$. In what sense are these problems equivalent? We will see that solving one of them solves the other. Let $x_0 \in C$ be a minimizer of (CP). We shall show that (x_0, t_0) with $t_0 = f(x_0)$ is a minimizer of (LCP). Observe that $(x_0, f(x_0))$ is a feasible point for (LCP). Now consider any (x, t) feasible to (LCP). Hence, $x \in C$, and thus $f(x_0) \leq f(x) \leq t$ which shows that $(x_0, f(x_0))$ is a solution to (LCP). For the converse, let us assume that (x_0, t_0) is a solution to (LCP). It is clear that $(x, f(x))$ is a feasible point of (LCP) for any $x \in C$. Hence, we have for any $x \in C$

$$f(x_0) \leq t_0 \leq f(x).$$

This shows that x_0 solves the problem (CP). Thus, without loss of generality, we may claim that in order to study convex optimization problems, it is enough to focus on the minimization of a linear function over a convex set. The problem (LCP) is indeed of that form. In fact, the most important classes of convex optimization problems actually fall under this particular format. Our focus is thus on the problem (CPL) given as

$$\min\langle c, x\rangle, \quad \text{subject to} \quad x \in X,$$

where $c \in \mathbb{R}^n$ and X is a convex set in $\mathbb{R}^n$. When in particular X is a polyhedral set, the problem (CPL) is called the linear optimization problem or the linear programming problem.

A study of linear programming cannot proceed without several important aspects of convex analysis. Among them are the separation theorems about convex sets, and furthermore, a few important notions and properties of polyhedral sets and cones should be known. Let us first start with separation theorems. In these theorems, we establish that two non-empty convex sets that do not intersect can be "separated" from each other by a hyperplane, which acts like a boundary wall between them.

Given two convex sets A and B such that $A \cap B = \emptyset$, the phrase *separation* of A and B implies that there exists a hyperplane

$$H = \{x \in \mathbb{R}^n : \langle a, x\rangle = b\}$$

such that

$$\langle a, y\rangle \geq b, \forall y \in A$$

and

$$\langle a, z\rangle \leq b, \forall z \in B.$$

Of course, the inequalities also hold in reverse order for A and B by changing the signs of a and b. Our next result will introduce the concept of strict separation of convex sets.

Theorem 2.2.1 *Let S be a closed convex set in $\mathbb{R}^n$ and $y \notin C$. Then there exits $a \in \mathbb{R}^n$, with $a \neq 0$ and $\alpha \in \mathbb{R}$ such that for all $x \in S$,*

$$\langle a, x\rangle < \alpha < \langle a, y\rangle.$$

This idea of separation, as we see above, is called strict separation. This strict separation can happen when one considers two non-intersecting convex sets, one of which is compact and the other closed. In finite dimensions, a compact set is equivalent to a closed and bounded set. Figure 2.4 pictorially illustrates the concept of strict separation.

Before we write down the result for strict separation of two convex sets, it is important for us to introduce the notion of Minkowski addition of two sets. Let A and B be two subsets of $\mathbb{R}^n$. Then the *Minkowski addition* of the sets is denoted as

Fig. 2.4 Strict separation

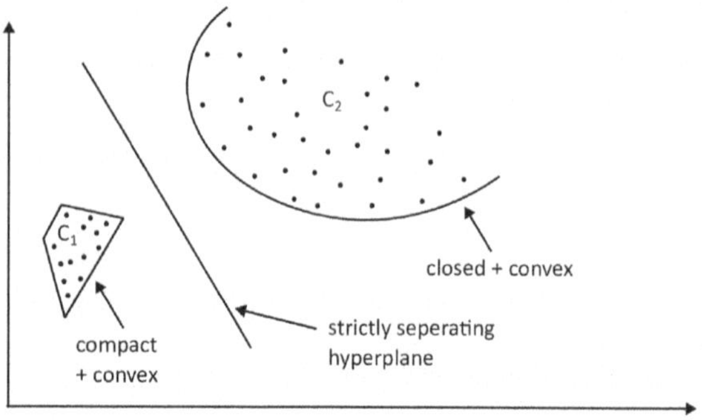

$A + B$, and is given as

$$A + B = \{z : z = x + y, x \in A, y \in B\}.$$

The set $A - B$ is similarly defined. Note that if A and B are closed, $A + B$ or $A - B$ need not be closed unless one of them is compact. This will be a crucial fact in establishing the strict separation between two disjoint convex sets.

Theorem 2.2.2 *Let C_1 be a closed convex set and C_2 be a compact convex set. Further, let $C_1 \cap C_2 = \emptyset$. Then there exists $a \in \mathbb{R}^n$, with $a \neq 0$ and $\beta \in \mathbb{R}$ such that*

$$\langle a, x\rangle < \beta < \langle a, y\rangle,$$

for all $x \in C_1$ and all $y \in C_2$.

Proof Observe that, as C_1 and C_2 do not intersect, it is clear that $0 \notin C_1 - C_2$. From the hypothesis of the theorem, it is clear that $C_1 - C_2$ is closed. This allows us to apply Theorem 2.2.1 to conclude that there exists $0 \neq a \in \mathbb{R}^n$ and $\alpha \in \mathbb{R}$ such that for all $v \in C_1 - C_2$ we have

$$\langle a, v\rangle < \alpha < 0.$$

Writing $v = x - y$, where $x \in C_1$ and $y \in C_2$, we have

$$\langle a, x\rangle < \alpha + \langle a, y\rangle. \tag{2.1}$$

Since C_2 is compact by the celebrated Weierstrass Theorem, there exists $\hat{y} \in C_2$ such that

$$\langle a, \hat{y}\rangle = \inf_{y \in C_2} \langle a, y\rangle.$$

Thus, from (2.1), we have for any $x \in C_1$

$$\langle a, x \rangle < \alpha + \langle a, \hat{y} \rangle.$$

As $\alpha < 0$, we can conclude from the above inequality that for all $x \in C_1$ and $y \in C_2$

$$\langle a, x \rangle < \alpha + \langle a, \hat{y} \rangle < \langle a, \hat{y} \rangle \leq \langle a, y \rangle.$$

Setting $\beta = \alpha + \langle a, \hat{y} \rangle$, the result is established. □

Following this is the supporting hyperplane theorem, which states that if we consider a closed convex set S and x be any point on the boundary, then there exists $a \neq 0$ such that $\langle a, x \rangle = b$ and $\langle a, y \rangle \leq b$ for all $y \in C$. This simply says that we can pass a hyperplane through each point of the boundary of a closed convex set such that the convex set always lies in only one of the half-spaces generated by the hyperplane.

Though this may not be important from the point of view of linear programming, it may be worthwhile for the reader to take a detour and take a look at a subclass of convex functions called *strongly convex functions*. A function $f : \mathbb{R}^n \to \mathbb{R}$ is called strongly convex if there exist $\rho > 0$ and a convex function $g : \mathbb{R}^n \to \mathbb{R}$ such that for any $x \in \mathbb{R}^n$ we have

$$f(x) = g(x) + \frac{\rho}{2}\|x\|^2.$$

This is equivalent to the fact that for any $x, y \in \mathbb{R}^n$ and $\lambda \in [0, 1]$, we have

$$f(\lambda y + (1 - \lambda)x) + \frac{\rho}{2}\lambda(1 - \lambda)\|y - x\|^2 \leq \lambda f(y) + (1 - \lambda) f(x).$$

The positive real ρ number is called the *modulus of strong convexity*. A strongly convex function with modulus $\rho > 0$ is often referred to as the ρ-strongly convex function. A key fact to note is that a linear function can never be strongly convex. This is pretty obvious, as linear functions do not have any quadratic parts. Strongly convex functions have a very important property that endears them to numerical optimizers. This is the fact that over a closed convex set, they always have a minimizer, and the minimizer is unique. However, this simple fact about strongly convex function leads us to some deeper geometrical facts about convex sets and also allows us to devise geometrical structures that carry with them the information about the Lagrange multipliers associated with optimization problems.

Consider the problem of computing the distance of a point $x \in \mathbb{R}^n$ to a closed convex set C. In the literature, one sees the notion of a distance function given as

$$d(x, C) = \inf_{y \in C} \|y - x\|.$$

Since C is convex, the distance function $d(x, C)$, a function of x, is also convex. To compute the distance of a point x to a closed convex set C, we can equivalently solve the following quadratic optimization problem:

$$\min_{y \in C} \frac{1}{2}\|y - x\|^2. \tag{2.2}$$

The above problem (2.2) always has a minimizer as the function $f(y) = \frac{1}{2}\|y - x\|^2$ is a strongly convex function, and thus the above problem has a unique minimizer say $\hat{y} \in C$, which is called the projection of x on the closed convex set C and is written as

$$\hat{y} = proj_C(x).$$

Further it has been shown in Theorem 3.1.1 of Hiriart-Urruty and Lemaraechal [11] that $\hat{y} \in C$ is a projection of a point x on the closed convex set C if and only if

$$\langle x - \hat{y}, y - \hat{y} \rangle \leq 0, \quad \text{for all} \quad y \in C.$$

The vector $v = x - \hat{y} = x - proj_C(x)$ is called the normal to the closed convex set C at the point $x \in C$. This idea is actually crucial in proving Theorem 2.2.1. Observe that if C is a closed convex set and $x \notin C$, then we have $a = x - proj_C(x) \neq 0$, and thus using the above inequality, we conclude that

$$\langle a, y - proj_C(x) \rangle \leq 0,$$

for all $y \in C$. Thus, we have

$$\begin{aligned} \langle a, y \rangle &\leq \langle a, proj_C(x) \rangle \\ &\leq \langle a, x - a \rangle \\ &\leq \langle a, x \rangle - \|a\|^2 \\ &< \langle a, x \rangle. \end{aligned}$$

We leave it to the reader to see that, from the above discussion, we can easily deduce the conclusions reached in Theorem 2.2.1. Actually, the set of normals at a point $x \in C$, where C is a closed convex set, forms a special kind of convex set that is also a cone. A set $S \subset \mathbb{R}^n$ is a *cone* if $\lambda x \in S$ whenever $x \in S$ and $\lambda \geq 0$. A cone is called convex if it is a convex set, and in other words, a cone is convex if and only if, for any x and y in C, the vector sum $x + y$ is also in C. The idea of a cone of normals can be extended to convex sets, and we can define it as follows.

Definition 2.2.3 Let C be a convex subset of $\mathbb{R}^n$ and let $x \in C$. Then the normal cone to C at x is defined as

$$N_C(x) = \{v \in \mathbb{R}^n : \langle v, y - x \rangle \leq 0, \quad \forall v \in C\}.$$

It is important to observe that if C is an open convex set, then $N_C(x) = \{0\}$ (Fig. 2.5). The normal cone is a useful geometric tool to express optimality conditions.

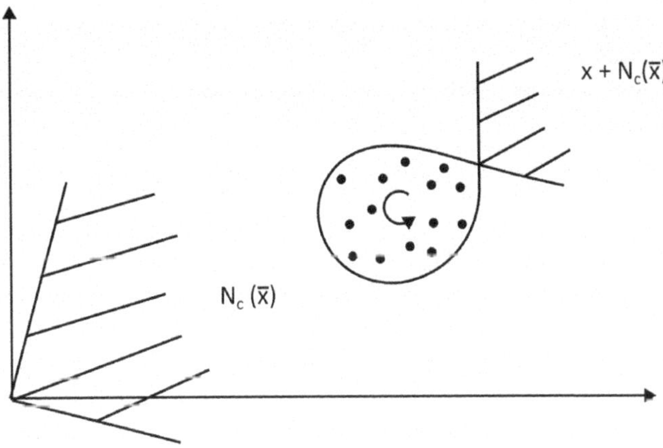

Fig. 2.5 Normal cone

We will also see that it will be the source of Lagrange multipliers, or more precisely, the Karush-Kunh-Tucker (KKT) multipliers themselves. To begin with, let us first consider the problem of minimizing a differentiable convex function over a convex set C. This is the basic format of a convex optimization problem. Can we develop an optimality condition for such a problem? The answer turns out to be surprisingly simple. If x^* is a minimizer of f over C, then

$$-\nabla f(x^*) \in N_C(x^*).$$

Observe that if $x^* \in C$ be a minimizer of f over C, then for any $x \in C$ and $\lambda \in (0, 1)$, we have, $x^* + \lambda(x - x^*) \in C$ as C is convex, and hence

$$f(x^* + \lambda(x - x^*)) \geq f(x^*).$$

The definition of differentiability leads to the following expression:

$$\langle \nabla f(x^*), x - x^* \rangle + \frac{o(\lambda)}{\lambda} \geq 0.$$

As $\lambda \to 0$, we have $\dfrac{o(\lambda)}{\lambda} \to 0$, proving that

$$\langle \nabla f(x^*), x - x^* \rangle \geq 0,$$

for all $x \subset C$. This simply proves that $-\nabla f(x^*) \in N_C(x^*)$. For the converse, let us note that a differentiable function is convex if and only if

$$f(y) - f(x) \geq \langle \nabla f(x), y - x \rangle, \tag{2.3}$$

for all $x, y \in \mathbb{R}^n$. Once this is known, the converse can be easily established, and we leave it to the reader.

Our focus will now turn to the convex optimization problem, where the feasible set C is given as

$$C = \{x \in \mathbb{R}^n : g_i(x) \le 0, \quad i = 1, \dots, m\},$$

and each g_i is a real-valued differentiable convex function over $\mathbb{R}^n$. If we want to minimize the function f over the above set C, then the key to writing the optimality condition lies in calculating the normal cone to the above set at the point of optimality. Let us consider x_0 to be a minimizer, and we shall denote by $I(x_0)$ the set of all indices i for which $g_i(x_0) = 0$. If we are considering a problem with only two variables, we can visualize that at x_0, the gradients are perpendicular to the surface, and thus one may feel that an element of the normal cone may be generated by a non-negative linear combination of the gradient vectors of g_i at x_0, where $i \in I(x_0)$. In other words, we are asking the question: Is the following true?

$$N_C(x^*) = \left\{ \sum_{i \in I(x_0)} \lambda_i \nabla g_i(x_0) : \lambda_i \ge 0, i \in I(x_0) \right\}. \tag{2.4}$$

It is not hard to see that using (2.3) and replacing f by g_i, we can conclude that the set on the right-hand side of the above equality is definitely contained in $N_C(x_0)$. The reverse is however true under the assumption that the *Slater condition* holds, i.e., there exists $\hat{x}$ such that $g_i(\hat{x}) < 0$ for all $i = 1, \dots, m$. Thus, if the Slater condition is satisfied, the equality (2.4) holds true. This leads to the following theorem.

Theorem 2.2.3 *Let us consider the problem of minimizing the real-valued convex function f with respect to the constraints $g_i(x) \le 0$, $i = 1, \dots, m$, where each g_i is also a real-valued convex function. Assume that f and each g_i, $i = 1, 2, \dots, m$, are differentiable. Assume that the Slater condition holds true, i.e. there exists $\hat{x}$ such that $g_i(\hat{x}) < 0$. Assume that x^* is the global minimizer of the above stated minimization problem. Then there exists scalars $\lambda_i \ge 0$ $i = 1, \dots, m$ such that*

(i) $\nabla f(x^*) + \sum_{i=1}^m \lambda_i \nabla g_i(x^*) = 0$
(ii) $\lambda_i g_i(x^*) = 0$, $i = 1, \dots, m$

Conversely if $(x^*, \lambda) \in \mathbb{R}^n \times \mathbb{R}^m_+$, with $\lambda = (\lambda_1, \dots, \lambda_m)$, is a vector which satisfies the above two conditions and x^* satisfies the constraints, then x^* is the global minimizer of f with respect to the constraints $g_i(x) \le 0$, $i = 1, \dots, m$.

We leave it to readers to patch up a proof of the above theorem with all the discussions preceding it. Graduate students who have never seen the above result should definitely try to write down the proof of this result, as that will be their first true step into convex optimization. This is the celebrated **Karush-Kuhn-Tucker** condition; in short, the **KKT-conditions** for a convex programming problem.

It is important to now gradually take steps toward writing down optimality conditions for optimization problems with linear functions or affine functions as objective and constraints since linear functions and affine functions are also convex functions. To begin with, let us look at the following convex optimization problem, which we call (CP) for convenience, in which we seek to minimize a real-valued and differentiable convex function f with respect to the constraints $g_i(x) \leq 0, i = 1, \ldots, m$ and $x \in X$, where X is a convex set. In most applications, X is usually a closed convex set. In proceeding with this, we will first introduce the Lagrangian function. The Lagrangian function $L : \mathbb{R}^n \times \mathbb{R}^m_+ \to \mathbb{R}$ is given as

$$L(x, \lambda) = f(x) + \sum_{i=1}^{m} \lambda_i g_i(x).$$

We shall now explore a key result in convex optimization. This is called the **saddle point theorem**.

Theorem 2.2.4 *Let x^* be the global minimizer of (CP). Further, observe that the generalized Slater condition holds true, i.e., there exists $\hat{x} \in X$ such that $g_i(\hat{x}) < 0$ for all $i = 1, \ldots, m$. Then there exists $\lambda^* \in \mathbb{R}^m_+$ such that*

(i) $L(x^*, \lambda) \leq L(x^*, \lambda^*) \leq L(x, \lambda^*), \forall x \in X$ *and* $\lambda \in \mathbb{R}^m_+$
(ii) $\lambda_i g_i(x^*) = 0, i = 1, \ldots, m.$

Conversely if there exists a vector $(x^*, \lambda) \in X \times \mathbb{R}^m_+$ such that the above two conditions are satisfied, then $x^* \in X$ is feasible to (CP) and also the global minimizer of (CP).

We shall pick up the threads of the above theorem in the last part of the next section. However, let us now focus on what we have learned in the above result. The key learnings are the following:

$$L(x^*, \lambda^*) = \min_{x \in X} L(x, \lambda^*) \tag{2.5}$$

and

$$L(x^*, \lambda^*) = \max_{\lambda \in \mathbb{R}^m_+} L(x^*, \lambda). \tag{2.6}$$

The crucial fact to observe here is that the saddle point conditions reveal that actually, as we try to minimize a convex optimization problem, a minimization problem and a maximization problem are simultaneously being solved. Using (2.5), we can conclude that

$$0 \in \nabla_x L(x^*, \lambda^*) + N_X(x^*).$$

This condition is actually equivalent to condition (2.5). Thus, the above conditions combined with the *complementary slackness condition*, i.e., $\lambda_i g_i(x^*) = 0$, for all $i = 1, \ldots, m$ are called the generalized version of the KKT-conditions for the convex optimization problem with inequality constraints and an abstract constraint like $x \in X$. The reason for calling the second condition in Theorem 2.2.1 a complementary slackness condition stems from the fact that both λ_i and $g_i(x^*)$ cannot hold with strict inequality at the same time. More formally, the generalized KKT-conditions for a convex programming problem are given as

(i) $0 \in \nabla_x L(x^*, \lambda^*) + N_X(x^*)$
(ii) $\lambda_i g_i(x^*) = 0, i = 1, \ldots, m.$

Conditions (i) and (ii) play an important role in taking the first step in writing down the KKT-conditions for linear programming, which we will use in Sect. 2.3. Let us now focus on Eq. (2.6). This says that for a given $\lambda \in \mathbb{R}^m_+$ we have

$$\min_{x \in X} L(x, \lambda) \leq L(x^*, \lambda^*).$$

Now, using the complementary slackness condition, we conclude that

$$\min_{x \in X} L(x, \lambda) \leq f(x^*).$$

If we set

$$\theta(\lambda) = \min_{x \in X} L(x, \lambda),$$

then we can conclude that for any $\lambda \in \mathbb{R}^m_+$

$$\theta(\lambda) \leq \min_{x \in X, g(x) \leq 0} f(x).$$

In fact, the interesting question that we can ask here is the following: Is it possible that

$$\max_{\lambda \in \mathbb{R}^m_+} \theta(\lambda) = \min_{x \in X, g(x) \leq 0} f(x)? \tag{2.7}$$

Note that in the above expressions $g(x) \leq 0$ means less than equal to in component-wise sense. Further from the discussion above, we are motivated to present what is known in the literature as the Lagrangian dual problem, denoted by (LDP), and given by

$$\max_{\lambda \geq 0} \theta(\lambda). \tag{2.8}$$

If the result in (2.7) holds, we say that the **strong duality** holds. Though we shall show in the next section that for linear programming problems, duals appears in a more natural way from a completely different context, it is actually the Lagrangian dual. We shall end this section by mentioning the result and stating the conditions under which strong duality holds.

Theorem 2.2.5 *Let us consider the problem of minimizing the convex function f subject to the constraints $g_i(x) \leq 0$, $i = 1, \ldots, m$, $x \in X$. Consider x^* to be a minimizer of this problem. Assume that the generalized version of the Slater condition holds. Then there exists $\lambda^* \geq 0$ such that it solves the Lagrangian dual problem (LDP) and $f(x^*) = \theta(\lambda^*)$.*

It is important to note that absence of the Slater's condition in the above theorem may lead to the failure of strong duality.

2.3 Linear Programming: Basic Structure and Results

The discussion in this section and the next ones deals with developing the basic framework of linear programming and, at the same time, developing a tableau-free method for linear programming, i.e., developing the simplex method more in the flavor of non-linear programming. In doing so, these two sections are largely based on Chaps. 3, 4, and 5 of the monograph by Padberg [17]. I openly admit my own intellectual debt to Manfred Padberg since, through his beautiful writing and tableau-free approach, I first appreciated the beauty and power of the simplex method. It is my attempt to discuss his approach in a more detailed fashion in these two sections. In this section and the next, we shall use the symbols $\langle c, x\rangle$ and $c^T x$ interchangeably since both denote the dot product or inner product of the vectors c and x in $\mathbb{R}^n$.

Linear programming involves minimizing a linear function over a polyhedral set. In other words,

$$\min c^T x = \langle c, x\rangle$$

subject to

$$\begin{aligned} &\langle a_j, x\rangle \geq b_j \qquad && j = 1, \ldots, m \\ &x_i \geq 0 && i = 1, \ldots, n \end{aligned}$$

where $c \in \mathbb{R}^n$, $a_j \in \mathbb{R}^n$, $\forall j$ and $b_j \in \mathbb{R}$, $\forall j$. The constraints $x_i > 0$, are called the non-negativity restrictions and can be compactly written as

$$\langle e_i, x\rangle \geq 0,\ i = 1, \ldots, n$$

where e_i is the ith column of the $n \times n$ identity matrix, which has zero at all places except the ith position where it has a 1. The non-negativity constraints can also be written as $x \geq 0$ where

$$x \geq 0 \Longleftrightarrow x_i \geq 0,\ \forall l = 1, \ldots, n.$$

Let A be an $m \times n$ matrix whose rows are the vectors a_j written as row vectors. Thus, the above linear programming problem can be written as

$$\min c^T x$$
$$\text{subject to}$$
$$Ax \geq b, x \geq 0.$$

This is often called the linear programming problem written in the *canonical form*.

Let $w \in \mathbb{R}^n$ with $w \geq 0$ be such that

$$Ax - w = b.$$

Then, we can again equivalently reformulate the above problem as

$$\min c^T x + 0^T w$$
$$\begin{bmatrix} A \vdots - I \end{bmatrix} \begin{bmatrix} x \\ w \end{bmatrix} = b$$
$$x \geq 0, \, w \geq 0.$$

This reformulates a few inequality constraints as equality constraints. We do not, however, touch the constraints $x \geq 0$. The non-negativity restrictions are often called **hard constraints**. Hence, in general, we can write the linear programming problem as

$$\min c^T x \qquad \text{(LP)}$$
$$Ax = b$$
$$x \geq 0$$

This problem, which we refer to as (LP), is often called the linear programming problem in the *standard form*. Our goal is to study the nature of this problem and methods for solving it.

A key fact about the linear programming problem is the following.

> If the problem (LP) is bounded below, it has a minimizer.

Remember: The problem (LP) has no local minimizers. All minimizers are global minimizers. So we have the following question:

> Under what condition will (LP) be bounded below?

But before that, let us answer the question, when will it not be bounded below? This is provided in the following result by Padberg [17].

Proposition 2.3.1 *If $b = 0$, then either $x = 0$ is a solution or the problem (LP) is unbounded below.*

Proof Let $b = 0$. Then, $x = 0$, is a trivial feasible solution of (LP). If x^* is a solution of (LP), we have $c^T x^* \leq 0$. If $x = 0$ is not a solution, then there exists a feasible point $\widetilde{x}$ of (LP) such that

$$c^T \widetilde{x} < 0. \tag{2.9}$$

Let $\lambda > 0$ be any real number. Since $b = 0$, $\lambda\widetilde{x}$ is also feasible for (LP), as

$$A(\lambda\tilde{x}) = \lambda A\tilde{x} = 0; \text{ and } \lambda\tilde{x} \geq 0.$$

Further, $\langle c, \lambda\widetilde{x}\rangle < 0$ from (2.9). Thus,

$$\lambda\langle c, \widetilde{x}\rangle < 0.$$

Now as $\langle c, \widetilde{x}\rangle < 0$ by (2.9), and $\lambda \to \infty$, we have

$$\lambda\langle c, \widetilde{x}\rangle \to -\infty.$$

Therefore, $\langle c, \widetilde{x}\rangle \to -\infty$ when $\lambda > 0$. Thus, moving along the ray passing through $\widetilde{x}$, the function value approaches $-\infty$. Thus, (LP) is unbounded below. This proves our result. □.

Now let us try to see when the problem (LP) will be bounded. Let x be feasible for (LP), then

$$Ax = b.$$

Then, for any $y \in \mathbb{R}^m$ one has

$$\langle Ax, y\rangle = \langle b, y\rangle$$
$$\text{or} \quad \langle A^T y, x\rangle = \langle b, y\rangle.$$

Let us choose $y \in \mathbb{R}^m$ in such a way that $c \geq A^T y$ (component wise). Thus, for any x which is feasible to (LP), we have

$$\langle c, x\rangle \geq \langle A^T y, x\rangle$$
$$\langle c, x\rangle \geq \langle b, y\rangle, \text{ where } y \text{ satisfies } A^T y \leq c.$$

This is true for any feasible set of (LP). We have

$$\operatorname*{Inf}_{\{x \geq 0: Ax = b\}} \langle c, x\rangle \geq \langle b, y\rangle, \text{ such that } A^T y \leq c.$$

If we write val(LP) = Inf$\{\langle c, x\rangle : x \geq 0, Ax = b\}$, then

$$\text{val(LP)} \geq \langle b, y \rangle, \text{ such that } A^T y \leq c \qquad \text{(Dual problem)}$$
$$\therefore \text{val(LP)} \geq \text{Sup}\{\langle b, y \rangle : A^T y \leq c\}.$$

If there is a y such that $A^T y \leq c$, then val(LP) is finite, and hence the (LP) has a minimizer.

The dual problem is given as

$$\max \langle b, y \rangle$$
$$A^T y \leq c.$$

An equivalent formulation in the standard form allows us to write the dual problem as

$$\begin{aligned} &\max \langle b, y \rangle \\ &A^T y + s = c \\ &s \geq 0 \end{aligned} \qquad \text{(DP)}$$

where y is the dual variable and s is the dual slack variable. The key fact is that the dual of the dual is indeed the primal in the case of a linear programming problem, though in the case of a convex programming problem, this may not hold true in general. Of course, for a convex programming problem, the dual problem associated with it is the Lagrangian dual problem, as stated in the previous section. In fact, it is natural to ask whether the dual problem that we have just constructed above for (LP) is actually its Lagrangian dual. This is exactly what we are going to demonstrate here. Let us do this for the linear programming problem written in the canonical form, i.e.

$$\min\langle c, x \rangle, \quad \text{subject to} \quad Ax \geq b, \quad x \geq 0.$$

The Lagrangian function in this case is defined as

$$L(x, y) = \langle c, x \rangle + \langle y, b - Ax \rangle,$$

where $x \in \mathbb{R}^n$ and $y \in \mathbb{R}^m_+$. The dual function in this case is written as

$$\begin{aligned} \theta(y) &= \min_{x \geq 0} L(x, y) \\ &= \min_{x \geq 0}\{\langle c, x \rangle + \langle y, b - Ax\} \\ &= \min_{x \geq 0}\{\langle c - A^T y, x \rangle + \langle b, y \rangle\}. \end{aligned}$$

We leave it to the reader to argue that if $c - A^T y \geq 0$, or equivalently $A^T y \leq c$, then $\theta(y) = \langle b, y \rangle$, or else $\theta(y) = -\infty$. Thus, our required dual problem is

$$\max\langle b, y\rangle, \quad \text{subject to} \quad A^T y \leq c, \quad y \geq 0.$$

We strongly encourage the reader to use this approach for (LP) by breaking up the constraints $Ax = b$ as $Ax \geq b$ and $-Ax \geq -b$. A key result follows.

Strong Duality: If both (LP) and (DP) have an optimal solution, then

$$opt(LP) = opt(DP)$$

where *opt* is used to denote the optimal value.

For a primal dual pair of linear programming problems, there can be several combinations that one might face. Strong duality, as mentioned above, is one of them. There are three more possibilities, which we list below.

- If the primal problem (LP) is infeasible, the dual (DP) is feasible and unbounded (above), i.e., $\langle b, y\rangle \to +\infty$.
- If the dual (DP) is infeasible, the primal (LP) is feasible and unbounded (below), i.e., $\langle c, x\rangle \to -\infty$.
- Both the primal and the dual do not have a feasible solution.

Let us end this section by providing a very simple approach to proving strong duality for the linear programming problem. This approach was first shown to me by my post-doctoral mentor, Prof. Juan Enrique Martinez-Legaz of the Autonomous University of Barcelona, Spain, We produce it here with his kind permission. The approach has a two-fold effect. For a linear programming problem, it provides its dual and at the same time establishes the strong duality theorem. The problem considered here is of the following form:

$$\max\langle c, x\rangle, \quad \text{subject to} \quad Ax \leq b,$$

where A is an $m \times n$ matrix and $b \in \mathbb{R}^m$ and $c \in \mathbb{R}^m$. We call this problem (LP1). The maximization problem is chosen since the whole proof is based on the so called *lub-axiom* (least upper bound axiom) of real analysis, i.e., the supremum is the minimum of the upper bounds.

$$\sup\{\langle c, x\rangle : Ax \leq b\} = \min\{k : Ax \leq b \implies \langle c, x\rangle \leq k\}.$$

To proceed further, we need to use the celebrated Farkas Lemma, which was introduced in 1902. This lemma says that the inequality $\langle c, x\rangle \leq k$ is a consequence of the inequalities $Ax \leq b$ if and only if there exists $w \in \mathbb{R}^m$ and $w \geq 0$ such that $A^T w = c$ and $\langle b, w\rangle \leq k$. Thus, the Farkas lemma tells us that

$$\sup\{\langle c, x\rangle : Ax \leq b\} = \min\{k : k \geq \langle b, w\rangle, A^T w = c, w \geq 0\}.$$

Now we would like to recall that in Sect. 2.2, we discussed how we can equivalently write a convex optimization problem as minimizing a linear function over a convex set. Using that equivalence, we can see that

$$\sup\{\langle c, x\rangle : Ax \leq b\} = \min\{\langle b, w\rangle : A^T w = c, w \geq 0\}.$$

Therefore, we have proved that if we maximize the problem (LP1), its maximum value is the minimum value of the linear program (DP2) given as

$$\min\langle b, w\rangle \quad \text{subject to} \quad A^T w = c, \quad w \geq 0.$$

This is exactly the strong duality theorem.

Readers may try to deduce the problem (DP) from the problem (LP) using the above approach, but instead using the fact the infimum is the maximum of the lower bounds, which is called the *glb-axiom* (greatest lower bound axiom) in real analysis.

2.4 Simplex Method Without Tableau

This section is a mathematician's point of view on linear programming problems and how to solve them using the celebrated simplex method without tableau. Manfred Padberg will serve as the guide for this, and I will attempt to convey to the reader the simplex method approach outlined in his book [17]. Denote the feasible set of (LP) by S, i.e.,

$$S = \{x \in \mathbb{R}^n : Ax = b, x \geq 0\}$$

Let us set $N = \{1, 2, \ldots, 3\}$. For any $\boldsymbol{x} = \begin{bmatrix} x_1 \\ \cdot \\ \cdot \\ x_n \end{bmatrix} \in S$, let us begin by considering the following index set:

$$\begin{aligned} I_x &= \{j \in N : x_j > 0\} \\ N \setminus I_x &= \{j \in N : x_j = 0\} \\ \boldsymbol{x}_{I_x} &= (x_i : i \in I_x) \\ \boldsymbol{x}_{N \setminus I_x} &= (x_i : i \in N \setminus I_x) \end{aligned}$$

Padberg shows us that we can rewrite the facts: $Ax = b$, with $\boldsymbol{x}_{N\setminus I_x} = 0$ as

$$\begin{bmatrix} A_{I_x} & A_{N\setminus I_x} \\ 0 & I_{n-k_x} \end{bmatrix} \begin{bmatrix} x_{I_x} \\ x_{N\setminus I_x} \end{bmatrix} = \begin{bmatrix} b \\ 0 \end{bmatrix} \tag{2.10}$$

where A_{I_x} consists of columns of A corresponding to $\boldsymbol{x}_{I_x}$, and $A_{N\setminus I_x}$ comprises of columns of A corresponding to $\boldsymbol{x}_{I\setminus N_x}$. Let $k_x = \text{card}(I_x)$. Here, A_{I_x} is an $m \times k_x$ submatrix of A, and $A_{N\setminus I_x}$ is an $m \times (n - k_x)$ submatrix of A. Note that to solve the above system, we need to focus on the rank of the matrix in (2.10). Observe that

$$\text{Rank}\begin{bmatrix} A_{I_x} & A_{N\setminus I_x} \\ 0 & I_{n-k_x} \end{bmatrix} = \text{Rank}(A_{I_x}) + (n - k_x).$$

The system (2.10) has a unique solution if and only if

$$\text{Rank}(A_{I_x}) + (n - k_x) = n,$$

i.e., if and only if $\text{Rank}(A_{I_x}) = k_x$, i.e., $\text{Rank}(A_{I_x}) = \text{card}(I_x)$. Note that for having a unique solution we should also have,

$$\begin{aligned} m + n - k_x &\geq n \\ \text{i.e.,}\quad k_x = \text{card}(I_x) &< m \end{aligned}$$

Here, $\boldsymbol{x}$ is called a *basic feasible solution* (BFS) if

$$\text{Rank}(A_{I_x}) = m.$$

And, $\boldsymbol{x}$ is called a *degenerate* BFS if

$$\text{Rank}(A_{I_x}) < m.$$

An $m \times m$ submatrix of A is called a *basis* if it has rank m. Therefore, if B is an $m \times m$ submatirx of A and $\text{Rank}(B) = m$, we call B a basis. Therefore, we can always write $Ax = b$ as

$$Bx_B + Rx_R = Ax = b.$$

In fact, the matrix B is the collection of all linearly independent columns in the matrix A. Since A is assumed to be an $m \times n$ matrix with $m \leq n$ and A is full-rank, i.e., A has rank m, the number of linearly independent columns of A, is m. Thus, the matrix B is an $m \times m$ matrix consisting of the linearly independent columns of A. One may question: are the first m columns of A linearly independent? Of course, that may not be the case, but by multiplying with elementary matrices, we can rearrange the columns in the above fashion.

If $x_R = \mathbf{0}$, then

$$x_B = B^{-1}b$$

and if $x_B \geq 0$, i.e., $B^{-1}b \geq 0$, we say "$\boldsymbol{x} = (x_B, x_R)^T$" is a BFS. Let us illustrate this operation. Let

$$A = \begin{array}{c} \begin{matrix} C_1 & C_2 & C_3 & C_4 \end{matrix} \\ \begin{bmatrix} a_{11} & a_{12} & a_{13} & a_{14} \\ a_{21} & a_{22} & a_{23} & a_{24} \\ a_{31} & a_{32} & a_{33} & a_{34} \end{bmatrix} \end{array}.$$

Here, C_1, C_2, C_3, C_4 denote the four columns of the matrix A. Let columns C_1, C_3 and C_4 be linearly independent, then

$$B = \begin{bmatrix} a_{11} & a_{13} & a_{14} \\ a_{21} & a_{23} & a_{24} \\ a_{31} & a_{33} & a_{34} \end{bmatrix}, \; R = \begin{bmatrix} a_{12} \\ a_{22} \\ a_{32} \end{bmatrix}$$

and Rank$(B) = 3$. Now, $Ax = b$ in this case implies

$$x_1 \begin{bmatrix} a_{11} \\ a_{21} \\ a_{31} \end{bmatrix} + x_2 \begin{bmatrix} a_{12} \\ a_{22} \\ a_{32} \end{bmatrix} + x_3 \begin{bmatrix} a_{13} \\ a_{23} \\ a_{33} \end{bmatrix} + x_4 \begin{bmatrix} a_{14} \\ a_{24} \\ a_{34} \end{bmatrix} = \begin{bmatrix} b_1 \\ b_2 \\ b_3 \end{bmatrix}$$

$$x_1 \begin{bmatrix} a_{11} \\ a_{21} \\ a_{31} \end{bmatrix} + x_3 \begin{bmatrix} a_{13} \\ a_{23} \\ a_{33} \end{bmatrix} + x_4 \begin{bmatrix} a_{14} \\ a_{24} \\ a_{34} \end{bmatrix} + x_2 \begin{bmatrix} a_{12} \\ a_{22} \\ a_{32} \end{bmatrix} = \begin{bmatrix} b_1 \\ b_2 \\ b_3 \end{bmatrix}$$

$$\begin{bmatrix} a_{11} & a_{13} & a_{14} \\ a_{21} & a_{23} & a_{24} \\ a_{31} & a_{33} & a_{34} \end{bmatrix} \begin{bmatrix} x_1 \\ x_3 \\ x_4 \end{bmatrix} + x_2 \begin{bmatrix} a_{12} \\ a_{22} \\ a_{32} \end{bmatrix} = b$$

$$\begin{aligned} \boldsymbol{x}^T &= \left[x_1,\, x_2,\, x_3,\, x_4\right] \\ \boldsymbol{x}_B^T &= \left[x_1,\, x_3,\, x_4\right] \\ &= \left[x_{l_1},\, x_{l_2},\, x_{l_3}\right] \quad \text{where } l_1 = 1, l_2 = 3, l_3 = 4 \\ \boldsymbol{x}_R &= \left[x_2\right]. \end{aligned}$$

Let us now state the following fundamental theorem by Padberg [17].

Theorem 2.4.1 *Let us consider the problem (LP) and let Rank$(A) = m$. Then the following holds,*

1. *If (LP) is feasible, i.e., if there is an $x \in \mathbb{R}^n$ feasible to (LP), then there exists a basic feasible solution.*
2. *If there is an optimal solution to (LP), then there exists an optimal solution to (LP) which is a basic feasible solution as well.*

Proof The proof here is purely linear algebraic in nature. We now outline the steps of the proof below.

(i) Let x be a feasible point of (LP). Let $I = \{j \in N : x_j > 0\}$. Thus, $x_j = 0$ $\forall j \in N \setminus I$. So the matrix A can be partitioned as A_I and $A_{N \setminus I}$. If Rank(A_I) = card(I), then x is a BFS by definition. If Rank$(A_I) <$ card(I), then the columns of A_I are linearly dependent. Hence, there exists scalars λ_j and not all zero such that

$$\sum_{j \in I} \lambda_j a_j = 0$$

where a_j, $j \in I$, are the columns of the matrix. Without loss of generality, we assume that $\lambda_j > 0$ for some $j \in I$.

Define

$$\frac{x_r}{\lambda_r} = \min \left\{ \frac{x_j}{\lambda_j} : \lambda_j > 0, j \in I \right\}.$$

Now, define $y \in \mathbb{R}^n$ as

$$y_j = \begin{cases} x_j - \lambda_j \dfrac{x_r}{\lambda_r}, & j \in I \\ 0, & j \in N \setminus I. \end{cases} \tag{2.11}$$

If $\lambda_j \leq 0$, then $y_j \geq 0$. If $\lambda_j > 0$, then also $y_j \geq 0$ with $y_r = 0$. Hence, y is feasible to (LP). Further, if $I_y = \{j \in N : y_j > 0\}$, then card$(I_y)$ <card(I). If Rank(A_{I_y}) =card(I_y), then we are done, or we can keep repeating the above argument, and in a finite number of steps we shall reach a basic feasible solution.

(ii) We can now look at the next part with solutions. Let $\boldsymbol{x}$ be a solution of (LP), i.e. $c^T x \leq c^T y$, for all feasible y of (LP).

Let $I = \{j \in N, x_j > 0\}$. Then $A_I x_I = b$ and $z_0 = c_I^T x_I$ where c_I is the vector containing the components of c corresponding to indices in I. If $Rank(A_I) = card(I)$, then we are done. If not, again we shall have scalars λ_j, $j \in J$ such that $\lambda_j^{'s}$ are not all zero and

$$\sum_{j \in I} \lambda_j a_j = 0$$

Without loss of generality, assume that $c_I^T \lambda_I \geq 0$. Here, we shall deal with two cases.

- Case (a): $\lambda_j \leq 0$, $\forall\ j \in I$ and $c_I^T \lambda_I > 0$.
- Case (b): $\lambda_j > 0$, for some $j \in I$.

Note that if $c_I^T \lambda_I = 0$, then we are in Case (b) always (the reader can think on it). Suppose we first consider Case (b). Then, construct $y \in \mathbb{R}^n$ as we have done in (2.11) Then,

$$\begin{aligned} c^T y &= c_I^T y_I \\ &= \sum_{j \in I} c_j (x_j - \lambda_j \theta), \{\text{where } \theta = \frac{x_r}{\lambda_r} > 0\} \\ &= \sum_{j \in I} c_j x_j - \theta \sum_{j \in I} c_j \lambda_j \\ &= z_0 - \theta c_I^T \lambda_I. \end{aligned}$$

If $c_I^T \lambda_I > 0 \Rightarrow z_0 - \theta c_I^T \lambda_I < z_0$, contradicting that z_0 is the optimal value. Thus, $c_I^T \lambda_I = 0$. Also, $c^T y = z_0$ proves that y is an optimal solution. We have already checked the feasibility of y. Further, $card(I_y) < card(I)$, so one can repeat the argument as given in part (i) of this proof.

Now consider Case (a). Define $y \in \mathbb{R}^n$ as follows

$$y_j = \begin{Bmatrix} x_j - \mu \lambda_j, & j \in I \\ 0, & j \in N \setminus I \end{Bmatrix} \tag{2.12}$$

where $\mu \geq 0$.

Now,

$$\begin{aligned} Ay &= \sum_{j \in I} y_j a_j + \sum_{j \in N \setminus I} y_j a_j \\ &= \sum_{j \in I} (x_j - \mu \lambda_j) a_j \\ &= \sum_{j \in I} x_j a_j - \mu \sum \lambda_j a_j \\ &= b. \end{aligned}$$

For any $\mu \geq 0$, we write $y = y(\mu) \in \mathbb{R}^n$, whose components are given by (2.12).

$$z(\mu) = c^T y(\mu) = z_0 - \mu c_I^T \lambda_I$$

As $\mu \to +\infty$, $z(\mu) \to -\infty$. This contradicts the fact that the problem (LP) has a solution. Hence, the proof is complete. □

It is a well known fact and always discussed in linear programming texts that basic feasible solutions correspond to the vertices of the feasible polyhedron of (LP). A point x in a polyhedron P is called a vertex if for any two points u and v in P and $\lambda \in (0, 1)$, $x = \lambda u + (1 - \lambda) v$ implies that $x = u = v$. It is important to note that a polyhedron has only finite number of vertices. The often used technical name for vertices is *extreme points*. We present the following important result from Guenin, Koneman and Tuncel [10].

Theorem 2.4.2 *Let us consider the problem (LP) where the rank of the matrix A is m, i.e., it has full row rank. Then any feasible point of (LP) is an extreme point (or vertex) of the feasible polyhedron*

$$S = \{x \in \mathbb{R}^n : Ax = b, x \geq 0\},$$

if and only if the point is a basic feasible solution of the system $Ax = b$.

In order to start solving (LP), we first focus on the expression

$$Bx_B + Rx_R = b$$

where B is an $m \times m$ matrix of rank m. If we set $x_R = 0$, we shall get

$$x_B = B^{-1}b.$$

If $B^{-1}b \geq 0$, we get a basic feasible solution x which has two vector components

$$x_B = B^{-1}b \text{ and } x_R = 0.$$

Let us denote by I_B the index set of B as

$$I_B = \{k_1, k_2, \ldots, k_m\}$$

where k_1 is the index of the first column of B as a column of A and so on. For our example, we have

$$k_1 = 1, k_2 = 3, k_3 = 4$$

Let us fix a few symbols following Padberg [17].

- $\bar{b} = B^{-1}b = (\bar{b}_1, \bar{b}_2, \ldots, \bar{b}_m)$—transformed right-hand side,
- $\bar{c} = c - c_B B^{-1} A$—reduced cost vector or reduced cost coefficient $\bar{c}_i$ of the components of $\bar{c}$, and
- $y_j = B^{-1}a_j$—transformed jth column of A.

The Basic Feasible Solution: (This is just too important so we repeat) The expression, $Ax = b$ can be written as,

$$Bx_B + Rx_R = b$$

where, $\text{Rank}(A) = \text{Rank}(B) = m$.

$$\therefore x_B + B^{-1}Rx_R = B^{-1}b$$
$$i.e.\ x_B + B^{-1}Rx_R = \bar{b}$$

So one can write,

$$x_{k_i} = \bar{b}_i,\ for\ 1 \le i \le m$$
$$x_l = 0,\ for\ l \in N \setminus I_B$$

where, I_B is the index set of x_B , i.e. $I_B = \{k_1, k_2, \ldots, k_m\}$.

Detecting Optimality: Let B be the matrix corresponding to the vector component x_B. We call B as the feasible basis (or feasible basis matrix)

$$z_B = c_B^T x_B = \langle c_B, B^{-1}b\rangle,$$
$$(x_B = B^{-1}b,\ \text{since}\ x_B\ \text{symbolizes the BFS})$$

In general, $x_B = \bar{b} - B^{-1}Rx_R$

$$\begin{aligned}
\langle c, x\rangle &= \langle c_B, x_B\rangle + \langle c_R, x_R\rangle \\
&= \langle c_B, \bar{b} - B^{-1}Rx_R\rangle + \langle c_R, x_R\rangle \\
\langle c, x\rangle &= \langle c_B, x_B\rangle + \langle c_R, x_R\rangle \\
&= \langle c_B, B^{-1}b - B^{-1}Rx_R\rangle + \langle c_R, x_R\rangle \\
&= \langle c_B, B^{-1}b\rangle + \langle c_R, x_R\rangle - \langle c_B, B^{-1}Rx_R\rangle \\
&= \langle c_B, B^{-1}b\rangle + \langle c_R, x_R\rangle - \langle (B^{-1}R)^T c_B, x_R\rangle \\
&= z_B + \langle c_R - (B^{-1}R)^T c_B, x_R\rangle
\end{aligned}$$

If $c_R - (B^{-1}R)^T c_B \ge 0 \Rightarrow \langle c_R - (B^{-1}R)^T c_B, x_R\rangle \ge 0 \Rightarrow \langle c, x\rangle \ge z_B$. Hence, $c_R - (B^{-1}R)^T c_B \ge 0$ is a sufficient condition for optimality, though not necessary.

$$c_R - (B^{-1}R)^T c_B = c_R - (c_B^T B^{-1}R)^T$$

Note that $c_B^T B^{-1}R$ is a row vector. In component wise form it means

$$c_j - c_B^T B^{-1}a_j \ge 0, \forall j \in N \setminus I_B$$

How to detect unboundedness: Let us consider B to be a feasible basis matrix. If there is no confusion, set $I_B = I$. consider that in the problem(LP), there exists $j \in N \setminus I$, such that

- $c_j - c_B^T B^{-1} a_j < 0$
- $B^{-1}\, a_j \leq 0$

Then (LP) is unbounded.

We shall now provide a demonstration of the fact. The proof is due to Padberg [17].
Consider for any real number μ, the vector $x(\mu)$ defined as,

$$\begin{aligned} x_B(\mu) &= B^{-1}b - \mu B^{-1}a_j \\ x_j(\mu) &= \mu \\ x_k(\mu) &= 0, \forall k \in N \setminus I, k \neq j \end{aligned}$$

Since $B^{-1}a_j \leq 0$, we have $x_B(\mu) \geq 0$, $x_j(\mu) \geq 0$ for all $\mu \geq 0 \Rightarrow x(\mu) \geq 0$, for all $\mu \geq 0$.

$$\begin{aligned} Ax(\mu) &= Bx_B(\mu) + Rx_R(\mu) \\ &= B(B^{-1}b - \mu B^{-1}a_j) + Rx_R(\mu) \\ &= BB^{-1}b - \mu BB^{-1}a_j + \mu a_j \\ &= b - \mu a_j + \mu a_j \\ &= b \end{aligned}$$

$\therefore$ $x(\mu)$ is feasible for (MP), for all $\mu \geq 0$. Since $c_j - c_B^T B^{-1}a_j < 0$, we have

$$\begin{aligned} c^T x(\mu) &= \langle c_B, x_B(\mu)\rangle + \langle c_R, x_R(\mu)\rangle \\ &= \langle c_B, B^{-1}b - \mu B^{-1}a_j\rangle + \mu c_j \\ &= \langle c_B, B^{-1}b\rangle - \mu\langle c_B, B^{-1}a_j\rangle \\ &= z_B - \mu c_B^T B^{-1}a_j + \mu c_j \end{aligned}$$

Since $(c_j - c_B^T B^{-1}a_j) < 0$, we have $c^T x(\mu) \to -\infty$ as $\mu \to +\infty$.

Hence, (LP) is unbounded.
Our next step is to know that if a basis B is not optimal, and if (LP) is unbounded, how can we find another basis B' such that $z_{B'} \leq z_B$. The move from B to B' is done through what is called a **rank one update**.

Let $u, v \in \mathbb{R}^m$, then uv^T is an $m \times m$ matrix given by

$$uv^T = \begin{bmatrix} u_1v_1 & u_1v_2 & & u_1v_m \\ u_2v_1 & u_2v_2 & \cdots & u_2v_m \\ \vdots & \vdots & \ddots & \vdots \\ u_mv_1 & u_2v_m & \cdots & u_mv_m \end{bmatrix}.$$

Let B be $m \times m$ matrix which has an inverse then

$$(B + uv^T)^{-1} = B^{-1} - \frac{1}{1 + v^T B^{-1} u}(B^{-1}u)(v^T B^{-1})$$

provided $1 + v^T B^{-1} u \neq 0$. Here $B + uv^T$ is called an rank one update of B as uv^T has a rank at most one.
If B is not an optimal basis and (LP) is not unbounded then we must have

- $\bar{c_j} = c_j - c_B^T B^{-1} a_j < 0$, for some $j \in N \setminus I$
- $B^{-1} a_j \nleq 0$, for all $j \in N \setminus I$ for which it holds.

One of the key steps in the simplex method is that of pivoting. This means removing one column of B and replacing it with one column of R. How to do it?
So replace a_l of B with some column a_j of R
As $B^{-1} a_j \nleq 0$, there exists one component strictly positive. Of course there can be more than one. Now using this information how can we detect which column in B should we replace.
Let,

$$\theta = \frac{\bar{b}_r}{y_j^r} = \min\left\{\frac{\bar{b}_i}{y_j^i} : y_j^i > 0, i = 1, 2, \ldots, m\right\}.$$

In the column vector $B^{-1}a_j$, find the positive components and compute θ. The basis column that has to leave is the rth one, i.e., $a_{k_r} = a_l$ (say). This is a convention. But this helps us in achieving $z_{B'} \leq z_B$.
Consider the rank one update

$$B^{-1} = B + (a_j - a_l)u_r^T$$
$$\text{where} \quad u_r^T = (0 \ldots 0, 1 \ldots 0) \in \mathbb{R}^m.$$

$\downarrow$
(rth position)

This simply means that u_r is the rth unit vector in $\mathbb{R}^m$. Now a key thing we need to compute $z_{B'}$ is B', i.e. $(B + (a_j - a_l)u_r^T)^{-1}$.

Note that,

$$\begin{aligned}
1 + u_r^T B^{-1}(a_j - a_l) &= 1 + u_r^T B^{-1} a_j - u_r^T B^{-1} a_l \\
&= 1 + u_r^T B^{-1} a_j - u_r^T u_r \\
&= 1 + u_r^T B^{-1} a_j - 1 \\
&= u_r^T y_j \\
&= y_j^r > 0
\end{aligned}$$

Hence,

$$
\begin{aligned}
(B')^{-1} &= B^{-1} - \frac{1}{1+u_r^T B^{-1}(a_j - a_l)} B^{-1}(a_j - a_l)u_r^T B^{-1} \\
&= B^{-1} - \frac{1}{y_j^r} B^{-1}(a_j - a_l)u_r^T B^{-1} \\
&= B^{-1} - \frac{1}{y_j^r}[B^{-1}a_j - B^{-1}a_l][u_r^T B^{-1}] \\
&= B^{-1} - \frac{1}{y_j^r}[B^{-1}a_j - u_r][u_r^T B^{-1}]
\end{aligned}
$$

Is $(B')^{-1}$ truly a basis matrix? In order to check this we have to see if $(B^{'})^{-1}$ $b \geq 0$.

$$
\begin{aligned}
(B')^{-1}b &= B^{-1}b - \frac{1}{y_j^r}(y_j - y_r)(u_r^T B^{-1}b) \\
&= B^{-1}b - \frac{1}{y_j^r}(y_j - y_r)(u_r^T \bar{b}) \\
&= B^{-1}b - \frac{1}{y_j^r}\bar{b}_r(y_j - y_r) \\
&= \begin{bmatrix} \bar{b}_1 \\ \vdots \\ \bar{b}_r \\ \vdots \\ \bar{b}_m \end{bmatrix} + \begin{bmatrix} -\frac{y_j^1}{y_j^r}\bar{b}_r \\ \vdots \\ -\bar{b}_r \\ \vdots \\ -\frac{y_j^m}{y_j^r}\bar{b}_r \end{bmatrix} + \begin{bmatrix} 0 \\ \vdots \\ \bar{b}_r \\ \vdots \\ 0 \end{bmatrix} = \begin{bmatrix} \bar{b}_1 - \theta y_j^1 \\ \vdots \\ \bar{b}_r \\ \vdots \\ \bar{b}_m - \theta y_j^m \end{bmatrix}, \text{ where } \theta = \frac{\bar{b}_r}{y_j^r}.
\end{aligned}
$$

By definition of θ, we have $(B^{'})^{-1}b \geq 0$. Hence, $B^{'}$ is a feasible basis. So this shows the importance of pivoting.

$$
\begin{aligned}
\text{Define } x_{B'} &= (B')^{-1}b \\
&= B^{-1}b - \frac{1}{y_j^r}\bar{b}_r(y_j - y_r) \\
&= B^{-1}b - \theta(y_j - y_r)
\end{aligned}
$$

$$
x_k = 0, \quad \forall\, k \in N \setminus I^{'}, (I^{'} = I_{B'})
$$

Now update c_B as follows

$$c_{B'} = c_B + (c_j - c_l)u_r$$

$$\begin{aligned}
z_{B'} &= \langle c_{B'}, x_{B'} \rangle \\
&= \langle c_B, B^{-1}b - \theta(y_j - y_r) \rangle + \langle (c_j - c_l)u_r^T, B^{-1}b - \theta(y_j - y_r) \rangle \\
&= z_B + -\theta \langle c_B, y_j - y_r \rangle \\
&= z_B - \theta c_B^T B^{-1} a_j + \theta c_l + (c_j - c_l)(\bar{b}_r - \theta(y_j^r - 1)) \\
&= z_B - \theta(c_j - c_B^T B^{-1} a_j) \\
&\le z_B \quad \left[\text{since } (c_j - c_B^T B^{-1} a_j) < 0 \right]
\end{aligned}$$

Computing B^{-1} is difficult for large scale problems. So it is better to start with an identity matrix as the starting basis matrix. This is done by introducing m artificial variables $x_{n+1}, x_{n+2}, \ldots, x_{n+m}$, all non-negative, i.e., (LP) would look like

$$\min c_1 x_1 + \cdots + c_n x_n + 0x_{n+1} + \cdots + 0x_{n+m}$$

subject to

$$\begin{aligned}
a_{11}x_1 + \cdots + a_{1n}x_n + x_{n+1} + 0 + \cdots + 0 \quad &= b_1 \\
\vdots \qquad\qquad\qquad\qquad &\quad \vdots \\
a_{m1}x_1 + \cdots + a_{mn}x_n + 0 + 0 + \cdots + x_{n+m} \quad &= b_m
\end{aligned}$$

Starting basis $= I$
If $B = I$, then

$$B' = \begin{bmatrix} 1 & \cdots a_j^1 & 0 \\ \vdots & \vdots & \vdots \\ 0 & \cdots a_j^m & 1 \end{bmatrix}.$$

So the lth column is replaced with the jth column
In fact if $B = I$, then

$$(B^{'})^{-1} = \begin{bmatrix} 1 & \cdots - \dfrac{y_j^1}{y_j^r} & \cdots 0 \\ \vdots & \vdots & \vdots \\ 0 & \cdots \dfrac{1}{y_j^r} & \cdots 0 \\ \vdots & \vdots & \vdots \\ 0 & \cdots - \dfrac{y_j^m}{y_j^r} & \cdots 1 \end{bmatrix}.$$

The vector leaving the basis has these new entries.

The main Simplex algorithm: (Padberg [17]) (Key data: m, n, A, b, c)

Step 0: Find a basis B, which is feasible. If none exists, declare (LP) infeasible. Else, compute $B^{-1}, \bar{b} = B^{-1}b$, and initialize C_B.

Step 1: Compute $\bar{c} - c - (c_B^T B^{-1} A)^T$ (reduced cost vector). Note that $c_B^T B^{-1} A$ is a row vector. If $\bar{c} \geq 0$, then

- Set $x_B = \bar{b}, x_R = 0$,
- Stop and declare these as optimal BFS. Else, choose $j \in \{k \in N : \bar{c_k} < 0\}$.

Step 2: Compute $y_j = B^{-1}a_j$. If $y_j \leq 0$, STOP and declare (LP) to be unbounded. Else,

- compute $\theta = \min\left\{\dfrac{\bar{b}_i}{y_j^i} : y_j^i > 0, 1 \leq i \leq m\right\}$,
- choose $l \in I$ such that $\dfrac{\bar{b}_{p_l}}{y_j^{p_l}} = \theta, 1 \leq p_l \leq m$,
- set $p_l = r$.

Step 3: Set

- $B := B + (a_j - a_l)u_r^T$,
- $c_B := c_B + (c_j - c_l)u_r$,
- $I := (I \setminus (\{l\}) \cup \{j\})$,
- $p_j = r$.

Step 4: Compute $B^{-1}, \bar{b} = B^{-1}b$, and go to Step 1.

Example. We provide an example to illustrate how the tableau-free method can be used. We consider a linear programming problem (LP), i.e.,

$$\min\langle c, x\rangle, \quad \text{subject to} \quad Ax = b, x \geq 0.$$

In our case there will be five decision variables and three equality constraints. We have

$$c = [-2, -3, 0, 0, 0]^T$$

$$A = \begin{bmatrix} -1 & 1 & 1 & 0 & 0 \\ 1 & -1 & 0 & 1 & 0 \\ 1 & 1 & 0 & 0 & 1 \end{bmatrix}$$

and $b = [3, 2, 7]^T$. We shall denote the columns of the matrix A as a_i, $i = 1, \ldots, 5$. Observe here that the columns $\{a_3, a_4, a_5\}$ can be considered as a starting basis which is the identity matrix. Here $c_B = [0, 0, 0]^T$. Thus the reduced cost vector $\overline{c}$ is given as

$$\overline{c} = [-2, -3, 0, 0, 0]^T.$$

Thus, our current basis is not optimal. Our starting basic feasible solution include variables x_3, x_4, x_5 and its value is given as

$$x_B = [3, 2, 7]^T.$$

Let us choose from the components of $\overline{c}$ the one which is most negative, i.e. -3, and thus compute $y_2 = B^{-1}a_2$. Thus, $y_2 = [1, -2, 1]^T$. Hence, a_2 will now enter the basis. Our next step is to compute θ and determine the vector leaving the basis. We have

$$\theta = \left\{\frac{3}{1}, -, \frac{7}{1}\right\}.$$

Thus, the first column of the basis matrix has to be replaced by a_2, and column a_3 will leave the basis. So the new basis, which we also mark as B, should look like

$$B_1 = \begin{bmatrix} 1 & 0 & 0 \\ -2 & 1 & 0 \\ 1 & 0 & 1 \end{bmatrix}.$$

What we have done is just placed column a_2 of A as the first column of the basis matrix where originally we had a_3. This can also be checked using the formula

$$B := B + (a_2 - a_3)u_1^T,$$

where $u_1 = [1, 0, 0]^T$. Further, the updated version of c_B, which we again mark by the same symbol, is

$$c_B := c_B + (c_2 - c_3)u_1.$$

This gives us $c_B = [-3, 0, 0]^T$. In the current case the inverse of the basis matrix is given as

$$B^{-1} = \begin{bmatrix} 1 & 0 & 0 \\ 2 & 1 & 0 \\ -1 & 0 & 1 \end{bmatrix}.$$

And in this basis, $\bar{b} = B^{-1}b$ is given as $\bar{b} = [3, 8, 4]^T$. Now the reduced cost coefficient in this case can be computed as

$$\bar{c} = [-5, 0, 3, 0, 0]^T.$$

We shall thus have to compute y_1 since a_1 will now enter the basis. So with this new basis we have

$$y_1 = B^{-1}a_1 = [-1, -1, 2]^T.$$

It is clear that $\theta = 2$ and the third column of the basis matrix leaves the basis matrix. Thus, a_5 will be replaced by a_1. The updated basis matrix, is given as

$$B = \begin{bmatrix} 1 & 0 & -1 \\ -2 & 1 & 1 \\ 1 & 0 & 1 \end{bmatrix}.$$

Now, we have

$$B^{-1} = \begin{bmatrix} \frac{1}{2} & 0 & \frac{1}{2} \\ \frac{3}{2} & 1 & \frac{1}{2} \\ \frac{-1}{2} & 0 & \frac{1}{2} \end{bmatrix}.$$

We leave it to the reader to compute that

$$c_B = [-3, 0, 2]^T$$

and

$$c_B^T B^{-1} A = \left[-2, -3, \frac{-1}{2}, 0, \frac{-5}{2}\right].$$

Thus, we have

$$\bar{c} = \left[0, 0, \frac{1}{2}, 0, \frac{5}{2}\right]^T.$$

Since $\bar{c} \geq 0$, the current basis is optimal, and the basic variables are x_1, x_2, and x_4. We hope the reader will figure out that the optimal solution is given as

$$x^* = [2, 5, 0, 10, 0].$$

Note that in order to arrive at this solution the reader must compute $B^{-1}b$ which will give $\bar{b} = x_B$.

2.5 KKT Conditions for Linear Programming

There are several ways one may approach the Karush-Kuhn-Tucker necessary and sufficient optimality conditions for linear programming. Unfortunately, the usual texts on Optimization do not focus on the important fact that the KKT conditions always hold for linear programming problem. To begin with, let us briefly state the KKT conditions for a general mathematical programming problem with smooth data. Consider the following problem (MP) given as

$$
\begin{aligned}
\min \quad & f(x) \\
\text{subject to} \quad & \\
& g_i(x) \leq 0 \qquad i = 1, \ldots, m \\
& h_j(x) = 0 \qquad j = 1, \ldots, k
\end{aligned}
$$

where $f : \mathbb{R}^n \to \mathbb{R}$, $g_i : \mathbb{R}^n \to \mathbb{R}$ for all $i = 1, \ldots, m$ and $h_j : \mathbb{R}^n \to \mathbb{R}$, for all $j = 1, \ldots, k$. We begin by assuming that all functions in the problem (MP) are differentiable. We have already learned about the KKT conditions for a convex programming problem with inequality constraints under the important assumption that the Slater condition or Slater constraint holds. The very basic optimality conditions for the problem (MP) where no assumptions other than continuous differentiability are needed are due to Fritz John, who in 1948 developed such necessary conditions for the problem (MP) with only inequality constraints. However, the full power of the Fritz John condition was revealed by Mangasarian and Fromovitz [14], who combined the method of Fritz John with the implicit function theorem in order to establish the following result.

Theorem 2.5.1 *Consider the problem (MP) and let x^* be a local minimizer. Then there exist real numbers $\lambda_0 \geq 0, \lambda_1 \geq 0, \ldots, \lambda_m \geq 0$ and $\mu_1, \ldots, \mu_n$ such that*

(i) $\lambda_0 \nabla f(x^*) + \sum_{i=1}^{m} \lambda_i \nabla g_i(x^*) + \sum_{i=1}^{k} \mu_j \nabla h_j(x^*) = 0$
(ii) $\lambda_i g_i(x^*) = 0$ *for all* $i = 1, \ldots, m$
(iii) $(\lambda_0, \lambda, \mu) \neq 0$

where $\lambda = (\lambda_1, \ldots, \lambda_m)$ *and* $\mu = (\mu_1, \ldots, \mu_k)$.

The vector $(\lambda_0, \lambda, \mu) \in \mathbb{R}_+ \times \mathbb{R}_+^m \times \mathbb{R}^k$ which satisfies the conditions of Theorem 2.5.1 are called **Fritz John Multipliers** of (MP) at the local minimizer x^*.

The case when $\lambda_0 > 0$ is crucial, and in such a case we can always consider $\lambda_0 = 1$ without loss of generality. When $\lambda_0 = 1$, then we say that the KKT conditions holds, and (λ, μ) is called the **KKT multiplier vector** and the individual components of

λ and μ are called *KKT multipliers*. It is simple to note that the problem (LP) is a special case of the problem (MP). Our aim in this section would be to prove that the linear programming problem (LP) has at least one KKT multiplier. The key idea is that in order to have a KKT multiplier we need not impose additional assumptions on the constraints of (LP). In order to derive the optimality conditions for (LP) we shall begin with deriving the optimality condition for the problem (LP2), which is given as

$$\min\langle c, x\rangle, \quad \text{subject to} \quad Ax \leq b,$$

where as before A is an $m \times n$ matrix and $b \in \mathbb{R}^m$. Our approach will closely follow along the lines of Pshenichny [18], where the author had developed the necessary and sufficient condition for minimization problem. The objective is a convex function and the constraints are same as that of (LP2). The approach of Pshenichny [18] is indeed so elegant that I succumb to the temptation of presenting it in this note tailoring it to the needs of (LP2).

Pshenichny [18] begins with the idea of a **cone of feasible directions**. Consider a convex subset C of $\mathbb{R}^n$. Then given $x_0 \in C$ the cone of feasible directions is given by

$$K_C(x_0) = \{d \in \mathbb{R}^n : d = \lambda(x - x_0), \lambda > 0, x \in C\}.$$

Simply put, $K_C(x_0)$ is the convex cone generated by the set $C - x_0$. However, let us first show that the set $K_C(x_0)$ is indeed a convex cone. The fact that it is a cone is simple to see and we leave it to the reader to verify it. Also observe that $0 \in K_C(x_0)$ and further note that if $d \in K_C(x_0)$, then $x_0 + \alpha d \in C$ if $\alpha \leq \frac{1}{\lambda}$, with $\alpha > 0$. Of course, here the λ comes from writing $d = \lambda(x - x_0)$. Observe that $x_0 + \alpha d = x + \alpha\lambda(x - x_0)$. If $\alpha \leq \frac{1}{\lambda}$ then $0 < \alpha\lambda \leq 1$. Hence, $x_0 + \alpha\lambda(x - x_0) \in C$ since C is convex and that shows that $x_0 + \alpha d \in C$.

We shall now introduce the notion of the dual cone of a given cone K in $\mathbb{R}^n$. The dual cone is given as

$$K^* = \{w \in \mathbb{R}^n : \langle w, v\rangle \geq 0, \forall v \in \mathbb{K}\}.$$

Using the separation theorem (Theorem 2.2.2) discussed in Sect. 2.2, we can show that if K is closed then, $K^{**} = K$, where K^{**} is the bi-dual cone which simply means the dual cone of K^*. Let us now begin by trying to derive a necessary and sufficient condition for minimizing the linear function $f(x) = \langle c, x\rangle, (c \in \mathbb{R}^n)$, over the convex set S in $\mathbb{R}^n$. We present below Theorem 1.2 from Pshenichny [18], but specialized to the case of linear objective.

Theorem 2.5.2 *A necessary and sufficient condition for a point x_0 to be a minimizer of $f(x) = \langle c, x\rangle$ over the convex set S is*

$$c \in K_S^*(x_0).$$

Proof Let x_0 be a minimizer of the linear function $\langle c, x \rangle$ over the convex set S. Then for any $x \in S$, we have

$$\langle c, (x - x_0) \rangle \geq 0.$$

For any $\lambda > 0$ we can show that

$$\langle c, \lambda(x - x_0) \rangle \geq 0$$

for any $x \in C$. This shows that

$$\langle c, d \rangle \geq 0,$$

for all $d \in K_S(x_0)$ This proves that $c \in K_S^*(x_0)$. For the converse, observe that $c \in K_S(x_0)$, shows that $\langle c, d \rangle \geq 0$ for all $d \in K_S(x_0)$. Thus, we have for any $x \in S$ and $\lambda > 0$

$$\langle c, \lambda(x - x_0) \rangle \geq 0.$$

This shows that $\langle c, x \rangle \geq \langle c, x_0 \rangle$ for all $x \in S$. Hence, proves the result. □

Let us consider

$$S = \{x \in \mathbb{R}^n : Ax \leq b\}.$$

If a_i, with $i = 1, \ldots, m$ denotes the rows of the matrix A then we can write S as

$$S = \{x \in \mathbb{R}^n : \langle a_i, x \rangle \leq b_i, i = 1, \ldots, m\}.$$

Given any $x_0 \in S$ as before let $I(x_0)$ denote the set of active constraints at x_0. Now we claim that

$$K_S(x_0) = \{d \in \mathbb{R}^n : \langle a_i, d \rangle \leq 0, i \in I(x_0)\}. \tag{2.13}$$

Let $d \in K_S(x_0)$, then $x_0 + \alpha d \in S$ for $\alpha > 0$ and sufficiently small. Thus, for $i \in I(x_0)$ we have

$$\langle a_i, x_0 + \alpha d \rangle \leq b_i.$$

Thus, we have

$$\langle a_i, x \rangle + \alpha \langle a_i, d \rangle \leq b_i.$$

Since $i \in I(x_0)$, we have $\langle a_i, x_0 \rangle = b_i$. Thus, from the above inequality we conclude that

$$\langle a_i, d \rangle \leq 0, \quad \forall i \in I(x_0).$$

This shows that

$$K_S(x_0) \subseteq \{d \in \mathbb{R}^n : \langle a_i, d \rangle \leq 0, i \in I(x_0)\}.$$

For proving the reverse inclusion, consider any d belonging to the set

$$\{d \in \mathbb{R}^n : \langle a_i, d \rangle \leq 0, i \in I(x_0)\}.$$

Consider the point $x_0 + \lambda d$ with $\alpha > 0$ and sufficiently small. Now for any $i \in \{1, 2, \ldots, m\}$

$$\langle a_i, x_0 + \alpha d \rangle = \langle a_i, x_0 \rangle + \alpha \langle a_i, d \rangle.$$

If $i \in I(x_0)$, we have $\langle a_i, x_0 \rangle = b_i$ and $\langle a_i, d \rangle \leq 0$. This shows that

$$\langle a_i, x_0 + \alpha d \rangle \leq b_i$$

If $i \notin I(x_0)$, then $\langle a_i, x_0 \rangle < b_i$. Thus, $\langle a_i, x_0 \rangle - b_i < 0$, then we can choose $\alpha > 0$ sufficiently small such that

$$\langle a_i, x_0 \rangle - b_i + \alpha \langle a_i, d \rangle < 0.$$

Hence, we can conclude that

$$\langle a_i, x_0 + \alpha d \rangle < b_i$$

for $\alpha > 0$ and sufficiently small. This shows that $d \in K_S(x_0)$, and (2.13) holds.

Our next job is to compute $K_S^*(x_0)$, so that we can directly apply Theorem 2.5.2 and deduce the optimality conditions. We shall show that

$$K_S^*(x_0) = \left\{ - \sum_{i \in I(x_0)} \lambda_i a_i : \lambda_i \geq 0 \quad \forall i \in I(x_0) \right\}. \tag{2.14}$$

Let us set the right hand side of the above equality as Q, i.e.

$$Q = \left\{ - \sum_{i \in I(x_0)} \lambda_i a_i : \lambda_i \geq 0 \quad \forall i \in I(x_0) \right\}.$$

Now, for any $d \in K_S(x_0)$ observe that

$$\left\langle -\sum_{i\in I(x_0)} \lambda_i a_i, d \right\rangle = -\sum_{i\in I(x_0)} \lambda_i \langle a_i, d\rangle.$$

Further, since $d \in K_S(x_0)$, we have $\langle a_i, d\rangle \leq 0$ for all $i \in I(x_0)$. Hence,

$$\left\langle -\sum_{i\in I(x_0)} \lambda_i a_i, d \right\rangle \leq 0.$$

This shows that

$$-\sum_{i\in I(x_0)} \lambda_i a_i \in K_S^*(x_0).$$

Also, $Q \subset K_S^*(x_0)$. We shall now show the opposite inclusion, i.e., $K_S^*(x_0) \subset Q$. Assume on the contrary that this is not true. Thus, there exists $v \in K_S^*(x_0)$, but $v \notin Q$. The cone Q is finitely generated, and hence closed. Thus, using the separation arguments, we conclude that there exists $0 \neq w \in \mathbb{R}^n$ and $\alpha \in \mathbb{R}$ such that for any $q \in Q$ we have

$$\langle w, v\rangle < \alpha < \langle w, q\rangle.$$

Since $0 \in Q$, we conclude that $\alpha < 0$, and hence $\langle w, v\rangle < 0$. We claim that $\langle w, q\rangle \geq 0$ for all $q \in Q$. On the contrary, let us assume that there exists $\hat{q} \in Q$ such that $\langle w, \hat{q}\rangle < 0$. Since Q is a cone, for any $\mu > 0$ we have $\mu q \in Q$. Thus, if $\mu > 0$ the above inequality leads us to conclude that

$$\langle w, v\rangle < \alpha < \langle w, \mu\hat{q}\rangle. \tag{2.15}$$

But as $\mu \to \infty$ we have $\langle w, \mu\hat{q}\rangle \to -\infty$, and hence we can find $\mu_0 > 0$, such that for all $\mu > \mu_0$,

$$\langle w, \mu\hat{q}\rangle \leq 0.$$

This contradicts (2.15). Thus, we have $\langle w, q\rangle \geq 0$ for all $q \in Q$. Hence, we have

$$-\sum_{i\in I(x_0)} \lambda_i \langle a_i, w\rangle \geq 0,$$

for any $\lambda_i \geq 0$, for each $i \in I(x_0)$. This shows that $\langle a_i, w\rangle \leq 0$ for all $i \in I(x_0)$. Thus, $w \in K_S(x_0)$. Since $v \in K_S^*(x_0)$, we have $\langle w, v\rangle \geq 0$. This contradicts the fact that $\langle w, v\rangle < 0$. Hence, the result, i.e. (2.14) holds. □

This will allow us to write the following theorem.

Theorem 2.5.3 *Let x_0 be a minimizer of (LP2). Then there exists $\lambda \in \mathbb{R}^m_+$ such that*

(i) $c + A^T\lambda = 0$
(ii) $\langle \lambda, Ax_0 - b\rangle = 0.$

Conversely if x_0 is feasible for (LP2) and the conditions (i) and (ii) as given above hold, then x_0 is a minimizer of (LP2).

Proof Since x_0 is a minimizer, we have from Theorem 2.5.2 that $c \in K_S^*(x_0)$. Hence, there exists $\lambda_i \geq 0$, $i \in I(x_0)$ such that

$$c = -\sum_{i \in I(x_0)} \lambda_i a_i.$$

Further, if we set $\lambda_i = 0$ for $i \notin I(x_0)$, then we can rewrite the above conditions as

(i) $c + \lambda_{i=1}^m a_i = 0$
(ii) $\lambda_i(\langle a_i, x_0\rangle - b) = 0.$

The result is established by noting that $a_i's$ are the rows of the matrix A in (LP2). We leave the converse to the reader. □

Let us now consider the standard linear programming problem (LP), i.e.

$$\min\langle c, x\rangle, \quad \text{subject to} \quad Ax = b, x \geq 0.$$

This can be re-written equivalently as

$$\begin{aligned} &\min\langle c, x\rangle, \\ &\text{subject to} \\ &\qquad Ax \leq b \\ &\qquad -Ax \leq -b \\ &\qquad x \geq 0. \end{aligned}$$

It can be further simplified as

$$\begin{aligned} &\min\langle c, x\rangle, \\ &\text{subject to} \\ &\qquad \langle a_i, x\rangle \leq b_i, \quad i = 1, \dots, m \\ &\qquad -\langle a_i, x\rangle \leq -b_i, \quad i = 1, \dots, m \\ &\qquad -\langle e_j, x\rangle \leq 0, \quad j = 1, \dots n, \end{aligned}$$

where $e_j \in \mathbb{R}^n$ is the column vector whose jth component is 1 and rest $(n-1)$ components are zero. Let x_0 be a solution of (LP). Then, using Theorem 2.5.2, we see that there exist scalars $\lambda_i \geq 0, i = 1, \ldots, m$, $\mu_i \geq 0, i = 1, \ldots, m$ and $s_j \in \mathbb{R}$, $j = 1, \ldots, n$ such that

(i) $c + \sum_{i=1}^{m} \lambda_i a_i - \sum_{i=1}^{m} \mu_i a_i - \sum_{j=1}^{n} s_j e_j = 0$
(ii) $\lambda_i(\langle a_i, x\rangle - b_i) = 0 \quad \forall i = 1, \ldots, m$
(iii) $\mu_i(\langle -a_i, x\rangle + b_i) = 0 \quad \forall i = 1, \ldots, m$
(iv) $s_j x_j = 0 \quad \forall j = 1, \ldots, n.$

Observe that (i) can be written as

$$c + \sum_{i=1}^{m} (\lambda_i - \mu_i) a_i = s,$$

where $s = (s_1, s_2, \ldots, s_n)^T$. Setting $y_i = \mu_i - \lambda_i$, we conclude from the above equality

$$c = s + \sum_{i=1}^{m} y_i a_i.$$

Putting this in matrix form shows that $c = s + A^T y$, where $y = (y_1, \ldots y_m)^T$. Thus, we can compactly put the above discussion in the form of the following theorem. Observe that, once we use the vector y which is unsigned, conditions (ii) and (iii) are no longer required as they are automatically satisfied.

Theorem 2.5.4 *Let us consider the problem (LP). Then $x \in \mathbb{R}^n$ is a global minimizer of (LP) if and only if there exists $y \in \mathbb{R}^m$ such that*

(i) $A^T y + s = c$
(ii) $Ax = b$
(iii) $\langle x, s\rangle = 0$
(iv) $x \geq 0, s \geq 0.$

For proving the converse, we can show using the above conditions that

$$\langle x, s\rangle = \langle b, y\rangle - \langle c, x\rangle.$$

The condition (iii) shows that $\langle b, y\rangle = \langle c, x\rangle$. From Sect. 2.3, we know that this leads to the fact—x solves (LP) and y solves (DP). The KKT-multiplier associated with (LP) is actually the dual solution. The KKT conditions will play a key role in studying the interior point method in the next section. Before we close this section, we shall demonstrate another approach to obtain the KKT conditions for the problem (LP) by directly applying the generalized version of the KKT conditions discussed in Sect. 2.2. Note that the problem (LP) can be written down as the generalized form of the problem (CP) with $g_i(x) = -x_i$ and $X = \{x \in \mathbb{R}^n : Ax = b\}$. Now, let us write down the feasible set of (LP) as follows

$$S = \{x \in \mathbb{R}^n : Ax = b, x \geq 0\}.$$

Its *relative interior* is given by

$$S^\circ = \{x \in \mathbb{R}^n : Ax = b, x > 0\}.$$

Readers who have not made a detailed study of convex analysis, the word relative interior may not be familiar to them. We will refer the reader to the classic book on *Convex Analysis* by Rockafellar [22]. However, to provide a crude idea of what a relative interior point means, let us begin by imagining a line on the two-dimensional plane. Consider any point on the line. If we draw an open circular disk (i.e., an open ball) with that point as the center, then such a disk cannot be contained in the line itself. However, the intersection of the open disk with the line is contained naturally in the line itself. Thus, the point is interior relative to the line and not to the two-dimensional plane. Further, it has been proven, for example, in Rockafellar [22], that every convex set in finite dimensions have a relative interior. Thus, in the case of (LP), if its feasible set S is non-empty, then so is its relative interior S°. This shows that the generalized Slater condition holds true. Hence, using the generalized version of the KKT conditions, we can show that if x is a solution of (LP), there exist scalars $s_i \geq 0$ such that

(i) $0 \in c - \sum_{i=1}^{n} s_i e_i + N_X(x)$,
(ii) $x_i s_i = 0, i = 1, \ldots, n$,

where e_i, $i = 1, \ldots, n$, is the ith unit vector.

We leave it to the reader to show that

$$N_X(x) = \mathrm{Im} A^T,$$

where Im denotes the image of a mapping. Thus, there exists $y \in \mathbb{R}^m$ such that

(i) $c - \sum_{i=1}^{n} s_i e_i + A^T y = 0$,
(ii) $x_i s_i = 0, i = 1, \ldots, n$.

The reader can easily show that the above two conditions, along with the feasibility conditions, actually provide the KKT conditions for (LP), as we have derived before in this section. This approach using the relative interior is only applicable in finite dimensions since the relative interior of convex sets need not always be non-empty in infinite dimensions. But the first approach that has been shown in this section is applicable even in infinite dimensions.

2.6 Interior-Point Methods

In contrast to the simplex method, which seeks to decrease the objective value by hopping from one vertex of the feasible polyhedron to the next, the interior point method in linear programming attempts to remain within the feasible polyhedron and decrease the value of a parameter known as the duality measure. Putting it more simply, the interior point methods use non-linear programming techniques to solve linear-programming problems. Though the seminal work of Karmarkar [12] in 1984 initiated the interior point revolution, its true impact was felt in the optimization community when it was realized that an interior point method for linear programming could be attained by employing the barrier function technique from non-linear programming. The barrier function technique at each step solves an auxiliary problem, where an additional function called the barrier function is added to the objective function, which is also multiplied by a barrier parameter. The barrier function has the property that its value tends to become very large as the decision variable approaches the boundary of the feasible region. The barrier function approach in non-linear programming was popularized through the celebrated monograph of Fiacco and Mc Cormick [8]. The interior point methods gained importance because theoretically, it could be demonstrated that they were all polynomial time algorithms, in contrast to the simplex method, whose worst-case complexity was shown to be exponential by Klee and Minty [13].

Let us now begin our discussion on the barrier function approach for linear programming. Given the problem (LP), for a given *barrier parameter* $\mu > 0$, the barrier problem (BP_μ) is given as

$$\min f_\mu(x) = \langle c, x\rangle - \mu \sum_{i=1}^{n} \ln x_i \quad \text{subject to} \quad Ax = b, x > 0.$$

There are two ways to approach the problem. The first is to try to find a minimizer of the above problem using Newton's method, while the other approach is to write down the KKT conditions for the above convex optimization problem (BP_μ) and solve that KKT system of equations using Newton's method for solving a system of equations. The first approach has been beautifully described in the survey by Margaret Wright [23] and also in Dantzig and Thapa [7]. The second approach has been championed by Stephen Wright [24]. We shall discuss the first approach from Margaret Wright [23] and the second approach from Stephen Wright [24].

Let us begin with the first approach. We shall develop Newton's method for minimizing (BP_μ). Observe that $x > 0$ represents inactive constraints, and hence the multipliers associated with them are zero. Thus, it is enough to solve the problem of minimizing $f_\mu(x)$ subject to the constraints $Ax = b$. Suppose we are at the kth iteration x_k. Then we shall consider the quadratic approximation of f_μ around x_k. This is given by the function

$$\phi(p) = f_\mu(x_k) + \langle \nabla f(x_k), p\rangle + \frac{1}{2}\langle p, \nabla^2 f(x_k)p\rangle.$$

Newton's method is obtained by simply writing down the first order optimality condition for the quadratic approximation. Thus, if p_k is the unconstrained minimizer of $\phi(p)$ then the standard optimality conditions tell us

$$\nabla^2 f_\mu(x) p_k + \nabla f(x_k) = 0.$$

This provides the Newton step given as

$$\nabla^2 f(x_k) p_k = -\nabla f(x_k).$$

The vector p_k is called the Newton direction or even Newton step, and the next iterate x_{k+1} is given as $x_{k+1} = x_k + p_k$. But in our particular case, we want to also have $Ax_{k+1} = b$, and thus we see that p_k must also satisfy,

$$Ap_k = b - Ax_k.$$

In effect, for the problem (QP_μ), given as,

$$\min \phi(p) \quad \text{subject to} \quad Ap = b - Ax_k,$$

the vector p_k is indeed a minimizer. The KKT conditions automatically hold as the constraints are given by affine equalities. Thus, there exists a vector y_k such that

(i) $\nabla^2 f_\mu(x_k) p_k + \nabla f_\mu(x_k) - A^T y_k = 0,$
(ii) $Ap_k = b - Ax_k.$

This is the Newton step for (QP_μ). Thus, in the matrix form we have

$$\begin{bmatrix} \nabla^2 f_\mu(x_k) & A^T \\ A & 0 \end{bmatrix} \begin{bmatrix} p_k \\ -y_k \end{bmatrix} = \begin{bmatrix} -\nabla f_\mu(x_k) \\ b - Ax_k \end{bmatrix}.$$

Let us note that

$$\nabla f_\mu(x) = c - \mu X^{-1} e,$$

where $X = \text{diag}(x)$, which means that X is a diagonal matrix whose diagonal entries are the components of the vector x, entered with the first component of x as the first entry in the diagonal, the second component as the second entry in the diagonal, and so on, and e is the vector in $\mathbb{R}^n$ whose all components are 1. Further, we also have,

$$\nabla^2 f(x) = \mu X^{-2}.$$

If we have $x_k > 0$ and $Ax_k = b$, the above Newton scheme becomes

$$\begin{bmatrix} \mu X_k^{-2} & A^T \\ A & 0 \end{bmatrix} \begin{bmatrix} p_k \\ -y_k \end{bmatrix} = \begin{bmatrix} -\nabla f_\mu(x_k) \\ 0 \end{bmatrix}.$$

Wright [23] provides an easy way, at least theoretically, to compute the Newton steps p_k and y_k. Explicitly writing down the first matrix equation we get,

$$\mu X_k^{-2} p_k - A^T y_k = -c + \mu X_k^{-1} e. \tag{2.16}$$

This can be rewritten as

$$c - \mu X_k^{-1} e + \mu X_k^{-2} p_k = A^T y_k.$$

Multiplying both sides of the above equation by Ax_k^2 and noting that $Ap_k = 0$ (since x_k is feasible), we have

$$AX_k^2 A^T y_k = AX_k(X_k c - \mu e).$$

Once y_k is known, we can use (2.16) to show that

$$p_k = X_k e + \frac{1}{\mu} X_k^2 (A^T y - c).$$

Then, we can take a pure Newton step and write

$$x_{k+1} = x_k + p_k.$$

In fact, the idea is to consider the sequential problems (BP_{μ_k}), where $\mu_k > 0$ and $\mu_k \to 0$. The sequence $\{x_k\}$ thus obtained can be viewed as a discrete version of the central path, the path along which one needs to move in the interior and toward the solution. However, implementing this method may not always be so simple if the matrix A has a high dimension or the data is huge. Nevertheless, from a theoretical point of view, the above approach is illuminating and shows the power of the KKT conditions.

Our next approach is to use Newton's method for solving a system of equations, and we would like to solve the system of KKT equations in order to compute the optimum. Let us begin by providing a brief overview of Newton's method. Consider the problem $F(x) = 0$, where $F : \mathbb{R}^n \to \mathbb{R}^n$ is a differentiable function. The key idea of Newton's method is as follows. Let x_0 be the initial point. If $F(x_0) = 0$, then we are done. Otherwise, we need to find a point x_1 that might be a solution. How to move to the new point x_1? In doing so, we consider the linear approximation of F around x_0, which is given by

$$L(x) = F(x_0) + JF(x_0)(x - x_0),$$

where $JF(x_0)$ is the Jacobian matrix of F at x_0. Let x_1 be a point such that $L(x_1) = 0$. Then, we have

$$JF(x_0)(x_1 - x_0) = -F(x_0).$$

This is the required Newton step, and if we start from the kth iteration, we shall get the Newton step as

$$JF(x_k)(x_{k+1} - x_k) = -F(x_k).$$

If we set $\Delta x_k = x_{k+1} - x_k$, then we call it the Newton step, and the Newton iteration can be compactly written as

$$JF(x_k)\Delta x_k = -F(x_k). \tag{2.17}$$

The primal-dual interior point methods that are so elegantly described in the celebrated book by Wright [24] are based on the above approach to solve the KKT conditions for linear programming problems. This might not be as easy as it sounds. Let us illustrate the issues here. Consider the problem (LP) and its dual problem (DP), which we recollect here again for the convenience of the reader. The primal problem (LP) is given as follows

$$\min\langle c, x\rangle, \quad \text{subject to} \quad Ax = b, \quad x \geq 0,$$

while the dual problem (DP) is given as

$$\max\langle b, y\rangle \quad \text{subject to} \quad A^T y + s = c, \quad s \geq 0.$$

Let us now introduce the primal-dual feasible set and its relative interior as given in Wright [24]. The primal dual feasible set $\mathcal{F}$ is given as

$$\mathcal{F} = \{(x, y, s) \in \mathbb{R}^n \times \mathbb{R}^m \times \mathbb{R}^n : A^T y + s = c, Ax = b, x \geq 0, s \geq 0\},$$

and its relative interior is given as

$$\mathcal{F}^\circ = \{(x, y, s) \in \mathbb{R}^n \times \mathbb{R}^m \times \mathbb{R}^n : A^T y + s = c, Ax = b, x > 0, s > 0\}.$$

It now goes without saying that if $\mathcal{F}$ is non-empty, then so is $\mathcal{F}^\circ$. Further, solving the KKT conditions for the problem (LP) involve solving the following equations for the vector (x, y, s),

$$\begin{aligned} A^T y + s &= c \\ Ax &= b \\ \langle x, s\rangle &= 0 \\ x &\geq 0 \\ s &\geq 0. \end{aligned}$$

If $(x, y, s) \in \mathcal{F}$, then except the equation $\langle x, s\rangle = 0$, everything else is automatically satisfied. In fact, one of the ways to solve the KKT system is to solve a system of equations with the following pattern.

$$
\begin{aligned}
A^T y + s &= c \\
Ax &= b \\
x_i s_i &= \mu, \quad \forall i = 1, \ldots, n \\
x &> 0 \\
s &> 0.
\end{aligned}
$$

The above system is directly linked with (BP_μ), as it follows from the KKT system of that problem. The KKT conditions associated with (BP_μ) tell us the existence of the vector $y \in \mathbb{R}^m$ such that

$$
\begin{aligned}
A^T y + \mu X^{-1} e &= c \\
Ax &= b \\
x &> 0.
\end{aligned}
$$

Now observe that

$$
\mu X^{-1} e = \left(\frac{\mu}{x_1}, \ldots, \frac{\mu}{x_n} \right)^T .
$$

If we set $\dfrac{\mu}{x_i} = s_i$ and $s = (s_1 \ldots, s_n)^T$, this leads to the system

$$
\begin{aligned}
A^T y + s &= c \\
Ax &= b \\
x_i s_i &= \mu, \quad \forall i = 1, \ldots, n \\
x &> 0 \\
s &> 0.
\end{aligned}
$$

Any solution to the above system using Newton's method will depend on the factor μ which is often called the duality measure. The key is to solve the above system with decreasing values of μ. The Newton scheme for the above system based on (2.17) is given as

$$
\begin{bmatrix} 0 & A^T & I \\ A & 0 & 0 \\ S & 0 & X \end{bmatrix}
\begin{bmatrix} \Delta x \\ \Delta y \\ \Delta s \end{bmatrix}
=
\begin{bmatrix} 0 \\ 0 \\ -\mu e \end{bmatrix} .
$$

Thus, the updated point (x', y', s') can now be given by a pure Newton iteration as

$$
\begin{aligned}
x' &= x + \Delta x \\
y' &= y + \Delta y \\
s' &= s + \Delta s.
\end{aligned}
$$

As μ is pushed toward zero, the trajectory of points $(x, y, s) \in \mathcal{F}^{\circ}$ obtained by solving the above KKT type system is called the *central path*. However, it is always not feasible from a practical point of view to compute the exact central path. Thus, a less precise version of the above KKT type condition is used to practically compute the solution of (LP) by staying in the interior. Actually, the tweaking that we need to make is in the constraints $x_i s_i = \mu$, $i = 1, \ldots, n$ or $XSe = \mu e$. These are called complementarity constraints and are more difficult to handle. So in the less precise version, we are not expecting that the complementarity constraints are satisfied exactly, but some leverage is allowed. With this framework in mind, let us now present the basic structure of the primal-dual interior point method. The idea is to start with a point $(x_0, y_0, z_0) \in \mathcal{F}^{\circ}$, and then for $k = 0, 1, 2, \ldots$ compute the Newton steps $(\Delta x_k, \Delta y_k, \delta s_k)^T$ by solving the following system, which we call the *Approx-KKT-Newton* system, i.e.

$$\begin{bmatrix} 0 & A^T & I \\ A & 0 & 0 \\ S_k & 0 & X_k \end{bmatrix} \begin{bmatrix} \Delta x_k \\ \Delta y_k \\ \Delta s_k \end{bmatrix} = \begin{bmatrix} 0 \\ 0 \\ -X_k S_k e + \theta_k \mu_k e \end{bmatrix}$$

where $\theta_k \in [0, 1]$ and

$$\mu_k = \frac{\langle x_k, s_k \rangle}{n}.$$

In general, instead of using a pure Newton step we shall take a damped Newton step and update as follows

$$\begin{aligned} x_{k+1} &= x_k + \alpha_k \Delta x_k \\ y_{k+1} &= y_k + \alpha_k \Delta y_k \\ s_{k+1} &= s_k + \alpha_k \Delta s_k. \end{aligned}$$

Moreover, we need to choose α_k in such a way that $x_{k+1} > 0$ and $s_{k+1} > 0$.

Before we proceed further, we shall take a more careful look into the matrix equation, which we call the Approx-KKT-Newton system. Observe the expression: $-X_k S_k e + \theta_k \mu_k e$. The key point here is that even if we choose $\theta_k = 1$, we need not have $-X_k S_k e + \mu_k e = 0$. Note that

$$X_k S_k e = (x_{k_1} s_{k_1}, \ldots, x_{k_n} s_{k_n}).$$

Recall that μ_k is the average value of the individual components of the vector $X_k S_k e$, but each individual component need not have the value μ_k. The parameter μ_k is called the duality measure. The idea of the above algorithm is based on this insight (Fig. 2.6).

Based on this basic structure, there are several different types of primal-dual interior point methods discussed in Wright [24]. Since we are not computing exact

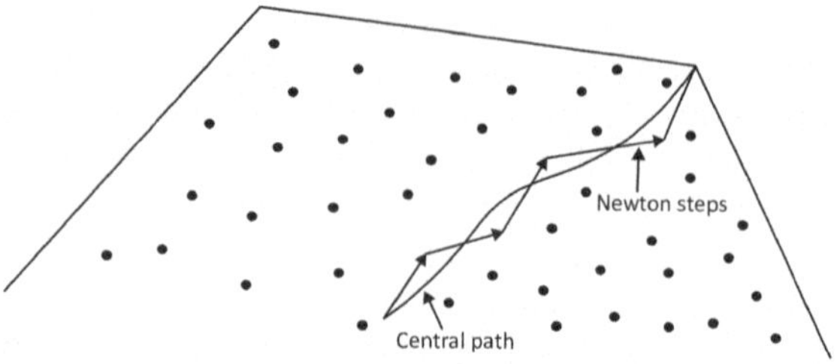

Fig. 2.6 Central path and Newton steps

points on the central path, but some approximations of it, it is important to know how close the computed points are to the actual central path. Thus, one needs to develop some way to measure the nearness of the computed points such that the points that are obtained by solving the Approx-KKT-Newton are forced to satisfy some proximity measure. Let us start by taking a look at one such measure as described in Wright [24]. Consider the following set for $\beta \in [0, 1)$ and $\mu > 0$, given as

$$\mathcal{N}_2(\beta) = \{(x, y, s) \in \mathcal{F}^\circ : \|XSe - \mu e\| \leq \beta\mu\}.$$

We would like to note that the norm used in the above definition is the usual Euclidean norm. One can guess that we have somehow built up a kind of proximity measure since points on the central path must satisfy $XSe - \mu e = 0$. But the key question is: if we have $(x, y, s) \in \mathcal{N}_2(\beta)$, what is the guarantee that the updated point is also in $\mathcal{N}_2(\beta)$. For the sake of simplicity, following Wright [24] consider the updated point as

$$\begin{aligned} x(\alpha) &= x + \alpha\Delta x, \\ y(\alpha) &= y + \alpha\Delta y, \\ s(\alpha) &= s + \alpha\Delta s. \end{aligned}$$

Theorem 5.6 in Wright [24] shows that—if $(x, y, s) \in \mathcal{N}_2(\beta)$, $(x(\alpha), y(\alpha), s(\alpha)) \in \mathcal{N}_2(\beta)$. Since this theorem is of such significance, we reproduce it here.

Theorem 2.6.1 *Consider $\beta \in (0, 1)$ and $\theta \in (0, 1)$ so chosen that*

$$\frac{\beta^2 + n(1-\theta)^2}{2^{\frac{3}{2}}(1-\beta)} \leq \theta\beta.$$

Then if $(x, y, s) \in \mathcal{N}_2(\beta)$, $(x(\alpha), y(\alpha), s(\alpha)) \in \mathcal{N}_2(\beta)$ for all $\alpha \in [0, 1]$.

In fact, Wright [24] shows that if we choose $\theta = 1 - \frac{0.4}{\sqrt{n}}$ and $\beta = 0.4$, the conditions of the above theorem hold and the conclusion of the theorem holds for $\alpha = 1$. If $\theta_k = \theta$ for all $k \in \mathbb{N}$, then what we get is called the *short-step path following method*, or, in short, SPF. However, this algorithm is not a very practical one to implement on a PC. I tried and failed on simple problems. Nevertheless, it is an important schematic algorithm that brings out the basic features of the primal-dual interior point method.

The next obvious question is: what will be the most reasonable stopping criteria for the short-step path following algorithm, or, in general for a primal dual method? The stopping criteria as given in Wright [24] tells us to stop the algorithm once the value of the duality measure goes below a predetermined threshold. This means that given the threshold $\varepsilon > 0$, stop the algorithm at the kth iteration if $\mu_k \leq \varepsilon$. The question is: can this be done in polynomial time, i.e., put more crudely can we reach the stopping criteria in a reasonable number of iterations and the required number of iterations not become exponentially large? The result that we shall present below (from Theorem 5.2 in Wright [24]) is called the *iteration complexity* of the short-step path following method.

Theorem 2.6.2 *Let us consider the short-step path following method. Let the starting point* $(x_0, y_0, s_0) \in \mathcal{N}_2(0.4)$. *Suppose we have*

$$\mu_0 \leq \frac{1}{\varepsilon^\gamma}$$

where $\varepsilon > 0$ *is the threshold parameter and* γ *is a positive constant. Then there is an index* K *with*

$$K = O\left(\sqrt{n}\ln\left(\frac{1}{\varepsilon}\right)\right),$$

such that $\mu_k \leq \varepsilon$, *for all* $k \geq K$.

The term $O\left(\sqrt{n}\ln(\frac{1}{\varepsilon})\right)$ implies a term that is a constant time $\sqrt{n}\ln(\frac{1}{\varepsilon})$. In fact, K provides the minimum number of steps required to reach the stopping criteria. Of course, the number of actual steps in this case can also be large, but not exponential.

Let us end our technical discussion on the interior points here, and we refer the reader to the wonderful monograph of Wright [24] for more on primal-dual interior point methods. The measure of proximity given here is not the only possible proximity measure. In fact, different proximity measures can give rise to different methods, like the long-step path following method, or the predictor-corrector method, and so on. We shall provide a few more references in the next section.

The most important question that can be asked is whether the simplex method, despite not being polynomial time, provides the vertex solution in the majority of important application cases, whereas the interior-point approach provides only an approximate solution that lies in the interior. But then, can one reach the vertex solution using interior point methods. To the best of our knowledge, Mehrotra [15] was the first to propose this method. In this paper, the author computes a basic feasible solution where the objective value is close to the optimal objective value. This is achieved using interior-point methods. Once such a vertex point has been computed, the actual solution can be obtained using the simplex method.

2.7 Further Reading, Another Story and a Dedication

Any author writing a survey on a mathematical topic is faced with two dilemmas. The first is to hold the attention of the reader by making the subject matter interesting while presenting as much material as possible within a reasonable number of pages. As the author of this survey on linear programming, I adopted the strategy of providing the reader with a view point of linear programming from the perspective of convex programming or convex optimization. If one compares textbooks written solely for linear programming with those written solely for convex programming, the approaches are usually very different. The texts on linear programming present the simplex method with the use of the tableau, and the approach has a strong combinatorial flavor. While any modern text on convex optimization naturally leans toward a presentation using the tools of real or functional analysis and linear algebra, thus giving it what one would call a continuous flavor. Many readers, like myself, would find it hard to appreciate the beauty and depth of the simplex method if we see the standard presentation. It was because of Padberg's [17] approach to the simplex method without the tableau that I could not only learn but also enjoy and appreciate its beauty. Actually, the approach taken by Padberg is quite naive in the sense that it shows that an optimal point must also be a basic feasible solution. Thus, if a basic feasible solution is not optimal, we must update it to obtain another basic feasible solution at which the objective value decreases. This is quite the standard way of algorithms in "non-linear programming" problems. Thus, Padberg may have made it possible for many individuals, such as myself, for whom the door to optimization was opened by the famous text on nonlinear programming by Bazaraa, Sherali and Shetty [3] to enjoy and appreciate the beauty of the simplex method. The reader must have observed that in Sect. 2.5 at the end, we have used the ideas from convex programming to provide a very simple way to show that the KKT conditions always hold for a linear programming problem. In fact, the debate in optimization about linearity and non-linearity has shifted to the debate between convexity and non-convexity, as powerful interior point methods can provide polynomial time algorithms for even non-linear convex problems. Rockafellar [21] has already envisaged this when he wrote that

> *The great watershed in optimization isn't between linearity and nonlinearity, but convexity and nonconvexity.*

Thus, keeping in line with the view that Rockafellar has mentioned, I have tried to present a view of linear programming in this survey through the lens of convex programming. For a more comprehensive survey of linear programming, I would like to mention the two-volume monographs of Dantzig and Thapa [6, 7]. There are several other great texts on linear programming that take the combinatorial view; however, I avoid mentioning them here. Karmarkar's method, though revolutionary, is not so easy to learn. Karmarkar, in fact, uses linear programming in the following format

$$\begin{aligned} &\min\langle c, x\rangle \\ &\text{subject to} \\ &\qquad Ax = 0 \\ &\qquad \langle e, x\rangle = n \\ &\qquad x \geq 0. \end{aligned}$$

I was lucky enough to speak to Narendra Karmarkar at the sidelines of a conference at IIT Kanpur. He told me that this particular format was chosen to eliminate the problem of unboundedness. According to Dantzig and Thapa [7], Karmarkar makes some additional assumptions that, at the minimizing point, the objective value is zero. Karmarkar's algorithm starts from a point in the relative interior, and the kth iteration x_k is considered optimal by Karmarkar if

$$\frac{z_k}{z_0} \leq \frac{e^{-\delta}}{h_0},$$

where $\delta > 0$, h_0 is the geometric mean of the components of the starting solution, z_0 is the objective value at the starting solution, and z_k is naturally the objective value at the kth iteration. We would also like to bring to the attention of the reader the beautiful paper by Bayer and Lagarias [2] on the geometry associated with Karmarkar's algorithm. Though the fascinating monograph by Wright [24] is a wonderful source for most of the well-known interior point methods, to the best of our knowledge, the first use of Newton's method for linear programming was done by Renegar [19]. I would also like to suggest to the reader the wonderful and short monograph by Renegar [20] on interior-point methods and the references therein. Of course, this monograph by Renegar is about convex programming, but it shows the power of the interior-point approach. A classic book on the subject of interior point methods in convex programming is by Nesterov and Nemirovskii [16], but it would be difficult for a beginner to read. Another beautiful survey on linear programming with a huge geometric flavor is by Gonzaga [9].

The interest in linear programming still continues, and areas that seemingly appear to be quite far from optimization are now being shown to have deep links with it. For example, the exploration of the connection between tropical geometry and linear programming. We strongly recommend the paper by Allamigeon et al. [1].

In the very first section of the chapter, I admitted my discomfort with the tableau approach for the simplex method. This strange uneasiness with linear programming translated into my dismal performance in the operations research course in the masters program, where the instructor possibly gave me a pass grade out of sympathy. It was only during my Ph.D. days when my supervisor understood that I was not interested in the problems on fixed point theory that he was suggesting and this brought me to optimization. One day he took me to the library and handed over the paper by Rockafellar [21] and told me that he had always wanted to study optimization, and since I was not interested in fixed points, perhaps I should look at this subject. I was taken aback, given my aversion to optimization at that time. However, I began reading that paper and could understand up to 10–11 pages. It was enough to convince me that optimization, and especially the idea of Lagrange multipliers, is indeed fascinating, and I remained firmly glued to the subject after that.

Many times in Barcelona, I took a walk with Professor Martinez-Legaz, my friend, collaborator, and post-doctoral mentor, in the beautiful avenue called Rambla de Catalunya. Once during such a walk, I told him if I were given the option to be a famous optimizer, I would choose to be R. T. Rockafellar. He looked at me and said that he would choose to be George Dantzig. While writing this survey, I understood what he meant. Linear programming is indeed one of the main pillars of optimization, whose beauty is unparalleled. Also, the simplex method is one of the greatest algorithms of all time. I now understand that my experience of optimization would have been much poorer without an appreciation of linear programming. Thus, in memory of our walks and his decades-long illustrious career as a famous convex analyst and optimizer, I dedicate this chapter to my friend Juan Enrique Martinez-Legaz.

Acknowledgements I thank Dr. Faiz Hamid of Department of IME at IIT Kanpur for helping with typing a part of the manuscript. I also thank Prof. J. E. Martinez-Legaz for his kind permission to reproduce his proof of the strong duality theorem for linear programming in this chapter.

References

1. Allamigeon X, Benchimol P, Gaubert S, Joswig M (2021) What tropical geometry tells us about the complexity of linear programming. SIAM Rev 63(1):123–164
2. Bayer DA, Lagarias JC (1989) The nonlinear geometry of linear programming. I. Affine and projective scaling trajectories. Trans Am Math Soc 314(2): 499–526
3. Bazaraa MS, Sherali HD, Shetty CM (2013) Nonlinear programming: Theory and algorithms. Wiley
4. Borwein J, Lewis A (2006) Convex analysis. Springer
5. Boyd SP, Vandenberghe L (2004) Convex optimization. Cambridge University Press
6. Dantzig GB, Thapa MN (2003) Linear programming: Theory and extensions, vol 1. Springer
7. Dantzig GB, Thapa MN (2003) Linear programming: Theory and extensions, vol 2. Springer
8. Fiacco AV, McCormick GP (1990) Nonlinear programming: Sequential unconstrained minimization techniques. SIAM
9. Gonzaga CC (1992) Path-following methods for linear programming. SIAM Rev 34(2):167–224
10. Guenin B, Könemann J, Tuncel L (2014) A gentle introduction to optimization. Cambridge University Press
11. Hiriart-Urruty J-B, Lemaréchal C (2013) Convex analysis and minimization algorithms I: Fundamentals, vol 305. Springer Science & Business Media
12. Karmarkar N (1984) A new polynomial-time algorithm for linear programming. In: Proceedings of the sixteenth annual ACM symposium on Theory of computing, pp 302–311
13. Klee V, Minty GJ (1972) How good is the simplex algorithm. In: Shisha O (ed) Inequalities-III. Academic Press, New York, pp 159–175
14. Mangasarian OL, Fromovitz S (1967) The Fritz John necessary optimality conditions in the presence of equality and inequality constraints. J Math Anal Appl 17(1):37–47
15. Mehrotra S (1991) On finding a vertex solution using interior point methods. Linear Algebra Appl 152:233–253

16. Nesterov Y, Nemirovskii A (1994) Interior-point polynomial algorithms in convex programming. SIAM
17. Padberg MW (1999) Linear optimization and extensions, vol 12. Springer Science & Business Media
18. Pshenichnyj BN (2012) The linearization method for constrained optimization, vol 22. Springer Science & Business Media
19. Renegar J (1988) A polynomial-time algorithm, based on Newton's method, for linear programming. Math Program 40(1–3):59–93
20. Renegar J (2001) A mathematical view of interior-point methods in convex optimization. SIAM
21. Rockafellar RT (1993) Lagrange multipliers and optimality. SIAM Rev 35(2):183–238
22. Rockafellar RT (1997) Convex analysis, vol 11. Princeton University Press
23. Wright M (2005) The interior-point revolution in optimization: History, recent developments, and lasting consequences. Bull Am Math Soc 42(1):39–56
24. Wright SJ (1997) Primal-dual interior-point methods. SIAM

Chapter 3
Mixed-Integer Programming Modeling Strategies for Scheduling-Based Optimization Problems

J. Cole Smith and Robert M. Curry

Abstract We provide a tutorial treatment of basic techniques for modeling scheduling optimization problems as mixed-integer linear programs, presented for readers having at least a limited background in linear programming formulations and algorithms for solving mathematical optimization problems. Although the scope of scheduling applications is far too broad to cover with a single chapter, many scheduling formulations employ a common set of modeling strategies that can be directly used, modified, or combined to formulate novel scheduling problems. We briefly cover these fundamental scheduling principles and illustrate how they can be adapted to model several modern applications arising in the field of scheduling optimization. These applications include optimizing radar pulse interleaving, identifying schedules that are robust to changes in data, and scheduling path flows in a dynamic flow optimization problem.

Keywords Scheduling · Mixed-integer programming · Optimization applications · Robust optimization

3.1 Introduction to Scheduling Problems

This chapter explores the use of mathematical optimization formulations with the aim of modeling several classes of scheduling problems. These problems determine how jobs should be scheduled or ordered on a set of machines, satisfying all given scheduling requirements, while optimizing a given objective. Specific details pertaining to the machines, jobs, objective, and schedule restrictions vary according to

J. C. Smith (✉)
223 Link Hall, Syracuse University, Syracuse, NY 13244, USA
e-mail: colesmit@syr.edu

R. M. Curry
4166 Bell Engineering Center, University of Arkansas, 800 W Dickson St., Fayetteville, AR 72701, USA
e-mail: rmcurry@uark.edu

F. Hamid (ed.), *Optimization Essentials*, International Series in Operations Research & Management Science 353, https://doi.org/10.1007/978-981-99-5491-9_3

the application area in which the scheduling problem is found. Because scheduling problems arise in so many practical settings, there are thousands of scheduling studies in the literature today.

Scheduling problems include a set J of jobs, usually with the requirement that each job must be processed by some machine. These problems may involve only a single machine, or may instead include multiple machines for processing the jobs. Different scheduling problems have various assumptions on the restrictions that govern each schedule. For instance, a schedule can usually process only one job at a time. Jobs might have an earliest time they can start (a release date) or the latest time they can finish (a due date). There may be precedence restrictions among the jobs, or machine setup times between jobs. In some problems, a job is not completed until it undergoes processing by each of the machines, as opposed to being completed by just one machine. The objectives are normally a function of the jobs' completion time, i.e., the vector of times at which each job finishes processing on a machine. For example, as we will describe below, one such objective is called *makespan*, denoted by $C_{\max}$, which is the time at which the last job finishes processing.

This chapter is designed to give the reader a sense of how to formulate scheduling problems using mixed-integer (linear) programs (MIPs) in particular. An exhaustive coverage of the topic would be impossible; however, there are a few key ideas that form the foundation of most practical formulations. We describe these foundations here. To set the stage for our discussions, we describe a general method for classifying scheduling problems in Sect. 3.1.1, along with some core concepts that will be used in this chapter. Section 3.1.2 briefly covers some of the most common algorithms to solve scheduling problems, and outlines the remainder of this chapter.

3.1.1 Categorizing Scheduling Problem Variations

To organize scheduling problem variations, Graham's notation [17] identifies problems according to three fields: $\alpha|\beta|\gamma$. The α-field describes the machine environment, the β-field describes the constraint environment, and the γ-field describes the objective environment. Unique characteristics of each environment are listed in their respective field. For example, if a single machine can complete the set of jobs, then a 1 would be listed in place of α. Similarly, when a job j carries a due date d_j by which it should be completed and our job schedule objective seeks to minimize makespan, then d_j would be listed in the β-field and C_{max} in the γ-field. If multiple characteristics exist simultaneously for a single field, then those terms are separated by a semicolon. For example, if a job j carries a due date d_j and a release date r_j before which it cannot be started, then d_j; r_j would be listed in the β-field. Bringing these together, a scheduling problem having a single machine, release and due dates for each job, and a makespan objective would be denoted as $1|r_j; d_j|C_{max}$ using Graham's notation. We next describe each field and introduce a non-exhaustive attribute list.

The machine environment, α, often describes the number of machines in the environment as well as the specific characteristics of those machines. In the general case, there exist m machines, and each job j has a unique processing time p_{jk} on each machine k. This is often referred to as the *related* (denoted by the letter R) case, since the time to process job j depends on the machine to which it has been assigned. In the uniform machine case (Q), some machines may be uniformly faster than the other machines. For the parallel (and identical) machine case (P), each machine is identical, so the processing time of job j is the same across all machines. Finally, the multi-purpose machine case occurs when a subset of identical machines is associated with a specific job operation.

In addition to completing each job j, the constraint environment β includes details about added restrictions on the job schedule. In many applications, each job j carries a due date (d_j) before which the job must be completed and/or a release date (r_j) before which a job cannot begin. Additionally, precedence constraints (*prec*) may enforce that some job j cannot be started until all jobs that are supposed to precede that job have been completed.

Finally, the objective field γ describes the objective or performance measure for a job assignment schedule. For each job $j \in J$, define C_j to be the completion time of j. As mentioned above, some applications will seek to minimize makespan C_{max}, which is formally defined as $\max_{j \in J}\{C_j\}$. Another objective associates importance weights w_j with each job $j \in J$, and minimizes the sum of weighted completion times ($\sum w_j C_j$). Given job due dates as described above, we sometimes permit a job to be completed after its due date, but at a penalty. In that case, there exist several possible objectives. One objective defines a binary parameter $U_j = 0$ if job j completes by the due date, and $U_j = 1$ if it is late, with an objective of minimizing the number of late jobs ($\sum U_j$). (These could be weighted sums as well, using the w_j-values as before.) We can also define the *tardiness* of job j as $\tau_j = \max\{0, C_j - d_j\}$ as a measure that gives zero penalty if a job is completed on time, and otherwise gives a linear penalty proportional to the amount of time by which the job is late. We can minimize the sum of weighted tardiness ($\sum w_j \tau_j$) or maximum tardiness $\tau_{max} = \max_{j \in J}\{\tau_j\}$, in addition to other measures.

Observe that the objectives given here are all said to be *regular*: The objectives are non-decreasing functions of the vector of completion times (see, e.g., [9]). For an example of a non-regular function, suppose that we minimize $\sum_{j \in J} |C_j - d_j|$, i.e., the sum of tardiness and the sum of earliness for each job. By delaying the completion time of jobs that completed early (if possible), the objective can decrease. The objectives we consider in this chapter, except where noted, are regular.

3.1.2 Solving Scheduling Problem Variations

Solving scheduling problems is often quite difficult due to the fact that, as the number of jobs in J increases, the number of possible schedules grows exponentially. For example, in a single-machine scheduling problem in which idleness is not allowed, there will exist up to $|J|!$ different solutions, depending on other problem constraints. Thus, scheduling problems are often NP-hard, and no general polynomial-time algorithms exist for their solution.

Fortunately, some scheduling problems do allow for efficient solution techniques. For these problems, solution approaches are tailored to the unique aspects of that problem. For instance, minimizing total weighted completion time on a single machine ($1||\sum w_j C_j$ in Graham's notation) without precedence constraints can be solved optimally using the weighted shortest processing time rule (WSPT) of Smith [35], which schedules jobs in non-decreasing order according to the ratio p_j/w_j, under the assumption that $w_j > 0,\ \forall j \in J$. One could modify the WSPT rule to optimally solve the variation of this problem having release dates and without job weights by iteratively scheduling an unfinished and available job j having the smallest processing time among the set of unfinished jobs. Modified versions of the earliest due date (EDD) rule, which schedules jobs in order of non-decreasing due dates, can optimally solve single-machine problems that minimize the maximum tardiness ($1||\tau_{max}$) or minimize the number of late jobs ($1||\sum U_j$).

For NP-hard scheduling problems, no specifically tailored polynomial-time algorithms exist unless P = NP. In such cases, worst-case exponential approaches like constraint programming (CP) may become particularly appropriate. For the job shop scheduling problem, where each job has a set of operations that must be completed in a specific order, Brailsford et al. [8] show that CP tends to be comparable to a MIP. Other works apply CP to a parallel machine scheduling problem with tool loading [16] and sequence-dependent constraints [15], among others. We also note that a subset of the literature employs heuristic algorithms (e.g., genetic algorithms [23]) as an alternative to exact approaches.

In addition to the large number of possible schedules, modifications and/or side constraints commonly exist in the problem environments that further complicate solving scheduling problems. The aforementioned approaches heavily rely on problem characteristics unique to that problem variation. Thus, introducing a new restriction or slightly amending another problem aspect could degrade the algorithm's effectiveness significantly or even render it unable to produce a feasible schedule. For this reason, MIPs are an attractive alternative to formulate scheduling problems due to their flexibility in accommodating unique problem characteristics.

While useful for modeling an array scheduling problems, one must assess the effectiveness of MIPs for *solving* these problems. Effectiveness is often (but not exclusively) measured computationally by the time to solve said problem to optimality. The computational time of the most general and widely used approach to solving MIPs, the branch-and-bound algorithm, relies in part on the *tightness* of a MIP's linear programming relaxation (LPR) formulated by relaxing MIP integer variables to be continuous. A branch-and-bound tree iteratively updates upper and lower bounds on the optimal objective function value, and the tightness of an LPR relates to the current performance of these bounds.

As is common in optimization modeling, multiple MIPs can be selected to represent the same scheduling problem. While some models may be smaller in terms of the number of variables or constraints, their LPR may not be as tight as an LPR for a larger model. Thus, there exists a trade-off between the size of a MIP and the tightness of its LPR.

We now present the organization of the rest of this chapter. We explore some of the basic MIP formulation strategies in Sect. 3.2. In Sects. 3.3–3.5, we present three illustrations of MIP modeling techniques for unique scheduling problems. Section 3.3 discusses a problem arising in the field of radar pulse interleaving, in which each task is performed in two phases, separated by a fixed window of time. In Sect. 3.4, we highlight a problem related to the field of robust optimization. Next, Sect. 3.5 presents a MIP model for a problem in which flows are routed non-simultaneously across a series of paths, and the scheduling of these routes affects the overall objective. Finally, we conclude this chapter in Sect. 3.6.

3.2 Scheduling Formulation Fundamentals

There are several general approaches for modeling scheduling problems that result from the initial choice of decision variables. We cover three such approaches here. In each subsection we present a different choice of decision variables, assuming that we are optimizing problem $1||\sum w_j C_j$: That is, there is a single machine, and we seek to minimize the sum of weighted completion times, where the weight for job $j \in J$ is given by w_j. At the end of each subsection, we discuss how the models can be extended to accommodate the presence of multiple machines, and how alternative scheduling objective functions can be modeled using the choice of decision variables in that subsection.

Before discussing these modeling techniques, we present a small numerical example to illustrate the $1||\sum w_j C_j$ problem. For this example, we have three jobs each having processing times and weights as detailed in Table 3.1.

For this example, we seek to minimize the sum of the three jobs' weighted completion times, where the completion time of each job is processing time of that job plus the time at which it began. A naive and suboptimal approach is to schedule the jobs according to their weights in decreasing order, and start each job as soon as the prior jobs end. Using this approach, job j_1 has the highest weight and would begin at time 0 and end at time 4, the second-highest weighted job j_3 would begin at time 4 and end at time 6, and the third-highest weighted job j_2 would begin at time 6 and end at time 9.

Table 3.1 Jobs and their processing times and weights

Job	p_j	w_j
j_1	4	5
j_2	3	2
j_3	2	3

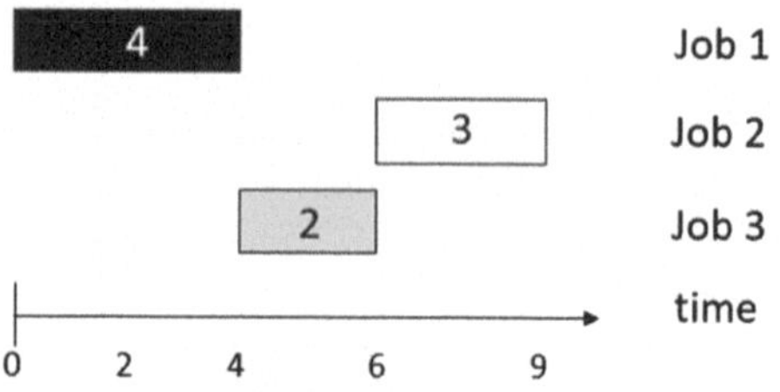

Fig. 3.1 Schedule (j_1, j_3, j_2) depicted on a single machine as a Gantt chart

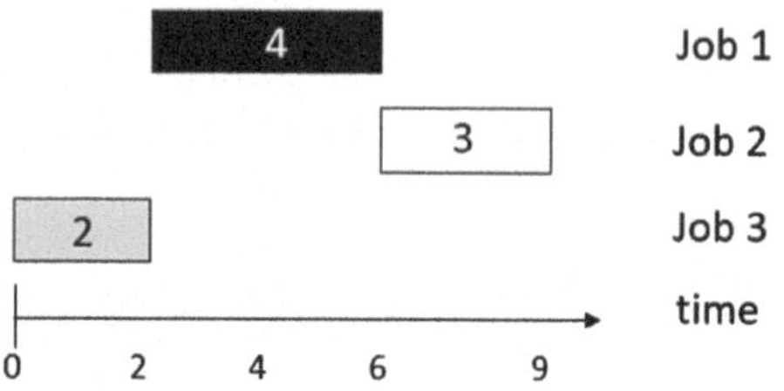

Fig. 3.2 Optimal schedule (j_3, j_1, j_2) depicted as a Gantt chart

We next use a Gantt chart to visually represent this solution. Figure 3.1 displays the interval of time for which each job is processed by the single machine when the jobs are scheduled in the order (j_1, j_3, j_2). The label on each interval in Fig. 3.1 displays the processing time for that job, and the length of each interval is proportional to its processing time. The machine is also never idle, so each interval begins as soon as the prior interval ends, with overlapping not allowed. The solution in Fig. 3.1 provides the following weighted completion times:

- Job j_1: $w_1(0 + p_1) = 5(0 + 4) = 20$
- Job j_3: $w_3(4 + p_3) = 3(4 + 2) = 18$
- Job j_2: $w_2(6 + p_2) = 2(6 + 3) = 18$
- Total weighted completion time: $20 + 18 + 18 = 56$.

While choosing jobs in their weighted order might provide an optimal solution in some cases, it is generally a heuristic approach. Figure 3.2 depicts the optimal schedule, (j_3, j_1, j_2), as a Gantt chart, in which job j_3 begins at time 0 and ends at time 2, job j_1 begins at time 2 and ends at time 6, and job j_2 begins at time 6 and ends at time 9. The solution in Table 3.1 provides the following weighted completion times:

- Job j_3: $w_3(0 + p_3) = 3(0 + 2) = 6$
- Job j_1: $w_1(2 + p_1) = 5(2 + 4) = 30$
- Job j_2: $w_2(0 + p_2) = 2(6 + 3) = 18$
- Total weighted completion time: $6 + 30 + 18 = 54$.

Thus, the total weighted completion time for the optimal schedule is 54.

3.2.1 Time-Indexed

Our first approach divides the time horizon over which scheduling is performed into individual segments called *time slots*. As a preliminary, a common assumption in the scheduling literature—and indeed in many optimization problems—is that all data is integer-valued. This assumption can be justified by anticipating that data will be input to a computer and should therefore be rational. By scaling as appropriate, it is possible to obtain all integer data.

Thus, suppose that the processing time of job $j \in J$ is given by a positive integer p_j. Assume for now that there is no idleness on the machine. The machine will begin processing one job at time zero, and because that job's processing time is an integer value, it will immediately begin processing another job at an integer starting time after the first job completes. Continuing in this manner, all jobs will have an integer starting time, and so we only need to consider integer start times for each job.

Based on these principles, suppose that we define a time horizon from 0 to T, where T is the maximum amount of time that the machine processes jobs. (With single-machine scheduling and no idleness, we simply set $T = \sum_{j \in J} p_j$.) Then, we *discretize* the time horizon by dividing it into T unit-length slots, where slot t represents the interval from time t to time $t + 1$, for each $t = 0, \ldots, T - 1$. In this chapter, $\boldsymbol{T} = \{0, \ldots, T - 1\}$ denotes the set of time slots.

We say that a job is scheduled in slot t if it begins processing at the beginning of the slot. Define binary variables x_{jt} equal to 1 if job $j \in J$ begins processing in slot $t \in \{0, \ldots, T - 1\}$, and 0 otherwise. As a result, if $x_{jt} = 1$ then slots $t, \ldots, t + p_j - 1$ are all occupied: The machine is occupied by job j from the beginning of slot t through the end of slot $t + p_j - 1$, i.e., from time t through time $t + p_j$. In fact, this necessarily means that x_{jt} must equal zero for $t \geq T - p_j + 1$, or else job j will not finish before time T. Thus, variables x_{jt} are defined just over $j \in J$ and $t \in \{0, \ldots, T - p_j\}$.

To ensure that every job must be scheduled on the machine, we state that every job must be scheduled to begin in some time slot:

$$\sum_{t=0}^{T-p_j} x_{jt} = 1 \qquad \forall j \in J. \tag{3.1}$$

Then, to prevent having overlapping jobs on the machine, we need to ensure that no two jobs simultaneously occupy the machine in any time slot. Perhaps the simplest way of achieving this is to define a set J_k, for each $k \in \boldsymbol{T}$, that represents all combinations of jobs j and time slots t that would occupy the machine at time k if job j is scheduled at time t. Formally:

$$J_k = \left\{(j, t) : j \in J \text{ and } t \in \{\max\{0, k - p_j + 1\}, \ldots, \min\{k, T - p_j\}\}\right\} \qquad \forall k \in \boldsymbol{T}. \tag{3.2}$$

In (3.2), the condition that t starts on or after $k - p_j + 1$ is present because starting job j at time $t - p_j$ or earlier means that the machine will finish processing that job before time t. The term $\max\{0, t - p_j + 1\}$ serves to ignore (j, t) pairs in which t is negative. Also, j cannot start later than k if it is to occupy slot k, and $k \leq T - p_j$ is necessary so that job j is not allowed to start too late to finish processing by time T.

With these definitions in place, stating that every time slot must be assigned to one job becomes straightforward:

$$\sum_{(j,t) \in J_k} x_{jt} = 1 \qquad \forall k \in \boldsymbol{T}. \tag{3.3}$$

Most objective functions that would be optimized are easy to formulate using these variables, because completion time is easy to ascertain: The completion time of job j is given by $\sum_{t=0}^{T-p_j} (t + p_j)x_{jt}$. Thus, minimizing weighted completion time is simply

$$\min \sum_{j \in J} \sum_{t=0}^{T-p_j} w_j(t + p_j)x_{jt}. \tag{3.4}$$

The overall problem is then stated as follows.

Optimize (3.4), s.t. constraints (3.1) and (3.3), with x binary.

As we will discuss in Sect. 3.3, this modeling structure affords flexibility to handle a range of variations on scheduling assumptions. The linear programming relaxation of these models tends to be tight, in part because the constraint coefficients are binary-valued.

However, there is a significant disadvantage to this approach. The discretization of the time horizon into T time slots potentially requires a very large number of decision variables and constraints. There are $O(|J|T)$ variables and $O(|J| + T)$ constraints in the formulation. The size of the single-machine scheduling problems that we consider is given by $|J|$ and the number of bits required to store the input data. Define $\nu = \lceil \log_2\{\max_{j \in J} p_j\} \rceil$ as the number of bits required to store the largest processing time, and note that, therefore, the input size of any scheduling problem having general integer processing times depends on ν (in addition to other elements). This means that T is at least 2^ν, and thus the number of variables and constraints grows as an exponential function of problem size. (That is, T is said to be *pseudo-polynomial*.) If the data is known to be limited to relatively small numbers, or perhaps bounded by a polynomial function of the number of jobs to be scheduled, then these models are polynomial in size and could be very practical in real applications. In other cases when T is very large, these discretized models could be too large to use in practice. Computational experimentation is often needed to determine whether a discretized model is appropriate for a certain application, given the problem's objective, additional constraints pertaining to the problem, and assumptions on the data.

3.2.1.1 Multiple Machines

Suppose now that there exists a set of m machines. In a single-machine scheduling model, the x_{jt}-variables indicate whether or not job j will be scheduled at time t. In a multiple-machine scheduling model, we revise these variables to include a third index. Now, x_{jtc} will equal 1 if job $j \in J$ is scheduled at time $t \in \{0, \ldots, T - p_j\}$ on machine $c \in \{1, \ldots, m\}$. We then revise (3.1) as

$$\sum_{t=0}^{T-p_j} \sum_{c=1}^{m} x_{jtc} = 1 \qquad \forall j \in J \tag{3.5}$$

to enforce that each job j should be scheduled at some time, on some machine. To avoid processing multiple jobs on the same machine simultaneously, we revise (3.3) as

$$\sum_{(j,t) \in J_k} x_{jtc} \le 1 \qquad \forall k \in \boldsymbol{T},\ c = 1, \ldots, m. \tag{3.6}$$

Note that these constraints are written as less-than-or-equal-to constraints to allow idleness (especially at the end of the time horizon). This is necessary because when multiple machines are used, we typically do not use each machine over the entire time horizon, nor do we know how long each machine will be utilized before solving the scheduling problem.

Again, the x-variables must be binary-valued. The weighted completion time objective would simply include a summation over the machine indices to become

$$\text{Min} \sum_{j \in J} \sum_{t=0}^{T-p_j} \sum_{c=1}^{m} w_j(t + p_j)x_{jtc}. \tag{3.7}$$

Remark 3.1 This discussion assumes that the machines are identical. In some applications, jobs cannot be placed on certain machines, or the processing time of a job depends on the machine. These extensions are very simple to handle. However, if the machines are truly identical, then another potential complication exists: Every solution has a factorial number of alternative optimal solutions, which can be obtained by simply re-indexing the machines. This *symmetry* among the machines can cause a branch-and-bound solver to explore an exponential number of isomorphic solutions, greatly (and unnecessarily) increasing computational effort required to solve the problem. Anti-symmetry constraints [31] can alleviate this problem by, for instance, insisting that the sum of completion times on machine c is at least as large as that for machine $c + 1$, for $c = 1, \ldots, m - 1$. This symmetry-breaking hierarchy would be enforced using the following constraints:

$$\sum_{j\in J}\sum_{t=0}^{T-p_j} w_j(t+p_j)x_{jtc} \geq \sum_{j\in J}\sum_{t=0}^{T-p_j} w_j(t+p_j)x_{jt(c+1)} \qquad \forall c=1,\ldots,m-1. \tag{3.8}$$

Many alternative hierarchies can be used; we refer the reader to [31] for a thorough discussion of symmetry-breaking considerations.

3.2.1.2 Alternative Objectives

Now, we extend the prior formulation strategy to model three different objectives: Minimizing weighted tardiness, minimizing the number of late jobs, and minimizing makespan (when multiple machines are present). For generality and consistency within this section, we assume the presence of multiple machines (and variables x indexed by job, time slot, and machine). The single-machine cases follow easily.

For *weighted tardiness*, recall that the completion time of job j is $(t+p_j)$ if $x_{jtc}=1$ for some machine c. In that case, the tardiness of job j would be given by $\max\{0, t+p_j-d_j\}$. Before solving the model, we can simply compute tardiness parameters $\tau_{jt}=\max\{0, t+p_j-d_j\}$ for each $j\in J$ and $t\in\{0,\ldots,T-p_j\}$. The objective would then be

$$\min \sum_{j\in J}\sum_{t=0}^{T-p_j}\sum_{c=1}^{m} w_j\tau_{jt}x_{jtc}. \tag{3.9}$$

A similar approach also handles the case in which we *minimize the number of late jobs*. Define binary parameters $\ell_{jt}=1$ if $(t+p_j)>d_j$ (and thus scheduling job j at time t would cause it to be late), and $\ell_{jt}=0$ otherwise, for all $j\in J$ and $t\in\{0,\ldots,T-p_j\}$. To minimize the number of late jobs, we would use the objective function:

$$\text{Min} \sum_{j\in J}\sum_{t=0}^{T-p_j}\sum_{c=1}^{m} \ell_{jt}x_{jtc}. \tag{3.10}$$

Minimizing makespan requires a slightly different approach. Define a variable, z, that represents the maximum completion time of any job. Because the completion time of job j is given by $\sum_{t=0}^{T-p_j}\sum_{c=1}^{m}(t+p_j)x_{jtc}$, the following set of constraints guarantees that z is *at least* as large as the makespan:

$$z \geq \sum_{t=0}^{T-p_j}\sum_{c=1}^{m}(t+p_j)x_{jtc} \qquad \forall j\in J. \tag{3.11}$$

Minimizing the makespan requires the following simple objective:

$$\text{Min } z. \tag{3.12}$$

Because z is being minimized, an optimal solution will exist in which z equals the completion time of the latest job to finish on a machine.

3.2.2 Permutation

A different approach avoids the discretization of the time horizon as done in Sect. 3.2.1, and instead implicitly determines job starting times by stating the order in which jobs are scheduled on the machine.

We first consider a model that specifies the order in which each job is processed on the (single) machine. Define binary variables x_{jq} that equal 1 if job $j \in J$ is the qth job processed on the machine, where $q \in \{1, \ldots, |J|\}$. Then, the following set of assignment constraints ensures that every job is assigned to one position in the permutation, and every position is assigned to one job:

$$\sum_{q=1}^{|J|} x_{jq} = 1 \qquad \forall j \in J \tag{3.13a}$$

$$\sum_{j \in J} x_{jq} = 1 \qquad \forall q = 1, \ldots, |J|. \tag{3.13b}$$

These constraints are totally unimodular, meaning that every square submatrix of the constraint set has a determinant of 1, 0, or −1, and so optimizing a linear program over these constraints yields an extreme-point optimal solution whose values are integer. Put simply, there would be no reason to restrict the x-variables to be binary-valued if optimizing over a linear function subject only to these constraints.

However, more constraints are almost always required in scheduling formulations of interest. Note that the completion time of job j is given by

$$p_j + \sum_{q=2}^{|J|} x_{jq} \left(\sum_{i \in J} p_i \sum_{h=1}^{q-1} x_{ih} \right). \tag{3.14}$$

That is, the completion time is p_j, plus the sum of job processing times that precede j in the schedule. If $x_{j1} = 0$, (3.14) simplifies to p_j. Otherwise, let $\hat{q}$ be such that $x_{j\hat{q}} = 1$. Job i precedes job j in the schedule if $\sum_{h=1}^{\hat{q}-1} x_{ih} = 1$. Expression (3.14) therefore adds p_i to the completion time for job j when i precedes j.

Unfortunately, (3.14) is a nonlinear function. Thus, minimizing (3.14) over the assignment constraints in (3.13) is very difficult. As an alternative, we instead define *precedence variables* I_{ij}, which determine whether job i precedes job j in a schedule, for $i, j \in J,\ i \neq j$. These constraints can be written in a number of different ways. For our purposes, we present them as

$$I_{ij} + I_{ji} = 1 \quad \forall 1 \leq i < j \leq |J| \tag{3.15a}$$

$$I_{ij} \geq x_{iq} + \sum_{h=q+1}^{|J|} x_{jh} - 1 \quad \forall i, j \in J,\ i \neq j,\ q \in \{1, \ldots, |J| - 1\} \tag{3.15b}$$

$$I_{ij} \geq 0 \quad \forall i, j \in J,\ i \neq j. \tag{3.15c}$$

Constraints (3.15a) state that either job i precedes job j, or vice versa. (Note that by declaring these constraints for all $1 \leq i < j \leq |J|$, we examine only unique combinations of i and j.) We then enumerate the conditions under which i would precede j, which is done by (3.15b). These constraints state that if job i is placed in position q, and job j is placed in a position after q (as determined by the summation on the right-hand side of (3.15b)), then the right-hand side of (3.15b) equals 1. These constraints, along with (3.15a), force $I_{ij} = 1$ and $I_{ji} = 0$.

We can now represent the completion time of job j by the following linear expression:

$$p_j + \left(\sum_{i \in J} p_i I_{ij} \right), \tag{3.16}$$

where I_{jj} is defined to be 0. This allows us to minimize weighted completion time using the objective

$$\min \sum_{j \in J} w_j \left(p_j + \left(\sum_{i \in J} p_i I_{ij} \right) \right). \tag{3.17}$$

The resulting optimization problem requires binary restrictions on the x-variables, because the presence of constraints (3.15a)–(3.15c) destroys the totally unimodular constraint structure and allows fractional solutions to exist at optimality. The overall problem is given by

Optimize (3.17), s.t. constraints (3.13) and (3.15), with x binary.

Remark 3.2 Note that (3.15b) could equivalently be written as

$$I_{ij} \geq x_{jq} + \sum_{h=1}^{q-1} x_{ih} - 1 \quad \forall i, j \in J,\ i \neq j,\ q \in \{2, \ldots, |J|\},$$

which, when $x_{jq} = 1$, forces $I_{ij} = 1$ whenever job i is scheduled in a position before q.

Also, constraints (3.15a) are not typically required, as their role simply forces $I_{ij} = 0$ whenever $I_{ji} = 1$. Hence, we can simply define $I_{ji} = 1 - I_{ij}$ and retain only variables I_{ij} in the formulation, $\forall 1 \leq i < j \leq |J|$.

Remark 3.3 A different approach that seems appealing at first glance is to define binary variables $x_{ij} = 1$ if job i immediately precedes job j in a sequence, and 0 otherwise, for all distinct job pairs i and j. (For simplicity in notation, we define these variables for all job pairs i and j, where x_{ii} is simply defined to be 0, $\forall i \in J$.) Observe that every job (except the last one scheduled) must precede exactly one job, and that every job (except the first one scheduled) must be preceded by exactly one job in the sequence. To state these conditions, we define a dummy start job, 0, and a dummy end job, e. Similar to job positioning constraints, we can impose the following assignment constraints:

$$\sum_{i \in J \cup \{0\}} x_{ij} = 1 \quad \forall j \in J \tag{3.18a}$$

$$\sum_{j \in J \cup \{e\}} x_{ij} = 1 \quad \forall i \in J. \tag{3.18b}$$

We would also require that the dummy start job immediately precedes one job:

$$\sum_{j \in J} x_{0j} = 1, \tag{3.18c}$$

and that one job precedes the dummy end job:

$$\sum_{i \in J} x_{ie} = 1. \tag{3.18d}$$

In fact, this is the same modeling strategy used to represent the traveling salesperson problem (TSP), or more precisely, the Hamiltonian Path problem with fixed starting and ending nodes. Just as in the TSP, the collection of constraints (3.18) are not sufficient to enforce a valid schedule, because of the presence of *subtours*. For instance, if $|J| = 5$, then one solution might yield the following:

$$x_{01} = x_{12} = x_{20} = 1, \text{ and } x_{34} = x_{45} = x_{53} = 1.$$

These values satisfy constraints (3.18), but do not form a sensible schedule, as job 3 precedes job 4, which precedes job 5, which precedes job 3. The cycle 3–4–5–3 is an example of a subtour. These subtours can indeed be eliminated either using an exponential set of *subtour elimination constraints*, or using an auxiliary set of variables that state the relative position of each job [25]. We refer the reader to the seminal work of [2] for a thorough study of TSP models. The strategy of [25] is particularly relevant to this chapter, as it relates to the idea of employing variables that determine at which time a job starts processing. Hence, rather than fully explore TSP-inspired scheduling models, we turn our attention to continuous-start-time models in Sect. 3.2.3.

3.2.2.1 Multiple Machines

We return to the job-positioning variable strategy explored in Eqs. (3.13)–(3.17), and consider how this strategy can be extended for the case of multiple machines. First, we expand the variable indices to include a machine index c. Thus, x_{jqc} equals 1 if job $j \in J$ is the qth job on machine c, and is 0 otherwise. The assignment constraints are adjusted to state that every job must be scheduled in some position on some machine, and every position on a machine must be assigned to *no more than* one job. Note that some machine-position combinations will have no jobs assigned to them now, because with multiple machines, it is impossible to ascertain how many jobs will exist on each machine before solving the problem:

$$\sum_{q=1}^{|J|}\sum_{c=1}^{m} x_{jqc} = 1 \quad \forall j \in J \tag{3.19a}$$

$$\sum_{j \in J} x_{jqc} \leq 1 \quad \forall q = 1, \ldots, |J|,\ c = 1, \ldots, m. \tag{3.19b}$$

Note that constraints (3.19) in fact allow a machine to have no assigned job to position q, while having a job assigned to position $q + 1$. This by itself will not cause the model to fail (this would be akin to having a zero-processing time "phantom" job sitting in position q); however, it does generate symmetric solutions that should be avoided. Hence, we instead eliminate these "phantom" jobs by insisting that if there are k jobs assigned to a machine, they are positioned $1, \ldots, k$ with no gaps in between. We can achieve this by revising (3.19b) as

$$\sum_{j \in J} x_{j1c} \leq 1 \quad \forall c = 1, \ldots, m \tag{3.19c}$$

$$\sum_{j \in J} x_{jqc} \leq \sum_{j \in J} x_{j(q-1)c} \quad \forall q = 2, \ldots, |J|,\ c = 1, \ldots, m. \tag{3.19d}$$

Constraints (3.19c) require that the first position on any machine c cannot be occupied by more than one job. For every subsequent position q, (3.19d) requires that one job can be placed in position q on machine c, only if position $q - 1$ was also occupied on the machine. (Otherwise, no job can be placed in position q on machine c.)

These new variable definitions now require a more specific definition of I_{ij}: Those variables should equal 1 if job i precedes job j on a *common* machine. Hence, if jobs i and j are scheduled on two different machines, then $I_{ij} = I_{ji} = 0$. With this revision in mind, we no longer impose (3.15a) and revise (3.15b) as

$$I_{ij} \geq x_{iqc} + \sum_{h=q+1}^{|J|} x_{jhc} - 1 \quad \forall i, j \in J,\ i \neq j,\ q = 1, \ldots, |J| - 1,\ c = 1, \ldots, m. \tag{3.20}$$

As referenced above, constraints (3.20) force $I_{ij} = 1$ when i precedes j on some machine. They do not force $I_{ij} = 0$ when i does not precede j on a machine; however, optimality will enforce this condition for regular objectives.

Because completion time is still given by (3.16), the objective (3.17) can still be used without modification.

3.2.2.2 Alternative Objectives

For *weighted tardiness*, the tardiness for job $j \in J$ is now given by

$$\max\left\{0,\ p_j + \left(\sum_{i \in J} p_i I_{ij}\right) - d_j\right\}.$$

However, this expression is nonlinear, and minimizing the sum of these terms over all job pairs leads to a nonlinear mixed-integer program. Fortunately, these terms are piecewise-linear and convex, and are thus amenable to a standard reformulation using linear constraints. Define variable τ_j as the tardiness for job $j \in J$, and note that the following constraints must hold for job j:

$$\tau_j \geq p_j + \left(\sum_{i \in J} p_i I_{ij}\right) - d_j \tag{3.21a}$$

$$\tau_j \geq 0. \tag{3.21b}$$

These constraints guarantee that $\tau_j \geq \max\{0, p_j + \left(\sum_{i \in J} p_i I_{ij}\right) - d_j\}$. Because we will seek to minimize weighted tardiness, τ_j equals this desired maximum value at optimality. The objective function simply becomes

$$\min \sum_{j \in J} w_j \tau_j. \tag{3.22}$$

To *minimize the number of late jobs*, we now need to define binary variables z_j, $\forall j \in J$, which equal 1 if job j will be scheduled late, and 0 otherwise. The objective function is given as

$$\min \sum_{j \in J} z_j. \tag{3.23}$$

To guarantee that $z_j = 1$ when job j is late, we state that the completion time for job j must be no later than d_j, unless job j is late (and $z_j = 1$). The following constraints enforce those conditions:

$$p_j + \left(\sum_{i \in J} p_i I_{ij}\right) \leq d_j + (T - d_j) z_j \quad \forall j \in J. \tag{3.24}$$

Note that if $z_j = 0$, then job j completes on or before the due date. If the job is late, then $z_j = 1$, and the completion time is only restricted to be no more than T.

The objective of *minimizing makespan* is similar to the approach used in Sect. 3.2.1. Letting z represent makespan, we define an objective variable z, and create the objective

$$\min z. \tag{3.25}$$

The constraints

$$z \geq p_j + \left(\sum_{i \in J} p_i I_{ij} \right) \quad \forall j \in J \tag{3.26}$$

guarantee that z is at least as large as the latest completion time; the objective (3.25) guarantees that z equals the largest completion time.

3.2.3 Continuous Start Times

Some of the prior modeling difficulties can be avoided by instead defining variables that represent the start time of each variable on the machine. Here, we let s_j represent the time at which job j begins processing. We know that the machine will be occupied during the time interval $[s_j, s_j + p_j]$. Therefore, every other job must finish before time s_j, or begin processing after time $s_j + p_j$.

To enforce this disjunction, we use binary variables I_{ij} for each pair of jobs $i, j \in J, i \neq j$, as defined in Sect. 3.2.2. Recall that $I_{ij} = 1$ if job i precedes job j, and $I_{ij} = 0$ otherwise. If job i precedes job j, then $s_i + p_i \leq s_j$, and if job j precedes job i, then $s_j + p_j \leq s_i$. These conditions are represented as follows:

$$s_i + p_i \leq s_j + T(1 - I_{ij}) \quad \forall i, j \in J, \ i \neq j. \tag{3.27}$$

To justify (3.27), note that when $I_{ij} = 1$, we have that $s_i + p_i \leq s_j$, and so job i precedes job j as desired. On the other hand, if $I_{ij} = 0$, then the right-hand side of (3.27) becomes $s_j + T$. However, we could never have job i complete later than T (the latest sensible completion time). Noting that $s_j \geq 0$, we have that (3.27) becomes redundant when $I_{ij} = 0$. In this way, constraints (3.27) are only active for the i, j combinations for which $I_{ij} = 1$.

The only remaining step is to enforce constraints (3.15a) (or simply enforce that I_{ij} is binary for each $i, j \in J, i < j$, while defining $I_{ji} = 1 - I_{ij}$ throughout). Given that the completion time of job j is $s_j + p_j$, the objective is

$$\min \sum_{j \in J} w_j(s_j + p_j). \tag{3.28}$$

These developments yield the following optimization problem.

Optimize (3.28), s.t. constraints (3.15a) and (3.27), with $s \geq 0$ and I binary.

This formulation has the advantage of using relatively few constraints and variables (there are $O(|J|^2)$ constraints, $O(|J|)$ continuous variables, and $O(|J|^2)$ binary variables). However, the linear programming relaxations are typically weak, as is common in formulations that have constraint-activation structures such as in (3.27). When I_{ij} (and I_{ji}) are both fractional for i, j, constraints (3.27) do not effectively separate the jobs on the machine, requiring a branch-and-bound solver to execute many branching operations before terminating.

3.2.3.1 Multiple Machines

To adapt the prior formulation strategy to the case of multiple machines, it is helpful to define job-to-machine assignment variables. Define binary variables $v_{jc} = 1$ if job $j \in J$ is assigned to machine $c \in \{1, \ldots, m\}$, and 0 otherwise. With these variables, it is necessary to ensure that every job is assigned to one machine:

$$\sum_{c=1}^{m} v_{jc} = 1 \qquad \forall j \in J. \tag{3.29}$$

As in Sect. 3.2.2, we now need a different mechanism to guarantee that either I_{ij} or I_{ji} must equal 1 when jobs i and j are on the same machine. We enforce this condition via the following constraints:

$$I_{ij} + I_{ji} \geq v_{ic} + v_{jc} - 1 \qquad \forall 1 \leq i < j \leq |J|,\ c = 1, \ldots, m. \tag{3.30}$$

These constraints allow $I_{ij} = I_{ji} = 0$ if $v_{ic} + v_{jc} - 1$ is never more than 0, for each possible c. If there is a c such that $v_{ic} + v_{jc} - 1 = 1$, then both jobs i and j are scheduled on machine c, and either job i precedes job j (and $I_{ij} = 1$) or job j precedes job i (and $I_{ji} = 1$). The same constraints that force job i to complete before job j starts if $I_{ij} = 1$ can be used as before:

$$s_i + p_i \leq s_j + T(1 - I_{ij}) \qquad \forall i, j \in J,\ i \neq j. \tag{3.31}$$

3.2.3.2 Alternative Objectives

We only briefly cover the alternative objectives in this subsection, as they are handled in a similar manner to those in Sect. 3.2.2.

For *weighted tardiness*, we use variables τ_j as the tardiness for job $j \in J$ once again. This allows us to use the objective

$$\min \sum_{j \in J} w_j \tau_j, \tag{3.32}$$

along with constraints

$$\tau_j \geq s_j + p_j - d_j \qquad \forall j \in J \tag{3.33a}$$
$$\tau_j \geq 0 \qquad \forall j \in J. \tag{3.33b}$$

We *minimize the number of late jobs* by again defining binary variables $z_j, \forall j \in J$, along with the objective

$$\min \sum_{j \in J} z_j \tag{3.34}$$

and constraints

$$s_j + p_j \leq d_j + (T - d_j)z_j \qquad \forall j \in J. \tag{3.35}$$

The objective of *minimizing makespan* follows the prior developments. We again let z represent the makespan, define the objective

$$\min z, \tag{3.36}$$

and use constraints

$$z \geq s_j + p_j \qquad \forall j \in J. \tag{3.37}$$

3.2.4 Which Model is Best?

The prior discussion presents a series of alternative formulations for single-machine scheduling problems, along with motivation for when each of these models might be effective. However, it can still be difficult to assess which model will solve a scheduling problem the fastest, because the answer to that question depends on several factors.

First, consider the computational hardware on which these models will be solved. If memory is limited, then it may not even be possible to load the time-indexed model, much less solve it. Software plays a key role as well: Since state-of-the-art solvers employ a different set of strategies to solve mixed-integer programs, the best modeling strategy may often depend simply on the choice of solver. A third factor depends on the scheduling problem data. As discussed before, when the processing times are very large, a time-indexed model is likely to be a poor choice. The continuous start-time models exhibit weak linear programming relaxations, leading to a large branch-and-bound tree, but contain fewer variables and constraints and can be preferable for a large number of jobs. A fourth consideration involves the specifics of the problem being solved, e.g., which objective is being used, and what side constraints or specific characteristics of the model exist. We demonstrate several examples of specific model characteristics in the following three sections.

An array of computational studies exists in the literature, and the conclusions drawn from those studies have evolved as hardware and software have changed over time. A sampling of excellent computational studies can be found in [21, 26, 27, 37].

3.3 Interleaving Two-Phase Jobs

We present the background on two-phase jobs in Sect. 3.3.1 before giving a formal problem definition and discretized model formulation in Sect. 3.3.2. Section 3.3.3 examines some of the weaknesses of the approach in Sect. 3.3.2 and provides an alternative formulation strategy for the problem.

3.3.1 Background

The first of our featured applications is a single-machine scheduling problem that arises in radar pulse interleaving. Here, each job j in the set J of jobs must be completed on a single machine as before, and jobs must be done in two *phases*, with a required exact waiting time in between the phases. Thus, once the first phase of job j begins processing, the machine must finish processing the first phase of job j (without preemption), and then after exactly the specified waiting time for job j, it must process the second phase for job j (also without preemption).

The unique aspect of this problem is the enforced waiting time between the two phases of each job. The key is to schedule phases of some jobs during the waiting times between phases of other jobs, thus interleaving the job phases in an efficient manner. Crafting such a schedule is challenging, because it is necessary to ensure that phases do not overlap. Figure 3.3 illustrates the situation with three jobs. Job 1 has a first-phase processing time of 2, a waiting time of 14 between phases, and a second-phase processing time of 1. For job 2, those times are 4, 13, and 2, respectively, and for job 3, those times are 3, 5, and 4, respectively. If these three jobs were scheduled consecutively, then the total amount of time to complete all three jobs would be (2 + 14 + 1) + (4 + 13 + 2) + (3 + 5 + 4) = 48 time units; moreover, the machine would be idle for 32 of those units. By contrast, the proposed schedule in Fig. 3.3 places job 1 on the machine first, and after 2 units of idle time, schedules the first phase of job 2, interleaving it between the two phases of job 1. Then, after 1 more unit of idle time after the first phase of job 2, we schedule job 3. Both phases of job 3 occur in the waiting time between job 2's phases, and the second phase of job 1 is processed between the two phases of job 3. The machine requires 23 units of time to process all three jobs in this schedule, and is idle for only 7 units of time.

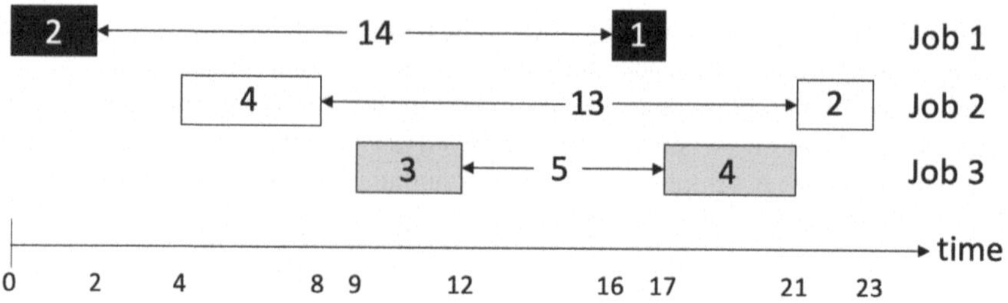

Fig. 3.3 Illustration of job interleaving

In the context of radar pulse interleaving, the machine is the radar. Tracking a target requires the emission of a pulse (which represents the first phase of the job), waiting a certain amount of time for the pulse to return (the enforced waiting time in between phases), and receiving the pulse (the second phase of the job). Any uncertainty on when the pulse will return to the radar is embedded within the duration of the job's second phase. In this setting, the ability to effectively interleave job phases reduces the time required to track a large array of targets. As such, high-quality solutions to the two-phase job scheduling problems are important to air-traffic controllers who manage a large volume of aircraft within tight airspace sectors. These solutions can also aid defense systems in identifying and tracking threats from missile attack. Details of the radar pulse interleaving problem are discussed in Elshafei et al. [12]; see Skolnik [34] for details pertaining to radar technology. A broader discussion of two-phase scheduling in this section is taken from Sherali and Smith [32].

There are two variations of the problem. The simplest, which we only briefly cover here for the sake of brevity, is to minimize *makespan*. That is, we seek to find the minimum amount of time, T^*, required to complete the scheduling of every job. Notably, Farina and Neri [13] contribute a constructive heuristic to address a simplified version of this problem. The other problem, which we focus on primarily, is a *reward maximization* problem. That problem involves a time limit, T, along with importance weights, w_j, for each job $j \in J$. Completing job j on or before time T yields a reward of w_j. The goal is to find a schedule that maximizes the sum of rewards due to completed jobs. We refer the reader to [19] for a neural network approach to this problem, [1, 18] for algorithms tailored for special cases of the problem, and [22] for a survey of literature on related problems.

3.3.2 *Problem Definition and Discretized Formulation*

We now formally state the problem. There exists a single machine on which each job $j \in J$ can be processed. Each job contains two phases. The processing time of job j is p_j^1 for phase one of the job and p_j^2 for phase two, and the waiting time in between those two phases is d_j for job j. Therefore, if we begin processing job j at an integer time t, then the machine will be occupied in the intervals $[t, t + p_j^1]$ and $[t + p_j^1 + d_j, t + p_j^1 + d_j + p_j^2]$, noting that job preemption is not allowed. We assume that p_j^1, p_j^2, and d_j are integer-valued for all $j \in J$. For the reward maximization problem, recall that each job $j \in J$ is associated with a reward weight w_j, which is earned if the job is completed before a specified time, T.

It is not clear how to formulate this problem using variables that simply state whether job i precedes job j. As shown in Fig. 3.3, it is necessary to accommodate idle periods on the machine, and it is also possible that job i could both start before job j and finish after job j. As a result, perhaps the most intuitive formulation (proposed in [12]) uses a discretized set of variables as presented in Sect. 3.2.1. First, define $\boldsymbol{T} = \{0, \ldots, T - 1\}$ as a set of time slots. We define binary variables x_{jt} such that x_{jt} equals 1 if and only if the first phase of job $j \in J$ begins at time $t \in \boldsymbol{T}$ (and 0 otherwise). Importantly, note the following consequences of these variable definitions.

- If $x_{jt} = 1$, then time slots $t, \ldots, t + p_j^1 - 1$ are occupied by job j, as are time slots $t + p_j^1 + d_j, \ldots, t + p_j^1 + d_j + p_j^2 - 1$. The "$-1$" adjustment on the right endpoints of those intervals is because each interval is one unit long. For instance, job 1 in Fig. 3.3 has $p_1^1 = 2$, $p_1^2 = 1$, and $d_1 = 14$; hence, if $x_{10} = 1$, then time slots 0, 1, and 16 are occupied by the job.
  - Note that some of the time slots defined above may not belong to $\boldsymbol{T}$, as they may be indexed at time T or above, which implies that the scheduled job would not complete on time. To avoid these undefined indices, we define $f_j \equiv T - p_j^1 - d_j - p_j^2 + 1$, $\forall j \in J$, as the final time slot in which job j can start processing in order to finish before the time limit, T.
- It is also useful to examine the inverse concept: For a given time slot $k \in \boldsymbol{T}$, when would scheduling job j occupy slot k? Note that the first phase of job j would occupy slot k, if j is scheduled to start at time $t \in \{k - p_j^1 + 1, \ldots, k\}$. The second phase of job j would occupy slot k if (the first phase of) job j starts at time $t \in \{k - p_j^1 - d_j - p_j^2 + 1, \ldots, k - p_j^1 - d_j\}$. Some of the time slots defined here may not belong to $\boldsymbol{T}$, because they may be indexed at a negative value. Such time slots will be ignored in our notation below by intersecting those intervals with the set of defined time slots, $\boldsymbol{T}$.

Based on the prior analysis, we define J_k to be the set of job/time pairs (j, t) such that scheduling job j at time t occupies the machine at time slot k, for each $k \in \boldsymbol{T}$. Formally,

$$\begin{aligned} J_k = \{(j, t) : j \in J, \text{ and either } & t \in \{k - p_j^1 + 1, \ldots, k\} \cap \boldsymbol{T} \text{ or} \\ & t \in \{k - p_j^1 - d_j - p_j^2 + 1, \ldots, k - p_j^1 - d_j\} \cap \boldsymbol{T}\} \qquad \text{for } k \in \boldsymbol{T} \end{aligned} \tag{3.38}$$

We are now ready to formulate the reward maximization problem using the discrete time slot concepts. To understand this formulation, we first insist that no job is scheduled more than once:

$$\sum_{t=0}^{f_j} x_{jt} \leq 1 \quad \forall j \in J.$$

Similarly, we insist that no time slot can be occupied by more than one job:

$$\sum_{(j,t)\in J_k} x_{jt} \leq 1 \quad \forall k \in \boldsymbol{T}.$$

Noting that job j is scheduled if and only if $\sum_{t=0}^{f_j} x_{jt} = 1$, the reward we accrue from job j is given by $\sum_{j\in J}\sum_{t=0}^{f_j} w_j x_{jt}$. Putting these ideas together, we obtain the following formulation for the reward maximization problem:

$$\text{Max} \sum_{j\in J}\sum_{t=0}^{f_j} w_j x_{jt} \tag{3.39a}$$

$$\text{s.t.} \sum_{t=0}^{f_j} x_{jt} \leq 1 \quad \forall j \in J \tag{3.39b}$$

$$\sum_{(j,t)\in J_k} x_{jt} \leq 1 \quad \forall k \in \boldsymbol{T} \tag{3.39c}$$

$$x_{jt} \in \{0, 1\} \quad \forall j \in J,\ t = 0, \ldots, f_j. \tag{3.39d}$$

As stated before, objective (3.39a) maximizes the sum of job rewards, constraints (3.39b) state that each job can be processed no more than once, and constraints (3.39c) state that each time slot can be occupied by at most one job. Constraints (3.39d) require the x-variables to be binary-valued.

As a side note, minimizing makespan can also be done using a similar strategy. First, we need an upper bound, T, on the makespan objective. (We can compute T by executing a heuristic scheme such as the one presented in [13].) We formulate the makespan minimization problem as follows, where the objective function variable z represents the latest time that the machine is used in the schedule:

$$\text{Min } z \tag{3.40a}$$

$$\text{s.t. } z \geq \sum_{t=0}^{f_j} (t + p_j^1 + d_j + p_j^2) x_{jt} \quad \forall j \in J \tag{3.40b}$$

$$\sum_{t=0}^{f_j} x_{jt} = 1 \quad \forall j \in J \tag{3.40c}$$

$$\sum_{(j,t)\in J_k} x_{jt} \leq 1 \quad \forall k \in \boldsymbol{T} \tag{3.40d}$$

$$x_{jt} \in \{0, 1\} \quad \forall j \in J,\ t = 0, \ldots, f_j. \tag{3.40e}$$

Here, constraints (3.40c)–(3.40e) are analogous to (3.39b)–(3.39d), except that (3.40c) now requires that every job is processed exactly once. Constraints (3.40b) state that z must be at least as large as the completion time for each job $j \in J$; therefore, z must be at least as large as the latest completion time over all jobs. By minimizing z in the objective, optimization forces z to be the smallest value it can feasibly take, which is precisely the time at which the last job completes.

3.3.3 *Alternative Formulation*

Returning to the revenue maximization formulation (3.39), note that the set-packing structure of constraints (3.39b) and (3.39c) tends to promote a tight linear programming relaxation. However, this formulation could require a very large set of variables and constraints. In particular, there are $O(|J|T)$ binary variables and $O(|J| + T)$ constraints. Because T is a function of the input data, this formulation is pseudo-polynomial in size.

Searching for a polynomial-size formulation is challenging because there is ambiguity in stating that one job starts before another, as discussed earlier. However, it is possible to specify, using continuous variables as introduced in Sect. 3.2.3, when jobs start. To approach such a revised formulation, we start by defining variables s_j as the time at which job $j \in J$ starts processing its first phase on the machine. This choice of variable presents two challenges:

1. If s_j is the starting time of job j, then it is implied that every job is actually scheduled at some time. Hence, we need to determine how to signify whether a job finishes before time T. (Toward the end of this section, we also impose rules governing s_j if it is designated as being late.)
2. Given two jobs $i, j \in J$, we must constrain s_i and s_j so that jobs i and j do not overlap.

To address the first challenge, let u_j be a binary decision variable that equals 1 if job j completes on time (i.e., the second phase of job j finishes processing on the machine before time T), and 0 otherwise. We therefore use the following objective function:

$$\text{Max} \sum_{j \in J} w_j u_j.$$

At optimality, u_j will equal 1 if possible; hence, we must force $u_j = 0$ if job j will be late. Noting that the completion time for job j is $s_j + p_j^1 + d_j + p_j^2$, we need a set of constraints stating that the completion time is no more than T if $u_j = 1$, but otherwise can be as large as desired. Letting M be a very large constant (the value of which we will address shortly), this condition can be stated as

$$s_j + p_j^1 + d_j + p_j^2 \leq T + M(1 - u_j) \quad \forall j \in J.$$

Note that the right-hand side of this constraint is T when $u_j = 1$ (and when the job is therefore completed before the time limit). When $u_j = 0$, the job can essentially complete at any convenient time. Our strategy is to find an upper bound, UB, on the maximum completion time of any job in an optimal solution. For instance, a simple bound is given by

$$UB = \sum_{i=1}^{|J|}(p_i^1 + d_i + p_i^2), \tag{3.41}$$

which would result from simply scheduling the jobs without interleaving. With this bound defined, we can take $M = (UB - T)$.

Next, we prevent overlapping jobs using a new set of binary variables. To understand their function, recall that it is not sufficient to simply say whether one job precedes another in the sequence. However, because there are only two phases for each job, it is possible to enumerate all different ways that the jobs can be scheduled relative to one another. That is, for jobs $i, j \in J$, there are up to five different ways that jobs i and j might be interleaved. Each of these five ways is called a *configuration*. For each unique pair of jobs i and j, the following configurations exist:

1. The second phase of job i finishes processing before the first phase of job j begins.
2. The first phase of job j (but not also its second phase) is scheduled during the waiting time between two phases for job i.
3. The first phase of job j begins processing after the first phase of job i is completed, and the second phase of job j is completed before the second phase of job i begins; i.e., both phases of job j are scheduled in the waiting time between the two phases for job i. Alternatively, both phases of job i are scheduled in the waiting time between the two phases for job j. (Note that either job j can be scheduled completely within job i's waiting time, or vice versa, or neither is possible.)
4. The first phase of job i (but not also its second phase) is scheduled during the waiting time between two phases for job j.
5. The second phase of job j finishes processing before the first phase of job i begins.

Note that configurations 5 and 4 are essentially the same as configurations 1 and 2, respectively, with the indices of i and j swapped. To illustrate these concepts, consider the example depicted in Fig. 3.4. The six configurations for each pair of jobs are given as follows.

- Between jobs 1 and 2 (with $i = 1$ and $j = 2$): Configuration 2.
- Between jobs 1 and 3: Configuration 2.
- Between jobs 1 and 4: Configuration 1.
- Between jobs 2 and 3: Configuration 3 (specifically, in which both phases of job 3 are scheduled in the waiting time between job 2's phases).
- Between jobs 2 and 4: Configuration 1.
- Between jobs 3 and 4: Configuration 1.

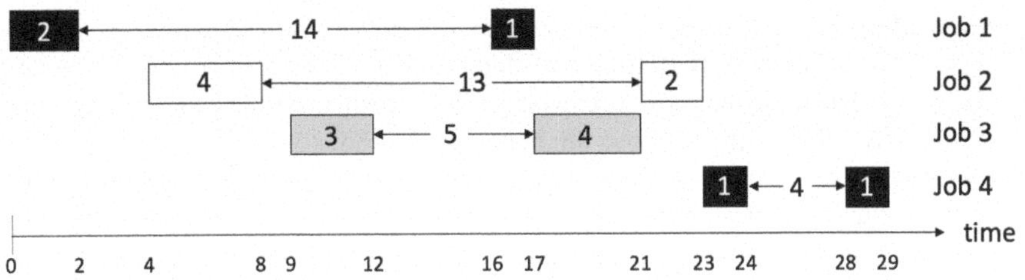

Fig. 3.4 Job configuration illustration

Indexing $J = \{1, \ldots, |J|\}$, we require that for every pair of jobs $i, j \in J, i < j$ (to guarantee that we examine each *unique* pair of jobs), jobs i and j must be scheduled in one of the five configurations listed above relative to one another. To achieve this, define binary variables $y_{ijk} = 1$ if jobs i and j in J $(i < j)$ are arranged relative to one another in configuration k, for $k = 1, \ldots, 5$, and $y_{ijk} = 0$, otherwise. We will require $\sum_{k=1}^{5} y_{ijk} = 1$ for each such pair of jobs $i, j \in J$.

For a given combination of jobs i, j and configuration k, the next challenge is to require s_i and s_j to be positioned in a certain way relative to one another depending on which y_{ijk} value is chosen to equal 1. First, if it is not possible for jobs i and j to exist in configuration k, then we simply fix $y_{ijk} = 0$ before solving the problem. (For instance, $y_{123} = 0$ because neither job can start and finish both its phases in the waiting time of the other.) Now, assuming that jobs i and j can exist in configuration k, we define ℓ_{ijk} and r_{ijk} to be the earliest and latest times, respectively, that job j can begin processing relative to job i. More precisely, if i and j exist in configuration k, then ℓ_{ijk} and r_{ijk} are the smallest and largest values, respectively, for which

$$\ell_{ijk} \leq s_j - s_i \leq r_{ijk}$$

holds true.

Given a job pair i, j, exactly one value of y_{ijk} will equal 1; we seek to bound $s_j - s_i$ in the $[\ell_{ijk}, r_{ijk}]$ interval corresponding to the configuration k for which $y_{ijk} = 1$. This bounding can be achieved by imposing the constraint:

$$\sum_{k=1}^{5} \ell_{ijk} y_{ijk} \leq s_j - s_i \leq \sum_{k=1}^{5} r_{ijk} y_{ijk}$$

for job pair i, j. If $y_{ijk'} = 1$ for some $k' \in \{1, \ldots, 5\}$, then all $y_{ijk} = 0$ for $k \neq k'$, and the constraint reduces to $\ell_{ijk'} \leq s_j - s_i \leq r_{ijk'}$ as desired.

To illustrate the concepts of the ℓ and r values, we examine the possible configurations between jobs 2 and 3 (noting that all five configurations are possible for this pair of jobs), treating $i = 2$ and $j = 3$. Suppose that $UB = 100$ (this value is necessary to compute the ℓ- and r-values, as we will see below):

Configuration 1. All of job 2 completes before job 3 starts. Because $p_2^1 + d_2 + p_2^2 = 19$, job 3 must start 19 time units later than job 2. Hence, $s_3 - s_2 \geq 19$, i.e., $\ell_{231} = 19$. The most that $s_3 - s_2$ can be is if job 2 begins at time 0 and job 3 begins at time $UB - (p_3^1 + d_3 + p_3^2) = 88$; hence, $r_{231} = 88$.

Configuration 2. The first phase of job 3 is scheduled within the waiting time between the job 2 phases (but the second phase of job 3 is not). To achieve this, job 3 cannot start earlier 11 time units after the start of job 2, or else the second phases of these jobs would overlap, and cannot start later than 14 time units after job 2, or else the first phase of job 3 overlaps with the second phase of job 2. Hence, $\ell_{232} = 11$ and $r_{232} = 14$.

Configuration 3. Both phases of job 3 are scheduled within the waiting time between the job 2 phases. This means that job 3 must start between 4 and 5 time units after job 2, in order to fit both phases within the 13-unit waiting time separating the two job 2 phases. Hence, $\ell_{233} = 4$ and $r_{233} = 5$.

Configuration 4. The first phase of job 2 is scheduled within the waiting time between the job 3 phases. This means that job 3 must start between 3 and 4 time units before job 2 starts, in order to fit the first phase of job 2 within the 5-unit waiting time separating the two job 3 phases. Note that in this case, $s_3 - s_2$ becomes negative: We obtain $\ell_{234} = -4$ and $r_{234} = -3$.

Configuration 5. All of job 3 completes before job 2 starts. Because $p_3^1 + d_3 + p_3^2 = 12$, job 2 must start 12 time units later than job 3. In this configuration, noting that $UB = 100$ (and, therefore, $s_2 \leq 81$), we must have $s_3 - s_2 \geq -81$, which occurs if $s_3 = 0$ and $s_2 = 81$. Thus, $\ell_{235} = -81$ and $r_{235} = -12$.

With this background, we can state the model formulation as follows:

$$\text{Max} \sum_{j \in J} w_j u_j \tag{3.42a}$$

$$\text{s.t. } s_j + p_j^1 + d_j + p_j^2 \leq UB - u_i(UB - T) \quad \forall j \in J \tag{3.42b}$$

$$\sum_{k=1}^{5} y_{ijk} = 1 \quad \forall 1 \leq i < j \leq |J| \tag{3.42c}$$

$$s_j - s_i \geq \sum_{k=1}^{5} \ell_{ijk} y_{ijk} \quad \forall 1 \leq i < j \leq |J| \tag{3.42d}$$

$$s_j - s_i \leq \sum_{k=1}^{5} r_{ijk} y_{ijk} \quad \forall 1 \leq i < j \leq |J| \tag{3.42e}$$

$$y \in Y, \ s \geq 0, \ u, y \text{ binary.} \tag{3.42f}$$

These constraints follow the developments in this section, where (3.42d) and (3.42e) combine to form the interval restrictions on $s_j - s_i$, and Y represents a list of constraints that force $y_{ijk} = 0$ for all job pairs $i < j$ that cannot be scheduled according to configuration k.

Model (3.42) contains a polynomial number of variables and constraints; in particular, the problem size no longer depends on the input data values. There are $O(|J|)$ (binary) u-variables, $O(|J|^2)$ (binary) y-variables, and $O(|J|)$ (continuous) s-variables. Also, there are $O(|J|^2)$ constraints. Thus, the size of this formulation becomes much smaller than that of (3.39) when the data input becomes large relative to $|J|$.

Despite the polynomial number of variables and constraints in model (3.42), this formulation may exhibit a weaker linear programming relaxation than (3.39) due to the presence of "big-M constraints" of the form (3.42b). Additionally, the y-variables tend to take fractional values, which in turn can allow the relative positioning given by $s_j - s_i$ to take on values that in fact result in overlapping jobs on the machine. We briefly describe a "placeholder strategy" to mitigate these difficulties.

Model (3.42) forces the solver to construct a feasible schedule of *all* jobs on the machine, even if the job will be completed after time T. In fact, the positioning of the jobs that finish after this due date is irrelevant, and an improved formulation would allow the solver to avoid having to place these jobs in a sequence. Accordingly, an alternative strategy designates a *placeholder* time for each job, such that if the job is scheduled late, then it will be scheduled at its placeholder time. To implement the idea, we first run a heuristic (e.g., the one of Farina and Neri [13]) to determine a feasible schedule of jobs that has a reasonably limited makespan. In this schedule, let $\hat{s}_j^c$ be the time at which job $j \in J$ starts, and let $\hat{z}^c$ be the makespan of the schedule. Our strategy sets $UB = T + \hat{z}^c$, leaving the first T time units to be used for on-time jobs (those finishing on or before T) and the last $\hat{z}^c$ time units to schedule any late jobs. In particular, for each job $j \in J$, we either schedule j so that it completes on time, or we force $s_j = T + \hat{s}_j^c$, i.e., we schedule the job in its placeholder. By scheduling all jobs in their placeholders, we guarantee a feasible schedule and relieve the solver from performing unnecessary optimization effort.

This strategy is implemented simply by replacing (3.42b) with the following two sets of constraints:

$$s_j + p_j^1 + d_j + p_j^2 \leq T + (\hat{s}_j^c + p_j^1 + d_j + p_j^2)(1 - u_j) \quad \forall j \in J \tag{3.43a}$$

$$s_j \geq (T + \hat{s}_j^c)(1 - u_j) \quad \forall j \in J. \tag{3.43b}$$

Note that whenever $u_j = 1$, and the job is to be completed on time, constraints (3.43a) and (3.43b) force $0 \leq s_j \leq T - p_j^1 - d_j - p_j^2$ as appropriate. Otherwise, when $u_j = 0$, the constraints force $T + \hat{s}_j^c \leq s_j \leq T + \hat{s}_j^c$, as desired.

3.4 Robust Scheduling

In this section, we turn to a problem in robust scheduling, which appears in the work of Tadayon and Smith [36]. Their study considers a single-machine scheduling problem in which all jobs in J must be processed without preemption. The key aspect of this paper is that some data pertaining to the problem is uncertain, and the schedule of jobs must be determined before ascertaining the actual data. Without knowing the data before solving the problem, it is impossible to state which schedule will be the best. Hence, before one can attempt to solve an optimization problem under uncertainty, it is vital to first determine precisely what "optimal" means when the data is uncertain.

For the sake of simplicity, we focus specifically on the problem of minimizing the sum of weighted completion times, when the processing times are uncertain. (Tadayon and Smith [36] consider four total objectives in their study, but a discussion of all four objectives is beyond the scope of this chapter.) Before beginning, we review the deterministic problem description, in which every job $j \in J$ has a processing time of p_j and a weight of w_j. For a given schedule, letting C_j be the completion time of job j, the objective is given by $\sum_{j \in J} w_j C_j$. When all data is known, the problem can be solved by the weighted shortest processing time rule (WSPT): Schedule the jobs in non-decreasing order of ratios p_j/w_j [35].

Section 3.4.1 introduces the robust scheduling problem, and demonstrates that the WSPT rule no longer applies in this case. In Sect. 3.4.2, we derive a model that optimizes this robust scheduling problem, using a combination of a classical scheduling formulation and an application of linear programming duality.

3.4.1 The Robust Weighted Completion Time Scheduling Problem

We consider the problem in which the jobs' processing times are uncertain. In particular, we now say that each job $j \in J$ has an *ideal processing time* of p_j. The ideal processing time is the shortest-possible processing time for a job, and can be thought of as the processing time for a job under normal circumstances. However, delays in processing a job can randomly occur, due for instance to a machine that occasionally malfunctions, or to a job that requires additional work to complete its processing. In that case, the processing time of job $j \in J$ is equal to its ideal processing time plus a *delay time*, δ_j.

For each job $j \in J$, we must specify a maximum delay time value, μ_j. In the absence of job-specific data, it is common to specify a constant, $K > 0$, and set $\mu_j = Kp_j, \forall j \in J$. Such a model assumes that delays do not exceed $K \cdot 100\%$ of the ideal processing times. That approach is used in [36]; however, we use the more general approach in which μ_j can be any (finite and nonnegative) bound on the delay for job j.

Given this characterization of uncertainty, there are alternative metrics that can be optimized. If one has a probability distribution for the δ-values, then it may be appropriate to find a schedule having the minimum *expected* objective function value as done, e.g., in stochastic programming formulations (see classic works by [7, 20]). A conditional value-at-risk (CVaR) approach contains a parameter $\beta \in (0, 1]$ that the user specifies. In the context of our problem, a CVaR model would minimize the average scheduling objective over the worst $\beta \cdot 100\%$ of data outcomes with respect to the chosen objective [29]. Note that the worst-case outcomes depend on the schedule chosen. If some job j appears first in some sequence, then delaying it will affect the completion time of every job in the sequence. On the other hand, if j appears last in the sequence, then delaying it only affects the completion time of job j itself. Hence, a delay in j would be more likely to be among the worst-case delay outcomes in the first sequence than in the second sequence. If β is close to 1, then the CVaR objective considers almost all of the possible outcomes, signifying that the user is risk-neutral. If β is close to 0, then the CVaR objective focuses only on those cases that are among the worst, revealing a risk-averse user.

Robust optimization models [4, 6] are appropriate for a risk-averse scheduler, but do not utilize the CVaR framework exactly. The goal of a robust optimization formulation is to provide a solution that performs as well as possible "in the worst case." However, it can be difficult to define what the "worst case" means. If there only exist bounds $0 \leq \delta_j \leq \mu_j$, $\forall j \in J$, then the worst case is easy and does not depend on the schedule: Simply set $\delta_j = \mu_j$, $\forall j$, and use the WSPT rule as usual. However, the probability that every job's delay value simultaneously takes its largest possible value is typically very small. This observation motivates the use of *budgeted uncertainty*.

In a budgeted uncertainty model, the δ-values have upper bounds as before, but the sum of delays is restricted by a positive parameter, Δ, i.e., $\sum_{j \in J} \delta_j \leq \Delta$. For small values of Δ, the delays are assumed to be very small, reflecting confidence that the processing times will in fact be close to their ideal values. As Δ approaches its largest meaningful value, $\sum_{j \in J} \mu_j$, the user assumes that all jobs may be maximally delayed. In real-world modeling situations, the user tunes this parameter based on risk tolerance and knowledge of the processing times. A deeper analysis regarding the validity of these objective bounds under assumptions regarding the underlying probability distributions that govern the δ-values can be found in [5].

We employ this concept in stating the robust weighted completion time (RWCT) scheduling problem. Summarizing, the RWCT considers a set of jobs J having ideal processing times p_j and weights w_j for all $j \in J$. The RWCT also specifies non-negative maximum delay values μ_j, $\forall j \in J$, and a delay budget $0 < \Delta \leq \sum_{j \in J} \mu_j$. The goal is to identify a minimum-cost job schedule, where the cost is given by the largest weighted completion time objective that could occur given values for μ_j, $\forall j \in J$, and Δ.

Before continuing, we provide an example of the RWCT, where $J = \{1, 2\}$.

- Job 1: $p_1 = 4$, $w_1 = 5$, and $\mu_1 = 4$
- Job 2: $p_2 = 1$, $w_2 = 1$, and $\mu_2 = 1$.

Because there are only two jobs, we can simply enumerate both sequences and determine their objective function values. Schedule A, which places job 1 first and job 2 second, is the WSPT order when the jobs' processing times are at their ideal values (with $\delta_1 = \delta_2 = 0$). This schedule has an objective function value of

$$obj(A) = (p_1 + \delta_1)w_1 + (p_1 + \delta_1 + p_2 + \delta_2)w_2 = 6(4 + \delta_1) + (1 + \delta_2) = 25 + 6\delta_1 + \delta_2. \tag{3.44}$$

Here, the worst-case delays will increase δ_1 as much as possible until $\delta_1 = \mu_1 = 4$, and will then increase δ_2 as much as possible (if $\Delta > 4$).

Schedule B places job 2 before job 1, and has an objective function value of

$$obj(B) = (p_2 + \delta_2)w_2 + (p_2 + \delta_2 + p_1 + \delta_1)w_1 = 6(1 + \delta_2) + 5(4 + \delta_1) = 26 + 6\delta_2 + 5\delta_1. \tag{3.45}$$

In this case, the worst-case delays will increase δ_2 as much as possible until $\delta_2 = \mu_2 = 1$, and will then increase δ_1 as much as possible (if $\Delta > 1$).

Case 1: Suppose that $\Delta = 1$. In schedule A, we would set $\delta_1 = 1$ and $\delta_2 = 0$, while in schedule B, we set $\delta_1 = 0$ and $\delta_2 = 1$. This yields *obj*(A) = 31 and *obj*(B) = 32, so schedule A is optimal.

Case 2: Now, suppose that $\Delta = 3$. In schedule A, we would set $\delta_1 = 3$ and $\delta_2 = 0$, while in schedule B, we set $\delta_1 = 2$ and $\delta_2 = 1$. This yields *obj*(A) = 43 and *obj*(B) = 42, so schedule B now becomes optimal.

It is easy to show that schedule A is optimal for $\Delta \in [0, 2]$ and schedule B is optimal for $\Delta \in [2, 5]$ (with both schedules being optimal when $\Delta = 2$). Therefore, simply sorting the jobs in WSPT order does not guarantee optimality.

3.4.2 Robust Model Formulation

This subsection begins with a conceptual formulation for the RWCT that contains a nonlinear function due to the presence of the robust element of the problem. The next part of this subsection models the robustness element via a linear programming formulation. Finally, we close by integrating the model into a single mixed-integer programming formulation.

3.4.2.1 Basic Formulation

Because the RWCT seeks to minimize (by selecting a schedule) the maximum objective value that can occur (by selecting delay values), this problem is referred to as a *minimax* problem. In this section, we establish a framework formulation around which we will build our model.

First, we adopt decision variables that specify a position for each job, as described in Sect. 3.2.2. Define decision variables $x_{jq} = 1$ if job $j \in J$ is the qth job in the schedule, and 0 otherwise. We include the following constraints:

$$\sum_{q=1}^{|J|} x_{jq} = 1 \quad \forall j \in J$$

$$\sum_{j \in J} x_{jq} = 1 \quad \forall q = 1, \ldots, |J|$$

$$x_{jq} \in \{0, 1\} \quad \forall j \in J,\ q = 1, \ldots, |J|.$$

The first set of constraints ensures that every job is scheduled exactly once, and the second set ensures that every position in the schedule corresponds to exactly one job. The final set of constraints states that the x-variables are binary. A challenge with this formulation is that the x-variables do not directly yield information about the job completion times. This information is necessary for the objective, and can be obtained by adding another set of variables. For each pair of jobs i and j in J, define I_{ij} as a binary variable that equals 1 if job i appears before job j in the schedule, and 0 otherwise. We use the following constraints to force I_{ij} to their correct values:

$$I_{ij} + I_{ji} = 1 \quad \forall i, j \in J,\ i < j$$

$$I_{ij} \geq x_{iq} + \sum_{s=q+1}^{|J|} x_{js} - 1 \quad \forall i, j \in J,\ i \neq j,\ 1 \leq q < |J|$$

$$I_{ij} \in \{0, 1\} \quad \forall i, j \in J,\ i \neq j.$$

The first set of constraints states that for every unique pair of jobs i and j, either i precedes j or j precedes i in the schedule. To understand the function of the second set of constraints, examine the right-hand side of that constraint and suppose that $x_{iq} = 1$. If job j is scheduled in position $q + 1$ or later, then $\sum_{s=q+1}^{|J|} x_{js} = 1$, and the right-hand side equals 1. In that case, I_{ij} is forced to 1 as desired. Hence, if i precedes j, then the constraint corresponding to jobs i and j and position q force $I_{ij} = 1$. If not, then the same argument shows that I_{ji} will be forced to 1 by these constraints; moreover, the condition that $I_{ij} + I_{ji} = 1$ would force $I_{ij} = 0$ as desired. Finally, the last set of constraints forces the I-variables to be binary-valued.

The (ideal) completion time of job $j \in J$ can be represented as $\sum_{i \in J} p_i I_{ij} + p_j$, i.e., the ideal processing time of all jobs scheduled before j, plus the ideal processing time of job j itself. The maximum possible delay, however, has not been accounted for at this point. Temporarily define $\theta(I)$ as the total increase in the objective function value caused by job delays, given variable values I. (We will return to this function in the following subsection.) Putting these concepts together, we have the following base formulation:

$$\text{Min} \sum_{j \in J} w_j \left(\sum_{i \in J} p_i I_{ij} + p_j \right) + \theta(I) \tag{3.46a}$$

$$\text{s.t.} \sum_{q=1}^{|J|} x_{jq} = 1 \quad \forall j \in J \tag{3.46b}$$

$$\sum_{j \in J} x_{jq} = 1 \quad \forall q = 1, \ldots, |J| \tag{3.46c}$$

$$I_{ij} + I_{ji} = 1 \quad \forall i, j \in J, \ i < j \tag{3.46d}$$

$$I_{ij} \geq x_{iq} + \sum_{s=q+1}^{|J|} x_{js} - 1 \quad \forall i, j \in J, \ i \neq j, \ 1 \leq q < |J| \tag{3.46e}$$

$$x_{jq} \in \{0, 1\} \quad \forall j \in J, \ q = 1, \ldots, |J| \tag{3.46f}$$

$$I_{ij} \in \{0, 1\} \quad \forall i, j \in J, \ i \neq j. \tag{3.46g}$$

The objective function sums the weighted (by w_j) completion times, and adds them to the maximum possible objective function increase caused by delays, $\theta(I)$. The constraints are exactly as described before.

3.4.2.2 The Worst Case

We now tackle the problem of anticipating what the worst-case data outcome will be, given a specified schedule, and formulating an expression that captures $\theta(I)$ in model (3.46). In the previous example, the worst-case data outcome used a greedy algorithm: Delay the first job in the schedule as much as possible, then the second job, and so on, until the sum of delays reaches the budget limit given by Δ. In fact, this greedy algorithm is optimal.

To see why, and to motivate the formulation that we will use subsequently, observe that delaying job j postpones the completion time for job j and every job scheduled after it. Define $\pi = \pi_1, \ldots, \pi_{|J|}$ as a given permutation schedule (i.e., every job in J appears exactly once in the sequence). Then the worst-case data outcome would be obtained by solving the following problem:

$$\theta(I) = \max \sum_{j=1}^{|J|} \left(\sum_{k=j}^{|J|} w_{\pi_k} \right) \delta_{\pi_j} \tag{3.47a}$$

$$\text{s.t.} \sum_{j \in J} \delta_j \leq \Delta \tag{3.47b}$$

$$0 \leq \delta_j \leq \mu_j \quad \forall j \in J. \tag{3.47c}$$

In this formulation, the objective (3.47a) notes that by delaying the jth job in the sequence (denoted by π_j), the completion times for job π_j and all jobs after π_j in the sequence are delayed. The term $\sum_{k=j}^{|J|} w_{\pi_k}$ gives the sum of weights over π_j and jobs following π_j in the schedule, and multiplies this sum by δ_{π_j}. Summing over every job position j gives the total delay. The two constraint sets are simpler: (3.47b) states the budgeted uncertainty constraint, and (3.47c) bounds each delay value for job $j \in J$ between zero and μ_j.

Formulation (3.47) is a knapsack problem with simple upper bounds on the variables, and is thus optimized by the greedy algorithm stated above. However, this formulation assumes knowledge of the schedule, π, which is not known ahead of time in the RWCT problem. Fortunately, the I-variables used in (3.46) yield a convenient tool to solve this problem. We first revise (3.47a) so that it does not depend on π, where $J(I, j)$ is the set of all jobs scheduled after job j given I:

$$\text{Max} \sum_{j \in J} \left(w_j + \sum_{k \in J(I,j)} w_k \right) \delta_j.$$

Note that w_j must be split out of the inner summation, because $J(I, j)$ refers to those jobs occurring strictly after job j, and delaying job j affects the completion time of those jobs in $J(I, j)$ and that of job j itself. Still, the presence of the set $J(I, j)$ presumes knowledge of this set before solving the problem. To avoid this problem, note that $\sum_{k \in J(I,j)} w_k = \sum_{k \in J} I_{jk} w_k$, where the second summation picks up only those weights corresponding to jobs k appearing after j in the schedule. Now, we can state a linear programming problem as a function of I:

$$\theta(I) = \max \sum_{j \in J} \left(w_j + \sum_{k \in J} w_k I_{jk} \right) \delta_j \tag{3.48a}$$

$$\text{s.t.} \sum_{j \in J} \delta_j \leq \Delta \tag{3.48b}$$

$$0 \leq \delta_j \leq \mu_j \quad \forall j \in J. \tag{3.48c}$$

3.4.2.3 Solving the Minimax Formulation

Finally, we show how to convert the minimax formulations given by (3.46) and (3.48) into a single mixed-integer program. First, note that by strong duality, the linear program in (3.48) can be converted to an equivalent dual formulation. Define u as the dual variable associated with constraint (3.48b), and v_j as the dual variable associated with bounding constraints (3.48c), for each $j \in J$. We obtain the following dual formulation:

$$\theta(I) = \min \ \Delta u + \sum_{j \in J} \mu_j v_j \tag{3.49a}$$

$$\text{s.t. } u + v_j \geq \left(w_j + \sum_{k \in J} w_k I_{jk} \right) \quad \forall j \in J \tag{3.49b}$$

$$u \geq 0, \ v_j \geq 0 \quad \forall j \in J. \tag{3.49c}$$

Now, because (3.46) and (3.49) are both minimization problems, they can be combined into one. To achieve this, we substitute $\theta(I)$ with $\Delta u + \sum_{j\in J} \mu_j v_j$ within the objective (3.46a), and add constraints (3.49b) and (3.49c) to (3.46). The resulting formulation becomes

$$\text{Min} \sum_{j\in J} w_j \left(\sum_{i\in J} p_i I_{ij} + p_j\right) + \Delta u + \sum_{j\in J} \mu_j v_j \tag{3.50a}$$

$$\text{s.t.} \sum_{q=1}^{|J|} x_{jq} = 1 \quad \forall j \in |J| \tag{3.50b}$$

$$\sum_{j\in J} x_{jq} = 1 \quad \forall q = 1, \ldots, |J| \tag{3.50c}$$

$$I_{ij} + I_{ji} = 1 \quad \forall i, j \in J,\ i < j \tag{3.50d}$$

$$I_{ij} \geq x_{iq} + \sum_{s=q+1}^{|J|} x_{js} - 1 \quad \forall i, j \in J,\ i \neq j,\ 1 \leq q < |J| \tag{3.50e}$$

$$u + v_j \geq \left(w_j + \sum_{k\in J} w_k I_{jk}\right) \quad \forall j \in J \tag{3.50f}$$

$$u \geq 0,\ v_j \geq 0 \quad \forall j \in J \tag{3.50g}$$

$$x_{jq} \in \{0, 1\} \quad \forall j \in J,\ q = 1, \ldots, |J| \tag{3.50h}$$

$$I_{ij} \in \{0, 1\} \quad \forall i, j \in J,\ i \neq j. \tag{3.50i}$$

As a final note to this section, observe in (3.50) that the I-variables appear only in the objective and in constraints (3.50d), (3.50e), (3.50f), and (3.50i). The objective forces the I-variables to be as small as possible, because they affect the objective in two ways. One, the objective coefficients on I-variables are positive, so larger I-values increase the objective. Two, in (3.50f), larger I-values force $u + v_j$ to be larger on the left-hand side of the equation, and the objective coefficients of u and v_j are both positive. Therefore, there is no need to include constraints (3.50d), as the I-values will simply take on the smallest value allowed by (3.50e). Furthermore, since that smallest value will be either 0 or 1, given binary x-variables, the binary restrictions on I given in (3.50i) can be relaxed.

Thus, a slightly simpler formulation for the RWCT is (3.50), modified by deleting constraints (3.50d) and replacing (3.50i) with $I_{ij} \geq 0,\ \forall i, j \in J,\ i \neq j$.

3.5 Scheduling Non-simultaneous Path-Flows

Next, we highlight a dynamic flow scheduling problem introduced by Curry and Smith [11] referred to as the minimum-cost network flow problem having arc activation costs (MCFA). The MCFA determines a minimum-cost network flow solution in which a single commodity is transmitted from a source to a sink according to an ordered sequence of consecutive path-flows. The cost of an MCFA solution includes standard per-unit flow costs, as in the general minimum cost flow problem, plus costs to *activate* arcs. An arc is activated when it is used to transmit flow on the current path in the schedule, but not on the prior path in the schedule. In this way, the activation cost for an arc must be paid in order to begin transmitting flow on that arc.

In addition to a network flow interpretation, one can view this problem as a variation of a job scheduling problem, in which each path-flow corresponds to a job, each arc corresponds to a machine, and jobs are scheduled consecutively without interruption. Each path-flow (job) is scheduled on a combination of arcs (machines) that form a source-to-sink path. One could then view the cost to activate an arc as the effort required to set up a machine to be used on a sequence of consecutive jobs. After being activated, if the machine goes unused for any positive period of time, then the cost to activate it again must be paid. As long as that machine continues working on consecutive jobs, it need not be activated again. The order in which path-flows/jobs are scheduled therefore affects the objective function value.

For the rest of the section, we predominantly discuss the MCFA in terms of its network flow representation. Section 3.5.1 introduces all relevant notation, formally defines the MCFA, and illustrates the problem with two examples. Next, in Sects. 3.5.2 and 3.5.3, we formulate two MIPs for solving the MCFA and a relaxed version of the MCFA, respectively. For both of these MIPs, we also present a series of inequalities for improving their solvability.

3.5.1 Defining the MCFA

Let graph $\mathcal{G}$ consist of node set $\mathcal{V}$ and arc set $\mathcal{A}$, in which $\mathcal{V}$ contains a source node s and a sink node t. As opposed to a general network flow in which all flows are assumed to be transmitted simultaneously, multiple paths cannot simultaneously transmit flow for the MCFA. Thus, a feasible MCFA solution consists of an ordered schedule P of s–t path-flows whose total flow equals a given integer value, $f > 0$. To characterize such a solution, let k be the number of paths in P, let P be composed of paths $p_1, \ldots, p_k$, and let each path have a respective flow value $f_1, \ldots, f_k$, in which $f_i \geq 1$, $\forall i = 1, \ldots, k$. Thus, the sum of all path-flows $\sum_{i=1}^{k} f_i$ must equal f, and flows are transmitted on the paths in the order listed in P.

Each arc $(i, j) \in \mathcal{A}$ has a maximum integer capacity $c_{ij} > 0$ and a per-unit flow cost $d_{ij} > 0$. (That is, the sum of flows sent on $(i, j) \in \mathcal{A}$ cannot exceed c_{ij}.) Given a path-flow schedule P, the activation cost $a_{ij} > 0$ for arc $(i, j) \in \mathcal{A}$ is incurred when (i, j) has positive flow on path p_i, in which $1 \leq i \leq k$ but not path p_{i-1}, where path p_0 represents an empty path (having no arcs or flow). Thus, an arc must be activated when first included on a subsequence of path-flows in P. Formally defined, the MCFA problem determines a schedule P of s–t path-flows minimizing the sum of activation and transmission costs required to send f units of flow from s to t.

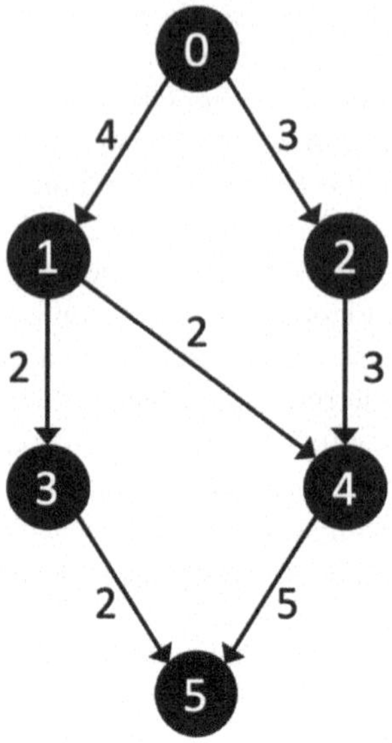

Fig. 3.5 MCFA example in which arc labels refer to arc capacities, $s = 0$, $t = 5$, and $f = 7$. All activation costs a equal 1 and all unit-flow costs d equal 1

To better understand the MCFA, consider the example MCFA instance depicted in Fig. 3.5, in which all unit-flow costs d and activation costs a are 1. An optimal MCFA schedule P for this instance is as follows:

- Send two units of flow on path p_1 : 0–1–3–5
  - Arcs $(0, 1)$, $(1, 3)$, and $(3, 5)$ are activated, total cost = 9.
- Send two units of flow on path p_2 : 0–1–4–5
  - Arcs $(1, 4)$ and $(4, 5)$ are activated, total cost = 8.
- Send three units of flow on path p_3 : 0–2–4–5
  - Arcs $(0, 2)$ and $(2, 4)$ are activated, total cost = 11.

Since each arc is activated exactly once, the total activation cost equals 7. Additionally, since each arc is fully saturated for this solution, the sum of flow costs equals 21. The total cost for this solution is thus 28.

It might be tempting to conjecture that any optimal path-flow schedule P activates each arc no more than once, but this is not necessarily true. The MCFA example in Fig. 3.6 details an example in which some arcs must be activated twice. Like the previous example, all feasible path-flow schedules saturate flow on every arc and all flow costs d and activation costs a equal 1. To see why an arc must be activated twice, suppose that the schedule begins by transmitting two units of flow on path 0–1–3–5. Arcs $(0, 1)$, $(1, 3)$, and $(3, 5)$ have now been activated. Although there is no more flow that can be sent on arc $(1, 3)$, two more units of flow must be sent on arc $(0, 1)$ and one more unit of flow must be sent on arc $(3, 5)$. In order to avoid

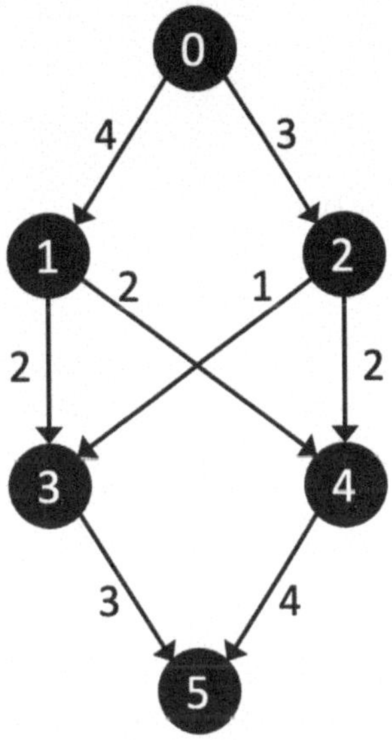

Fig. 3.6 MCFA instance in which arc labels refer to arc capacities, $s = 0$, $t = 5$, and $f = 7$

activating those arcs twice, the next path in the solution must send flow on arcs (0, 1) and (3, 5). This is impossible, because no path exists that uses these two arcs, but not the saturated arc (1, 3). Therefore, in order to avoid activating either arc (0, 1) or (3, 5) twice, this path cannot be the first one scheduled. However, an identical analysis can be performed for *every* path from node 0 to node 5 in this network, which means that no matter which path is scheduled first, at least one arc on this path must be activated twice.

One optimal solution to the instance depicted in Fig. 3.6 schedules flows as follows:

- 0–1–3–5 (two units)
- 0–1–4–5 (two units)
- 0–2–4–5 (two units)
- 0–2–3–5 (one unit).

This solution activates arc (3, 5) twice: Once at the beginning of the schedule, and then again at the end. An alternative representation of the solution can be given as a Gantt chart. Representing each arc as a machine, and equating the path flows with the time required to schedule the job, Fig. 3.7 displays the intervals of time during which the arc is used to transmit flows. For instance, this figure shows that for arc (0, 1), flows are transmitted during the interval of time from 0 to 4, without interruption. As such, that arc is activated once. A similar observation holds for every arc in this diagram except for arc (3, 5), where flows are transmitted during the interval of time from 0 to 2, and then again from 6 to 7. The fact that there are two non-consecutive intervals of time during which arc (3, 5) transmits flows means that the arc must be

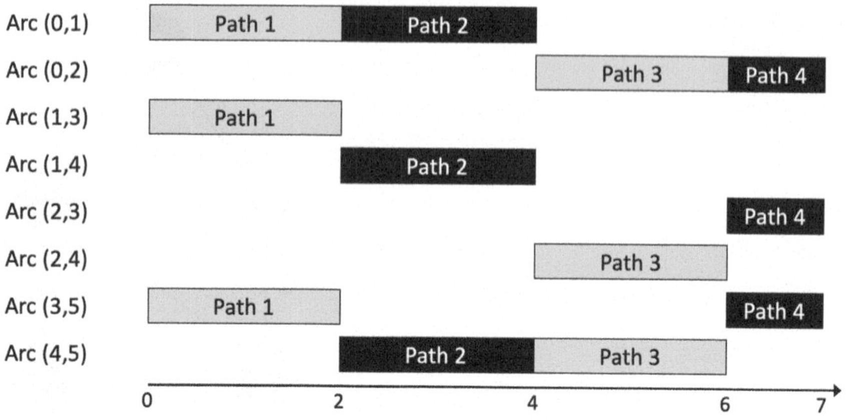

Fig. 3.7 MCFA solution envisioned as a Gantt chart

activated twice. This machine-scheduling representation of the problem informs our modeling strategy for the MCFA.

3.5.2 Mixed-Integer Program for the MCFA

For modeling the MCFA as a MIP, we adopt here the basic permutation modeling strategy in Sect. 3.2.2: Our model will prescribe which "job" (path) is assigned to be in position p of the flow schedule. However, for the MCFA, we need to expand our approach in Sect. 3.2.2 in two ways. First, we must specify for each path (job) the set of arcs (machines) that it will use: This information is also not known ahead of time. Second, the number of jobs to be scheduled is unknown, as the MCFA does not specify how many paths must be sent across the network.

Thus, we choose our decision variables with the goal of determining which arcs have a positive flow on each path, which arcs are activated for each path, and the amount of flow on each path. Thus, we include the following variables:

- Continuous variable x_{ij}^p equals the amount of flow on arc $(i, j) \in \mathcal{A}$ on the pth scheduled path.
- Binary variable $y_{ij}^p \in \{0, 1\}$ equals 1 when arc $(i, j) \in \mathcal{A}$ is activated on the pth path, and 0 otherwise.
- Binary variable $z_{ij}^p \in \{0, 1\}$ equals 1 when arc $(i, j) \in \mathcal{A}$ has positive flow on the pth path, and 0 otherwise.

Additionally, Table 3.2 summarizes all sets and parameters in our formulation. Since the maximum number of positive-flow paths in any optimal MCFA schedule is f, under the restriction that all positive path-flows are at least 1, we have that the maximum number of paths required in an optimal MCFA schedule is no greater than

Table 3.2 Sets and parameters for modeling the MCFA

Sets	$\mathcal{V}$	Nodes
	$\mathcal{A}$	Arcs
	P	Ordered sequence of paths to be used in the solution
Parameters	c_{ij}	Arc (i, j)'s integer capacity
	a_{ij}	Arc (i, j)'s activation cost
	d_{ij}	Arc (i, j)'s per-unit flow cost
	$f > 0$	Amount of flow sent from s to t

f. With this information, we construct a MIP for solving the MCFA. We begin by detailing various linear inequalities required to appropriately model the MCFA.

First, an MCFA solution must transmit f units of flow from s to t over a sequence of flows. We include a constraint requiring f units of total flow to be transmitted out of s among all scheduled path-flows, and constraints enforcing the balance of flow at each node for each path. Together, these constraints enforce that the f total units of flow beginning at s also terminate at t:

$$\sum_{p=1}^{f} \left(\sum_{i:(s,i)\in\mathcal{A}} x_{si}^{p} - \sum_{j:(j,s)\in\mathcal{A}} x_{js}^{p} \right) = f$$

$$\sum_{j:(i,j)\in\mathcal{A}} x_{ij}^{p} - \sum_{h:(h,i)\in\mathcal{A}} x_{hi}^{p} = 0 \quad \forall i \in \mathcal{V} \setminus \{s, t\},\ p = 1, \ldots, f.$$

Next, we require that each flow should occur over a simple s–t path, with no repeated vertices. To do this, we introduce a constraint ensuring that only one arc (s, i) emanating from s has positive flow for each path, and we include additional constraints to enforce that flows do not split at any intermediate node. Specifically, when some path includes positive flow on arc (i, j), then only a single arc (j, k) can also have positive flow:

$$\sum_{i:(s,i)\in\mathcal{A}} z_{si}^{p} \le 1 \quad \forall p = 1, \ldots, f,$$

$$\sum_{j:(i,j)\in\mathcal{A}} z_{ij}^{p} - \sum_{h:(h,i)\in\mathcal{A}} z_{hi}^{p} = 0 \quad \forall i \in \mathcal{V} \setminus \{s, t\},\ p = 1, \ldots, f.$$

Next, a MIP solution must appropriately account for all costs related to an MCFA solution. To do so, some set of inequalities must enforce the relationship between variables x, y, and z. When $x_{ij}^{p} > 0$ for some $(i, j) \in \mathcal{A}$ and $p = 1, \ldots, f$, then z_{ij}^{p} must equal 1 by definition. The following constraints enforce this relationship:

$$z_{ij}^p \leq x_{ij}^p \leq c_{ij} z_{ij}^p \quad \forall (i, j) \in \mathcal{A},\ p = 1, \ldots, f.$$

Thus, if $z_{ij}^p = 0$, then $x_{ij}^p = 0$ as well, and if $z_{ij} = 1$, then $x_{ij}^p \in [1, c_{ij}]$. This constraint by itself does not completely enforce the arc capacity constraints. To achieve this, we must require that the sum of all flows using arc $(i, j) \in \mathcal{A}$ does not exceed c_{ij}. We enforce those capacity restrictions as follows:

$$\sum_{p=1}^{f} x_{ij}^p \leq c_{ij} \quad \forall (i, j) \in \mathcal{A}.$$

Since the y-variables indicate whether an arc is activated or not for each path, constraints need to enforce $y_{ij}^p = 1$ when arc (i, j) is activated on the pth path:

$$y_{ij}^p \geq z_{ij}^p - z_{ij}^{p-1} \quad \forall (i, j) \in \mathcal{A},\ p = 1, \ldots, f.$$

These constraints force $y_{ij}^p = 1$ when $z_{ij}^p = 1$ and $z_{ij}^{p-1} = 0$, and 0 otherwise. Whenever arc (i, j) is included (or not included) on both paths p and $p - 1$, then y_{ij}^p is allowed to take on a value of 0. Since the activation costs a_{ij} are all positive, optimization forces $y_{ij}^p = 0$ in such cases. Because z-variables are binary, then $z_{ij}^p - z_{ij}^{p-1}$ either equals $-1, 0$, or 1. Thus, at optimality, $y_{ij}^p = 0$ when $z_{ij}^p - z_{ij}^{p-1} \leq 0$, and $y_{ij}^p = 1$ when $z_{ij}^p - z_{ij}^{p-1} = 1$, for all $(i, j) \in \mathcal{A}$ and $p = 1, \ldots, f$. We can therefore relax binary restrictions on y_{ij}^p for all $(i, j) \in \mathcal{A}$ and $p = 1, \ldots, f$.

Finally, we include the nonnegativity restrictions on x and y and the binary restrictions on z, respectively:

$$x_{ij}^p,\ y_{ij}^p \geq 0 \quad \forall (i, j) \in \mathcal{A},\ p = 1, \ldots, f$$
$$z_{ij}^p \in \{0, 1\} \quad \forall (i, j) \in \mathcal{A},\ p = 1, \ldots, f.$$

When aggregated, the proposed constraints, along with the objective function

$$\min \sum_{p=1}^{f} \sum_{(i,j)\in\mathcal{A}} \left(d_{ij} x_{ij}^p + a_{ij} y_{ij}^p \right),$$

correctly model the MCFA. The computational performance of solving this MIP using traditional branch-and-bound methods depends on the size of the resulting branch-and-bound tree. We next explore a few ways to reduce the size and symmetry of the branch-and-bound tree.

First, we note that the number of unique path-flow schedules producing the same total cost could be exponentially large. This can occur when the linear programming relaxation of the MCFA MIP produces a path-flow schedule in which zero-flow paths are interleaved among subsequences of positive-flow paths. Such an occurrence would significantly and unnecessarily increase computational time as the branch-and-bound tree would need to examine very many identical solutions of this type [31]. However, when zero-flow paths are interleaved among positive-flow paths in an optimal solution, there must exist an alternative optimal solution in which all positive-flow paths are scheduled consecutively at the beginning of the sequence. Using this observation, we thus add the following constraints:

$$\sum_{i:(s,i)\in\mathcal{A}} z^1_{si} = 1$$

$$\sum_{i:(s,i)\in\mathcal{A}} z^{p+1}_{si} \leq \sum_{i:(s,i)\in\mathcal{A}} z^p_{si} \quad \forall p = 1,\ldots,f-1.$$

The first constraint ensures that the first scheduled path must have a positive flow. (If an optimal solution produces a path-flow schedule having zero flow on the first path, then an alternative optimal solution is guaranteed to exist in which all positive-flow paths are scheduled earlier, so that the first scheduled path-flow has positive flow.) Similarly, the second set of constraints ensures that if the $(p+1)$st path has positive flow, then so does the pth path, for all paths $p = 1,\ldots,f-1$. The inclusion of these constraints ensures that no zero-flow paths are scheduled in the middle of a path-flow schedule. Additionally, these constraints also imply that $\sum_{i:(s,i)\in\mathcal{A}} z^p_{si} \leq 1$ for each $p = 1,\ldots,f$, and so we omit those restrictions moving forward.

With the addition of these constraints, and by defining $z^0_{ij} = 0,\ \forall (i,j) \in \mathcal{A}$, the MCFA can be modeled by the following MIP:

$$\text{Min} \sum_{p=1}^{f} \sum_{(i,j)\in\mathcal{A}} \left(d_{ij}x^p_{ij} + a_{ij}y^p_{ij}\right) \tag{3.51a}$$

$$\text{s.t.} \sum_{p=1}^{f} \left(\sum_{i:(s,i)\in\mathcal{A}} x^p_{si} - \sum_{j:(j,s)\in\mathcal{A}} x^p_{js} \right) = f \tag{3.51b}$$

$$\sum_{j:(i,j)\in\mathcal{A}} x^p_{ij} - \sum_{h:(h,i)\in\mathcal{A}} x^p_{hi} = 0 \quad \forall i \in \mathcal{V}\setminus\{s,t\},\ p = 1,\ldots,f \tag{3.51c}$$

$$\sum_{p=1}^{f} x^p_{ij} \leq c_{ij} \quad \forall (i,j) \in \mathcal{A} \tag{3.51d}$$

$$\sum_{i:(s,i)\in\mathcal{A}} z^1_{si} = 1 \tag{3.51e}$$

$$\sum_{i:(s,i)\in\mathcal{A}} z^{p+1}_{si} \leq \sum_{i:(s,i)\in\mathcal{A}} z^p_{si} \quad \forall p = 1,\ldots,f-1 \tag{3.51f}$$

$$\sum_{j:(i,j)\in\mathcal{A}} z^p_{ij} - \sum_{h:(h,i)\in\mathcal{A}} z^p_{hi} = 0 \quad \forall i \in \mathcal{V}\setminus\{s,t\},\ p = 1,\ldots,f \tag{3.51g}$$

$$z^p_{ij} \leq x^p_{ij} \leq c_{ij}z^p_{ij} \quad \forall (i,j) \in \mathcal{A},\ p = 1,\ldots,f \tag{3.51h}$$

$$y_{ij}^p \geq z_{ij}^p - z_{ij}^{p-1} \quad \forall (i, j) \in \mathcal{A},\ p = 1, \ldots, f \tag{3.51i}$$

$$x_{ij}^p,\ y_{ij}^p \geq 0 \quad \forall (i, j) \in \mathcal{A},\ p = 1, \ldots, f \tag{3.51j}$$

$$z_{ij}^p \in \{0, 1\} \quad \forall (i, j) \in \mathcal{A},\ p = 1, \ldots, f. \tag{3.51k}$$

Model (3.51) can be solved using commercial MIP solvers. In this case, the model size heavily depends on the number of possible path-flows f; however, f is input data. Thus, model (3.51) is pseudo-polynomial in size, and the number of constraints and variables in (3.51) could become very large whenever a significant amount of flow must be transmitted from s to t. Alternatively, the number of positive-flow paths in an optimal solution will typically be far fewer than f, since each path-flow is *likely* to transmit more than one unit of flow from s to t. This observation motivates the exploration of an MCFA relaxation that requires fewer variables and constraints, as we discuss in the next subsection.

3.5.3 Alternative MIP for an MCFA Relaxation

We now create a relaxation of MCFA, i.e., a version of MCFA in which some of the restrictions are removed. We refer to the relaxed MCFA as problem MCFA-R, and describe it as follows.

Rather than scheduling flows over the (potentially very large) set of indices $1, \ldots, f$, the MCFA-R permits only $\lambda < f$ path-flows. However, any remaining flows not scheduled on the first λ path-flows can then be sent *simultaneously* (rather than consecutively) on a final flow pattern. Because these final flows are transmitted simultaneously, they need not correspond to an s–t path. We refer to this final flow as an *aggregate* flow, since all remaining path-flows are aggregated into a general network flow that conserves flow on all intermediate nodes in $\mathcal{V} \setminus \{s, t\}$. The aggregate flow occurs in position $\lambda + 1$ of the flow schedule.

Because the aggregate flow does not necessarily occur over a path, a schedule that uses an aggregate flow is not guaranteed to be feasible to the MCFA. It will, however, provide a lower bound on the optimal MCFA objective. (We refer the reader to [11] for a full justification of this claim, but the intuition is that we can take any optimal solution to the MCFA and convert it into a feasible solution to the MCFA-R as follows: Use the same solution over paths $1, \ldots, \lambda$, and then let the aggregate flow in MCFA-R be the sum of flows over paths $\lambda + 1, \ldots, f$ in the MCFA solution. The resulting MCFA-R solution has an objective function value less than or equal to that for the MCFA problem, by possibly avoiding activation costs in the aggregate flow. Moreover, the optimal objective to MCFA-R may be even smaller than the one generated by aggregating the flows $\lambda + 1, \ldots, f$ from the MCFA solution.) Additionally, we force the aggregate flow to be zero unless there is positive flow on the first λ paths, or else all flow in an MCFA-R solution could be sent through the aggregate flow to minimize activation costs. Formally defined, MCFA-R determines a schedule of λ path flows, possibly followed by a single aggregate flow, minimizing the sum of activation and transmission costs required to send f units of flow from s to t.

We now show how to build a MIP for solving the MCFA-R. As with the MIP model for the MCFA, variables x, y, and z and parameters c, d, and a are defined as before, but over $p = 1, \ldots, \lambda + 1$. The objective is exactly as for MCFA, except that the first summation goes to $\lambda + 1$ instead of f. That is, rather than minimizing the total cost of scheduling path-flows over all f paths, the MCFA-R objective minimizes the total cost of scheduling the first λ paths plus the cost of the aggregate flow in position $\lambda + 1$. The objective function still includes the cost to activate arcs that are included in the aggregate flow (in position $\lambda + 1$) when those arcs are not included in the path-flow in position λ.

Most of the MCFA-R constraints are analogous to those in the MCFA MIP. These include changing the flow index limit f to $\lambda + 1$ with the following exceptions:

- The right-hand side of (3.51b) remains f, as this constraint states the required flow that exits node s.
- Symmetry-breaking constraints (3.51f) should now be defined from $p = 1, \ldots, \lambda - 1$, not from $p = 1, \ldots, \lambda$, as we explain below.

The symmetry-breaking constraints (3.51f) now insist that the $(p + 1)$st path-flow is positive only when the pth path-flow is also positive, for $p = 1, \ldots, \lambda - 1$. To ensure that the aggregate flow is positive only when the first λ flows are positive (or, equivalently, when the flow on path λ in particular is positive), we state the following constraints:

$$z_{sj}^{\lambda+1} \leq \sum_{i:(s,i)\in\mathcal{A}} z_{si}^{\lambda} \qquad \forall j : (s, j) \in \mathcal{A}.$$

This revised form is necessary because the aggregate flow is not necessarily a path.

Next, observe that without the addition of any other constraints, the bound provided by the model developed so far may be poor. For example, the presented constraints allow for a solution in which the first λ paths send a single unit of flow, each along an identical path. In doing so, the activation cost for each arc on this path is paid only a single time. Then, the remaining flow, $f - \sum_{i:(s,i)\in\mathcal{A}} \sum_{p-1}^{\lambda} x_{si}^{p}$, is transmitted as an aggregate flow. The activation cost on each arc having positive flow on the aggregate flow, if not included on path-flow λ, is paid only once as well. For this given solution, the activation cost on every arc is paid no more than once, and we have already shown that some optimal solutions may require some arcs to be activated more than once.

Thus, to improve the bound provided by solving the MCFA-R, we seek to prohibit flow schedules having consecutive flows on identical paths. To accomplish this, we propose a combination of constraints. First, we add a set of constraints to ensure that at least one arc is activated for each positive-flow path:

$$\sum_{(u,v)\in\mathcal{A}} y^p_{uv} \geq \sum_{i:(s,i)\in\mathcal{A}} z^p_{si} \quad \forall p = 1, \ldots, \lambda + 1.$$

Note that these constraints have a different implication for the aggregate flow: They state that the number of arcs activated on the aggregate flow must be at least the number of arcs incident to node s that are used to transmit flow. To see why this is constraint is valid, consider an aggregate flow that uses, say, three arcs emanating from node s. At least two of those arcs were not used in path λ (because one arc is used in that path that emanates from node s). If all three of the arcs were not used in path λ, then the constraint is immediately seen to be valid. If not, then consider any s–t path in the aggregate flow that uses a common first arc. That path must differ from path λ (or else more flow should have been transmitted on λ), and it must differ on an arc that is not incident to s. Thus, a third arc would be activated in this case as well.

The addition of these constraints, though, does not strictly enforce that each positive-flow path $p = 2, \ldots, \lambda$ is distinct from path-flow $p - 1$. For example, some subsequence of positions could still use the same path while paying the activation cost for some arc not actually on that path. Therefore, we propose the following constraints, in addition to $y^p_{ij} \geq z^p_{ij} - z^{p-1}_{ij}$, $\forall (i, j) \in \mathcal{A},\ p = 1, \ldots, \lambda + 1$, to enforce that $y^p_{ij} = 1$ if and only if $z^p_{ij} = 1$ and $z^{p-1}_{ij} = 0$:

$$\begin{aligned} &y^p_{ij} \leq z^p_{ij} && \forall (i, j) \in \mathcal{A},\ p = 1, \ldots, \lambda + 1 \\ &y^p_{ij} \leq 1 - z^{p-1}_{ij} && \forall (i, j) \in \mathcal{A},\ p = 1, \ldots, \lambda + 1. \end{aligned}$$

These constraints ensure that the activation cost for some arc must be paid only in the case that it fits the definition of arc activation. The inclusion of these constraints again allows the binary constraints on the y-variables to be relaxed.

The MCFA-R can be modeled by the following MIP:

$$\text{Min} \sum_{p=1}^{\lambda+1} \sum_{(i,j)\in\mathcal{A}} \left(d_{ij} x^p_{ij} + a_{ij} y^p_{ij} \right) \tag{3.52a}$$

$$\text{s.t.} \sum_{p=1}^{\lambda+1} \left(\sum_{i:(s,i)\in\mathcal{A}} x^p_{si} - \sum_{j:(j,s)\in\mathcal{A}} x^p_{js} \right) = f \tag{3.52b}$$

$$\sum_{j:(i,j)\in\mathcal{A}} x^p_{ij} - \sum_{h:(h,i)\in\mathcal{A}} x^p_{hi} = 0 \quad \forall i \in \mathcal{V} \setminus \{s, t\},\ p = 1, \ldots, \lambda + 1 \tag{3.52c}$$

$$\sum_{j:(i,j)\in\mathcal{A}} z^p_{ij} - \sum_{h:(h,i)\in\mathcal{A}} z^p_{hi} = 0 \quad \forall i \in \mathcal{V} \setminus \{s, t\},\ p = 1, \ldots, \lambda \tag{3.52d}$$

$$z^p_{ij} \leq x^p_{ij} \leq c_{ij} z^p_{ij} \quad \forall (i, j) \in \mathcal{A},\ p = 1, \ldots, \lambda + 1 \tag{3.52e}$$

$$\sum_{p=1}^{\lambda+1} x_{ij}^p \le c_{ij} \qquad \forall (i,j) \in \mathcal{A} \tag{3.52f}$$

$$y_{ij}^p \ge z_{ij}^p - z_{ij}^{p-1} \qquad \forall (i,j) \in \mathcal{A},\ p = 1,\dots,\lambda+1 \tag{3.52g}$$

$$y_{ij}^p \le z_{ij}^p \qquad \forall (i,j) \in \mathcal{A},\ p = 1,\dots,\lambda+1 \tag{3.52h}$$

$$y_{ij}^p \le 1 - z_{ij}^{p-1} \qquad \forall (i,j) \in \mathcal{A},\ p = 1,\dots,\lambda+1 \tag{3.52i}$$

$$\sum_{(u,v)\in\mathcal{A}} y_{uv}^p \ge \sum_{i:(s,i)\in\mathcal{A}} z_{si}^p \qquad \forall p = 1,\dots,\lambda+1 \tag{3.52j}$$

$$\sum_{i:(s,i)\in\mathcal{A}} z_{si}^1 = 1 \tag{3.52k}$$

$$\sum_{i:(s,i)\in\mathcal{A}} z_{si}^{p+1} \le \sum_{i:(s,i)\in\mathcal{A}} z_{si}^p \qquad \forall p = 1,\dots,\lambda-1 \tag{3.52l}$$

$$z_{sj}^{\lambda+1} \le \sum_{i:(s,i)\in\mathcal{A}} z_{si}^{\lambda} \qquad \forall j : (s,j) \in \mathcal{A} \tag{3.52m}$$

$$x_{ij}^p,\ y_{ij}^p \ge 0 \qquad \forall (i,j) \in \mathcal{A},\ p = 1,\dots,\lambda+1 \tag{3.52n}$$

$$z_{ij}^p \in \{0,1\} \qquad \forall (i,j) \in \mathcal{A},\ p = 1,\dots,\lambda+1. \tag{3.52o}$$

Observe that constraints (3.52h)–(3.52j) are included in the MCFA-R MIP but not in the MCFA MIP. In the MCFA MIP, an optimal solution exists having all unique positive-flow paths, so that (3.52j) is unnecessary for the MCFA MIP. As discussed in Sect. 3.5.2, the upper bound on y enforced by (3.52h) and (3.52i) is also unnecessary for the MCFA MIP. However, including these constraints in the MCFA MIP may tighten its relaxation. This tightening comes at the cost of a slight increase in the size of the MIP. For more explanation on how these inequalities strengthen the relaxation of the MCFA MIP, see [11].

Recall that an optimal solution to model (3.52) only provides a lower bound on the cost of an optimal MCFA path-flow schedule. However, in practice, solving model (3.52) often provides a feasible solution to the MCFA. An optimal MCFA-R solution is feasible to the MCFA if the aggregate flow forms a positive-flow path, or if there is no aggregate flow at all. In either case, the path flows in the MCFA-R solution form a feasible solution to the MCFA, and any feasible solution to the MCFA provides a lower bound to that problem. Therefore, if our MCFA-R solution yields a feasible MCFA solution, then that solution must be optimal to the MCFA.

3.6 Conclusion

The chapter discusses some modeling principles that can be applied to formulate a vast array of practical scheduling optimization problems. It would be too ambitious a task to provide a taxonomy of all viable models for all problems in the literature, and attempting to do so for as-yet unresearched problems is hopeless. However, our experience is that many (perhaps most) scheduling problems can be approached either using variables that discretize the time horizon over which scheduling takes place, using variables that describe the relative position of jobs in a schedule (or perhaps using immediate precedence relations among the jobs), or variables that prescribe the time at which a job begins processing on the machine. Each approach has its corresponding advantages and disadvantages in terms of its resulting size and solvability using branch-and-bound techniques, depending perhaps on the objective function being used.

Formulating scheduling problems is only part of the science behind computing optimal schedules. Fortunately, there now exist powerful MIP solvers that can quickly solve surprisingly large instances of these optimization problems. However, there will always be limits to the size of scheduling problems that can be optimized exactly within the required computational limits. As a result, recent research has identified various modeling and solution approaches, such as heuristic algorithms [28, 30, 38], decomposition methods [3, 10, 39], and constraint programming (CP) approaches [14, 16, 24, 33] that prove to be viable alternatives, and indeed are sometimes the only practical way of addressing these problems.

In particular, promising work highlights the efficacy of CP approaches, especially when combined to create a hybrid CP/MIP solution technique. Constraint programming models are well-known for quickly finding quality feasible solutions to small- or large-scale scheduling problems and tend to be smaller than MIPs for small instances [15]. Thus, hybrid CP/MIP approaches often use CPs to generate initial feasible solutions that serve as warm-start solutions for a MIP in order to decrease computational effort. Recent research shows that this hybrid approach can outperform stand-alone MIP or CP models [24]. Future research trends may continue to explore various ways for combining multiple modeling approaches to scheduling problems. We encourage any reader with an interest in scheduling optimization to include MIP modeling within their toolkit, but never at the expense of considering or combining alternative approaches as well.

References

1. Ageev AA, Baburin AE (2007) Approximation algorithms for UET scheduling problems with exact delays. Oper Res Lett 35(4):533–540
2. Applegate DL, Bixby RE, Chvátal V, Cook WJ (2007) The traveling salesman problem: A computational study. Princeton University Press, Princeton, NJ
3. Barzanji R, Naderi B, Begen MA (2020) Decomposition algorithms for the integrated process planning and scheduling problem. Omega 93:102025
4. Ben-Tal A, El-Ghaoui L, Nemirovski A (2009) Robust optimization. Princeton University Press, Princeton, NJ
5. Bertsimas D, Sim M (2004) The price of robustness. Oper Res 52(1):35–53
6. Bertsimas D, Thiele A (2006) Robust and data-driven optimization: Modern decision-making under uncertainty. In: Johnson MP, Norman B, Secomandi N (eds) Tutorials in operations research: Models, methods, and applications for innovative decision making. INFORMS, pp 95–122

7. Birge JR, Louveaux FV (1997) Introduction to stochastic programming. Springer, New York
8. Brailsford SC, Potts CN, Smith BM (1999) Constraint satisfaction problems: Algorithms and applications. Eur J Oper Res 119(3):557–581
9. Brucker P, Knust S (2012) Complex scheduling. Springer, Berlin Heidelberg
10. Bruni M, Di Puglia Pugliese L, Beraldi P, Guerriero F (2018) A computational study of exact approaches for the adjustable robust resource-constrained project scheduling problem. Comput Oper Res 99:178–190 (2018)
11. Curry RM, Smith JC (2022) Minimum-cost flow problems having arc-activation costs. Naval Res Logistics 69:320–335
12. Elshafei M, Sherali HD, Smith JC (2004) Radar pulse interleaving for multi-target tracking. Naval Res Logistics 51(4):72–94
13. Farina A, Neri P (1980) Multitarget interleaved tracking for phased-array radar. IEEE Proc Part F: Commun Radar Signal Process 127(4):312–318
14. Gedik R, Kalathia D, Egilmez G, Kirac E (2018) A constraint programming approach for solving unrelated parallel machine scheduling problem. Comput Ind Eng 121:139–149
15. Gedik R, Rainwater C, Nachtmann H, Pohl EA (2016) Analysis of a parallel machine scheduling problem with sequence dependent setup times and job availability intervals. Eur J Oper Res 251(2):640–650
16. Gökgür B, Hnich B, Özpeynirci S (2018) Parallel machine scheduling with tool loading: A constraint programming approach. Int J Prod Res 56(16):5541–5557
17. Graham RL, Lawler EL, Lenstra J, Rinnooy Kan AHG (1979) Optimization and approximation in deterministic sequencing and scheduling: A survey. Ann Discret Math 5:287–326
18. Hwang F, Lin BM (2011) Coupled-task scheduling on a single machine subject to a fixed-job-sequence. Comput Ind Eng 60(4):690–698
19. Izquierdo-Fuente A, Casar-Corredera JR (1994) Optimal radar pulse scheduling using a neural network. In IEEE Int Conf Neural Netw 7:4588–4591
20. Kall P, Wallace SW (1994) Stochastic Programming. Wiley, Chichester, UK
21. Keha AB, Khowala K, Fowler JW (2009) Mixed integer programming formulations for single machine scheduling problems. Comput Ind Eng 56(1):357–367
22. Khatami M, Salehipour A, Cheng TCE (2020) Coupled task scheduling with exact delays: Literature review and models. Eur J Oper Res 282(1):19–39
23. Ławrynowicz A (2011) Advanced scheduling with genetic algorithms in supply networks. J Manuf Technol Manag 22(6):748–769
24. Pour SM, Drake JH, Ejlertsen LS, Rasmussen KM, Burke EK (2018) A hybrid constraint programming/mixed integer programming framework for the preventive signaling maintenance crew scheduling problem. Eur J Oper Res 269(1):341–352
25. Miller CE, Tucker AW, Zemlin RA (1960) Integer programming formulations and traveling salesman problems. J Assoc Comput Mach 7:326–329
26. Nogueira TH, de Carvalho CRV, Ravetti MG, de Souza MC (2019) Analysis of mixed integer programming formulations for single machine scheduling problems with sequence dependent setup times and release dates. Pesquisa Operacional 39(1):109–154
27. Pan C-H (1997) A study of integer programming formulations for scheduling problems. Int J Syst Sci 28(1):33–41
28. Pellerin R, Perrier N, Berthaut F (2020) A survey of hybrid metaheuristics for the resource-constrained project scheduling problem. Eur J Oper Res 280(2):395–416
29. Rockafellar RT, Uryasev SP (2000) Optimization of conditional value-at-risk. Journal of Risk 2:21–41
30. Rohaninejad M, Tavakkoli-Moghaddam R, Vahedi-Nouri B, Hanzálek Z, Shirazian S (2022) A hybrid learning-based meta-heuristic algorithm for scheduling of an additive manufacturing system consisting of parallel SLM machines. Int J Prod Res 60(20):6205–6225
31. Sherali HD, Smith JC (2001) Improving zero-one model representations via symmetry considerations. Manag Sci 47(10):1396–1407
32. Sherali HD, Smith JC (2005) Interleaving two-phased jobs on a single machine with application to radar pulse interleaving discrete optimization. Discret Optim 2(4):348–361

33. Shi G, Yang Z, Xu Y, Quan Y (2022) Solving the integrated process planning and scheduling problem using an enhanced constraint programming-based approach. Int J Prod Res 60(18):5505–5522
34. Skolnik MI (ed) Radar handbook. McGraw Hill, New York, NY (1990)
35. Smith WE (1956) Various optimizers for single-stage production. Naval Res Logistics Q 3(1–2):59–66
36. Tadayon B, Smith JC (2015) Algorithms and complexity analysis for robust single-machine scheduling problems. J Sched 18(6):575–592
37. Ying K-C, Cheng C-Y, Lin S-W, Hung C-Y (2019) Comparative analysis of mixed integer programming formulations for single-machine and parallel-machine scheduling problems. IEEE Access 7:152998–153011
38. Zhang B, Pan Q-K, Gao L, Li X-Y, Meng L-L, Peng K-K (2019) A multiobjective evolutionary algorithm based on decomposition for hybrid flowshop green scheduling problem. Comput Ind Eng 136:325–344
39. Zhang B, Pan Q-K, Gao L, Meng L-L, Li X-Y, Peng K-K (2020) A three-stage multiobjective approach based on decomposition for an energy-efficient hybrid flow shop scheduling problem. IEEE Trans Syst Man Cybern: Syst 50(12):4984–4999

Chapter 4
Converting Weak to Strong MIP Formulations: A Practitioner's Guide

Ed Klotz and Richard Oberdieck

Abstract Enormous progress in Mixed Integer Programming (MIP) solvers has come from synergies involving the increasingly powerful computing environments on which they run and the algorithms they implement. Essential to the algorithmic improvements are mathematical techniques to strengthen weak formulations. The literature contains sophisticated mathematical theory to generate cutting planes that make the branch and bound algorithm most solvers use more effective by tightening the formulation of the model the solver receives. Nonetheless, numerous MIP models remain difficult or impossible to solve to optimality, as evidenced by the hard and open models in the MIPLIB test set. In addition, as larger and more difficult MIPs have become easier to solve, practitioners have responded by creating larger, more granular, and more difficult MIPs that pose new challenges for the solvers. The cuts implemented in the MIP solvers tend to be generic, looking for common structures and characteristics in MIP models from diverse sources. However, some models have more uncommon characteristics that elude the implemented cuts. Leveraging knowledge specific to their models can help practitioners create more customized cuts that strengthen the formulation in ways that significantly improve solver performance. Optimization practitioners with limited familiarity with the theoretical constructs regarding the strength of formulations may conclude that they lack the prerequisites to effectively strengthen their models. In this chapter the authors will argue that, while such knowledge is an extremely valuable asset, it is not always needed to effectively tighten formulations. After reviewing some previous work in this area and discussing some easily remedied common pitfalls that result in slow solver times, the authors will consider how to identify weak formulations based on the contrast between the physical systems modeled by the MIP and that of the LP relaxation that initiates the branch and bound algorithm. Both the algebraic representation of the constraints and a more visual approach will be considered. Practitioners can use an

E. Klotz (✉) · R. Oberdieck
Gurobi Optimization, LLC, Beaverton, OR, USA
e-mail: klotz@gurobi.com

R. Oberdieck
e-mail: r.oberdieck@gmail.com

F. Hamid (ed.), *Optimization Essentials*, International Series in Operations Research & Management Science 353, https://doi.org/10.1007/978-981-99-5491-9_4

understanding of the weakness of a formulation to strengthen it by customizing cuts tailored to the characteristics of the model at hand. The authors will illustrate this approach by considering some examples involving challenging models drawn from practical business applications.

Keywords Mixed integer programming · Weak formulations · Strong formulations · Branch and bound · Run time · Tight formulations · Cuts

4.1 Introduction

Enormous progress in Mixed Integer Programming (MIP) solvers has come from synergies involving the increasingly powerful computing environments on which they run and the algorithms they implement. Models that were completely unsolvable 10 or more years ago are now solved easily. This has had a transformative effect within various industries (see, e.g., Carlson et al. [6]) Essential to the algorithmic improvements are mathematical techniques to strengthen weak formulations. The literature contains sophisticated mathematical theory to generate cuts that make the branch and bound algorithm most solvers use more effective by tightening the formulation of the model the solver receives. Polyhedral and graph theory are two common strategies used to generate cuts.

Nonetheless, numerous MIP models remain difficult or impossible to solve to optimality, as evidenced by the hard and open models in the MIPLIB test set [12]. This should come as no surprise, as the mathematical models underpinning many of these models are NP hard or NP complete, meaning that while MIP solvers can effectively solve many instances of these models, it is unlikely that they will be able to solve all instances unless $P = NP$. In addition, as the solvers have thinned the herd of difficult models, practitioners have formulated larger, more complex, and more granular models, posing new challenges for the MIP solvers.

Optimization practitioners with limited familiarity with the theoretical constructs regarding the strength of formulations may conclude that they lack the prerequisites to effectively strengthen their models. In this chapter the authors will argue that, while such knowledge is an extremely valuable asset, it is not always needed to effectively tighten formulations. This chapter seeks to empower practitioners on how to effectively tighten their formulation. We encourage them to begin by using their knowledge of the model and a set of generic, easily understandable strategies to understand the sources of weakness in a MIP formulation. They can then use that understanding to tighten the formulation. Emphasis will be given to a more visual approach for distinguishing weak formulations from strong ones. The authors hope that this material will also provide useful background material for those interested in learning the more advanced theory used to strengthen formulations.

After reviewing some previous work in this area and discussing easily remedied common pitfalls that result in slow solver times, the authors will consider how to strengthen weak formulations based on the contrast between the physical systems modeled by the MIP and that of the LP relaxation that initiates the branch and bound algorithm. Both the algebraic representation of the constraints and a more visual approach will be considered. Practitioners can use an understanding of the weakness of a formulation to strengthen it by customizing cuts tailored to the characteristics of the model at hand. The authors will illustrate this approach by considering some examples involving challenging models drawn from practical business applications.

The chapter is organized as follows.

- Section 4.3 considers previous efforts in the literature on tightening MIP formulations.
- Section 4.4 provides a checklist of common problems in MIP formulations that can be addressed by looking for the problematic characteristic and making the suitable correction. While these may involve some notion of weakness in the formulation, the practitioner can just correct the problems listed without any detailed knowledge about the underlying issue.
- Section 4.5 describes different methods to identify weak formulations. Each method may be better suited for certain types of MIPs, so familiarity with all of them can help the practitioner immensely.
- Section 4.6 describes three fundamental methods for strengthening formulations. These can be effective on almost any MIP model, although applying them effectively is not always easy.
- Section 4.7 considers some additional strengthening tactics that are less generic but are sufficiently common that they can help on a significant number of MIPs.
- Section 4.8 provides a summary of the cuts found in most state-of-the-art MIP solvers. This serves to help the practitioner avoid spending time deriving cuts that the solver will identify without any assistance.
- Section 4.9 considers some examples from industrial applications where the authors apply the developed strategies.
- Section 4.10 describes some techniques to help the practitioner validate the derived cuts.
- Section 4.11 summarizes with some key takeaways and also considers future work in this area.

Regarding background, familiarity with linear and integer programming and the basic algorithms used to solve them in practice is assumed. Advanced knowledge of the literature regarding theoretical results of integer programming is not required. Nonetheless, those with strong theoretical backgrounds in the theory of integer programming may find the chapter helpful for explaining the concepts of weak and strong formulations to audiences learning about them for the first time.

As both authors worked together at Gurobi Optimization, LLC while writing this document, all solver logs and other output will come from the Gurobi Optimizer. However, most of the concepts discussed can be used effectively, independent of the solver being used.

Remark 4.1 This book chapter focuses on the notion of strengthening the existing formulation through cutting planes that improve the performance of the branch and bound algorithm. Other methodologies for solving MIPs more efficiently exist. However, as these methods would merit their own chapters, they will not be covered in this chapter in any detail. Summarizing briefly, other avenues to pursue performance gains include:

- **Equivalent formulations on which the branch and bound algorithm runs faster**. This involves choosing a different set of variables, which then will result in a different formulation that may be easier for a branch and bound solver to handle.
- **Reformulations that require decomposition methods rather than branch and bound to solve**. This involves reformulating the model with a much larger number of constraints or variables that makes it too large for the branch and bound algorithm. However methods such as Dantzig-Wolfe (Lübbecke and Desrosiers [19] and Santos and Nepomuceno [24]) and Benders' (Benders [5] and Rahmaniani et al. [23]) Decomposition can solve these significantly larger models by incrementally augmenting the set of variables or constraints that are actually involved in an optimal solution without ever creating the full model.
- **Other algorithms that use different relaxations**. For example, algorithms using surrogate constraints Gavish and Pirkul [10] and Gavish et al. [11] create a relaxation from a linear combination of the original constraints. The algorithm then systematically modifies the multipliers associated with the linear combination until the optimal relaxation solution satisfies all the original constraints and is therefore optimal for the original MIP as well.

4.2 Notation and Terminology

Let x be a vector of variables in $\mathbb{R}^n$, A be an m by n matrix, c, l and u be vectors in $\mathbb{R}^n$ and b a vector in $\mathbb{R}^m$. Let C and I be index sets that partition $1, \ldots, n$ and identify continuous (x_C) and integer variables (x_I), respectively. In its broadest sense, a Mixed Integer Program (MIP) can then be defined as follows:

$$
\begin{aligned}
(P_{MIP}): \quad & \underset{x}{\text{minimize}} \quad c_C^T x_C + c_I^T x_I + .5 x^T Q x \\
& \text{subject to} \quad A_C x_C + A_I x_I \sim b \\
& \qquad\qquad\qquad l \le x \le u, \quad x_I \ \text{integer}
\end{aligned}
$$

If $Q = 0$, the MIP is a Mixed Integer Linear Program (MILP); otherwise it is a Mixed Integer Quadratic Program (MIQP); the MIP designation subsumes both problem types.[1] The $\sim$ notion for the linear constraints denotes that they can be an arbitrary mixture of equality and inequality constraints. No assumption is made that

[1] In some circumstances, the abbreviation "MIP" is used synonymously with "MILP". Please note that this is not the case in this chapter.

C or I is nonempty, so P_{MIP} subsumes the cases where the problem involves purely integer variables (an IP) or purely continuous variables (an LP or QP). The variable bounds in the vectors l and u are arbitrary real numbers, but most MIPs have lower bounds of 0 and positive (finite or infinite) upper bounds. In the special case where $C = \emptyset, l = 0$ and $u = 1$, the MIP is a binary integer program (BIP). However, all of these special cases are included in the MIP definition, which will be used throughout the chapter unless the context requires a more specific problem type to be identified. P_{MIP} is expressed as a minimization problem, and all discussions of the algorithm make that assumption. But maximization problems can be expressed by minimizing the negative of the objective to be maximized.

4.3 Previous Work

Most of the MIP literature involving weak and strong formulations focuses on more theoretical advances. The MIPs considered tend to have only one or two categories of constraints, enabling characterizations of both the convex hull of the set of integer feasible solutions, as well as the polytope associated with the LP relaxation. This approach has been very effective for the development of valid inequalities (i.e., cuts) that define facets of the LP relaxation polytope, resulting in the sharpest cuts possible. Chapter 9 of Wolsey [28] provides an excellent description of numerous different methods to do this. The papers in the literature on this topic are too numerous to mention here, but Grötschel [14] provides a centralized location with many examples. These theoretical advancements have made huge contributions to the cuts implemented in MIP solvers. However, practitioners often encounter MIPs with more groups of constraints than those considered in the literature, as well as continuous variables that make the establishment of facet defining inequalities more challenging. Furthermore, practitioners may have limited familiarity with the concepts needed for this approach.

Nonetheless, the literature does include some guidance for strengthening MIP formulations for practitioners less familiar with the theoretical concepts. Williams [27] focuses primarily on techniques for formulating models, but in doing so it implicitly discusses modeling practices that are more likely to yield strong formulations. Klotz and Newman [17] instead assume that the MIP has been formulated and focuses on an organized set of concepts and guidelines to strengthen arbitrary MIP formulations. This section of the chapter will summarize some of the core ideas described therein, and Sects. 4.5, 4.6 and 4.7 build upon and extend some of the guidelines in that previous work.

In its broadest terms, Klotz and Newman [17] can be summarized as follows when a difficult MIP is encountered.

1. Use the solver node log to identify the source of the slow performance.
2. Try adjusting solver parameters that are particularly well suited to addressing the source of the slow performance. A MIP solver is a collection of configurable

algorithms and procedures. Default values are chosen to provide good average performance, but many solver parameters can be set to improve performance on models with specific characteristics.

3. If adjusting parameters in this way does not yield a sufficient performance improvement, use an organized set of guidelines to first identify the weakness in the formulation, then use the knowledge of the weakness to strengthen the formulation by adding cuts to the MIP that are too specific to be derived by the more generic cut generators of the MIP solvers.

The solver node log essentially provides a scoreboard of solver performance, and it frequently sheds light on common sources of MIP performance problems:

A. There is insufficient node throughput due to slow node relaxation solves, as none of the solver's LP algorithms efficiently solve the node LPs.
B. There is a lack of progress in the best integer solution (the upper or primal bound in the node log).
C. There is a lack of progress in the lower bound on the optimal objective value of the MIP (the best or dual bound in the node log).
D. There is insufficient node throughput due to numerical instability or ill conditioning arising from the problem data.
E. There is insufficient node throughput due to excessive memory usage.

A closer look at the individual columns of a solver node log can help identify which, if any, of these performance bottlenecks are present.

Figure 4.1 contains a Gurobi node log with annotations that illustrate the connections between the algorithm and the potential performance bottlenecks listed above. Here are some more details on the individual columns of the node log.

Nodes: The first two columns (explored (`Expl`) and unexplored (`Unexpl`) nodes) provide the total number of nodes processed and the number of unexplored nodes remaining in the branch and bound tree. This can be useful in assessing progress in the dual bound and the associated tightness of the current formulation. Rapid growth in the number of unexplored (also known as active) nodes indicates that the solver is unlikely to reach an optimal solution soon, even if the MIP gap is relatively small. Note that when `Expl` = `Unexpl` = 0, then Gurobi is processing the root node.

Current Node: The third through fifth columns (objective (`Obj`), depth (`Depth`) and integer infeasibilities (`IntInf`)) provide information about the current node relaxation after it has been solved. The `Obj` column lists the optimal objective value of the node relaxation, or a solve status of infeasible or cutoff in the event of a node relaxation that is infeasible or exceeds the cutoff value established by the incumbent integer feasible solution. The `IntInf` column lists the number of variables in the node relaxation that violate their integrality restrictions in the MIP to be solved. The `Depth` column lists the depth in the tree of the current node. These columns of the node log are useful for both assessing potential progress in the incumbent and best bound values. They can help the practitioner assess whether the solver is exploring nodes deep in the search tree of the branch and

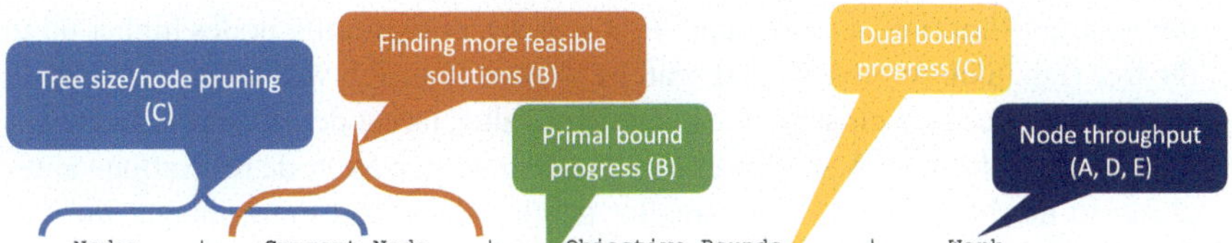


```
    Nodes    |    Current Node    |     Objective Bounds      |     Work
 Expl Unexpl |  Obj  Depth IntInf | Incumbent    BestBd   Gap | It/Node Time

     0     0  27.61749    0  112   57.00000   27.61749  51.5%     -    0s
H    0     0                      53.0000000   27.61749  47.9%     -    0s
H    0     0                      47.0000000   27.61749  41.2%     -    0s
     0     0  27.93121    0  119   47.00000   27.93121  40.6%     -    0s
H    0     0                      46.0000000   27.93121  39.3%     -    0s
     0     0  27.93121    0  120   46.00000   27.93121  39.3%     -    0s
     0     0  27.93121    0  120   46.00000   27.93121  39.3%     -    0s
     0     0  27.94960    0  115   46.00000   27.94960  39.2%     -    0s
     0     0  28.11503    0  121   46.00000   28.11503  38.9%     -    1s
     0     0  28.11503    0  117   46.00000   28.11503  38.9%     -    1s
     0     0  28.53708    0  120   46.00000   28.53708  38.0%     -    1s
     0     0  28.53708    0  120   46.00000   28.53708  38.0%     -    1s
     0     0  28.65875    0  121   46.00000   28.65875  37.7%     -    1s
     0     0  28.65875    0  121   46.00000   28.65875  37.7%     -    1s
     0     0  28.71799    0  120   46.00000   28.71799  37.6%     -    1s
H    0     0                      45.0000000   28.71799  36.2%     -    2s
     0     2  28.71799    0  120   45.00000   28.71799  36.2%     -    2s
H   28    32                      44.0000000   28.71799  34.7%   147    2s
H   93    97                      43.0000000   28.71799  33.2%   148    3s
   265   245  30.45442    8  112   43.00000   28.74524  33.2%   113    5s
   752   603  40.87728   30   56   43.00000   29.60949  31.1%  94.9   10s
  1065   792  39.16136   27  120   43.00000   29.68942  31.0%  91.6   15s
  1076   802  29.68942   10  116   43.00000   29.68942  31.0%  92.6   20s
  1385  1003  30.10596   15  109   43.00000   29.68942  31.0%  95.6   25s
  1636  1106  34.72416   28   92   43.00000   29.68942  31.0%  92.5   30s
  2213  1373  31.46757   17  108   43.00000   29.90665  30.4%  88.0   35s
  2891  1594    cutoff   51        43.00000   30.33320  29.5%  84.4   40s
  3529  1921  31.69003   19  115   43.00000   30.58867  28.9%  82.7   45s
  4061  2349  37.56288   31   91   43.00000   30.73984  28.5%  82.6   50s
  4838  2882  33.23450   20  108   43.00000   30.87811  28.2%  81.3   56s
  5498  3434  41.30303   48   77   43.00000   30.97873  28.0%  80.2   61s

...

H18910 11343                      41.0000000   31.95165  22.1%  76.0  158s
 19077 11579  32.96948   20  104   41.00000   31.95737  22.1%  75.9  160s
 19865 12101    cutoff   38        41.00000   31.99372  22.0%  76.0  166s
 20651 12379  37.06513   34  120   41.00000   32.02031  21.9%  76.1  218s

...

 8075464 4048188    cutoff   51        41.00000   36.07945  12.0%  62.1 44521s
 8076746 4048773  39.01331   45   79   41.00000   36.07956  12.0%  62.1 44527s
 8077384 4049081  38.44205   38   76   41.00000   36.07963  12.0%  62.1 44530s
 8078566 4049652  39.55251   47   83   41.00000   36.07972  12.0%  62.1 44536s
 8079877 4050180  37.57895   37   86   41.00000   36.07984  12.0%  62.1 44542s
 8080502 4050423  37.02682   40   88   41.00000   36.07988  12.0%  62.1 44545s
 8081573 4050996  38.56802   48   71   41.00000   36.07995  12.0%  62.1 44551s
 8082789 4051442  37.28332   39   86   41.00000   36.08007  12.0%  62.1 44557s
```

Fig. 4.1 Gurobi node log

bound algorithm (i.e., a more depth first search), or exploring nodes higher up in the tree (i.e., a more breadth first search). A deeper search will tend to have node relaxation objectives closer to the incumbent value, larger depth values, and fewer integer infeasibilities. A shallower search will tend to have node relaxation values closer to the best bound value, smaller depth values, and more integer infeasibilities. The former tends to be better at finding more integer feasible solutions, while the latter tends to move the best bound value more effectively. Thus, node exploration deep in the tree when the log indicates lack of progress in the lower bound, or node exploration high in the tree when the log indicates lack of progress in the incumbent value, are both indications that adjusting the solver parameters that affect the depth of the tree search should be adjusted to better align with the information about the solve progress in the node log. In other words, do not use strategies that will make more progress in the dual bound when the performance issue involves the primal bound, and vice versa.

Objective Bounds: The sixth and seventh columns contain the values of the primal bound (`Incumbent`) associated with the current integer feasible solution and the dual bound (`BestBd`) of the optimization. These can help the practitioner identify whether the performance bottlenecks involve the primal bound, the dual bound, or both. In particular, little or no progress in the dual bound, combined with growth in the active node list, indicates that strengthening the model formulation, either by setting solver parameters or using strategies customized to the model, is in order. The eighth column contains the `Gap`, which measures the relative difference between the primal and dual bound, thus providing a percentage within the optimality of the current incumbent solution. The branch and bound algorithm distinguishes itself from many methodologies used to solve MIPs that do not provide any measure of the objective function quality of the current solution. This distinction is particularly useful because for some MIPs and associated physical systems they represent, a provably optimal solution with a MIP gap of 0% is not needed, and practitioners can save solver time by accepting a solution with a positive percentage (e.g., 5 or 10%) of optimality.

Work: The ninth and tenth columns contain the average relaxation solve iterations per node (`It/Node`) and the total run time (`Time`), respectively. This can help identify whether node relaxation solve time is a performance bottleneck, in which case adjusting the choice of linear programming algorithm to solve the relaxations may help. Otherwise, it may also help if the solver offers functionality to move the primal or dual bounds without solving node relaxations for the entire MIP (e.g., Gurobi's No Relaxation heuristic).

By using the node log to identify the specific source of slow performance, the practitioner can identify the appropriate solver parameters to use to improve performance. However, in some cases a manual inspection, even with a deep knowledge of the connections between the columns of the node log and the branch and bound algorithm, may overlook more subtle information about the solve. For example, manual inspection of the incremental changes in the total time between consecutive lines in the node log is tedious at best, and error prone at worst. To that end, the Gurobi

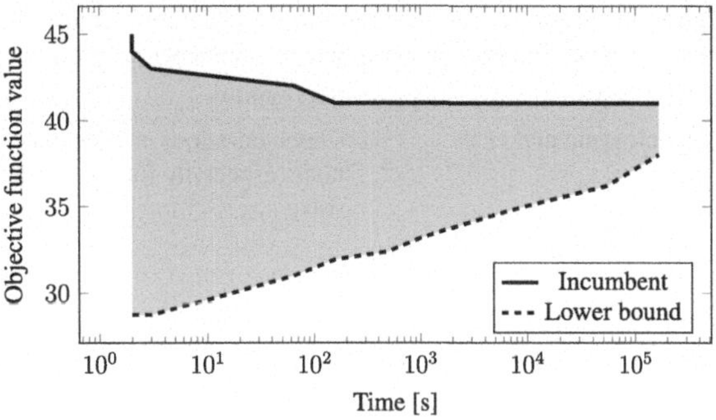

Fig. 4.2 A visual representation of the log file shown in Fig. 4.1

Experts team has created an open source Python package for analyzing and plotting node log data. This plot can reveal aspects of the solve that might otherwise go overlooked. For example, the tool can convert the log file in Fig. 4.1 into a visual representation of the progress of the primal and dual bound, as shown in Fig. 4.2.

The package is available at https://github.com/Gurobi/grblogtools, and a recording that describes its usage can be found at https://www.gurobi.com/resource/tech-talk-visualization-tools/.

The node log in Fig. 4.1 and the corresponding plot in Fig. 4.2 both indicate a lack of progress in the primal and dual bounds, with no new progress in the former and very slow progress in the latter in over 10 h. While one could try strategies to improve both, a closer examination suggests that lack of progress in the dual bound represents the real issue. The integer objective value suggests that the objective function consists of all integer variables. Examination of the objective of the model (`v150d30-2hopcds`, a difficult model from the MIPLIB 2017 test set) confirms this. Therefore, even if an optimal solution exists with objective 37, the slow progress in the dual bound, combined with the still growing list of over 4 million unexplored nodes, suggests that significant performance improvements can be attained via faster progress in the dual bound. That is indeed the case, as in fact the optimal solution has objective value of 41 and thus has already been found.

Once any performance bottlenecks have been identified, adjusting optimizer parameters based on the nature of the bottlenecks provides the easiest way to improve performance. Some solver parameters are effective at addressing certain performance problems but are totally ineffective for others. For example, in the node log for `v150d30-2hopcds` shown in Fig. 4.1, increasing the intensity of the node heuristics is likely to degrade performance since Gurobi has found the optimal solution within 3 min. On the other hand, reducing the intensity of the node heuristics might speed up the rate of node throughput without hurting the primal bound, provided that the optimal solution can still be found reasonably quickly. Klotz and Newman [17] provide a detailed discussion of different solver parameters and strategies, along with their strengths and weaknesses. As this book chapter focuses on strengthening formulations, it will not provide the same level of detail. However, an updated version

Table 4.1 Under various circumstances, different formulations and algorithmic settings have a greater chance of faster solution time on an integer programming problem instance

Characteristic	Recognition	Suggested tactic(s)
Troublesome node LPs	Check iterations or time per node, especially for root node solve	• Try alternate LP algorithms including barrier algorithm if advanced starts do not help simplex • Try avoiding node LPs completely (e.g., use Gurobi's No Relaxation heuristic)
Lack of progress in best integer	Little or no change in best integer solution in log even after many explored nodes	• Use best estimate or depth first search • Apply heuristics more frequently • Supply an initial solution, regardless of objective value • Solve with zero or other alternate objective • Branch up or down to resolve integer infeasibilities • Use branching priorities or variable hints
Lack of progress in best bound	Little or no change in best bound in log after even after many explored nodes	• Use breadth first search • Use aggressive probing • Use aggressive algorithmic cut generation • Apply strong branching • Increase intensity of symmetry detection • Derive cuts *a priori* • Reformulate with different variables
Data problems	Limited solution accuracy	• Avoid large differences in size of data • Reformulate "big M" constraints
Memory problems	Out of memory error Node throughput degrades as memory usage grows	• Apply memory focus setting • Buy more memory

of a summary table from Klotz and Newman [17] that connects useful parameters to specific performance bottlenecks is provided in Table 4.1.

Practitioners with limited familiarity of the branch and bound algorithm and solver parameters may still be able to obtain performance improvements by using automated parameter tuning procedures available in some solvers. These procedures may consume significant amounts of time performing multiple tuning tests, but they can be very helpful when solving similar instances of the model where good parameter settings for one instance will be effective on other instances as well.

4.4 Common Formulation Pitfalls and Oversights

If setting solver parameters[2] by using the analysis described in Table 4.1 does not provide the necessary performance improvements, more refined strategies are called for. While the focus of this chapter involves strengthening MIP formulations by using specific knowledge about the model at hand, there are some characteristics of formulations that occur frequently enough that they are worth summarizing. The practitioner can view this section as a checklist to examine whenever seeking to improve performance without drawing on the more customized approaches described in subsequent sections of this chapter.

- **Do not assume fewer integer variables result in a better formulation**. While more integer variables conceptually imply a potentially larger search tree, they can also enable the solver to perform additional MIP preprocessing, cuts, and other computations that tighten the formulation and hence improve performance. The node log can shed light on whether more integer variables results in better performance. The practitioner should not spend inordinate amounts of time trying to minimize the number of integer variables in the initial formulation.
- **Do not assume a more compact formulation is better**. The strength or weakness of the formulation is more important than the size. Additional constraints in the model that strengthen the formulation can yield significant performance improvements that can far exceed any decline in speed solving the node relaxations due to the resulting larger problem size. This will be discussed in greater detail in subsequent sections, especially in Sect. 4.7.1.
- **Do not ignore slow node relaxation solve times**. Node relaxation solutions provide a treasure trove of information for MIP solvers to exploit. However, excessive node relaxation solve times can compromise the performance gains obtained from that information. The authors have encountered a significant number of MIP models containing node relaxations so time consuming as to render the tree search ineffective, regardless of the actual branching decisions made by the solver. In such cases setting parameters that improve the tree search will not pay off (with the possible exception of when the parameters somehow involve solving LPs, such as with pseudo cost or strong branching variable selection). Examination of the solver node logs will reveal this problematic behavior. In such cases one should always try alternate LP algorithms and parameter settings to speed up the node throughput. If none of the solver's LP algorithms (typically the primal and dual simplex and barrier algorithms, but possibly also specialized network and sifting algorithms) provide a sufficient speedup in node throughput, then a more careful analysis of the node relaxation solve times is in order. See Klotz and Newman [18] for more details on improving LP performance.

[2] The specific parameter names differ between the various solvers, and the authors recommend that the reader consults the documentation of the respective solver to find the most suitable fit.

- **Consider reducing as well as increasing the computational effort associated with the solver parameters**. As MIP solvers have evolved over more than 50 years, many of the improvements have come from the addition of new features that extract more information via additional computations. Many of these features are controlled by solver parameters available to the practitioner. When the solver's node log reveals a lack of progress in the primal or dual bound, a natural tendency is to raise the intensity of parameters (i.e., perform more of the associated calculations) that will address the source of the slow progress, as recommended in Table 4.1. This approach is often effective. However, as solvers have matured, multiple features may essentially yield the same improvements. For example, node heuristics may find good solutions, but recent progress in variable branching such as pseudo costs may find equally good solutions. In that case, instructing the solver to apply the heuristics less aggressively (or even disabling heuristics completely) may pay off, resulting in equally good solutions via branching and better ones due to processing more nodes per unit time. Similarly, reducing the level of computational effort with which the solver applies cuts may result in equally good progress in the dual bound by faster node throughput combined with powerful branching methods.
- **Do not ignore numerical issues in the model**. Issues of ill conditioning and numerical instability in the model formulation can slow down performance by creating time consuming node relaxation solves or by compromising the accuracy of various algorithmic improvements to the branch and bound algorithm. This lengthy topic will not be discussed here. The reader is referred to Klotz [16] for a discussion that focuses on numerical issues in linear and integer programming, and Higham [15] for a thorough general discussion of the details of finite precision floating point arithmetic.
- **Beware of multi-objective models in disguise**. Recently, Gurobi and several other solvers have added the ability to formulate MIPs (and LPs as well) with multiple objectives directly. Whether using this functionality or instead taking advantage of the flexible solver APIs to implement their own multi-objective procedure, practitioners must choose between a hierarchical approach in which multiple MIP subproblems are solved, or a single optimization involving a blended objective that is a linear combination of individual objectives. While one's intuition suggests that a single optimization will perform much better, the associated blended objective can often result in an unfocused search process that takes longer than the multiple solves in the hierarchical approach. Or, putting it less diplomatically, sometimes a blended objective appears more like a blurred objective to the solver. By contrast, the multiple solves in the hierarchical approach take advantage of the optimal solution and objective value from the preceding solve. Assessment of the trade-offs between objectives may help select the best multi-objective approach to use. Benchmarking both approaches on a suitable test set of models can also help.

  One particularly common type of multi-objective model involves soft constraints and their associated penalty variables. These models typically have an objective component associated with the soft constraints and a separate one associated with the rest of the model's activities. Particularly when the objective component asso-

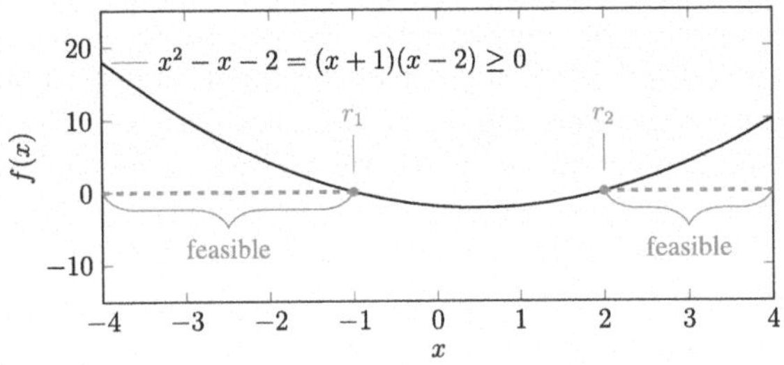

Fig. 4.3 The nonlinear constraint $x^2 - x - 2 \geq 0$ can be converted into a pair of logical constraints by recognizing that $x^2 - x - 2$ has the real roots -1 and 2. Therefore we can substitute $x^2 - x - 2 \geq 0$ with the logical constraint $x \leq -1$ or $x \geq 2$

ciated with the soft constraints is much larger than the component associated with the rest of the model, using a hierarchical objective is likely to work better, as using a blended objective is more likely to result in an unfocused tree search during the branch and bound algorithm.

- **Consider simplifying constraints involving polynomials**. While not as developed as MIP solvers, advances in mixed-integer nonlinear programming (MINLP) solvers have gained momentum. As a result, more challenging industrial problems are available, and more strategies for solving these problems can be useful. In particular, practitioners should keep in mind that some nonconvex constraints involving polynomials can instead be expressed using logical constraints involving the factors of the polynomials. For example, consider a nonconvex constraint involving arbitrary second degree polynomial (with real roots) $ax^2 + bx + c \geq 0$. Let r_1 and r_2 be the roots of the polynomial and suppose $r_1 < 0 < r_2$. If x has a lower bound of 0, this seemingly nonconvex constraint can be replaced with the equivalent simple linear bound constraint $x \geq r_2$. If x is a free variable with infinite lower and upper bounds, one can replace the nonconvex quadratic constraint with the logical constraint $x \geq r_2$ or $x \leq r_1$. Figure 4.3 illustrates this type of reformulation for the nonconvex quadratic constraint $x^2 - x - 2 \geq 0$. Some solvers support such logical constraints explicitly. Others require the practitioner to formulate the logical constraint using binary variables or special ordered sets; see Williams [27] for details and examples.

4.5 Identification of Weak MIP Formulations

In order to strengthen a weak MIP formulation, the practitioner must first identify the aspects of the model that make it weak, which may be difficult as well. We draw on and augment a set of guidelines from Klotz and Newman [17].

- **Simplify the model if necessary**. Filter out any constraints or integrality restrictions by systematically removing them and checking if the model remains difficult to solve. If so, focus on the filtered model. Otherwise, restore the removed constraints and proceed with the next group of constraints or restrictions. Try to filter groups of similar constraints and variables rather than operating on individual ones. This filtering process can help identify the parts of the model that create the weakness in the formulation, and they should be examined for potential strengthenings. If filtering is not effective, try to reproduce the performance problem on a smaller model.

- **Identify the constraints that prevent the objective from improving**. With a cost minimization problem, this typically means identifying the constraints that force activities to be performed. More generally, consider the solution obtained by retaining the objective function and variable bounds but removing the linear constraints. Then identify the constraints that prevent the variables from reaching those values. These components of the model almost certainly will be involved in any tightening of the model formulation.
- **Determine how removing integrality restrictions allows the root node relaxation objective to improve**. Weak formulations typically involve root node relaxations with optimal objective values significantly better than that of the associated MIP. The variables with fractional solutions in the root node relaxation help identify the constraints and variables that are most likely to motivate additional cuts that will tighten the formulation. Globally valid cuts obtained by simply examining the MIP model without considering the relaxation solutions and the objective function are less likely to help tighten the formulation.

Section 4.9 will illustrate all three of these guidelines. However, the third guideline merits a more detailed discussion. After verifying from the node log that little or no progress occurs in the dual bound (e.g., refer to Fig. 4.1), one can then try to locate the specific sources of weakness in the formulation. Note that while such weakness has a theoretical basis involving differences between the convex hull of the integer feasible solutions and the polyhedron associated with the LP relaxation, it is primarily a symptom of the branch and bound algorithm. A formulation that is considered weak for the branch and bound algorithm may work very well for other solution methods. In its broadest sense, the branch and bound algorithm relaxes all the discrete objects to continuous ones, then gradually moves them to discrete values. If the information in the LP relaxation solution provides unrealistically favorable objective values, the effort required during the tree search phase of the branch and bound algorithm will be much greater than if the LP relaxation has a stronger connection to the associated MIP model. In other words, a weak formulation is characterized by a high level of disconnectedness between the MIP model and its LP relaxation.

Here are three methods the practitioner can use to assess the disconnect between the MIP model and its associated LP relaxation.

4.5.1 The Polyhedral Approach

This method uses polyhedral theory to compare the convex hull of the MIP's integer feasible solutions with the polyhedron of its LP relaxation. It makes use of a formal definition of the strength of the formulation based on facet defining valid inequalities. Proving that a particular valid inequality defines a facet of the convex hull of the integer feasible solutions of a MIP establishes that no stronger version of the inequality exists. Thus, the cut is as strong as possible regarding separating fractional solutions from the LP relaxation. This formal approach has led to the discovery of a numerous facet defining inequalities that have contributed greatly to the cuts implemented in solvers like Gurobi.

This chapter will not focus on this methodology. The literature (including other chapters in this book) is full of work on this subject matter. The authors recommend Wolsey [28] to get started, and Grötschel [14] and its references for more advanced study. While this material is extremely important, it tends to be mathematically advanced. Practitioners unfamiliar with the concepts and underlying mathematics may conclude that tightening MIP formulations is impossible without this type of knowledge. The authors seek to dispel this misconception by focusing on the algebraic and visual approaches in the next two subsections. These can be used effectively with a less advanced background comprised of more basic linear algebra and familiarity with the meaning of the MIP model at hand. Get started with what you know, and then systematically expand your knowledge base to include the more advanced mathematical concepts if they are currently unfamiliar. More advanced readers familiar with the polyhedral approach may still find this material helpful for explaining these concepts to others.

4.5.2 *The Algebraic Approach*

This method involves examining the solution values of the LP relaxation to understand if the objective value is unrealistically favorable compared to that of feasible solutions for the MIP model. This may require simplifying the model or looking at smaller instances. Here is a simple example where this method sheds light on unrealistic and deceptive information in the LP relaxation solution.

$$
\begin{aligned}
(P_1): \quad & \underset{x,t}{\text{minimize}} \quad t && && \text{(4.1a)}\\
& \text{subject to} \quad t \geq x_j && j = 1, \ldots, n && \text{(4.1b)}\\
& \sum_{j=1}^{n} x_j = 1 && && \text{(4.1c)}\\
& t \geq 0 && && \text{(4.1d)}\\
& x_j \in \{0, 1\} && j = 1, \ldots, n && \text{(4.1e)}
\end{aligned}
$$

One can see that setting $x_j = 1$ for any index j forces $t = 1$, and yields one of the n optimal solutions. However, by removing the integrality restrictions on the x variables, the optimal solution for the LP relaxation consists of $x_j = \frac{1}{n}, j = 1, \ldots, n,$

yielding an objective value of $t = \frac{1}{n}$. Thus, both the nature of the optimal solutions and optimal objective function value in the LP relaxation are quite disconnected from the MIP, and the situation only worsens as n increases. For even modest values of n, the LP relaxation values are much closer to the variable lower bounds of 0 than the upper bound of 1. Little information is available to distinguish one branching variable candidate from another, and the variable values suggest branching down on the selected variable (i.e., setting it to 0 for the resulting child node). Branching down will reduce the number of integer infeasibilities by 1 and result in an LP relaxation solution with both t and the remaining $n-1$ fractional x variables set to $\frac{1}{n-1}$. Thus, the same problem can occur repeatedly if branching down continues. By contrast, constraint (4.1c) ensures that branching up on a single x variable in the initial relaxation immediately fixes all the remaining binary variables to 0 and immediately yields a feasible solution (that turns out to be optimal).

Even though the weakness of the formulation manifests itself primarily for values of $n \gg 2$, the graphical representation of the MIP and its relaxation for $n = 2$ does shed light on a way to tighten the formulation using a polyhedral approach to incorporate the convex hull of the integer feasible solutions into the formulation. In Fig. 4.4, the gray area represents the feasible solution space for the LP relaxation of problem (P_1). Note that it allows for the solution $(x_1, x_2, t) = \left(\frac{1}{2}, \frac{1}{2}, \frac{1}{2}\right)$, which is clearly not on the convex hull of the MIP, denoted by the thick black line. One can then cut off the additional feasible space by constraining any feasible solution to lie on the convex hull of the integer feasible solutions. This is accomplished by constraints (4.2b) and (4.2c) in the following reformulation of (P_1).

$$(P_1 2D): \quad \underset{x,t,\lambda}{\text{minimize}} \quad t \tag{4.2a}$$

$$\text{subject to} \quad (x_1, x_2, t) = \lambda_1(1, 0, 1) + \lambda_2(0, 1, 1) \tag{4.2b}$$

$$\lambda_1 + \lambda_2 = 1 \tag{4.2c}$$

$$t \geq 0 \tag{4.2d}$$

$$\lambda_j \geq 0, \quad x_j \in \{0, 1\} \; j = 1, 2 \tag{4.2e}$$

This approach of enumerating integer feasible solutions works for arbitrary values of n as well. However, it only works because (P_1) is simple enough that one can enumerate the integer feasible points of the model in polynomial (and in this case linear) time. For more elaborate MIP models, such enumeration is typically not possible. Nonetheless, enumerating integer feasible solutions for individual or small groups of constraints within a larger model and then replacing said constraints with their convex hull representation can help tighten a formulation.

Note also that (P_1) is simple enough that most MIP solvers will have no trouble tightening the existing formulation without performing the reformulation described by $(P_1 2D)$. While the continuous variable t in (P_1) can assume fractional values such as 1.5 in feasible solutions, in any optimal solution it can only take on integral values because of the integrality of the x variables. Constraints (4.1b) ensure that

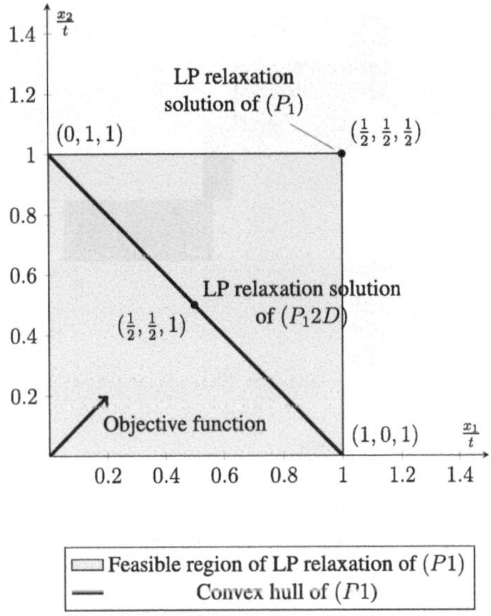

Fig. 4.4 A visualization of the strengthening of the original model (P_1) by explicitly specifying the convex hull, denoted by the thick line from (1, 0, 1) to (0, 1, 1), of the integer feasible solutions through the introduction of the auxiliary variable λ in problem $(P_1 2D)$. Note the projection onto the $\frac{x_1}{t}$ and $\frac{x_2}{t}$ space in order to enable a 2D representation; the coordinates added refer to the corresponding values of (x_1, x_2, t)

$t \geq \max_{1 \leq j \leq n} x_j$. Since t appears in no other constraints, this inequality will be satisfied at equality in any optimal solution. Having established the integrality of t in any optimal solution, one can now deduce the lower bound $t \geq 1$ either by probing or combining constraints and applying Chvátal-Gomory rounding [28]. Regarding probing, fixing $t = 0$ and propagating this upper bound into constraints (4.1b) forces $x_j = 0$ for $j = 1, \ldots, n$. This implies constraint (4.1c) is infeasible. The integrality of t then implies $t \geq 1$ in any feasible solution. While this tighter bound on t still allows fractional solutions such as $t = 1, x_j = \frac{1}{n}, j = 1, \ldots, n$, most MIP solvers will have heuristics that quickly identify feasible solutions consisting of $x_j = 1, x_k = 0, k \neq j$. And even if not, setting the solver branching direction to consider the up branch first will do so. The primal and dual bounds of 1 are then equal, so the solver has a provably optimal solution. Similarly, having established the integrality of t, one can derive a mixed integer rounding cut by summing up the constraints (4.1b) and using (4.1c) to get $n \cdot t \geq \sum_{j=1}^{n} x_j = 1$. Hence $t \geq \frac{1}{n}$, together with the newly established integrality of t enables the integer rounding to obtain $t \geq 1$.

4.5.3 The Visual Approach

By comparing the physical systems associated with a MIP model and the associated LP relaxation, practitioners can frequently identify the source of the disconnect between them. This strategy may be more effective than the previously discussed polyhedral or algebraic approaches. Practitioners typically are familiar with the MIP model formulation since they created (or at least worked on) it. However, they can

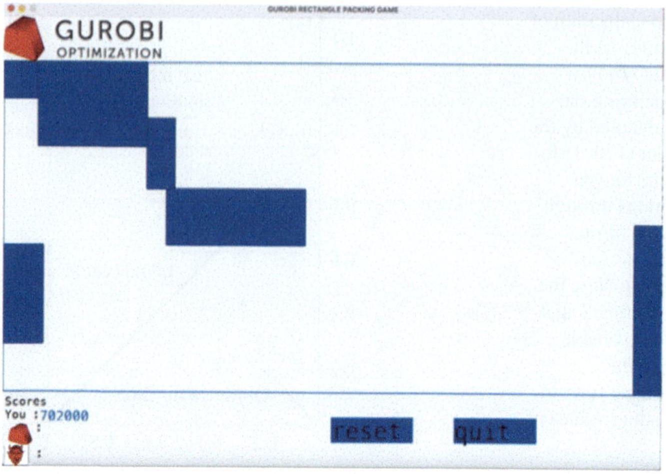

Fig. 4.5 Rectangle packing game: initial game layout

easily overlook the information available from the physical system associated with the LP relaxation of the MIP formulation. Two examples, both of which are drawn from MIPs with meaningful industrial applications, illustrate how this visual approach not only identifies the weakness in the formulation but sheds light on how to strengthen it.

4.5.3.1 The Rectangle Game

The first example involves a rectangle packing game where the player's objective is to place various rectangles next to each other without overlap so that the area of the enclosing rectangle (i.e., the product of the maximum height and width of the rectangular bin created by the packed rectangles) is as small as possible. The game can be downloaded at https://github.com/Gurobi/techtalks, and the authors encourage the reader to download the game and play it a few times before continuing. Figure 4.5 shows the initial game layout for a dataset involving 6 rectangles. At the start, the rectangular bin enclosing the 6 rectangles is the whole playing surface defined by the thin blue line. The initial area of the bin is 702000 units, which is reflected in the player score on the bottom left of the game screen. The player can click on the individual rectangles, moving them so that they reduce the size of the enclosing rectangular bin, as illustrated in Fig. 4.6, where the player has reduced the score from 702000 to 113625. The player can continue moving the rectangles, to try to make additional reductions, or he can click on the red Gurobi logo to instruct Gurobi to show the best possible configuration for the associated MIP model. In addition, he can examine the best possible configuration when overlap is permitted by clicking on the devil button below the Gurobi logo to examine the best possible configuration in that case. Figure 4.7 illustrates the areas of the enclosing bins for the player configuration

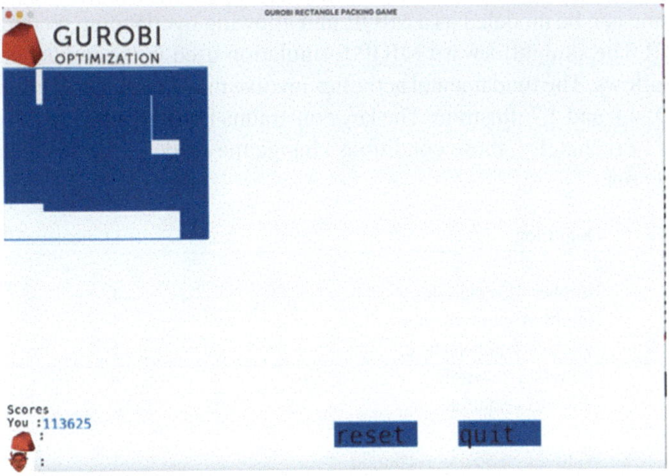

Fig. 4.6 Rectangle packing game: player score

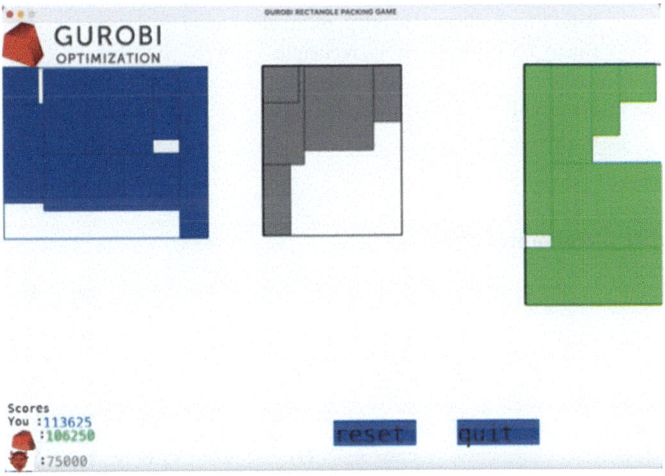

Fig. 4.7 Rectangle packing game: player, Gurobi and devil scores

in blue on the left, the best possible configuration in green on the right, and the best possible configuration when overlap is permitted in gray in the middle.

This game can be modeled as a MIQP, and allowing overlapping rectangles turns it into a QP. The straightforward MIQP formulation used in the implementation of the game follows. The fundamental activities involve the placement of the rectangles, as Figs. 4.5, 4.6 and 4.7 illustrate. The key constraints involve forbidding the overlap of any pair of rectangles; those conditions change the model from a continuous one to a discrete one.

Name	Definition	Dimensionality
Data		
n	Number of rectangles	1
H	Height of game surface	1
W	Width of game surface	1
h_j	Height of rectangle j	n
w_j	Width of rectangle j	n
Variables		
$\hat{x}$	Width of enclosing rectangle	1
$\hat{y}$	Height of enclosing rectangle	1
x_j	Upper left corner horizontal position of rectangle j	n
y_j	Upper left corner vertical position of rectangle j	n
u_{ij}	Binary for horizontal overlap, rectangles i and j	n^2
v_{ij}	Binary for vertical overlap, rectangles i and j	n^2
Sets		
N	$\{1, \ldots, n\}$	
N_i	$\{i+1, \ldots, n\}$	

$$(Pgame): \quad \underset{x,y,u,v}{\text{minimize}} \quad \hat{x}\hat{y} \tag{4.3a}$$

$$\text{subject to} \quad x_j + w_j \leq \hat{x} \qquad \forall j \in N \tag{4.3b}$$

$$y_j + h_j \leq \hat{y} \qquad \forall j \in N \tag{4.3c}$$

$$x_j - x_i \geq w_i - W u_{ij} \qquad \forall i \in N, \forall j \in N_i \tag{4.3d}$$

$$x_i - x_j \geq w_j - W u_{ji} \qquad \forall i \in N, \forall j \in N_i \tag{4.3e}$$

$$y_j - y_i \geq h_i - H v_{ij} \qquad \forall i \in N, \forall j \in N_i \tag{4.3f}$$

$$y_i - y_j \geq h_j - H v_{ji} \qquad \forall i \in N, \forall j \in N_i \tag{4.3g}$$

$$u_{ij} + u_{ji} + v_{ij} + v_{ji} \leq 3 \qquad \forall i \in N, \forall j \in N_i \tag{4.3h}$$

$$x, y \geq 0,\ u, v \in \{0, 1\} \tag{4.3i}$$

Figure 4.8 connects the activities in the model to a feasible configuration of rectangles. The objective minimizes the score in the game, or equivalently the area of the enclosing rectangle, which is the product of its height and width. Constraints (4.3b) and (4.3c) define the width and height of the enclosing rectangle. For example, the n constraints (4.3b) ensure that $\hat{x}$ is at least the rightmost horizontal coordinate of the n rectangles for any configuration. Since $\hat{x}$ has a lower bound of 0 and appears only in the objective to be minimized with a positive coefficient, it will never exceed this

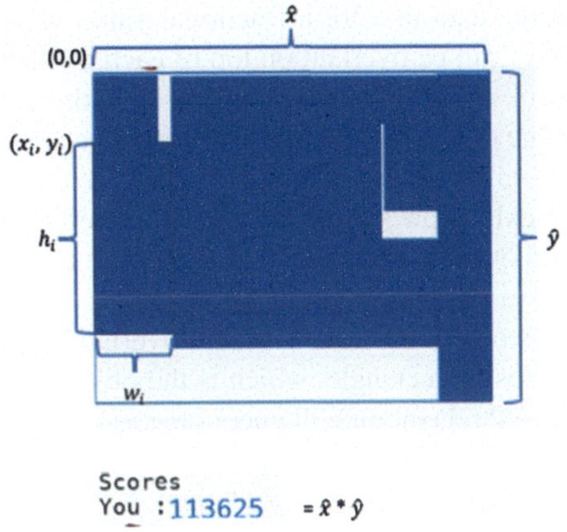

Fig. 4.8 Rectangle packing game model

coordinate in any node relaxation solutions, or any optimal integer feasible solutions. A similar argument shows that $\hat{y}$ is the height of the enclosing rectangle. Furthermore, the nature of the objective ensures that one of the rectangles will have its corner placed at the origin in any optimal solution. The game is implemented using the Pygame package (https://www.pygame.org/news), in which the origin is the upper left corner of the playing region, so the x and y variables are defined accordingly in model $Pgame$. However, this is only relevant for the game display in the figures, not the underlying model formulation. Constraints (4.3d)–(4.3g) and (4.3h) ensure that no overlap occurs in a feasible configuration of the rectangles. Consider constraint (4.3d) for a pair of rectangles i and j. If the binary variable $u_{ij} = 0$, this constraint simplifies to $x_j \geq x_i + w_i$, which forces rectangle j to be positioned to the right of rectangle i in any feasible configuration. However, if $u_{ij} = 1$, note that $w_i \leq W$, so the constraint allows the two rectangles to have the same x coordinate. Similar reasoning applies to constraints (4.3e), (4.3f), and (4.3g), which require that rectangle j be positioned to the left, above or below rectangle i, respectively. As long as at least one of these non-overlap conditions holds, no overlap occurs for rectangles i and j. Equivalently, at least one of u_{ij}, u_{ji}, v_{ij} and v_{ji} must be 0, which is specified by constraint (4.3h).

Is this formulation weak? And if so, what is the source of the weakness? Constraints (4.3d)–(4.3g) and (4.3h) are the only ones that involve any discrete variables, so they merit a closer look.

Using the algebraic approach, again consider constraint (4.3d) for a pair of rectangles i and j, and the effect of fractional solution values for the binary variable u_{ij}. The binary values $\{0, 1\}$ ensure that either overlap is forbidden or the constraint is nonbinding, respectively. However, as soon as u_{ij} starts to increase above 0, the non-overlap requirement is lost. For example, in the game instance with 6 rectangles, the height and width of the game surface (H and W in the model formulation) are 800 and 1200 respectively, while the individual rectangle heights and widths in Fig. 4.5 all have much smaller values, in fact ranging from 50 to 300. This means that for constraints such as $x_j - x_i \geq w_i - W u_{ij}$ in (4.3d), even for the maximum width in the data of 250, a fractional value of $u_{ij} = \frac{250}{1200} = \frac{5}{24}$ implies that rectangles i and j can be overlaid on top of each other in the horizontal direction. The situation worsens for smaller rectangles; a width of 50 implies that a much smaller fractional value of $u_{ij} = \frac{50}{1200} = \frac{1}{24}$ allows the rectangles to completely overlap. Thus, 0 values for u_{ij} forbid horizontal overlap of the rectangles, but small values in the relaxation quickly allow significant levels of overlap. The same reasoning applies to the other non-overlap constraints (4.3e), (4.3f) and (4.3g), and constraint (4.3h) does nothing to prevent this since small values for u variables do not threaten to violate this constraint. Furthermore, more overlap among the rectangles reduces the area of the enclosing rectangle, which is the objective to be minimized. Therefore, optimizing the QP relaxation will encourage as much overlap as possible.

With the algebraic approach having established how overlap occurs, the visual approach sheds additional light on the weakness in the formulation. Figure 4.7 provides, from left to right, the player's configuration, the optimal QP relaxation configuration, and the optimal MIQP configuration. In the optimal QP relaxation configuration, complete overlap occurs, as all 6 rectangles have their upper left corner positioned at the origin (Pygame, the tool used to implement the game, uses the upper left corner as the origin (0, 0)). Fractional values for the binary variables u_{ij} allow pairwise overlap; inductively applying this reasoning to all pairs enables the complete overlap seen in the figure. At first glance, this does not appear to be particularly problematic. For the 6 rectangle instance in Fig. 4.7, the difference in objectives of the optimal QP relaxation and MIQP configurations is not that extreme, and Gurobi can find the optimal MIQP configuration in a fraction of a second. Figure 4.9 illustrates now the situation quickly worsens as the number of rectangles increases. For this 12 rectangle instance, the differences between the objectives of the fractional solution of the QP relaxation and the integer feasible solutions found by the player and Gurobi are much more pronounced than for the 6 rectangle example in Fig. 4.7. Furthermore, given that every rectangle in the relaxation solution in gray has its upper left corner at position (0, 0), one can see that adding more rectangles to this instance of the model will not increase the objective of the relaxation solution unless the new rectangles have width or height that exceeds that of the maximum width (w_{max}) and height (h_{max}) on the existing set of rectangles. Thus, one could add an infinite number of rectangles within the maximum width and height to the model without increasing the QP relaxation objective, yet the increase in cumulative area of all the rectangles would dramatically increase the objective associated with integer feasible solutions that do not allow overlap. This means that the MIQP model and associated QP relaxation model are almost completely disconnected, with the QP relaxation solutions providing unrealistic and deceptive solution information that can be problematic for the branch and bound algorithm.

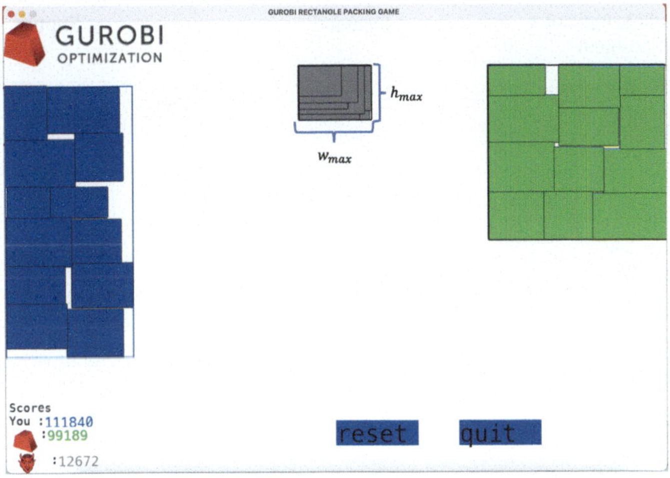

Fig. 4.9 Rectangle packing game, harder instance: player, Gurobi, and devil scores

In addition, the solution displayed by Gurobi has not been proven to be optimal, as seen in the following node log fragment of the associated MIQP after 3 min of run time. Note that the problem size for the MIQP remains small for modest values of n. From the table describing the model data and variables, an instance with n rectangles will have $2n^2 + 2n + 2$ variables, of which $2n^2$ are binary. Similarly, it will have $5(n \cdot \frac{n+1}{2} - n) + 2n$ constraints. For $n = 12$, this translates to 314 variables and 354 constraints.

```
    Nodes    |    Current Node    |     Objective Bounds      |     Work
 Expl Unexpl |  Obj  Depth IntInf | Incumbent    BestBd   Gap | It/Node Time
...
 160801 83880 95976.0000   46    24    99189.0000 42120.0000  57.5%    66.9  165s
 164492 85779 79182.0000   38    41    99189.0000 42143.0790  57.5%    67.2  170s
 168696 87795 infeasible   53          99189.0000 42246.1507  57.4%    67.3  175s
 173455 90297 68607.0000   42    67    99189.0000 42325.0601  57.3%    67.5  180s
```

In fact, optimality was not proven even after 3 h of run time. Note also that the list of unexplored notes in column 2 of the node log is still increasing, indicating that this run will probably run for many more hours.

```
    Nodes    |    Current Node    |     Objective Bounds      |     Work
 Expl Unexpl |  Obj  Depth IntInf | Incumbent    BestBd   Gap | It/Node Time
...
 8833626 2707084 91827.0000   46   18 99189.0000 56943.0000  42.6%   61.2 10785s
 8838517 2708131 68580.0000   49   36 99189.0000 56943.0000  42.6%   61.2 10790s
 8843233 2709326 58441.5000   41   94 99189.0000 56943.0000  42.6%   61.1 10796s
 8846747 2710147 infeasible   47      99189.0000 56943.0000  42.6%   61.1 10800s
```

The weakness of the formulation has now been established. A discussion of strengthening strategies for this type of model will appear in Sect. 4.9.

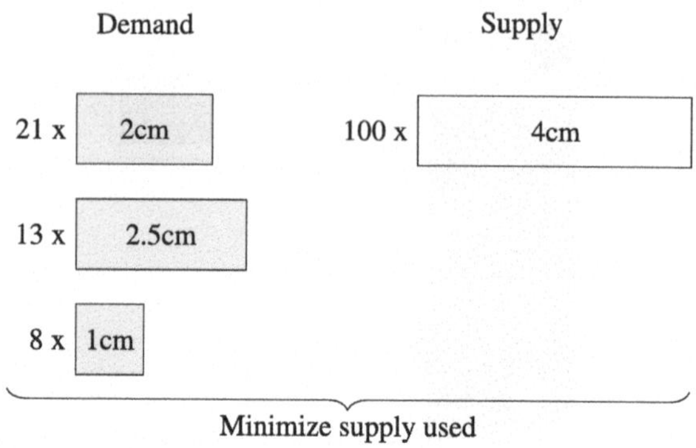

Fig. 4.10 An example of a cutting stock problem

4.5.3.2 Visualizing the Weakness of the Cutting Stock Problem

In the cutting stock problem, we aim to find the lowest number of master rolls of length c that are needed to cut n different elements with a specified length w_i for each element i. A common application for this problem is to cut wood planks into a set of smaller pieces of pre-specified length. This example is shown in Fig. 4.10.

One can define the following variables:

$x_{i,j}$: Is element i assigned to master roll j?.
y_j: Is master roll j being used?

Using these variables, the cutting stock problem can be described as follows:

$$(P_{CS}): \quad \underset{x,y}{\text{minimize}} \quad \sum_{j \in J} y_j \tag{4.4a}$$

$$\text{subject to} \quad \sum_{j \in J} x_{i,j} = 1, \qquad \forall i \in I \tag{4.4b}$$

$$\sum_{i \in I} w_i x_{i,j} \leq c y_j, \qquad \forall j \in J \tag{4.4c}$$

$$x_{i,j} \in \{0, 1\}, y_j \in \{0, 1\} \qquad \forall i \in I, \forall j \in J \tag{4.4d}$$

This formulation is known to be weak, but it may not be obvious why. However, it becomes clear quite quickly when considering Fig. 4.11.

Specifically, notice that a portion of the master roll may go to waste, as the elements must attain specific sizes. Therefore, during the LP relaxation phase, it is possible to obtain the same assignments of elements to master rolls while maintaining a fractional value for y. This fractional value is beneficial since the objective is to minimize the sum over all y.

In other words: if one could shrink the master roll by a certain amount, this would not impact the overall number of master rolls needed. This slack (or waste) in the

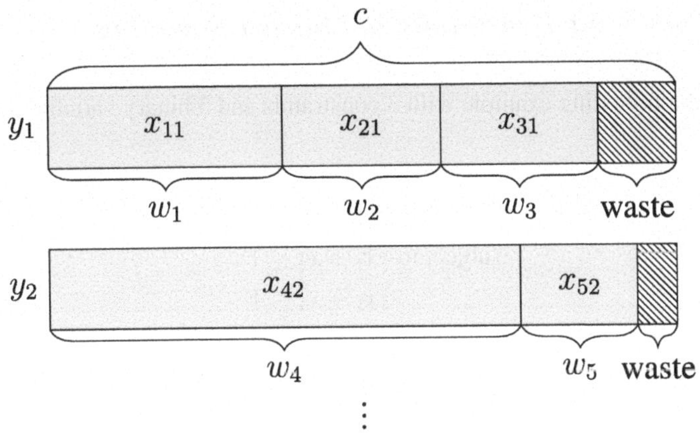

Fig. 4.11 Visualization of the weakness of the classical cutting stock formulation P_{CS}: when we fit the cut patterns, there is almost always an unused amount in the master roll. This waste means that the corresponding binary can be fractional without impacting the distribution of the cut patterns. This requires the solver to force the fixing of these fractional values through branching

master roll length is used by the LP relaxation and the solver has to use the search tree to successively fix these fractional y values.

The most common way to address this weakness is by pre-calculating all possible cut patterns, which is an application of a change of variable that we will discuss in Sect. 4.7. In case this cut pattern generation becomes computationally prohibitively expensive, column generation is known to be a very effective strategy to tighten the formulation. However, an in-depth discussion around the merits and challenges around column generation goes beyond the scope of this chapter, and the interested reader is referred to Lübbecke and Desrosiers [19] and references therein.

4.6 Fundamental Strategies to Strengthen MIP Formulations

By understanding the weakness of the formulation, the practitioner can then focus on the aspects of the model that need strengthening. Klotz and Newman [17] provide an extensive list of strengthening strategies from which much of the discussion here is drawn. This section will focus on broader strategies that not only form the foundation of many of the established cuts in the literature used by most MIP solvers, but also can be effectively applied to derive cuts that elude the solvers. Section 4.7 describes strategies that are less conceptual and require more customization to the specific characteristics of the model of interest.

4.6.1 Combining Constraints and Integer Rounding

Consider the following example with 3 constraints and 3 binary variables.

$$
\begin{aligned}
(P_2): \quad \underset{x}{\text{maximize}} \quad & x_1 + x_2 + x_3 \\
\text{subject to} \quad & x_1 + x_2 \le 1 \\
& x_1 + x_3 \le 1 \\
& x_2 + x_3 \le 1 \\
& x_j \in \{0, 1\}, \quad j = 1, 2, 3
\end{aligned}
$$

The optimal solution for the continuous relaxation of P_2 consists of setting $x_1 = x_2 = x_3 = \frac{1}{2}$, attaining an objective of $\frac{3}{2}$. Combining the 3 constraints in P_2 yields the inequality $2x_1 + 2x_2 + 2x_3 \le 3$. Dividing by 2 yields $x_1 + x_2 + x_3 \le \frac{3}{2}$. The left-hand side is integral, so the fractional right-hand side value can be rounded down to its greatest integer, yielding the cut $x_1 + x_2 + x_3 \le 1$. No integer feasible solutions were harmed by this cut; the preservation of all such solutions makes it globally valid. The cut is also useful in that it separates fractional solutions in the continuous relaxation. It cuts off the optimal relaxation solution of $x_1 = x_2 = x_3 = \frac{1}{2}$, which sums to $\frac{3}{2}$. Separation of fractional solutions and global validity are two fundamental conditions for adding cuts that strengthen a MIP formulation. In some cases the requirement of global validity can be relaxed, provided that the cut in question guarantees that at least one optimal solution is preserved in the tightened formulation. Such cuts that involve the objective function will be considered in future sections.

Problem P_2 is a simple example of the more general Chvátal-Gomory integer rounding procedure. In addition to summing constraints and rounding, as was done with P_2, one can look for more general linear combinations of constraints that yield an integral left-hand side and a fractional right-hand side. Consider a system of m linear inequalities in n variables involving all integer variables.

$$
\begin{aligned}
(P_3): \quad \underset{x}{\text{minimize}} \quad & 0 \\
\text{subject to} \quad & Ax \le b \\
& x \ge 0 \\
& x \text{ integer}
\end{aligned}
$$

Nonnegative multipliers of each $\le$ constraint in P_3 yield equivalent constraints that can be combined into a single base inequality to which integer rounding can be applied. Mathematically, given $\lambda \in R^m_+$, the base inequality $\lambda^T Ax \le \lambda^T b$ remains valid. Let $v = A^T\lambda$ and $b' = \lambda^T b$, and rewrite the base inequality as $\sum_{j=1}^n v_j x_j \le b'$. One can then round down the coefficients of the left-hand side without compromising the validity of integer feasible solutions. The resulting inequality $\sum_{j=1}^n \lfloor v_j \rfloor x_j \le b'$

has an integral left hand side, so the right-hand side can be rounded down, yielding the valid inequality $\sum_{j=1}^{n} \lfloor v_j \rfloor x_j \leq \lfloor b' \rfloor$.

Chvátal-Gomory rounding can be extended to arbitrary mixtures of constraint types. Non-positive multipliers λ can be used to transform $\geq$ constraints into $\leq$ constraints; the same rounding procedure then applies. Multipliers of any sign can be used with equality constraints. It can also be extended to constraints that include continuous as well as integer variables; see Wolsey [28] for more information.

While many MIP solvers have powerful Mixed Integer Rounding (MIR) cut generators, they rely on heuristics to identify constraints to try to combine to derive a suitable base inequality upon which to apply the Chvátal-Gomory (C-G) rounding. Therefore, the practitioner may be able to take advantage of specific model characteristics to derive base inequalities for C-G rounding that eludes the more generic rounding procedures of the solvers.

4.6.2 Graph-Based Cut Generation

Associating graphs with constraints in the MIP model is a well-established technique for strengthening formulations. Analogous to combining constraints in Sect. 4.6.1, commercial MIP solvers make use of some graphs to identify model structures that yield useful cuts. However, once again, the practitioner may be able to identify specific graphs associated with less common types of constraints, and use them to derive customized cuts.

The conflict graph, which is used by many MIP solvers and discussed extensively in the literature, provides an illustration. Each node of the graph is associated with a binary variable, and two nodes are connected by an edge if one or more constraints imply that both variables cannot equal one in any integer feasible solution. For model P_2 from the previous subsection (4.6.1), this results in the conflict graph in Fig. 4.12.

The graph forms a clique, i.e., every node is connected to every other node. Therefore, setting the binary variable associated with one node to 1 forces all other variables in the clique to 0 in any integer feasible solution. Hence, the sum of the binary variables in any clique in the conflict graph can be at most 1. For P_2, this derives the same globally valid cut $x_1 + x_2 + x_3 \leq 1$ that was derived by combining the constraints and rounding. However, a slightly more elaborate example shows

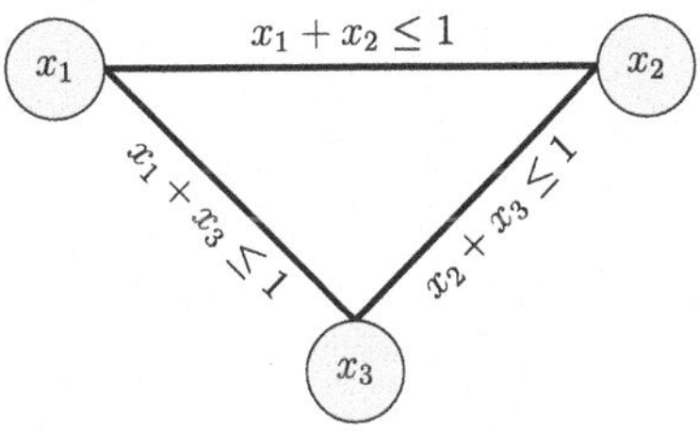

Fig. 4.12 The Conflict Graph for Model P_2: each node represents a binary variable, and each edge represents a conflict, i.e., a constraint that does not allow both binary variables to be one at the same time

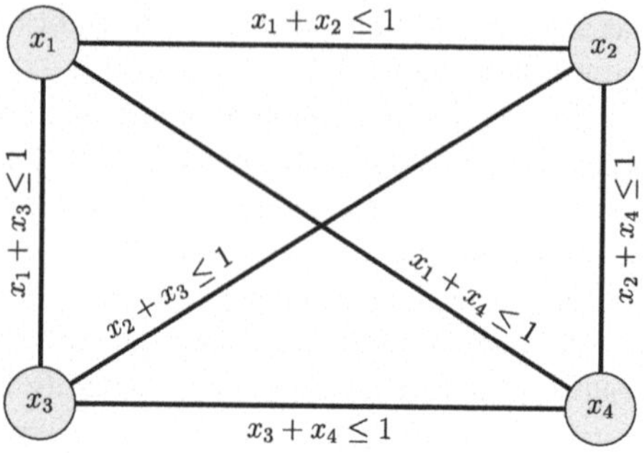

Fig. 4.13 The Conflict Graph for Model P_4: each node represents a binary variable, and each edge represents a conflict, i.e., a constraint that does not allow both binary variables to be one at the same time

that the conflict graph can yield cuts that cannot be deduced directly by combining constraints in the model. The following MIP consists of 4 binary variables with constraints ensuring that no pairs of variables can be 1.

$$
\begin{aligned}
(P_4): \quad \underset{x}{\text{maximize}} \quad & x_1 + x_2 + x_3 + x_4 \\
\text{subject to} \quad & x_1 + x_2 \leq 1 \\
& x_1 + x_3 \leq 1 \\
& x_1 + x_4 \leq 1 \\
& x_2 + x_3 \leq 1 \\
& x_2 + x_4 \leq 1 \\
& x_3 + x_4 \leq 1 \\
& x_j \in \{0, 1\}, \quad j = 1, 2, 3, 4
\end{aligned}
$$

Figure 4.13 shows the conflict graph. Since all nodes in the graph are part of a clique, the cut $x_1 + x_2 + x_3 + x_4 \leq 1$ is globally valid and cuts off the LP relaxation solution $x_1 = x_2 = x_3 = x_4 = \frac{1}{2}$. However, this cut cannot be derived by simply summing the 6 constraints in P_4, in contrast to the way summing the 3 constraints in P_2 yielded the clique cut. For P_4, the base inequality resulting from summing the constraints in the model is $3x_1 + 3x_2 + 3x_3 + 3x_4 \leq 6$, from which no integer rounding can be performed.

Note that one could derive the cut by first generating the 4 cuts associated with the cliques of size 3 in the conflict graph. The cuts are

$$x_1 + x_2 + x_3 \leq 1 \tag{4.5a}$$

$$x_1 + x_2 + x_4 \leq 1 \tag{4.5b}$$

$$x_1 + x_3 + x_4 \leq 1 \tag{4.5c}$$

$$x_2 + x_3 + x_4 \leq 1 \tag{4.5d}$$

The combined constraint is then $3(x_1 + x_2 + x_3 + x_4) \leq 4$, which yields $x_1 + x_2 + x_3 + x_4 \leq 1$ after dividing by 3 and performing the integer rounding. This provides an example of cuts of higher rank, as the constraints that were combined to derive the cut did not come from the original model, but rather from cuts derived from the original model. Cuts derived from only the original model have rank 1, while the cuts that also use rank 1 cuts in their derivation have rank 2. Cuts that include cuts of rank $k - 1$ in their derivation have rank k. Nemhauser and Wolsey [20] contains more details on cut rank. Using the conflict graph to derive clique cuts is more direct and efficient, but both approaches ultimately rely on heuristics. Finding all maximal cliques in a graph currently has exponential worst case complexity, so commercial solvers use heuristics to find the cliques used to generate clique cuts. Atamturk et al. [2] provide more details on the effective use of the conflict graph to strengthen MIP formulations.

Effective use of graphs to tighten a MIP formulation may involve using graphs that solvers like Gurobi don't use, or it may make use of graphs they do use, but in different ways specific to individual models. The p-dispersion-sum problem provides a good example of the latter approach. This model is discussed in Klotz and Newman [17] and also in a blog by Kalvelagen at http://yetanothermathprogrammingconsultant.blogspot.com/2019/06/maximum-dispersion.html. The problem involves a collection of n objects with some measure of distance between them. The objective is to select $k < n$ of the objects that maximize the cumulative distance between the objects. The objective thus provides a metric of diversity among the objects. Data science applications in which the objects are data points in a vector space provide one common use case. The most direct model formulation involves binary variables and a quadratic objective. Let $d_{ij} \geq 0$ be the known distance between object i and object j, and let x_i be a binary variable that is 1 if object i is selected, and 0 otherwise. The model can be formulated directly as a binary quadratic program with a single cardinality constraint.

$$
\begin{aligned}
(P_{BQP}): \quad & \underset{x}{\text{maximize}} && \sum_{i=1}^{n} \sum_{j=i+1}^{n} d_{ij} x_i x_j \\
& \text{subject to} && \sum_{j=1}^{n} x_j \leq k \\
& && x_j \in \{0, 1\}, \quad \forall j = \{1, \ldots n\}
\end{aligned}
$$

The quadratic objective is nonconvex since all diagonal terms are zero. P_{BQP} can be solved directly by spatial branch and bound algorithms that branch on the nonconvexity. However, solver performance on even fairly small instances of this formulation with $n = 50$ and $k = 10$ in Kalvelagen's blog cannot be solved to optimality in any reasonable amount of time. Part of the difficulty is that the cardinality constraint that comprises the feasible region for the QP relaxation is the convex hull of the integer feasible points; any solution with fractional values can be written as a convex combination of two integer feasible points and is therefore not an extreme point. A convex hull formulation implies that an optimal LP relaxation solution is optimal for the associated MILP, but the same does not hold for QP relaxations and MIQPs, regardless of the convexity of the quadratic objective. None of the polyhedral cut theory based on facets of the convex hull of integer feasible solutions applies to P_{BQP}.

Fortunately, linearized formulations exist that introduce additional binary variables to model the products of binaries [26] in the objective of P_{BQP}. Such linearized formulations can potentially be strengthened. Numerous different linearizations are discussed in the literature, but the so-called standard linearization suffices here. Each product of binaries $x_i x_j$ in P_{BQP} is replaced with a new binary variable z_{ij} and associated linear constraints that ensure that $z_{ij} = x_i x_j$ in any integer feasible solution. P_{BQP} can thus be reformulated as an equivalent Binary Integer Program (BIP).

$$
\begin{aligned}
(P_{BIP}): \quad & \underset{x,z}{\text{maximize}} && \sum_{j=1}^{n} \sum_{\substack{i=1 \\ i<j}}^{n} d_{ij} z_{ij} && && (4.6a) \\
& \text{subject to} && \sum_{j=1}^{n} x_j \leq k && && (4.6b) \\
& && z_{ij} \leq x_i && \forall i,j \in \{1, ..., n\} && (4.6c) \\
& && z_{ij} \leq x_j && \forall i,j \in \{1, ..., n\} && (4.6d) \\
& && x_i + x_j \leq 1 + z_{ij} && \forall i,j \in \{1, ..., n\} && (4.6e) \\
& && x_j, z_{ij} \in \{0, 1\} && \forall i,j \in \{1, ..., n\} && (4.6f)
\end{aligned}
$$

The constraints (4.6c), (4.6d) and (4.6e) need only be created for indices (i, j), $i < j$ because the selection of both i and j is equivalent to the selection of both j and i. Hence, the model only need define z_{ij} variables with $i < j$. Note that if x_i or $x_j = 0$, then constraints (4.6c) and (4.6d) force z_{ij} to 0, while (4.6e) imposes no restriction on z_{ij}. Similarly, if both x_i and $x_j = 1$, (4.6c) and (4.6d) impose no restriction on z_{ij}, while (4.6e) forces z_{ij} to 1. So, regardless of the values of x_i, and x_j, $z_{ij} = x_i x_j$, and occurrences of $x_i x_j$ can be replaced with z_{ij} to obtain the linearized reformulation above.

Unlike P_{BQP}, the relaxation feasible region for the linearized formulation P_{BIP} is not the convex hull of integer feasible solutions. The formulation can be strengthened with globally valid cuts. Specifically, one can add the globally valid cut

$$
\sum_{i=1}^{n} \sum_{j=i+1}^{n} z_{ij} \leq k(k-1)/2
$$

Klotz and Newman [17] and Kalvelagen use different approaches to derive this cut, but both draw on techniques previously discussed in this chapter. Klotz and Newman [17] start by looking at a small instance with $n = 3$ and $k = 2$, then use the algebraic approach to contrast the BIP and its LP relaxation. The resulting disconnect is manifested by all 3 z variables taking on positive values of $\frac{2}{3}$ in the optimal LP relaxation solution, but only one z variable taking on a positive value in the BIP. This raises the question of whether any feasible solutions exist in the BIP that have two z variables set to 1, and examination of the constraints reveals no such solutions exist. The authors then generalize this to arbitrary n and k by assuming, without loss of generality, that $x_1 = x_2 = \cdots = x_k = 1$, and $x_{k+1} = \cdots = x_n = 0$. They then use a counting argument to deduce the cut. Kalvelagen instead starts with the cardinality constraint (4.6b), then creates n quadratic constraints by multiplying the cardinality constraint by each of the n x variables. He then replaces each $x_i x_j$ term with the corresponding z_{ij} variable and adds up these n linearized quadratic constraints. Performing some simplifications and substitutions yields the cut. He does not discuss the weakness in the formulation.

Both of these approaches work, but a graph-based approach offers a very easy way to derive the cut directly from P_{BQP}. Let $G = (V, E)$ be a graph with nodes V corresponding to the n x variables in P_{BIP}. Each z variable identifies an edge in the graph connecting the nodes associated with the product of the two x variables that the z variable linearizes, i.e., z_{ij} defines an edge between the nodes for x_i and x_j. This graph is frequently called the product graph, and it can also be derived directly from the original formulation P_{BQP} by connecting edges for pairs of variables with a nonzero quadratic term. Note that the product graph is a complete graph for the p-dispersion-sum problem discussed here, but it may not be for other models. Padberg [21] pioneered the use of this graph in unconstrained binary QPs. He used it to derive facet defining inequalities based on cliques in the product graph. Gurobi's BQP cuts are derived from such cliques.

While Padberg used the product graph based purely on the objective, it can also be used in conjunction with linearized constrained BQPs, and the p-dispersion-sum problem provides a useful example. Figure 4.14 illustrates how the graph provides a simple way to visualize the counting argument in Klotz and Newman [17] that determines the maximum number of z variables that can be 1.

The number of z variables set to 1 will be the same regardless of which k of the x variables are set to 1. Without loss of generality, assume that variables $x_1, x_2, \ldots, x_k$ are set to 1, as denoted by the gray nodes in the figure and the solid edges connecting them. Dashed edges correspond to the z variables associated with pairs of x variables of which at least one x variable is set to 0, and hence the corresponding z variable is 0. Every pair of objects is assumed to have a positive distance, so the gray nodes associated with $x_1, x_2, \ldots, x_k$ and their associated edges with solid lines comprise a complete subgraph of the original product graph. The number of edges in the subgraph is $(k-1) + (k-2) + \cdots + 1 = k(k-1)/2$. As all the edges with dashed lines correspond to z variables that are 0, the number of edges in this complete subgraph provides the upper bound on the number of z variables that can be 1, establishing the cut. The authors contend that this graph-based approach which takes

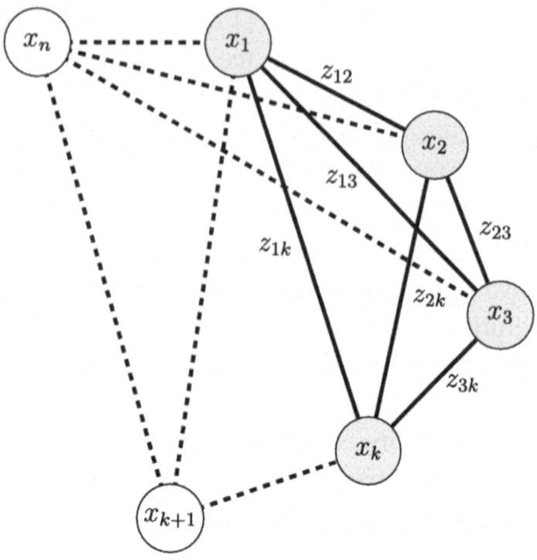

Fig. 4.14 The product graph for the P-dispersion-sum problem

the standard product graph and augments it with constraint information specific to the model provides the easiest way to derive a cut that is essential to strengthening the formulation.

Readers interested in exploring linearizations of binary variables in greater detail can read Glover [13] and Sherali and Smith [25], which describe additional linearizations beyond the standard linearization used here. Forrester and Hunt-Isaak [9] provides a detailed computational comparison of these various linearizations. Readers interested in the p-dispersion-sum problem and the product graph can download Python programs to generate model instances and product graphs for arbitrary quadratic expressions of binary variables at https://github.com/Gurobi/techtalks.

4.6.3 Disjunctive Arguments

Disjunctions are an extremely powerful method for deriving general cuts mathematically using polyhedral theory as well as more specialized cuts that strengthen particular MIP formulations. However, effectively harnessing their power can be more challenging than the constraint combining and graph oriented strategies discussed in the previous two subsections. This subsection attempts to provide some guidance on how to use them.

Before considering some examples, consider the following proposition that provides the foundation of basic disjunctive arguments. Chapter 8 of Wolsey [28] contains more details.

Proposition 4.1 *Let X_1 and X_2 be subsets of R^n_+. Let $X = X_1 \cup X_2$. For $x \in R^n_+$ and $u, v \in R^n$, suppose $u^T x \leq b_1$ is valid for X_1 and $v^T x \leq b_2$ is valid for X_2. Let $w \in R^n$ be less than or equal to the component-wise minimum of u and v, i.e., $w_j \leq$*

$\min(u_j, v_j), j \in \{1, \ldots n\}$, *and let* $b \geq \max(b_1, b_2)$. *Then* $w^T x \leq b$ *is valid for all* $x \in X$.

Proof Since $X = X_1 \cup X_2$, for any $x \in X$, $x \in X_1$ or $x \in X_2$. Since $w_j \leq \min(u_j, v_j)$, then $w \leq u$ and $w \leq v$. The nonnegativity of $x \in X$ implies that $w^T x \leq u^T x$ and $w^T x \leq v^T x$. Since $u^T x \leq b_1 \leq b$ and $v^T x \leq b_2 \leq b$, chaining these inequalities implies that $w^T x \leq b$, establishing its validity for $x \in X$. □

Some discussion is in order before moving on to some simple examples that illustrate how to use this result to derive useful disjunctive cuts. First, note that X_1 and X_2 are arbitrary subsets of R^n_+. They could be linear or nonlinear, discrete or continuous, convex or nonconvex, and the proposition still holds. So this result may be useful in modeling more complex problem types than the MILPs and MIQPs under consideration in the chapter. Second, an equivalent proposition exists for $\geq$ constraints.

Proposition 4.2 *Let* X_1 *and* X_2 *be subsets of* R^n_+. *Let* $X = X_1 \cup X_2$. *For* $x \in R^n_+$ *and* $u, v \in R^n$, *suppose* $u^T x \geq b_1$ *is valid for* X_1 *and* $v^T x \geq b_2$ *is valid for* X_2. *Let* $w \in R^n$ *be greater than or equal to the component-wise maximum of u and v, i.e.,* $w_j \geq \max(u_j, v_j), j \in \{1, \ldots n\}$, *and let* $b \leq \min(b_1, b_2)$. *Then* $w^T x \geq b$ *is valid for all* $x \in X$.

The proof is omitted since one can multiply the $\geq$ constraints by -1 and apply Proposition 4.1 and its proof. Third, the relative simplicity of the proof (which simplifies even more when looking at small examples) may prompt the reader to doubt that this result could lead to any useful cuts that will strengthen a formulation. A small example is in order to address this doubt.

$$(P_{abs}): \quad \underset{x,y}{\text{minimize}} \quad x + y \tag{4.7a}$$

$$\text{subject to} \quad |x - y| \geq 1 \tag{4.7b}$$

$$x, y \in [0, 5] \tag{4.7c}$$

Note that x and y are continuous variables in this model. Constraint (4.7b) can be equivalently expressed as $x - y \geq 1$ or $y - x \geq 1$. The lower bounds of 0 on x and y simplify this disjunctive condition to $x \geq 1$ or $y \geq 1$, which in turn implies $x + y \geq 1$. This disjunctive inequality could also have been derived using Proposition 4.2 by taking $X_1 = \{(x, y) \in R^n_+ : x - y \geq 1\}$, $X_2 = \{(x, y) \in R^n_+ : y - x \geq 1\}$, $u = (1, -1)$, $v = (-1, 1)$, and $b_1 = b_2 = 1$. Then $w = (1, 1)$ and $b = 1$, yielding the desired inequality.

But does this inequality actually tighten the formulation of (P_{abs})? Figure 4.15 illustrates the feasible region for (P_{abs}) and the disjunctive inequality derived by Proposition 4.2. From the figure one can see that (P_{abs}) is a discrete optimization problem involving a nonconvex feasible region.

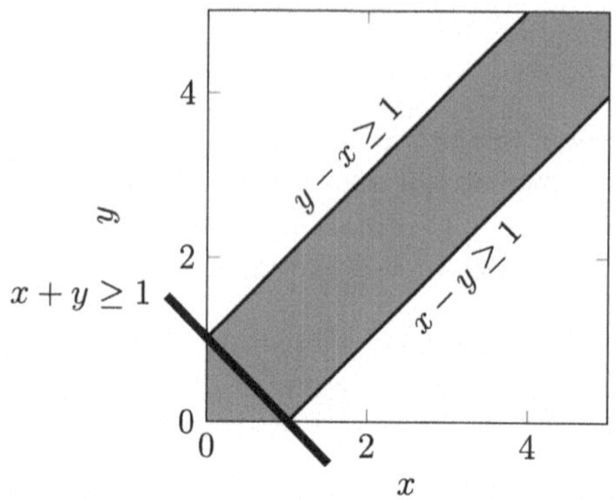

Fig. 4.15 Feasible region and disjunctive inequality for model (P_{abs}). The region colored in gray is made infeasible by the disjunction $|x - y| \geq 1$. Based on this image it is not clear whether the disjunctive cut $x + y \geq 1$ actually tightens the formulation

However, (P_{abs}) does not fit the description of a MIP given by (P_{MIP}), and thus does not have as direct a notion of a convex relaxation analogous to the continuous relaxation of (P_{MIP}). Furthermore, while the figure shows that the disjunctive inequality $x + y \geq 1$ is globally valid both algebraically and visually, one cannot discern whether the inequality actually tightens the formulation in any way. In the space of variables x and y in Fig. 4.15, it does not do so. (P_{abs}) has no discrete variables, and the absolute value constraint does not fit the requirement of linear constraints in (P_{MIP}). A linearized formulation that uses a binary variable and linear constraints to represent the nonconvex absolute value constraint as described in (P_{abs}) is in order:

$$(P_{abslin}): \quad \underset{x,y}{\text{minimize}} \quad x + y \tag{4.8a}$$

$$\text{subject to} \quad x - y \geq 1 - 6z \tag{4.8b}$$

$$y - x \geq 1 - 6(1 - z) \tag{4.8c}$$

$$x, y \in [0, 5], z \in \{0, 1\} \tag{4.8d}$$

Here, the linear constraints (4.8b) and (4.8c) model the explicit absolute value constraint in (P_{abs}). If the binary variable $z = 0$, then constraint (4.8b) is enforced, while constraint (4.8c) is always satisfied for $x, y \in [0, 5]$. Similarly, if $z = 1$, constraint (4.8c) is enforced, while constraint (4.8b) is not. The choice of coefficient 6 for the binary variable is based on the upper bound of 5 on x and y; if that upper bound was U, the coefficient would be $U + 1$. Note that these two constraints resemble the ones in model (P_{game}) described in Sect. 4.5.3 used to forbid overlap of the rectangles in the game. However, the latter constraints forbade overlap of two-dimensional rectangles; here the absolute value constraint in (P_{abs}) or the linearized counterparts in (P_{abslin}) forbid overlap of one-dimensional line segments of length 1. The two variables x and y represent one-dimensional coordinates on a vertical or horizontal axis starting at the origin of line segments of length 1. The two-line segments must be placed on the axis so that they do not overlap. As with the rectangle game, relaxing the integrality conditions allows complete overlap. Algebraically, this means that

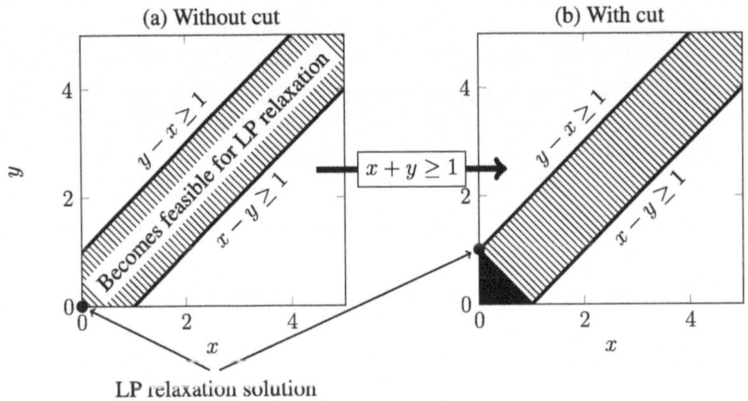

Fig. 4.16 Feasible region and disjunctive inequality for model P_{abslin}: by introducing the disjunctive cut $x + y \geq 1$, we can discard the previous LP relaxation solution and attain an integer feasible solution directly from the LP relaxation

in the LP relaxation of (P_{abslin}), the solution $x = y = 0, z = \frac{1}{2}$ is optimal with an objective of 0. However, the disjunctive inequality $x + y \geq 1$ cuts off this solution, tightening the formulation.

Figure 4.16 illustrates the feasible regions for the MILP P_{abslin} and its LP relaxation. In Fig. 4.16a, the LP relaxation feasible region is given by the complete 5×5 square. The feasible region for P_{abslin} is the same as the feasible region of P_{abs}, namely the union of the two white triangles with vertices $\{(1, 0), (5, 0), (5, 4)\}$ and $\{(0, 1), (0, 5), (4, 5)\}$, respectively. However, by introducting the disjunctive cut in Fig. 4.16b, the feasible region for the LP relaxation is reduced by the black region, which represents the portion of the LP relaxation feasible region separated by the disjunctive cut. In this case the cut defines a facet of the convex hull of the set of integer feasible solutions for the MILP.

While (P_{abs}) and (P_{abslin}) illustrate the power of disjunctions, determining the appropriate linear expressions on which to apply the disjunction can be challenging. For (P_{abs}) and its linearized formulation (P_{abslin}), the absolute value in the $|x - y| \geq 1$ constraint suggested a natural disjunction, as absolute value constraints have an equivalent representation as logical *or* constraints. But such a natural disjunction may not present itself so clearly in other cases. However, using the algebraic approach to identify the weakness in the formulation can help identify relevant constraints and the associated variables to use in the disjunction. Klotz and Newman [17] contain a one constraint, pathological example that still proves challenging for many solvers in its original form. However, deriving a suitable disjunction by identifying the weakness in the formulation results in two cuts that enable the model to solve immediately.

The late Egon Balas was the pioneer of the theory of disjunctions, so there is no better source of further reading than his comprehensive, recently published book "Disjunctive Programming" [3]. The book describes not only the original concepts developed in the 1970s, but traces the evolution and contributions of others all the way to more recent developments such as Lift-and-Project cuts, which have been implemented in several MIP solvers, including Gurobi.

4.7 Additional Strategies to Strengthen MIP Formulations

Combining constraints, graphs, and disjunction are foundational strategies to strengthen MIP formulations in the sense that they are used heavily in the literature yet are also very useful for the less theoretical approaches to derive cuts on individual models discussed here. This section describes some additional methods that can be useful on individual models despite appearing much less frequently in the literature, expanding on a similar discussion in Klotz and Newman [17].

4.7.1 Disaggregation

There is the notion that the number of variables and constraints is correlated with the time it takes to solve a model. This is often not accurate, as model structure and tightness of the formulation are significantly more important than size.

An example of this is disaggregation, i.e., to decompose a single constraint into several, new constraints. For example, let $x_1, x_2, x_3, y \in \{0, 1\}$. Then a disaggregation can look as follows:

$$x_1 + x_2 + x_3 \leq 3y \Rightarrow \begin{cases} x_1 \leq y \\ x_2 \leq y \\ x_3 \leq y \end{cases} \tag{4.9}$$

While this disaggregation increases the number of constraints, the disaggregated formulation is tighter than the aggregated form. In fact, Wolsey [28] demonstrates through various proofs that this particular disaggregation actually forms the convex hull.

Note though that in constraint (4.9), the feasible region for the original MIP does not change, but rather the feasible region for the corresponding LP relaxation. For example, the solution $x_1 = 1, x_2 = x_3 = 0, y = \frac{1}{3}$ is feasible for the original relaxation but cut off for the disaggregated form. This demonstrates that disaggregation can be used to strengthen a weak dual bound, however at the cost of generating a potentially very large number of constraints.

4.7.1.1 Disaggregating a Facility Location Problem

To illustrate this further, consider the facility location problem, where we aim to place facilities to serve customers so as to minimize cost. Each customer is served by at least one factory, however there is a cost associated with opening a factory.

Using the following binary variables, with a value of 1 indicating the associated action is performed:

$x_{i,j}$: Does factory i deliver to customer j?
y_i: Is factory i being built?

We can now describe the problem as follows:

$$(P_{FLP}): \quad \underset{x,y}{\text{minimize}} \quad \sum_{i \in I} b_i y_i + \sum_{j \in J} c_{i,j} x_{i,j} \qquad \text{(minimize cost)} \tag{4.10a}$$

$$\text{subject to} \quad \sum_{i \in I} x_{i,j} \geq 1 \qquad \forall j \in J \qquad \text{(customers served)} \tag{4.10b}$$

$$\sum_{j \in J} x_{i,j} \leq |J|\, y_i \quad \forall i \in I \quad \text{(open to serve)} \tag{4.10c}$$

where b_i is the cost of building factory i, $c_{i,j}$ is the cost of delivering from factory i to customer j, I and J refer to the set of factories and customers respectively, and $|J|$ denotes the cardinality of the set J, i.e., how many customers there are.

In this example, the "open to serve" constraint can in fact be disaggregated:

$$\sum_{j \in J} x_{i,j} \leq |J|\, y_i \Rightarrow \begin{cases} x_{i,1} \leq y_i \\ x_{i,2} \leq y_i \\ \vdots \\ x_{i,|J|} \leq y_i \end{cases} \tag{4.11}$$

This can be visualized as shown in Fig. 4.17.

Note that this disaggregation is proven to produce a tighter LP relaxation, however at the cost of adding $|I| \cdot (|J| - 1)$ more constraints. The addition of these extra constraints might impact the time it takes to solve the root relaxation, or the quality of the first solution.

Remark 4.2 Since Gurobi, like most other solvers, has a limit for the number of cuts it will add, it does not perform this disaggregation automatically.

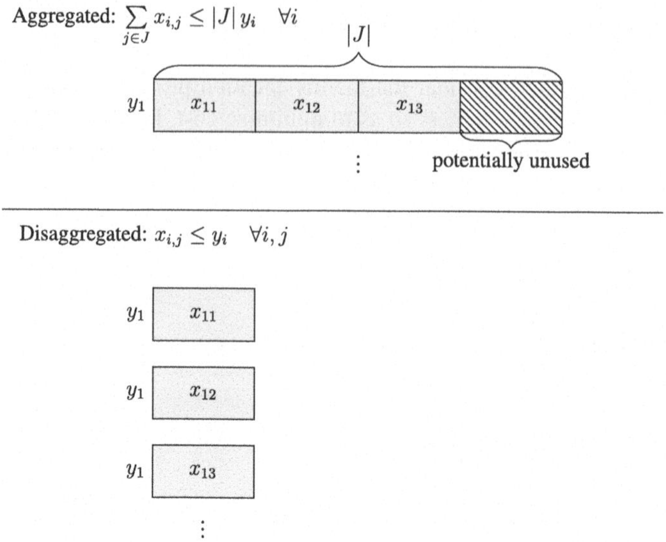

Fig. 4.17 Visual representation of the difference between the aggregated and disaggregated constraint(s). Similar to the cutting stock example (see Fig. 4.11), the weakness of the formulation is apparent as the maximum allowed number of customers need not be assigned. This creates unused capacity, which enables a fractional solution for y without changing the assignment

4.7.2 *Solving Related Models to Derive Cuts*

Solving a related model can provide information in the form of cuts or heuristic solutions. This is relevant to the discussion of multi-objective models in disguise in Sect. 4.4, particularly in the presence of soft constraints and their associated penalty variables. If minimizing the sum of penalty variables results in a positive objective value, constraining the sum of penalty variables to be greater than or equal to this value can strengthen the original model formulation. Specifically, a tighter formulation occurs when the initial dual bound of this related problem is less than the optimal minimum sum. In addition, the optimal solution obtained from minimizing the sum of penalty variables provides a feasible starting solution to the original model, regardless of the value of the minimum sum.

4.7.3 *Using Infeasibility to Derive Cuts*

If one can derive an infeasible condition from the MIP to be tightened, the reverse of that infeasible condition is valid for all integer feasible solutions. In particular, for any linear expression involving integer variables with integer coefficients a and an integer right-hand side b, if $a^T x \leq b$ can be proven infeasible, then $a^T x \geq b + 1$

provides a valid cut. Non-integral variables or problem data create complications by requiring the use of a suitable tolerance to enforce the reverse inequality. In that case, if $a^T x \leq b$ can be proven infeasible then $a^T x \geq b + \epsilon$ is a valid cut. The proper choice of the tolerance ϵ depends on factors including the number of decimal places in the non-integral problem data, the solver feasibility and integrality tolerances, and condition numbers of the optimal bases of the node relaxations during the solve. The use of disjunctive arguments can effectively find such cuts. The disjunctions discussed in Sect. 4.6.3 described two-sided disjunctions in which both sides of the disjunction (i.e., the logical *or* condition) implied a constraint to be satisfied. However, for some disjunctions, one side of the disjunction implies the associated model is infeasible, in which case the other side of the disjunction provides a valid cut. Klotz and Newman [17] provide examples of both one- and two-sided disjunctions. As with disjunctive arguments, the challenge to derive cuts by using infeasibility arguments consists of finding suitable related models or one-sided disjunctions that yield cuts that strengthen the formulation.

4.7.4 Dual-Based Reductions

Globally valid cuts are based on the constraints of a MIP. Therefore, they will remain globally valid regardless of the choice of objective function. However, the strength they add to the formulation may vary depending on the objective. This can be seen in Fig. 4.16 and the associated model P_{abslin} in Sect. 4.6.3. The figure reveals two globally valid cuts, $x + y \geq 1$ and $x + y \leq 9$. The former strengthens the formulation for the objective of $x + y$ to be minimized in P_{abslin}; the latter is of no use in that model. However, if the objective sense of the model were changed to maximize, the latter cut would strengthen the formulation while the former would not.

Remark 4.3 The cut $x + y \leq 9$ can be derived in an analogous fashion to $x + y \leq 1$. However, as it does not contribute to the improvement in the LP relaxation solution, it was omitted from Fig. 4.16 for clarity.

Additional strengthenings can be obtained by sacrificing the requirement of global validity, but doing so in a way that ensures that at least one optimal solution to the MIP is preserved. This is the foundation of so-called dual tightenings that feature prominently in many commercial solvers. A simple example illustrates. Consider a MILP model with a single soft constraint, and an arbitrary number of other linear constraints. The penalty variable s on the soft constraint is continuous. For simplicity, assume all of the other variables are in $\mathbb{Z}_n^+$ (the set of nonnegative integers), but the conclusions are the same regardless of the mix of continuous and integer variables or their bounds.

$$
\begin{aligned}
(P_{soft}): \quad & \underset{x,s}{\text{minimize}} \quad c^T x + \alpha s \\
& \text{subject to } \hat{a}^T x + s \geq \hat{b} \\
& Ax \sim b \\
& x, s \geq 0 \\
& x \in \mathbb{Z}_n^+
\end{aligned}
$$

The penalty variable s allows for shortfalls in the associated hard constraint $\hat{a}^T x \geq \hat{b}$, at a per unit cost of $\alpha > 0$. Given that s only appears in the objective with a positive value and a single constraint that it relaxes, a dual argument can be used to tighten the soft constraint from the inequality $\hat{a}^T x + s \geq \hat{b}$ to the equality $\hat{a}^T x + s = \hat{b}$. For any integer feasible solution $(\tilde{x}, \tilde{s})$ with $\hat{a}^T \tilde{x} + \tilde{s} > \hat{b}$, any value δ in $[0, \hat{a}^T \tilde{x} + \tilde{s} - \hat{b})$ results in another integer feasible solution, $(\tilde{x}, \tilde{s} - \delta)$ with a smaller objective than the previous solution $(\tilde{x}, \tilde{s})$. Since the penalty variable s appears only in the objective and the soft constraint, it can be reduced until $\delta = 0$, or equivalently, $\hat{a}^T x + s = \hat{b}$. Note that this tightening is not a globally valid cut; it removes the set of integer feasible solutions $(\tilde{x}, \tilde{s} - \delta)$ for all positive values of δ in $(0, \hat{a}^T \tilde{x} + \tilde{s} - \hat{b})$. However, it does not remove any optimal solutions. More aggressive dual tightenings allow for removal of optimal solutions, as long as at least one optimal solution remains.

Strengthening a model formulation typically is made possible by understanding which variable values are feasible in the original problem regardless of the objective function, as seen in the Facility Location example in Sect. 4.7.1. However, in example P_{soft}, the argument for strengthening the formulation is based on the objective function. This involves considering dual feasible solutions, along with complementary slackness conditions, to deduce the primal variable values that are optimal in the LP relaxation of the MIP.[3] Since these reductions hold in the root node of the MIP, they will also hold in any child node, including the ones that yield optimal integer solutions. Since these reductions depend implicitly on both dual feasibility and strong duality, this model strengthening is referred to as "dual-based" reductions. However, as illustrated by example P_{soft} above, one can often derive such reductions more easily by looking at the primal objective and constraints rather than explicitly deriving reductions to the dual or the LP relaxation of the MIP based on feasibility and complementary slackness, then translating this reduction back to the primal LP relaxation.

A common dual tightening strategy involves making use of incumbent solution objective values or other known cutoff values that characterize good solutions. For example, in a model with soft constraints with some penalty variables with extremely high penalty costs, the presence of a solution with an objective value that rules out incurring any penalties may enable fixing the costly penalty variables to 0.

The lotsizing problem in Sect. 4.9 will discuss dual tightening in more detail.

[3] Note that the notion of a dual is not well-defined for a MIP due to presence of discrete components.

4.8 Table of Solver Cuts

Commercial MIP solvers generate a wide variety of cuts that are sufficiently generic to apply to a significant percentage of the MIPs they solve. They may be based on polyhedral theory (e.g., Gomory cuts), combining constraints (e.g., MIR cuts), graph theory (e.g., clique cuts), disjunctions (e.g., lift-and-project cuts), or special structure (e.g., cover cuts). When looking to strengthen a MIP formulation, knowledge of these standard cuts can help the practitioner avoid deriving cuts that the solver will derive as well. By understanding the types of constraints that generate the various solver cuts, the practitioner is better equipped to focus on aspects of the model that are more likely to yield cuts that the solver will miss. A detailed description of all the different cuts generated by state-of-the-art solvers would require a book of its own. However, Table 4.2 provides a summary of solver cuts and the associated model characteristics that generate them. This table was taken from Klotz and Newman [17] and augmented with cuts that have recently found their way into commercial solvers.

Note that lift-and-project cuts are a special case of disjunctive cuts in which the disjunction is based on the up and down branches associated with the branching

Table 4.2 Different types of cuts and their characteristics, where z is binary unless otherwise noted, and x is continuous; * denotes based on general polyhedral theory; † denotes based on specific, commonly occurring problem structure

Cut name	Mathematical description of cut	Structure of original MILP that generates the cut
Boolean quadric polytope† (BQP)	$\sum_{j=1}^{n} x_j - \sum_{i=1}^{n} \sum_{k=i+1}^{n} z_{ik} \leq 1$	Quadratic expressions with binary variables
Clique†	$\sum_i z_i \leq 1$	Packing constraints
Cover†	$\sum_i z_i \leq b$, b integer	Knapsack constraints
Disjunctive*	Constraint derived from an LP solution	$\sum_i a'_i x_i \geq b'$ or $\sum_i a''_i x_i \geq b''$
Flow cover†	Linear combination of flow and binary variables	Fixed charge network
Gomory*	Mixed integer rounding applied to simplex tableau associated with optimal node LP basis	$\bar{a}_C x_C + \bar{a}_{I/k} x_{I/k} + x_k = \bar{b}$, x_k integer, $x \geq 0$
Generalized upper bound†	$\sum_i x_i \leq b$, b integer	Knapsack constraints with precedence or packing
Implied bound†	$x_i \leq \frac{b}{a_i}$	$\sum_i a_i x_i \leq bz$, $x \geq 0$
Mixed integer rounding*	Use floors and ceilings of coefficients together with original variables	$\sum_i a_i z_i + \sum_j a_j x_j \leq b$, z integer, $x \geq 0$
Zero-half*	$\lambda^T A x \leq \lfloor \lambda^T b \rfloor$, $\lambda_i \in \{0, 1/2\}$	Constraints containing integer variables and coefficients

variable selected by the branch and bound algorithm. Also, while BQP cuts are listed as specific to quadratic expressions with binary variables, they can also be applied to MILPs with constraints similar to those that linearize the product of a pair of binary variables [26], e.g., (4.6c), (4.6d) and (4.6e) in (P_{BIP}).

4.9 Industrial Strength Examples

4.9.1 Laying Cables for Offshore Wind Farms

The principle of an offshore wind farm is simple: place wind turbines on water and bring the generated electricity to shore to be delivered to customers. This "bringing to shore" typically has two phases:

Array cables: The array cables connect all the turbines to one or more offshore substations, where the power is gathered and converted into a higher voltage to reduce power loss.

Export cables: The export cables connect the offshore substations to the onshore substations, where the power is fed into the national grid of the respective country.

A visual overview over this setup is given in Fig. 4.18.

How should these cables be laid? In this section, the focus will be on the array cable placement, which has received some attention in the literature, most recently with the work by Fischetti and Pisinger (see [7, 8] and references therein). The interested reader is referred to their work for a thorough literature review.

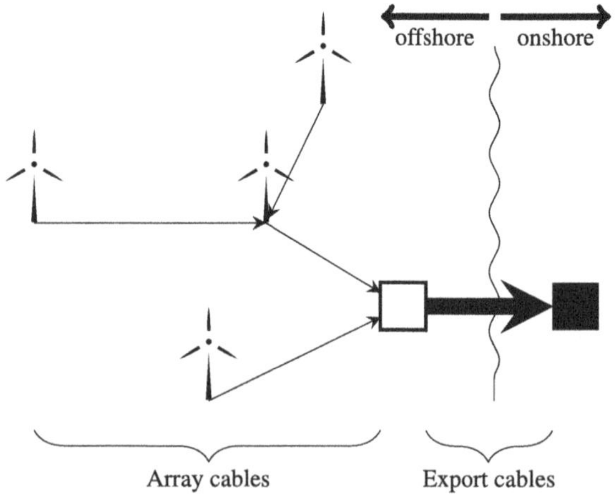

Fig. 4.18 A high-level overview over cabling on offshore wind farms. The wind turbines produce the power, which is carried through relatively small cables to an offshore substation (□). Then, one or more large export cables carry the power to the onshore substation (■), from where it is distributed to the respective national grid

4.9.1.1 Problem and Model Description

In its simplest form, the problem can be summarized as follows:

> Minimize the installation cost for the array cable layout such that all turbines are connected, and the maximum capacity of the cable, called rating, is not exceeded.

Mathematically, this model can be expressed using the following elements:

$x_{i,j} \in \{0, 1\}$: This binary variable denotes whether a cable is laid between turbine i and turbine or offshore substation j.
$f_{i,j} \geq 0$: This continuous variable denotes how much power flows between turbine i and turbine or substation j.
T: The set of all turbines
O: The set of all substations
N: The set of all turbines and substations, i.e., $N = T \cup O$

Based on these building blocks, we can now describe the following mathematical programming problem:

$$(P_{\text{array}}): \quad \underset{x}{\text{minimize}} \quad \sum_{\substack{i \in T \\ j \in N}} c_{i,j} x_{i,j} \tag{4.12a}$$

$$\text{subject to} \quad \sum_{j \in N} x_{i,j} = 1 \qquad \forall i \in T \tag{4.12b}$$

$$\sum_{j \in N} f_{t,j} = \sum_{i \in T} f_{i,t} + P \qquad \forall t \in T \tag{4.12c}$$

$$f_{i,j} \leq P^{\max} x_{i,j} \qquad \forall i \in T, j \in N \tag{4.12d}$$

$$x_{i,j} \in \{0, 1\}, \; f_{i,j} \geq 0 \qquad \forall i \in T, j \in N \tag{4.12e}$$

In this model, $c_{i,j}$ denotes the cost of building a connection between i and j, P denotes the power produced at a turbine[4] and $P^{\max}$ denotes the maximum power the cable can carry.

4.9.1.2 Where is the Weakness?

Although model (P_{array}) is straightforward, its performance is underwhelming, even on small datasets (see Fig. 4.19).

In fact, although we find the optimal solution after 25 s, Gurobi is unable to prove optimality even after 20000 s.

---

[4] P is identical for all turbines, because typically all turbines in a wind farm produce the same amount of power.

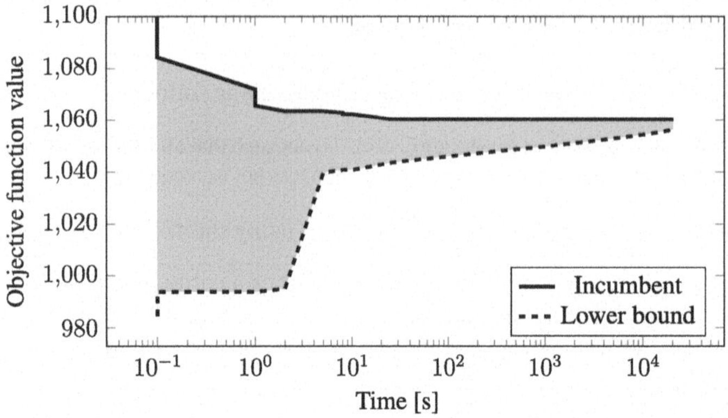

Fig. 4.19 Although the optimal solution is found quickly, the dual bound does not close easily, even for small datasets

As discussed in Sect. 4.3, the inability to close the dual bound points toward a weakness in the mathematical formulation. But where is it? Let us consider the flow balance constraint for a given turbine t:

$$\underbrace{\sum_j f_{t,j}}_{\text{Power out}} = \underbrace{\sum_i f_{i,t}}_{\text{Power in}} + \underbrace{P}_{\text{Power produced}} \tag{4.13}$$

where the power produced P is a constant scalar. This means that the value of any flow variable $f_{i,j}$ will be a multiple of P, since for any turbine the power produced consists of a set of predecessor turbines. This means that a turbine with k predecessors has kP units of powers flowing into it, and $(k + 1)P$ power flowing out. Considering this fact, let us consider the cable rating constraint for a given connection between i and j:

$$f_{i,j} \leq P^{\max} x_{i,j} \tag{4.14}$$

In words, this constraint states that the flow can only be nonzero if $x_{i,j} = 1$, i.e., if a cable is being built. However, even when a cable is built, the flow cannot exceed the maximum rating of the cable, denoted by $P^{\max}$. Unless $P^{\max}$ also is a multiple of P, there is a discrepancy between the values that $f_{i,j}$ can attain and the limit imposed by $P^{\max}$. This discrepancy is visualized in Fig. 4.20.

This weakness can also be understood in terms of the discrepancy between the solution of the MIP and LP relaxation. In the LP relaxation of the problem, $x_{i,j}$ can adhere to a fractional value such that $P_{\max} x_{i,j}$ is a multiple of P without impacting the flow variable $f_{i,j}$. Since we are aiming at minimizing the cost incurred by building a cable, which is related to $x_{i,j}$, this is the solution that the LP relaxation will and does attain. Therefore, the solver must use the branch and bound algorithm to fix these

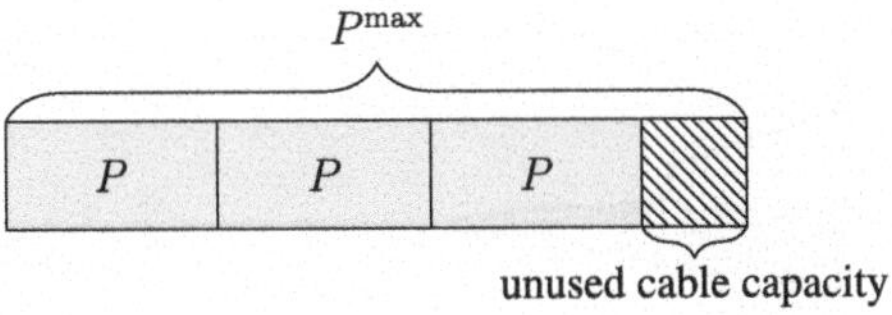

Fig. 4.20 A visual representation of the weakness of the formulation in problem (P_{array}): the cable fits a fractional amount of power produced by a turbine, which allows the corresponding binary variable x to take a fractional value without the amount of power transmitted through the cable

binaries, which results in a run time of over 3 hours without proving optimality, as illustrated in Fig. 4.19.

4.9.1.3 How Can This Formulation Be Strengthened?

Once we understand the source of the weakness, the solution becomes quite obvious: by simply rounding P^{max} down to the nearest multiple P, the discrepancy between the MIP and LP relaxation solution is removed. More concretely, this means that we define

$$\hat{P}^{max} = \lfloor \frac{P^{max}}{P} \rfloor P \quad (4.15)$$

and modify the cable rating constraint accordingly to:

$$f_{i,j} \le \hat{P}^{max} x_{i,j} \quad (4.16)$$

4.9.1.4 Computational Results and Conclusion

By this simple rounding operation,[5] the performance changes drastically, as the problem is solved to optimality within 17 s. A visual representation of the difference can be seen in Fig. 4.21.

This example emphasizes that it is crucial to understand why a certain formulation is weak before any action can be taken. Without understanding this, it is very challenging to find out that this one coefficient can be rounded. More importantly even, it also means understanding when this strengthening will no longer work: for example, if there were multiple types of turbines with different power productions P_1, P_2, etc., then it is not as obvious how to avoid this type of weakness.

An implementation of this offshore wind farming model is available at https://github.com/Gurobi/modeling-examples/tree/master/offshore_wind_farming.

[5] In fact, one can consider this a mixed-integer rounding cut in disguise, as it is possible to sum up the corresponding cable rating and combine them with the respective flow balance constraint.

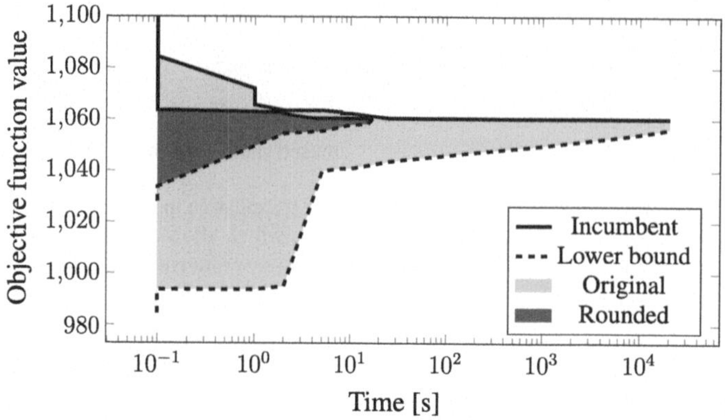

Fig. 4.21 By rounding the coefficient, we obtain a drastically better performance

4.9.2 A Lot Sizing Example

Lot sizing models involve production planning for a collection of products over a series of time periods. Customer demand for the products must be satisfied. Resources for producing the products are available but with costs. Usually products can be inventoried and sold in time periods subsequent to their production. Additional characteristics may be in play as well, such as perishability of inventory, time windows for production, and backordering. The objective typically involves minimizing the cost of operations, which includes cost associated with production, inventory, and possibly additional activities.

Lot sizing models tend to be challenging for MIP solvers, and the MIPLIB test set contains at least 3 lot sizing models that are all very modest in size, yet difficult to solve to optimality. One of those models is `b2c1s1`. In 2012, Gurobi 4.6.1 solved the model in 32.38 h. Gurobi 9.5, the current version as of this writing, solves it in 3 h with 8 threads on a modest machine (Intel Xeon E3-1240 3.50Ghz CPUs), 50 and 18.5 min with 8 and 32 threads respectively on a high end machine (one 16 core AMD 7313P processor with 32 logical processors). The model is very modest in size, with 3904 constraints, 3872 variables, and 11408 nonzeros, so even with these major improvements in the solver, the question of whether additional performance improvements are possible remains. Parameter tuning did not improve performance. Examination of the node log can help answer this question. The GRBlogtools (https://github.com/Gurobi/grblogtools) plot of the primal and dual bounds for the 32 thread run with default parameter settings appears in Fig. 4.22.

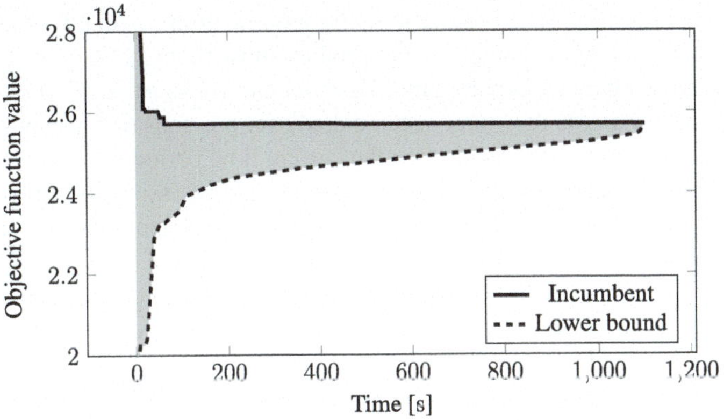

Fig. 4.22 GRBlogtools plot of primal and dual bounds for model `b2c1s1`

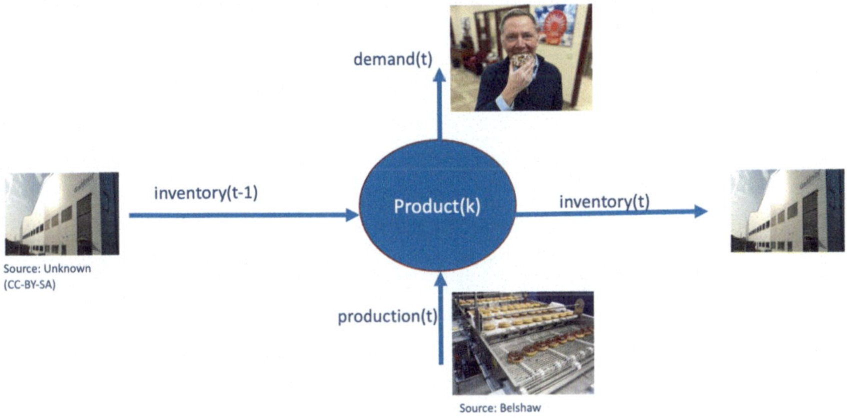

Fig. 4.23 Fundamental lot sizing process, single product and time period

The solid line shows progress in the incumbent solution objective (primal bound), while the dashed line shows progress in the best possible solution objective value (dual bound). Gurobi has found an optimal solution during the first 10% of the run, while the rest of the run time is spent very slowly improving the dual bound until it matches the primal bound, at which point optimality is proven. This suggests that the model formulation has some weakness that results in slow progress in the dual bound. If the formulation could be strengthened so that the dual bound moves much faster, performance gains of up to 80% seem possible even without any improvements in primal bound progress.

For completeness, a full description of the model formulation will follow. However, examination of the fundamental model activities for a single product and time period can make the complete model description easier to digest, as well as facilitate any subsequent assessment of the weakness of the formulation.

Figure 4.23 illustrates this basic activity for a single product (in this case, donuts) and single time period. Demand for the product must be met; the model does not permit any backorders or backlogging. Demand can be satisfied by drawing on a combination of available inventory accumulated over previous time periods and production from operating the machines in the current time period. Excess product can be forwarded to the next time period as inventory. Costs are associated with inventory and production. In this model, the production cost is represented as a fixed charge associated with turning on a production machine in a particular time period. Other lot sizing models may also include a per unit production cost. Let $x_{p,t}$, $s_{p,t}$ and $d_{p,t}$ represent production, inventory, and demand for product p in time period t, and let $u_{m,t}$ be a binary variable used to model the fixed charge associated with turning on machine m in time period t. The model has two fundamental elements. The first involves balancing the available product with the current demand data and the amount to be forwarded to the next time period. As illustrated by the arcs in Fig. 4.23, this can be represented as conservation of flow in a network. The second element represents the fixed cost of operating the machine.

$$s_{p,t-1} + x_{p,t} = s_{p,t} + d_{p,t} \quad \text{(flow balance)} \tag{4.17a}$$

$$x_{p,t} \leq M u_{m,t} \quad \text{(fixed charge)} \tag{4.17b}$$

A full description of the variables and constraints in model `b2c1s1` follows. The hurried reader can skip this, as the subsequent assessment of the weakness in the formulation depends only on the aforementioned flow balance and fixed charge constraints. The primary extension beyond the basic lot sizing model involves handling of component products that have no direct demand but go into the production of either the final product or another component product. In addition, the model has some constraints that limit production of selected products in individual time periods, and some production limitations on selected groups of machines and products, also in individual time periods. Downloading the column-oriented MPS file for model `b2c1s1` and using a MIP solver to write the associated row-oriented LP format file should aid in the digestion of the formulation.

Name	Definition	Dimensionality
Data		
N	Number of final products	1
R	Number of component products	1
P	Total number of products $N+R$	1
M	Number of machines	1
T	Number of time periods	1
$\hat{N}$	Set of all final products	
$\hat{P}$	Set of all products	
$\hat{M}$	Set of all machines	
$\hat{T}$	Set of all time periods	
S_1, S_2	Disjoint sets of final product	$(N/2, N/2)$
S_3	Set of limited production components	$\|S_3\|$
K	Set of machines that limit component production	$\|K\|$
S_K	Set of component products associated with set K	$\|S_K\|$
C_p	Set of component product dependencies, product p	$\sum_{p=1}^{P} \|C_p\|$
$d_{p,t}$	Demand for final product p, time period t	$N \cdot T$
$c_{p,t}$	Inventory cost for product p, time period t	$P \cdot T$
$f_{k,t}$	Fixed cost for using machine k, time period t	$M \cdot T$
$\beta_{k,t}$	Component consumption, machine k, time period t	$\|K\|$
μ_p	Production machine associated with product p	N
Variables		
$x_{p,t}$	Production for product p, time period t	$P \cdot T$
$s_{p,t}$	Inventory for product p, time period t	$P \cdot T$
$u_{k,t}$	Binary variable using machine k, time period t	$M \cdot T$

$$
\begin{aligned}
(b_{2c1s1}):\quad & \underset{x,s,u}{\text{minimize}} \quad \sum_{p=1}^{P}\sum_{t=1}^{T} c_{p,t}s_{p,t} + \sum_{k=1}^{M}\sum_{t=1}^{T} u_{k,t}f_{k,t} & & \\
& \text{subject to} \quad s_{p,t-1} + x_{p,t} = s_{p,t} + d_{p,t} & \forall p \in \hat{N}, \forall t \in \hat{T} \quad & \text{FLOW}(p,t) \\
& s_{i,t-1} + x_{i,t} - s_{i,t} = x_{p,t} & \forall p \in \hat{P}, \forall, i \in C_p, \forall t \in \hat{T} \quad & \text{FLO}(i,p,t) \\
& x_{p,t} \le 480 u_{\mu_p,t} & \forall p \in \hat{N}, t \in \hat{T} \quad & \text{VUB}(p,t) \\
& \sum_{p \in S_3} x_{p,t} \le 7500 & \forall t \in \hat{T} \quad & \text{COMPLIM}(t) \\
& \sum_{p \in S_1} x_{p,t} \le 120 & \forall t \in \hat{T} \quad & \text{PRODLIM}(t,1) \\
& \sum_{p \in S_2} x_{p,t} \le 120 & \forall t \in \hat{T} \quad & \text{PRODLIM}(t,2) \\
& \sum_{k \in K} \beta_{k,t}u_{k,t} + \sum_{p \in S_K} x_{p,t} \le 400 & \forall t \in \hat{T} \quad & \text{CAPLIM}(t) \\
& x_{p,t}, s_{p,t} \ge 0 & \forall p \in \hat{N}, t \in \hat{T} \quad & \\
& u_{k,t} \in \{0,1\} & \forall k \in \hat{M}, t \in \hat{T} \quad &
\end{aligned}
$$

where the descriptors denote the following aspects:

FLOW	Flow conservation, final product
FLO	Flow conservation, component products
VUB	Fixed charge
COMPLIM	Component production limits
PRODLIM	Final production limits
CAPLIM	Machine limits on component production

Note that these descriptors are used inside of the LP file.

4.9.2.1 Where is the Weakness?

As discussed in Sect. 4.5, identifying the constraints that prevent the objective from improving can help focus the assessment of the weakness in the formulation. To more easily digest the model formulation, consider a single product across the $T = 16$ time periods in the model, and consider the fundamental flow balance and fixed charge constraints. The model involves minimizing costs, so understanding why the zero cost option of keeping all the machines off and producing nothing is infeasible can help identify the activities that must occur. In some cases the practitioner may need to fix the cost function to 0 and invoke Gurobi's infeasibility finder to obtain a minimal explanation of the infeasibility. However, for this model, the fundamental flow balance and fixed cost constraints answer this question. One can use the LP Format file to examine the flow balance constraint for a single final product (in this case product 1, indexed in the LP file by the first two digits in the inventory and production variable names) across all 16 time periods.

```
FLOW0101: - s___0101 + x___01101 = 0
FLOW0102: s___0101 - s___0102 + x___01102 = 0
FLOW0103: s___0102 - s___0103 + x___01103 = 0
FLOW0104: s___0103 - s___0104 + x___01104 = 0
FLOW0105: s___0104 - s___0105 + x___01105 = 0
FLOW0106: s___0105 - s___0106 + x___01106 = 0
FLOW0107: s___0106 - s___0107 + x___01107 = 3
FLOW0108: s___0107 - s___0108 + x___01108 = 3
FLOW0109: s___0108 - s___0109 + x___01109 = 4
FLOW0110: s___0109 - s___0110 + x___01110 = 1
FLOW0111: s___0110 - s___0111 + x___01111 = 4
FLOW0112: s___0111 - s___0112 + x___01112 = 4
FLOW0113: s___0112 - s___0113 + x___01113 = 2
FLOW0114: s___0113 - s___0114 + x___01114 = 5
FLOW0115: s___0114 - s___0115 + x___01115 = 0
FLOW0116: s___0115 - s___0116 + x___01116 = 2
```

The right-hand sides of these constraints are the demands for product 01 across the 16 time periods. There is no initial inventory available in the first time period, so the first time period with positive demand requires a positive level of production. For the 16 FLOW constraints associated with product 01, that occurs in time period 7. Adding up the first 7 FLOW constraints for the first 7 time periods,

$$\sum_{t=1}^{7} x_{1,t} - s_{1,7} = 3 \tag{4.18}$$

Like all the other variables in the model, inventory variable s_{0107} has a lower bound of 0, making it effectively a slack variable. Therefore, the constraint (4.18) implies a production level of at least 3 across the production variables for time periods 1 through 7. But each production variable has a fixed charge for turning on the machine that can produce product 1 (in this case machine 19):

$$x_{1,t} \leq 480 u_{19,t} \quad t = 1, \ldots, 7 \tag{4.19}$$

Combining constraints (4.18) and (4.19),

$$3 \leq \sum_{t=1}^{7} x_{1,t} \leq 480 \sum_{t=1}^{7} u_{19,t} \tag{4.20}$$

In other words, machine 19 must be turned on in at least one of time periods 1 through 7, and the associated fixed charges will be incurred.

The weakness of the formulation can be assessed by extracting the following inequality from (4.20):

$$3 \leq 480 \sum_{t=1}^{7} u_{19,t} \tag{4.21}$$

Algebraically, the weakness is apparent by comparing the MIP and LP versions of constraint (4.21). Rearranging (4.21) results in $\sum_{t=1}^{7} u_{19,t} \geq \frac{3}{480} = \frac{1}{160}$. Enforcing the binary restrictions on the u variables requires rounding the fractional value of $\frac{1}{160}$ up to 1. Thus, in the MIP formulation at least one machine must be turned on during the first 7 time periods to meet the demand for product one in time period 7. By contrast, for the physical system when the binary restrictions are removed, only $\frac{1}{160}$th of the production machine operating cost is enough to satisfy all of the demand.

Figure 4.24 illustrates the weakness of the formulation by contrasting the physical systems associated with the MIP and LP formulations of (4.21). For the MIP formulation, the full cost of the machine to produce the product is borne. In the LP formulation, a tiny replica of the machine still has the ability to meet all demand, despite having only $\frac{1}{160}$th of the operating cost.

Fig. 4.24 Contrast of LP and MIP systems illustrates weakness in formulation

4.9.2.2 How Can This Formulation Be Strengthened?

The previous discussion illustrated the discrepancy between the physical systems associated with the MIP and its LP relaxation, both algebraically and visually. Understanding the source of the weakness in the formulation is the first step toward strengthening it. The weakness came from the value of 480 in the inequality $\sum_{t=1}^{7} u_{19,t} \geq \frac{3}{480}$ in the LP relaxation. As the value of 480 decreases, so does the weakness, as the right-hand side of this combined constraint increases toward 1. So where does this value of 480 come from? The model has 16 time periods, and examination of the demand values in the FLOW constraints reveals that the largest demand in any single time period for any product is 25. The value of 30 is used as the maximum possible demand in any single time period. The value of 480 is then based on the assumption of a demand of 30 for every product in all 16 time periods. Examination of the flow constraints reveals an average demand of 10.09. Since all the operating costs in the model are nonnegative, there is no benefit to having inventory on hand in the final time period. This raises the question of whether the value of 480 used to model the fixed charges for operating the machines can be reduced. To assess this, consider FLOW and VUB (i.e., fixed charge) constraints for the final 3 time periods for product 1.

```
FLOW0111: s____0110 - s____0111 + x__01111 = 4
FLOW0112: s____0111 - s____0112 + x__01112 = 4
FLOW0113: s____0112 - s____0113 + x__01113 = 2
FLOW0114: s____0113 - s____0114 + x__01114 = 5
FLOW0115: s____0114 - s____0115 + x__01115 = 0
FLOW0116: s____0115 - s____0116 + x__01116 = 2

VUB01114: x__01114 - 480 u____1914 <= 0
VUB01115: x__01115 - 480 u____1915 <= 0
VUB01116: x__01116 - 480 u____1916 <= 0
```

At time period 14, all demand from the previous 13 time periods has already been met since this formulation doesn't permit backlogging. Since there is no benefit to having inventory in the final time period, at most $5+0+2=7$ units need to be

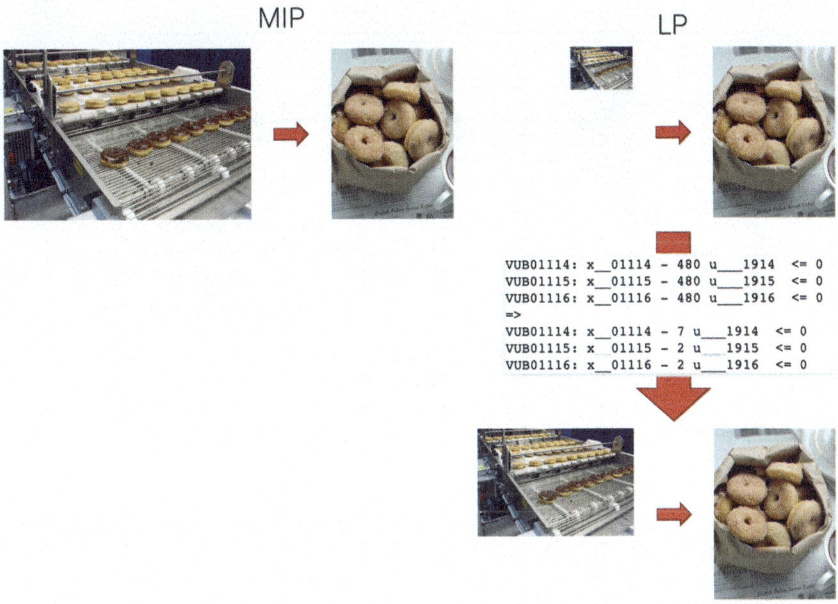

Fig. 4.25 Reduced coefficients decrease contrast in LP and MIP systems

produced in time periods 14–16, 2 in periods 15–16, and 2 in period 16. The worst case value of 480 for the fixed charge constraints for the machine that is used for product 1 can be reduced to 7, 2, and 2, respectively.

```
VUB01114: x__01114 - 7 u___1914 <= 0
VUB01115: x__01115 - 2 u___1915 <= 0
VUB01116: x__01116 - 2 u___1916 <= 0
```

Similar reasoning can be applied to all products in the model, final or component. Note that these cuts are not globally valid. They do cut off integer feasible solutions, but only those that will never be optimal. Solutions with inventory in the final time period are integer feasible, but since all optimal solutions have no such inventory, cuts that remove such sub-optimal solutions can be used even though they are not globally valid. For example, returning to the last 3 time periods for product 1, producing 100 units in time period 14 is a feasible option despite the total remaining demand of 7 over periods 14–16. This cannot be optimal since the model has positive inventory costs over periods 14–16. Furthermore, even if the inventory costs were zero, there is no gain to be had producing the extra 93 units, so the added cuts still ensure that at least one optimal solution exists.

Figure 4.25 illustrates how the reduced coefficients in the cuts strengthen the formulation in terms of the physical systems. No longer does the LP relaxation enable demand to be met with a tiny replica of the production machine with associated tiny

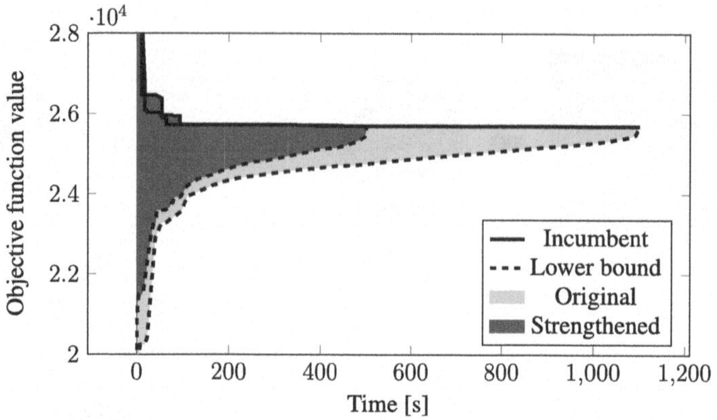

Fig. 4.26 GRBlogtools plot of primal and dual bound for strengthened model `b2c1s1`

costs. Now, the machine in the LP relaxation is much closer in size and cost to the one in the MIP.

Remark 4.4 The reader may notice that the integer rounding of constraint (4.21) was obtained by combining constraints and doing integer rounding, which was discussed in Sect. 4.6. This is true, but Gurobi's Mixed Integer Rounding cut separator already found those cuts, and they did not significantly strengthen the formulation.

4.9.2.3 Computational Results and Conclusion

Figure 4.26 compares Gurobi 9.5 performance on the original and strengthened formulation. The solid and dashed lines surrounding the black region show the primal and dual bounds for the strengthened formulation overlaid against the original results from Fig. 4.22. The optimal solution is not found any faster, but the dual bound is immediately better, and it progresses faster. The net result is a reduction in run time by 52%. While the plot shows results for a single run, comparisons were done for 5 different settings of Gurobi's Seed parameter on both formulations. Different settings of this parameter typically result in different solution paths (as was the case for these runs) and provide a useful tool for determining if a performance improvement was due to luck on models with significant performance variability. For this model the reduction in run time was consistent.

What are the key takeaways from this model? As with the offshore wind farm example that preceded it, understanding the weakness of the formulation is a prerequisite to effectively strengthen the formulation. Also similar to the preceding example, understanding when the strengthening will no longer work is important. In this lot sizing model, the cuts were derived by using a dual-based reduction that exploited the nonnegativity of the production and inventory variable costs. Other lot sizing models with the same characteristics can benefit from this reduction, provided that the maximum production level (480 in this case) exceeds the total product demand across all time periods. Had the cost structure been different (e.g., if inventories were not perishable, and production was subject to seasonable variability that created value for stockpiling inventory), the dual-based reduction may no longer have worked. More likely, it would have required a modified approach to reducing the values associated with the maximum possible demand. Finally, this example illustrated the boundary between globally valid cuts and dual-based cuts. The former are based only on the primal constraints in the model. The latter can remove integer feasible solutions but still can be used to tighten the formulation provided they allow at least one optimal solution to remain.

For additional discussions of lot sizing models, Pochet and Vyve [22] and Akartunali and Miller [1] describe heuristic approaches to improve the primal bound, while Belvaux and Wolsey [4] focuses more on strengthening the formulation to get better dual bounds.

4.9.3 Revisiting the Difficult Rectangle Game Model

The 12 rectangle game instance in Fig. 4.9 is a small yet challenging MIP model that illustrated how to visually assess the weakness of a MIP formulation. After 3 h, the MIP gap was still only at 42.6% and a dual bound of 56943, with little indication of significant progress to follow soon thereafter. While the model was described in the context of a game, the MIP formulation has applications in semiconductor design involving the efficient placement of rectangular components on a chip. In addition, the forbidding of overlap that necessitated the use of binary variables is common in other models, including bin packing and scheduling.

4.9.3.1 How Can This Formulation Be Strengthened?

The weakness of the formulation has been discussed in Sect. 4.5 and will not be repeated here. However, for convenience, Fig. 4.9 is repeated here, as it facilitates strengthening the formulation as well as understanding the weakness.

Examination of the player and Gurobi configurations and scores in Fig. 4.27 indicates very little wasted space in either of those configurations. The best possible configuration is one with no whitespace in the enclosing rectangle that defines the score. In such a configuration, the area of the enclosing rectangle will match the sum of the areas of the (in this instance 12) individual rectangles. For this instance, the sum of the areas of the individual rectangles is 96000, far tighter than the dual bound of 56943 Gurobi reported after 3 h of run time. Note that a configuration with no whitespace may not be attainable as a feasible arrangement of rectangles; the individual rectangle heights and widths will determine whether such an arrangement exists. But the sum of the areas of the rectangles will always provide a valid dual bound that can be used in the calculation of the MIP gap.

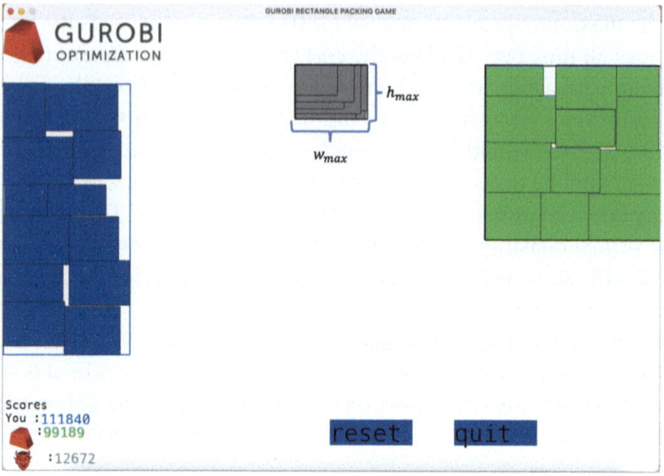

Fig. 4.27 Rectangle packing game, harder instance: player, Gurobi and devil scores

The issue here is that the algebraic model formulation in Sect. 4.5 provided to Gurobi does not include information about rectangles. The constraints are formulated using rectangle widths and heights, not with areas. So the information about rectangle areas that is clear from the visual representation of the model in Fig. 4.27 is lost in translation. Gurobi is finding much better solutions than is indicated by the dual bound and associated MIP gap, and it does so within a matter of minutes. In fact, Gurobi is not aware that the coefficients in the constraints of the model involve rectangle heights and widths. They are just values in linear expressions. Gurobi cannot "see" the game representation in Fig. 4.27, and valuable information about the model is lost. The information about the best possible area can be added to the model in the form of the nonconvex quadratic constraint $\hat{x} \cdot \hat{y} \geq 96000$, establishing a gap of 3.32% for the solution with objective value 99189 that Gurobi found within the first 3 min of the run. More work remains, as this tightening of the formulation does not enable Gurobi to prove optimality for this model of modest size. Nonetheless, with this additional information about the model, the practitioner is in a better position to assess the quality of the solution regarding optimality, and also pursue other strategies to strengthen the formulation.

Remark 4.5 Due to various details about pseudo cost calculations in branch and bound algorithms, the practice of adding constraints based on the objective function and a lower or upper bound on the objective value can degrade solver performance. This is less likely in the rectangle game model with a quadratic objective than in models with a linear objective. Nonetheless, the authors urge caution regarding this practice. A strong dual bound based on model knowledge may be better used by maintaining it outside of the optimization (e.g., by using a callback function to compare the best solution to the stronger dual bound).

4.9.3.2 Conclusion for Rectangle Game Models

As with the previous examples, understanding the weakness in the formulation was essential to discerning how to strengthen it. The visual approach was far more effective than the algebraic approach for the rectangle game models. While one can understand the weakness of the formulation algebraically as well as visually, the algebraic formulation lacks valuable information available in the visual representation of the model.

4.10 Cut Validation Methods

Much of this chapter has been devoted to strengthening MIP formulations by adding cuts. Two questions emerge. First, are the cuts correct? Cuts that are too tight may result in a model whose optimal solution is actually sub-optimal for the original model of interest. Second, are the cuts correct but weaker than intended? Besides careful examination of the cuts for correctness, validation strategies are limited, but some do exist. The following tactics can help.

- **Examine any infeasibilities created by the cuts**. If adding the cuts result in an infeasible model, they clearly have cut off too much. Gurobi and several other solvers offer an infeasibility finder that provides a minimal explanation of the infeasibility, and at least one cut will be involved.
- **Examine any declines in the optimal objective created by the cuts**. The optimal objective value of the strengthened model should always match that of the original model (within suitable tolerances). In the event of a mismatch, add a constraint on the objective function that it is at least as good as the optimal objective for the original model of interest to the strengthened model. This will make the strengthened model infeasible, and once again an infeasibility finder can help identify the problem with the cuts. If the optimal objective value is not known, use a smaller instance of the model for which the optimal solution can be found. For example, with the rectangle game models, use a 6 rectangle model rather than one of the unsolvable 12 rectangle ones.
- **Reverse the direction of the cuts**. For a globally valid cut of the form $a^T x \leq b$, reversing means replacing it with $a^T x \geq b + \epsilon$. With integer variables and data coefficients, the tolerance ϵ can be set to 1. Otherwise, the proper choice of the tolerance ϵ depends on factors including the number of decimal places in the non-integral problem data, the solver feasibility and integrality tolerances, and condition numbers of the optimal bases of the node relaxations during the solve. Since globally valid cuts preserve all integer feasible solutions, reversal should result in an infeasible model. If the model with the reversed constraint has any feasible solutions, the cut is too tight.

- **Examine the effect of the cuts on the initial root relaxation**. If the cuts have little or no impact on the initial root relaxation objective, they may not be as strong as intended. This is particularly true for globally valid cuts but may also apply to dual-based cuts.

4.11 Conclusions and Future Work

Here are some key takeaways from this chapter.

- **Understanding the weakness in a MIP formulation is essential to strengthening it**. In all the examples discussed, assessing the weakness in the formulation sheds light on how to strengthen it. This can be accomplished in numerous ways, but the polyhedral, algebraic and visual approaches repeatedly showed their effectiveness in the examples in this chapter. With the polyhedral approach, weakness is typically assessed by comparing the polyhedron of the LP relaxation with the convex hull of the integer feasible solutions of the MIP. For algebraic and visual approaches, comparing the physical systems associated with the MIP and LP, and looking for major disconnects or discrepancies between the two systems usually sheds light on the weakness in the formulation, and the remedies needed to strengthen it.
- **The solver may already do a lot of this for you**. Generally, modern MIP solvers have implemented a lot of features which automatically strengthen a weak formulation without the practitioner realizing it. Therefore, the authors recommend that the practitioners start by implementing the most intuitive model formulation before analyzing and applying the techniques described in this chapter.
- **Weakness in the formulation can be addressed in different ways**. While this chapter focused on cuts to strengthen the existing formulation, other strategies include reformulating the model or using algorithms other than branch and bound, such as decomposition.
- **Consider dual-based cuts as well as globally valid cuts based only on the primal constraints**. While globally valid cuts dominate the literature, dual-based cuts have gained momentum in recent years. They can exclude integer feasible solutions, provided that they guarantee that at least one optimal solution remains.
- **The visual approach can identify information in the model that is no longer available in the algebraic formulation**. The rectangle game example illustrated this particularly well. The authors believe this is the most underutilized practical method to identify weakness in the formulation.
- **Make good use of solver features to improve performance first**. State-of-the-art MIP solvers have numerous parameters and other features to facilitate faster performance. Make good use of the node log, either by direct examination or plotting packages like GRBlogtools to determine if the performance problem involves the primal bound, dual bound, or both. Then set parameters that are suited to address the source of the performance problem (e.g., do not use more aggressive heuristics

on models where the optimal solution is found quickly but the dual bound moves slowly). Some solvers also have automatic tuning features that will search for good parameter configurations. If parameter tuning does not help, then make use of the methods discussed in this chapter.

4.11.1 Future Work

The MIPLIB 2017 test set contains numerous models that either remain unsolved to optimality or take days to reach an optimal solution (sometimes using specialized algorithms or supercomputers). Many of the models, while challenging, are relatively simple, consisting of just a few different types of constraints and variables, often with references to papers that provide more information about the model. If this guide is as effective as the authors hope, techniques discussed herein should lead to improved performance on a significant number of these models. Three models in particular are relevant to the examples here. The models `bg512142` and `dg012142` are two significantly more challenging lot sizing models. While not dramatically larger in size, they have additional conditions that complicate the models, resulting in run times of days rather than hours. Furthermore, the dual-based cuts discussed in the `b2c1s1` example are included in formulation of these models, so other strategies are needed to improve performance. The other model relevant to this chapter is the open model `liu`, which is a rectangle game model with a linearized objective involving 33 rectangles, with the additional option of rotating the rectangles by 90°. The MPS file of the model can be downloaded from the MIPLIB 2017 website. However, as discussed in Sect. 4.9.3, valuable information about the rectangles has already been lost when looking at the algebraic formulation in an MPS or LP format file. Fortunately, the original rectangle data can be obtained by downloading the file at https://miplib.zib.de/downloads/models.zip, in the `miplib2010_liu` directory.

Another category of future work involves extending the strengthening strategies discussed here to work for the spatial branch and bound algorithm used in global solvers for models with nonconvex quadratic constraints and objectives. Branch and bound is actually a special case of this more general algorithm. Both make use of convex relaxations to handle nonconvex constraints, followed by branching to resolve violations of the nonconvexity conditions. However, for MIPs, the specific nonconvex condition of integrality enables numerous specializations that simplify the comparison of the physical systems associated with the MIP model and its continuous relaxation. The methods discussed in this chapter (particularly the visual approach) may be difficult to extend to the more general nonconvexities handled in the spatial branch and bound algorithm.

Another aspect of future work concerns marketing, rather than the more technical topics are discussed here. With the recent increased interest in data science, machine learning, and artificial intelligence, scientists from other disciplines are encountering an increasing number of MIPs, many of which involve NP-hard problems, in their work. Some scientists with limited familiarity of the state of the art in MIP solvers tend to assume that if a problem is NP-hard, no instance of significant size can be solved to optimality, and heuristic approaches must be tried. Optimization practitioners have an opportunity to challenge this misconception by demonstrating that meaningful instances of NP-hard problems can be solved to optimality using MIP solvers. The potential advances in different industries is quite significant. Wolsey [28] summarizes this quite well ($conv(X)$ in this quote refers to the convex hull of the integer feasible solutions of a MIP):

> If the optimization problem is NP-hard, there is no hope (unless P = NP) of obtaining an explicit description of $conv(X)$, but this should not deter us from looking for families of strong valid inequalities.

The authors might update this to mention dual-based inequalities as well. Far too many scientists from other disciplines stop reading after "there is no hope" in this quotation, and optimization practitioners have the opportunity to demonstrate the full potential of MIP solver technology. Hopefully this chapter provided some additional techniques to help them in this effort.

4.11.2 Additional Reading

The literature in integer programming is endless, but here are a few recommendations. "Integer Programming" by L. Wolsey has already been cited numerous times, and provides a good blend of theoretical and practical insights, with a comprehensive discussion of many of the globally valid cuts used in state-of-the-art MIP solvers. More theoretical discussions can be found in "Optimization over Integers" by Bertsimas and Weismantel, and "The Sharpest Cut: The Impact of Manfred Padberg and his Work" (edited by Martin Groetschel). On the opposite end of the spectrum, "The Woolsey Papers" by R.E.D. Woolsey focuses totally on the successful deployment of optimization applications and how to gain acceptance from stakeholders unfamiliar with the technical concepts. While the articles have a somewhat outdated view on the kinds of integer programs that can be solved in a reasonable amount of time, the lessons stand the test of time despite the major advances in MIP solver technology that have occurred since the book's publication. "The Traveling Salesman Problem" by Applegate, Bixby, Chvátal, and Cook provides insight customized to a popular but specific optimization problem, with many of the insights being applicable to other problem classes as well. Similarly, "Constraint programming and maths programming" by Puget and Lustig provide both useful insights on scheduling models that resemble the rectangle game model discussed in Sect. 4.5, as well as a broader discussion of improving the effectiveness of branch and bound and constraint programming (CP) algorithms on challenging industrial optimization problems. In particular, they describe how breaking symmetries in MIP and CP formulations can improve performance. "Constraint Integer Programming" by Achterberg provides an encyclopedic discussion of many of the features that go into an MIP solver. It provided the foundation for the SCIP solver, which is the most advanced MIP solver from the academic community as of this writing.

Finally, one work outside the realm of mixed-integer programming influenced the material in this chapter. The visualization of the physical systems associated with the MIPs and their relaxations drew significantly on the concepts in the video course "The Power of Mathematical Visualization", by James Tanton (https://www.thegreatcourses.com/courses/the-power-of-mathematical-visualization). While this course seems to focus on teaching pre-college mathematics, it can also be used effectively for more advanced math as well.

Acknowledgements Much of the material in this chapter was drawn from two Gurobi Tech Talks entitled "Converting Weak to Strong MIP Formulations." Recordings of the Tech Talks are available at https://www.gurobi.com/resource/tech-talk-chat-converting-weak-to-strong-mip-formulations/ and https://www.gurobi.com/event/tech-talk-converting-weak-to-strong-mip-formulations-part-ii/. The authors wish to thank Dr. Kostja Siefen and Dr. Steven Edwards, the other contributors to these presentations. In addition, they wish to thank Dr. Alexandra Newman and her students at the Colorado School of Mines, who agreed to be a test audience for these and other Tech Talks, thus helping with refining the material before the final presentations.

The material also benefited from feedback from numerous readers. The authors wish to thank our Gurobi colleagues Ronald van der Velden, Drs. Rodrigo Fuentes, Greg Glockner, David Torres Sanchez, Steven Edwards, Edward Rothberg, Maliheh Aramon, and Mario Ruthmair for numerous useful comments that have improved the document. We also wish to thank Drs. Alexandra Newman and John Cox of the Colorado School of Mines, Dr. Irv Lustig of the Princeton Consultants Inc., Dr. Paul Rubin of Michigan State University, and an anonymous reviewer for additional improvements. Of course, any errors or omissions are the sole responsibility of the authors.

References

1. Akartunali K, Miller A (2009) A heuristic approach for big bucket multi-level production planning problems. Eur J Oper Res 193(2):396–411. https://doi.org/10.1016/j.ejor.2007.11.033
2. Atamturk A, Nemhauser G, Savelsbergh M (2000) Conflict graphs in solving integer programming problems. Eur J Oper Res 121(1):40–55. https://doi.org/10.1016/S0377-2217(99)00015-6
3. Balas E (2018) Disjunctive programming. Springer.
4. Belvaux G, Wolsey L (2001) Modelling practical lot-sizing problems as mixed-integer programs. Manag Sci 47(7):993–1007. https://doi.org/10.1287/mnsc.47.7.993.9800
5. Benders JF (1962) Partitioning procedures for solving mixed-variables programming problems. Numerische Mathematik 4(3):238–252. https://doi.org/10.1007/BF01386316
6. Carlson B, Chen Y, Hong M, Jones R, Larson K, Ma X, Nieuwesteeg P, Song H, Sperry K, Tackett M, Taylor D, Wan J, Zak E (2012) MISO unlocks billions in savings through the application of operations research for energy and ancillary services markets. INFORMS J Appl Anal 42(1):58–73. https://doi.org/10.1287/inte.1110.0601
7. Fischetti M (2017) Mathematical Programming Models and Algorithms for Offshore Wind Park Design. PhD thesis, DTU Management Engineering.
8. Fischetti M, Pisinger D (2019) Mathematical optimization and algorithms for offshore wind farm design: An overview. Bus Inf Syst Eng 61:469–485. https://doi.org/10.1007/s12599-018-0538-0

9. Forrester RJ, Hunt-Isaak N (2020) Computational comparison of exact solution methods for 0-1 quadratic programs: Recommendations for practitioners. J Appl Math 2020:1–21. https://doi.org/10.1155/2020/5974820
10. Gavish B, Pirkul H (1985) Efficient algorithms for solving multiconstraint zero-one knapsack problems to optimality. Math Program 31:78–105. https://doi.org/10.1007/BF02591863
11. Gavish B, Glover F, Pirkul H (1991) Surrogate constraints in integer programming. J Inf Optim Sci 12(2):219–228. https://doi.org/10.1080/02522667.1991.10699064
12. Gleixner A, Hendel G, Gamrath G (2021) MIPLIB 2017: Data-driven compilation of the 6th mixed-integer programming library. Math Program Comput 13:443–490. https://doi.org/10.1007/s12532-020-00194-3
13. Glover F (1975) Improved linear integer programming formulations of nonlinear integer problems. Manag Sci 22(4):455–460. https://doi.org/10.1287/mnsc.22.4.455
14. Grötschel (ed) (2004) The sharpest cut: The impact of Manfred Padberg and his work. SIAM, Philadelphia
15. Higham N (2002) Accuracy and stability of numerical algorithms. SIAM
16. Klotz E (2014) Identification, assessment, and correction of ill-conditioning and numerical instability in linear and integer programs. In: INFORMS tutorials in operations research, pp 54–108. https://doi.org/10.1287/educ.2014.0130
17. Klotz E, Newman AM (2013a) Practical guidelines for solving difficult mixed integer linear programs. Surv Oper Res Manag Sci 18(1–2):18–32. https://doi.org/10.1016/j.sorms.2012.12.001
18. Klotz E, Newman AM (2013b) Practical guidelines for solving difficult linear programs. Surv Oper Res Manag Sci 18(1–2):1–17. https://doi.org/10.1016/j.sorms.2012.11.001
19. Lübbecke ME, Desrosiers J (2005) Selected topics in column generation. Oper Res 53(6):1007–1023. https://doi.org/10.1287/opre.1050.0234
20. Nemhauser G, Wolsey L (1988) Integer and combinatorial optimization. Wiley, New York
21. Padberg M (1989) The boolean quadric polytope: Some characteristics, facets and relatives. Math Program 45(1–3):139–172. https://doi.org/10.1007/BF01589101
22. Pochet Y, Van Vyve M (2004) A general heuristic for production planning problems. INFORMS J Comput 16(3):316–327. https://doi.org/10.1287/ijoc.1030.0042
23. Rahmaniani R, Crainic TG, Gendreau M, Re W (2017) The Benders decomposition algorithm: A literature review. Eur J Oper Res 259(3):801–817. https://doi.org/10.1016/j.ejor.2016.12.005
24. Santos JV, Nepomuceno N (2022) Computational performance evaluation of column generation and generate-and-solve techniques for the one-dimensional cutting stock problem. Algorithms 15(11):1–14. https://doi.org/10.3390/a15110394
25. Sherali H, Smith C (2007) An improved linearization strategy for zero-one quadratic programming problems. Optim Lett 1:33–47. https://doi.org/10.1007/s11590-006-0019-0
26. Watters L (1967) Reduction of integer polynomial programming to zero-one linear programming problems. Oper Res 15(6):1171–1174 December
27. Williams HP (1990) Model building in mathematical programming, 3rd edn. Wiley, West Sussex
28. Wolsey L (1998) Integer programming. Wiley, New York

Chapter 5
Valid Inequalities for the Telecommunications Network Design Problem

Yogesh K. Agarwal

Abstract This chapter addresses the problem of designing a backbone telecommunications network using transmission facilities (e.g., optical fibers) of a fixed capacity. Given a set of nodes, traffic demands among the nodes, and the costs of installing the facilities between the nodes, we need to design a minimum-cost network that can carry all the traffic demands. The problem can be modeled as a mixed integer program using a multicommodity network flow formulation. The chapter briefly introduces the concept of valid inequalities and reviews integer rounding and Chvátal-Gomory methods generating valid inequalities. It then identified the p-partition-based substructures for p = 2, 3, and 4, from which valid inequalities can be generated by applying the Chvatál-Gomory method. Several families of such inequalities are derived, and the effect of adding these inequalities is demonstrated through simple examples. A simple but effective shrinking heuristic is developed to find the p-partitions which are likely to result in violated valid inequalities. Some computational results are presented to demonstrate the effectiveness of the proposed approach.

5.1 Introduction and Formulation

Consider the problem of designing a telecommunications network using optical fibers or similar other transmission facilities of a given capacity. We are given a graph $G = (N, E)$, where N is the set of nodes representing locations among which a given set of telecom requirements have to be satisfied, and E is the set of candidate edges on which we may install one or more transmission facilities to carry these demands. Let q_{ij} be the traffic demand between nodes i and j, usually expressed in appropriate units of bandwidth, e.g., megabits/s or gigabits/s. This traffic will flow along one or more paths in the designed network that may pass through one or more intermediate nodes. Each transmission facility has capacity C, usually expressed in the same units as the traffic, and it costs c_{ij} to install one facility between nodes i and

Y. K. Agarwal (✉)
Jaipuria Institute of Management, Lucknow, India
e-mail: yka@iiml.ac.in

F. Hamid (ed.), *Optimization Essentials*, International Series in Operations Research & Management Science 353, https://doi.org/10.1007/978-981-99-5491-9_5

j. The traffic and the transmission facilities may be directed of undirected, resulting in two different versions of the problem. In general, the flow of information on the telecom network is always directed. But it may be considered as undirected when between any given pair of nodes, the traffic demand is the same in both directions, and the same path must be used to carry the traffic in both directions. A typical example of this is when the network is carrying mainly either voice traffic, or data traffic of different point-to-point applications where the bandwidth required in both directions is identical for each application. Here we address the undirected version of the problem.

This problem belongs to a class or problems generally known as *multicommodity flow* problems. The classical transportation problem, where goods have to be shipped from multiple plants to multiple customers, is an example of a single commodity problem, because it is assumed that all plants are producing the same product, and the demand of any customer can be satisfied from any plant. By analogy, in a telecom network, if the traffic needs to go from location A to location B, we may view location A as a plant, location B as the customer, and data as the product being shipped. However, the analogy ends there, because the demand of location B cannot be satisfied by sending data from a location other than A. Therefore, every origin-destination (OD) pair of traffic demands must be treated as a separate product or commodity. Hence the name multicommodity flow.

This would suggest that in a network with n nodes, there may be as many as ${}^nC_2$ commodities, one for each possible pair of nodes. However, this number can be reduced to just $n - 1$, if all traffic originating at a node is treated as the same commodity. Thus, a commodity will have a single origin and multiple destinations. The traffic from node i to any node $j > i$ will be part of commodity i, if we identify the commodity with its origin node. This method of defining the commodities is called the aggregate commodity definition and leads to an enormous reduction in the number of variables needed to formulate the problem. As we need two flows variables for each commodity for each edge (one for each direction), the number of variables needed with ${}^nC_2$ commodities is $mn(n - 1)$, while using aggregate commodity definition only $2m(n - 1)$ variables are needed, where $m = |E|$. Thus, if $n = 50$ and $m = 100$, without the aggregation of commodities we need 245,000 flow variables, but with aggregate commodity definition this number comes down to only 9800.

This problem is known by several names in the network design literature, e.g., Network Loading Problem (NLP) and Single Facility Network Design Problem. An early paper on a closely related version of this problem is by [6]. Further significant work on the single-facility version was done by [1–4]. Subsequent papers on the two-facility version of this problem are [5, 7–9]. This chapter is primarily based on the work in Agarwal [1].

We introduce the following notation in order to formulate the problem as a mixed integer program using the aggregate commodity definition described above.

δ_i : set of nodes adjacent to node i in G.
k : index used for commodities, same as the origin node of the commodity.
K : set of all commodities, usually $|K| = n - 1$.

O_k : origin node of commodity k.
c_{ij} : cost of installing one facility on edge (i, j).
d_i^k : demand of commodity k at node i. $d_i^k = q_{ki}$ for $i > k$, and $d_k^k = -\sum_{i>k} q_{ki}$.
x_{ij} : number of facilities installed on edge (i, j).
f_{ij}^k : flow of commodity k on edge (i, j) in the direction from i to j.

Using the above notation, NLP may be formulated as the following mixed integer programming problem.

$$\text{Minimize} \sum_{(i,j)\in E} c_{ij}x_{ij}$$

subject to:

$$\sum_{j\in\delta_i} f_{ji}^k - \sum_{j\in\delta_i} f_{ij}^k = d_i^k \qquad \forall i \in N, k \in K \qquad (1)$$

$$\sum_k (f_{ij}^k + f_{ji}^k) \le Cx_{ij} \qquad \forall (i, j) \in E \qquad (2)$$

$$x_{ij} \ge 0 \text{ and integer} \qquad \forall (i, j) \in E$$

$$f_{ij}^k,\ f_{ji}^k \ge 0 \qquad \forall (i, j) \in E, \forall k \in K$$

The objective function minimizes the total cost of facilities installed. Constraints (1) are the flow conservation constraints for each commodity at each node. The first term on left-hand side (LHS) of the constraint is the total inflow of the commodity into node i, and the second term is the total outflow of the commodity from node i. Thus the LHS represents the net inflow of the commodity into node i, which must be equal to the RHS, i.e., the demand of the commodity at that node. Note that k is the origin node of commodity k, and d_k^k is a negative number equal to the total demand of commodity k at other nodes, the flow conservation constraint of node k for commodity k ensures that the net quantity flowing out of node k is equal to the total demand of commodity k. Note that the demand between nodes i and j is part of commodity i if $i < j$. Facility installation variables x_{ij}, also known as the design or capacity variables, are integer variables, while the flow variables f_{ij}^k are continuous variables. Thus, the problem is to find a minimum cost installation of facilities on the edges of G, which permits a feasible flow of all commodities.

5.2 Japanese Theorem and the Capacity Formulation

The mixed integer formulation described in Sect. 5.1 has a fairly modest number of capacity variables, but a large number of flow variables. Note that for the case of $n = 50$ and $m = 100$ discussed earlier, the number of capacity variables is only 100 while the number of flow variables is 9800. In view of this, it is tempting to look for a formulation of the problem involving only the capacity variables, and completely eliminating the flow variables. Such a formulation is indeed possible, but it comes at the cost of adding an exponential number of constraints involving the capacity variables. The question is what conditions must the capacity variables satisfy so that a solution satisfying these conditions is guaranteed to permit a feasible flow of a given set of demands.

This question is answered by the so-called Japanese Theorem proposed by [11]. This theorem specifies a set of linear constraints in capacity variables so that any solution satisfying these constraints is guaranteed to permit a feasible flow of all commodities. These constraints are of the form $\mu c \geq \mu d$, where μ, c and d, respectively are vectors of size ${}^nC_2$. Note that in a n-node problem there are ${}^nC_2$ node pairs. For each node-pair (i, j), the demand is d_{ij} and the capacity is c_{ij}. If edge (i, j) is not present in G, then c_{ij} is taken as zero. The vector μ, called a metric vector, is a vector of values μ_{ij} satisfying the triangle inequality, i.e., for every $i, j, k \in N$ it satisfies $\mu_{ij} + \mu_{jk} \geq \mu_{ik}$.

The theorem states that if a solution x satisfies these metric inequalities for *all* metric vectors, it is guaranteed to permit a feasible flow of all commodities. Although there are an infinite number of metric vectors, these vectors form a convex cone which has a finite number of extreme rays, which are called primitive metrics. If the metric inequalities corresponding to the primitive metrics are satisfied, all others automatically get satisfied because they can be expressed as convex combinations of primitive metrics. Let Ω be the set of all primitive metrics. Then the NDP can be reformulated as the following pure integer program.

$$\begin{aligned}
&\text{Minimize} \sum_{(i,j)\in E} c_{ij}x_{ij} \\
&\text{subject to:} \\
&\mu x \geq \mu d \qquad \forall \mu \in \Omega \\
&x_{ij} \geq 0 \text{ and integer} \quad \forall (i, j) \in E
\end{aligned}$$

5.2.1 The Cut Metric

An important class of metric inequalities is the so-called "cut" metrics, also called a 2-partition metric. Partition the nodes of the graph into two subsets S and S', so that $S \cap S' = \phi$ and $S \cup S' = V$. Then, let $\mu_{ij} = 1$ if $i \in S$ and $j \in S'$ (or $i \in S'$ and $j \in S$), and zero otherwise. Let $D(S, S') = \sum_{i\in S}\sum_{j\in S'} d_{ij}$ be the total demand across the cut (S, S'). Then the resulting metric inequality $\mu x \geq \mu d$ takes the following form:

$$\sum_{i\in S}\sum_{j\in S'} Cx_{ij} \geq \sum_{i\in S}\sum_{j\in S'} d_{ij}$$

This inequality has a natural interpretation. The LHS is the total capacity available across the cut (S, S'), and the RHS is the total demand across the cut. Obviously, in any feasible solution, the capacity across the cut cannot be less than the required flow across the cut.

This inequality is also known as the "cut"-inequality in the network design literature and has been extensively used in solving almost all versions of the network design problem. An important result in this context is that for problems with 4 or fewer nodes, the cut-metrics constitute the complete set of metrics. In other words, if a capacity vector satisfies all the cut inequality for such a problem, then a feasible flow of all demands is assured for any such capacity vector. This point is illustrated with the following example.

5.2.2 *Example 1 (3-Nodes)*

Consider a 3-node NLP with $C = 10$ and $d_{ij} = 4\ \forall(i, j)$. A 3-node graph has only three cuts, each with one of the three nodes in S, and the remaining two in S'. Consider the cut $\{1{:}23\}$, i.e., the cut with node 1 in S, and nodes 2 and 3 in S'. The two edges crossing the cut are: $(1, 2)$ and $(1, 3)$, and the total traffic demand across the cut is 8. Thus, the metric inequality for this cut is

$$10x_{12} + 10x_{13} \geq 8$$

All three cut inequalities for this problem are given in Table 5.1.

If an integer program is solved with the above constraints, it will produce the optimal solution of the problem that guarantees a feasible flow of all traffic, even though the flow variables are not included in the constraints or the objective function at all, because the Japanese Theorem guarantees this. As only the capacity variables are involved, this formulation is called the capacity formulation. If the facility installation cost $c_{ij} = 5$ for all edges, the optimal solution for this integer program is $x_{12} = x_{13} = 1$ with $Z_{IP} = 10$. It is easy to verify that all traffic can be routed in this solution.

It is worth noting that the LP relaxation of this integer program produces a very weak bound. The LP relaxation solution is $x_{ij} = 0.4\ \forall(i, j)$ with $Z_{LP} = 6$, which is only 60% of the optimal solution. The difference between the values of the LP relaxation solution and the IP solution is called the *optimality gap* of the integer program, which is 40% in this case.

Table 5.1 Cut inequalities of the 3-node example

Sl. no.	Cut	x_{12}	x_{13}	x_{23}		RHS
1	1:23	10	10	0	$\geq$	8
2	2:13	10	0	10	$\geq$	8
3	3:12	0	10	10	$\geq$	8

5.2.3 Example 2 (4-Nodes)

Consider a 4-node NLP with $C = 10$ and $d_{ij} = 3\ \forall(i, j)$. A 4-node graph has two types of cuts, those with $|S| = 1$ and those with $|S| = 2$. Clearly, there are 4 cuts with $|S| = 1$, and three with $|S| = 2$. Consider the cut {1:234}, i.e., the cut in which node 1 is on one side, and the remaining three nodes are on the other side. The three edges crossing the cut are: $(1, 2)$, $(1, 3)$, and $(1, 4)$, and the total traffic demand across the cut is 9. Thus, the metric inequality corresponding to this cut is

$$10x_{12} + 10x_{13} + 10x_{14} \geq 9$$

On the other hand, consider a cut with $|S| = 2$, e.g., the cut $\{1, 2 : 3, 4\}$. For this cut, the edges crossing the cut are $(1, 3)$, $(1, 4)$, $(2, 3)$, and $(2, 4)$, and the total traffic demand across the cut is 12 units. Thus, the metric inequality corresponding to this cut is

$$10x_{13} + 10x_{14} + 10x_{23} + 10x_{24} \geq 12$$

The seven cut inequalities corresponding to these cuts for the problem at hand are given in Table 5.2.

Once again, if an integer program is solved with the above constraints, it will produce the optimal solution of the problem that guarantees a feasible flow of all traffic, even though the flow variables are not included in the constraints or the objective function at all, because the Japanese Theorem guarantees this. If the facility installation cost $c_{ij} = 5$ for all edges, the optimal solution for this integer program is $x_{12} = x_{13} = x_{14} = 1$ with $Z_{IP} = 15$. Again, it is easy to verify that all traffic can be routed in this solution.

For this example also, the LP relaxation produces a very weak bound. The LP relaxation solution is $x_{ij} = 0.3\ \forall(i, j)$ with $Z_{LP} = 9$, which is again 60% of the optimal solution. In the rest of this chapter, we shall develop valid inequalities for this problem, which will help to substantially improve the LP lower bound resulting in a much more efficient solution of the integer program via branch and bound.

Table 5.2 Cut inequalities of the 4-node example

Sl. no.	Cut	x_{12}	x_{13}	x_{14}	x_{23}	x_{24}	x_{34}		RHS
1	1:234	10	10	10	0	0	0	$\geq$	9
2	2:134	10	0	0	10	10	0	$\geq$	9
3	3:124	0	10	0	10	0	10	$\geq$	9
4	4:123	0	0	10	0	10	10	$\geq$	9
5	12:34	0	10	10	10	10	0	$\geq$	12
6	13:24	10	0	10	10	0	10	$\geq$	12
7	14:23	10	10	0	0	10	10	$\geq$	12

5.3 Review of Valid Inequalities, Integer Rounding and the Chvátal-Gomory (C-G) Cuts

Although most readers of this book are likely to be familiar with cutting planes and valid inequalities, we very briefly review these concepts for completeness, and for the benefit of those who may not be too familiar with these concepts. Readers are referred to any good integer programming book, such as [12] or [10] for a more detailed discussion.

Figure 5.1 shows the graphical representation of an integer programming problem. The solid lines represent the constraints of the problem, and the shaded polygon enclosed by these lines is a feasible region (R) of the LP relaxation of the integer program. The dots within this region are the intersection points of the dotted grid lines corresponding to the integer values of the variables, and constitute the set of feasible solutions of the integer program. Consider the convex hull of these points shown by dashed lines. If the inequalities corresponding to these dashed lines could be known, they will define another linear program with the feasible region (R') shown in the figure, which is called the *integer polyhedron*. Please note that at the extreme points of R' all variables have integer values. Therefore, if a linear program is solved over R', the solution will automatically be integral. In other words, for a given integer program, if we could find the equations of the constraint lines of R', it would be trivially easy to solve such an integer program, because solving a linear program over these constraints will automatically produce the integer optimal solution. However, finding the equations of these constraint lines, called the *facet inequalities* of the integer program, is a very challenging task, except for very few

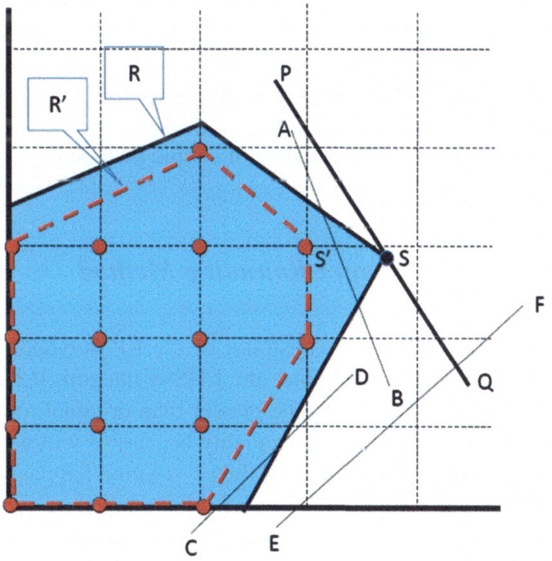

Fig. 5.1 Valid inequalities and cutting planes

problems with special characteristics. Note that facet inequalities are the strongest possible cutting planes.

If line P-Q represents the objective function, the optimal solution of the LP relaxation is point S, while the optimal solution of the integer program is the point S'.

5.3.1 Valid Inequalities and Cutting Planes

It is intuitively clear that if the region between the polygons R and R' is small, generally it will be somewhat easier to solve the integer program via branch and bound method than otherwise. Consider the possibility of adding additional constraint to the original LP formulation represented by R so that some part of this region is cut off. For the example in Fig. 5.1, line A-B represents such a constraint. These constraints are called *cutting planes* (or simply a *cut*) because they cut off some part of R (including the LP relaxation solution) *without* cutting of any part of R'. It is well-known that by adding such cutting planes, the integer programs can generally be solved much faster via the branch and bound method than without them. A large body of research in the integer programming literature is devoted to finding these cutting planes for various integer programming problems.

A *valid inequality* is very similar to a cutting plane, but there is small difference. Any inequality which does *not* cut off any part of R' is called a valid inequality of the integer program. Thus, all cutting planes are clearly valid inequalities, but the converse is not always true. An inequality which does not cut off the current LP relaxation solution is clearly a valid inequality, but not a cutting plane. Therefore, for solving an integer program, a valid inequality is actually useful only if it is also a cutting plane. In Fig. 5.1, line C-D is a valid inequality but *not* a cutting plane because although it cuts off some part of R, but does not cut off the optimal solution S. By definition, any inequality that does not cut off any part of R' is called valid. Therefore, an inequality represented by line E-F is also valid even though it does not cut off any part of R.

Next we describe a couple of simple methods of generating valid inequalities.

5.3.2 The Integer Rounding Method

Let $\sum a_i x_i \geq b$ be a constraint of the LP polyhedron S. Suppose that all a_i are integral, but b is not. Since the LHS is integral, if it is greater than or equal to b, then clearly it must also be greater than or equal to $\lceil b \rceil$. Therefore, the constraint $\sum a_i x_i \geq \lceil b \rceil$ is a valid inequality because it will not cut off any part of S', while cutting off some part of S. If the LP solution happens to be in the cut-off portion of S, then it is also a cutting plane. This method of generating valid inequalities and cutting planes is called the integer rounding method. If the inequality is of $\leq$ type, the RHS is rounded downward instead of upward using the same logic.

We can see an application of this method in Example 1. The first constraint $10x_{12} + 10x_{13} \geq 8$ is the cut inequality from the cut {1:23}. Let us divide the constraint by 10 yielding $x_{12} + x_{13} \geq 0.8$, which does not change it in any way. Now the stated condition is satisfied, i.e., the coefficients of all integer variables on the LHS are integral, but the RHS is not. Applying the integer rounding method to this constraint yields the valid inequality $x_{12} + x_{13} \geq 1$. Such an inequality is called the *rounded cut inequality*. If integer rounding is applied to all three inequalities and they are added to the linear program, the LP solution changes to $x_{ij} = 0.5 \ \forall (i, j)$, with the objective function value increasing from 6.0 to 7.5, thus reducing the optimality gap from 40% to 25%. If this gap is small, the branch and bound method can, in general, solve problem more quickly.

This method can also be applied to the constraints of Example 2, and it will increase the LP solution value from 9.0 all the way to 15.0, which is the IP optimal solution value, thus reducing the optimality gap to zero.

5.3.3 The Chvátal-Gomory (C-G) Method

The opportunities for applying integer rounding idea to the original constraints of the problem may be rather limited and by themselves would not be very impactful in solving the problem. However, new opportunities for applying this idea can be created via the Chvátal-Gomory method [12]. Consider taking a non-negative linear combination of two or more valid inequalities so that the resulting inequalities have the necessary conditions for applying integer rounding. By non-negative linear combination of several inequalities, we mean multiplying each of these inequalities by a non-negative number and adding them to get a new inequality. An inequality obtained in this way is implied by the original inequalities, i.e., its addition does not at all affect the LP feasible region R. However, if this inequalities satisfied the necessary conditions, i.e., the coefficients of all variables on the LHS are integral, and the RHS is fractional, then applying integer rounding to this inequality will result in a valid inequality that cuts off some parts of the original feasible region.

Consider the following two variable integer programs.

$$\begin{aligned}
&\text{Maximize } x_1 + x_2 \\
&\text{subject to} \\
&5x_1 + 3x_2 \leq 15 \\
&3x_1 + 5x_2 \leq 15 \\
&x_1, x_2 \geq 0, \text{ integer}
\end{aligned}$$

Figure 5.2 shows the two constraints (lines AD and EC), which intersect at point B. Thus, the feasible region of the LP relaxation of this IP is the shaded area OABC, and B is the LP optimal solution with $x_1 = x_2 = 1.875$ and $Z = 3.75$. Now multiplying each constraint with 0.125 and adding them yields the inequality $x_1 + x_2 \leq 3.75$,

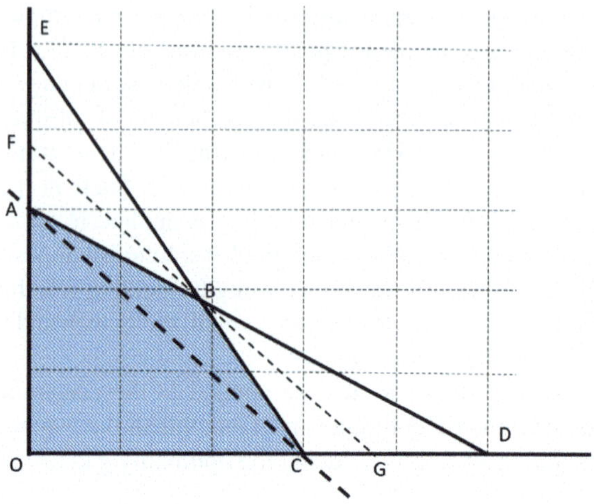

Fig. 5.2 The Chvátal-Gomory cutting plane

which is shown as the thin dotted line FG in the diagram. Note that this line does not cut off any part of the feasible region OABC at all. Applying integer rounding to this inequality, we get $x_1 + x_2 \leq 3$, which is shown as the heavy dotted line AC in the diagram. This inequality cuts off a significant portion of the region OABC. Note that after adding this inequality, the feasible region reduces to the triangle OAC, and all of its extreme points are integral.

Now let us apply this idea to the 3-node problem of Example 1. After applying integer rounding to the three constraints of the problem, we get the following rounded inequalities.

$$x_{12} + x_{13} \geq 1$$
$$x_{12} + x_{23} \geq 1$$
$$x_{13} + x_{23} \geq 1$$

Recall that after adding these inequalities, the LP lower bound of the problem had improved from 6.0 to 7.5. Now, apply the C-G procedure to these three inequalities by adding all of them and dividing the resulting inequality by 2, yielding the inequality $x_{12} + x_{13} + x_{23} \geq 1.5$. As the LHS is integral, we may round up the RHS, yielding the C-G inequality $x_{12} + x_{13} + x_{23} \geq 2$, i.e., the RHS has been *strengthened* to larger number. It is this strengthening which shifts the line from position FG to position AC, cutting off some part of the original feasible region. Without this strengthening, the inequality is quite useless as it would not cut off any part of the feasible region.

If the linear program is solved after adding this inequality, the resulting solution is $x_{12} = x_{13} = 1, x_{23} = 0$ and $Z = 10$. As the solution is integral, it is the optimal solution of the integer program. In the rest of this chapter we show how to apply the Chvátal-Gomory procedure to obtain valid inequalities for the NLP from the so-called p-partitions of the underlying graph. First we discuss the idea of p-partition and shrinking.

5.4 p-Partitions of the NLP Graph and the Shrunk Problem

Partition the nodes of the NLP graph into p subsets: $S_1, S_2.., S_p$, so that $S_i \cap S_j = \phi \; \forall i, j$, and $\bigcup_{i=1}^{p} S_i = N$. Now shrink the nodes in each S_i into a single node, creating a shrunk graph $G' = (N', E')$, where $N' = \{1, 2, .., p\}$, and $(i, j) \in E'$ if there is at least one edge between S_i and S_j in G. This shrinking process is shown in Fig. 5.3 for $p = 3$. Define a shrunk NLP on this graph so that for $i, j \in N'$, the aggregate traffic demand from S_i to S_j is $Q_{ij} = \sum_{u \in S_i} \sum_{v \in S_j} q_{uv}$. Similarly, we aggregate the original capacity variables x_{ij} into aggregate capacity variables associated with the edges of G', i.e., for $i, j \in E'$, $X_{ij} = \sum_{u \in S_i} \sum_{v \in S_j} x_{ij}$.

This shrinking process converts a larger problem into a smaller problem. It can be shown that a valid inequality of the smaller problem is also valid for the larger problem after the appropriate substitution of variables. The reader is referred to Agarwal [1] for details. In order to illustrate this point, consider the cut {1:23} of the shrunk graph. The rounded cut inequality corresponding to this cut for the shrunk problem is $X_{12} + X_{13} \geq \lceil (Q_{12} + Q_{13})/C \rceil$. By substituting $X_{12} = x_{24} + x_{26} + x_{36}$, etc., we get the rounded cut inequality for the original problem for the cut {123:456789} of the original graph. It is much easier to find valid inequalities using the Chvátal

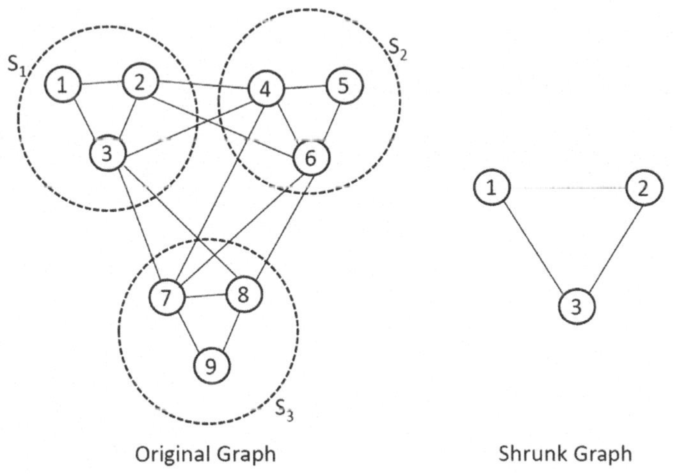

Fig. 5.3 Shrinking of graph G

Gomory procedure for the smaller size problems than for the larger ones. We already derived such a valid inequality for the 3-node problem, and in the rest of this paper we derive several such inequalities for the 4-node problem.

Given the above background, we may take the following approach for solving the problem. For $p = 2, 3$ or 4, find a $p-$partition of the graph that is likely to produce a violated valid inequality. Keep adding such inequalities as long as possible, and then solve the problem by branch and bound. Note that there are a very large number of possible p-partitions in a larger graph, and a large number of violated valid inequalities can be found and added via this approach. Addition of these inequalities leads to a significant improvement in the LP relaxation bound for the problem, and a substantial reduction is the solution time via branch and bound is obtained due to the addition of these inequalities. The computational results given in Sect. 5.8 amply demonstrate this effect.

5.5 p-Partition Inequalities for $p = 2, 3$ and 4

5.5.1 *Two-Partition Inequality*

When the shrinking is done with $p = 2$, the resulting shrunk problem is quite trivial, as there is only a single inequality: $X_{12} \geq Q_{12}$. Applying integer rounding we get the rounded cut inequality: $X_{12} \geq \lceil Q_{12}/C \rceil$. In the rest of this chapter we shall refer to this inequality as the *Two-Partition Inequality*. We also call them bi-partition inequalities or simply B-inequalities. These inequalities have been extensively used in the network design literature. Note that there are many ways to partition the graph into two subsets, resulting in a large number of such inequalities.

5.5.2 *Three-Partition Inequality*

For the case of $p = 3$, the shrunk problem has 3 nodes. There are three 2-partitions of a 3-node graph, resulting in three rounded cut inequalities given in Table 5.3.

Here $b_1 = \lceil (Q_{12} + Q_{13})/C \rceil$, $b_2 = \lceil (Q_{12} + Q_{23})/C \rceil$ and $b_3 = \lceil (Q_{13} + Q_{23})/C \rceil$. If $b_1 + b_2 + b_3$ is odd, than we can derive a Chvátal-Gomory cut by

Table 5.3 Cut inequalities of the 3-node graph

Sl. no.	Cut	X_{12}	X_{13}	X_{23}		RHS
1	1:23	1	1	0	$\geq$	b_1
2	2:13	1	0	1	$\geq$	b_2
3	3:12	0	1	1	$\geq$	b_3

adding the three inequalities and dividing by 2. As already seen in Sect. 5.2.3, if $b_1 + b_2 + b_3$ is odd, this will result in fractional RHS, which can be rounded up to produce the following inequality:

$$X_{12} + X_{13} + X_{23} \geq \lceil (b_1 + b_2 + b_3)/2 \rceil$$

We call this inequality the *Three-Partition Inequality* or T-inequality. Please note that if $(b_1 + b_2 + b_3)$ is even, then rounding up does not lead to any strengthening of the inequality, and will be of no use. It is well-known that for a 3-node NLP, the above four inequalities provide a *complete description* of the problem (see Agarwal [1] for details). This means that with these inequalities added, the LP solution will always be integral.

5.5.3 4-Partition Inequalities

As mentioned earlier, the 4-node NLP has seven cut inequalities. The rounded versions of these inequalities are given in Table 5.2. Note that the original inequality becomes redundant after its rounded version has been included in the formulation, because the former is dominated by the latter. Thus, the seven inequalities given in Table 5.4 provide a valid formulation of the 4-node problem.

Here $b_1 = \lceil (q_{12} + q_{13} + q_{14})/C \rceil$, and b_2, b_3 and b_4 are similarly defined. $b_{12} = \lceil (q_{13} + q_{14} + q_{23} + q_{24})/C \rceil$, and b_{13} and b_{14} are similarly defined, as already described earlier. As these inequalities are based on Bi-partitions, we call them as B-inequalities and label them as B_1, B_2, etc. as shown in the last column of the table. Note that the subscript of the label indicates the nodes which are on one side of the bi partition.

If we shrink any two nodes of a 4-node graph, we get 3-node shrunk graph, from which we can derive a T-inequality. Thus there are six possible T-inequalities for a 4-node problem. These are listed in the table below.

Table 5.4 B inequalities of a 4-node graph

Sl. no.	Cut	x_{12}	x_{13}	x_{14}	x_{23}	x_{24}	x_{34}		RHS	Label
1	1:234	1	1	1	0	0	0	$\geq$	b_1	B_1
2	2:134	1	0	0	1	1	0	$\geq$	b_2	B_2
3	3:124	0	1	0	1	0	1	$\geq$	b_3	B_3
4	4:123	0	0	1	0	1	1	$\geq$	b_4	B_4
5	12:34	0	1	1	1	1	0	$\geq$	b_{12}	B_{12}
6	13:24	1	0	1	1	0	1	$\geq$	b_{13}	B_{13}
7	14:23	1	1	0	0	1	1	$\geq$	b_{14}	B_{14}

Note that when nodes 1 and 2 are shrunk, the resulting 3-node graph has 3 B-inequalities, which are exactly the same as inequalities B_3, B_4 and B_{12} given in Table 5.1. Thus, inequality T_{12} is obtained by adding the B-inequalities B_3, B_4 and B_{12} dividing the resulting inequality by 2 and rounding up the RHS. Therefore, $t_{12} = \lceil (b_3 + b_4 + b_{12})/2 \rceil$.

We shall now derive three new classes of Chvátal-Gomory inequalities by combining the B- and T-inequalities given in Tables 5.1 and 5.2. As these inequalities are obtained from a tetRa-partition of a graph, we call them R-inequalities.

5.5.4 *Inequality R1*

Consider adding the six T-inequalities given in Table 5.3 and dividing the resulting inequality by 5 and rounding up the RHS if it is fractional. This results in the following inequality.

$$x_{12} + x_{13} + x_{14} + x_{23} + x_{24} + x_{34} \geq \lceil (t_{12} + t_{13} + t_{14} + t_{23} + t_{24} + t_{34})/5 \rceil = \alpha \quad (R1)$$

Note that an inequality with the same LHS can be obtained in two other ways. Add inequalities B_1, B_2, B_3 and B_4, divide the resulting inequality by 2 and round up the RHS. This results in the following inequality.

$$x_{12} + x_{13} + x_{14} + x_{23} + x_{24} + x_{34} \geq \lceil (b_1 + b_2 + b_3 + b_4)/2 \rceil \quad (R1A)$$

Similarly, add inequalities B_{12}, B_{13} and B_{14}, divide the resulting inequality by 2 and round up the RHS, yielding the following inequality.

$$x_{12} + x_{13} + x_{14} + x_{23} + x_{24} + x_{34} \geq \lceil (b_{12} + b_{13} + b_{14})/2 \rceil \quad (R1B)$$

Since all three inequalities have the same LHS, clearly whichever one has the largest RHS will be most effective, making the other *redundant*. A redundant inequality is one whose addition does not cut off any part of the current LP feasible region obtained after adding other inequalities. This introduces us to the notion of *dominance* of valid inequalities. If two inequalities of $\geq$ type have the same LHS, then the one with larger RHS dominates the other, making it redundant. For $\leq$ type inequalities, if LHS is identical, then one with smaller RHS dominates the other.

However, in the present case, it can be proven that inequality $R1$ always dominates $R1A$ and $R1B$, and there is no need to compare the RHS values. For details the reader is referred to Agarwal [1].

5.5.4.1 Example 3

Consider a 4-node NLP with $C = 5$, and $q_{ij} = 3\,\forall(i, j)$. Then $b_1 = b_2 = b_3 = b_4 = 2$, and $b_{12} = b_{13} = b_{14} = 3$. Computing the RHS of T-inequalities, we get $t_{12} = \lceil(b_3 + b_4 + b_{12})/2\rceil = \lceil(2 + 2 + 3)/2\rceil = 4$. In fact, the RHS values of all other T-inequalities turn out to be 4 due to symmetry of demands. Therefore, the RHS of R1 is $\alpha = \lceil(24/5)\rceil = 5$. Let $c_{ij} = 5\,\forall(i, j)$. Then, the LP solution value with only the B-inequalities is Z = 22.5. When the T-inequalities are added, it increases to 24, but the solution is still fractional. When inequality R1 is also added, it increases to 25.0, and the solution turns integral.

Note that for this example the RHS or inequality $R1A$ is 4, and that or $R1B$ is 5. Thus, $R1A$ is dominated by $R1$, and $R1B$ is same as $R1$. As mentioned before, it can be shown that $R1$ always dominates $R1A$ and $R1B$, i.e., the RHS of $R1A$ or $R1B$ can never be greater than that or $R1$.

5.5.5 *Inequality R2*

Consider adding the inequalities B_1, B_2, B_{13} and B_{14}, and dividing the resulting inequality by 2. If $b_1 + b_2 + b_{13} + b_{14}$ is odd, the RHS will be fractional and can be rounded up to get the following inequality.

$$2x_{12} + x_{13} + x_{14} + x_{23} + x_{24} + x_{34} \geq \lceil(b_1 + b_2 + b_{13} + b_{14})/2\rceil = \beta$$

Notice that all six variables are present in this inequality, but one of them has coefficient 2, while all others have coefficient 1. This clearly suggests that there are six possible versions of this inequality, each with a different variable with coefficient 2. In Table 5.2, each variable appears in exactly 4 out of 7 inequalities. In the C-G inequality derived by adding these inequalities, the corresponding variable will have coefficient 2, while all others will have coefficient 1. For example, if we use the inequalities B_1, B_3, B_{12}, and B_{14}, in the resulting inequality variable x_{13} will have coefficient 2. Thus, this inequality represents a *class* of 6 inequalities which are *isomorphic* to each other. Two inequalities are considered isomorphic to each other if one turns into the other by re-indexing the nodes. Such isomorphic classes of inequalities are quite common and arise frequently in many problems. This observation is important because identifying large classes of such isomorphic inequalities can be computationally very impactful in solving the problem. In order to distinguish these six inequalities, we identify each inequality with the edge whose coefficient is 2 on the LHS. Thus, the above inequality is labeled $R2_{12}$.

5.5.5.1 Example 4

Consider the problem described in Example 3, and increase the value of q_{12} from 3 to 5, and increase c_{12} from 5 to 8. The value of C and all other q_{ij} values remain unchanged. Then the RHS values of the six inequalities will be as follows:

$$b_1 = 3, b_2 = b_3 = b_4 = 2, b_{12} = b_{13} = b_{14} = 3$$

Note that in this case $b_1 + b_2 + b_{13} + b_{14}$ is an odd number. Therefore, we can derive inequality $R2_{12}$ as described above, and it will have the RHS value of 6. With only the B-inequalities, we get $Z = 22.5$. When the six T-inequalities with RHS = 4 are added, Z increases to 27.5, but the solution is fractional. When the R1 inequality (RHS = 5) there is no change in the solution or the Z value. The solution at this stage is $x_{12} = 0$, $x_{13} = x_{14} = 1.5$, $x_{23} = x_{24} = 1$, and $x_{34} = 0.5$. Now, we add the $R1_{12}$ with RHS 6, Z value increases to 28 and the solution turns integral with $x_{23} = 0$ and all other $x_{ij} = 1$.

5.5.6 *Inequality R3*

Consider adding inequalities T_{23}, T_{24}, T_{34} and B_1, and dividing the resulting inequality by 2. If $t_{23} + t_{24} + t_{34} + b_1$ is odd, then the RHS will be fractional and can be rounded up to obtain the following C-G inequality.

$$2x_{12} + 2x_{13} + 2x_{14} + x_{23} + x_{24} + x_{34} \geq \lceil (t_{23} + t_{24} + t_{34} + b_1)/2 \rceil = \gamma$$

Notice that the edges corresponding to the three variables with coefficient 2 on the LHS are all incidents on node 1. As in the case of R2, this again suggests that there are four possible versions of this inequality. In each one, the edges corresponding to the three variables with coefficient 2 are incident on a different node. As explained earlier, these four inequalities are isomorphic to each other under re-indexing of nodes and belong to a single class.

We call this inequality $R3$ and distinguish the four versions by adding the common node of the three edges as a subscript. For example, the above inequality will be labeled $R3_1$.

5.5.6.1 Example 5

With $C = 5$, let the demand vector be

q_{12}	q_{13}	q_{14}	q_{23}	q_{24}	q_{34}
3	4	3	2	5	4

Then the RHS values of the B-inequalities are

b_1	b_2	b_2	b_4	b_{12}	b_{13}	b_{14}
2	2	2	3	3	3	4

The RHS values of the T-inequalities are

t_{12}	t_{13}	t_{14}	t_{23}	t_{24}	t_{34}
4	4	4	5	4	4

Note that $t_{23} + t_{24} + t_{34} + b_1 = 15$ is odd, and the inequality $R3_1$ obtained has the RHS of $\gamma_1 = 8$.

Let us take the following cost vector:

c_{12}	c_{13}	c_{14}	c_{23}	c_{24}	c_{34}
8	8	8	5	5	5

When the problem is solved with this cost vector, and only the B-inequalities are included, we get $Z = 31$, and the solution is highly fractional. When the T-inequalities are also added, Z increases to 33.5, but the solution is still fractional as shown below.

x_{12}	x_{13}	x_{14}	x_{23}	x_{24}	x_{34}
0.5	1.0	0.5	0.5	1.5	1.5

Inequality R1, which has the RHS $\alpha = 5$, is already satisfied by this solution. RHS values of the six R2 inequalities for this problem are given below.

β_{12}	β_{13}	β_{14}	β_{23}	β_{24}	β_{34}
6	6	6	5	6	6

Addition of these inequalities also does not change the solution, as all of them are already satisfied by the solution. But when the inequality $R3_1$ derived above is added, the Z value increases to 34, and the solution turns integral with $x_{23} = 0$, and $x_{ij} = 1$ for all other (i, j).

5.6 Complete Polyhedral Description

Recall the discussion about valid inequalities and facet inequalities from Sect. 5.3. We know that the inequalities describing the convex hull of integer solutions of an integer program are called facet inequalities. As mentioned earlier, if all facet inequalities of an integer program are known, we say that a complete polyhedral description of that integer program is known. Such an integer program is easily solved as a linear program.

It has been proved that the inequalities R1, R2, and R3 are actually facet-defining for the 4-node NLP under appropriate conditions (see Agarwal [1]). Moreover, it was proved in Agarwal [2] that these inequalities constitute a complete polyhedral description of the 4-node NLP. Also, according to a key theorem proved in Agarwal [1], if a valid inequality derived from a k-partition of an NLP is facet-defining for the k-node problem resulting from the k-partition, then it is also facet defining for the original NLP under a fairly mild condition, i.e., the subgraph induced by each of the k subsets is connected. Thus, identifying the facets of 2-, 3- and 4-node NLPs yields large classes of facet inequalities for the larger NLPs. These inequalities constitute very large classes of inequalities for larger problems because there are many ways to partition the original problem into k subsets, and each one gives rise to several valid inequalities of the original problem.

The importance of identifying large classes of facet inequalities of a problem lies in the fact that using these inequalities leads to substantial improvement in the LP relaxation bounds, and the branch and bound search becomes very efficient. On the other hand, if the inequalities used are not facet-defining, then, in spite of a significant improvement in the lower bound, the branch and bound search may still take a long time. The computational efficiency of the inequalities described above is demonstrated in the computational results given in Sect. 5.8.

5.7 Separation Heuristics

Separation is an important issue in the implementation of valid inequalities. Generally, a given class of valid inequalities (e.g., B, T, R1, R2 etc.) contains a very large number of valid inequalities as there are many ways to partition the graph. Given an LP relaxation solution, some of these inequalities may cut off that solution, and therefore are useful, while the others are not useful. The challenge is to find a suitable partition which is more likely to result in a violated inequality. We describe a simple heuristic to obtain such a p-partition (for $p = 2, 3, 4$) from which to derive these inequalities.

Given a 2-partition (S, S') of G, we had earlier defined the demand the capacity across the cut as $D(S, S')$ and $C(S, S')$. It is intuitively obvious that a partition with the smallest value of $C(S, S') - D(S, S')$ is more likely to result in a violated B-inequality. We call this value the slack of the 2-partition, and use a simple shrinking heuristic to obtain a partition with the least slack.

We start with a p-partition of G with $p = n$, i.e., each subset contains a single node. For each pair of subsets S_u and S_v, we compute

$$D(S_u, S_v) = \sum_{i \in S_u} \sum_{j \in S_v} d_{ij}$$

$$C(S_u, S_v) = C \sum_{i \in S_u} \sum_{j \in S_v} \overline{x}_{ij}$$

where $\overline{x}_{ij}$ is the value of variable x_{ij} in the current fractional solution. Then the pair of subsets indexed with u^* and v^* is selected that maximizes the value $C(S_u, S_v) - D(S_u, S_v)$. These two subsets are then shrunk into a single subset, thus reducing the value of p by one. This process is continued until p reaches the desired value of 2, 3, or 4. All inequalities associated with the p-partition are then examined, and if one or more violated inequalities are found, they are added to the problem, which is solved again to obtain a new $\overline{x}$ solution. The shrinking heuristic is then applied all over again with respect to the new solution, leading to another p-partition that may yield more valid inequalities. The process is continued until no further violated inequalities can be found.

5.8 Computational Results

While a detailed computational study of the inequalities presented is outside the scope of this chapter, we demonstrate the effectiveness of these inequalities with the example of a 10-node network with 27 edges in the supply graph. The facility costs are computed using a fixed and a distance based cost, and range from \$1800 to \$5400. The facility capacity is 25 units. Three demand sets were randomly generated with the number of demands for each node-pair following a uniform distribution between 0 and 5 units of demand.

For each instance a lower bound was obtained for the following three cases: (1) with 2-partition inequalities only, (2) with 2- and 3-partition inequalities, and (3) with 2-, 3- and 4-partition inequalities. The purpose was to examine the incremental change in the bound value with the addition of 3- and 4-partition inequalities.

The results listed in Table 5.5 demonstrate that the addition of 3- and 4-partition inequalities led to a substantial reduction in the gap between the lower bound and the optimal solution. The computational advantage of the addition of these inequalities can be seen in the reduction in the number of nodes needed in the branch and bound tree to solve the problem optimally. Note that the effect is most significant in the case of Prob. 2 for which the addition of 3- and 4-partition inequalities decreased the number of nodes in the search tree from 913 to only 147 (Table 5.5).

Table 5.5 T-inequalities of a 4-node graph

Prob.	LB1	LB2	LB3	OPT	GAP1	GAP2	GAP3	BBN1	BBN2	BBN3
1	27686	30365	31546	32378	14.5%	6.2%	2.6%	>1000	241	83
2	31165	32870	33899	34976	10.9%	6.0%	3.1%	>1000	913	147
3	28773	30623	32031	32770	12.2%	6.6%	2.3%	>1000	181	109

5.9 Conclusion

This chapter has introduced the topic of valid inequalities and their application to the telecommunications network design problem. After reviewing the basics of valid inequalities and the Chvátal-Gomory method, we introduced a methodology of shrinking to obtain small subproblems to which the C-G method was applied to obtain effective valid inequalities for the problem. The computational efficacy of these inequalities is evident from the computational results.

The NLP and the related classes of problems constitute a very challenging set of combinatorial optimization problem. In spite of significant advances in the methodology of solving these problems, it is still not feasible to solve problems larger than 25–30 nodes optimally in most instances. Future work on these problems should focus on generating valid inequalities from 5 and higher partitions, and applying the presented methodology to other versions of the problem. It is also worth exploring to develop good heuristic methods for solving larger instances that utilize the valid inequalities presented here.

References

1. Agarwal YK (2006) k-Partition-based facets of the network design problem. Netw Int J 47(3):123–139
2. Agarwal YK (2009) Polyhedral structure of the 4-node network design problem. Netw Int J 54(3):139–149
3. Agarwal YK (2018) Network loading problem: Valid inequalities from 5-and higher partitions. Comput Oper Res 99:123–134
4. Avella P, Mattia S, Sassano A (2007) Metric inequalities and the network loading problem. Discret Optim 4(1):103–114
5. Bienstock D, Günlük O (1996) Capacitated network design—polyhedral structure and computation. Inform Journal Comput 8(3):243–259
6. Gavish B, Altinkemer K (1990) Backbone network design tools with economic tradeoffs. ORSA J Comput 2(3):236–252
7. Hamid F, Agarwal YK (2015) Solving the two-facility network design problem with 3-partition facets. Networks 66(1):11–32
8. Magnanti TL, Mirchandani P, Vachani R (1993) The convex hull of two core capacitated network design problems. Math Program 60(1):233–250
9. Magnanti TL, Mirchandani P, Vachani R (1995) Modeling and solving the two-facility capacitated network loading problem. Oper Res 43(1):142–157
10. Nemhauser G, Wolsey L (1988) Integer and combinatorial optimization. Wiley, New York
11. Onaga K , Kakusho O (1971) On feasibility conditions of multicommodity flows in networks. IEEE Trans Circ Theor 18.4:425–429
12. Wolsey L (1998) Integer programming. Wiley, New York

Chapter 6
Modeling and Solution Approaches for Resource Assignment with Deployment Restrictions

Anantaram Balakrishnan and Prakash Mirchandani

Abstract Since personnel and equipment costs constitute a large portion of the operating expenses for most organizations, judiciously using these resources to perform the organization's activities is important to ensure cost-effective operations. This paper addresses a class of resource management problems that entail assigning resources to perform a given set of ordered tasks while meeting policies and restrictions governing the deployment of these resources. We consider two particular kinds of restrictions, called span limits and work content restrictions, that specify upper and lower limits on pairwise and cumulative metrics based on the tasks assigned to each resource. We refer to this problem as the Resource Assignment with Deployment Restrictions (RADR) problem. We discuss several alternative ways to model the problem as a network-based integer program, and contrast these models in terms of the tightness of their linear programming relaxations and model size. This discussion serves to highlight the benefits of developing extended formulations and embedding constraints within the network representation, underscoring the importance of using strong formulations to effectively solve integer programs such as the RADR problems. We also outline three additional strategies—problem reduction, tightening the model with valid inequalities and coefficient lifting, and optimization-based heuristics—to further accelerate solution procedures for the RADR problem. Results from a prior study confirm the computational benefits of incorporating these strategies jointly in the solution method.

A. Balakrishnan (✉)
McCombs School of Business, University of Texas at Austin, Austin, TX, USA
e-mail: anantb@mail.utexas.edu

P. Mirchandani
Katz Graduate School of Business, University of Pittsburgh, Pittsburgh, PA, USA
e-mail: pmirchan@katz.pitt.edu

F. Hamid (ed.), *Optimization Essentials*, International Series in Operations Research & Management Science 353, https://doi.org/10.1007/978-981-99-5491-9_6

6.1 Introduction

Effective utilization of resources, including people and equipment, to perform scheduled tasks is vital to ensure cost-efficient and responsive operations and gain competitive advantage in many service and manufacturing settings. This chapter compares modeling alternatives and outlines solution approaches for a class of resource assignment and scheduling problems that entails assigning a limited number of resources to complete a given set of ordered tasks at minimum total cost while meeting resource deployment policies and restrictions. Tasks are ordered in the sense that, for any resource, the sequence in which it must perform each of its assigned tasks is given (e.g., in increasing order of scheduled start times). Resources may have different capabilities that limit the subset of tasks they can perform. Among resource deployment restrictions, we particularly consider two requirements: (i) *span limits* that constrain the tasks assigned to any resource relative to its first task, such as the total deployment time between the first and last tasks and (ii) the total effort or *work content* of the tasks assigned to each resource. To motivate the problem, let us consider the following example.

A home healthcare service provider offers skilled at-home health services such as administering nutrition therapy or injections, monitoring illness, wound care, and patient education. The organization has a group of staff members (nurses, medical assistants) with varying skills who visit and can serve multiple clients in a day. Every day, given the set of client appointments, the organization's planner needs to assign one staff member to visit each home at the scheduled time, complete the service at this home, and travel to the next home assigned to them or return to the office after their last visit. For each client, the planner knows the location, type of service needed, scheduled time of the visit, and total time needed at this household. Also known are the distances and travel times between the households and from/to the office. Staff members can begin their workday at different times during the day but must end all their assigned work and return to their assigned office within 8 hours after leaving the office. The span of time between leaving the office for the first visit and returning to the office is sometimes called the staff member's *duty time*. Moreover, the organization has a policy that all clients assigned to a staff member must be within a 30 miles radius from the first client.

In our resource assignment framework, the resources in this example are the organization's staff members, and the tasks are the home visits scheduled for the day. The basic daily decisions facing the planner are how many staff members of each type (defined by their skills and assigned office) to dispatch for client visits and what clients to assign to each staff member. For any given set of clients assigned to a staff member, the visit sequence is determined by the appointment times, i.e., the task sequencing rule is in order of increasing task start times. Moreover, this assignment must satisfy several deployment requirements and restrictions: (i) the staff member should have the skills needed for each assigned client; (ii) for any two consecutive visits, say, to clients A and B, the appointment or visit start time for client B must not be earlier than the visit end time for client A plus the travel time from A to B;

(iii) the total duty time, from leaving the office to returning back, must not exceed 8 hours; and (iv) the farthest distance of any client from the first client must be no more than 30 miles. In addition to these assignment rules, the organization may also have other auxiliary requirements such as ensuring that the total idle time between visits is not excessive and the workload (number of visits, total time at client locations) is balanced across staff members. The organization's goal is to minimize the total cost of its operations, including the daily wage for each staff member dispatched for visits and their travel costs.

Figure 6.1a depicts a home healthcare example with five clients and two resource types. The nodes represent clients, with node numbers representing client indices; the time window shown next to each node specify the scheduled visit start and end times. Green (solid fill) nodes represent clients who require only a medical assistant (regular skill level), whereas red (hatched fill) nodes require nurses (higher skill level). The two resource types, nurses and medical assistants, operate from the same office shown at the center. For simplicity, assume that the travel time between all locations is 1 hour, except between clients 1 and 3 (30 min). The figure shows a feasible assignment: green (solid) arrows show the sequence of visits (to clients 1 and 4) for a medical assistant, and red (dashed) arrows show a nurse's task assignment and sequence covering clients 2, 3, and 5.

We next briefly outline other contexts in which similar resource assignment problems arise:

- **Urban transit operations**: An urban transit (e.g., bus, train) service provider needs to assign drivers (or other personnel) to operate its scheduled transit services between various terminals so that no driver works for more than 8 hours, each driver has at least 1 hour rest time between trips, duty times are balanced across drivers, and the driver's utilization (total work time divided by duty time) exceeds some threshold (e.g., 80%), with the goal of minimizing the number of drivers used. Unlike the home healthcare example, where each task has a single service

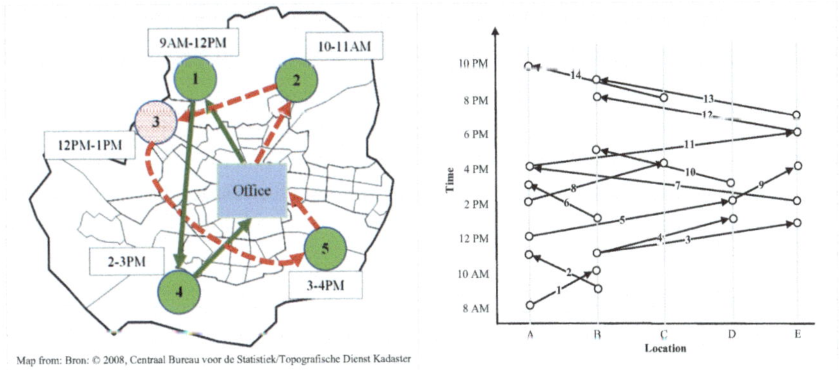

Map from: Bron: © 2008, Centraal Bureau voor de Statistiek/Topografische Dienst Kadaster

(a) Home healthcare example **(b)** Urban transit example

Fig. 6.1 Resource assignment examples

location, in this application, a task starts in one location and can end at another. Figure 6.1b depicts a time–space representation of a problem instance with 14 scheduled trips between five locations, with trip numbers shown on the arcs. For simplicity, each trip is 2 hours; for example, trip 1 starts at 8 AM from location A and ends at location B at 10 AM. Crew members who work in 8-h shifts need to be assigned to operate each trip. Union rules mandate a rest time of at least 1 hour between each crew member's assigned trips, and deadheading is not allowed. The crew scheduler seeks a cost minimizing solution that covers all trips while satisfying duty time and other policy restrictions.

- **Aircraft assignment**: A regional airline needs to decide which aircraft to assign to its scheduled flights each week to minimize its total operating costs. Aircraft vary in their seating capacities and operating costs, and each flight requires some minimum capacity based on anticipated demand. Every aircraft needs to return every week to its base for inspection.
- **Railroad maintenance**: A railway company has to periodically inspect and maintain track segments in its rail network. These geographically dispersed tasks require appropriate equipment and operating personnel, e.g., geometry car for track inspection and rail grinder for maintenance that are based at certain locations. Given the schedule for inspection and maintenance of various rail segments, the planner needs to assign the jobs to available teams (equipment and personnel) and route these teams so that each team returns to its home base within a certain period.
- **Heat treatment**: A manufacturer of aluminum extrusions produces items to order for a variety of applications. Each production batch requires heat treatment for a specified time at a given temperature. The facility has multiple heat treatment ovens with varying capacities. The production planner must decide which oven to use for each batch so as to minimize the total time to complete all batches. Each oven processes its assigned batches in order of increasing temperature (to reduce energy usage). The oven changeover time from one batch to the next depends on the temperature difference between the batches. The facility wants to assign no more than five batches to each oven, and limits the range of temperature differences between the first batch and subsequent batches processed in each oven.

Similar applications arise in other service and manufacturing contexts such as scheduling automobile painting operations for multiple painting bays where the preferred sequence for each bay is light to dark color, dispatching drones to deliver products to multiple customers, scheduling technicians for residential service calls, and assigning couriers to pick up food from restaurants and deliver to one or more homes.

These examples have the following common elements:

- **Resources**. Resources (people, equipment, teams) perform the needed work and have several *attributes* that distinguish them: skills or capabilities, base or starting/ending location, fixed usage cost per planning period (e.g., daily) if a resource is used, task operating and changeover costs, number available at each location, and work rules.

- **Tasks**: Tasks (patient visits, scheduled trips, production jobs) represent work to be done and can have the following attributes: starting time, processing time, starting and/or ending location, skills needed, due dates, priorities for sequencing, and other relevant metrics (e.g., temperature, color, etc.).

In assigning resources to tasks, we need to account for several resource deployment policies and constraints. We can broadly classify these requirements into the following three categories:

- **Task ordering**: The sequence in which each resource performs its assigned tasks is governed by some rule based on an operational requirement or scheduling policy such as increasing order of scheduled times, due dates, or task times. For example, in the urban transit context, the resource can operate a trip j after trip i only if the starting time for j equals or exceeds the completion time of i plus any needed transfer and rest time. Moreover, a policy that limits the maximum idle time between tasks may preclude this transition. Similarly, in the home healthcare context, a staff member who is qualified to provide service to client i but not client j cannot transition to j after completing i. In general, the permitted task-to-task transitions may vary by resource.
- **Span limits**: For any resource, the span of its assigned tasks is the largest difference in some attribute(s) between the first task and each subsequent task. For instance, the maximum (or minimum) duty time is an upper (or lower) limit on the time difference between the starting time of the first task and the ending time of the last task plus any travel time to/from the base. Similarly, span may be defined in terms of the radial distance or temperature difference between the resource's first task and all remaining tasks assigned to that resource. Resource deployment restrictions may impose multiple span limits (defined in terms of different attributes) that must all be satisfied for every resource that is used.
- **Work content** or **effort limits**: Resource deployment rules may impose upper or lower limits on the total work content, total time spent working on tasks or total distance traveled, summed over all the tasks assigned to a resource. Again, the assignments may be subject to several such limits.

Note that the task ordering, transitions, and limits on span and work content are resource-specific requirements, i.e., they must hold for the tasks assigned to each resource.

The general resource deployment problem that we study entails assigning one resource to each task subject to resource availability and capability, task ordering and permitted transitions, span limits, and work content restrictions for each resource, so as to minimize the total cost of the operation including costs for resource usage, resource-to-task assignment, task processing, and task-to-task transitions. Both the constraints and costs may vary by resource type, defined by the capabilities, deployment rules, and location of the resources. We refer to this problem as the *Resource Assignment with Deployment Restrictions* (RADR) problem. The literature has mainly studied a particular version of this problem in the context of scheduling crews to operate transport services. For crew scheduling problems (CSP), each task corresponds to a scheduled trip that starts at a given time and location and ends after the trip time at a possibly different location. Resource types are distinguished by their depot or base location, and span limits correspond to the maximum permissible duty time for each crew member, from the time they leave their base or depot to the time they return after the last trip. As our initial examples suggest, the general RADR problem is more broadly applicable, and can have span and work content limits as well as task sequencing and transition constraints that are governed by other metrics besides time and distance.

Resource deployment decisions have significant economic importance. For instance, for urban transport organizations, crew costs comprise a large proportion of their operating costs. Making resource assignment decisions manually can be time-consuming and may result in solutions with high costs. So, optimally solving the RADR problem can yield significant cost savings, improve service and responsiveness, and ensure balanced workload across resources. However, solving the problem is computationally challenging since, except for a few special cases, even simple versions of the RADR problem are NP-hard (see Sect. 6.2.1). In this chapter, we discuss modeling issues and algorithmic approaches to solve the RADR problem effectively. Like most large-scale integer optimization problems, the RADR problem can be modeled in alternative ways. The algorithmic approach and its effectiveness, in terms of the computational time needed to find optimal or near-optimal solutions, depend on the how the problem is modeled. Specifically, the modeling framework affects how strong its linear programming (LP) relaxation is and also determines how we can tighten the formulation to improve the bounds. The chapter highlights five complementary ideas to accelerate the RADR solution procedure: (i) variable disaggregation (using multi-commodity flows), (ii) capturing span limits within the network structure to define a strong problem formulation, (iii) problem reduction via variable elimination, (iv) valid inequalities and constraint tightening, and (v) optimization-based heuristic methods. These general approaches also apply to and can be adapted for other large-scale IP problems. Some of the applications discussed in this section and the general modeling framework presented in Sect. 6.2 are new; later parts of this chapter draw heavily from the work reported in [3]. We henceforth use the abbreviation *BML*22 when citing this paper.

Section 6.2 defines the RADR problem in more detail, sets up the notation and modeling ingredients, and describes computational complexity results for the RADR problem and some of its special cases. Section 6.3 describes four different ways to model the problem and Sect. 6.4 compares the tightness of the LP relaxations of these models for RADR problems with only span limits. Section 6.5 outlines three strategies for accelerating the solution procedure for general RADR problems—problem reduction, model tightening with valid inequalities and coefficient lifting, and optimization-based heuristic procedures. Section 6.6 presents illustrative computational results and Sect. 6.7 concludes the chapter.

6.2 Problem Ingredients and Model Development

To simplify the exposition, including notation, concepts, and model formulation, we will henceforth focus on RADR problems that are defined over spatially dispersed tasks to which resources travel to provide the needed service. Further, for these problems we will consider two main metrics relevant to tasks and deployment restrictions: time, and location or distance. All our initial illustrative examples, except the last application (heat treatment), fall in this category of problems. However, the model extends to other problem contexts where performing the tasks does not involve spatial movements, and may entail other relevant task and deployment metrics and attributes. Next, we discuss the problem ingredients.

Tasks: Let I denote the set of tasks to be completed within the planning horizon. Each task $i \in I$ has associated "attributes" that vary by context. For example, attributes for a task $i \in I$ may include its scheduled start time s_i (e.g., for trips, production jobs), physical location(s) (e.g., a single location for home health care or different starting and ending locations for transport operations), processing time τ_i, and needed resource capabilities or skills.

Resources: Resources vary in their: (a) capabilities (e.g., capacity, skills) which, in turn, determine the tasks that they can perform; (b) costs for usage, task completion, transitioning or changing over from one task to the next, and so on; (c) deployment restrictions such as span limits, and work content rules; and, (e) initial and final states, e.g., the resource's location before it begins and after it ends its assigned tasks, its available start time and required end time, or starting and ending temperatures (for heat treatment). We refer to the joint specification of its starting and ending states as its *base*. The above attributes of each resource characterize its operation. We refer to the set of available resources that have the same set of attributes as a resource *type*. Let R denote the set of *resource types*, and let μ_r, for $r \in R$, be the number of type-r resources available for deployment. Denote the base for resource type-r as b_r, and define $I^r \subseteq I$ as the set of tasks that this resource type can perform. We group together resources into types because, for the RADR problem, instead of defining task assignment variables for each individual resource, we only need to define these variables for each resource type (since all resources within a type have identical attributes, and we are not concerned with which particular member of this set is used). This approach not only yields a model with far fewer variables but also reduces computational effort by avoiding symmetry (i.e., when we have assignment variables for each resource of a given type, these assignments are interchangeable with other resources of the same type). For convenience, when obvious from the context, we use the index r to denote both a resource type and a specific resource of this type.

Task sequence: Each resource that is used performs its assigned tasks in sequence, changing over from one task to the next as it completes each task. The permissible task changeovers, which we call transitions, depend on operational rules and policies, and can vary by resource type. For instance, in transportation and production scheduling applications, a resource can transition from i to j only if the starting time for j is at or after the ending time for i plus any needed transition time δ_{ij} (including any travel time from the ending location of task i to the starting location of task j), i.e., only if $s_j \geq s_i + \tau_i + \delta_{ij}$. Moreover, some transitions may be prohibited for operational reasons, e.g., to avoid excessive idle time between tasks or long travel distances. Note that these transition rules impose precedence relationships among the tasks that are assigned to the same resource, but do not apply when the tasks are performed by different resources (as is typically the interpretation of precedence constraints in traditional project management problems). For resource type-r and any task $i \in I^r$, let $J^r(i) \subseteq I^r$ denote the subset of next tasks j that the resource is permitted to perform after completing task i. Define $J(i) = \cup_{r \in R} J^r(i)$ as the set of possible transitions from task i for any resource type. We will assume that, for any subset of l tasks assigned to a resource r, the corresponding transition rules define a consistent and complete ordering of these tasks, i.e., we can label the nodes as $i_1, i_2, \ldots, i_l$ such that i_q precedes i_{q+j} for $j = 1, \ldots, l-q$, and $q = 1, 2, \ldots, l-1$. The sequencing of tasks that are assigned to a resource (and the permitted transitions) may stem from operational reasons (e.g., if starting times of tasks are fixed) or other policy considerations (e.g., in order of increasing due dates or processing times). Note that when a resource r is assigned a task sequence, its first action is to transition or move from its base b_r to the first task i_1, and its last action is to return to the base from the last task i_l. These initial and final transitions may also be subject to some transition policies. For instance, the organization may limit the distance that a resource r has to travel from its base b_r to the location of its first task, as well as the return distance from its last task to the base. Let $IF^r \subseteq I^r$ and $IL^r \subseteq I^r$, respectively, denote the set of permitted *first* tasks and *last* tasks for resource r.

Span Limits: Given the ordered set of tasks $I' = \{i_1, i_2, \ldots, i_l\}$ assigned to a resource, we refer to the maximum difference in some attribute between that of the first task i_1 and any other task $i_q, q = 2, \ldots l$, as the *span* of this assignment. Span limits can be either upper or lower limits on the span of an assignment, based on the resource's deployment rules or organizational policies. For instance, suppose the rules specify that every task that a resource r is assigned must lie within a radius of D^r miles from the first task. Then, the attribute or metric of interest is the distance between task locations, and the span limit specifies that $\max_{q=2,\ldots,l} d_{i_1,i_q}$ must be less than or equal to D^r, where d_{ij} is the distance between the locations of tasks i and j. Similarly, if the work rules require that each type-r resource's duty time must not exceed T^r, then the metric is time. Given an assignment (and corresponding sequence) of tasks assigned to this resource, the actual duty time only depends on the starting time s_{i_1} of the first task $i_1 \in IF^r$ and the ending time $s_{i_l} + \tau_{i_l}$ of the last task. If duty time also includes the time, say, t_{b_r,i_1} for the resource to travel from the base b_r to the first task i_1 and the time t_{i_l,b_r} from the last task $i_l \in IL^r$ to return to the base, then the span limit translates to the requirement that: $\left(s_{i_l} + \tau_{i_l}\right) - s_{i_1} + t_{b_r,i_1} + t_{i_l,b_r} \leq T^r$. In general, let $\mathbf{\Gamma}^r_{\max}$ and $\mathbf{\Gamma}^r_{\min}$ be the vectors of upper and lower span limits over various metrics for resource r, and let $\mathbf{\Gamma}^r(I')$ denote the corresponding vector of actual values of the metrics for a given task assignment I' to this resource, then collectively the Span Limit (*SL*) restrictions can be expressed as $\mathbf{\Gamma}^r_{\min} \leq \mathbf{\Gamma}^r\left(I'\right) \leq \mathbf{\Gamma}^r_{\max}$. These SL constraints capture the requirement that the assignment must *simultaneously* satisfy all the span limits.

Work content restrictions: In addition to span limits, task assignments are also constrained by limits on the "work content." Unlike span limits that apply to the maximum difference in some metric between the first and each of the later tasks, the work content restrictions apply to the sum of a metric over all the tasks and transitions in a resource's work sequence. For instance, work rules in an urban transit organization may require that a driver (with resource type-r) may not operate a vehicle for more than W hours during their duty time. Given a set of tasks I' assigned to this resource, the working time constraint specifies that $\sum_{i \in I'} \tau_i \le W^r$. Similarly, in a home health service context, another work rule may specify that the total travel distance must not exceed L^r miles, which implies the constraint $d_{b_r, i_1} + \sum_{q=1}^{l-1} d_{i_q, i_{q+1}} + d_{i_l, b_r} \le L^r$. In general, let $\mathbf{\Omega}^r_{\max}$ and $\mathbf{\Omega}^r_{\min}$ be the vectors of upper and lower limits on work content for resource type-r defined in terms of various metrics, and let $\mathbf{\Omega}^r(I')$ denote the corresponding vector of actual values of the metrics (sum over all tasks assigned to the resource) for a given task assignment I'. Then, the Work Content (WC) restrictions can be represented as $\mathbf{\Omega}^r_{\min} \le \mathbf{\Omega}^r(I') \le \mathbf{\Omega}^r_{\max}$.

Costs: Assigning resources to perform tasks entails the following three types of costs that may vary by resource type:

- Resource *usage* or deployment cost: Let f^r be the cost of using one resource of type-r. In the home healthcare context, this cost may represent the daily wage paid to the staff member, and can vary depending on staff member's skill level. For resources that incur only a fixed periodic (e.g., monthly) cost regardless of whether or not they are deployed, there are no separate usage costs.
- Resource-to-task *assignment* cost: Let g_i^r be the cost for assigning a type-r resource to perform task $i \in I^r$. This cost may depend on the work requirements (e.g., time) of the task as well as the resource type. If all resource types have the same assignment cost, then we need not separately consider this cost since every task needs to be performed and so the total assignment cost is a fixed constant.
- *Transition* or changeover cost: Let h_{ij}^r denote a type-r resource's cost to transition from a task $i \in I^r$ to task $j \subset J^r(i)$. For instance, this cost can include the travel or equipment repositioning cost from i to j, and the cost of idle time before task j begins. $h_{b_r,i}^r$ and h_{j,b_r}^r are the costs for transitions from a resource's base b_r to its first task i, and from its last task j back to the base.

Problem definition: Based on these problem ingredients, the RADR optimization problem is as follows: Given the tasks that need to be completed and the available resources of each type, decide which resource type to assign to each task so as to minimize total cost while meeting the following requirements: (a) each task must be assigned one resource; (b) the number of resources used must not exceed the number of available resources of each type; (c) the tasks assigned to each resource must meet the span limits; (d) these assignments must also meet work content requirements; and (v) any additional policy constraints.

6.2.1 Network Flow Representation

To model the RADR problem, it is convenient to introduce a graphical representation of the problem, defined over the following directed network $G:(N, A)$. The network's node set N contains one node for each task $i \in I$, and two nodes—a base-starting node o_r and a base-ending node e_r—for each resource type $r \in R$. These latter two nodes represent the resource's base, with the first serving as the origin of the resource before its first task, and the latter the destination after completing its last task. For every resource type-r, the arc set A contains an arc from its base-starting node o_r to each of its possible first task nodes $i \in IF^r$, and from every possible last node $i \in IL^r$ to the base-ending node e_r. The set also includes an arc from each task node i to every task node $j \in J(i)$. Due to the complete ordering of tasks implied by the transition rules for a resource, the graph G is acyclic. In this network, the assignment of tasks to a resource type-r will correspond to unit flow along a "path" from o_r to e_r that passes through each of the assigned tasks in sequence. We account for the usage, assignment, and transition costs for each resource type by defining appropriate arc costs. For a resource type $r \in R$ and any arc (i, j) from task node $i \in I^r$ to task node $j \in J^r(i)$, define $c^r_{ij} = g^r_j + h^r_{ij}$. For arcs (o_r, i), with $i \in IF^r$, emanating from a type-r resource's starting-base node o_r, set $c_{o_r,i} = f^r + g^r_i + h^r_{o_r,i}$. Similarly, for base-return arcs (j, e_r), for $j \in IL^r$, set $c_{j,e_r} = h^r_{j,e_r}$. With these definitions of arc costs, the total cost of a path from o_r to e_r exactly equals the sum of the resource usage, assignment, and transition costs for a resource that is assigned the tasks on this path. (Some of the models from the literature that we later use for comparison cannot incorporate resource-dependent transition rules or costs; these models effectively assume that the costs do not vary by resource type.) To capture the total work content of the tasks and transitions assigned to a resource, for each metric or attribute m (e.g., working time, travel distance), we define arc "weights" w^{rm}_{ij} for all $(i, j) \in A$ and resource types $r \in R$. For $j \neq e_r$, we can set w^{rm}_{ij} equal to the value of metric m for task j plus its value for the transition from i to j. And, for $j \in IL^r$, the weight w^{rm}_{j,e_r} for the return arc is the value of metric m for the transition from task j to e_r. With these definitions, the sum of arc weights on any o_r to e_r path for metric m is this metric's total work content for the task assignment corresponding to the path. The RADR problem thus translates to selecting the least cost set of o_r-to-e_r paths (or equivalently routing unit arc flows) for different resource types so that: (i) the chosen paths cover all the task nodes exactly once, (ii) no more than μ_r paths emanate from o_r (and terminate at e_r), (iii) all the tasks on a path satisfy the SL constraints, and (iv) the total weight of each path satisfies the WC constraints for the corresponding resource type and metric.

Figures 6.2 (a) and (b) depict the networks for the home health care and the urban transit examples of Fig. 6.1. In the home healthcare example, since staff members have two possible skill levels, regular and high, we have two starting-bases, o_R and o_H, and two corresponding ending-bases. Staff members leave from and return to their respective offices. Observe although the travel time between clients 1 and 3 is low, they cannot both be served by the same staff member of either type due to the overlapping visit time intervals for these two clients. Therefore, arc (1,3) is omitted from the network. The figure also shows the flow paths corresponding to the solution shown in Fig. 6.1a (with solid green arrows for tasks assigned to a regular resource and dashed red arrows for a high-skilled resource).

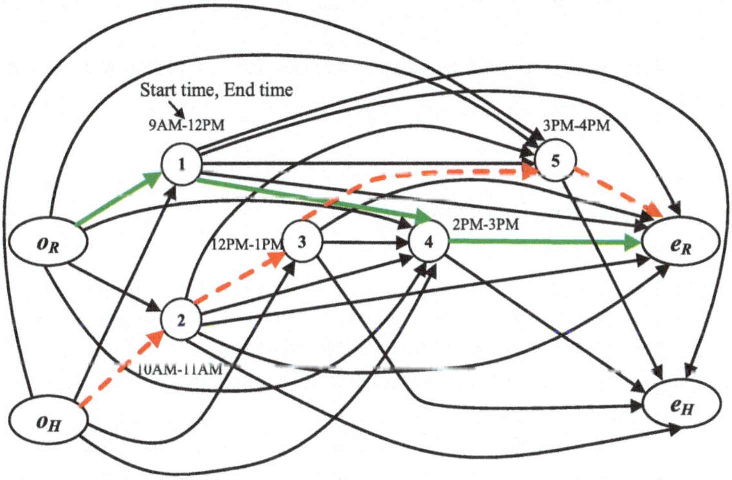

(a) Network for home healthcare example.

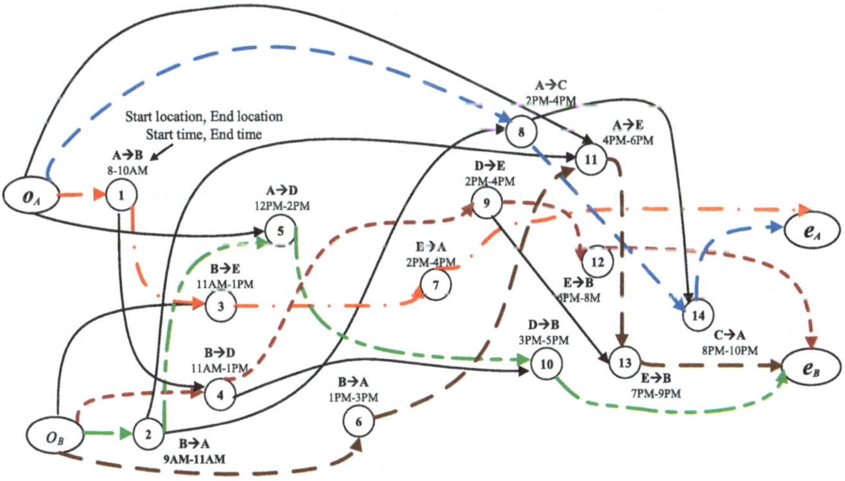

(b) Network for urban transit example.

Fig. 6.2 Network flow representation for assignment examples

For the urban transit example, Fig. 6.2b shows the corresponding network representation, with nodes assigned the same index as the trips in Fig. 6.1b, and arcs corresponding to permitted trip-to-trip transitions (without deadheading). Crews based at two locations, A and B, represent the two resource types, both with the same trip-to-trip transition rules. The colored (dashed, dotted) arcs show a feasible solution, using five crew members; the black (solid) arrows are feasible transitions that are not used.

We next discuss the computational complexity of this problem's variants and special cases.

6.2.2 Computational Complexity

The RADR problem with multiple resources is NP-hard, even with just one duty time (SL) constraint or just one WC constraint. ([7, 8] proved these results in the context of job scheduling.) This section explores some special cases that are polynomially solvable. There are three problem features that distinguish these special cases: (i) the number of resource types (one versus more than one); (ii) the number of resources of each type; and (iii) whether the problem imposes only SL constraints, only WC constraints, neither or both types of constraints. Based on these dimensions, this section explores the computational complexity for a sequence of special cases of the RADR problem to determine which dimensions are most influential in making the problem computationally intractable.

Let us first consider the special case with a single resource type, but no SL or WC restrictions. This problem is solvable in strongly polynomial time by solving a minimum cost flow problem over the following augmented network. Modify the network $G:(N, A)$ over which the RADR problem is defined by splitting each task node $i \in I$ into two nodes i' and i'', adding a zero-cost arc $\left(i', i''\right)$ with minimum and maximum (capacity) flow limits of one. Also, add a zero-cost arc (o_r, e_r) from the starting-base node to the ending-base node, with capacity μ_r. All other arcs of the network have unit capacity. Node o_r has a supply of μ_r, and node e_r has a demand of μ_r units. If the (single commodity) minimum cost flow problem on this network has a feasible solution, then so does the original RADR problem without SL and WC constraints. In this case, the minimum cost flow problem must have an integer (i.e., 0–1) optimal solution that can be identified in $O(|A|\log(|N|)(|A| + |N|\log(|N|)))$ time [12]. Decomposing the arcs flows in this solution into unique o_r-to-e_r paths (that are node-disjoint in terms of task nodes) gives the optimal task assignment for each resource that is used. The flow on arc (o_r, e_r) equals the number of unused resources.

Having multiple resource types together with SL and WC constraints increases the computational difficulty of the RADR problems. Further, there is an interesting distinction between problems that have only SL constraints and those with WC constraints (with or without SL constraints). In general, WC constraints seem to make the problem harder. BML22 showed, using a constructive proof, that a

RADR problem instance with two resources of the same type and one duty time (SL) constraint but no WC constraints is solvable in $O(|N|(|A| + |N|\log|N|))$ time. However, when the problem has one or more WC constraints but no SL constraints, then even the two-resource single resource-type version of the problem is NP-hard (BML22).

Three special cases that limit the number of tasks each resource can perform are easy to solve even with WC and SL constraints (we can interpret these special cases as having another WC constraint where each task has a weight of one, and there is a limit on the total weight of tasks assigned to any resource). Propositions 1 and 2 show that problems with multiple resource types as well as WC and SL constraints are strongly polynomial when each resource is permitted to perform no more than one or two tasks. For problem feasibility, we assume that $\mu_r \geq |I|$ for all $r \in R$, i.e., resource availability is not limited (but the solution may use far fewer resources due to the fixed resource usage costs).

Proposition 1: The RADR problem with multiple resource types and SL and WC constraints, but requiring each resource to perform no more than one task, can be solved in strongly polynomial time.

Proof. Construct a transportation problem with $|R|$ supply nodes, one for each resource type-r with a supply of μ_r units, and $|I|$ demand nodes corresponding to each task with a demand of one unit each. If resource type-r can perform task i without violating the WC and SL constraints, add the arc (r, i) connecting supply node r to demand node i with cost $\gamma_{ri} = f^r + g_i^r + h^r_{o_r,i} + h^r_{i,e_r}$. Then, the minimum cost transportation solution gives the optimal RADR solution. The computational complexity of this method is $O\big(|I||R|^2(\log|I| + |R|\log|R|)\big)$ [5]. ◆

Proposition 2: The RADR problem with multiple resource types and SL and WC constraints, but requiring each resource to perform no more than two tasks, can be solved in strongly polynomial time.

Proof. Construct a network with two nodes, i and i' for each original task i. For every pair of tasks i, j, let R_{ij} denote the subset of resource types for which a single resource can perform both i and j without violating any WC or SL constraints. Assume without loss of generality that i precedes j for the resource. For each resource type $r \in R_{ij}$, compute the cost $\gamma^r_{ij} = f^r + g^r_i + g^r_j + h^r_{ij} + h^r_{o_r,i} + h^r_{j,e_r}$, and let $\gamma^*_{ij} = \min_{r \in R_{ij}} \gamma^r_{ij}$. In the graph, connect nodes i and j with an edge $\{i, j\}$ having cost γ^*_{ij}, and also connect nodes i' and j' with a zero-cost edge. For each $r \in R$, let $\gamma^r_{ii'} = f^r + g^r_i + h^r_{o_r,i} + h^r_{i,e_r}$, with $g^r_i = \infty$ if resource r cannot perform task i, and define $\gamma^*_{ii'} = \min_{r \in R} \gamma^r_{ii'}$. For every $i \in I$, connect nodes i and i' using edge (i, i') with cost $\gamma^*_{ii'}$. The minimum cost perfect matching on this network gives the optimal RADR solution with no more than two tasks assigned to each resource. The matching problem can be solved in $O\big(|I|\big(|A'| + |I|\log|I|\big)\big)$ time, where $|A'| \leq |I|(|I| - 1)$ is the number of arcs in the above graph [10]. ◆

Now, consider the problem with one resource type-r and *limited* availability, $\mu_r < |I|$. If each resource can be assigned to at most two tasks, Proposition 3 shows that the problem can also be solved efficiently.

Proposition 3: The RADR problem with SL and WC constraints, one resource type-r with $\mu_r < |I|$, where the resource can perform no more than two tasks, can be solved in strongly polynomial time.

Proof. Construct a network with task nodes i for each original task i. Let i, j denote two tasks that the resource can perform without violating the SL and WC constraints. Assume without loss of generality that i precedes j. Connect nodes i and j with an edge $\{i, j\}$ having cost $\gamma_{ij} = f^r + g_i^r + g_j^r + h_{ij}^r + h_{o_r,i}^r + h_{j,e_r}^r$. Now, add $2\mu_r - |I|$ (which is ≥ 0 since feasibility requires $\mu_r \geq (|I| + 1)/2$) dummy nodes to the graph, and connect each task node i with each dummy node j using an edge of cost $\gamma_{ij}^r = f^r + g_i^r + h_{o_r,i}^r + h_{i,e_r}^r$. If $2\mu_r - |I|$ equals zero or one no additional edges are needed. If $2\mu_r - |I| \geq 2$, index the dummy nodes arbitrarily as 1, 2, ..., $2\mu_r - |I|$, and connect alternate pairs of nodes 1 and 2, 3 and 4, and so on, with zero-cost edges. If $2\mu_r - |I|$ is odd, then node $2\mu_r - |I|$ is not connected to any other dummy node. Find the minimum cost perfect matching in this network. For every pair of matched task nodes, the optimal RADR solution assigns a single resource to both the corresponding tasks; individual resources are assigned to the tasks corresponding to the remaining task nodes. Every pair of matched dummy nodes in the minimum cost perfect matching corresponds to an unused resource. Based on the time complexity of the weighted matching problem on general graphs [10], this RADR matching problem can be solved in $O\big(|I|\big(|A'| + |I|\log|I|\big)\big)$ time, where $|A'|$ denotes the number of arcs in the network constructed above. We can also formulate this problem as a minimum cost model. ♦

Although RADR problems with at most two tasks per resource are polynomially solvable, we next prove that increasing the maximum number of tasks to three makes the problem intractable.

Proposition 4: When there is one resource type and each resource can be assigned to at most three tasks, the RADR problem with WC constraints is NP-hard even if there are no SL constraints.

Proof. We reduce the NP-hard 3-Partition Problem (3PP) in polynomial time to an instance of this RADR problem. Given $3q$ integers $S_1, S_2, \ldots, S_{3q}$ and a positive integer T such that $T/4 < S_i < T/2$ for $i = 1, 2, \ldots 3q$ and $\sum_{i=1}^{3q} S_i = qT$, 3PP seeks a partition of the integers into q disjoint sets σ_j, with three elements each, such that $\sum_{i\in\sigma_j} S_i = T\ \forall j$. To reduce 3PP to the RADR problem, define a RADR instance with $3q$ tasks, indexed in the same order as the given $3q$ integers. The start time s_1 of task 1 is zero, and the start time of task $i \in [2, 3q]$ is $s_i = \sum_{j=1}^{i-1} S_j$. The work content of task i is S_i. The arc set consists of: (i) arcs from origin o_r to all the task nodes, (ii) arcs from each task node to the destination e_r, and (iii) for $i = 1, 2, \ldots, 3q - 1$, arcs from i to all $j > i$. All transition times, including to and from the base, are zero, and all arcs have zero costs. There are q resources, all of the same type, and the working time limit for every resource is T. Then, any feasible solution to the RADR problems assigns a total work content of exactly T to each of the q resources. The subsets of integers corresponding to the tasks in these q assignments are therefore a solution to 3PP. Since the 3PP problem is NP-hard, solving (or even finding a feasible solution to) the RADR problem is NP-hard. ♦

Table 6.1 summarizes the complexity results discussed in this section.

6.3 Model Formulations for the RADR Problem with SL Constraints

Based on the RADR problem's network representation from Sect. 2.1, a natural problem formulation consists of modeling resource assignments and task transitions as flows over the arcs or paths of this network. In this section, we present four different RADR model formulations when the problem imposes only SL constraints (no WC constraints), and compare their sizes and tightness. The first model is based on path flows, and the other three are arc-flow formulations. The path-flow model and the first two arc-flow formulations were previously proposed for crew and vehicle scheduling problems. The fourth model, an extended arc-flow formulation, was developed by BML22 for crew assignment with duty time (SL) constraints.

6.3.1 Path-Flow Formulation

This formulation selects a subset of paths of G that satisfy the SL requirements, and together span all the task nodes. Each path represents a sequence of tasks assigned to one resource. For each resource type $r \in R$, let Π^r denote the set of all the o_r-to-e_r paths that are *feasible*, i.e., satisfy the SL restrictions, for this resource type. Let $c_p^r = \sum_{(ij)\in p} c_{ij}^r$ denote the total cost of path $p \in \Pi^r$. By our definition of arc costs c_{ij}^r (including the costs when $i = o_r$ or $j = e_r$), this path cost c_p^r equals the total cost incurred for resource usage, assignment, and transitions when a type-r resource is assigned all the tasks associated with the task nodes on path p. For any task $i \in I$, let P_i^r denote the set of all feasible paths $p \in \Pi^r$ for resource type-r that visit task node i (some of these sets may be empty, e.g., if $i \notin I^r$). Let z_p^r denote a binary variable for each path $p \in \Pi^r$ that equals one if resource r completes tasks on path p, and is zero otherwise. Selecting path p by setting z_p^r to one is equivalent to sending a flow of one resource from o_r to e_r on path p. With these decision variables, the path-flow-based formulation for the RADR problem with SL constraints is

Table 6.1 Summary of computational complexity of RADR special cases

# Resource types	# Resources of each type	Max. # of tasks per resource	SL constraints	WC constraints	Time complexity	Reference
1	Arbitrary	Arbitrary	None	None	$O(\lvert A\rvert\log(\lvert N\rvert)(\lvert A\rvert + \lvert N\rvert\log(\lvert N\rvert)))$	This paper (before Proposition 1)
1	2	Arbitrary	Yes (duty time limit)	None	$O(\lvert N\rvert(\lvert A\rvert + \lvert N\rvert\log\lvert N\rvert))$	BML22
1	2	Arbitrary	None	Yes	NP-hard	BML22
Arbitrary	Arbitrary	≤ 1	Yes	Yes	$O(\lvert I\rvert\lvert R\rvert^2(\log\lvert I\rvert + \lvert R\rvert\log\lvert R\rvert))$	Proposition 1
Arbitrary	High avail. $\mu_r \geq \lvert I\rvert$	≤ 2	Yes	Yes	$O(\lvert I\rvert(\lvert A'\rvert + \lvert I\rvert\log\lvert I\rvert))$	Proposition 2
1	Limited avail. $\mu_r < \lvert I\rvert$	≤ 2	Yes	Yes	$O(\lvert I\rvert(\lvert A'\rvert + \lvert I\rvert\log\lvert I\rvert))$	Proposition 3
1	Arbitrary	≤ 3	No	Yes	NP-hard	Proposition 4
1	Arbitrary	Arbitrary	Yes	No	NP-hard	[7]

[*Path-flow*] Minimize

$$\sum_{r \in R} \sum_{p \in \Pi^r} c_p^r z_p^r \tag{6.3.1}$$

subject to:

Task completion:

$$\sum_{r \in R} \sum_{p \in P_i^r} z_p^r = 1 \qquad \forall i \in I \tag{6.3.2}$$

Resource availability:

$$\sum_{p \in \Pi^r} z_p^r \leq \mu_r \qquad \forall r \in R \tag{6.3.3}$$

Integrality:

$$z_p^r = 0 \text{ or } 1 \qquad \forall p \in \Pi^r, r \in R \tag{6.3.4}$$

Constraints (6.3.2) ensure that every task is assigned on resource, and constraints (6.3.3) specify that no more than μ_r type-r resources can be deployed. This formulation can incorporate other resource-based requirements such as WC constraints more easily than arc-flow-based formulations can, and as we will see later it also has a tight LP relaxation. However, the number of variables in the model is exponential in the number of tasks. So, even for small problem instances, we have to resort to specialized column generation methods (see, for instance, [4]) to iteratively identify feasible paths that can potentially belong to the optimal solution, and resolve the master problem after adding the corresponding path variables. We note that identifying suitable paths (that are feasible and have negative reduced costs) to add to the model can itself be computationally intractable (i.e., NP-hard) for general RADR problems (e.g., with WC constraints).

6.3.2 *Arc-Flow Formulations*

Arc-flow formulations define variables corresponding to the flows on each arc, and so have only polynomial number of variables. The flow on any arc $(i, j) \in A$ is a binary variable that takes the value one if a crew member makes the transition from a base or task i to a task or base j, and is zero otherwise. Flow conservation requirements ensure that the flows correspond to paths from o_r to e_r for different resource types. Given a solution for either the path-flow or arc-flow formulation, we can obtain an equal-cost solution for the other. By the flow decomposition property [1], any path-flow solution can be converted into a unique arc-flow solution, and any given arc-flow solution can be decomposed into path flows. (In general, the path-flow solution that we obtain from an arc-flow solution may not be unique. But, for the RADR problem, the decomposition of integer arc-flow solutions is unique since every task node has exactly one incoming and one outgoing arc.) Although arc-flow models are more parsimonious, they require side constraints to enforce some of the resource deployment restrictions, rendering them weaker (in terms of LP relaxation) than the path-flow model. For arc-flow formulations, there are multiple ways of defining commodities, these alternative versions vary in the strength of the LP relaxation, measured by the relative gap between the IP optimal value and the LP lower bound. We discuss three different arc-flow formulations, starting with a weak single-commodity model (Sect. 3.2.1) followed by a somewhat stronger but limited multi-commodity flow model (Sect. 3.2.2) that were both developed mainly in the context of vehicle scheduling problems (VSP); for our discussion, we also consider the added SL restrictions. We then introduce an extended multi-commodity flow model (Sect. 3.2.3) that is markedly stronger and very effective for capturing SL upper and lower bounds. These three models differ in the way they define commodities and impose the SL constraints. The first two models have fewer variables but also a weaker LP relaxation. The definition of commodities and corresponding flow variables also influences the granularity of the costs that can be captured by the model. This discussion serves to highlight the modeling choices for difficult integer programs such as the RADR problem that entail trade-offs between formulation size, model strength, and versatility.

6.3.2.1 Single Commodity Arc-Flow Formulation

For the single-commodity model, the set of base-starting nodes $O = \{o_r\}_{r \in R}$ are the origin nodes for the commodity, with node o_r having a supply of at most μ_r units. The set of base-ending nodes $E = \{e_r\}_{r \in R}$ are the flow destinations. This formulation does not permit the costs to vary by resource type, and so we drop the superscript r from the arc cost parameters c^r_{ij}. It imposes SL and other crew deployment requirements using an exponential number of constraints as follows. Let Π_{infeas} denote the set of *infeasible* paths in $G(N, A)$ from nodes in O to nodes in E. A path may be infeasible for three reasons: (i) it does not start and end at the same base, i.e., the path starts at node o_{r1} and ends at e_{r2}, $r1 \neq r2$; (ii) it entails task-to-task transitions that are not permitted for the particular resource; or (iii) it violates the SL constraints. Let $|p|$ denote the number of arcs on any path p. Using binary decision variables x_{ij} to represent the flow on each arc $(i, j) \in A$, the single-commodity formulation for the RADR problem, denoted as [*Arc-flow*-SC], is as follows:

[*Arc-flow*-SC] Minimize

$$\sum\nolimits_{(i,j)\in A} c_{ij} x_{ij} \tag{6.3.5}$$

subject to:

Flow conservation:

$$\sum_{(g,i)\in A} x_{gi} = \sum_{(i,j)\in A} x_{ij} \qquad \forall i \in I \qquad (6.3.6a)$$

$$\sum_{i\in IF^r} x_{o_r i} = \sum_{i\in IL^r} x_{ie_r} \qquad \forall r \in R \qquad (6.3.6b)$$

Resource availability:

$$\sum_{i\in IF^r} x_{o_r i} \leq \mu_r \qquad \forall r \in R \qquad (6.3.7)$$

Task coverage:

$$\sum_{(i,j)\in A} x_{ij} = 1 \qquad \forall j \in I \qquad (6.3.8)$$

Infeasible path:

$$\sum_{(i,j)\in p} x_{ij} \leq |p| - 1 \qquad \forall p \in \Pi_{infeas} \qquad (6.3.9)$$

Binary:

$$x_{ij} = 0 \text{ or } 1 \qquad \forall (i,j) \in A \qquad (6.3.10)$$

The objective function minimizes the total cost. Constraints (6.3.6a, 6.3.6b) are the flow conservation constraints. Constraints (6.3.7) ensure that the number of resources of each type used from the corresponding base do not exceed the number available. Constraints (6.3.8) require each task to be assigned exactly one resource, and constraints (6.3.10) specify that the flow variables can only take values 0 or 1. To ensure that the commodity flows only on feasible o_r-to-e_r paths, constraints (6.3.9) preclude infeasible paths p by specifying that no more than $(|p|-1)$ arcs of such paths can be chosen. Importantly, the full model contains exponentially many constraints (6.3.9) since the number of infeasible paths in a network can be exponential in the number of nodes in the network. Therefore, solving this model requires developing specialized cutting-plane procedures that iteratively add violated inequalities from (6.3.9) that the current LP solution does not satisfy and resolving the problem (see, for instance, [6] which applied a cutting-plane procedure to solve this single-commodity model formulation in the context of crew and vehicle scheduling problems). We refer to this RADR model as the *SC model*.

We note that that constraints (6.3.9) do not prevent LP solutions from routing fractional flows on infeasible paths. As a simple example, any LP solution that routes

$(|p| - 1)/|p|$ or less units of flow on all the arcs of an infeasible path $p \in \Pi_{infeas}$ satisfies the corresponding constraint (6.3.9). So, one way to strengthen the model is to define flow variables so that some or all of the three reasons for a path to be infeasible do not apply. This strategy will also reduce the number of infeasible path constraints needed. We can achieve this goal by replacing the single-commodity representation with multiple commodities, an approach that has the added advantage of permitting more general cost structures. Essentially, this tactic "disaggregates" the single-commodity flow variables into multi-commodity flows. The next two models progressively refine the SC model using this variable disaggregation approach.

6.3.2.2 Resource-Indexed (Weak) Multi-Commodity Arc-Flow Formulation

To overcome the single-commodity model's shortcoming that some of the constraints in (6.3.9) are needed to eliminate infeasible paths that begin from one base and end at another, we can expand the arc-flow model by defining one commodity for each resource type-r. For brevity, instead of presenting the full multi-commodity model formulation, we discuss its main distinguishing features vis-à-vis the formulation [*Arc-flow*-SC]. (Hajdar et al. [9] adopted this modeling approach in the context of vehicle scheduling problems for which the resources only differ in their base locations but not in their capabilities, costs, or deployment rules, and were not constrained by span limits or other work restrictions.) In this model, commodity r, for $r \in R$, originates from the base-starting node o_r and terminates at the base-ending node e_r. No more than μ_r units of this commodity are available at o_r. To distinguish the flows of this commodity, we replace the single-commodity variables x_{ij} with the resource-specific flow variables x^r_{ij}. These flows are permitted only on arcs in the resource-specific arc subset $A^r = \{(i, j) : i, j \in I^r, j \in J^r(i)\} \cup \{(o_r, i) : i \in IF^r\} \cup \{(j, e_r) : j \in IL^r\}$. With resource-specific cost parameters c^r_{ij} for each arc in A^r, the model can now account for usage, assignment, and transition costs that vary by resource (unlike the SC model). Define $G^r : (N^r, A^r)$ as the subgraph over which commodity r can flow (where N^r is the subset of nodes spanned by the arcs in A^r). When the arc flows x^r_{ij} are decomposed into path flows, the task nodes traversed by each path for commodity r represent the tasks assigned to a type-r resource. Since each commodity corresponds to a resource type, we refer to this model as the "Resource-type-indexed" multi-commodity model, abbreviated as the *RMC model*. Let [*Arc-flow-RMC*] denote its formulation. The RMC model's objective function minimizes total cost, summed over all arc flows and commodities. Flow conservation constraints analogous to constraints (6.3.6a), at every node $i \in I^r$, (6.3.6b), and (6.3.7), but now for each commodity r (i.e., using x^r_{ij} instead of x_{ij} variables), ensure that flows of each resource type-r start at o_r and end at e_r and no more than μ_r resources are used. For the task coverage constraints, in (6.3.8), we replace x_{ij} with the total flow $\sum_{r \in R} x^r_{ij}$ across all commodities. Compared to the SC model, the number of variables and flow conservation constraints for this resource-type-indexed model have increased by a factor of $|R|$. However, its flow paths no longer violate the requirement that they must start and end at the base nodes corresponding to the same resource. Moreover, by the definition of the arc subsets A^r over which each commodity r is permitted to flow, the model already incorporates resource-specific deployment rules such as the permitted task-to-task, base-to-task, and task-to-base transitions. So, the flows in this model avoid the first two causes of path infeasibility for the SC model. However, they may not satisfy the SL constraints. So, we need to add infeasible path constraints analogous to (6.3.9), but far fewer, for every path $p \in \Pi^r_{infeas}$, where Π^r_{infeas} denotes all the o_r-to-e_r paths that violate the SL constraints. We note that LP solutions to the RMC model can split and send fractional flows over paths that do not meet the SL constraints; our next model enhancement prevents such infeasibility.

6.3.2.3 Strong Multi-commodity Arc-Flow Formulation

BML22 recently proposed a new extended multi-commodity arc-flow formulation, with very tight LP relaxation, for crew assignment problems that also applies to the more general RADR problem. This new approach overcomes the last cause of path infeasibility, namely, violation of SL constraints, by further disaggregating the commodities. Specifically, for every resource type $r \in R$, we now define $|IF^r|$ commodities, indexed as *commodity* $\langle r, k\rangle$ for every possible *first task* $k \in IF^r$. The origin of this commodity is resource r's base-starting node o_r and its destination is node e_r. The only outbound arc from the starting-base node o_r on which commodity $\langle r, k\rangle$ is permitted to flow is on arc (o_r, k), implying that task k must necessarily be the first task for this commodity. A RADR solution routes one unit of this commodity from o_r to e_r via k if and only if the solution deploys one type-r resource to begin its assigned work at task k. For brevity, we refer this resource as resource r-k. As before, the task nodes spanned by any o_r-to-e_r path for commodity $\langle r, k\rangle$ in the RADR solution (after path decomposition from arc flows) correspond to an assignment of the tasks on this path to a type-r resource. By limiting the flow of commodity $\langle r, k\rangle$ to the arcs of A^r we can ensure, as before, that its flow paths start and return to resource r's base and conform to the transition rules for this resource type. We now discuss how to further restrict the network on which the commodity can flow so that its paths also satisfy the span limits. Recall that each span limit specifies the maximum (or minimum) permitted difference in the value of some attribute between the first task that a type-r resource performs and its subsequent tasks. We say that a task i is "compatible" with the first task k for resource type-r if i can be performed after k by the same resource (i.e., subject to the transition rules for this resource type) and the difference between the attribute values for this task and those of the first task k meet *all* the span limits. For instance, for a span limit corresponding to the maximum permitted duty time T^r of a resource, a task $i \in N^r$ is compatible with first task k only if $(s_i + \tau_i) - s_k + t_{o_r,k} + t_{i,e_r} \leq T^r$ (see Sect. 6.2). Further, if resource r's deployment policy specifies that no subsequent task must be more than D^r miles from the first task, then task i must also satisfy the relationship $d_{k,i} \leq D^r$ for it to be compatible with k. Let $I^{rk} \subseteq I^r$ be the set of all tasks (including task k itself) that are compatible with the first task k for resource type-r, and define $N^{rk} = I^{rk} \cup \{o_r, e_r\}$. Let $A^{rk} \subseteq A^r$

be the arcs of the subgraph induced by the nodes N^{rk}. Let G^{rk}:(N^{rk}, A^{rk}) denote the subnetwork consisting of these node and arc subsets.

By permitting commodity $\langle r, k\rangle$ to only flow on network G^{rk}, every o_r-to-e_r path for this commodity not only conforms to the transition rules for a resource type-r (or resource r-k), but all the task nodes on this path also satisfy the resource's span limits with respect to the first task k. So, with this disaggregation of commodities and definition of the corresponding flow networks, we no longer require any infeasible path constraints in the model. Another benefit of this approach is that it can now permit both costs and resource deployment rules to vary by the first task that a resource performs. Let c_{ij}^{rk} denote the total cost (usage, assignment, and transition) incurred to traverse arc (i, j) by resource r-k. If transition rules and other resource deployment restrictions also vary both by resource type and first task, we can readily incorporate these rules and restrictions by further pruning the node and arc sets of G^{rk}. This feature is useful, for instance, when resources are people (e.g., home health care, crew scheduling) and their daily wage and work rules depend on the time they begin their work shift (early morning, late night), which in turn depends on the starting time s_k of the first task k assigned to this resource. To formulate the model, let $RK = \{\langle r, k\rangle : k \in IF^r, r \in R\}$ denote the set of all disaggregated commodities. For each commodity $\langle r, k\rangle$, define the binary commodity flow variables x_{ij}^{rk} for all arcs $(i, j) \in A^{rk}$. Then, the multi-commodity formulation for the RADR problem with commodities indexed by resource type-first task (RF) pairs, which we denote as formulation [*Arc-flow-RFMC*] and refer to as the *RFMC model*, is

$$[Arc - flow - RFMC] \quad \text{Minimize} \sum_{\langle r,k\rangle \in RK} \Big(\sum_{(i,j)\in A^{rk}} c_{ij}^{rk} x_{ij}^{rk}\Big) \tag{6.3.11}$$

subject to:

Flow conservation:

$$\sum_{(i,j)\in A^{rk}} x_{ij}^{rk} - \sum_{(j,i)\in A^{rk}} x_{ji}^{rk} = 0 \qquad \forall i \in I^{rk}, \langle r, k\rangle \in RK \tag{6.3.12a}$$

$$x_{o_r,k}^{rk} = \sum_{(j,e_r)\in A^{rk}} x_{j,e_r}^{rk} \qquad \forall \langle r, k\rangle \in RK \tag{6.3.12b}$$

Resource availability:

$$\sum_{k\in IF^r} x_{o_r,k}^{rk} \le \mu_r \qquad \forall r \in R \tag{6.3.13}$$

Task coverage:

$$\sum_{\langle r,k\rangle \in RK} \sum_{(j,i)\in A^{rk}} x_{ji}^{rk} = 1 \qquad \forall i \in I^{rk} \tag{6.3.14}$$

Binary:

$$x_{ij}^{rk} = 0 \text{ or } 1 \qquad \forall (i, j) \in A^{rk}, \langle r, k\rangle \in RK \qquad (6.3.15)$$

The objective function (6.3.11) minimizes the total cost for resource deployment or usage, resource-to-task assignment, and changeovers. Recall that the arc set A^{rk} contains the arc (o_r, k) from the starting-base o_r to the first task k whose cost includes the resource usage cost. Also, the solution can route at most one unit of a commodity $\langle r, k\rangle$ over the network since this commodity first visits node k and no more than one unit can enter and flow through node k. Constraints (6.3.12a) are the standard flow conservation constraints for every commodity $\langle r, k\rangle$ at each task node $i \in I^{rk}$ through which this commodity can flow. Constraint (6.3.12b) specifies that, if one unit of commodity $\langle r, k\rangle$ leaves the starting-base node o_r (on its only incident outbound arc for this commodity), then that unit of flow must return to the ending-base node e_r. The total number of type-r resources used in any RFMC solution equals the total flow $\sum_{k\in IF^r} x^{rk}_{o_r,k}$ emanating from the starting-base node o_r (or returning to the ending-base node e_r). Constraints (6.3.13) ensure that this number does not exceed the number of available resources of each type. By setting the total flow entering each task node to one, the task coverage constraints (6.3.14) ensure that each task is performed exactly once. Constraints (6.3.15) enforce integrality (binary) for all variables. Finally, as noted earlier, we do not need any infeasible path constraints for this RFMC model since all flow paths implied by its commodity arc flows meet the resource deployment and SL constraints. Nevertheless, its LP solutions can be fractional due to the resource availability and task coverage side constraints (6.3.13) and (6.3.14). Unlike the previous arc-flow models, this formulation is polynomial in size, with $O(|I||R||A|)$ variables and $O(|I|^2|R|)$ constraints.

To conclude, we have outlined several alternative IP model formulations for the RADR problem, starting with the path-flow formulation with an exponential number of variables, a single-commodity arc flow model, and two increasingly disaggregate multi-commodity arc-flow versions. Among the arc-flow models, we have seen that "extended" models obtained by disaggregating commodities permit eliminating some causes for infeasible paths, but at the expense of adding more variables. The RFMC model does not require any "infeasible path" constraints because these requirements are embedded in the restricted networks G^{rk} over which the respective disaggregate commodities flow. This formulation illustrates one way to avoid imposing side constraints by suitably defining commodities and constructing their flow networks. As we show next, this modeling strategy makes the LP relaxation much tighter.

6.4 Tightness of RADR Models

We now compare the previous four formulations in terms of the tightness of their LP relaxations. The results in this section apply to RADR problems with multiple resource types and resource-type-dependent transition rules and costs, and deployment constraints (except for Proposition 6 where we omit WC restrictions). Recall that, for any given problem instance, a tighter model yields a higher (or equal) LP value compared to weaker formulations, which in turn reduces the enumeration needed for exact methods such as branch-and-bound that rely on LP-based lower bounds. The overall time to solve a problem also depends on the computational time per iteration (e.g., per branch) which depends on the size of the model and the solution method (e.g., standard solver or specialized iterative methods such as cutting-plane or column generation methods). As is common for difficult IP problems, the different models for the RADR problem present trade-offs between LP tightness and model size. We first compare the tightness of the three arc-flow models, before comparing with the path-flow version.

6.4.1 Comparison of Arc-Flow Models

Let $LP(\text{Model}X)$ and $Z^{LP}(\text{Model}X)$, respectively, denote the LP relaxation of model X and its optimal value. The following result establishes a hierarchy, in terms of tightness, of the three arc-flow models.

Proposition 5:

(a) $Z^{LP}(SC) \leq Z^{LP}(RMC) \leq Z^{LP}(RFMC)$;
(b) the ratio $Z^{LP}(RMC)/Z^{LP}(SC)$ can be arbitrarily large; and,
(c) the ratio $Z^{LP}(RFMC)/Z^{LP}(RMC)$ can be arbitrarily large.

Proof. Given any solution $\left\{x_{ij}^r\right\}_{(i,j)\in A^r, r\in R}$ for the LP relaxation of the RMC model, the solution $x_{ij} = \sum_{r\in R} x_{ij}^r$ for all $(i, j) \in A$ is feasible (and has the same cost) for the LP relaxation of the SC model. Hence, $Z^{LP}(SC) \leq Z^{LP}(RMC)$. Similarly, given any LP solution $\left\{x_{ij}^{rk}\right\}_{(i,j)\in A^{rk}, \langle r,k\rangle \in RK}$ for the RFMC model, the solution $x_{ij}^r = \sum_{k\in I F^r} x_{ij}^{rk}$ for all $r \in R$, $(i, j) \in A^r$ is a feasible, equal-cost solution for the LP relaxation of the RFMC model, implying that $Z^{LP}(RMC) \leq Z^{LP}(RFMC)$. The following simple example establishes parts (ii) and (iii), and also shows that the inequalities in part (i) can be strict.

Figure 6.3a displays a problem instance with two resource types, labeled as $r1$ and $r2$, and 12 tasks (and their corresponding nodes) indexed as 1A, 1B, ..., 6A, 6B. The numbers on the nodes represent the starting and ending times of the corresponding task. The arcs of the network show the permitted transitions between tasks and the bases for the two resource types. The numbers on each arc are the transition time and cost, which are the same for all resources, where ϵ_1 and ϵ_2 are small positive quantities. In this example, there are no WC constraints on task assignments, and each resource has one SL constraint, namely, a duty time limit of ten from the time a resource leaves the starting-base and returns to the ending-base. So, for instance, assigning tasks 1A and 5A to a type $r1$ resource requires the resource to leave its starting-base o_{r1} at time 0 and return to its ending-base e_{r1} only at time 11, so this assignment does not satisfy the SL limit of ten. Figure 6.3b presents an optimal solution to *LP*(*SC*). The value on each arc is the flow of the commodity on that arc in this solution; the figure also depicts the unique path-flow decomposition of this arc-flow solution. Observe that, although the SC flow solution satisfies the infeasible path constraints (6.3.14) in aggregate, it routes 2/3 units on path $o_{r1} - 1A - 5A - e_{r2}$, which is infeasible because it starts at one base and ends at another. The total cost of the LP(SC) solution is $2\epsilon_1$. Figure 6.3c shows the LP solution to the RMC model with total cost of $2\epsilon_1 + \epsilon_2$. In this solution, all paths in the path-flow decomposition start and end at the same base, and the fractional solution satisfies the infeasible path inequalities (to comply with SL constraints) in aggregate. However, in one path-flow decomposition of the RMC solution, 1/2 unit flows on path $0_{r1} - 1A - 5A - e_{r2}$; as noted earlier, this path violates the SL limit of ten. Finally, Fig. 6.3d depicts the optimal LP solution for the RFMC model. This solution, with cost $4+2\epsilon_2$, is integer-valued and is therefore optimal for the original IP model. For this example, we see that $Z^{LP}(RMC)/Z^{LP}(SC) = (2\epsilon_1 + \epsilon_2)/2\epsilon_1$ which increases without bound as ϵ_1 tends to zero. And, $Z^{LP}(RFMC)/Z^{LP}(RMC) = (4 + 2\epsilon_2)/(2\epsilon_1 + \epsilon_2)$ which can become arbitrarily large as ϵ_1 and ϵ_2 tend to zero.♦

This result demonstrates that, even though the RFMC model has the highest number of variables among the three arc-flow models (but only polynomial number of constraints versus exponential constraints in the SC and RMC models), it yields

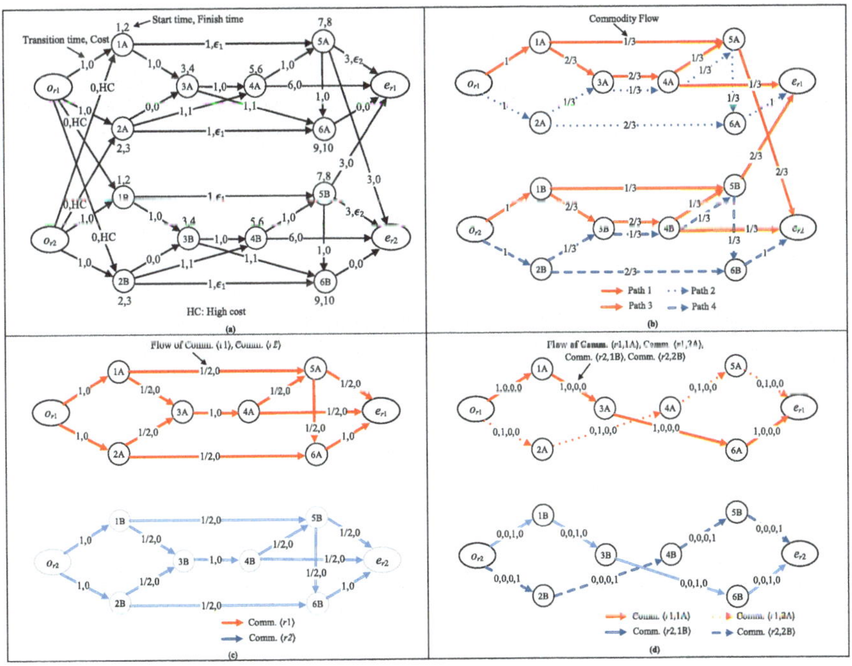

Fig. 6.3 Comparison of arc-flow models

the tightest LP lower bound. So, we also refer to this model as the *Strong* arc-flow model.

6.4.2 Strong (RFMC) Arc-Flow Model Versus Path-Flow Model

The following proposition establishes the LP-equivalence of the path-flow model and the strong RFMC model for the class of RADR problems with only SL restrictions.

Proposition 6: For the RADR problem with only SL restrictions, the LP relaxations of the path-flow model and the RFMC model have the same optimal value.

Proof. Given any LP solution LP solution $\left\{x_{ij}^{rk}\right\}_{(i,j)\in A^{rk},\langle r,k\rangle\in RK}$ to the RFMC model, for each commodity $\langle r, k\rangle \in RK$, decompose its flow from o_r to e_r, via first task node k, into flows along L_{rk} paths $P_{LP}^{rk} = \{p_1^{rk}, p_2^{rk}, \ldots, p_{L_{rk}}^{rk}\}$, and let $y(p)$, $p \in P_{LP}^{rk}$ denote the flow value on each of these paths. For every $\langle r, k\rangle \in RK$, by construction of the network G^{rk}, each path in P_{LP}^{rk} starts from the o_r and ends at e_r, satisfies all the transition rules for type-r resources, and meets all their SL requirements. So, each of these paths also belongs to the set of feasible paths Π^{r} for resource type-r in the path-flow model. Moreover, the RFMC solution routes exactly one unit of flow into each task node. Hence, setting $z_p^r = y(p)$ for all $p \in P_{LP}^{rk} \subseteq \Pi^r$ and every $\langle r, k\rangle \in RK$ gives an equal-cost solution to the path-flow model. Conversely, consider any path p that carries positive flow in the LP solution to the path-flow model, and suppose path p connects o_r to e_r and passes through task node k first after leaving o_r. All arcs of p must belong to the arc set A^{rk} in the RFMC model. So, by setting $x_{ij}^{rk} = z_p^r$ for every arc $(i, j) \in p$ for every path p carrying positive flow in the path-flow solution, we get a feasible and equal-cost solution to the RFMC model. Hence, the two models have the same optimal LP value. ♦

So, despite having only polynomial number of variables and constraints, the strong RFMC arc-flow model is as tight as the path-flow model (which has an exponential number of variables) when the RADR problem imposes only SL constraints.

6.4.3 Summary

Among the arc-flow models, the "extended" RFMC model with "disaggregated" commodities that are indexed by both resource type and first task is tighter than the RMC model with fewer resource-type-indexed commodities. In turn, the RMC model is tighter than the single-commodity SC model. This comparison shows how disaggregating the decision variables can lead to stronger formulations, but at the expense of having more variables in the model. Although the RFMC model has the highest number of variables, by embedding the resource deployment rules and SL limits within the definition of the restricted networks G^{rk}, it does not require any separate side constraints to enforce these requirements. Moreover, since this model has only a polynomial number of constraints, we can solve it using standard solvers, whereas the exponential number of constraints in the RMC and SC models necessitate a tailored cutting-plane solution approach. The path-flow model is as tight as the RFMC model, and can be tighter if the problem also imposes WC constraints. However, it contains an exponential number of variables, requiring column generation method to solve the problem. Thus, the new RFMC model strikes an excellent compromise between model size, tightness of its LP, and solution approach. In summary, the discussions in this section highlight the importance of selecting the "right" model formulation from among alternative models, and the impact this choice has on both the solution methods employed and computational time needed to obtain optimal or near-optimal solutions. The next section outlines three strategies to further accelerate the solution procedure.

6.5 Strategies to Accelerate the Arc-Flow Solution Procedure

The previous two sections have discussed two inter-related ways to strengthen the arc-flow model in order to reduce computational time—by disaggregating the flow variables into multi-commodity flows, and by capturing the SL constraints via restrictions on the arcs over which each commodity flows. These strategies yield the polynomial-sized strong RFMC model that is LP-equivalent to the exponential path-flow model for the RADR problem with SL constraints. This section outlines three additional strategies to reduce the time to solve the RFMC model—problem reduction, model tightening using valid inequalities and coefficient lifting, and optimization-based heuristics. The broad principles underlying these strategies apply and have been successfully used to solve many other difficult IP problems. We consider the more general RADR problem with both SL and WC constraints.

The work content (WC) of a particular assignment may be expressed in terms of one or more metrics, most commonly (but not limited to) time and distance. For instance, the deployment restrictions may impose upper and/or lower limits on the total time a resource spends working on the assigned tasks, the total distance that the resource travels (possibly including distance from and to the base), or the resource's total idle time between assigned tasks. They may stem from union rules, technological restrictions (e.g., operating time for equipment), or policies to ensure equitable assignment of workload across resources (e.g., lower/upper limits on workload). For each metric, work content is its total value, summed over all (or a subset of) the tasks assigned to a resource and the transitions needed to perform these tasks. The RADR model can accommodate multiple WC constraints that must simultaneously apply. For notational simplicity, in the following discussions, we will focus on problems with a single metric for WC upper and lower limits, but the principles we discuss here also extend to situations with multiple WC metrics. One of the advantages of the RFMC model over the more aggregate SC and RMC models is that the value of the metric and its limits can vary both by resource type and the first task that the resource performs (e.g., the model can incorporate metrics that vary with the time or work shift at which the resource begins working). Let v_i^{rk} denote the work content of task i for resource r-k (type-r resource with task k as first task), and ψ_{ij}^{rk} the work content for this resource to transitioning from i to j (possibly including the starting- and ending-bases). If a task i or a transition from i to j is not relevant to calculating a particular metric, then the corresponding value is zero. For instance, if the work content of an assignment of tasks to a particular resources is measured by the resource's total idle time, then $v_i^{rk} = 0$ for all tasks i, and $\psi_{ij}^{rk} = s_j - (s_i + \tau_i) - \delta_{ij}$, where δ_{ij} is the i-to-j transition time. Let $W_{\max}^{kr}$ and $W_{\min}^{kr}$, respectively, denote maximum and minimum permissible values of the work content for any assignment of tasks to the resource. Then, for the RFMC model, the maximum WC constraint is

$$\sum_{i \in N^{rk}} v_i^{rk} \sum_{(h,i) \in A^{rk}} x_{hi}^{rk} + \sum_{(i,j) \in A^{rk}} \psi_{ij}^{rk} x_{ij}^{rk} \leq W_{\max}^{rk}, \ \forall \langle r, k \rangle \in RK$$

where the second summation in the first term of the left-hand side (LHS) is the total flow entering node i. Since this total flow is one if and only if commodity $\langle r, k \rangle$ passes through node i, the first LHS term captures the work content v_i^{rk} if resource-type r-k performs task i. To simplify this expression, for each arc $(i, j) \in A^{rk}$, define $w_{ij}^{rk} = v_j^{rk} + \psi_{ij}^{rk}$ as the "weight" of arc (i, j). With this terminology, we refer to $W_{\max}^{rk}$ as the maximum weight. Then, the maximum WC constraint can be simplified as

$$\sum_{(i,j) \in A^{rk}} w_{ij}^{rk} x_{ij}^{rk} \leq W_{\max}^{rk} \ \forall \langle r, k \rangle \in RK. \tag{6.5.1}$$

Similarly, the minimum WC constraint is

$$\sum_{(i,j) \in A^{rk}} w_{ij}^{rk} x_{ij}^{rk} \geq W_{\min}^{rk} \ \forall \langle r, k \rangle \in RK. \tag{6.5.2}$$

6.5.1 Problem Reduction

Problem reduction refers to methods for eliminating (or fixing the values of) some variables and constraints in an optimization model prior to solving it. These methods identify and exploit properties of optimal and feasible solutions to determine if some variables can be fixed a priori at zero or some other value and if certain constraints can be omitted without raising the optimal value or rendering the solution infeasible. This tactic decreases solution time by reducing the size of the formulation, and can also increase the lower bound, i.e., make the model tighter. We illustrate this principle for the RADR with the WC upper and lower limits (6.5.1) and (6.5.2). For any pair of nodes $s, t \in N^{rk}$, we refer to the shortest (longest) path from s to t using the arc weights w_{ij}^{rk} as arc lengths as the shortest weight path (or longest weight path). Let LS_{st}^{rk} and LL_{st}^{rk}, respectively, denote the total weight of the shortest weight and longest weight paths from s to t in graph G^{rk}. Since the network is acyclic, we can identify the longest path between any two nodes in this network in polynomial time. Suppose, for some arc $(i, j) \in A^{rk}$, $LS_{o_r,i}^{rk} + w_{ij}^{rk} + LS_{j,e_r}^{rk} > W_{\max}^{rk}$, i.e., the total weight of the shortest weight e_r-to-o_r path containing arc (i, j) exceeds the weight limit $W_{\max}^{rk}$. Then, we can drop arc (i, j) from A^{rk} and omit the flow variable x_{ij}^{rk} from the model. Similarly, if $LL_{o_r,i}^{rk} + w_{ij}^{rk} + LL_{j,e_r}^{rk} < W_{\min}^{rk}$, we can drop arc arc (i, j) and the corresponding flow variable x_{ij}^{rk} since the expression in the left-hand side of this condition is the total weight of the longest weight e_r-to-o_r path containing arc (i, j). When the reduction procedure drops arc (i, j), we may be able to identify and drop other arcs that cannot belong to any feasible solution since dropping (i, j) can raise (reduce) the length of the shortest (longest) weight origin-to-destination path containing these other arcs. That is, the arc elimination tests can be applied iteratively until no further reduction is possible. Further, if this procedure eliminates all incoming (or outgoing) arcs at a node $i \in N^{rk}$, then we can also delete node i from the network. Note that these reductions not only decrease model size but may also increase the LP lower bound since the LP solution can no longer route (partial) flows on the high or low weight (and hence infeasible) paths in the original network.

6.5.2 *Model Tightening: Valid Inequalities and Coefficient Lifting*

Although the RFMC model is strong, we can further tighten it and raise the LP lower bound, particularly with the added WC constraints, by adding valid inequalities and enhancing the model's current constraints. The LP relaxation of the RFMC model for the RADR problem can have fractional solutions for two reasons: (a) due to the requirement that the total flow, summed over all commodities, entering each task node must be exactly one and (b) due to the WC constraints on the total weight of the o_r-to-e_r path for any commodity $\langle r, k\rangle$ that is routed on the network. We note that the first reason applies even if the model does not include WC constraints (or SL constraints). We illustrate how to strengthen the model by adding valid inequalities to eliminate some fractional solutions due to the first reason, and by enhancing (e.g., coefficient lifting) the WC constraint to address the second cause.

(a) *Adding valid inequalities.* Valid inequalities are constraints that raise the LP lower bound by cutting off some fractional LP solutions without eliminating any feasible solution to the original problem. Such inequalities are very problem-specific;

developing them requires insights about the structure of feasible or optimal solutions to the problem being solved. They have proven to be very effective for solving difficult IP problems, including the celebrated Traveling Salesman problem. In the RFMC model, consider the task coverage constraints (6.3.14). Instead of assigning one resource fully to perform each task, i.e., sending just one commodity (with unit flow) through each task node, the LP solution may satisfy this constraint by assigning fractional resources of multiple types to the task in order to reduce total cost while meeting all constraints. As an example, consider a problem with three tasks and two resource types, $r1$ and $r2$, and suppose two resources are available for each type. In this problem's network, Fig. 6.4 shows arcs with positive flows for three commodities, with task 1 or task 3 as the first tasks, in an LP solution of the RFMC model. All arcs shown have zero cost, except for arc $(o_{r1}, 1)$ which has a cost of 1, all flow paths satisfy SL constraints, and the problem does not impose any WC constraints. Each arc carries ½ unit of one of the three commodities, $\langle r1, 1\rangle$, $\langle r1, 3\rangle$, and $\langle r2, 1\rangle$, except arc $(2, e_{r1})$ which carries ½ unit each of both commodities $\langle r1, 1\rangle$ and $\langle r1, 3\rangle$. Let us develop a valid inequality to cut off this solution. Since no more than one unit can flow out of node 1, every feasible solution must satisfy $x_{12}^{r1,1} + x_{13}^{r2,1} \leq 1$. This type of requirement is called a *conflict* relationship since it specifies that only one of the two flows in the LHS can be positive in an integer solution. Similarly, the inequalities $x_{12}^{r1,1} + x_{32}^{r1,3} \leq 1$ and $x_{32}^{r1,3} + x_{13}^{r2,1} \leq 1$ are conflict inequalities at nodes 2 and 3. Adding these three inequalities, dividing both sides by two and rounding down the right-hand side (since the left-hand side must be integer in an IP solution) gives the Chvátal–Gomory cut $x_{12}^{r1,1} + x_{32}^{r1,3} + x_{13}^{r2,1} \leq 1$. The current solution violates this inequality, and so adding it will cut off the current LP solution, increasing the LP value from ½ to 1. The Chvátal–Gomory procedure can yield many other valid inequalities by adding more conflict inequalities with more (odd number) variables. Further, we can develop broader families of Chvátal–Gomory cuts by also considering inequalities based on *conditional* relationships. BML22 discuss one such generalization, called the extended Eulerian subgraph inequality, obtained by combining conflict and conditional constraints on Eulerian subgraphs of the LP flow solution for the RFMC model. They also develop another general class of inequalities, called multi-commodity partition inequalities.

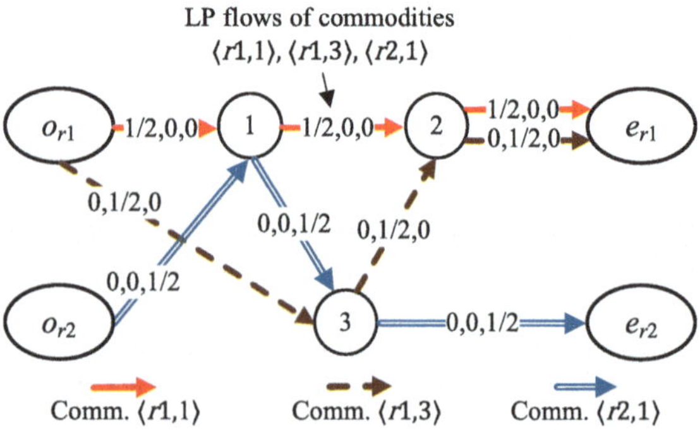

Fig. 6.4 Subgraph to illustrate valid inequality for RFMC model

(b) *Coefficient lifting.* To illustrate opportunities for strengthening the model by enhancing its existing constraints, consider the maximum WC constraint (6.5.1). First, since the constraint only applies if we deploy a type r-k resource, i.e., only if the flow $x^{rk}_{o_r,k}$ on arc (o_r, k) is one, we can replace the constant $W^{rk}_{\max}$ in the RHS with $W^{rk}_{\max}x^{rk}_{o_r,k}$. For every LP solution, this new RHS value is no larger than before, and so the modified constraint can be tighter. Similarly, we can tighten the minimum WC constraint by replacing $W^{rk}_{\min}$ with $W^{rk}_{\min}x^{rk}_{o_r,k}$ in the RHS. Second, the following observation provides another way to tighten the WC constraints. Suppose an integer solution to the RFMC model selects a resource r-k and assigns task h as its last task. Let $U^{rk}_{\max}(h)$ and $U^{rk}_{\min}(h)$ denote the maximum and minimum possible work content of this resource, given that h is its last task. On the network G^{rk}, we can calculate these bounds by solving longest and shortest weight problems from node k to node h and setting $U^{rk}_{\max}(h) = w^{rk}_{o_r,k} + LL^{rk}_{kh} + w^{rk}_{h,e_r}$ and $U^{rk}_{\min}(h) = w^{rk}_{o_r,k} + LS^{rk}_{kh} + w^{rk}_{h,e_r}$. Then, if $U^{rk}_{\max}(h) < W^{rk}_{\max}$, i.e., the every path with task k as first task and task h as last task has work content less than the maximum permitted work content $W^{rk}_{\max}$, we can "lift" the coefficient of the last arc (h, e_r) from w^{rk}_{h,e_r} to $w^{rk}_{h,e_r} + (W^{rk}_{\max} - U^{rk}_{\max}(h))$ in the maximum WC constraint (6.5.1) without eliminating any feasible integer solution. More generally, define $\overline{w}^{rk}_{h,e_r} = w^{rk}_{h,e_r} + \left[W^{rk}_{\max} - U^{rk}_{\max}(h)\right]^+$, where $[a]^+ =$ Max$\{a, 0\}$. Then the lifted inequality

$$\sum_{(i,j)\in A^{rk}, j\neq e_r} w^{rk}_{ij}x^{rk}_{ij} + \sum_{(i,e_r)\in A^{rk}} \overline{w}^{rk}_{h,e_r}x^{rk}_{h,e_r} \leq W^{rk}_{\max}\ \forall\,\langle r,k\rangle \in RK \tag{6.5.2}$$

is valid for the RFMC model and tighter than constraints (6.5.1) since $\overline{w}^{rk}_{h,e_r} \geq w^{rk}_{h,e_r}$ and $x^{rk}_{h,e_r} \geq 0$ for all h. Similarly, replacing w^{rk}_{h,e_r} with $\underline{w}^{rk}_{h,e_r} = w^{rk}_{h,e_r} + \left[U^{rk}_{\min}(h) - W^{rk}_{\min}\right]^+$ in the minimum WC constraint (6.5.2), for every possible last node h for commodity $\langle r, k\rangle$, we can strengthen this constraint. Besides these model-tightening approaches, by exploiting the knapsack-like structure of the WC constraints, we can develop and add valid inequalities to eliminate fractional LP solutions that arise because of these constraints. ([2] identify several families of valid inequalities for a broader class of problems that also apply to the RADR problem with WC constraints.)

In summary, tailored valid inequalities and constraint tightening methods that exploit problem structure can raise the LP lower bound, thereby reducing enumeration in exact solution methods. However, adding valid inequalities can also pose challenges. Typically, there are an exponential number of inequalities in each family of cuts, and so they need to be added iteratively via cutting-plane procedures instead of adding them all a priori. Such an implementation requires developing and applying "separation" procedures (see, for instance, [11]) that can identify violated inequalities for any given intermediate LP solution.

6.5.3 Optimization-Based Heuristic Method

Solving the full model (after problem reduction) can be time-consuming when the problem instance is large. This can happen even when the integrality gap of the model is very low due to difficulty in finding feasible integer solutions that satisfy all the problem requirements. So, to complement the previous strategies, generating good feasible solutions and tight upper bounds using optimization-based heuristic algorithms, before applying exact solution methods such as branch-and-bound, can also significantly reduce total computational time to solve the problem. For the RADR problem, we propose a heuristic method that solves a *restricted* problem, obtained by omitting flow variables for high-cost arcs, to generate good starting solutions. Interestingly, this approach also enables further problem reduction by eliminating variables and can also determine if the heuristic solution is optimal. The rationale underlying the restricted problem approach is that arcs with high cost are unlikely to carry flow in the optimal solution. Dropping these arcs reduces problem size, permitting solvers to find the optimal solution much faster. To use this approach, we need to identify an effective rule to decide which arcs to drop. One approach is to set a cost threshold, and drop any arc whose cost exceeds this threshold value. This method entails the following trade-off. Setting too high a threshold will not reduce the problem size much, while setting a very low threshold yields small, restricted problems but may eliminate the optimal solution or even make the restricted problem infeasible. As an alternate approach, for each arc (i, j) and commodity $\langle r, k\rangle$, if we can show that setting a flow variable x_{ij}^{rk} to one raises the cost of the solution (or an estimate of this cost) above the cost of a neighboring solution that does not use this arc, then we can drop this variable. We refer to this approach as the estimated incremental cost method for omitting flow variables. It is based on the following principle. For arc (i, j) and resource r-k, suppose the resource is assigned to perform tasks i and j consecutively. Hence, $x_{ij}^{rk} = 1$. Let $I1$ and $I2$ denote the tasks the resource performs before i and after j, respectively. Now consider a modified solution that assigns only tasks in $I1 \cup \{i\}$ to this resource (so the resource returns to the base after completing task i), and assigns a different resource to performs the remaining tasks in $\{j\} \cup I2$. The resource type for this second resource is chosen to minimize the cost of performing these remaining tasks. This modified solution has $x_{ij}^{rk} = 0$. If its cost is no more than c_{ij}^{rk} higher than the cost of the first solution, then, we drop the variable x_{ij}^{rk} from the restricted problem. BML22 describes the method in more detail. The solution to the restricted problem provides an initial upper bound and warm-start to solve the full problem. Let Z_{LP}^{R} and $\widehat{Z}$, respectively, denote the optimal LP value and a feasible (or optimal) solution value of the restricted problem. The following result, established by BML22, demonstrates how to use these objective function values as well as the optimal dual values from the LP relaxation of the restricted model to identify omitted variables x_{ij}^{rk} in the restricted model that can also be dropped from the full unrestricted model, leading to a sufficient condition for the restricted solution to be optimal for the original problem.

Proposition 7: If the reduced cost (using the optimal dual values for the restricted LP) for an arc-flow variable x_{ij}^{rk} that is omitted in the restricted model is greater than or equal to $(\widehat{Z} - Z_{LP}^{R})$, then x_{ij}^{rk} must also be zero in the optimal solution to the original problem. And, if all variables that were omitted in the restricted model satisfy this reduced cost condition, then the optimal solution to the restricted problem is also optimal for the original problem.

Clearly, the smaller the gap $(\widehat{Z} - Z_{LP}^{R})$, the more effective is this optimality test. So, solving the restricted problem optimally and using its value Z^{R*} as $\widehat{Z}$ can lead to more variable fixing.

We note that this principle of solving restricted problems by dropping high-cost variables both to obtain good feasible solutions quickly (and possibly use dual information to perform problem reduction) apply to other difficult IP problems including the Traveling Salesman problem and many network design problems. In these problems, we can omit arc or flow variables that have high cost, determined using either a threshold cost or the estimated incremental cost method to obtain smaller restricted problems that can be solved quickly. Similarly, even for the simple knapsack problem, we can omit items that have low value-to-weight ratio to obtain good "restricted" solutions, and use properties of these solutions to show that including the omitted items in the consideration set does not improve the objective function value.

6.6 Summary

This section has discussed three strategies—problem reduction, model tightening using valid inequalities and coefficient lifting, and optimization-based heuristics—to complement strong model formulations for accelerating solution procedures. An important learning from these discussions is that these enhancements require a good understanding of the problem structure and insights about the properties of solutions, and must be tailored to exploit these characteristics. Composite solution methods that combine strong problem formulations with the above three strategies have proven very effective to reduce computational time for other difficult IP problems, and have the same improvement potential for RADR problems.

6.7 Computational Results

To demonstrate the effectiveness of the RFMC model and solution approach for RADR problems with SL and WC limits, BML22 applied the approach to a set of large crew and vehicle scheduling problems from the literature. These problems are based on real-world multi-depot vehicle assignment problems from an Italian bus company (http://www.or.deis.unibo.it/ORinstances/) and were originally provided by [6], abbreviated as FLMT. In these problems, each task represents a scheduled trip, resources are crew members, and multiple resource types correspond to crews that are located at different depots. [4], abbreviated as BMR, modified the problems to include upper limits on SL and WC. This section presents a summary of the results obtained by BML22, using CPLEX as the LP and IP solver on a desktop computer (Intel i7 processor running at 3.4 GHz). The tightness of the RFMC model is measured by the *Initial LP* gap, which is the gap between the optimal IP value (or the best upper bound at termination) and the LP value of the model, expressed as a percentage of the upper bound.

The FLMT problem instances have 80 available resources, with a fixed cost of 5,000 for using any resource. The number of tasks vary from 184 to 580 (five values as shown in Table 6.2), with number of resource types equal to the number of depots (2, 3, or 5), yielding a total of 15 instances. FLMT solved these problems as vehicle scheduling problems, without any SL or WC constraints, but BMR tested their column generation approach after adding upper limits of 800 on the span (duty time) and 600 on work content for every resource. For the *full* (unrestricted) problem, the RFMC model for the largest problem instance (with 5 bases and 580 tasks) contains more than 10 million variables and 63,000 constraints.

The LP relaxation of the RFMC model provides very tight lower bounds for the FLMT problem instances (as modified by BMR). Table 6.2 shows the average Initial LP gaps for the full model (without any valid inequalities) and the LP solution times (in seconds), where the results are averaged over the three instances (with varying number of bases) for each of the five values for the number of tasks. Remarkably, the average initial LP gap for the RFMC model is only 0.005% even without adding any of the valid inequalities discussed in Sect. 6.5, confirming the tightness of our formulation. (Due to these very low initial gaps, the inequalities were not added when solving the IP.) Table 6.2 also reports the average IP solution times (in seconds) and average number of branch-and-bound (B&B) nodes to reach optimality, when the full problems are solved using CPLEX, without running our optimization-based heuristic to generate initial upper bounds or apply the optimality test. Due to the tight lower bounds, only one out of the 15 problem instances required branching; CPLEX solved all other problem instances at the root node.

Table 6.2 Result summary for FLMT instances: full unrestricted problem

#Tasks	Initial LP gap (%)	LP time (secs)*	#B&B node*	Total time (secs)*
184	0.005	3.7	0	11.6
285	0.007	20.4	0	71.0
352	0.005	43.9	3	178.6
463	0.007	183.8	0	616.9
580	0.000	648.7	0	1109.8
Overall average	0.005	180.1	0.6	397.6

* all reported values are averages over the three instances with 2, 3, and 5 resource types

Table 6.3 compares the computational results for our method using the *enhanced* method that includes the initial optimization-based heuristic approach, i.e., solving the restricted problem and applying the optimality test, with the performance of the BMR algorithm. Applying the initial heuristic significantly reduces (by 60%) the overall average LP solution time and average total computational time.

Column 5 shows that the total computational times averaged between 7 seconds for problems with 184 tasks to 7 min for problems with 580 tasks. Columns 6 and 7 contain the results that BMR obtained using their method (FLMT did not test these problem instances with SL or WC limits). With two exceptions out of 15 problem instances (results not presented in this table), BMR's best lower bound gaps are higher than the RFMC model's initial LP gaps; on average, their gap is twice the RFMC model's gap, but still only 0.01%. BMR's computational time is, on average, 3.32 times the computational time needed for our approach, but using (slower) different hardware and software (Pentium II 400 MHz computer). Comparing the trends in computational times, the computational times for the BMR method increase much more rapidly with the number of tasks than for our method. The decrease in the percentage gap as the number of tasks increases may partly stem from the larger value (solution cost) of the denominator for problems with more tasks.

To illustrate the versatility and value of the RFMC model, we next report results after adding minimum span (duty time) requirements. (BMR's problem instances did not include minimum duty time limits.) Imposing and parametrically varying these lower limits can yield crew assignment solutions that are more equitable (in terms of duty times) across crew members, but at higher total cost. To assess this trade-off, Fig. 6.5 presents results for one illustrative real-life FLMT problem instance with two bases, 184 tasks, and 80 available crew members, for minimum span limits varying from 0 to 700; all other parameter values are the same as those for the original instance.

Figure 6.5 shows how total cost, average span, and variance of span across crew members change as the minimum span, say $SL_{\min}$, increases (along the x-axis). To facilitate the comparison, for each of the three metrics and every value of $SL_{\min}$,

Table 6.3 Results using enhanced method for FLMT instances

# tasks	RFMC model average				BMR averages	
	Initial LP Gap (%)	LP time (secs)	# B&B nodes	Total time (secs)	LB gap G_3 (%)	Total time (secs)
184	0.005	2.2	0.0	7.1	0.017	34.6
285	0.008	12.7	0.0	43.1	0.018	115.4
352	0.003	24.7	7.3	88.4	0.008	305.6
463	0.008	85.5	0.0	230.8	0.005	644.1
580	0.000	239.5	0.0	422.5	0.001	1650.3
Overall average	0.005	72.9	1.5	158.4	0.010	550.0

* all reported values are averages over the three instances with 2, 3, and 5 resource types

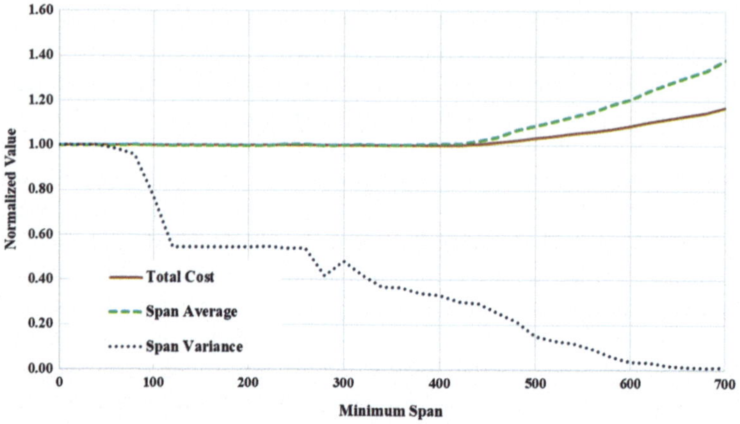

Fig. 6.5 Effect of changing minimum span (duty time) limit (from BML22)

Fig. 6.5 plots the normalized value obtained by dividing the actual value of the metric for $SL_{\min}$ by the metric's value when there is no minimum span constraint. As expected, the variance in the span decreases as $SL_{\min}$ increases, reflecting more balanced task assignments. Interestingly, the total cost and average span are both relatively flat until $SL_{\min}$ increases to about 400. This analysis shows that the organization can substantially reduce the variance of the spans (by over 60%) across crew members, and hence their potential dissatisfaction or stress, without increasing average span or total cost. In fact, imposing the minimum span constraints provides the added benefit of reducing the spans for more crew members (not shown in this figure).

6.8 Concluding Remarks

This chapter has proposed a general modeling framework for resource assignment to precedence-constrained tasks when the deployment of resources needs to meet some restrictions such as limits on the span and work content of each resource. All but the simplest versions of this problem are computationally intractable, i.e., NP-hard. Using this problem class as a platform, we have illustrated the various alternative ways in which integer programming problems can be modeled, and highlighted the trade-offs between formulation size and LP tightness (or strength) of the model. Interestingly, for problems with only SL constraints, the extended RFMC multi-commodity model for RADR problems achieves the same LP value as the path-flow model (which has exponential number of variables) despite having only polynomial number of variables and constraints. Besides using a strong model formulation, our discussions have also emphasized the need to develop composite solution procedures combining problem reduction, valid inequalities, approximate methods, and optimality tests in order to effectively solve difficult IP problems. The RADR is a general class of problems that has applications in many different classical and contemporary domains. It offers many opportunities for further methodological research, algorithmic development, computational testing, and application to real-life problems. Future research opportunities include the following: (a) applying this model to other emerging and practical problems (e.g., drone deployment, courier assignment for food delivery services); (b) testing the model and method for larger problem instances; (c) benchmarking this method with other classical and decomposition approaches such as column generation and cutting-plane procedures; (d) understanding why the initial LP (integrality) gaps are low, and identifying the problem features that are most influential in determining these gaps and problem difficulty; and (e) exploring other types of constraints, besides SL constraints, that can be captured within the network structure for an extended formulation.

Acknowledgements We thank the editor for his careful review, detailed comments, and thoughtful suggestions on an earlier version of this chapter that have led to significant improvements in content and exposition.

References

1. Ahuja RK, Magnanti TL, Orlin JB (1995) Network flows: Theory, algorithms and applications. Prentice Hall
2. Balakrishnan A, Li G, Mirchandani P (2017) Optimal network design with end-to-end service requirements. Oper Res 65:729–750
3. Balakrishnan A, Mirchandani P, Lin S (2022) Crew assignment with duty time limits for transport services: Tight multi-commodity models. Oper Res 70:690–714
4. Boschetti MA, Mingozzi A, Ricciardelli S (2004) An exact algorithm for the simplified multiple depot crew scheduling problem. Ann Oper Res 127:177–201
5. Brenner U (2008) A faster polynomial algorithm for the unbalanced Hitchcock transportation problem. Oper Res Lett 36:408–413
6. Fischetti M, Lodi A, Martello S, Toth P (2001) A polyhedral approach to simplified crew scheduling and vehicle scheduling problems. Manag Sci 47:833–850
7. Fischetti M, Martello S, Toth P (1987) The fixed job schedule problem with spread-time constraints. Oper Res 35:849–858
8. Fischetti M, Martello S, Toth P (1989) The fixed job schedule problem with working-time constraints. Oper Res 37:395–403
9. Hadjar A, Marcotte O, Soumis F (2006) A branch-and-cut algorithm for the multiple depot vehicle scheduling problem. Oper Res 54:130–149
10. Gabow HN (2018) Data structures for weighted matching and extensions to b-matching and f-factors. ACM Trans Algorithms 14:1–80
11. Nemhauser GL, Wolsey LA (1988) Integer and combinatorial optimization. Wiley
12. Orlin JB (1993) A faster strongly polynomial minimum cost flow algorithm. Oper Res 41:338–350

Chapter 7
Modeling and Exact Solution Approaches for the Distance-Based Critical Node and Edge Detection Problems

Glory Uche Alozie, Kerem Akartunali, and Ashwin Arulselvan

Abstract The performance of many networked systems including energy, telecommunication, and transport networks is dependent on the functionality of a few components of these systems whose malfunction compromises the optimal performance of the system. With respect to network connectivity as a performance metric, such components are termed critical nodes and edges. The optimization problem associated with identifying critical nodes of a network is termed the critical node detection problem (CNDP). The CNDP has gained a significant amount of research owing to its applicability in diverse real-life problems including disaster management, social network analysis, and disease epidemiology, as well as its computational complexity. However, traditional models, whose underlying objective is to maximize network fragmentation fail to capture cohesiveness and extent of connectivity within the resultant network. Therefore, a new variant of the problem termed the distance-based critical node detection problem (DCNDP) was proposed to address this gap. The DCNDP takes into account pairwise distances between nodes as part of its network connectivity objective, which is modeled by pre-defined distance-based connectivity measures. Distance-based connectivity plays an important part in everyday life. For instance, our choice of route of travel and the cost of a flight ticket are influenced by the duration of travel and number of stopovers involved. Therefore, while a source-destination route might exist, if the duration of a trip via the route precludes the attainment of a time-bound activity, then such is a practical disconnection. Similarly, in communication and telecommunication networks, speed and coverage are key operational issues for assessing connectivity which are both related to pairwise distances in the network. In this chapter, we study a generalization of the DCNDP on weighted networks, where the distance between any source-destination (s-t) pair is not limited to hop distance (number of edges along an s-t path). We present a new

G. U. Alozie
Department of Biomedical Engineering, University of Strathclyde, 106 Rottenrow East, Glasgow, G4 0NW, United Kingdom

K. Akartunali (✉) · A. Arulselvan
Department of Management Science, University of Strathclyde, 199 Cathedral Street, Glasgow, G4 0QU, United Kingdom
e-mail: kerem.akartunali@strath.ac.uk

F. Hamid (ed.), *Optimization Essentials*, International Series in Operations Research & Management Science 353, https://doi.org/10.1007/978-981-99-5491-9_7

model with fewer entities than the models in previous studies. Moreover, we show that the proposed model admits different distance-based connectivity measures, hence is valid for all existing classes of the distance-based critical node detection problem. We introduce a new version of the problem, in which edges rather than nodes are to be deleted. This version is useful for application contexts where it is impractical or too expensive to delete nodes. Furthermore, we study social and transportation networks, where we also demonstrate practical aspects of the problem. Some computational experiments on instances of different real-world networks are presented for the different application contexts studied using the proposed models and algorithm. The Chapter concludes with directions for future research.

Keywords Critical node detection problem · Distance-based connectivity · Mixed integer programming · Branch-and-cut · Networks

7.1 Introduction

A large volume of social and economic activities take place daily across the globe. These activities are heavily dependent on networked infrastructures such as transportation (air, road, and rail), telecommunication, and energy networks. Therefore, ensuring that these infrastructures operate in the most efficient way possible is a critical objective. One important strategy to achieving this is to assess the vulnerability of the networks to component (nodes and edges) failures. In other words, to what extent would the failure of one or more components of a networked system degrade the performance of the entire network, and which components are the most critical to maintaining the network performance? Having connectivity as a measure of network performance and nodes (or vertices) as the network component whose failure is of interest, the related problem as investigated by the optimization community is named the critical node detection problem (CNDP), as introduced by [4]. The connectivity in a network relates to the extent to which nodes can reach other nodes within the network. Given an undirected graph $G = (V, E)$ with $n = |V|$ nodes and $m = |E|$ edges, the goal of the CNDP is to find a subset of nodes with limited cardinality B such that, when deleted (that is, failed), the resulting subgraph has maximum disconnectivity based on a connectivity measure that is defined in advance. Inspired by the contagion control problem in the human population, the authors defined a connectivity metric which measures the number of pairs of nodes in the resultant subgraph for which there exists a path connecting them. Thus, in relation to the contagion control problem, having fewer connected nodes means that the spread of the contagion can be minimized. Following the pioneering study of [5], subsequent studies on the critical node detection problem are hinged on two fundamental issues, namely, appropriate connectivity metrics for different application contexts and stronger formulation to improve computational tractability.

7.1.1 Connectivity Metrics for Different Application Contexts

Indeed, the study of the critical node detection problem is motivated by the varied areas in which it finds application. Different connectivity measures are more suitable for specific application areas. For example, the edge deletion variant of the classical CNDP which minimizes the number of connected node pairs has been applied as a method for controlling the spread of infection [28]. However, in the context of security defense operations, where the goal is to dismantle communication of a terrorist network, the topology of such networks requires a connectivity metric that would delete (neutralize) key members of the network such that no giant connected component survives. This would limit the chances of the small groups to launch a large-scale coordinated attack [5]. The CNDP variants that stem from this idea have as objective to maximize the number of connected components and minimize the largest component size which are often referred in literature as the "MaxNum" problem and the "MinMaxC" problem, respectively [32, 33]. A recent paradigm in the study of the critical node detection problem is motivated by the limitation of previous definitions of connectivity metrics which overlook internal cohesiveness of networks and the extent of connectivity in the network. This is primarily due to the fact that the underlying goal of these connectivity metrics is fragmentation. A graph is fragmented if it has disconnected components, i.e., some nodes cannot be reached from others due to the lack of a path. Thus, fragmentation-based connectivity metrics focus only on the existence of a path between two nodes regardless of its length thereby undermining the contribution of the length of the paths to the connectivity objective. One can immediately see the drawback of these metrics from many real-world situations. Indeed, daily real-life experiences reveal that cost (distance, duration, bandwidth) of a path plays an important role in determining whether or not a level of service is acceptable and the price that service users are willing to pay for different service levels. For example, long haul air trips with many overlays and connecting flights usually have lower ticket price than direct flight. Moreover, a passenger who has an important appointment does not consider such flights a practical means of connection between his origin and destination locations.

Another related concept is information diffusion in the area of communication networks. A key performance measure in this context is communication efficiency which is assumed to be inversely related to distances (length of shortest path) between nodes in the network. Therefore, individuals that have direct link (relationship) with sources of information are more likely to access information quicker and more reliably than those individuals with many intermediaries in between them and the information sources. Finally, in telecommunication network design problems, the hop distance (number of edges) of a path is used to measure the level of service (such as bandwidth, delay, etc.) specified for each commodity which is a key performance indicator for service providers (see, e.g., [7, 11, 17]). These practical issues led to the introduction of the distance-based critical node detection problem [35], which is a family of the critical node detection problem that uses distance-based connectivity functions to model connectivity. For an empirical illustration of the relevance

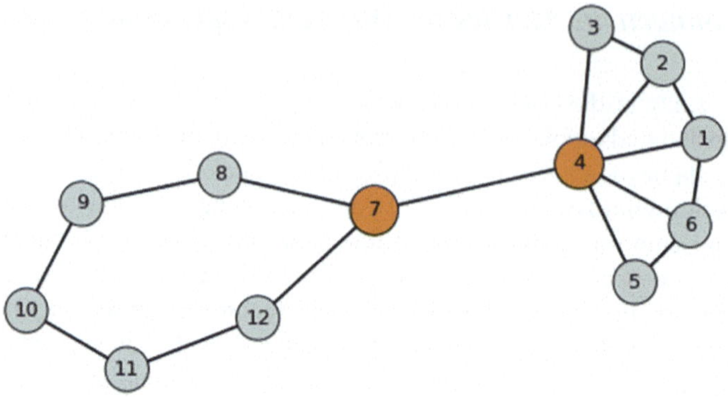

Fig. 7.1 Graphical illustration of where a single node deletion based on fragmentation objectives could lead to sub-optimal solution for a distance-based application (Adapted from [36])

of distance-based connectivity metrics, consider the graph in Fig. 7.1 in which a maximum of one node is to be deleted. Based on the goal of network fragmentation, deletion of either node 4 or 7 achieves equal value on all fragmentation-based objectives. In particular, deletion of either node 4 or 7 results in a subgraph consisting of 2 connected components, 25 connected node pairs, and largest component size of 6. Consider, however, a case where service level requirement for specified commodities is hop distance at most 3. It is easy to see that the deletion of node 4 compromises this quality of service for commodity (3, 5) whereas the deletion of node 7 does not. Similarly with respect to commodity (8, 12), node 7 is critical whereas node 4 is not. Hence, in such applications, use of fragmentation-based metrics would result in sub-optimal solutions which could have grave consequence for the decision-maker. The case study of this chapter is focused on the distance-based critical node detection problem (DCNDP) where interesting application areas are considered.

7.1.2 *Strong MIP Formulations*

The critical node detection problem like many combinatorial optimization problems is "hard". This leaves open the need for the development of more efficient techniques for the problem. Compact formulations are known to be theoretically appealing in terms of providing better dual bounds. However, computationally, this is not usually the case as shown in a recent study where a compact formulation and a lazy constraint formulation were compared for two classes of DCNDP [2]. In addition to using lazy constraint formulations, computational intractable problems such as the case in hand can be improved by exploiting the problem structure to arrive at a stronger reformulation. In this chapter, a reformulation of DCNDP for edge-weighted distance is considered in addition to the original lazy constraint formulation with some computational experiments comparing both. To further improve computational time for solving such challenging problems, approximate solutions from heuristic methods are often used in conjunction with exact methods either as warm start solution or to routinely guide the optimization process. In a related study [31], the authors reported significant improvements in runtime required to solve the integer programming models by starting the optimization solver with heuristic solutions. In the implementation of the formulations presented in this chapter, similar customization techniques are also used to improve the solution time. For a detailed review on the critical node detection problem and related problems including real-world examples, please refer to [1, 24].

7.1.3 Branch-and-Cut Method

The underlying optimization technique that is used to solve the combinatorial optimization problem discussed in this chapter is the branch-and-cut method [8, 29]. The branch-and-cut method is an exact algorithm that combines a branch-and-bound algorithm [25] and a cutting plane method [15]. The branch-and-bound algorithm is one of the foremost optimization techniques for solving integer programming (IP) and mixed integer programming (MIP) problems. The branch-and-bound algorithm is based on the divide and conquer principle wherein a given problem is partitioned into smaller subproblems which when solved and aggregated together provide a solution to the original problem. The basic idea of the branch-and-bound algorithm is as follows:

(i) partition the feasible set into smaller convex sets;
(ii) find lower and upper bounds for each subset;
(iii) form global lower and upper bounds;
(iv) if the global lower and upper bounds are close enough, then quit; else, refine the partition and repeat.

For a minimization IP or MIP problem, a feasible solution which can be obtained through a heuristic provides an upper bound to the problem since an optimal solution must have a smaller objective value. The linear programming relaxation provides a lower bound to the problem. The branch-and-bound algorithm relies on the computation of good lower and upper bounds since the bounds provide information to prune the branches thus reduce the computational cost of enumerating the tree. Bounds obtained from the linear programming (LP) relaxation solution might be far from the optimal objective value which makes the branch-and-bound tree slower. The cutting plane method provides a way to strengthen the bounds of LPs through an iterative addition of valid inequalities (or cuts) that are satisfied by all integer feasible solutions but are violated by the current LP-relaxation solution. The idea is that the addition of generated inequalities cuts off many fractional solutions. This is with the hope that the feasible region of the LP-relaxation is reduced to one whose extreme points are integer and hence solving the resultant linear program solves the integer program. However, it could require an exponential number of steps to encounter an integral extreme point, thus the cutting plane method on its own is not a very efficient way to solve difficult integer programs in practice. The branch-and-cut algorithm exploits the strength of both branch-and-bound and cutting plane methods. The underlying idea of the branch-and-cut algorithm is to work toward getting tighter dual bounds in each node of the tree by adding appropriate cuts thereby reducing the number of subproblems (nodes) in the search tree. The cuts are generated dynamically through a separation routine. The efficiency of the branch-and-cut method for solving difficult combinatorial optimization problems in practice requires an intelligent combination of techniques including fast feasible solutions, effective cutting planes, etc. We refer the reader to the recent survey by [27] for different mechanisms used in the implementation of the branch-and-bound algorithm. For a deeper understanding of the branch-and-cut technique, the reader is referred to [26, 30].

The next section presents a formal definition of the distance-based critical node detection problem on edge-weighted networks. A description of five distance connectivity metrics that exist in literature is presented alongside problem contexts where each metric finds interesting application.

7.2 Problem Definition

We first define a simple undirected edge-weighted graph $G = (V, E)$ with the set of nodes V and the set of edges $E \subseteq V \times V$, both of finite sizes, where $n := |V|$ and $m := |E|$. For every edge $(i, j) \in E$, denote by w_{ij} a positive weight on (i, j) interpreted as the length of the edge (i, j). For any given node, say i, a node j is said to be a neighbor of node i if there exists an edge between i and j. We denote by $\mathcal{N}_G(i)$ the set of all neighbors of i, that is, $\mathcal{N}_G(i) = \{j \in V : (i, j) \in E\}$. Node i is said to be connected to another node j if there is a path between them where a path P_{ij} is defined as a sequence of nodes and edges (with no node or edge repeated) starting from node i and ending in node j. For any pair of nodes i and j, the length of any given path between i and j is the sum of all edge weights along that path. We define $d_G(i, j)$, the distance between i and j, which is the length of the shortest path between i and j. Then, define the diameter of the graph G as the maximum distance between any pair of nodes in the graph G. We assume that all edge weights w_{ij} are positive integers which is not too restrictive for most real-world applications. However, where edge weights are rational numbers, the procedure proposed in [37] can be used to modify rational weights. The goal of DCNDP is to identify a subset of nodes $R \subset V$ satisfying $|R| \leq B$, such that, when removed from the graph, a distance-based connectivity measure $f(.)$ is minimized. The problem can be formally stated as

$$\textbf{DCNDP} : \min_{R \subset V} \sum_{i,j \in V : i < j} f(d_{G^R}(ij)) : |R| \leq B$$

where $d_{G^R}(ij)$ is the distance between nodes i and j in the node-deleted subgraph $G^R = G[V \setminus R]$, $f : \mathbb{Z}_{>0} \cup \{\infty\} \to \mathbb{R}$ denotes a distance-based connectivity function and B is the available budget on the set of critical nodes. The distance connectivity function $f(.)$ is assumed to be non-increasing which implies that paths with larger lengths contribute less in the objective function. Also, for any pair of node $i, j \in V \setminus R$, if i and j are not connected, then $d_{G^R}(ij) = \infty$ and assume that $f(\infty) = 0$. For the sake of simpler exposure, $f(d)$ is used instead of $f(d_{G^R}(ij))$ in the remainder of the chapter. For introductory discussion on concepts and topics in graph theory and combinatorial optimization, the reader is referred to [6, 16].

7.2.1 Classes of the Distance-Based Critical Node Detection Problem

The concept of the connectivity metrics used in the definition of DCNDP comes from notions of cohesion and fragmentation in network analysis including the characteristic path length and efficiency [38]. In [37], five different classes of DCNDP were introduced. Each class corresponds to a different distance connectivity function $f(.)$, nevertheless the underlying idea is that pairwise distances in the induced subgraph are maximized.

Class 1. *minimize the number of node pairs connected by a path of length at most k*:

$$f(d) = \begin{cases} 1, & \text{if } d \leq k \\ 0, & \text{if } d > k \end{cases} \tag{7.1}$$

where k is a given positive integer representing the cut-off distance. For unweighted graphs, where distances are hop lengths, the special case where $k \geq n - 1$ is the classical critical node detection problem version which minimizes the number of connected node pairs. Instances of interest for this class of DCNDP are graphs with a small diameter, that is, most nodes are connected to each other within small distances. The reader is referred to [2, 3, 31, 37] for computational studies on this class of DCNDP. The DCNDP version for this class of distance function is referred to as DCNDP-1.

Class 2. *Minimize the efficiency of the graph*

$$f(d) = \begin{cases} d^{-1}, & \text{if } d < \infty \\ 0, & \text{if } d = \infty \end{cases}$$

This metric stems from the communication network analysis, where communication efficiency between node pairs is naturally inversely proportional to the distance [12]. Two nodes that are disconnected from each other have a distance of ∞. However, in many real-world applications such as telecommunication and social networks, two nodes that are far enough from each other cannot effectively communicate. To address this, one may consider a threshold model, which will not allow communication for two nodes separated by a distance that is more than some specified threshold value L, and this will modify the distance function as follows:

$$f(d) = \begin{cases} d^{-1}, & \text{if } d \leq L \\ 0, & \text{otherwise} \end{cases} \tag{7.2}$$

The DCNDP version corresponding to this connectivity function is denoted by DCNDP-2 and the reader is referred to [2, 35] for computational studies on the DCNDP-2.

Class 3. *Minimize the sum of power functions of distances:*

$$f(d) = \begin{cases} p^{d}, & \text{if } d < \infty \\ 0, & \text{if } d = \infty \end{cases} \tag{7.3}$$

where p is a fixed parameter such that $0 < p < 1$. The distance function is also known as the Hosoya polynomial [21] in chemistry and can be applied in the context of virus or rumor propagation. The parameter p here models the transmission probability of the phenomenon between two adjacent nodes. For non-adjacent node pairs, the transmission probability estimated from the power function would decrease with increase in the distance (d) between the nodes. Hence, class 3 of the DCNDP which is denoted hereafter by DCNDP-3 can be understood as identifying a subset of nodes whose removal maximally reduces the network's propensity to spread a virus.

Class 4. *maximize the characteristic path length of the graph:*

$$f(d) = \begin{cases} -d, & \text{if } d < \infty \\ -M, & \text{if } d = \infty \end{cases} \tag{7.4}$$

where M is a sufficiently large constant such that $M > n - 1$.

The distance function $f(.)$ is given here as negative since the DCNDP has been defined as a minimization problem. One can easily reformulate this as a maximization problem, where the sum of all pairs shortest path distances in the induced subgraph are maximized as in [20].

Class 5. *maximize the length of the shortest path between two specific nodes s and t:*

$$f(d) = \begin{cases} 0, & \text{if } i \neq s, \text{ or } j \neq t \\ -M, & \text{if } i = s, j = t, \text{ and } d = \infty \\ -d, & \text{otherwise} \end{cases} \tag{7.5}$$

where $s, t \in V \setminus R$.

This is simply the node deletion version of the defender's problem in the classical shortest path network interdiction problem [22] which maximizes the distance between node (commodity) pair (s, t). This can also be modified to handle non-singleton set of commodity pairs.

7.3 Integer Programming Formulations

The motivation behind the distance-based connectivity metrics is that paths with smaller length contribute more to the connectivity objective than paths with larger length. Since only paths with a length shorter than or equal to a specified threshold L are relevant, by tracking such paths, there is a guarantee that any given node pair (i, j) is L-distance disconnected iff at least one node along all candidate paths $\mathcal{P}_L(i, j)$ between i and j is deleted. Based on this idea, an integer programming model with constraint set corresponding to the paths connecting pairs of nodes in the graph can be formulated. Since path-based formulations are dependent on the number of paths connecting all pairs of nodes in a graph, the number of constraints in them will exponentially grow. However, only a few of these constraints would eventually be active in the formulation hence they can be treated as lazy constraints. These constraints can be separated in polynomial time by solving a shortest path problem.

For each node $i \in V$, there is a node deletion variable x_i defined as

$$x_i = \begin{cases} 1, & \text{if node } i \text{ is deleted,} \\ 0, & \text{otherwise.} \end{cases} \tag{7.6}$$

Likewise, for every pair of nodes i, j and $\ell \in \{1, 2, \ldots, L\}$ where $L \leq (n-1) * \max_{(i,j)\in E} \{w_{ij}\}$, defined pairwise connectivity variables y^l_{ij} by

$$y^l_{ij} = \begin{cases} 1, & \text{if } (i, j) \text{ are connected by a path of length } \leq l \text{ in } G^R \\ 0, & \text{otherwise.} \end{cases} \tag{7.7}$$

Using the decision variables x and y defined in (7.6) and (7.7), a valid integer programming formulation for DCNDP for edge-weighted distances is given as follows:
DCNDP-L

$$\text{minimize} \quad \sum_{i,j \in V: i<j} \left(f(1) y_{ij}^1 + \sum_{\ell=2}^{L} f(\ell) \left(y_{ij}^{\ell} - y_{ij}^{\ell-1} \right) \right) \tag{7.8}$$

$$+ \sum_{i,j \in V: i<j} f(\infty) \left(1 - y_{ij}^{L} \right) \tag{7.9}$$

$$\text{s.t.} \quad \sum_{r \in V(P)} x_r + y_{ij}^{d} \geq 1, \quad \forall\, P \in \mathcal{P}_L(i,j),\ i,j \in V,\ i<j \tag{7.10}$$

$$y_{ij}^{\ell} \leq y_{ij}^{\ell+1}, \quad \forall\, (i,j) \notin E,\ i<j,\ \ell \in \{2,\ldots,L-1\} \tag{7.11}$$

$$y_{ij}^{\ell} = y_{ij}^{(w_{ij})}, \quad \forall\, (i,j) \in E,\ i<j,\ \ell \in \left\{ w_{ij}+1,\ldots,L \right\} \tag{7.12}$$

$$y_{ij}^{\ell} = 0, \quad \forall\, (i,j) \in E,\ i<j,\ \ell \in \left\{ 1,\ldots, \min_{t \neq j:(i,t) \in E} \left\{ w_{it}, w_{ij} - 1 \right\} \right\} \tag{7.13}$$

$$y_{ij}^{\ell} = 0, \quad \forall\, (i,j) \notin E,\ i<j,\ \ell \in \left\{ 1,\ldots, \min_{t:(i,t) \in E} \left\{ w_{it} \right\} \right\} \tag{7.14}$$

$$\sum_{i \in V} x_i \leq B \tag{7.15}$$

$$y_{ij}^{\ell} \in \{0,1\}, \quad \forall\, (i,j) \in V,\ i<j,\ \ell \in \{1,\ldots,L\} \tag{7.16}$$

$$x_i \in \{0,1\}, \quad \forall\, i \in V \tag{7.17}$$

Objective terms (7.8)–(7.9) minimize some distance-based connectivity objective $f(.)$. In objective (7.8), each term of the form $(y_{ij}^{\ell} - y_{ij}^{\ell-1})$, $\ell \geq 2$ is equal to 1 iff $y_{ij}^{\ell} = 1$ and $y_{ij}^{\ell-1} = 0$. This implies that ℓ is the shortest path length between i and j in G^R. Moreover, the term $1 - y_{ij}^{L}$ in objective (7.9) is equal to 1 iff nodes i and j are disconnected. Constraints (7.10) ensure that a node pair (i, j) can be classified as L-distance disconnected if and only if there is at least one node deleted on each path connecting i and j with a length not greater than L. As we have exponentially many such constraints, where many are also redundant, we only add the constraints for edges

$$x_i + x_j + y_{ij}^{(w_{ij})} \geq 1, \quad \forall\, (i,j) \in E,\ i<j$$

and leave the remaining constraints to be added dynamically in a separation routine as presented in Sect. 7.4.1. Constraints (7.11)–(7.12) guarantee that there is a path of length $\ell + 1$, if there is a path of length ℓ. Constraints (7.13)–(7.14) provide enhancement to the formulation by considering shortest path length between pairs of node. In particular, constraints (7.13)–(7.14) fix y_{ij}^{ℓ} variables to zero for all ℓ less than the minimum distance from node i to its neighbors t since the length of any shortest path from node i to any other node j must be greater than this minimum. The cardinality of the critical node set is limited by the budget constraint as given by (7.15) whereas constraints (7.16)–(7.17) specify the domains of the decision variables. Following the same logic presented in [37], the integrality constraints can also be relaxed for the connectivity variables y_{ij}^{l} in this formulation. Observe that for arbitrary edge weights, the number of variables in the formulation (7.8)–(7.17) becomes pseudo-polynomial growing in the order of $L \leq (n-1) * \max_{(i,j)\in E}\{w_{ij}\}$. To overcome this, we aggregate the y variables into a new set of continuous variables y_{ij} which is independent of ℓ. This new variable y_{ij} is interpreted as the value of the distance function between nodes i and j. Using these aggregated continuous y_{ij} variables, we arrive at an alternative formulation for the weighted distance-based critical node detection problem defined as follows:

DCNDP-Agg

$$\text{minimize} \quad \sum_{i,j\in V, i<j} y_{ij} \tag{7.18}$$

$$\text{s.t.} \quad \sum_{r\in V(P)} f(d)x_r + y_{ij} \geq f(d), \quad \forall P \in \mathcal{P}_L(i,j),\ (i,j) \in V,\ i<j \tag{7.19}$$

$$\sum_{i\in V} x_i \leq B \tag{7.20}$$

$$x_i \in \{0,1\}, \quad \forall i \in V \tag{7.21}$$

$$y_{ij} \geq 0, \quad \forall (i,j) \in V,\ i<j \tag{7.22}$$

Objective (7.18) minimizes the connectivity of the input graph with respect to a specified distance-based connectivity function $f(.) \geq 0$. Constraints (7.19) basically indicate that for any pair of nodes i and j connected by a path of length at most L, if none of the nodes along the paths connecting i and j is deleted, then the distance-based connectivity between i and j is at least $f(d)$ where d is the length of path P. Since the objective is a minimization and the distance function is non-decreasing, we see that y_{ij} would be equal to $f(\tilde{d})$ which is the value of the distance function corresponding to the length of the shortest path between i and j. Thus, the non-redundant inequalities in constraint set (7.19) are those of the shortest path between each pair of nodes which can be separated through an appropriate separation routine.

Formulation (7.18)–(7.22) above naturally admits any non-negative non-increasing distance function such as the first 3 classes of the distance-based critical node detection problem defined in Sect. 7.2. To see this, consider the first DCNDP class with distance function $f(.)$ defined as in Eq. (7.1). Since $f(d) = 1\ if\ d < L$, the set of constraints (7.19) reduces to

$$\sum_{r\in V(P)} x_r + y_{ij} \geq 1,\ \forall\ P \in \mathcal{P}_L(i,j),\ i,j \in V,\ i<j$$

Observe that although the y variables are defined in the formulation (constraints (7.22)) to be non-negative, indeed, they eventually will only take on binary values 0 or 1 based on the minimization objective and the definition of the distance function of DCNDP-1. A similar logic follows for the other two DCNDP classes (DCNDP-2 & DCNDP-3) whose objectives respectively minimize the efficiency of the graph and minimize the sum of power functions of distances in the graph. Recall that the distance functions for these DCNDP objectives are defined respectively as $f(d) = d^{-1}$ and $f(d) = p^d, \quad 0 < p < 1$ if d is less than or equal to the specified threshold L and 0 elsewhere. By substituting the corresponding values of each of these distance functions into constraints (7.19), one realizes that the y variables would eventually take on the value of the distance function of the shortest path between any node pair if none of the nodes along that path is deleted. That is, for any pair of nodes i and j, if the length of the shortest path say $\hat{\mathcal{P}}_{ij}$ between i and j is denoted by $\hat{d}$, and $x_r = 0 \; \forall \; r \in \hat{\mathcal{P}}_{ij}$ then in the optimal solution to DCNDP-2 and DCNDP-3, y_{ij} will be $\hat{d}^{-1}$ and $p^{\hat{d}}$ respectively.

Furthermore, for the last two DCNDP classes whose objectives have a maximization sense, the enhanced formulation (7.18)–(7.22), can easily be adapted to accommodate their distance functions. For example, for class 4 of the DCNDP whose objective maximizes the Wiener index, we redefine the distance function similar to that presented by [20]:

Class 4. *maximize the generalized Wiener index:*

$$f(d) = \begin{cases} d, & \text{if } d \leq L \\ M, & \text{if } d > L \end{cases} \tag{7.23}$$

where M is a sufficiently large constant such that $M > L$. Then, objective (7.18) changes to maximization sense and constraint (7.19) is modified to

$$y_{ij} \leq d + M \sum_{r \in V(P)} x_r, \quad \forall \; P \in \mathcal{P}_L(i, j), \; i, j \in V, \; i < j \tag{7.24}$$

In addition, the y variables are bounded by M, that is

$$y_{ij} \leq M \; \forall \; i, j \in V, \; i < j$$

The variables y_{ij} measure the shortest path length between nodes i and j in the residual graph. Thus, y_{ij} is simply the length of this shortest path in the residual graph if i and j are connected; otherwise, it is set to M. The ideas behind constraints (7.24) are the same with those of constraints (7.19). They indicate that if none of the nodes along a candidate path P connecting a pair of nodes i and j is deleted, then, the length of the shortest path between them is at most d which is the length of the current path P.

7.3.1 *Models for Edge Deletion Problem*

In many real-life situations, deletion of nodes reflects an appealing case, nevertheless, it might be more practical or less expensive to delete edges in some situations. For instance, in a situation where there is an outbreak of disease, it is more practical to temporarily ban flights from one location to another than it would be to completely shut down an airport. For this reason, this section concludes the discussion on model formulation with a brief presentation on how the ideas for the node deletion case can be extended to solve the variant of the problem where the network elements to be deleted are edges. The resultant problem is termed the distance-based critical edge detection problem (DCEDP). From an undirected graph $G = (V, E)$, one can construct an auxiliary graph $G' = (V + V', E')$ as follows:

- For each edge $(i, j) \in E$, add a new node k and replace $(i, j) \in E$ with (i, k) and (k, j). Thus the new graph G' has $n + m$ nodes and $2m$ edges.
- For edge-weighted distances, the weight w_{ij} of edge $(i, j) \subset E$ would be propagated equally to the corresponding edges of the auxiliary graph. Thus, the weights for edges (i, k) and (k, j) would be w_{ij} and the distance threshold would be modified to $2L$.

Let us partition the nodes in the auxiliary graph G' into two disjoint sets V and V'. Let set V be the set of all nodes present in the initial input graph G while V' be the set of all new nodes added for each edge of G. Observe that the deletion of nodes $v' \in V'$ from the auxiliary graph G' is equivalent to the deletion of the corresponding edge of G. Thus, the distance-based critical edge detection problem on the input graph G is equivalent to DCNDP on the constructed auxiliary graph G' formulated as follows:
DCEDP-mod

$$\text{minimize} \quad \sum_{i,j \in V: i<j} y_{ij} \tag{7.25}$$

$$\text{s.t.} \quad \sum_{r \in V'(P)} f(d)x_r + y_{ij} \geq f(d), \quad \forall\, P \in \mathcal{P}_L(i, j),\ i, j \in V,\ i < j \tag{7.26}$$

$$\sum_{i \in V'} x_i \leq B \tag{7.27}$$

$$x_i = 0, \quad y_{ij} \geq 0, \quad \forall\, (i, j) \in V,\ i < j \tag{7.28}$$

$$x_i \in \{0, 1\}, \quad \forall\, i \in V' \tag{7.29}$$

The construction of the auxiliary graph doubles the graph size. However, we can reduce the variable and constraint size by fixing the x variables of the original nodes to zero and defining y_{ij} variables for only node pairs (i, j) of the original graph as in constraint (7.28).

We end the section by presenting an alternative model (formulation (7.31)–(7.35)) for the edge-weighted distance-based critical edge detection problem which uses the input graph directly. Given an edge weighted graph $G = (V, E)$ with positive integer edge weights w_{ij}. We define a new set of variables for edges as follows:

$$x_{ij} = \begin{cases} 1, & \text{if edge } (i, j) \text{ is deleted,} \\ 0, & \text{otherwise.} \end{cases} \tag{7.30}$$

We keep the aggregated continuous variables y_{ij} as defined for the node deletion version. DCEDP on the input graph G admits the formulation:

DCEDP

$$\text{minimize} \quad \sum_{i,j \in V: i<j} y_{ij} \tag{7.31}$$

$$\text{s.t.} \quad \sum_{(r,t) \in E(P)} f(d) x_{rt} + y_{ij} \geq f(d), \quad \forall\, P \in \mathcal{P}_L(i, j),\ i, j \in V,\ i < j \tag{7.32}$$

$$\sum_{i,j \in E: i<j} x_{ij} \leq B \tag{7.33}$$

$$x_{ij} \in \{0, 1\}, \quad \forall\, (i, j) \in E \tag{7.34}$$

$$y_{ij} \geq 0, \quad \forall (i, j) \in V,\ i < j \tag{7.35}$$

7.4 Implementation

Next, we elaborate on the separation routine for the lazy constraints in the path-based formulations presented in Sect. 7.3 for DCNDP. A simple heuristic for the generation of good solution to warm start the implementation of the mixed integer programming models is also presented.

7.4.1 Separation Algorithm

The separation routine for the lazy constraints of inequalities (7.10) and (7.19) involves in generating shortest path trees by employing a suitable shortest path algorithm while keeping a record of the shortest paths and their lengths. A pseudocode of the separation routine for the enhanced model **DCNDP-Agg** is given in Algorithm 1. We remark that in any incumbent (integer) solution, the variables take on values 0 or 1. Therefore, constraints (7.10) and (7.19) will potentially be violated by an integer solution when the first term of the left-hand-side sum of each inequality is zero. Consequently, only nodes with zero values for the node deletion variables x_i can yield the desired paths. To this end, the shortest path trees will be generated from a residual graph constructed by removing from the input graph all the nodes whose $x-$variable value in the current integer solution is equal to one along with all edges incident on them. Given an integer solution $(\bar{x}, \bar{y})$, let $\mathcal{C}$ denote the set of nodes whose node deletion variables x have values equal to one. That is, $\mathcal{C} = \{i \in V : \bar{x}_i = 1\}$.

Then, the auxiliary graph corresponding to the current integer solution is defined by the subgraph $G[V \setminus C]$ whose node and edge sets are defined respectively by $V' = \{v \in V : \bar{x}_v = 0\}$ and $E' = \{(u, v) \in E : u, v \in V'\}$. Using this new graph $G[V \setminus C]$, the separation routine entails generating a shortest path tree T_r rooted at each $r \in V'$ (line 4). For every t in the shortest path tree T_r whose distance d from the root r is within the specified threshold L, the corresponding constraint (7.19) is checked for violation (lines 6–7). More specifically, if $\bar{y}_{rt} < f(d)$, then the violated inequality $\sum_{i \in V(P_{rt})} f(d)x_i + y_{rt} \geq f(d)$ is added to the formulation, where P_{rt} is the shortest path from node t to the root r.

Algorithm 1: Separation algorithm for the enhanced model **DCNDP-Agg**

```
1 Input  : Graph G = (V, E), Integer solution (x̄, ȳ)
2 Output: Violated inequalities
3 C ← {i ∈ V : x̄_i = 1}
4 for r ∈ V \ C do
5     find a shortest path tree T_r of the subgraph G[V \ C] rooted at r;
6     For any node t ∈ T_r having shortest path P connecting to the root node r with distance
      d ≤ L;
7     if ȳ_rt < f(d) then
          add Σ_{i∈V(P_rt)} f(d) x_i + y_rt ≥ f(d)
      end
  end
```

For the original path-based model **DCNDP-L** (formulation (7.8)–(7.17)), the same ideas follow and one needs only to modify line 7 in Algorithm 1 to check that $\bar{y}^d_{rt} < 1$ for every node t at distance d, $(1 < d \leq L)$ from root node r and then to add the inequality $\sum_{i \in V(P_{rt})} x_i + y^d_{rt} \geq 1$. The separation routine is also used in the implementation as a heuristic in order to separate fractional solutions. The slight modification to Algorithm 1 is as follows:

- The input solution in line 1 is a fractional solution $(\hat{x}, \hat{y})$.
- In line 4, the set of candidate nodes $V \setminus C$ for generation of shortest path trees is ordered in increasing order of the fractional solution values $\hat{x}$ of the nodes following the same motivation presented in [2] for the hop-based DCNDP.
- Since the nodes that feature in the shortest path trees do not all have zero values for the node deletion variables x, the left-hand side of the inequality for checking for cut violation in line 7 must be modified. One must check that $\sum_{i \in V(P_{rt})} f(d)\hat{x}_i + \hat{y}_{rt} < f(d)$.

For the edge deletion variant, the separation routine for formulation (7.31)–(7.35) is presented in Algorithm 2. The ideas are similar to the node deletion problem however in this case, the input graph is retained for the generation of shortest path trees rooted at each node in the graph.

Note that when distances are hop-based, the shortest path trees will eventually be breadth first search trees, the reader can refer to the study by [2] for the case of hop-

Algorithm 2: Separation algorithm for edge deletion model **DCEDP**

```
1 Input  : Graph G = (V, E), Integer solution (x̄, ȳ)
2 Output: Violated inequalities
3 for r ∈ V do
4     find a shortest path tree T_r of the input graph G = (V, E) rooted at r;
5     For any node t ∈ T_r having shortest path P connecting to the root node r with distance
      d ≤ L;
6     if Σ_{(i,j)∈E(P_rt)} f(d) x̄_{i,j} + ȳ_rt < f(d) then
          add Σ_{(i,j)∈E(P_rt)} f(d) x_{i,j} + y_rt ≥ f(d)
      end
  end
```

based distances. For arbitrary distances, the exact separation of a fractional solution would require us to solve a resource constrained shortest path problem, which is NP-hard (see [14]).

7.4.2 Heuristic

The implementation of the generic models can be enhanced by using warm start solution generated by a simple heuristic approach based on node centrality measures in a graph [13]. A feasible solution to the distance-based critical node detection problem consists of any set of B nodes, where B is the specified budget (cardinality of the set of critical nodes). The heuristic begins by calculating the centrality of all nodes in the graph using an appropriate node centrality measure and then selects the top ranking B nodes. The node deletion variables x for the top B nodes are fixed to 1 while the x variables for the rest of the nodes in the input graph are fixed to zero. The betweenness centrality measure can be used for the implementation motivated by results presented in [31] for the first class of the distance-based critical node detection problem. However, due to the cost of evaluating objective function value and the modifications required for some of the application problems considered in this chapter, we do not implement the two-stage approach presented in their paper.

7.5 Computational Experiment

In this section, a computational study of the mixed integer linear programming models for the distance-based critical node detection problem is presented. First, the two models presented in Sect. 7.3 are compared based on the second class of the DCNDP on some social and communication network instances. Following the results, the enhanced formulation which is shown to be more computationally efficient is used to study other applications in contagion control and transportation network problems.

For each of these problems, the necessary modifications to the generic model and associated separation routine are elaborated on.

7.5.1 Hardware and Software

The experiments were run on a Dell PC with Windows 10 64-bit OS, Intel Core i7 processor (CPU 2.80 GHz) and RAM 16GB. For the implementation, Anaconda Python 3.9 using Gurobi 9.5.1 [18] was employed as an optimization suite, and NetworkX [19] for manipulating of the graphs.

7.5.2 Comparing the Two-Path Based Formulations

From a theoretical perspective, the original model **DCNDP-L** with the disaggregated y_{ij}^{ℓ} variables (formulation (7.8)–(7.17)) seems to be a tighter formulation (i.e., has better lower bound) than the aggregated reformulation **DCNDP-Agg** (i.e., formulation (7.18)–(7.22)). This can be seen from the basic fact that for every solution of the disaggregated model, one can create a solution of the aggregated version with equal objective value by fixing $y_{ij} = \max(f(l)y_{ij}^{l})$. However, in terms of scalability, the disaggregated formulation is not likely to scale well computationally for arbitrary edge weights. This is because for arbitrary edge weights, the disaggregated y variables grow in the order of $L \leq (n-1) \times \max_{(i,j)\in E} \{w_{ij}\}$.

We investigate the computational efficiency of the two path formulations using their respective models for the second class of the distance-based critical node detection problem (DCNDP-2), which minimizes the efficiency in the graph. The comparisons are based on moderately sized instances of some real world networks of the Pajek dataset [9] as the original disaggregated model cannot cope with larger instances. As the networks are originally unweighted, positive integer weights are assigned on edges of the network based on their edge betweenness centralities [10]. Following the same assumption made by [37], smaller edge weights are assigned to edges with large edge betweenness centrality values. Specifically, given the normalized edge betweenness b_{ij} of edge (i, j), set $w_{ij} = min(b, \bar{b})$, where $\bar{b}$ is the nearest positive integer gotten from approximating the quotient of 0.1 and b_{ij} while b is some specified positive integer that provides an upper bound on the edge weights. For our experiment, we used values of b in the range $\{1, 2, 3, 4, 5, 6\}$ for the **SmallWorld** network where $b = 1$ is equivalent to the hop-based distance. We set $b = 6$ for the rest of the real-world networks. Results are summarized in Tables 7.1 and 7.2.

The columns labeled **PBMt(s)** contain the running time of the aggregated path formulation (**DCNDP-Agg**) while the columns labeled **PBMLt(s)** show the running time for the original disaggregated path formulation (**DCNDP-L**). Running times are given in seconds. The objective values before and after node deletions are respectively displayed in columns labeled **InitObj** and **FinObj**. For both budget settings of 5%

Table 7.1 Comparing aggregated and disaggregated path-based models on some real-world instances of DCNDP-2 where edge weights are in the range {1, 2, 3, 4, 5, 6}

Graph	n	m	diam	InitObj (%)	B = 0.05n FinObj (%)	PBMt (s)	PBMLt (s)	B = 0.1n FinObj (%)	PBMt (s)	PBMLt (s)
Hi-tech	33	91	14	16.2	13.0	**0.27**	0.34	8.5	**0.7**	1.12
Karate	34	78	11	21.8	10.5	0.13	**0.11**	2.7	**0.12**	0.13
Mexican	35	117	16	13.6	10.4	**0.28**	0.55	5.7	**0.36**	0.62
Chesapeake	39	170	16	12.0	9.3	**0.36**	0.56	4.8	0.35	**0.31**
Lesmiserable	77	254	21	11.0	3.2	**2.22**	5.54	1.3	**0.92**	1.38
Sawmill	36	62	12	23.7	10.1	0.25	**0.12**	5.3	0.28	**0.27**
Dolphins	62	159	21	12.0	5.5	**1.38**	2.5	4.0	**3.39**	10.32
Santafe	118	200	20	8.4	0.6	**1.42**	2.73	0.2	**1.56**	4.84
Sanjuansur	75	155	24	10.5	5.8	**3.93**	17.68	2.6	**2.35**	11.76
Attiro	59	128	22	11.8	8.2	**1.45**	5.93	4.3	**1.68**	5.8
SmallWorld	233	994	24	8.3	1.6	**136.5**	633.25	0.7	**314.42**	1727.73

Table 7.2 Comparing aggregated and disaggregated path-based models on weighted instances of the SmallWorld network using DCNDP-2 objective

Edge weight Range	diam	InitObj (%)	B = 0.05n FinObj (%)	PBMt (s)	PBMLt (s)	B = 0.1n FinObj (%)	PBMt (s)	PBMLt (s)
{1}	4	45.44	9.28	**74.6**	178.2	4.10	**231.7**	656.7
{1, 2}	8	23.44	4.64	**57.8**	214.1	2.01	**243.4**	721.5
{1, 2, 3}	12	16.01	3.09	**69.0**	322.0	1.34	**196.5**	1202.15
{1, 2, 3, 4}	16	12.18	2.32	**70.8**	412.1	1.01	**260.9**	1021.2
{1, 2, 3, 4, 5}	20	9.83	1.86	**69.4**	514.1	0.80	**440.3**	1334.6
{1, 2, 3, 4, 5, 6}	24	8.28	1.55	**136.5**	633.25	0.67	**314.42**	1727.73

and 10% of input network size, one can see from Table 7.1 that the disaggregated model **DCNDP-L** is computationally more expensive than the aggregated model across different network types. This is equally true when the range of the edge weights is varied for the **SmallWorld** network which possesses characteristics of many real-world network having a small diameter of 4 (see Table 7.2). Indeed, the number of variables and constraints in the disaggregated model increases rapidly as the range of edge weight increases even for such a moderately sized graph (233 nodes and 994 edges). For example, for the instance with maximum edge weight equal to 6, the final disaggregated model after Gurobi presolve had 627, 192 variables and 718, 360 constraints which amounts to 96 and 85% reduction in comparison to $27,241$ variables and $107,367$ constraints respectively in the aggregated model. These results are in line with the computational study of the first class of the DCNDP across different instances of unweighted networks [31]. Therefore, we conclude that the aggregated path model (formulation (7.18)–(7.22)) is computationally more efficient than the original disaggregated model (formulation (7.8)–(7.17)).

7.5.3 *Application problems*

We now demonstrate the application of the underlying ideas of the models and algorithms described in Sects. 7.3–7.4 to solve practical problems. In each of the application problems discussed, the aggregated formulation is the formulation implemented.

7.5.3.1 Contagion Control Problem

The problem of contagion control aims to control the spread of a malicious phenomenon such as rumors, infectious disease, and computer virus through a social, transportation, or computer network. The nature of many contagions indicates that the farther away that elements of a network are, the less likely (the slower) the contagion can spread within that network. The critical node detection problem provides a strategy for controlling contagion in a network since reducing the connectivity of a network through the deletion of nodes and edges is equivalent to minimizing or slowing down the spread of infections in the network through the immunization of individuals and preventing individual-individual contacts [28]. The third class of DCNDP (that is, distance connectivity function (7.3)) is most suited for the contagion control problems. The necessary definitions as well as details of the modifications to the formulation and algorithm to solve an instance of the contagion control problem using DCNDP-3 connectivity metric are discussed in the following. Using the susceptible-infected (SI) disease spread model, the input network $G(V, E)$ consists of the set of infected nodes I and the set of susceptible nodes $S = V \setminus I$. Given the transmission probability p, $0 < p < 1$ of a contagion from an infected node $i \in I$ to a neighboring node $s \in S$, then for non-adjacent pair of susceptible-infected nodes (i, j), the transmission probability can be estimated from the power function (p^d) of DCNDP-3. Therefore, susceptible nodes that are sufficiently far from infected nodes (that is, a large distance d) are less likely to contact the contagion. The goal is to identify nodes (members of the network) whose removal (isolation or vaccination) would maximally reduce the propensity of the spread of the contagion. Based on the same idea as in [28], the edges between two infected nodes are removed from the graph. Let Ω be defined as the set of distinct pairs of nodes such that at least one of them is susceptible, that is, $\Omega = \{(i, j) : (i, j) \in V \times V,\ i < j,\ (i, j) \notin I \times I\}$. Also, we redefine the set $\mathcal{P}_L$ to be the set of transmission paths where a transmission path is a path between two nodes $(i, j) \in \Omega$ such that none of the intermediary nodes in the path is an infected node. A susceptible node $s \in S$ is at risk of infection if there is at least one transmission path connecting it to an infected node. Therefore, the objective function takes into account only pairs of susceptible and infected nodes. Following these modifications, the resultant mixed integer programming model is given by formulation (7.36)–(7.40).

DCNDP-PBM-App

$$\text{minimize} \quad \sum_{i\in S\,,\, j\in I} y_{ij} \tag{7.36}$$

$$\text{s.t.} \quad \sum_{r\in V(P)} f(d)\, x_r + y_{ij} \geq f(d), \quad \forall\, P \in \mathcal{P}_L(i,j),\ (i,j) \in \Omega \tag{7.37}$$

$$\sum_{i\in V} x_i \leq B \tag{7.38}$$

$$x_i \in \{0,\, 1\}\,, \quad \forall\, i \in V \tag{7.39}$$

$$y_{ij} \geq 0, \quad \forall\, (i,j) \in \Omega \tag{7.40}$$

Consequently, the separation algorithm for the path-based constraints (Algorithm 1) can be slightly modified by starting each shortest path tree generation in line 4 from a susceptible node. Furthermore, in lines 6–7, the shortest paths from the root of each generated tree for which we consider for cut violation check must be a transmission path. That is, we skip line 7 for any node t in a shortest path tree T_r rooted at a susceptible node r, if the corresponding shortest path connecting node t from the root r contains an infected node. We use an instance of a social network instances named **Hi-tech** accessed from the Pajek datasets [9]. The **Hi-tech** network is a friendship network of employees in a small hi-tech computer firm. The network consists of 33 nodes (employees who responded to a question on the colleagues they consider to be a personal friend) and 91 edges. An edge between a pair of nodes indicates that both nodes (employees) consider each other to be a personal friend. The parameters used in implementing the model are critical node budget B equals $0.1n$ and transmission probability p equals 0.15. 15% of nodes were infected with at least one infected node selected from each of the four-employee classifications. Nodes $6,\ 18,\ 19$ and 29 were the infected nodes resulting to an initial objective value of 3.21. Deleting nodes $6,\ 19$ and 29 reduces the objective value to 0.72.

7.5.3.2 Assessing Vulnerability of Transportation Networks

Another application problem where the models and algorithms proposed in this chapter can be employed in order to solve is in assessing the vulnerability of transportation networks. We apply the first class of DCNDP to an air transport network instance named **USAir97**. Recall that the objective of the DCNDP-1 is to minimize the number of pairs of nodes connected by a path of length less than or equal to some threshold k. The **USAir97** instance is an air transportation network of United

States of America flights [9]. The instance consists of 332 nodes and 2126 edges where edges correspond to direct flights while paths of length 2 and 3 correspond to one and two layover flights. The longest pairwise connectivity in the network is five layover flights (that is, diameter of the network is 6). We set the value of k to 3 considering the fact that many pairwise connections in an air transportation network would be at most two layover flights (that is by paths of length less than or equal to 3). Indeed, this is the case with the **USAir97** network where 84.8% of node pairs connectivity are not more than two layover flights. Deleting just 10 nodes from this network reduces the connectivity of the network to 34.9% which is less than half the initial connectivity of the network.

Finally, we apply the first class of distance-based critical node detection problem (DCNDP-1) to assess the vulnerability of a road network to node and edge deletion. The instance used for this application is the **Albany** network [34] which is a road transportation network around Albany, New York, and its nearby highway (Fig 7.2).

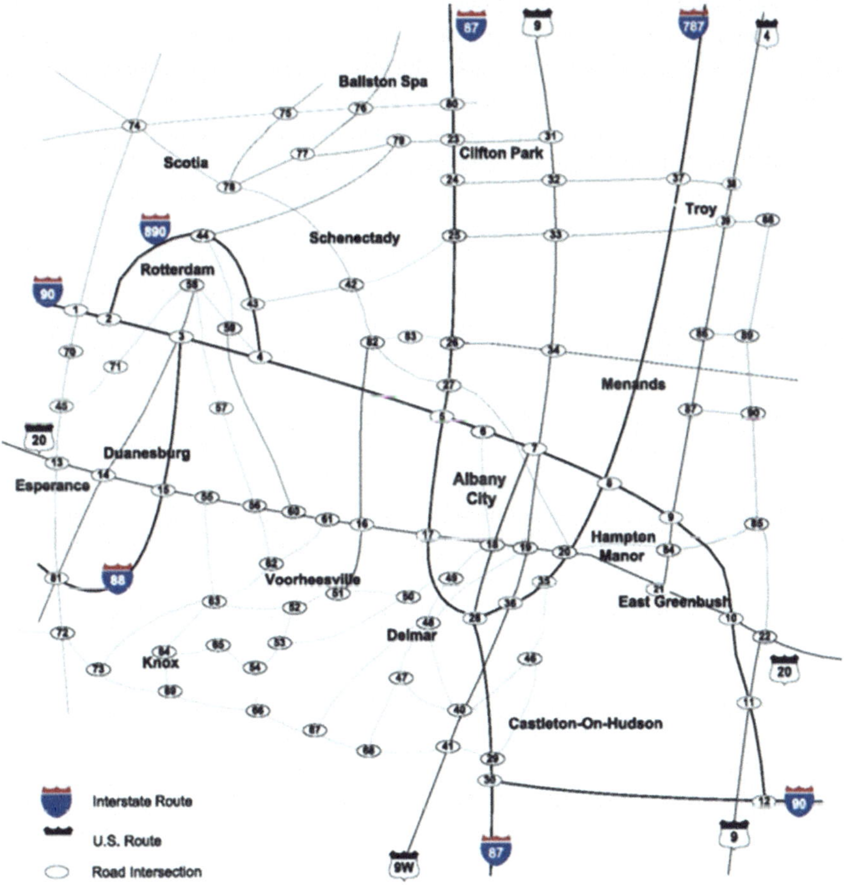

Fig. 7.2 Albany area road network (*Source* [23])

The **Albany** network consists of 90 nodes (road intersections) and 148 edges (road segments). The length on the edges is rounded up to the nearest integer and the threshold distance k is selected such that the proportion of node pairs with distance less than or equal to k is approximately 15% which results in $k = 4$. Deleting just 5% of nodes reduces the objective value by more than three-quarter, from 616 (15.4%) node pairs to 109 (2.7%) node pairs. As expected, deleting the same proportion (5%) of edges results in a higher objective value of 234 (5.84%) node pairs. This is intuitive since deleting a node means deleting every edge incident on it whereas the converse does not necessarily hold.

7.6 Conclusion and Future Research

In this chapter, a branch-and-cut technique is considered to address an important network optimization problem termed the distance-based critical node detection problem. Two mixed integer linear programming formulations were presented to model the problem on edge-weighted networks. Both models admit all five distance-based connectivity functions present in literature. The latter model which is an enhanced reformulation of the former overcomes the computational challenge posed by the pseudo-polynomial number of variables present in the original formulation for arbitrary edge weights. The ideas behind the formulation are extended to consider an edge deletion variant of the problem which is termed the distance-based critical edge deletion problem. A separation algorithm that is based on the shortest path tree is presented to separate the path-based constraints present in the proposed formulations. The chapter concludes with some computational study where the models and algorithms presented in the chapter were applied to some real-world network problems. In particular, the chapter addresses application problems within contexts where the set of nodes in the network need not be homogeneous, an assumption to which existing studies are limited.

An assumption of the models presented in this chapter for DCNDP and its edge deletion variant is that the decision-maker knows the required budget on the total cost of deleting the critical nodes or edges. However, this is not always the case. Sometimes, the decision-maker wants to uncover the extent of vulnerability of the network to different levels of damage (node or edge deletions). This would provide some insights as to how much cost he needs to incur to achieve the desired level of robustness of his networked infrastructure against component (node or edge) failures. In this sense, the relevant problem becomes the β-connectivity distance-based critical node detection problem where β, $0 < \beta < 1$ is a bound on the proportion of original distance-based connectivity left after node deletion. Also, it would be interesting to consider application contexts in which the costs associated with node or edge deletion are heterogeneous. One of the limitations of studies on the critical node detection problem in general is that they assume the knowledge about the real-world networks' parameters is known. Hence, future research should aim to address this.

Finally, the methods presented in this chapter might be useful to study an important network problem in social networks where the goal is to maximize the speed of diffusion of information within a communication network. We intend to perform experiments on instances in this application domain.

7.7 Data Availability

The Python codes and data can be accessed at https://github.com/galozie/dcndp-models.git.

Acknowledgements This material is based upon work supported by the Air Force Office of Scientific Research under award number FA9550-17-1-0029.

References

1. Alozie GU (2021) The distance-based critical node detection problem: Models and algorithms. University of Strathclyde, Glasgow. https://stax.strath.ac.uk/concern/theses/8k71nh57t
2. Alozie GU, Arulselvan A, Akartunalı K, Pasiliao EL Jr (2021) Efficient methods for the distance-based critical node detection problem in complex networks. Comput Oper Res 131:105254
3. Alozie GU, Arulselvan A, Akartunalı K, Pasiliao Jr EL (2022) A heuristic approach for the distance-based critical node detection problem in complex networks. J Oper Res Soc 73(6):1347–1361
4. Arulselvan A, Commander CW, Elefteriadou L, Pardalos PM (2009) Detecting critical nodes in sparse graphs. Comput Oper Res 36(7):2193–2200
5. Arulselvan A, Commander CW, Pardalos PM, Shylo O (2007) Managing network risk via critical node identification. In: Risk management in telecommunication networks. Springer, Berlin
6. Arumugam S, Brandstädt A, Nishizeki T, Thulasiraman K (2016) Handbook of graph theory, combinatorial optimization, and algorithms, vol 34. CRC Press
7. Balakrishnan A, Altinkemer K (1992) Using a hop-constrained model to generate alternative communication network design. ORSA J Comput 4(2):192–205
8. Balas E, Ceria S, Cornuéjols G (1996) Mixed 0–1 programming by lift-and-project in a branch-and-cut framework. Manag Sci 42(9):1229–1246
9. Batagelj V, Mrvar A (2006) Pajek datasets. http://vlado.fmf.uni-lj.si/pub/networks/data/. Accessed 2 Nov 2018
10. Brandes U (2008) On variants of shortest-path betweenness centrality and their generic computation. Soc Netw 30(2):136–145
11. Chemodanov D, Esposito F, Calyam P, Sukhov A (2018) A constrained shortest path scheme for virtual network service management. IEEE Trans Netw Serv Manag 16(1):127–142
12. Crucitti P, Latora V, Marchiori M, Rapisarda A (2003) Efficiency of scale-free networks: Error and attack tolerance. Physica A: Stat Mech Appl 320:622 642
13. Freeman LC (1978) Centrality in social networks conceptual clarification. Soc Netw 1(3):215–239
14. Garey MR, Johnson DS (1979) Computers and intractability: A guide to the theory of NP-completeness (Series of Books in the Mathematical Sciences). Freeman, W.H

15. Gomory RE (2010) Outline of an algorithm for integer solutions to linear programs and an algorithm for the mixed integer problem. In: 50 Years of Integer Programming 1958-2008. Springer, Berlin, pp 77–103
16. Gross JL, Yellen J, Anderson M (2018) Graph theory and its applications. Chapman and Hall/CRC
17. Guérin R, Orda A (2002) Computing shortest paths for any number of hops. IEEE/ACM Trans Netw 10(5):613–620
18. Gurobi Optimization I (2018) Gurobi optimizer reference manual. http://www.gurobi.com
19. Hagberg A, Swart P, S Chult D (2008) Exploring network structure, dynamics, and function using NetworkX, Technical report, Los Alamos National Lab.(LANL), Los Alamos, NM (United States)
20. Hooshmand F, Mirarabrazi F, MirHassani S (2020) Efficient Benders decomposition for distance-based critical node detection problem. Omega 93:102037
21. Hosoya H (1988) On some counting polynomials in chemistry. Discret Appl Math 19(1–3):239–257
22. Israeli E, Wood RK (2002) Shortest-path network interdiction. Networks 40(2):97–111
23. Kang Y, Batta R, Kwon C (2014) Value-at-risk model for hazardous material transportation. Ann Oper Res 222(1):361–387
24. Lalou M, Tahraoui MA, Kheddouci H (2018) The critical node detection problem in networks: A survey. Comput Sci Rev 28:92–117
25. Land A, Doig A (1960) An automatic method of solving discrete programming problems. Econometrica 28(3):497–520
26. Mitchell JE (2002) Branch-and-cut algorithms for combinatorial optimization problems. Handbook Appl Optim 1(1):65–77
27. Morrison DR, Jacobson SH, Sauppe JJ, Sewell EC (2016) Branch-and-bound algorithms: A survey of recent advances in searching, branching, and pruning. Discret Optim 19:79–102
28. Nandi AK, Medal HR (2016) Methods for removing links in a network to minimize the spread of infections. Comput Oper Res 69:10–24
29. Padberg M, Rinaldi G (1991) A branch-and-cut algorithm for the resolution of large-scale symmetric traveling salesman problems. SIAM Rev 33(1):60–100
30. Rebennack S, Reinelt G, Pardalos PM (2012) A tutorial on branch and cut algorithms for the maximum stable set problem. Int Trans Oper Res 19(1–2):161–199
31. Salemi H, Buchanan A (2022) Solving the distance-based critical node problem. INFORMS J Comput
32. Shen S, Smith JC (2012) Polynomial-time algorithms for solving a class of critical node problems on trees and series-parallel graphs. Networks 60(2):103–119
33. Shen S, Smith JC, Goli R (2012) Exact interdiction models and algorithms for disconnecting networks via node deletions. Discret Optim 9(3):172–188
34. STOM-Group (2019) Hazmat network data. https://github.com/STOM-Group/Hazmat-Network-Data. Accessed 19 April 2022
35. Veremyev A, Boginski V, Pasiliao EL (2014) Exact identification of critical nodes in sparse networks via new compact formulations. Optim Lett 8(4):1245–1259
36. Veremyev A, Prokopyev OA, Pasiliao EL (2014) An integer programming framework for critical elements detection in graphs. J Comb Optim 28(1):233–273
37. Veremyev A, Prokopyev OA, Pasiliao EL (2015) Critical nodes for distance-based connectivity and related problems in graphs. Networks 66(3):170–195
38. Zhou B, Cai X, Trinajstić N (2008) On harary index. J Math Chem 44(2):611–618

Chapter 8
Benders Decomposition: A Humanitarian Operation with Stochastic Programming Framework

Jyotirmoy Dalal and Faiz Hamid

Abstract Benders Decomposition (BD) is a popular large-scale optimization technique. Within an iterative framework, it attempts to solve difficult problems by fixing complicating variables. Researchers have successfully applied this exact method to a variety of challenging, large-scale optimization problems in diverse domains such as telecommunication, transportation and logistics, emergency management, sustainable energy management, healthcare, etc. Although this decomposition framework is quite general, the structural aspect of the two-stage stochastic programming (SP) models that represent data uncertainties by scenarios makes this class of problem particularly amenable to BD. The wide popularity of SP as an optimization tool for decision-making under uncertainties and the possibility of successfully applying BD to solve large-scale instances makes this technique attractive. This chapter begins with a mathematical foundation for the BD framework, followed by a step-by-step guide to iteratively solve a mixed-integer programming model by decomposing the same into a relaxed master problem and one or more subproblems. We demonstrate an application of this well-known large-scale optimization technique in stochastic settings and illustrate the concept through an example inspired by a research problem. We conclude the chapter by discussing a variety of implementation-related challenges and some contemporary implementation strategies utilizing the state-of-the-art CPLEX solver.

Keywords Large-scale optimization · Exact algorithm · Benders decomposition · Stochastic programming · Uncertainty · Nonprofit operations

---

J. Dalal (✉)
Sheffield University Management School (University of Sheffield),
Sheffield, United Kingdom
e-mail: j.dalal@sheffield.ac.uk

F. Hamid
Indian Institute of Technology Kanpur, Kanpur, India
e-mail: fhamid@iitk.ac.in

F. Hamid (ed.), *Optimization Essentials*, International Series in Operations Research & Management Science 353, https://doi.org/10.1007/978-981-99-5491-9_8

8.1 Need for an Exact Optimization Method

Optimization models have been successfully applied to numerous problems from diverse domains such as production planning, scheduling, transportation and logistics, healthcare, emergency management, supply chain network design, and management of various scarce resources, to name a few. The *INFORMS Journal on Applied Analytics* (formerly known as *Interfaces*) regularly publishes real-world applications of Operations Research (OR)/Management Science (MS) in "commerce, industry, government, or education" [11]. In many of these practical applications, large-scale problems involving hundreds and thousands of decision variables and constraints arise naturally. Although the linear programming (LP) models (even the huge ones) can be efficiently solved by state-of-the-art optimization packages (commercial: CPLEX, Gurobi, GAMS, etc.; open-source: COIN-OR), the frequent and unavoidable presence of a large number of binary and integer variables in various practical problems often pose severe computational challenges. Moreover, uncertainty regarding parameter estimation requires serious attention for various real-world optimization models, as the solution obtained by solving a problem instance with inaccurately estimated parameters can have severe negative practical consequences. To this end, stochastic programming (SP) is a popular approach for addressing data uncertainty in mathematical programming. However, as a result of the scenario-based uncertainty representation in SP, the number of constraints and decision variables can increase substantially, necessitating some computationally efficient methodology for solving large-scale problem instances.

Practitioners generally have two broad choices, given the real-world complexities and practical applicability of a solution obtained from a mathematical model. They can use simulation tools to replicate the system of their interest, understand the impact of specific changes made to the system, and choose an improvement strategy. Alternately, the decision-maker can adopt an optimization-based approach to obtain the *mathematically best* solution; however, to a simplified representation of the real system (since modeling the real one can be incredibly difficult—solving it optimally being out of the question). The latter approach can be further divided into two categories: *heuristics* and *exact methods*. Although many industry problems must resort to heuristics due to their sheer size, complexity, and a compelling practical requirement of obtaining a real time or "*on-the-fly*" solution (which is both feasible and intuitively reasonable), the drawback of this approach lies in the fact that in the absence of a benchmark, there is no way to evaluate the quality of a solution produced by a heuristic. On the other hand, an exact algorithm produces an optimal solution to a mathematical programming problem, albeit potentially at the expense of significant computational resources. Therefore, despite the fact that a practitioner may be in favor of developing and applying heuristics for genuine practical reasons, research and development of the exact approach are critically essential since only an exact method can produce *the optimal solution* to the problem. Spending a bit of computational time and resources is advisable for the problems that intend to make strategic decisions, as the decision-maker would prefer getting a high-quality solution at the expense of longer computational time. Since such mission-critical strategic decisions may only need to be made once per month, quarter, year, or even in 5 years, the exact method would be the most effective approach. Moreover, an optimal solution (or near-optimal solution with a clearly stated optimality gap) provided by an exact method can be used as a benchmark to evaluate the efficiency of various heuristics that practitioners would like to implement.

Over the past few decades, numerous integer and mixed-integer (IP/MIP) models have been formulated and solved to address challenging problems. The computational challenges have motivated many OR/MS researchers to develop theoretically elegant exact algorithms to efficiently solve many classes of large-scale IP/MIP and combinatorial optimization problems by exploiting the problem's structure. In this chapter, we present the widely applied exact approach known as the Benders Decomposition (BD) [1]. The strength of BD lies in solving a complex model within an iterative framework by decomposing it into a relaxed master problem and one or multiple subproblems that are computationally simpler to solve.

In this chapter, we first introduce the BD framework with its mathematical foundation. Then, we illustrate an elaborate BD implementation by considering a simplified version of a research problem involving the strategic decision-making of a nonprofit organization working on food insecurity of underprivileged children in India [6]. Specifically, in the presence of both supply and demand related uncertainties, through this problem, we also illustrate the application of BD in solving a stochastic programming (SP) problem.

This chapter is organized as follows. We describe the BD framework in the next section. Then, we discuss a SP problem motivated by a real-life example under uncertainty and apply BD as the solution methodology. After which, using a small numerical example, we illustrate the BD scheme in detail, demonstrating the computations iteration-by-iteration. The chapter concludes with a discussion on some recent advances in BD implementation.

8.2 Benders Decomposition Framework

Many researchers have extensively applied BD to solve complex MIP problems. The underlying principle of BD is to fix certain *complicating decision variables* in the mathematical model so that the remaining problem becomes computationally simpler. However, since these complicating variables are *fixed*, and not actually *solved* for their optimal values, the quality of the feasible solution to the overall problem can be quite low. To this end, BD offers a mathematically rigorous iterative approach to gradually improve the overall problem's solution. For an in-depth understanding of the state-of-the-art BD methodology and its many nuances, we refer the reader to [9, 16]. Moreover, illustrations of BD application on facility location problems can be found in [7]. In this section, we provide a brief introduction to BD for beginners.

We present the classical BD algorithm below with a generic math model with a minimization objective. All boldfaced notations represent vectors. For any maximization problem, the objective function should be converted to a minimization type (by taking its negative). Consider the following MIP problem with continuous decision vector $\boldsymbol{y}$ and integer decision vector $\boldsymbol{x}$:

$$\begin{aligned} \text{Minimize} \quad & \boldsymbol{c}^T\boldsymbol{x} + \boldsymbol{d}^T\boldsymbol{y} \\ \text{subject to} \quad & \\ & \boldsymbol{G}^T\boldsymbol{x} = \boldsymbol{q} \\ & \boldsymbol{A}\boldsymbol{x} + \boldsymbol{B}\boldsymbol{y} \geq \boldsymbol{b} \\ & \boldsymbol{x} \geq \boldsymbol{0} \text{ and integer;} \quad \boldsymbol{y} \geq \boldsymbol{0}. \end{aligned}$$

The presence of $\boldsymbol{x}$ variables is likely to make the MIP computationally hard. Nonetheless, if we can fix $\boldsymbol{x}$ by the feasible values of its components, i.e., $\widehat{\boldsymbol{x}} \in \mathcal{F} = \{\boldsymbol{x} \in Z_+ : \boldsymbol{G}^T\boldsymbol{x} = \boldsymbol{q}\}$, the remaining model becomes a tractable LP (with only unknown $\boldsymbol{y}$) as follows:

$$\begin{aligned} \text{Minimize}_{\widehat{x}\in\mathcal{F}} \quad & \boldsymbol{c}^T\widehat{\boldsymbol{x}} + \boldsymbol{d}^T\boldsymbol{y} \\ \text{subject to} \quad & \\ & \boldsymbol{B}\boldsymbol{y} \geq \boldsymbol{b} - \boldsymbol{A}\widehat{\boldsymbol{x}} \\ & \boldsymbol{y} \geq \boldsymbol{0}. \end{aligned}$$

Note that the first term of the objective function becomes constant, and the remaining problem turns into an LP. We would call it the primal subproblem (PSP). By introducing dual vector $\boldsymbol{w}$, we obtain the dual of this PSP as a maximization problem (referred to as the dual subproblem or DSP) as follows:

$$\begin{aligned} \text{(DSP)} \quad \text{Maximize} \quad & \boldsymbol{w}^T(\boldsymbol{b} - \boldsymbol{A}\widehat{\boldsymbol{x}}) \\ \text{subject to} \quad & \\ & \boldsymbol{w}^T\boldsymbol{B} \leq \boldsymbol{d} \\ & \boldsymbol{w} \geq \boldsymbol{0}. \end{aligned}$$

Then, in accordance with the duality theory, we substitute the inner minimization problem with its dual to obtain the equivalent formulation:

$$\begin{aligned} \text{Minimize}_{\widehat{x}\in\mathcal{F}} \quad & \boldsymbol{c}^T\widehat{\boldsymbol{x}} + \text{Max}\{\boldsymbol{w}^T(\boldsymbol{b} - \boldsymbol{A}\widehat{\boldsymbol{x}})\} \\ \text{subject to} \quad & \\ & \boldsymbol{w}^T\boldsymbol{B} \leq \boldsymbol{d} \\ & \boldsymbol{w} \geq \boldsymbol{0}. \end{aligned}$$

The above formulation is a nonlinear min-max problem. To obtain its linear reformulation, consider $\mathcal{P}$ to be the set of all extreme points in the feasible region of DSP, and $\mathcal{D}$ to be the set of extreme directions. Note that the feasible region of DSP is independent of the choice of $\boldsymbol{x} \in \mathcal{F}$; however, it impacts the objective function of DSP. Therefore, fixing $\boldsymbol{x}$ can result in one of the two distinct outcomes: (a) an optimal solution at an extreme point $\boldsymbol{w}^{(p)}$, with $p \in \mathcal{P}$, or (b) unbounded solution along an extreme direction $\boldsymbol{r}_d$, where $d \in \mathcal{D}$. Since we wish to restrict all movements along the directions of unboundedness, we should theoretically add one cut for each extreme ray r_d, i.e., the following cut set to the overall (i.e., the outer minimization) problem: $0 \geq \boldsymbol{r}_d^T(\boldsymbol{b} - \boldsymbol{A}\widehat{\boldsymbol{x}})$, $\forall d \in \mathcal{D}$. Then only case (a) is possible for the inner problem. Therefore, the optimal solution must be obtained at an extreme point $p \in \mathcal{P}$. Thus, the overall problem becomes

$$\min_{\widehat{\boldsymbol{x}} \in \mathcal{F}} \boldsymbol{c}^T\widehat{\boldsymbol{x}} + \max_{p \in \mathcal{P}} \{\boldsymbol{w}^{(p)T}(\boldsymbol{b} - \boldsymbol{A}\widehat{\boldsymbol{x}})\}$$

subject to

$$0 \geq \boldsymbol{r}_d^T(\boldsymbol{b} - \boldsymbol{A}\widehat{\boldsymbol{x}}) \qquad \forall d \in \mathcal{D}$$

Since the objective function is not linear due to the presence of the "*minimax*" term, we linearize it by introducing an auxiliary continuous variable $\rho \geq 0$, and adding an additional set of cuts for each extreme point of the DSP polyhedron. Finally, the overall problem with the original constraint involving $\boldsymbol{x}$, all the cuts corresponding to the extreme points $p \in \mathcal{P}$ and extreme directions $d \in \mathcal{D}$, is represented as follows:

$$\begin{aligned}
&\text{Minimize } \boldsymbol{c}^T\boldsymbol{x} + \rho \\
&\text{subject to} \\
&\qquad \boldsymbol{G}^T\boldsymbol{x} = \boldsymbol{q} \\
&\qquad \rho \geq \boldsymbol{w}^{(p)T}(\boldsymbol{b} - \boldsymbol{A}\boldsymbol{x}) && \forall p \in \mathcal{P} \\
&\qquad 0 \geq \boldsymbol{r}_d^T(\boldsymbol{b} - \boldsymbol{A}\boldsymbol{x}) && \forall d \in \mathcal{D} \\
&\qquad \rho > 0, \boldsymbol{x} \geq \boldsymbol{0} \text{ and integer.}
\end{aligned}$$

The above representation is known as the Benders Master Problem (BMP). The two sets of cuts corresponding to all the extreme points and extreme directions are known as the *optimality* and *feasibility* cuts, respectively. Theoretically, these cut sets can be exponentially large, making this BMP representation impractical for implementation. Therefore, in the BD algorithm, we begin with a *Relaxed Master Problem* (RMP) which does not contain any of these cuts. Then, using an iterative scheme, we gradually add these cuts to the RMP. Fortunately, in many real-world problems, the algorithm converges to the optimal solution only after adding a small subset of these optimality and feasibility cuts. We discuss next the iterative approach in BD using the above general model and explain the convergence mechanism. In the subsequent section, we apply BD on a specific problem.

8.2.1 The Iterative Approach of BD (Minimization-Type Problem)

- **Initialization**: Every iteration of BD requires the calculation of bounds. In order to accomplish this, we define four terms: *currentLB, currentUB, best LB*, and *bestUB*. Initially, we set: *currentLB* = *bestLB* = $-\infty$, and *currentUB* = *bestUB* = $+\infty$. In addition, the RMP has empty sets of extreme points and extreme directions, i.e., $\mathcal{P} = \mathcal{D} = \phi$.
- **Solving master problem and obtaining lower bound**: At the beginning, RMP provides a *relaxed* version of the original model with only $\boldsymbol{G}^T\boldsymbol{x} = \boldsymbol{q}$ constraints, integer decision vector $\boldsymbol{x}$, and a single continuous variable ρ. Therefore, upon solving the *minimization* problem RMP, its objective function value Z_{MP} provides the lower bound for the overall problem. Therefore, we update *currentLB* $\leftarrow Z_{MP}$. Observe that due to the addition of new cut(s) to the RMP, the *currentLB* value will never get worse than the previous in any BD iteration. Thus, *bestLB* $\leftarrow$ *currentLB*.
- **Solving subproblem and obtaining upper bound**: Now, we plug the RMP solution into DSP as parameter $\widehat{\boldsymbol{x}}$ and solve. Depending on whether we obtain a finite optimum at an extreme point or some extreme direction, we generate the respective optimality or feasibility cut. This cut would be added to the RMP in the next iteration. However, prior to the next BD iteration, the current and best upper bound values are compared. Note that an upper bound can be obtained from a feasible solution to the overall problem. Thus, we obtain: *currentUB* = $\boldsymbol{c}^T\widehat{\boldsymbol{x}} + Z_{DSP}(\boldsymbol{w}|\widehat{\boldsymbol{x}})$. If *currentUB* < *bestUB*, then update: *bestUB* $\leftarrow$ *currentUB*. In contrast to *bestLB*, the value *bestUB* may not be updated on every iteration.
- **Calculating optimality gap**: After updating these bounds, we determine: Optimality gap $= 100 \times$ (*bestUB* $-$ *bestLB*)/*bestLB*.
- **Terminating condition**: The BD iterations are terminated based on any (or some combination) of the following conditions:

  1. *MaxIter*, the maximum number of iterations, or
  2. Optimality gap % $\leq \epsilon$, predetermined threshold ($\epsilon \geq 0$).

8.3 An Application

In this section, we present a realistic problem adapted from [6] where mathematical programming is used to address a strategic problem faced by a nonprofit organization (NPO) working to alleviate food insecurity issues among children living in poverty. We begin with a discussion of the NPO's decision-making problem, followed by a presentation of the decomposition-based solution approach and an explanation of the step-by-step details using a small numerical example.

8.3.1 Problem Description

The NPO envisages a project of social awareness building among school-aged children regarding the serious impacts of food insecurity and malnutrition on the poor children of society. The project aims at strategically connecting a set of pre-identified slums to several upper primary and high schools, generally attended by students from affluent families. The project requires each participating school to adopt at least one slum for a full academic year, with students donating a certain amount of uncooked food items periodically. Then, student representatives accompanied by teaching staff would visit the adopted slum(s) to disburse the collected food. The NPO expects that through this *experiential learning exercise*, students will be exposed to the pressing issue of food insecurity, children in the adopted slums will receive regular nourishment, and participating schools will receive positive exposure that will strengthen their goodwill. A three-tier network comprising schools (donor nodes), slums (beneficiary nodes), and an intermediate tier of candidate warehouses has been considered to model this problem. One or more of these candidate warehouses may get "activated" if a need arises to temporarily store the donated food before distributing.

Various practical considerations
The NPO must pay attention to various practical concerns for the success of their project. The first condition is obvious: every participating school must be assigned at least one slum for its students to adopt; otherwise, why would they participate? In fact, a school with large student enrollment may adopt multiple slums. However, having multiple donors can invite complexities for a slum. Research on nonprofit operations management (e.g., Berenguer and Shen [2]) advises strongly against engaging multiple donors to a single beneficiary or a single project to avoid potential conflicts among donors regarding credit sharing, accounting complexities, etc. Therefore, while a large school may adopt multiple slums, a slum cannot be adopted by more than one school.

Consider the repercussions of this "one sponsor for one slum" restriction. For certain combinations of donor capacity and beneficiary demand values, a situation can arise where several slums may not be allocated to any school, despite the fact that several schools have leftover donation capacities. Such (in)action would further aggravate the system-wide shortage, which could otherwise be avoided. Since food insecurity alleviation is the primary goal of this project, to prevent avoidable shortages, the intermediate warehouse tier plays an important role. The warehouse(s), by serving as a *common donation pool*, allows schools to send excess quantities there. In other words, originally the donation from each school node was "earmarked" for some specific slum(s). Now, a portion of these outflows can be allocated to these pools as "non-earmarked" donations. The accumulated quantities can be supplied from these warehouses to multiple slums without conflicting with the "one donor" requirement. This way, the warehouse layer provides the much-needed flexibility to reduce any avoidable shortage across the system.

We present a small illustration of 3-school and 4-slum network in Fig. 8.1. Note that total supply = 70 + 130 + 150 = 350 meals (or meal equivalent) and total

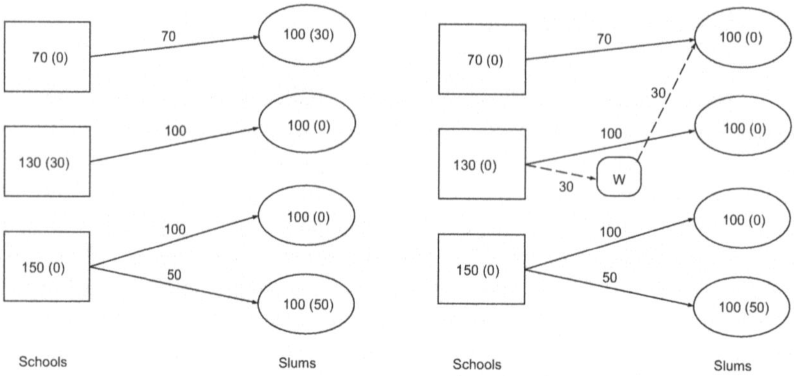

(a) No warehouse: shortage = 80 meals. (b) With warehouse: shortage = 50 meals.

Fig. 8.1 Donation flows in systems without and with warehouse. Numbers in parentheses are excess for schools and shortage for slums

demand = 100 + 100 + 100 + 100 = 400 meals, thus, having an unavoidable shortage of (400–350) = 50 meals in the system. However, the system without a warehouse (see Fig. 8.1a) experiences a shortage of 80 meals, even as school 2 has 30 meals worth capacity remaining (shown in parenthesis), which cannot be donated to any slum as it would violate the "one donor" condition. Observe in Fig. 8.1b, addition of a single warehouse (with adequate receiving capacity) provides a way of receiving that excess 30 units from school 2, thereby preventing the avoidable shortage.

Figure 8.2 is a schematic representation of the entire problem, including the parameters and decision variables that we describe in our explanation. The sets of participating donor schools, beneficiary slums, and candidate warehouses are represented by I, J, and K, respectively. We take into account the real-world uncertainties that may arise with regard to the donation capacities of individual schools and demands at the slums. To address these uncertainties, Dalal [6] adopts a two-stage stochastic programming (SP) framework in which a discrete scenario set Ω is considered. The occurrence probability of a scenario ω is p_ω, such that $\sum_{\omega \in \Omega} p_\omega = 1$. Now, in a specific scenario $\omega \in \Omega$, a school $i \in I$ assumes a supply capacity of $S_{i\omega}$ (number of meal equivalent) and a slum $j \in J$ realizes a demand of $D_{j\omega}$ (in terms of meal equivalent). We assume Q_k, the storage capacity of a warehouse $k \in K$ to be unaffected by scenarios. Except for $S_{i\omega}$ and $D_{j\omega}$, all other model parameters are deterministic. Since a detailed discussion on stochastic programming is beyond the scope of this chapter, we refer the reader to any related reference book on SP (e.g., Prékopa [15], Kall et al. [12], Birge and Louveaux [3]).

We consider g_k to be the fixed cost of opening and operating a warehouse at $k \in K$. As strategic connection costs, the model accounts for the time and effort spent engaging donor schools and maintaining donation collection, stockpiling, and distribution activities throughout the year. In particular, for $i \in I$, $j \in J$ and $k \in K$, we use f_{ij}, f_{ik} and f_{kj} to represent the annual strategic connection costs between

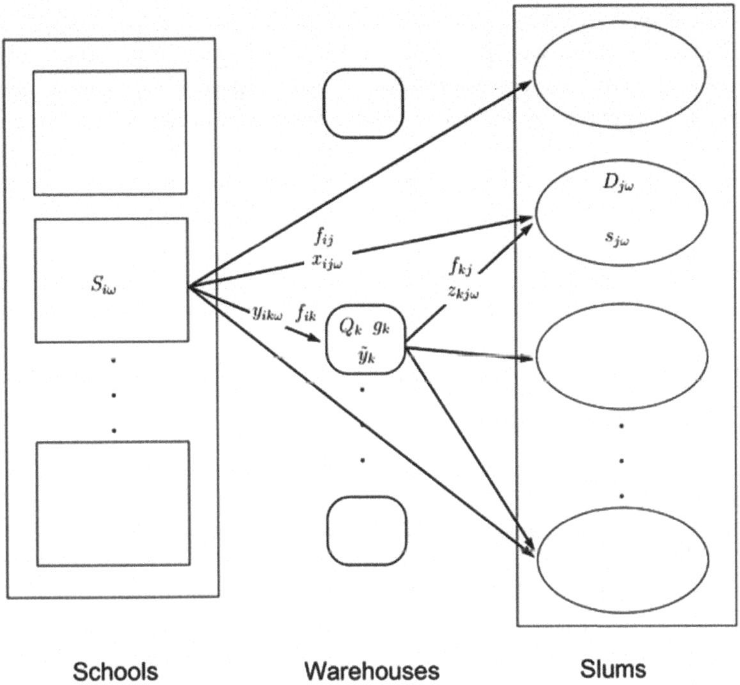

Fig. 8.2 Schematic diagram of the network

node pairs (i, j), (i, k), and (k, j). Note that these costs differ from the unit costs used in the classical transportation problem, as they are incurred regardless of the quantity of donations transported between tiers as long as a strategic connection is established. Thus, the distribution cost becomes a fixed cost in this problem. In the real world, total demand often overwhelms total supply, so shortages are inevitable in such a system. We consider a unit shortage cost Π_j, charged as a penalty for failing to meet unit demand at $j \in J$.

The indirect connection costs f_{ik} and f_{kj} are set to a higher value than f_{ij} to discourage indirect flows via warehouse(s), since the overarching motivation of the project is to involve the school students. In a similar vein, the following restriction has been imposed on how a slum's demand can be satisfied: we want to avoid a situation where a slum is not adopted by a school, but its demand is still being met by indirect flows from warehouse(s). Allowing it would defeat the primary purpose of engaging the school students in this program. Hence, we place a hard constraint to prevent such flows. Consequently, a slum faces one of the following three outcomes: (i) it receives only direct flow from one school, (ii) it receives a mix of one direct flow (from a school that adopts it) and one or more indirect flows from warehouse(s), and (iii) no donation inflow.

The initial stage of SP represents decisions that are independent of scenarios. In practice, such decisions are made prior to the unfolding of an uncertain scenario. These decisions are therefore referred to as "here and now" and can be strategic in nature. The second stage involves scenario-dependent decisions, i.e., "recourse" actions (tactical or operational decisions) corresponding to the scenario that is realized. The objective function of our SP model comprises sum of the first-stage strategic cost and the expected value of the second-stage cost. We consider sum of the warehouse opening and strategic connection costs (between (i, j), (i, k), and (k, j) pairs) as the first-stage cost. For the second stage, we take into account the expected value of the shortage penalty. The mathematical model with all constraints is presented below.

8.3.2 Model Formulation

Decision variables

A list of all decision variables for the MIP model is presented below.

$\tilde{y}_k$ 1, if warehouse is opened at $k \in K$, 0 otherwise
u_{ij} 1, if school $i \in I$ adopts slum $j \in J$ (i.e., direct supply along (i, j)), 0 otherwise
v_{ik} 1, if school $i \in I$ sends donation to warehouse $k \in K$, 0 otherwise
w_{kj} 1, if warehouse $k \in K$ sends donation to slum $j \in J$, 0 otherwise
$x_{ij\omega}$ donation flow from school $i \in I$ to slum $j \in J$ in scenario $\omega \in \Omega$
$y_{ik\omega}$ donation flow from school $i \in I$ to warehouse $k \in K$ (if open) in scenario $\omega \in \Omega$
$z_{kj\omega}$ flow from warehouse $k \in K$ (if open) to slum $j \in J$ in scenario $\omega \in \Omega$
$s_{j\omega}$ shortage at slum $j \in J$ in scenario $\omega \in \Omega$

We formulate a two-stage stochastic programming model as follows:

$$\min Z = \left[\sum_{k \in K} g_k \tilde{y}_k + \sum_{i \in I} \sum_{j \in J} f_{ij} u_{ij} + \sum_{i \in I} \sum_{k \in K} f_{ik} v_{ik} + \sum_{k \in K} \sum_{j \in J} f_{kj} w_{kj} \right] + \sum_{\omega \in \Omega} p_\omega \left[\sum_{j \in J} \Pi_j s_{j\omega} \right] \tag{8.1}$$

subject to

$$\sum_{j \in J} u_{ij} \geq 1 \quad \forall i \in I \tag{8.2}$$

$$\sum_{i \in I} u_{ij} \leq 1 \quad \forall j \in J \tag{8.3}$$

$$\sum_{j \in J} z_{kj\omega} = \sum_{i \in I} y_{ik\omega} \quad \forall k \in K, \omega \in \Omega \tag{8.4}$$

$$\sum_{j\in J} x_{ij\omega} + \sum_{k\in K} y_{ik\omega} \le S_{i\omega} \qquad \forall i \in I, \omega \in \Omega \tag{8.5}$$

$$\sum_{i\in I} x_{ij\omega} + \sum_{k\in K} z_{kj\omega} + s_{j\omega} \ge D_{j\omega} \qquad \forall j \in J, \omega \in \Omega \tag{8.6}$$

$$x_{ij\omega} \le M_{ij}^{(1)} u_{ij} \qquad \forall i \in I, j \in J, \omega \in \Omega \tag{8.7}$$

$$y_{ik\omega} \le M_{ik}^{(2)} v_{ik} \qquad \forall i \in I, k \in K, \omega \in \Omega \tag{8.8}$$

$$z_{kj\omega} \le M_{kj}^{(3)} w_{kj} \qquad \forall k \in K, j \in J, \omega \in \Omega \tag{8.9}$$

$$v_{ik} \le \tilde{y}_k \qquad \forall i \in I, k \in K \tag{8.10}$$

$$w_{kj} \le \tilde{y}_k \qquad \forall k \in K, j \in J \tag{8.11}$$

$$\sum_{i\in I} y_{ik\omega} \le Q_k \tilde{y}_k \qquad \forall k \in K, \omega \in \Omega \tag{8.12}$$

$$\sum_{k\in K} w_{kj} \le |K| \sum_{i\in I} u_{ij} \qquad \forall j \in J \tag{8.13}$$

$$\tilde{y}_k, u_{ij}, v_{ik}, w_{kj} \in \{0, 1\};\ x_{ij\omega}, y_{ik\omega}, z_{kj\omega}, s_{j\omega} \ge 0 \quad \forall i \in I, j \in J, k \in K, \omega \in \Omega. \tag{8.14}$$

The big-M terms are replaced as follows: $M_{ij}^{(1)} = \min\{D_{j\omega}, S_{i\omega}\}$, $M_{ik}^{(2)} = \min\{Q_k, S_{i\omega}\}$, and $M_{kj}^{(3)} = \min\{Q_k, D_{j\omega}\}$.

The objective function (8.1) minimizes the sum of first-stage strategic cost and the expected second-stage cost due to shortage in the system. Constraints (8.2) ensure that at least one slum is assigned to each participating school. The condition of each slum having no more than one adopting school is captured by constraints (8.3). Flow balance at warehouses (if opened) in each scenario is established by (8.4). Supply and demand requirements under each scenario $\omega \in \Omega$ are presented by (8.5) and (8.6), respectively. Constraints (8.7)–(8.9) connect the flow variables $x_{ij\omega}$, $y_{ij\omega}$, and $z_{ij\omega}$, with their corresponding binary variables using big M's. Similarly, forcing constraints (8.10)–(8.11) ensure that any school-to-warehouse and warehouse-to-slum connection can only be established if the warehouse itself is opened. Constraint (8.12) represents warehouse capacity. Constraints (8.13) capture an important condition underlying goal of this project. For any slum j, if there is no direct flow from any school i (i.e., $u_{ij} = 0$), then the right-hand side of (8.13) becomes zero, dictating all $w_{kj} = 0$ for that j. In other words, a slum cannot obtain its supply solely from one or more warehouses unless there is a direct donor. Note that if $w_{kj} = 0$, constraints (8.4) and (8.9) together ensure the zero indirect flow.

8.3.3 The Benders Decomposition Scheme

Large instances of the MIP model (8.1)–(8.14) are computationally difficult to solve due to the presence of binary decision variables ($\tilde{y}_k, u_{ij}, v_{ik}, w_{kj}$). Therefore, an efficient solution approach is required to solve large-scale instances. Observe that if we obtain feasible values for $\tilde{y}_k, u_{ij}, v_{ik}, w_{kj}$, the remaining model becomes an LP problem that can be solved efficiently by a standard solver. Therefore, to gain the computational advantage, we exploit the problem structure by decomposing the model into easier subproblems. In the line of our explanation for the BD algorithm in the previous section, we directly write below the master problem (MP) with all binaries and auxiliary variables $\rho_\omega \geq 0, \forall \omega \in \Omega$.

$$\text{Min } Z_{MP} = \sum_{k \in K} g_k \tilde{y}_k + \sum_{i \in I} \sum_{j \in J} f_{ij} u_{ij} + \sum_{i \in I} \sum_{k \in K} f_{ik} v_{ik} + \sum_{k \in K} \sum_{j \in J} f_{kj} w_{kj} + \sum_{\omega \in \Omega} \rho_\omega \tag{8.15}$$

subject to

$$\sum_{j \in J} u_{ij} \geq 1 \qquad \forall i \in I \tag{8.16}$$

$$\sum_{i \in I} u_{ij} \leq 1 \qquad \forall j \in J \tag{8.17}$$

$$v_{ik} \leq \tilde{y}_k \qquad \forall i \in I, k \in K \tag{8.18}$$

$$w_{kj} \leq \tilde{y}_k \qquad \forall k \in K, j \in J \tag{8.19}$$

$$\sum_{k \in K} w_{kj} \leq |K| \sum_{i \in I} u_{ij} \qquad \forall j \in J \tag{8.20}$$

(All the Benders optimality and feasibility cuts)

$$\tilde{y}_k, u_{ij}, v_{ik}, w_{kj} \in \{0, 1\};\ \rho_\omega \geq 0 \qquad \forall i \in I, j \in J, k \in K, \omega \in \Omega. \tag{8.21}$$

The subproblem is *separable* over scenarios $\omega \in \Omega$ as follows:

$$[\text{PSP}]\ \text{Min} \quad Z^{(\omega)}_{PSP}(x_{ij(\omega)}, y_{ik(\omega)}, z_{kj(\omega)}, s_{j(\omega)} | \widehat{\tilde{y}}_k, \widehat{u}_{ij}, \widehat{v}_{ik}, \widehat{w}_{kj}) = p_{(\omega)} \sum_{j \in J} \Pi_j s_{j(\omega)} \tag{8.22}$$

subject to

$$\sum_{j \in J} z_{kj(\omega)} = \sum_{i \in I} y_{ik(\omega)} \qquad \forall k \in K \tag{8.23}$$

$$\sum_{j \in J} x_{ij(\omega)} + \sum_{k \in K} y_{ik(\omega)} \leq S_{i(\omega)} \qquad \forall i \in I \tag{8.24}$$

$$\sum_{i \in I} x_{ij(\omega)} + \sum_{k \in K} z_{kj(\omega)} + s_{j(\omega)} \geq D_{j(\omega)} \qquad \forall j \in J \tag{8.25}$$

$$x_{ij(\omega)} \leq M^{(1)}_{ij} \widehat{u}_{ij} \qquad \forall i \in I, j \in J \tag{8.26}$$

$$y_{ik(\omega)} \leq M^{(2)}_{ik} \widehat{v}_{ik} \qquad \forall i \in I, k \in K \tag{8.27}$$

$$z_{kj(\omega)} \leq M^{(3)}_{kj} \widehat{w}_{kj} \qquad \forall k \in K, j \in J \tag{8.28}$$

$$\sum_{i \in I} y_{ik(\omega)} \leq Q_k \widehat{\tilde{y}}_k \qquad \forall k \in K \tag{8.29}$$

$$x_{ij(\omega)}, y_{ik(\omega)}, z_{kj\omega}, s_{j(\omega)} \geq 0 \qquad \forall i \in I, j \in J, k \in K, \omega \in \Omega. \tag{8.30}$$

With dual variables $\alpha_k^{(\omega)}$ (unrestricted), $\beta_i^{(\omega)}$ (≤ 0), $\gamma_j^{(\omega)}$ (≥ 0), $\delta_{ij}^{(\omega)}$ (≤ 0), $\phi_{ik}^{(\omega)}$ (≤ 0), $\psi_{kj}^{(\omega)}$ (≤ 0), and $\mu_k^{(\omega)}$ (≤ 0) corresponding to the PSP constraints, we write the dual subproblem as

$$\begin{aligned}
\text{[DSP] Max} Z_{DSP}^{(\omega)}(\alpha_k^{(\omega)}, \beta_i^{(\omega)}, \gamma_j^{(\omega)}, \delta_{ij}^{(\omega)}, \phi_{ik}^{(\omega)}, \psi_{kj}^{(\omega)}, \mu_k^{(\omega)} | \widehat{\tilde{y}}_k, \widehat{u}_{ij}, \widehat{v}_{ik}, \widehat{w}_{kj}) \\
-\sum_{i \in I} S_{i(\omega)} \beta_i^{(\omega)} \\
+\sum_{j \in J} D_{j(\omega)} \gamma_j^{(\omega)} + \sum_{i \in I}\sum_{j \in J} M_{ij}^{(1)} \widehat{u}_{ij} \delta_{ij}^{(\omega)} + \sum_{i \in I}\sum_{k \in K} M_{ik}^{(2)} \widehat{v}_{ik} \phi_{ik}^{(\omega)} \\
+\sum_{k \in K}\sum_{j \in J} M_{kj}^{(3)} \widehat{w}_{kj} \psi_{kj}^{(\omega)} + \sum_{k \in K} Q_k \widehat{\tilde{y}}_k \mu_k^{(\omega)}
\end{aligned} \tag{8.31}$$

subject to

$$\alpha_k^{(\omega)} + \gamma_j^{(\omega)} + \psi_{kj}^{(\omega)} \leq 0 \qquad \forall k \in K, j \in J \tag{8.32}$$

$$-\alpha_k^{(\omega)} + \beta_i^{(\omega)} + \phi_{ik}^{(\omega)} + \mu_k^{(\omega)} \leq 0 \qquad \forall i \in I, k \in K \tag{8.33}$$

$$\beta_i^{(\omega)} + \gamma_j^{(\omega)} + \delta_{ij}^{(\omega)} + \kappa_{ij}^{(\omega)} \leq 0 \qquad \forall i \in I, j \in J \tag{8.34}$$

$$\gamma_j^{(\omega)} \leq p_{(\omega)} \Pi_j \qquad \forall j \in J \tag{8.35}$$

$$\beta_i^{(\omega)}, \delta_{ij}^{(\omega)}, \phi_{ik}^{(\omega)}, \psi_{kj}^{(\omega)}, \mu_k^{(\omega)} \leq 0; \quad \gamma_j^{(\omega)} \geq 0; \quad \alpha_k^{(\omega)} \text{ unrestricted} \qquad \forall i \in I, j \in J, k \in K. \tag{8.36}$$

Thus, the Benders optimality cut resulting from solving DSP(ω) is

$$\begin{aligned}
\rho_\omega \geq \sum_{i \in I} S_{i(\omega)} \widehat{\beta}_i^{(\omega)} + \sum_{j \in J} D_{j(\omega)} \widehat{\gamma}_j^{(\omega)} + \sum_{i \in I}\sum_{j \in J} M_{ij}^{(1)} \widehat{\delta}_{ij}^{(\omega)} u_{ij} + \sum_{i \in I}\sum_{k \in K} M_{ik}^{(2)} \widehat{\phi}_{ik}^{(\omega)} v_{ik} \\
+\sum_{k \in K}\sum_{j \in J} M_{kj}^{(3)} \widehat{\psi}_{kj}^{(\omega)} w_{kj} + \sum_{k \in K} Q_k \widehat{\mu}_k^{(\omega)} \tilde{y}_k, \quad \forall \omega \in \Omega.
\end{aligned}$$

Note that in our model, for any feasible MP solution vector $(\tilde{y}_k, u_{ij}, v_{ik}, w_{kj})$, the PSP does not become infeasible. This is because the PSP is feasible even if all variables $\tilde{y}_k, u_{ij}, v_{ik}, w_{kj}$ are zero, as shortage is permitted in constraints (8.25). Thus, Benders feasibility cuts are not required in this problem. Nevertheless, for the sake of completeness, we present the feasibility cut below:

$$0 \geq \sum_{i \in I} S_{i(\omega)} \widehat{\beta}_i^{(\omega)} + \sum_{j \in J} D_{j(\omega)} \widehat{\gamma}_j^{(\omega)} + \sum_{i \in I}\sum_{j \in J} M_{ij}^{(1)} \widehat{\delta}_{ij}^{(\omega)} u_{ij} + \sum_{i \in I}\sum_{k \in K} M_{ik}^{(2)} \widehat{\phi}_{ik}^{(\omega)} v_{ik}$$

$$+\sum_{k\in K}\sum_{j\in J} M_{kj}^{(3)}\widehat{\psi}_{kj}^{(\omega)} w_{kj} + \sum_{k\in K} Q_k\widehat{\mu}_k^{(\omega)}\tilde{y}_k, \quad \forall\omega\in\Omega.$$

In our current implementation of BD, cuts are generated for each scenario. The cuts generated by this method are known as the *disaggregated cuts*. On the other hand, aggregated cuts are those that are generated by combining all the scenarios. Whether to generate aggregated or disaggregated cuts, must be decided judiciously. Disaggregated cuts are typically too many and may slow down solving the RMP. On the contrary, aggregated cuts may be weak. For our MIP model (8.1)–(8.14), we summarize the Benders decomposition approach as a flowchart in Fig. 8.3.

8.3.4 An Illustrative Example

We present a small instance of our model (8.1)–(8.14) as an illustration. Various parameters related to donor nodes (schools), beneficiary nodes (slums), and candidate warehouse nodes are presented in Tables 8.1, 8.2 and 8.3, respectively.

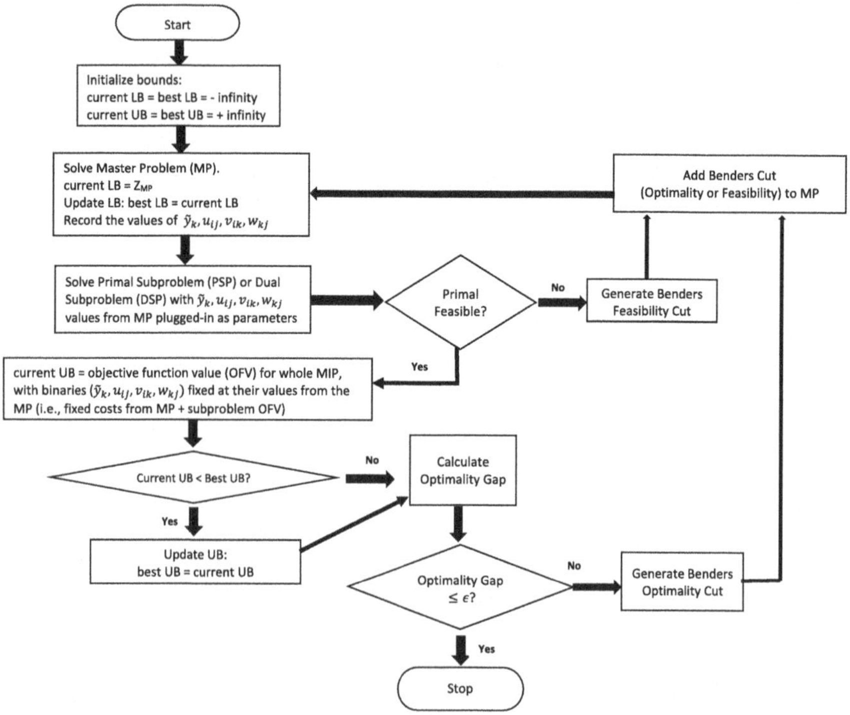

Fig. 8.3 Flow chart of the classic Benders decomposition for MIP model (8.1)–(8.14)

Table 8.1 School node data

School ID	x	y	Capacity ($\omega = 1$)	Capacity ($\omega = 2$)
1	12	26	15000	9900
2	78	37	18000	9500

Table 8.2 Slum node data

Slum ID	x	y	Demand ($\omega = 1$)	Demand ($\omega = 2$)
1	57	80	18000	11000
2	52	63	14400	7000

Table 8.3 Candidate warehouse node data

Warehouse ID	x	y	Fixed cost	Capacity
1	23	56	5000	2000
2	64	32	15000	5000

Using the aforementioned parameters, and considering direct and indirect unit transportation costs as 100 and 150, respectively, we construct the MIP model as presented below.

$$\begin{aligned}\text{Min } Z = {} & 7100u_{1,1} + 4800u_{2,1} + 5500u_{1,2} + 3700u_{2,2} + 50s_{1,1} + 50s_{1,2} + 50s_{2,1} \\ & + 50s_{2,2} + 4800v_{1,1} + 7950v_{1,2} + 8850v_{2,1} + 2250v_{2,2} + 6300w_{1,1} \\ & + 4500w_{1,2} + 7350w_{2,1} + 5100w_{2,2} + 5000\tilde{y}_1 + 15000\tilde{y}_2\end{aligned}$$

subject to

$$
\begin{aligned}
& u_{1,1} + u_{2,1} \leq 1 \\
& u_{1,2} + u_{2,2} \leq 1 \\
& u_{1,1} + u_{1,2} \geq 1 \\
& u_{2,1} + u_{2,2} \geq 1 \\
& z_{1,1,1} + z_{1,2,1} - y_{1,1,1} - y_{2,1,1} = 0 \\
& z_{1,1,2} + z_{1,2,2} - y_{1,1,2} - y_{2,1,2} = 0 \\
& z_{2,1,1} + z_{2,2,1} - y_{1,2,1} - y_{2,2,1} = 0 \\
& z_{2,1,2} + z_{2,2,2} - y_{1,2,2} - y_{2,2,2} = 0 \\
& y_{1,1,1} + y_{1,2,1} + x_{1,1,1} + x_{1,2,1} \leq 15000 \\
& y_{1,1,2} + y_{1,2,2} + x_{1,1,2} + x_{1,2,2} \leq 9900 \\
& y_{2,1,1} + y_{2,2,1} + x_{2,1,1} + x_{2,2,1} \leq 18000 \\
& y_{2,1,2} + y_{2,2,2} + x_{2,1,2} + x_{2,2,2} \leq 9500 \\
& z_{1,1,1} + z_{2,1,1} + x_{1,1,1} + x_{2,1,1} + s_{1,1} \geq 18000 \\
& z_{1,1,2} + z_{2,1,2} + x_{1,1,2} + x_{2,1,2} + s_{1,2} \geq 11000 \\
& z_{1,2,1} + z_{2,2,1} + x_{1,2,1} + x_{2,2,1} + s_{2,1} \geq 14400 \\
& z_{1,2,2} + z_{2,2,2} + x_{1,2,2} + x_{2,2,2} + s_{2,2} \geq 7000 \\
& x_{1,1,1} \leq 15000u_{1,1} \\
& x_{1,1,2} \leq 9900u_{1,1} \\
& x_{1,2,1} \leq 14400u_{1,2} \\
& x_{1,2,2} \leq 7000u_{1,2} \\
& v_{1,1} \leq \tilde{y}_1 \\
& v_{1,2} \leq \tilde{y}_2 \\
& v_{2,1} \leq \tilde{y}_1 \\
& v_{2,2} \leq \tilde{y}_2 \\
& w_{1,1} \leq \tilde{y}_1 \\
& w_{1,2} \leq \tilde{y}_1 \\
& w_{2,1} \leq \tilde{y}_2
\end{aligned}
$$

$$
\begin{aligned}
& x_{2,1,1} \leq 18000u_{2,1} \\
& x_{2,1,2} \leq 9500u_{2,1} \\
& x_{2,2,1} \leq 14400u_{2,2} \\
& x_{2,2,2} \leq 7000u_{2,2} \\
& y_{1,1,1} \leq 2000v_{1,1} \\
& y_{1,1,2} \leq 2000v_{1,1} \\
& y_{1,2,1} \leq 5000v_{1,2} \\
& y_{1,2,2} \leq 5000v_{1,2} \\
& y_{2,1,1} \leq 2000v_{2,1} \\
& y_{2,1,2} \leq 2000v_{2,1} \\
& y_{2,2,1} \leq 5000v_{2,2} \\
& y_{2,2,2} \leq 5000v_{2,2} \\
& z_{1,1,1} \leq 2000w_{1,1} \\
& z_{1,1,2} \leq 2000w_{1,1} \\
& z_{1,2,1} \leq 2000w_{1,2} \\
& z_{1,2,2} \leq 2000w_{1,2} \\
& z_{2,1,1} \leq 5000w_{2,1} \\
& z_{2,1,2} \leq 5000w_{2,1} \\
& z_{2,2,1} \leq 5000w_{2,2} \\
& z_{2,2,2} \leq 5000w_{2,2} \\
& w_{2,2} \leq \tilde{y}_2 \\
& y_{1,1,1} + y_{2,1,1} \leq 2000\tilde{y}_1 \\
& y_{1,1,2} + y_{2,1,2} \leq 2000\tilde{y}_1 \\
& y_{1,2,1} + y_{2,2,1} \leq 5000\tilde{y}_2 \\
& y_{1,2,2} + y_{2,2,2} \leq 5000\tilde{y}_2 \\
& w_{1,1} + w_{2,1} \leq 2u_{1,1} + 2u_{2,1} \\
& w_{1,2} + w_{2,2} \leq 2u_{1,2} + 2u_{2,2}
\end{aligned}
$$

In Table 8.4, we present our BD implementation's iterative progress on this small instance.

8.4 Modern Approaches of BD Implementation

We conclude this chapter by highlighting different approaches that researchers have taken over time to effectively apply BD in solving their research problems. While the decomposition scheme discussed in this chapter has theoretical significance for understanding the essence of this exact method, the implementation is fraught with numerous practical challenges. We discuss here a few acceleration schemes suggested in the literature for overcoming some of these challenges. Thereafter, we discuss a few modern approaches of implementing BD by employing the state-of-the-art CPLEX solver [5]. Due to the rapid evolution of implementation techniques in professional solvers, we strongly advise readers unfamiliar with hands-on BD implementation to consult the most recent user guidelines for the solver they intend to use.

Table 8.4 Iteration-wise progress of BD algorithm

Iter #	Best LB	Best UB	Gap (%)	Benders cuts added (two per iteration, one for each ω)
1	10300 ($u_{1,2} = 1, u_{2,1} = 1$)	85300	87.92	$750000\, u_{1,1} + 720000\, u_{1,2} + 900000\, u_{2,1} + 720000\, u_{2,2} + 100000\, w_{1,1} + 100000\, w_{1,2} + 250000\, w_{2,1} + 250000\, w_{2,2} + \rho_1 \geq 1620000$
				$495000\, u_{1,1} + 350000\, u_{1,2} + 475000\, u_{2,1} + 350000\, u_{2,2} + 100000\, w_{1,1} + 100000\, w_{1,2} + 250000\, w_{2,1} + 250000\, w_{2,2} + \rho_2 \geq 900000$
2	19800 ($\tilde{y}_1 = 1, u_{1,2} = 1,$	85300	76.79	$750000\, u_{1,1} + 900000\, u_{2,1} + 100000\, w_{1,1} + 250000\, w_{2,1} + \rho_1 \geq 900000$
	$u_{2,1} = 1, w_{1,2} = 1$)			$495000\, u_{1,1} + 475000\, u_{2,1} + 100000\, w_{1,1} + 250000 w_{2,1} + \rho_2 \geq 550000$
3	21600 ($\tilde{y}_1 = 1, u_{1,2} = 1, u_{2,1} = 1, w_{1,1} = 1$)	85300	74.68	$\rho_1 \geq 0$ $495000\, u_{1,1} + 350000\, u_{1,2} + 100000\, v_{1,1} + 250000\, w_{2,1} + 250000\, w_{2,2} + \rho_2 \geq 425000$
4	26400 ($\tilde{y}_1 = 1, u_{1,2} = 1, u_{2,1} = 1, v_{1,1} = 1, w_{1,1} = 1$)	26400	0	

The BD algorithm for this application problem has been implemented using Visual C++, and the callable library of CPLEX 22.1.0.0 is used to solve the LPs and IPs. The code and dataset for the project can be accessed from https://github.com/faizhamid0705/Benders_FoodProblem.

8.4.1 *Various Acceleration Schemes in BD*

In the previous section we have demonstrated how the traditional BD implementation solves the RMP and DSP sequentially in an iterative fashion. This approach often does not perform satisfactorily since the RMP must be constructed and solved from scratch in each iteration. Moreover, as the RMP gradually grows in size due to the accumulation of cuts (which can be of optimality or feasibility type), the computational time typically increases as the iterations progress. To address this issue, researchers have proposed several algorithmic improvements. Since a comprehensive discussion of all such methods is beyond the scope of this chapter, the reader is encouraged to consult the state-of-the-art literature (such as Rahmaniani et al. [16]) for an in-depth understanding. Here, we provide only a brief overview of various acceleration schemes.

Solving the RMP at each iteration can be very expensive if the original problem has a large number of integer variables. Several alternatives have been proposed in the literature to reduce the computational time to solve the RMP. Geoffrion and Graves [10] do not solve the RMP to optimality, rather stop once an incumbent solution is found which has value below ($bestUB - \epsilon$). Since the RMP contains insufficient information about SP costs in the early iterations, it is not worthwhile to strictly optimize the RMP. The BD procedure terminates if the current master problem cannot produce an integer solution with value below ($bestUB - \epsilon$). In their study, McDaniel and Devine [14] propose a modification to the classical BD in which the LP relaxation of the RMP is solved in a number of initial iterations. The resulting solution to RMP (possibly fractional) is then utilized in SP to generate a new cut for RMP. This cut is valid for the RMP as it corresponds to one of the extreme points of DSP. The idea is to generate many necessary cuts to the RMP by solving LPs rather than IPs. The RMP is solved as LP until no further iterations are possible (in which case the LP optimal solution to RMP has been found via Benders' partitioning), and is then solved as IP.

The example problem considered in Sect. 8.3.4 exhibits degeneracy. The authors obtained alternate optimal solutions for the DSPs within a few iterations while testing the problem instance separately in CPLEX interactive optimizer and Concert Technologies (Java-CPLEX API). Many network optimization problems have inherent degeneracy leading to alternate optimal solutions. Although these solutions have the same objective value of the DSP, they have the potential to generate different cuts with varying strengths. The reader may consult Van Roy [21], Wentges [22], Üster and Agrahari [19], and Üster and Dalal [20] for various cut strengthening schemes. One may be interested in identifying the solution among the many alternate optimal solutions of the DSP that will produce the cut not dominated by other cuts. The reader may refer to Magnanti and Wong [13] to find a method to judiciously select the optimal solution of the DSP that generates the Pareto-optimal cuts. Obviously, this will incur overhead of solving an additional LP in every iteration. Conforti and Wolsey [4] proposed a method to generate facet-defining BD cuts. The aforementioned techniques strengthen the optimality cuts. Rei et al. [18] develop a local branching procedure to strengthen feasibility cuts and accelerate the classical BD algorithm. Fischetti et al. [8] propose a cut selection criterion when both feasibility and optimality violated cuts exist. A recent approach, called adaptive Benders cuts, combines single-cuts and multicuts strategies to achieve faster initial iterations and overall convergence [17].

8.4.2 BD Implementation with CPLEX Solver

The following discussion pertains to CPLEX, the state-of-the-art optimization package from IBM [5]. CPLEX has evolved its BD-related functionalities and the Concert Technologies (APIs), specifically from version 12.4. While engaging CPLEX, we first categorize (possibly not exhaustive) different BD-implementation strategies.

- **Cutting Plane Benders**: This is the standard implementation, as outlined in the previous section (see Fig. 8.3). Here, the programmer is responsible for every aspect of implementation, e.g., modeling the master and subproblems, ensuring correct looping, cut generation, and termination conditions, etc. Thus, this is a fully manual approach.

- **Branch-and-cut Benders**: In this approach, the RMP is not fully optimized from scratch in every iteration. There are two methods to implement this approach.

  1. **Lazy constraint-based callback**: This CPLEX support, possibly introduced with version 12.4 of CPLEX, was very effective at addressing the typical computational challenges encountered during manual BD implementation. Now, the master problem is solved just once. As soon as an incumbent solution is found for the MP, the corresponding Benders cut is generated, added to the MP, and the branch-and-bound tree-search resumes thereafter. A possible issue with the computational performance of this legacy callback of CPLEX may be that it does not permit the use of multiple threads, even on multi-core computers. It also disables some CPLEX features such as dynamic search and presolve. Due to this, the MP may not be solved efficiently. In this feature, the Benders cut-generation part of the code is invoked automatically.
  2. **Generic callback**: CPLEX introduced generic callback in version 12.7, allowing the BD implementation to take advantage of multiple threads. Similar to the lazy callback, the MP and SP must be meticulously modeled, and the programmer must also ensure that multiple threads do not collide at runtime; therefore, this strategy would require advanced programming skills.

- **Built-in BD support**: Beginning with CPLEX 12.7, a built-in Benders support has been introduced. According to the CPLEX documentation: "*Given a formulation of a problem, CPLEX can decompose the model into a single master and (possibly multiple) subproblems. To do so, CPLEX makes use of annotations that you supply for your model. This approach can be applied to mixed-integer linear programs (MILP). For certain types of problems, this approach can offer significant performance improvements as subproblems can be solved in parallel.*" Considering the echoes in the CPLEX log and the solver's evolution over the past years, it may be safe to assume that this built-in BD can take advantage of multi-threading, despite the fact that the authors did not find it explicitly stated in writing.
  Moreover, this BD-implementation feature allows the user different modes to seek CPLEX's support at varying levels, depending on the amount of effort required by CPLEX. The user can communicate with CPLEX in two ways: (i) through a specially structured XML file - known as the "*annotation*" file, and (ii) through appropriate syntax of the programming language (this is API-dependent). Here are two important modes for using this built-in support:

  1. In the USER mode, we would decide the decomposition scheme ourselves by explicitly specifying which variables would go to MP and which ones would go to SP. We would then instruct CPLEX to adhere to this USER-specified scheme.
  2. In the FULL mode, the user can avoid this decomposition effort by allowing CPLEX to perform all of the work. CPLEX puts all integer variables into the master and all continuous variables into a subproblem. Obviously, the FULL

mode is the easiest to implement (one need only add one or two lines of code), but may be challenging for CPLEX.

8.5 Conclusions

The chapter demonstrates classical BD procedure using a real-life problem. Several practical implementation schemes and acceleration techniques have been discussed. It will be interesting to examine the various acceleration methods discussed in Sect. 8.4.1 on the example problem. To improve the efficiency of the solution procedure, it is essential to exploit the structure of the SP and solve it quickly using tailored algorithms. For the example problem, if the probabilities associated with each scenario are difficult to estimate, one may use Robust Optimization to deal with uncertainties.

References

1. Benders Jacques F (1962) Partitioning procedures for solving mixed-variables programming problems. Numerische mathematik 4(1):238–252
2. Berenguer G, Shen Z (2020) Challenges and strategies in managing nonprofit operations: an operations management perspective. Manuf Serv Oper Manag 22(5):888–905
3. Birge JR, Louveaux F (2011) Introduction to stochastic programming. Springer Science & Business Media
4. Conforti M, Wolsey Laurence A (2019) "Facet" separation with one linear program. Math Program 178:361–380
5. CPLEX (2022). https://www.ibm.com/analytics/cplex-optimizer. Accessed online: 19 February 2022
6. Dalal J (2022) A strategic donor-beneficiary assignment problem under supply and demand uncertainties. Int J Oper Res 43(1–2):188–207
7. Daskin Mark S (2011) Network and discrete location: Models, algorithms, and applications. Wiley, New York
8. Fischetti M, Salvagnin D, Zanette A (2010) A note on the selection of Benders' cuts. Math Program 124:175–182
9. Floudas Christodoulos A (1995) Nonlinear and mixed-integer optimization: fundamentals and applications. Oxford University Press
10. Geoffrion Arthur M, Graves Glenn W (1974) Multicommodity distribution system design by Benders decomposition. Manag Sci 20(5):822–844
11. INFORMS (2022) https://www.informs.org/Publications/INFORMS-Journals/INFORMS-Journal-on-Applied-Analytics. Accessed online: 19 February 2022
12. Kall P, Wallace SW, Kall P (1994) Stochastic programming, vol 6. Springer, Berlin
13. Magnanti Thomas L, Wong Richard T (1981) Accelerating Benders decomposition: algorithmic enhancement and model selection criteria. Oper Res 29(3):464–484
14. McDaniel D, Devine M (1977) A modified Benders' partitioning algorithm for mixed integer programming. Manag Sci 24(3):312–319
15. Prékopa A (2013) Stochastic programming, vol 324. Springer Science & Business Media
16. Ragheb R, Gabriel CT, Michel G, Walter R (2017) The Benders decomposition algorithm: a literature review. Eur J Oper Res 259(3):801–817
17. Ramírez-Pico C, Ljubić I, Moreno E (2023) Benders adaptive-cuts method for twostage stochastic programs. Transportation Science, 57(5):1252–1275

18. Rei W, Cordeau JF, Gendreau M, Soriano P (2009) Accelerating Benders decomposition by local branching. INFORMS J Comput 21(2):333–345
19. Üster H, Agrahari H (2011) A Benders decomposition approach for a distribution network design problem with consolidation and capacity considerations. Oper Res Lett 39(2):138–143
20. Üster H, Dalal J (2017) Strategic emergency preparedness network design integrating supply and demand sides in a multi-objective approach. IISE Trans 49(4):395–413
21. Van Roy TJ (1986) A cross decomposition algorithm for capacitated facility location. Oper Res 34(1):145–163
22. Wentges P (1996) Accelerating Benders' decomposition for the capacitated facility location problem. Math Methods Oper Res 44:267–290

Chapter 9
Delayed Column Generation: Solving Large-Scale Optimization Models From the Airline Industry

Adithya Patil and Milind G. Sohoni

Abstract Over the past several decades, the airline industry has widely used sophisticated large-scale optimization models and algorithms to improve operational efficiency, increase revenue, and improve profitability. This chapter discusses one such large-scale optimization technique called *Delayed Column Generation*. The approach is commonly used to solve instances of several challenging optimization models involving strategic business and operational processes. For example, the classical aircraft rotation set partitioning optimization model is routinely solved using delayed column generation. This chapter delves into the details of the delayed column generation and branch-and-price computational procedures. To illustrate their applicability in real-world airline models, we describe a simple application to the aircraft rotation model. Additionally, we provide a few references to articles describing other applications in the airline industry.

Keywords Large-scale optimization · Airline optimization · Column-generation · Set-partitioning models

9.1 Introduction

Managing airlines is a complex task and maintaining high efficiency during daily operations plays a major role in ensuring profitability, punctuality, and good service quality. The high demand for airline services, competition from rivals, and the expenses involved in operating an airline call for the implementation of sophisticated planning/operational models – for short-, medium-, and long-term decision-making. Since the 1950s, airline operators have been using operations research models to solve these complex problems. Over the past several decades these models have become much more sophisticated and with businesses growing rapidly the models have grown in scale.

A. Patil · M. G. Sohoni (✉)
Indian School of Business, Hyderabad 500111, India
e-mail: milind_sohoni@isb.edu

A. Patil
e-mail: adithya_patil@isb.edu

F. Hamid (ed.), *Optimization Essentials*, International Series in Operations Research & Management Science 353, https://doi.org/10.1007/978-981-99-5491-9_9

In this chapter, we present a commonly used approach for solving large-scale optimization problems. This algorithmic procedure is called the *delayed column generation* technique. We demonstrate the use of this algorithm in solving an important problem—the *aircraft routing and scheduling problem*. However, several other scheduling problems, such as the crew scheduling problem, are routinely solved using this technique.

We begin with a discussion of the set covering and partitioning models in Sect. 9.2. In the same section, we also describe the naive version of the branch-and-bound algorithm. Algorithms based on the branch-and-bound approach are routinely used to solve mixed integer programming formulations. We then present the delayed column generation technique to solve the set partitioning and covering models. The manner in which the delayed column generation technique is to be applied depends on the problem structure considered. We present two approaches for applying the column generation technique in Sect. 9.3. In Sect. 9.4, we present the Branch and Price algorithm that builds on the branch-and-bound and column generation approach to solve instances of large set covering and set partitioning problems. In particular, we discuss the Ryan-and-Foster branching strategy used in many aircraft routing and scheduling applications of Branch and Price. In Sect. 9.5, we discuss the application of column generation for the problem of optimal aircraft routing and scheduling. We conclude the chapter in Sect. 9.6.

9.2 Preliminaries

In this section we present the set partitioning and set covering formulations. We then describe the naive branch-and-bound algorithm.

9.2.1 Set Partitioning

A *partition* of a set is a grouping of the elements of the set into non-empty subsets such that each element only belongs to one subset. Thus, if P is a partition of a set X, the elements of P are a set of subsets of X that are mutually exclusive and collectively exhaustive. Many business problems can be modeled as set partitioning problems. For instance, consider the problem of assigning routes to delivery trucks where trucks need to pick up loads from a central warehouse, deliver them to customers around the vicinity of the warehouse, and return to the warehouse after the deliveries are completed. In this setting, the set X is the set of customers, and each element of the set P is the subset of customers serviced by a truck, i.e., the "routing" of a truck. The elements of set P are mutually exclusive, i.e., no truck shares a route (or customers) with another truck, and collectively exhaustive, i.e., all customers are serviced. Such a problem is an instance of a Set Partitioning Problem (SPP). The business problem is known as the Vehicle Routing Problem (VRP), and identifying a minimum-cost routing is equivalent to identifying a minimum-cost partition of the set X.

Formally, let X be a finite and non-empty set. Let P be a partition of X. Let S_j be the jth element (a subset of X) of P. Let c_j be the cost associated with the S_j and Ω be a set of all possible partitions of X. The SPP is to identify a partition P^*, where $P^* \in \Omega$, such that

$$P^* = \arg\min_{P \in \Omega} \left\{ \sum_{j \in P} c_j \right\}. \tag{9.1}$$

Note that the term $\sum_{j \in P} c_j$ is the cost associated with partition P. In matrix notation (9.1) can be expressed as follows. Consider a matrix $\mathbf{A}$ where column j represents subset j and row i represents element i. Matrix entry (a_{ij}) indicates if element i is included in subset j or not, i.e.,

$$a_{ij} = \left\{ \begin{matrix} 1: & \text{if } i \in j, \\ 0: & \text{Otherwise.} \end{matrix} \right\}. \tag{9.2}$$

$$\textbf{SPP:} \qquad \min \sum_j c_j x_j, \tag{9.3}$$

$$\text{s.t.} \qquad \mathbf{Ax} = \mathbf{1}, \qquad x_j \in \{0, 1\} \quad \forall\, j. \tag{9.4}$$

It is important to note that the column entries for each j, i.e., the column vector $\mathbf{A}_j$, and the number of columns depend on the problem modeled. We describe examples from the airline industry later in this chapter.

9.2.2 *Set Covering*

A closely related problem to the SPP is the Set Covering Problem (SCP). The SCP is as follows:

$$\textbf{SCP:} \qquad \min \sum_j c_j x_j, \tag{9.5}$$

$$\text{s.t.} \qquad \mathbf{Ax} \geq \mathbf{1}, \qquad x_j \in \{0, 1\} \quad \forall\, j. \tag{9.6}$$

Many business applications, such as scheduling problems and assignment problems, can be modeled as set partitioning or set covering problems.

9.2.3 Branch and Bound Algorithm with Simple Branching on Variables

The elementary version of the branch-and-bound (BB) essentially involves a *divide and conquer* procedure. Essentially, the BB procedure partitions the feasible region of the original problem into subregions and computes the optimal solution over each subregion separately. Suppose we are given the following mixed integer linear program (MILP) with a feasible region G.

$$\min_{x \in G} \mathbf{c}^T \mathbf{x}. \tag{9.7}$$

Consider a partition $G_1, \ldots G_k$ of G. Then (9.7) can be expressed as follows.

$$\min_{x \in G} \mathbf{c}^T \mathbf{x} = \min_{1 \leq i \leq k} \left\{ \min_{x \in G_i} \mathbf{c}^T \mathbf{x} \right\}.$$

In other words, the BB procedure optimizes over each partitioned subset separately. It builds on the fact that if the original problem cannot be solved directly because of its complexity, it may be possible to solve the divided subproblems recursively. This division into smaller subproblems is essentially called *branching*. It is noteworthy that a simple implementation of the BB may require enumerating all feasible integer solutions and can take arbitrarily long to solve even small instances of MILPs.

To reduce the computational effort, BB implementations employ a procedure called *pruning*. Suppose all decision variables are bounded above and below. Then any feasible solution to the MILP is an upper bound $U(G)$ on the optimal MILP objective value. Often, heuristics are employed in practice to obtain $U(G)$ efficiently. It is well-known that the feasible region of the linear program (LP) relaxation of a MILP is larger than that of the original MILP. Consequently, the LP relaxation optimal solution provides a lower bound on the optimal MILP objective value. Notice that, branching allows the BB procedure to compute a lower bound $l(G_i)$ on the optimal solution for every subregion $G(i)$. If it happens that for a particular subregion $l(G_i) \geq U(G)$, then the BB procedure doesn't need to explore the subproblem i any further (by branching) because it already has a better solution than any in partition G_i. Essentially, the LP relaxation of the subproblem provides a lower bound $l(G_i)$. The elimination of exploring possible solutions in a partition by creating an appropriate bound using the LP relaxation is called *Pruning*. *Bounding* iteratively finds *tighter* values of $l(G_i)$ and $U(G)$.

Branching. An elementary branching strategy is to select a fractional-valued variable k in the LP relaxation of a subproblem. The BB process then creates two further subproblems: in one branch (subproblem), the constraint $x_k \geq \lceil x_k \rceil$ is imposed, and in the other subproblem, the constraint $x_k \leq \lfloor x_k \rfloor$ is enforced. The selection of such a fractional-valued variable is called a *branching rule*. It is important to note that while branching creates two subproblems, it does not remove feasible integer solutions to the original problem. Essentially, the constraints generated during the branching process are *valid inequalities*.

Pruning. After branching the BB procedure recursively solves the two subproblems. If the LP relaxation optimal objective value is greater than the incumbent feasible upper bound solution to the MILP, the subproblem is pruned.
In summary, the basic steps of the BB algorithm are as follows.

1. Compute an upper bound U using a heuristic.
2. The original problem is put on a list of "candidates" to be solved.
3. In every iteration a problem P is selected from the candidate list and its LP relaxation is solved to compute a lower bound $l(P)$.
   (a) Prune P if (a) the LP relaxation results in an infeasible solution, or (b) $l(P) \geq U$.
   (b) Update the upper bound U if $l(P) < U$ and the solution is feasible to the original MILP.
   (c) Branch on P, i.e., create two subproblems using the branching rule and add them to the candidate list.
4. If the list of candidates is non-empty, go to step 3. Else, the procedure terminates.

It is noteworthy that there are multiple procedures to select a candidate in step 3. The *best-first* technique chooses a candidate with the lowest lower bound (optimal LP relaxation solution). Other techniques used include the*depth-first* (the most common technique) and the *breadth-first* technique. Both these techniques help in generating a feasible MILP solution relatively quickly. The reader is referred to [10] for details on the BB procedures.

Finally, we note that sometimes solving the LP relaxations of subproblems (which are done iteratively) on the candidate list can be challenging. For example, if the number of binary variables for an SPP-based formulation is extremely large, then solving the LP relaxation (multiple times in a BB framework) can be onerous. Many practical implementations of the BB procedure employ a technique called delayed column generation to solve such LP relaxations efficiently. Next, we discuss the delayed column generation technique.

9.3 Delayed Column Generation

Solving linear programs with a large number of variables (of exponential order or higher) poses many practical challenges like large memory requirements and long computational times. For such problem instances, the delayed column generation technique is an approach that is often used independently, or in conjunction with other algorithms.

Consider the following linear program, MP, where I and J are finite sets of indices, and $|J|$ is extremely large.

$$\textbf{MP:} \quad \min \sum_{j \in J} c_j \lambda_j, \tag{9.8}$$

$$\text{s.t.} \quad \sum_{j \in J} a_{ij} \lambda_j \geq b_i \quad \forall i \in I \quad \cdots (\pi_i), \tag{9.9}$$

$$\lambda_j \geq 0 \quad \forall j \in J. \tag{9.10}$$

Define $\Lambda = \langle \lambda_j,\ j \in J \rangle$ as the vector of decision variables $\lambda_j,\ j \in J$. The parameters $a_{ij},\ i \in I,\ j \in J$ represent the coefficients of the constraints of the linear program, while $b_i,\ i \in I$ represent the parameters on the right-hand side of the linear program. The variables $\pi_i,\ i \in I$ represent the dual multipliers of constraints (9.9). Constraints (9.10) represent the nonnegativity constraints. For notational ease, let $\mathbf{A}$ be a $|I| \times |J|$ matrix whose (i, j)th entry is a_{ij}, where $i \in I,\ j \in J$. Let $\mathbf{A}_j,\ j \in J$ refer to the jth column of matrix $\mathbf{A}$. Let $\mathbf{b} = \langle b_i,\ i \in I \rangle$ be the vector of the right-hand side parameters. Let $\Pi = \langle \pi_i,\ i \in I \rangle$ represent the vector of the dual multipliers $\pi_i,\ i \in I$.

The delayed column generation technique is essentially an iterative solution procedure to solve linear programs efficiently. The procedure begins with a restricted subset of columns of $\mathbf{A}$—called the *restricted* master problem. After solving this restricted master problem the procedure involves solving a suitable *subproblem* that identifies additional columns with lowest reduced cost with respect to the dual values generated in the master problem. If no additional columns with negative reduced costs are found (generated) then the procedure terminates with the optimal value of the primal and dual variables. Otherwise, the restricted master problem matrix $\mathbf{A}$ is appended with the newly generated columns, and the process is iterated with new primal and dual values. It is noteworthy that when the restricted master problem has too many columns after several iterations, columns with large reduced costs may be removed to keep future instances small. Below we detail the two main steps of the iterative delayed column generation procedure.

Solving the Restricted Master Problem. Consider the following restricted master problem (structurally identical to MP) which includes columns $J_{res} \subseteq J$ instead of J.

$$\textbf{RMP:} \quad \min \sum_{j \in J_{res}} c_j \lambda_j, \tag{9.11}$$

$$\text{s.t.} \quad \sum_{j \in J_{res}} a_{ij} \lambda_j \geq b_i \quad \forall i \in I \quad \cdots (\pi_i), \tag{9.12}$$

$$\lambda_j \geq 0 \quad \forall j \in J_{res}. \tag{9.13}$$

Suppose Λ^* and Π^* are RMP's optimal primal and dual solutions (vectors), respectively. Note that the primal optimal value for each $\lambda_j,\ j \in J - J_{res}$ is zero. However, we need to check if the solution at hand is also the optimal solution to MP, i.e., if the reduced costs of all variables $\lambda_j,\ j \in J$ are non-negative. $\bar{c}_j,\ j \in J$, the reduced cost of variable $\lambda_j,\ j \in J$ is computed as follows.

$$\bar{c}_j = c_j - \sum_{i \in I} \pi_i^* a_{ij} \geq 0,\ \forall j \in J_{res}. \tag{9.14}$$

If a solution to the RMP is not optimal to MP, then there exist some columns $j \in J - J_{res}$ such that $\bar{c}_j = c_j - \sum_{i \in I} \pi_i^* a_{ij} < 0,\ j \in J - J_{res}$. We need a systematic procedure to identify such a column $\mathbf{A}_j$, i.e,. values of $a_{ij}\ \ \forall\, i$.

Solving the Subproblem. Recall that our objective is minimization. Consequently, we have to identify a column $\mathbf{A}_j$ in matrix $\mathbf{A}$ such that the reduced cost of variable λ_j, i.e., $\bar{c}_j$, is minimal with respect to the optimal dual vector of the RMP, Π^*. That is, we have to solve the following problem of identifying z where

$$z = \min\ \ \{\bar{c}_j,\ j \in J\}. \tag{9.15}$$

Optimization problem (9.15) is referred to as a *subproblem*. In other words, we wish to identify j^* such that

$$j^* = \arg\min\ \ c_j - \sum_{i \in I} \pi_i^* a_{ij}. \tag{9.16}$$

Notice that our decision variables are entries of the $j*$th column, i.e., $|I|$ elements of $\mathbf{A}_{j^*}$. The entries of column $\mathbf{A}_{j^*}$ can be identified by solving the following optimization problem called SUB.

$$\textbf{SUB:} \qquad z = \min c_{j^*} - \sum_{i \in I} \pi_i^* a_{ij^*} \tag{9.17}$$

$$\text{subject to constraints on } a_{ij^*}, \forall\, i \in I, \text{ relevant to the problem context.} \tag{9.18}$$

Generating the minimal reduced cost column depends on the existence of an efficient algorithm to solve problem SUB. Often, this is application specific because of the constraints on y_i. We will look at some applications later where it is possible to solve for $\mathbf{A}_{j^*}$ efficiently. Solving SUB yields an optimal vector $\mathbf{A}_{j^*}$ with entries $a_{ij^*},\ i \in I$. If $z \geq 0$, then there is no column j^* such that $\bar{c}_{j^*} < 0$. This guarantees that we have an optimal solution to MP with respect to the dual vector Π^*. However, if $z < 0$ then we just generated a column with negative reduced cost and append it to the $\mathbf{A}$ matrix of RMP and iterate.

As mentioned earlier, the structure of SUB plays an important role in the manner by which the master problem is formulated. A well-known problem is a cutting stock problem [2]. In the case of the cutting stock problem, the subproblem is an integer knapsack problem. This is a well-studied problem and can be solved in pseudo-polynomial time using dynamic programming. Another class of problems is the capacitated vehicle routing problem with time windows for which the subproblem is a resource-constrained shortest path problem [5, 6].

9.3.1 Column Generation Using Dantzig-Wolfe Decomposition

Often the constraint structure of the master problem is more complex than the formulations described in Sect. 9.3. In such cases, the Datnzig-Wolfe decomposition approach to solve the LP may be useful. The column generation technique can then be applied to this reformulation. Consider a linear program (ORG) with two sets of constraints and decision variable vector $\mathbf{x}$. One set of constraints is the "complicating" set, while the other is the "easy" set.

$$\textbf{ORG:} \quad \min \ \mathbf{c}^{\mathrm{T}}\mathbf{x}, \tag{9.19}$$

$$\text{s.t.} \quad \mathbf{Ax} \geq \mathbf{b}, \quad \text{(Complicating constraints)} \tag{9.20}$$

$$\mathbf{Dx} \geq \mathbf{d}, \quad \text{(Easy constraints)} \tag{9.21}$$

$$\mathbf{x} \geq \mathbf{0}. \tag{9.22}$$

We assume that constraints (9.20) form the "complicating" set of constraints, and that constraints (9.21) form the "easy" constraints. Constraints (9.21) (the "easy" constraints) and constraints (9.22) represent a polyhedron. First, consider the "easy" constraints. Define the polyhedral feasible region $X = \{\mathbf{x} \geq \mathbf{0} \mid \mathbf{Dx} \geq \mathbf{d}\}$. Let Q be the set of extreme points of X, and let R be the set of extreme rays of X. Since X is a convex polyhedron, every element $\mathbf{x} \in X$ can be written in the following manner.

$$\mathbf{x} = \sum_{q \in Q} u_q \mathbf{x}_q + \sum_{r \in R} u_r \mathbf{x}_r, \tag{9.23}$$

where $\sum_{q \in Q} u_q = 1,\ u_q \geq 0,\ q \in Q$, and $u_r \geq 0,\ r \in R$. $\mathbf{x}_q,\ q \in Q$, is the qth extreme point of X and $\mathbf{x}_r,\ r \in R$ is the rth extreme ray of X.

Upon substituting (9.23) in the objective (9.19) and the "complicating" constraints (9.20), we get the following optimization problem DWRMP (or the Dantzig-Wolfe Reformulation Master Problem).

$$\textbf{DWRMP:} \quad \min \ \mathbf{c}^{\mathrm{T}}\left[\sum_{q \in Q} u_q \mathbf{x}_q + \sum_{r \in R} u_r \mathbf{x}_r\right], \tag{9.24}$$

$$\text{s.t.} \quad \mathbf{A}\left[\sum_{q\in Q} u_q\mathbf{x}_q + \sum_{r\in R} u_r\mathbf{x}_r\right] \geq \mathbf{b}, \tag{9.25}$$

$$\sum_{q\in Q} u_q = 1, \tag{9.26}$$

$$u_q \geq 0 \;\; \forall\, q \in Q, \tag{9.27}$$

$$u_r \geq 0 \;\; \forall\, r \in R. \tag{9.28}$$

Expanding the terms and grouping them together we get the following equivalent formulation.

$$\textbf{DWRMP:} \quad \min \sum_{q\in Q} u_q \underbrace{\mathbf{c}^T\mathbf{x}_q}_{c_q} + \sum_{r\in R} u_r \underbrace{\mathbf{c}^T\mathbf{x}_r}_{c_r}, \tag{9.29}$$

$$\text{s.t.} \quad \sum_{q\in Q} u_q \underbrace{\mathbf{A}\mathbf{x}_q}_{\mathbf{a}_q} + \sum_{r\in R} u_r \underbrace{\mathbf{A}\mathbf{x}_r}_{\mathbf{a}_r} \geq \mathbf{b}, \tag{9.30}$$

$$\sum_{q\in Q} u_q = 1, \tag{9.31}$$

$$u_q \geq 0 \;\; \forall\, q \in Q, \tag{9.32}$$

$$u_r \geq 0 \;\; \forall\, r \in R. \tag{9.33}$$

Thus, the final DWRMP model is as follows.

$$\textbf{DWRMP:} \quad \min \sum_{q\in Q} u_q c_q + \sum_{r\in R} u_r c_r, \tag{9.34}$$

$$\text{s.t.} \quad \sum_{q\in Q} u_q\mathbf{a}_q + \sum_{r\in R} u_r\mathbf{a}_r \geq \mathbf{b}, \qquad \cdots(\Pi) \tag{9.35}$$

$$\sum_{q\in Q} u_q = 1, \qquad \cdots(\pi_0) \tag{9.36}$$

$$u_q \geq 0 \;\; \forall\, q \in Q, \tag{9.37}$$

$$u_r \geq 0 \;\; \forall\, r \in R. \tag{9.38}$$

$u_q,\ q \in Q$ and $u_r,\ r \in R$ are the decision variables in DWRMP. Let Π be the dual multipliers (vector) associated with the constraint (9.35) and let π_0 be the dual multiplier associated with constraint (9.36). Note that DWRMP is equivalent to linear program ORG but has a large number of variables. Specifically, it has $|Q| + |R|$ variables.

Let DWRReMP be the restricted master problem of DWRMP where we consider a set $Q_{res} \subseteq Q$ and $R_{res} \subseteq R$ instead of Q and R respectively.

$$\textbf{DWRReMP:} \qquad \min \sum_{q \in Q_{res}} u_q c_q + \sum_{r \in R_{res}} u_r c_r, \tag{9.39}$$

$$\text{s.t.} \quad \sum_{q \in Q_{res}} u_q \mathbf{a}_q + \sum_{r \in R_{res}} u_r \mathbf{a}_r \geq \mathbf{b}, \qquad \cdots (\Pi) \tag{9.40}$$

$$\sum_{q \in Q_{res}} u_q = 1, \qquad \cdots (\pi_0) \tag{9.41}$$

$$u_q \geq 0 \;\; \forall \, q \in Q_{res}, \tag{9.42}$$

$$u_r \geq 0 \;\; \forall \, r \in R_{res}. \tag{9.43}$$

Solving DWRReMP gives us the optimal primal values of $u_q, \; q \in Q_{res}$ and $u_r, \; r \in R_{res}$, and the optimal dual values of Π and π_0. Note that $u_q = 0, \; q \in Q - Q_{res}$ and $u_r = 0, \; r \in R - R_{res}$ at optimality.

The subproblem is to identify either an extreme point $\mathbf{x}_q$ or an extreme ray $\mathbf{x}_r$ such that the reduced cost of the variables u_q or u_r is the lowest. The reduced cost expression for variable $u_q, \; q \in Q$ is as follows.

$$\bar{c}_q = c_q - \left[\Pi^T \; \pi_0\right] \begin{bmatrix} \mathbf{a}_q \\ 1 \end{bmatrix} = c_q - \Pi^T \mathbf{a}_q - \pi_0 = \mathbf{c}^T \mathbf{x}_q - \Pi^T \mathbf{A} \mathbf{x}_q - \pi_0. \tag{9.44}$$

Likewise, the reduced cost expression for variable $u_r, \; r \in R$ is as follows:

$$\bar{c}_r = c_r - \left[\Pi^T \; \pi_0\right] \begin{bmatrix} \mathbf{a}_r \\ 0 \end{bmatrix} = c_r - \Pi^T \mathbf{a}_r = \mathbf{c}^T \mathbf{x}_r - \Pi^T \mathbf{A} \mathbf{x}_r. \tag{9.45}$$

To identify an extreme point q or an extreme ray r with the lowest reduced cost, we find $\bar{c}^*$ where

$$\bar{c}^* = \min \left\{ \min_{q \in Q} \bar{c}_q, \min_{r \in R} \bar{c}_r \right\}. \tag{9.46}$$

Consequently, we have the following optimization problem (DWRSUB) as our subproblem.

$$\textbf{DWRSUB:} \qquad \min_{j \in Q \cup R} \; \mathbf{c}^T \mathbf{x}_j - \Pi^T \mathbf{A} \mathbf{x}_j. \tag{9.47}$$

Since Q and R contain the extreme points and rays of X, the optimization problem DWRSUB can be expressed as follows.

$$\textbf{DWRSUB:} \quad z_P^* = \min \left[\mathbf{c}^T - \Pi^T \mathbf{A}\right]\mathbf{y}, \tag{9.48}$$

$$\text{s.t.} \quad \mathbf{D}\mathbf{y} \geq \mathbf{d}, \tag{9.49}$$

$$\mathbf{y} \geq \mathbf{0}. \tag{9.50}$$

Here, DWRSUB is a linear program with "easy" constraints. The optimal vector $\mathbf{y}$ obtained after solving DWRSUB is either an extreme point or an extreme ray. After solving DWRSUB, we obtain z_P^*.

1. If $z_P^* = -\infty$, we identify an extreme ray $r^* \in R$ with $\bar{c}_{r^*} < 0$. We add variable u_{r^*} to the DWRReMP with cost coefficient $\mathbf{c}^T \mathbf{x}_{r^*}$ and column coefficient $\begin{bmatrix} \mathbf{a}_{r^*} \\ 0 \end{bmatrix} = \begin{bmatrix} \mathbf{A}\mathbf{x}_{r^*} \\ 0 \end{bmatrix}$ and solve DWRReMP.
2. If $-\infty < z_P^* - \pi_0 < 0$, we identify an extreme point $q^* \in Q$ with $\bar{c}_{q^*} < 0$. We add variable u_{q^*} to DWRReMP with cost coefficient $\mathbf{c}^T \mathbf{x}_{q^*}$ and column coefficients $\begin{bmatrix} \mathbf{a}_{q^*} \\ 0 \end{bmatrix} = \begin{bmatrix} \mathbf{A}\mathbf{x}_{q^*} \\ 1 \end{bmatrix}$ and solve DWRReMP.
3. If $z_P^* - \pi_0 \geq 0$, then there is no $j \in Q \cup R$ with $\bar{c}_j < 0$. This implies we have an optimal solution.

9.4 Branch and Price Algorithm

In this section we describe the Branch and Price (BP) algorithm that is often used to solve large-scale binary integer programming models. BP is essentially a branch-and-bound algorithm where the linear programming relaxation at every node is solved using delayed column generation. For example, for the SPP/SCP formulations described in Sect. 9.2, typical implementations of the BP algorithm solve the LP relaxation at each node using delayed column generation. Branching occurs when no new columns can be found, and the solution to the LP relaxation is not integral. An excellent review on BP is included in [1].

We describe the BP procedure for solving large SPP/SCP models. We focus on the important branching step because the bounding and pruning procedures are similar to routine implementations of the branch-and-bound algorithm. For SPP/SCP models branching on variables (by examining variables that take on fractional values after solving the LP relaxation of the SPP/SCP models at every) is not an efficient strategy because of the extremely large number of possible branches. Instead, we discuss another practically appealing branching rule described in [11]. This branching technique does not branch on variables, but instead, branches on the constraint set.

The Ryan-and-Foster Branching Strategy [11]. Consider the following LP relaxation of the SPP/SCP formulation described in Sect. 9.2.

$$\min \sum_{j \in J} c_j \lambda_j + M y_i, \tag{9.51}$$

$$\text{s.t.} \sum_{j \in J} a_{ij} \lambda_j + y_i = b_i \quad \forall i \in I, \tag{9.52}$$

$$0 \leq \lambda_j \leq 1 \quad \forall j \in J, \tag{9.53}$$

$$y_i \geq 0 \quad \forall i \in I. \tag{9.54}$$

Here, $c_j,\ j \in J$ is the cost associated with variable j, and variable y_i is a slack variable to ensure feasibility. The term $My_i,\ i \in I$, where M is a large value, ensures that $y_i,\ i \in I$ is kept as low as possible at optimality. The parameters $a_{ij},\ i \in I,\ j \in J$ are either 0 or 1. The parameters $b_i,\ i \in I$ are the right-hand sides of the first set of constraints. The decision variables $\lambda_j,\ j \in J$ are bounded above and below by 1 and 0 respectively.

It is noteworthy that in any fractional solution to the linear program there must exist at least one pair of constraints (say i_1 and i_2) such that

$$0 < \sum_{j \in J(i_1, i_2)} \lambda_j < 1, \tag{9.55}$$

where $J(i_1, i_2) \subseteq J$ are the set of variables that cover constraints i_1 and i_2 simultaneously. This allows us to partition the feasible region into two branches by introducing the following valid constraints.

$$\sum_{j \in J(i_1, i_2)} \lambda_j \leq 0 \qquad \text{(0-branch), and} \tag{9.56}$$

$$\sum_{j \in J(i_1, i_2)} \lambda_j \geq 1 \qquad \text{(1-branch).} \tag{9.57}$$

The 0-branch states that i_1 and i_2 must not be covered together (one does not follow the other), while the 1-branch states that i_1 and i_2 must be covered together (one follows the other). Notice that these two constraints describe a partition and are added to the linear programs for each of the branches in the branch-and-bound tree. The linear programs are then solved using delayed column generation.

It is also noteworthy that since the problem has a finite number of constraints, there exist only finitely many combinations of two constraints; hence, the BP algorithm terminates after a finite number of branch-and-bound nodes.

9.5 Aircraft Routing and Scheduling for Airline Operations

To understand how the column generation algorithm is used to develop optimal schedules for aircraft we consider a simplified version of the problem described in [4]. Specifically, consider the following simplified version of the daily aircraft routing/scheduling problem. Suppose there is only one fleet type, and all aircraft must return to a specific station (airport) at the end of the day. For example, such a station could be a maintenance station and all aircraft undergo inspection at the end of the day. The objective is to maximize the airline's profit by operating these n aircraft over a network of N daily flights providing service to multiple destinations. Essentially, we need to develop schedules for individual aircraft that maximize the airline's profitability and ensure that every flight in the (daily) network is assigned a unique aircraft. An aircraft schedule (also called rotation) is a feasible sequence of flights operated by an aircraft such that the rotation begins and ends at the "base" station. Daily rotations can then be repeated every day of the week.

Figure 9.1 is a schematic representation of a daily flight network. Flights $(f_1, \ldots, f_N)$ are represented as nodes and directed arcs as feasible connections, i.e., an arc connects flight i to j if (a) the destination airport of flight i is the same as the origin airport of j, and (b) the departure time of j is after the scheduled arrival of flight i and accounts for the aircraft turn-around time. There are two special nodes designated as s and t. The set of directed arcs originating at s connect to flights that depart from the base station. Similarly, directed arcs connect all flights that terminate at the base station to node t. Notice that a s-t path in this simple flight network represents a feasible aircraft schedule (rotation), originating and terminating at the base station. Even in a reasonably sized flight network (say a few hundred flights), the number of feasible rotations can be extremely large. Let $\mathcal{R}$ represent the set of all feasible aircraft rotations.

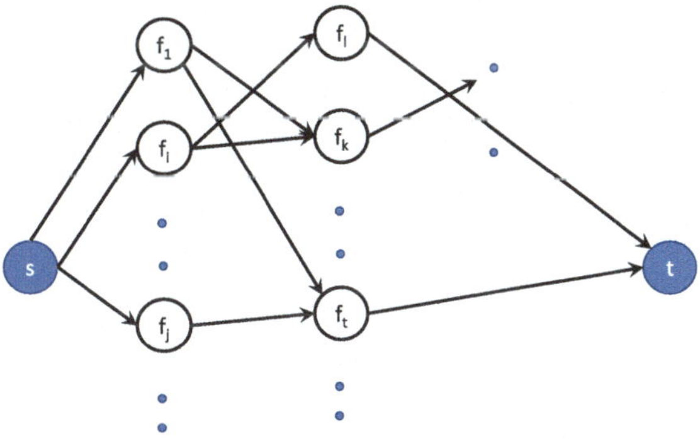

Fig. 9.1 A representative flight network

The SPP Formulation for the Routing Problem. Consider any feasible rotation R_j, i.e., a path in the daily flight network. Define

$$a_{ij} = \begin{cases} 1 : \text{if flight } i \in R_j, \text{ and} \\ 0 : \text{if flight } i \notin R_j. \end{cases} \tag{9.58}$$

Let p_i denote the operating profit of flight i, and c_j represent the profitability of rotation R_j. Then

$$c_j = \sum_{i=1}^{N} p_i a_{ij} \quad \forall\, j \in \mathcal{R}. \tag{9.59}$$

Let us define the binary decision variables x_j such that they take a value 1 if rotation j is chosen, and 0 otherwise. The simplified SPP version of the daily aircraft routing model (ARM) is expressed as follows.

$$\text{ARM:} \qquad \max \sum_{j \in \mathcal{R}} c_j x_j, \tag{9.60}$$

$$\text{s.t.} \quad \sum_{j \in \mathcal{R}} a_{ij} x_j = 1, \ \forall\, i = 1, \ldots, N, \tag{9.61}$$

$$\sum_{j \in \mathcal{R}} x_j = n, \tag{9.62}$$

$$x_j \in \{0, 1\} \quad \forall\, j \in \mathcal{R}. \tag{9.63}$$

Cover constraints (9.61) ensure that every flight is covered by exactly one feasible rotation, and the count constraint (9.62) makes sure that exactly n rotations are chosen from all the possible feasible rotations. Constraint (9.63) restricts our decision variables to binary values. The values α_i, $i = 1, \ldots, N$, and β are the dual values corresponding to the cover and count constraints when the LP relaxation of ARM is solved.

As mentioned the set $\mathcal{R}$ can be arbitrarily large even for medium-sized flight networks. In many real-world applications, instances of ARM are solved using the BP approach. Doing so further necessitates using column generation to solve the LP relaxations at each branch-and-bound node. We first consider a restricted master problem (R-ARM) structurally similar to ARM but has much fewer rotations, i.e., columns. Let the subset of the rotations considered in R-ARM be denoted by $\mathcal{R}_r$. The LP relaxation of R-ARM is modeled as follows.

$$\text{R-ARM:} \quad \max \sum_{j \in \mathcal{R}_r} c_j x_j, \tag{9.64}$$

$$\text{s.t.} \quad \sum_{j \in \mathcal{R}_r} a_{ij} x_j = 1, \ \forall i = 1, \ldots, N \quad \cdots (\alpha_i) \tag{9.65}$$

$$\sum_{j \in \mathcal{R}_r} x_j = n, \quad \cdots (\beta) \tag{9.66}$$

$$x_j \in [0, 1] \ \ \forall j \in \mathcal{R}_r. \tag{9.67}$$

The values $\alpha_i, i = 1, \ldots, N$, and β are the dual values corresponding to the cover and count constraints. The marginal profit (reduced cost) of any feasible rotation j is

$$\bar{c}_j = c_j - \sum_{i=1}^{N} \alpha_i a_{ij} - \beta \quad j \in \mathcal{R}. \tag{9.68}$$

Using column generation to solve R-ARM requires the subproblem to generate a new column in $\mathcal{R}$ (rotation in this case) that has the largest value of $\bar{c}_j$. To do so, consider the network in Fig. 9.1 with the following costs: (i) every directed arc from s to a flight node f_i is assigned cost $-\beta$, and (ii) every outbound directed arc from the flight node i is assigned a cost $-\alpha_i$. Recollect that, from equation (9.59), c_j is the sum of the individual profits of flights included in rotation j. It is readily observable that computing the marginal profit corresponding to any rotation j, $j \in \mathcal{R}$, with respect to the optimal dual values of R-ARM, is the same as finding the *longest path*, in the flight network. Given that the flight network is a directed acyclic graph, this problem can be transformed to a shortest path problem, which is solvable in polynomial time. If the longest path has a positive marginal profit, we have a new column to append to R-ARM. However, if the longest path has negative marginal profit, we have the optimal solution to the LP relaxation.

It is noteworthy that if more resource constraints (such as time) are added to ARM (consequently, to R-ARM) then the subproblem in the column generation procedure is modeled as a shortest path problem with resource constraints [5, 6]. The classical crew scheduling problem in airline operations [5, 8] is one such instance of a scheduling problem where the subproblem is a shortest path problem with resource constraints (SPPRC). The critical resource is the crew duty period (time) that needs to be computed along the paths generated on the flight network. It is important to note that SPRRC is NP-Hard. However, there are efficient algorithms to find good solutions in reasonable time [for example, see 7, 9].

9.6 Conclusions

In this chapter, we reviewed algorithmic techniques to solve large-scale mixed integer formulations frequently encountered in the airline industry. The column generation and Branch and Price algorithms are two techniques, among several others, frequently used to solve large and complex problem instances encountered in airline operations. To illustrate the solution mechanism, we introduced the aircraft rotation problem and demonstrated how to apply the technique to solve the model. The literature in this area is vast and with computational power increasing many folds, very large problem instances are routinely solved close to optimality. The algorithmic procedures in these, and related areas, have evolved considerably [8], and today large airlines solve integrated models to get better and more robust operational solutions. However, airline operations remain a vibrant area of research.

9.7 Exercise

A small regional airline operates 11 daily flights. The flights operate between stations A, B, C, D, and E. The airline has 3 small aircraft. Table 9.1 lists the daily profit (₹) associated with each flight in the network.

Figure 9.2 shows the possible connections between the daily flights. Direct arcs between the flight nodes represent the connections. The two special nodes (colored) represent each daily rotation's origin and termination stations (in this case station A). A directed path from the origin node (A) to the end node (A) represents an aircraft rotation.

Table 9.1 Flight profits

Flight	Expected profit in lacs (Rupee)
A–B	2
A–C	1.5
A–D	2
B–E	2
B–A	1
E–C	1.5
E–D	2
C–A	1.5
C–E	1.5
D–A	2
D–B	3

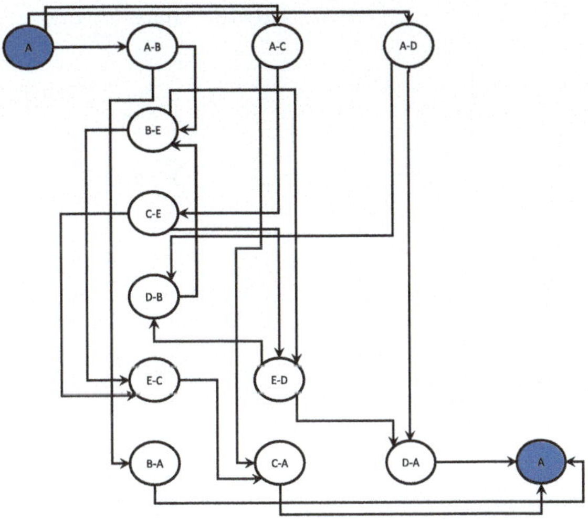

Fig. 9.2 The flight connection network

Table 9.2 Initial rotations for the restricted master problem

Flight	A_1	A_2	A_3	A_4	A_5	A_6	A_7
A–B	1	0	0	1	0	0	1
A–C	0	1	0	0	1	0	0
A–D	0	0	1	0	0	1	0
B–E	0	0	0	1	0	0	1
B–A	1	0	0	0	0	1	0
E–C	0	0	0	1	1	0	0
E–D	0	0	0	0	0	0	1
C–A	0	1	0	1	1	0	0
C–E	0	0	0	0	1	0	0
D–A	0	0	1	0	0	0	1
D–B	0	0	0	0	0	1	0

The goal is to choose exactly 3 rotations (paths) in the connection network to maximize the total profit. The profit of a path is the sum of the profits of all flights included along the path. Suppose we start with the 7 columns (rotations) $A_1, \ldots, A_7$ shown in Table 9.2. A '1' in the column represents the rotation covering the particular flight.

Let x_j represent the binary decision of whether we choose rotation j, c_i be the profit corresponding to operating rotation j, and a_{ij} indicate if flight i is covered by rotation j. The restricted master problem (ERMP) can be expressed as follows.

$$\text{ERMP:} \quad \max \sum_j c_j x_j, \tag{9.69}$$

$$\text{s.t.} \quad \sum_j x_j = 3, \tag{9.70}$$

$$\sum_j a_{ij} x_j = 1, \tag{9.71}$$

$$x_j \in \{0, 1\}. \tag{9.72}$$

Write down the subproblem to generate a new column (rotation) for the LP relaxation of ERMP. You may use any optimization software or MS Excel to solve for the relevant costs.

References

1. Barnhart C, Johnson EL, Nemhauser GL, Savelsbergh MW, Vance PH (1998) Branch-and-price: Column generation for solving huge integer programs. Oper Res 46(3):316–329
2. Ben Amor H, Valério de Carvalho J (2005) Cutting stock problems. In: Column generation. Springer, Berlin, pp 131–161
3. Desaulniers G, Desrosiers J, Dumas Y, Marc S, Rioux B, Solomon MM, Soumis F (1997) Crew pairing at Air France. Eur J Oper Res 97(2):245–259
4. Desaulniers G, Desrosiers J, Dumas Y, Solomon MM, Soumis F (1997) Daily aircraft routing and scheduling. Manag Sci 43(6):841–855
5. Desaulniers G, Desrosiers J, Ioachim I, Solomon MM, Soumis F, Villeneuve D (1998) A unified framework for deterministic time constrained vehicle routing and crew scheduling problems. Springer, Berlin
6. Desrosiers J, Dumas Y, Solomon MM, Soumis F (1995) Time constrained routing and scheduling. Handbooks Oper Res Manag Sci 8:35–139
7. Irnich S, Villeneuve D (2006) The shortest-path problem with resource constraints and k-cycle elimination for k ≥ 3. INFORMS J Comput 18(3):391–406
8. Klabjan D (2005) Large-scale models in the airline industry. In: Column generation. Springer, Berlin, pp 163–195
9. Lozano L, Duque D, Medaglia AL (2016) An exact algorithm for the elementary shortest path problem with resource constraints. Transp Sci 50(1):348–357
10. Nemhauser GL, Wolsey LA (1988) Integer and combinatorial optimization. Wiley
11. Ryan DM, Foster BA (1981) An integer programming approach to scheduling. In: Computer scheduling of public transport urban passenger vehicle and crew scheduling, pp 269–280
12. Sohoni MG (2019) Introduction to optimization. In: Essentials of business analytics. Springer, Berlin, pp 337–380

Chapter 10
Column Generation and Lagrangian Relaxation: Solving Real-World Crew (Re)Scheduling Problems

Twan Dollevoet and Dennis Huisman

Abstract In this chapter, we describe the combination of column generation and Lagrangian relaxation to solve a large-scale optimization problem. In the setting of a minimization problem, Lagrangian relaxation is used to compute lower bounds. Column generation is applied to deal with the huge number of columns. We apply these techniques to the crew scheduling and the crew rescheduling problem. Crew scheduling is an important planning problem for public transport companies and airlines. It has been studied by many Operations Researchers over the last few decades. The most complex problems arise at large public transport companies where 10000s of tasks have to be assigned to 1000s of crew members. Usually, the crew scheduling problem is solved once a year. During the year and in particular on the day of operations when disruptions occur, duties are rescheduled. We look at both the tactical planning phase as well as real-time crew rescheduling on the day of operations. In the tactical planning phase, a computation time of several hours is acceptable and it is important to find near-optimal solutions. We discuss the combination of Lagrangian relaxation and column generation to find a lower bound on the optimal objective value, and a Lagrangian heuristic to find good feasible solutions. When a disruption occurs, it is important to react quickly. Therefore, computation times should be in the order of a few minutes. Thus, heuristic approaches are often used for the real-time crew rescheduling problem. We discuss the combination of large neighborhood search with column generation and Lagrangian relaxation to solve a real-world crew rescheduling problem.

T. Dollevoet · D. Huisman (✉)
Econometric Institute and Erasmus Center for Optimization in Public Transport, Erasmus University Rotterdam, Rotterdam, The Netherlands
e-mail: huisman@ese.eur.nl

T. Dollevoet
e-mail: dollevoet@ese.eur.nl

D. Huisman
Process Quality and Innovation, Netherlands Railways, Utrecht, The Netherlands

F. Hamid (ed.), *Optimization Essentials*, International Series in Operations Research & Management Science 353, https://doi.org/10.1007/978-981-99-5491-9_10

10.1 Introduction

In this chapter, we discuss solution approaches to solve real-world crew scheduling and rescheduling problems. Crew (re)scheduling is an important topic in different modes of transportation and has been studied in the Operations Research literature for decades. Heil et al. [5] provide an excellent overview of the literature on railway crew scheduling.

At airlines, railway companies or mass transit operators, first a timetable is constructed. A timetable specifies the trips that are operated to move passengers from A to B. Afterward, the vehicle scheduling problem is solved, where vehicles are assigned to the different trips. This vehicle schedule can also contain some *deadhead* trips to move a vehicle empty from one place to another, thus without transporting passengers. The next step is crew scheduling. The main input to this step is a set of tasks, where a task is the smallest piece of work that has to be assigned to one single crew member. The output of the crew scheduling problem is a set of *duties* (sometimes called shifts or pairings).

In practice, the crew schedule is often constructed once a year from scratch. During the year, changes are made in the timetable or vehicle schedule. Then the crew schedule needs to be adjusted as well. This is called *crew rescheduling*. Moreover, crew rescheduling takes place on the day of operations where disruptions can render the original schedule infeasible and changes need to be made.

Both crew scheduling and rescheduling are commonly solved using column generation. We first describe the combination of Lagrangian relaxation with column generation for the tactical crew scheduling problem. In order to find solutions more quickly, we then embed this solution method in a large neighborhood search algorithm for the crew rescheduling problem.

The remainder of this chapter is set up as follows. In Sect. 10.2, we describe in detail the crew scheduling and crew rescheduling problem. In Sect. 10.3, we present a mathematical formulation of the crew scheduling problem. Moreover, we discuss the solution approach where we combine Lagrangian relaxation and column generation to solve the problem. We illustrate the details of the method with a simple example and show some general results. In a similar way, we present the crew rescheduling problem in Sect. 10.4. There we will particularly focus on crew rescheduling in railway disruption management. Finally, we conclude this chapter in Sect. 10.5.

10.2 Problem Description

We now describe the crew scheduling and crew rescheduling problems in more detail.

10.2.1 The Crew Scheduling Problem

The crew scheduling problem deals with assigning tasks to duties such that each task is performed, each duty satisfies all working rules and the total costs of the duties are minimized. We assume that there are several depots spread over the network and that each duty must be assigned to one of the depots.

We consider a basic version of the crew scheduling problem where there are only crew rules defined for individual duties. In many practical applications, there are also constraints on a subset of the duties, e.g. to limit the number of duties in a certain depot or to put a constraint on the average working time of all duties. We consider the following labor rules that each duty needs to satisfy.

1. Each duty needs to start and end in the same depot.
2. A maximum duty length must be respected, where the duty length is defined as the end time of the duty minus the start time of the duty.
3. Each duty needs to have a meal break in the duty, where there is a minimum time required for this break. To ensure that the meal break is roughly in the middle of the duty, we also impose a maximum working time before and after the break.
4. Between each pair of tasks, there is a minimum connection time if the two tasks are on a different vehicle.

10.2.2 The Crew Rescheduling Problem

In this chapter, we consider a crew rescheduling problem arising in railway disruption management. Crew rescheduling problems due to planned maintenance and/or events are discussed in [6]. Those problems have similar characteristics as the crew scheduling problem and can therefore be solved in a similar way. This is in contrast with crew rescheduling during disruptions, where feasibility is the main goal and computation time is very limited.

Railway operations are often rendered infeasible by so-called disruptions. These disruptions can be caused by accidents on the railway network, malfunctioning of the infrastructure or the rolling stock, or crew being absent. When a disruption occurs, the timetable and rolling stock schedule are adjusted. This results in a modified set of tasks that have to be assigned to railway crew members. As a consequence, their duties need to be rescheduled as well.

A complicating factor in crew rescheduling is that many crew members have already started their duty when the disruption occurs. Thus, part of their duties have already been executed. Moreover, it takes some time to find a solution to the crew rescheduling problem and to communicate the adjusted duties to the crew members. For that reason, one often incorporates some time during which the duties cannot be adjusted. For example, if the disruption starts at 11:00, changes can be made to the duties from 11:30 onwards.

There are often two types of crew members: drivers and guards. In practice, trains can often be operated without a guard. Therefore, we focus on the rescheduling of drivers. Finally, to deal with disruptions, railway operators commonly have standby drivers at some of the larger stations, to ensure that disruptions can be dealt with.

The objective in crew rescheduling consists of four parts:

1. Minimize cancelations of trains due to the fact that there is no driver assigned to a particular train.
2. Minimize the number of changes in the schedule such that the communication to the drivers is limited.
3. Minimize the number of taxis used to transport drivers from one side of the disruption to the other. The reason is not only the costs of the taxis, but also the uncertainty about the time the taxi trip will take and thus the risk of having an infeasible schedule at the end.
4. Minimize the use of standby drivers, because another disruption could occur where these standby drivers are more needed.

The objective function is usually a linear combination of these four parts, where minimizing the number of cancelations is by far the most important aspect.

Moreover, each adjusted duty needs to satisfy the labor rules. For the remainder of this chapter, we assume that the following rules apply: the duration of each duty can be extended by at most 30 min, each duty needs to end at the original depot, and there should be a meal break with a minimal duration. Finally, the solution should be found in a few minutes.

10.3 Solving the Crew Scheduling Problem

The aim in crew scheduling is to define a set of duties that together cover the work to be performed. In order to explain a solution methodology for crew scheduling, we now first introduce the required notation in Sect. 10.3.1. We present a mathematical formulation in Sect. 10.3.2, and Lagrangian relaxation in Sect. 10.3.3. Then, a column generation approach is described in Sect. 10.3.4. Finally, some numerical experiments are presented in Sect. 10.3.5.

10.3.1 Notation

The work to be scheduled is given as a set of tasks, which is denoted by I. Each task $i \in I$ is specified by a start location l_i^{S}, an end location l_i^{E}, a start time t_i^{S}, and an end time t_i^{E}. A task corresponds to being a driver or guard on a vehicle that departs from l_i^{E} at time t_i^{S} and arrives at l_i^{E} at time t_i^{E}. Two tasks $i \in I$ and $j \in I$ can be performed consecutively by the same crew member if $l_i^{\mathrm{E}} = l_j^{\mathrm{S}}$ and $t_j^{\mathrm{E}} \geq t_i^{\mathrm{S}} + \tau$, where $\tau > 0$ is a minimum connection time. The duration of a task $i \in I$ is computed as $t_i^E - t_i^S$. Table 10.1 provides a set of tasks as an example.

The tasks must be covered by so-called duties. The set of all possible duties is defined as K. Each duty $k \in K$ is a sequence of tasks that can be performed by one crew member. The duration of the duty, defined as the difference between the end time of the last task and the start time of the first task in the duty, can be at most $\max_{\text{dur}}$. Moreover, each duty should contain a break of a specified minimum duration. Finally, the working time before and after the break should be at most $\max_{\text{break}}$. The cost of a duty $k \in K$ is denoted by c_k. The cost contains a fixed part c_F and a variable part c_V per hour of duty duration. The binary parameter a_{ik} indicates whether duty $k \in K$ contains task $i \in I$ or not.

Each crew member works at one of the crew depots of the operator. The set of depots is denoted by D. We denote by K_d the set of duties that can be performed by a crew member from depot $d \in D$. In particular, this means that the duty starts and ends at depot d. Table 10.2 gives examples of duties covering the tasks in Table 10.1. The first duty covers tasks 1, 4, 7, and 9. It starts from A at 8:00 and ends there at 16:30. Its duration is 8.5 h. It contains a break of 45 min from 12:15 to 13:00.

Table 10.1 Examples of tasks

Task i	t_i^S	t_i^E	l_i^S	l_i^E
1	8:00	10:45	A	C
2	8:30	11:15	A	C
3	10:00	12:00	B	C
4	11:00	12:15	C	B
5	11:30	13:00	C	D
6	12:15	14:30	C	D
7	13:00	15:00	B	D
8	14:30	16:00	D	A
9	15:15	16:30	D	A
10	16:15	18:00	D	B

Table 10.2 Examples of duties. For the costs, we assume that $c_F = 100$ and $c_V = 10$

Duty k	Depot	Tasks	Start	End	Duration	Cost c_k
1	A	$\{1, 4, 7, 9\}$	8:00	16:30	8:30	185
2	A	$\{2, 5, 9\}$	8:30	16:30	8:00	180
3	A	$\{1, 5, 8\}$	8:00	16:00	8:00	180
4	B	$\{3, 6, 10\}$	10:00	18:00	8:00	180
5	B	$\{7, 10\}$	13:00	18:00	5:00	150

10.3.2 Mathematical Formulation

The crew scheduling problem can be formulated as a set covering model. Define a binary decision variable x_k for each duty $k \in K$ which is equal to 1 if duty k is selected, and 0 otherwise. The crew scheduling problem can be formulated as the following integer linear program (ILP).

$$\min \sum_{k \in K} c_k x_k \tag{10.1}$$

$$\text{s.t.} \sum_{k \in K} a_{ik} x_k \geq 1 \qquad \forall i \in I, \tag{10.2}$$

$$x_k \in \{0, 1\} \qquad \forall k \in K. \tag{10.3}$$

The objective (10.1) is to minimize the cost of the duties that are selected in the solution. The covering constraints (10.2) ensure that each task $i \in I$ is covered by at least one duty. If a task is covered multiple times, one of the crew members actually performs the task, while the other travels on the task as a passenger. This is referred to as deadheading and may be used to reposition crew. Finally, constraints (10.3) ensure that the variables are binary.

The crew scheduling problem can also be formulated as a set partitioning problem, obtained by replacing (10.2) with equalities. However, the set partitioning problem is computationally more difficult to solve. Moreover, the set covering formulation implicitly allows deadheading. For the set partition formulation, deadheading should be included explicitly.

Note that the number of duties grows exponentially in the number of tasks. For that reason, the set of duties K is generally huge. In such cases, column generation can be applied. We explain this in more detail in Sect. 10.3.4.

10.3.3 Lagrangian Relaxation

The crew scheduling problem can be solved using Lagrangian relaxation in combination with column generation. Using Lagrangian relaxation instead of the LP-relaxation provides several benefits. First, the solution to Lagrangian subproblems are generally integer valued, and allow to be used in heuristics that aim to find feasible solutions. Second, the Lagrangian multipliers are generally more stable than dual variables obtained via the LP-relaxation. This is advantageous when using column generation. We now first define the Lagrangian relaxation of model (10.1)–(10.3).

10.3.3.1 The Lagrangian Subproblem

Define Lagrangian multipliers $\lambda_i \geq 0$ for all tasks $i \in I$. The multiplier λ_i can be interpreted as a penalty for not covering task $i \in I$. For a given vector λ of Lagrangian multipliers, the Lagrangian subproblem reads as follows.

$$\theta(\lambda) = \min \sum_{k \in K} c_k x_k + \sum_{i \in I} \lambda_i \left(1 - \sum_{k \in K} a_{ik} x_k\right) \tag{10.4}$$

$$= \min \sum_{i \in I} \lambda_i + \sum_{k \in K} \left(c_k - \sum_{i \in I} \lambda_i a_{ik}\right) x_k \tag{10.5}$$

$$\text{s.t. } x_k \in \{0, 1\} \quad \forall k \in K. \tag{10.6}$$

The optimal solution of the Lagrangian subproblem provides a lower bound on the optimal solution of model (10.1)–(10.3) for every vector λ of Lagrangian multipliers. Moreover, the Lagrangian subproblem can easily be solved. Its solution is given by

$$x_k = \begin{cases} 1 \text{ if } c_k - \sum_{i \in I} \lambda_i a_{ik} < 0, \\ 0 \text{ otherwise.} \end{cases}$$

It follows that

$$\theta(\lambda) = \sum_{i \in I} \lambda_i + \sum_{k \in K} \min\left(0, c_k - \sum_{i \in I} \lambda_i a_{ik}\right).$$

In the remainder of this chapter, we define the expression

$$\bar{c}_k(\lambda) = c_k - \sum_{i \in I} \lambda_i a_{ik} \tag{10.7}$$

as the reduced cost of duty $k \in K$, given the Lagrangian multipliers λ.

10.3.3.2 Finding Upper Bounds

We now describe a simple heuristic procedure that computes an upper bound using a given vector λ of Lagrangian multipliers. This greedy heuristic is based on the one introduced by [3] for the Set Covering Problem. The idea of the algorithm is to iteratively select a duty with lowest reduced cost until all tasks are covered. When computing the reduced costs of the duties, we only consider the tasks that are not covered by the duties that have been selected in earlier iterations. This can be implemented by updating the Lagrangian multiplier λ_i to 0 when a duty is selected that covers task i. A detailed description of this method can be found in Algorithm 1.

Algorithm 1 Greedy heuristic procedure for finding an upper bound

```
function GreedyHeuristic(λ, K)
    Initialize: λ̂ = λ, x_k = 0 for all k ∈ K.
    while some tasks are not covered do
        Select a duty k ∈ K with lowest reduced cost c̄_k(λ̂).
        Set x_k = 1.
        for i ∈ I with a_ik = 1 do
            Set λ̂_i = 0.
```

10.3.3.3 The Lagrangian Dual Problem

Our aim is now to find the best (thus, the highest) lower bound: We try to find Lagrangian multipliers that solve

$$\theta^* = \max_{\lambda \geq 0} \theta(\lambda).$$

This optimization problem is called the Lagrangian dual problem. Substituting the solution of the Lagrangian subproblem, the Lagrangian dual problem can be rewritten as

$$\theta^* = \max_{\lambda \geq 0} \left(\sum_{i \in I} \lambda_i + \sum_{k \in K} \min \left(0, c_k - \sum_{i \in I} \lambda_i a_{ik} \right) \right).$$

Observing that the optimal dual variables of the LP-relaxation of model (10.1)–(10.3) provide a feasible solution to the Lagrangian dual problem, one can show that the lower bound θ^* of the Lagrangian Relaxation is at least as strong as the linear programming bound.

10.3.3.4 Solving the Lagrangian Dual Problem

The Lagrangian dual problem is commonly solved by a subgradient optimization algorithm. A subgradient is the generalization of a gradient to functions that are convex but not necessarily differentiable. Subgradient optimization algorithms can be applied to convex minimization problems and are then similar to steepest descent methods for minimizing differentiable functions. In the appendix, we illustrate the subgradient optimization algorithm by applying it to the Lagrangian dual problem for the tasks and duties of Tables 10.1 and 10.2, respectively. We refer to [9] for an even more detailed discussion of the subgradient algorithm.

In every iteration of the subgradient optimization algorithm, we have a vector of Lagrangian multipliers. For the n-th iteration, we denote that vector by λ^n. Solving the Lagrangian subproblem then provides a lower bound $\theta(\lambda^n)$ and a solution x^n.

Moreover, the vector d^n with coordinates

$$d_i^n = 1 - \sum_{k \in K} a_{ik} x_i^n \tag{10.8}$$

is a subgradient of the function θ at λ^n. We update the Lagrangian multipliers as follows: For all $i \in I$, we define

$$\lambda_i^{n+1} = \max \left\{ \lambda_i^n + \left(\alpha^n \frac{UB - \theta(\lambda^n)}{\|d^n\|^2} \right) d_i^n, 0 \right\}. \tag{10.9}$$

Here $\alpha^n > 0$ determines a step size for iteration n. We will later explain how to define these step sizes. Observe that if task $i \in I$ is not covered in the current solution x^n, then $d_i^n = 1$ and the Lagrangian multiplier λ_i is increased. If the task is covered exactly once, the Lagrangian multiplier does not change. If the task is covered multiple times, the Lagrangian multiplier is decreased.

By repeating this procedure, the Lagrangian multipliers λ^n converge to the optimal Lagrangian multipliers, if the step sizes α^n are chosen carefully (see [9]). A complete description of the subgradient optimization algorithm can be found in Algorithm 2. Here, the step size parameter α^n is halved whenever the lower bound has not improved in the last γ iterations. The number of iterations without improvement is counted using the parameter m.

Algorithm 2 Subgradient optimization algorithm

function SUBGRADIENTOPTIMIZATION(λ, α, $n_{\max}$, γ, ϵ)
  Initialize: $m = 0$, $n = 0$, $\lambda^0 = \lambda$, $\alpha^0 = \alpha$, $UB = \infty$, $LB = -\infty$.
  while $n < n_{\max}$ and $UB - LB > \epsilon$ and $\alpha^n > \epsilon$ **do**
    Compute a lower bound $LB^n = \theta(\lambda^n)$ and solution x^n.
    Compute an upper bound UB^n by Algorithm 1.
    Set $LB = \max\{LB, LB^n\}$ and $UB = \min\{UB, UB^n\}$.
    Compute the subgradient d^n using (10.8).
    Update the Lagrangian multipliers using (10.9).
    if $\theta(\lambda^n)$ is the best lower bound **then**
      $m = 0$.
    else
      $m = m + 1$.
    if $m = \gamma$ **then**
      $\alpha^{n+1} = \frac{1}{2}\alpha^n$
    else
      $\alpha^{n+1} = \alpha^n$
    $n = n + 1$.

10.3.4 Column Generation

Crew scheduling problems arising in practice commonly have 10000s of tasks to be covered. For such instances, the set of duties K is huge. Column generation can then be applied to solve the Lagrangian dual problem. In a column generation algorithm, a so-called restricted master problem is solved, that considers only a subset K' of the duties. For Lagrangian relaxation, the master problem is the Lagrangian dual problem. After solving the restricted master problem, the pricing problem determines whether any columns not yet in the set K' can improve the objective value of the current Lagrangian subproblem. If so, some of these columns are added to the set K' and the process continues. Otherwise, the master problem has been solved to optimality.

10.3.4.1 The Restricted Master Problem

We assume that an initial subset $K' \subset K$ of duties is given. To ensure that the Lagrangian dual problem is bounded, the set of duties in K' must cover each task at least once. We then define the restricted master problem as follows.

$$\theta^*(K') = \max_{\lambda \geq 0} \left(\sum_{i \in I} \lambda_i + \sum_{k \in K'} \min \left(0, c_k - \sum_{i \in I} \lambda_i a_{ik} \right) \right).$$

Applying the subgradient optimization algorithm, we obtain approximately optimal Lagrangian multipliers $\bar{\lambda}$ (for the set of duties K') and a solution $\bar{x}$.

Consider again the tasks in Table 10.1, and assume that the set of duties K' contains all duties of Table 10.2. Optimal Lagrangian multipliers are $\lambda_2 = 180$, $\lambda_3 = \lambda_4 = \lambda_6 = 185$, $\lambda_8 = 480$, and $\lambda_1 = \lambda_5 = \lambda_7 = \lambda_9 = \lambda_{10} = 0$. Note that there can be multiple vectors of optimal Lagrangian multipliers. We have chosen this specific vector for illustrative purposes. It holds that $\bar{c}_1(\lambda) = \bar{c}_2(\lambda) = 0$, $\bar{c}_3(\lambda) = -300$, $\bar{c}_4(\lambda) = -190$, and $\bar{c}_5(\lambda) = 150$. For example, the reduced cost of duty 3 is given by

$$\bar{c}_3(\lambda) = c_3 - \lambda_1 - \lambda_5 - \lambda_8 = 180 - 0 - 0 - 480 = -300.$$

Given that $\bar{c}_3(\lambda) < 0$ and that $\bar{c}_4(\lambda) < 0$, the solution of the Lagrangian subproblem is $x_3 = x_4 = 1$ and $x_1 = x_2 = x_5 = 0$. The Lagrangian relaxation provides a lower bound of

$$\theta(\lambda) = \sum_{i \in I} \lambda_i + \sum_{k \in K} \min(0, \bar{c}_k(\lambda)) = 1215 + 0 + 0 - 300 - 190 + 0 = 725$$

for the restricted master problem.

We now want to determine whether these Lagrangian multipliers are also (approximately) optimal if the full set of duties K is considered. The pricing problem decides whether this is the case.

10.3.4.2 Updating the Lagrangian Multipliers

In the pricing problem, we determine whether there are any duties that improve the solution of the Lagrangian subproblem $\theta(\bar{\lambda})$. Thus, we look for a duty $k \in K \setminus K'$ that satisfies $\bar{c}_k(\bar{\lambda}) < 0$. If such a duty exists, the Lagrangian subproblem $\theta(\bar{\lambda})$ can be improved, and the lower bound might not be valid. If no such duty can be found, the Lagrangian subproblem has been solved to optimality.

We aim to formulate the pricing problem as a shortest path problem. The length of a path should correspond to the reduced cost of the corresponding duty. In order to be able to solve the pricing problem as a shortest path problem, we must ensure that the shortest path corresponds to a duty $k \notin K'$. Indeed, if a duty $k \in K'$ would correspond to the shortest path, it would be much more difficult to determine whether any paths $K \setminus K'$ have negative reduced cost. We ensure that the shortest path corresponds to a duty $k \notin K'$ by adjusting the Lagrangian multipliers so that $\bar{c}_k(\bar{\lambda}) \geq 0$ for all $k \in K'$. By doing so, it holds that

$$\min_{k \in K \setminus K'} \bar{c}_k(\bar{\lambda}) < 0$$

if and only if

$$\min_{k \in K} \bar{c}_k(\bar{\lambda}) < 0.$$

We now explain how to update the Lagrangian multipliers to achieve this. When doing so, the Lagrangian multipliers should not become negative, and the objective value should not decrease.

The procedure to update the Lagrangian multipliers is given in Algorithm 3. We iterate over all duties $k \in K'$ and perform an update only if $\epsilon = \bar{c}_k(\bar{\lambda}) < 0$. In that case, we increase the reduced cost of duty k by subtracting a certain amount from the Lagrangian multipliers of all tasks that the duty covers. When decreasing the multipliers, we ensure that the multipliers remain non-negative and that the increase of the reduced cost of duty k is equal to $-\epsilon$. Thus, after the update, it holds that $\bar{c}_k(\bar{\lambda}) = 0$. In order to do so, we add

$$\epsilon \frac{\bar{\lambda}_j a_{jk}}{\sum_{i \in I} \bar{\lambda}_i a_{ik}} = \bar{\lambda}_j a_{jk} \frac{\epsilon}{\sum_{i \in I} \bar{\lambda}_i a_{ik}}$$

to the multiplier $\bar{\lambda}_j$ for all tasks $j \in I$. Note first that the multiplier is updated only if $a_{jk} = 1$, so if duty k covers task j. By definition of the reduced cost in (10.7), the total increase of the reduced cost adds up to $-\epsilon$. Finally, given that

$$-\epsilon = -\bar{c}_k(\bar{\lambda}) = -c_k + \sum_{i\in I} \bar{\lambda}_i a_{ik} < \sum_{i\in I} \bar{\lambda}_i a_{ik},$$

the Lagrangian multiplier does not become negative.

If there is a task $i \in I$ that is covered multiple times and for which $\bar{\lambda}_i > 0$ before the update, then the lower bound $\theta(\bar{\lambda})$ increases due to the update. This means that the lower bound becomes stronger. Moreover, after the procedure has been applied, all duties $k \in K'$ have non-negative reduced cost.

For our example, it holds that $\bar{c}_3(\bar{\lambda}) < 0$ and that $\bar{c}_4(\bar{\lambda}) < 0$. In particular, for $k = 3$, we obtain $\epsilon = -300$. Given that $a_{13} = a_{53} = a_{83} = 1$, $\bar{\lambda}_1$, $\bar{\lambda}_5$, and $\bar{\lambda}_8$ are updated when considering duty $k = 3$. In particular, $\bar{\lambda}_8$ is reduced by 300 to 180. Similarly, when considering duty $k = 4$, $\bar{\lambda}_3$ and $\bar{\lambda}_6$ are updated to 90. For the updated Lagrangian multipliers, $\theta(\bar{\lambda}) = 725$, and $\bar{c}_k(\bar{\lambda}) \geq 0$ for all $k \in K'$.

Algorithm 3 Updating the Lagrangian multipliers

```
function UPDATELAGRANGIANMULTIPLIERS(λ̄, K')
    for k ∈ K' do
        if ε = c̄_k(λ̄) < 0 then
            for j ∈ I do
                λ̄_j = λ̄_j + ε (λ̄_j a_jk / Σ_{i∈I} λ̄_i a_ik) = λ̄_j (1 + a_jk ε / Σ_{i∈I} λ̄_i a_ik)
            Set x̄_k = 0.
```

10.3.4.3 The Pricing Problem

In order to solve the pricing problem, we define a directed graph $G = (V, A)$ for each depot $d \in D$ separately. The set of nodes V contains a source s, a node v_i for each task $i \in I$, and a sink t:

$$V = \{s\} \cup \{v_i : i \in I\} \cup \{t\}.$$

We define an arc (s, v_i) for each task $i \in I$ that a duty from the given depot can start with. We define an arc (v_i, v_j) from task $i \in I$ to $j \in I$ if the two tasks can be performed consecutively by a crew member from the depot. Finally, we define an arc (v_j, t) for each task $j \in I$ that a duty from the depot can end with. Noting that two task nodes v_i and v_j can be connected only if v_j starts later than v_i, this graph is acyclic and allows a topological ordering. We define the following costs for the arcs.

$$c(v_1, v_2) = \begin{cases} c_F + c_V(t_j^{\mathrm{E}} - t_j^{\mathrm{S}}) - \bar{\lambda}_j & \text{if } v_1 = s \text{ and } v_2 = v_j, \\ 0 & \text{if } v_2 = t, \\ c_V(t_j^{\mathrm{E}} - t_i^{\mathrm{E}}) - \bar{\lambda}_j & \text{otherwise, with } v_1 = v_i, v_2 = v_j. \end{cases}$$

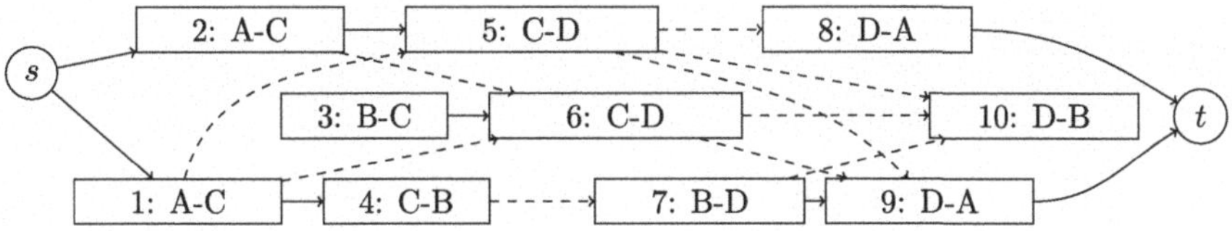

Fig. 10.1 Example of a pricing graph for depot A

Each feasible duty for the given depot is now a path from s to t in this graph. The reduced cost of a duty corresponds to the cost of this path in the graph. To illustrate this, consider a path $s \to v_1 \to v_2 \to v_3 \to \cdots \to v_m \to t$. The cost of this path is given by

$$c(s, v_1) + \sum_{j=2}^{m} c(v_{j-1}, v_j) + c(v_m, t)$$

$$= c_F + c_V(t_1^{\mathrm{E}} - t_1^{\mathrm{S}}) - \bar{\lambda}_1 + \sum_{j=2}^{m} \left(c_V(t_j^{\mathrm{E}} - t_{j-1}^{\mathrm{E}}) - \bar{\lambda}_j\right) + 0$$

$$= c_F + c_V(t_m^{\mathrm{E}} - t_1^{\mathrm{S}}) - \sum_{j=1}^{m} \bar{\lambda}_j.$$

This indeed equals the reduced cost of the path. Moreover, each path that satisfies the requirements on the duty duration and the break corresponds to a feasible duty. For the example of Table 10.1, the pricing graph for a depot located at A is depicted in Fig. 10.1. Here, dashed arcs indicate that a break is possible at the arc, where we use a minimum break duration of 45 min.

We now explain how to find a duty with most negative reduced cost. We consider each depot separately. Recall that each duty for the given depot corresponds to a path in the graph G. We thus look for the shortest path in G that respects the maximum duty duration and contains a break. These requirements could be incorporated by solving a resource-constrained shortest path problem (see [7]). However, we describe a simpler algorithm here.

Note that the duty duration is the difference of the end time of the last task in the duty and the start time of the first task. As a consequence, if we fix the first task in the duty, we can determine upfront which tasks the duty can possibly end with. By considering each possible task a duty can start with separately, we are sure that we still find a duty with lowest reduced cost. Similarly, we fix the first task after the break. By doing so, we can incorporate the maximum working time without a break.

To define the algorithm more formally, we select a start task i_1 of the duty and we select the first task i_2 after the break. We first remove from the graph G all arcs from s to other nodes than i_1. This ensures that the duty starts with task i_1.

We then remove all arcs (v_j, t) for which task j ends too late, which holds if $t_j^{\mathrm{E}} - t_{i_1}^{\mathrm{S}} > \max_{\mathrm{dur}}$ or for which the working time after the break is too long, which

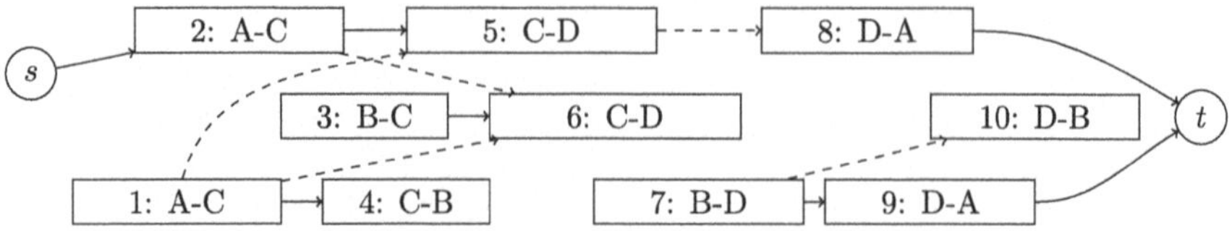

Fig. 10.2 Example of a pricing graph for depot A, for $i_1 = 2$ and $i_2 = 8$

holds if $t_j^{\mathrm{E}} - t_{i_2}^{\mathrm{S}} > \max_{\mathrm{break}}$. We also prevent that the duty ends too early, by removing all arcs (v_j, t) for which task j ends too early, which holds if $t_j^{\mathrm{E}} < t_{i_2}^{\mathrm{E}}$.

Finally, to ensure that task i_2 is the first task after the break, we remove all arcs (v_{j_1}, v_{j_2}) that satisfy the following properties.

- $t_{j_1}^{\mathrm{E}} \leq t_{i_2}^{\mathrm{S}} \leq t_{j_2}^{\mathrm{E}}$, and
- $j_2 \neq i_2$ or (v_{j_1}, v_{j_2}) does not contain a break or $t_{j_1}^{\mathrm{E}} - t_{i_1}^{\mathrm{S}} > \max_{\mathrm{break}}$.

Intuitively, removing these arcs ensures that at time $t_{i_2}^{\mathrm{S}}$, the employee must start with task i_2 after having had a break. Thus, just before time $t_{i_2}^{\mathrm{S}}$, the employee must be on an arc toward i_2 that allows a break and satisfies the maximum working time before the break. The first requirement selects all the arcs for which the employee is active at time $t_{i_2}^{\mathrm{S}}$. The second requirement then prevents the employee from arriving at another task than i_2, from arriving at i_2 without having had a break directly before task i_2, and from spending too much time before the break starts.

We can now find the duty that starts with i_1 and has a break just before i_2 by solving an ordinary shortest path problem. Given that the graph G is acyclic, this can be done in a computation time that is linear in the number of arcs in the graph. By considering all possible pairs of tasks (i_1, i_2), we obtain a polynomial algorithm for the pricing problem. This running time can be further improved by considering the parts of the duty before and after the break separately, as proposed by [6].

Returning to our example, we observe that the duty covering tasks 2, 5, and 8 has negative reduced cost of –185 and should be added to the set K'. Selecting task 2 as i_1 and task 8 as i_2, we obtain the graph that is shown in Fig. 10.2. In particular, arcs from the source to tasks other than i_1 are removed, as well as all arcs for which the employee is active just before time $t_{i_2}^{\mathrm{S}}$, but not taking a break, or which are not toward i_2. Given that i_1 starts at 8:00, the duty should end at the latest at 8:00 + $\max_{\mathrm{dur}}$. Moreover, given that i_2 starts at 14:30, the duty should end at the latest at 14:30 + $\max_{\mathrm{break}}$. If there would be tasks that end later than the earliest of these times, the arcs from those tasks to the sink would be removed as well. Running any ordinary shortest path algorithm on this graph, the duty covering tasks 2, 5, and 8 is found as the one with lowest reduced cost.

10.3.5 Some Numerical Results

Netherlands Railways (Nederlandse Spoorwegen, NS), the largest passenger railway operator in the Netherlands, have been using decision support to schedule the crew since the early 2000s. The algorithms for crew scheduling have been based on a combination of column generation and Lagrangian relaxation, see [1, 2]. We now summarize the main findings that are presented in [2].

NS also incorporates additional constraints in the set covering model. First of all, a maximum number of duties is incorporated for each of the 29 depots throughout the country. Moreover, the average duty duration of the duties assigned to each depot should be at most 8 hours. Finally, to ensure a fair distribution of work, desirable work must be evenly divided over the depots. This rule, referred to as the Sharing-Sweet-and-Sour rules, requires, for example, that the percentage of work on preferred trains is between 40 and 70% for each depot.

In the Netherlands, only few trains are running during the night. As a result, the crew scheduling problem can naturally be decomposed per day. We note that such a decomposition might not be possible if trains are operated throughout the night as well. Nonetheless, in [2], an algorithm is developed that can solve the crew scheduling problem for all days of the week at once. Such instances for a complete week contain 10000s of tasks and consider 1000s of crew members. The benefit of solving one big instance for a complete week instead of solving smaller instances for each day separately is the flexibility regarding the additional constraints. Consider, for example, the average duty duration which must be below 8 hours. Solving an instance for a complete week allows the solution to have a higher average duration in the weekends, if this is compensated for by a lower average duration for the weekdays.

The results of the computational experiments of [2] are summarized in Table 10.3. If there are no additional constraints, it is more efficient to schedule the seven days of the week individually. However, if constraints are included that consider duties of different days, solving the instance for one week reduces the number of duties significantly. In these experiments, it took several days of computation time to find a solution for the week instances.

10.4 Solving the Crew Rescheduling Problem

As stated in Sect. 10.2.2, the crew rescheduling problem needs to be solved quickly. Therefore, heuristics are often applied. In this section, we present a method proposed by [8] to solve this problem.

First, we discuss a simple example. This example contains the same 10 tasks from Table 10.1 and consists of 3 duties, as visualized in Fig. 10.3. In this example, duty A1 starts and ends in depot A and consists of the tasks 2, 5, and 8. Duty A2 consists of tasks 1, 4, 7, and 9 and is also from depot A. Finally, duty B1 from depot B contains

Table 10.3 Some computational results from [2]

Crew	Additional constraints	Number of duties	
		7 day instances	1 week instance
Drivers	No	5198	5221
	Avg. Duration	5393	5324
	Avg. Dur. + Pref. Trains	5489	5400
Guards	No	6388	6428
	Avg. Duration	6588	6552
	Avg. Dur. + Pref. Trains	6715	6636

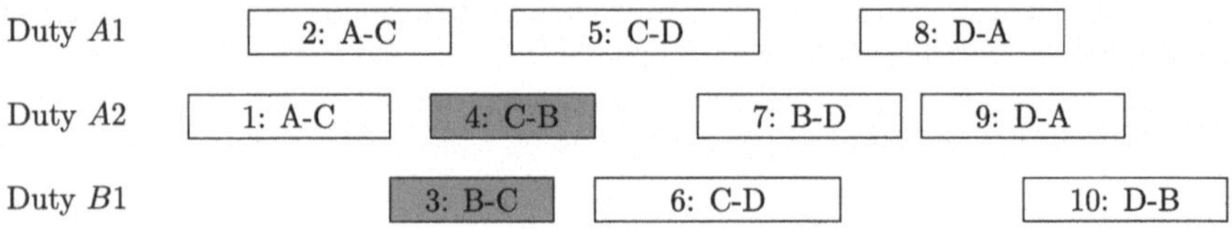

Fig. 10.3 Original duties A1, A2, and B1. Note that tasks 3 and 4 will be canceled due to the disruption

tasks 3, 6, and 10. Now, assume that at 9:00 it becomes clear that tasks 3 and 4 have to be canceled due to a disruption between B and C. As a result, duty A2 and duty B1 are infeasible and need to be repaired, where duty A2 has already started with task 1. A possible solution is to swap tasks 6 and 7 between duties A2 and B1. Then duty A2 will perform tasks 6 and 9, and duty B1 will start later and only execute tasks 7 and 10. As a result, all remaining tasks are covered and all duties end at their own depot at the original time, so there is no overtime.

In the remainder of this section, we present a mathematical formulation in Sect. 10.4.1, then discuss a solution approach in Sect. 10.4.2, and finally we present in Sect. 10.4.3 some numerical results on real-world instances.

10.4.1 Notation and Mathematical Formulation

Each original duty needs to be replaced by an adjusted duty. Note that this adjusted duty could be the same as the original duty, so without changes. This holds in particular for duties that have already been finished. Basically, we need to decide about the extension of the original duty from the start of the disruption. In the remainder of this section, we call this an extension. Note that an extension can also be empty, i.e. the original duty will not contain any tasks after the start of the disruption. Let $x_k^\delta = 1$ if we select extension $k \in K^\delta$ corresponding to original duty δ and 0 otherwise. Here, the set K^δ is defined as the set of all feasible extensions for original duty δ. Decision variables y_i are 1 if task i is canceled, and 0 otherwise. The crew rescheduling problem can then be formulated as follows.

$$\min \sum_{\delta\in\Delta}\sum_{k\in K^\delta} c_k^\delta x_k^\delta + \sum_{i\in I} f_i y_i \tag{10.10}$$

$$\text{s.t.} \sum_{\delta\in\Delta}\sum_{k\in K^\delta} a_{ik}^\delta x_k^\delta + y_i \geq 1 \qquad \forall i \in I, \tag{10.11}$$

$$\sum_{k\in K^\delta} x_k^\delta = 1 \qquad \forall \delta \in \Delta, \tag{10.12}$$

$$x_k^\delta \in \{0, 1\} \qquad \forall \delta \in \Delta,\ k \in K^\delta, \tag{10.13}$$

$$y_i \in \{0, 1\} \qquad \forall i \in I. \tag{10.14}$$

In the objective function (10.10), we minimize the costs of cancelations and the costs of the selected extensions in the solution. The cancelation costs of a task are denoted by parameters f_i. The costs of the extensions c_k^δ consist of the costs of taxis and overtime and a penalty for changing the original duty. This penalty could be larger if more tasks in the duty are changed, or it could be a constant if the company just wants to minimize the number of changed duties. The first set of constraints (10.11) ensures that each task $i \in I$ is covered by at least one duty or it is canceled. Constraints (10.12) ensure that each original duty gets exactly one extension. Finally, constraints (10.13) and (10.14) ensure that the variables are binary.

Note again that the number of variables is extremely large and that the problem needs to be solved quickly. Therefore, a heuristic approach is required, which will be explained next.

10.4.2 Solution Approach

The problem is solved iteratively.[1] In the first step, the core problem consists of the driver's duties that need to be rescheduled since they are directly affected by the disruption. In the example, the core problem consists of duties B1 and A2. All tasks in these duties are included as well (in the example, tasks 1, 3, 4, 6, 7, 9, and 10). In this way, an optimization problem with a relatively small number of tasks and duties can be solved. If tasks remain uncovered, additional tasks and drivers are added to the core problem, and the core problem is solved again. These additional drivers are selected by defining a neighborhood for each uncovered task. For example, suppose that we cannot cover a task starting at station A. It could be expected that a driver who is in station A around the start time of the uncovered task would be a good candidate to cover this uncovered task. Similarly, drivers which have a similar duty

[1] The content of the remainder of this section draws heavily from [8].

as the one we have just added could be good candidates as well. Think, for instance, of drivers having the same depot and visiting more or less the same stations in their duty.

This process repeats itself until all tasks are covered or a maximum computation time has been reached. In the latter case, trains need to be canceled because there is no driver.

10.4.2.1 Formulating the Core Problems

$\bar{\Delta}$ is the set of original duties in the core problem. Recall that this set is in general only a subset of all original duties. The set of all tasks covered by at least one $\delta \in \bar{\Delta}$ plus the tasks uncovered in the current solution is defined by $\bar{I}$. Finally, all feasible extensions in the core problem form the set $\bar{K}^\delta$. The core problem can then be formulated as follows.

$$\min \sum_{\delta \in \bar{\Delta}} \sum_{k \in \bar{K}^\delta} c_k^\delta x_k^\delta + \sum_{i \in \bar{I}} f_i y_i \tag{10.15}$$

$$\text{s.t.} \quad \sum_{\delta \in \bar{\Delta}} \sum_{k \in \bar{K}^\delta} a_{ik}^\delta x_k^\delta + y_i \geq 1 \quad \forall i \in \bar{I}, \tag{10.16}$$

$$\sum_{k \in \bar{K}^\delta} x_k^\delta = 1 \quad \forall \delta \in \bar{\Delta}, \tag{10.17}$$

$$x_k^\delta \in \{0, 1\} \quad \forall \delta \in \bar{\Delta},\ k \in \bar{K}^\delta, \tag{10.18}$$

$$y_i \in \{0, 1\} \quad \forall i \in \bar{I}. \tag{10.19}$$

10.4.2.2 The Lagrangian Subproblem

To quickly find a lower bound and a good feasible solution, we relax the covering constraints (10.16) in a Lagrangian way by introducing Lagrangian multipliers λ_i, $i \in \bar{I}$. The Lagrangian subproblem is then equal to

$$\Theta(\lambda) = \min \sum_{\delta \in \bar{\Delta}} \sum_{k \in \bar{K}^\delta} c_k^\delta x_k^\delta + \sum_{i \in \bar{I}} f_i y_i + \sum_{i \in \bar{I}} \lambda_i \left(1 - \sum_{\delta \in \bar{\Delta}} \sum_{k \in \bar{K}^\delta} a_{ik}^\delta x_k^\delta - y_i \right)$$

$$\text{s.t.} \quad (17), (18), \text{ and } (19).$$

We can rewrite this as:

$$\Theta(\lambda) = \min \sum_{i \in \bar{I}} \lambda_i + \sum_{\delta \in \bar{\Delta}} \sum_{k \in \bar{K}^{\delta}} \left(c_k^{\delta} - \sum_{i \in \bar{I}} \lambda_i a_{ik}^{\delta} \right) x_k^{\delta} + \sum_{i \in \bar{I}} (f_i - \lambda_i) y_i \quad (10.20)$$

$$\text{s.t.} \quad (17), (18), \text{ and } (19).$$

The optimal solution to the Lagrangian subproblem can be found as follows. For each $\delta \in \bar{\Delta}$, we set x_k^{δ} to 1 for one $k \in \bar{K}^{\delta}$, namely for any

$$k \in \arg\min\{\bar{c}_k^{\delta}(\lambda) : k \in \bar{K}^{\delta}\}.$$

The variables x_k^{δ} for the other $k \in \bar{K}^{\delta}$ are set to 0. Doing so, exactly one extension is selected for each original duty $\delta \in \bar{\Delta}$. Thus, (10.17) is satisfied. Notice that

$$\bar{c}_k^{\delta}(\lambda) = c_k^{\delta} - \sum_{i \in \bar{I}} \lambda_i a_{ik}^{\delta}$$

is equal to the reduced cost of feasible extension k, similar as in Sect. 10.3.3. Furthermore we set $y_i = 1$ if $f_i - \lambda_i < 0$ and $y_i = 0$ otherwise.

10.4.2.3 Finding Upper Bounds

To obtain a feasible solution to the core problem, we apply a greedy algorithm, which is presented in Algorithm 4. First (Line 1), the original duties are ordered randomly. Next (Lines 2–5), we choose the feasible extension with the lowest reduced cost for every original duty. Finally (Lines 8–12), all remaining uncovered tasks are covered by idle reserve duties. A reserve duty being idle means here that the feasible extension of the reserved duty does not cover any tasks.

10.4.2.4 The Lagrangian Dual Problem

The Lagrangian dual problem is now defined as

$$\Theta^* := \max_{\lambda \geq 0} \Theta(\lambda).$$

It can be solved by a subgradient algorithm similar to the one described in Algorithm 2.

10.4.2.5 Column Generation

Due to the fact that the number of feasible extensions grows exponentially large in the input size, we will apply column generation to solve the Lagrangian dual problem.

Thus, we only consider a subset of the x_k^δ variables in the so-called *restricted master problem (RMP)*.

We denote $\bar{K}_n^\delta \subseteq \bar{K}^\delta$ for the subset of feasible extensions that is considered in iteration n of the column generation algorithm. The variables x_k^δ that are included in the RMP are those in $\bigcup_{\delta\in\bar{\Delta}}\{x_k^\delta : k \in \bar{K}_n^\delta\}$. Let Θ_n^* be the value of the associated Lagrangian dual problem and define λ_n^* as the corresponding multiplier vector.

For every $\delta \in \bar{\Delta}$, we solve a pricing problem to determine whether there are feasible extensions that can improve the value of Θ_n^*. Each pricing problem is solved as a resource-constrained shortest path problem (see Sect. 10.4.2.7). Define $\bar{z}_n^\delta(\lambda_n^*)$ as $\min\{\bar{c}_k^\delta(\lambda_n^*) : k \in \bar{K}_n^\delta\}$. Then $\bar{z}_n^\delta(\lambda_n^*)$ equals the lowest reduced cost over all extensions of original duty $\delta \in \bar{\Delta}$ that are currently included in the RMP. Note that this is the reduced cost of the extension that is selected in the solution of the Lagrangian subproblem. Moreover, define $\bar{c}_*^\delta(\lambda_n^*) := \min\{\bar{c}_k^\delta(\lambda_n^*) : k \in \bar{K}^\delta\}$ as the lowest reduced costs over all extensions, including those that are not yet considered in the RMP. Then the feasible extension x_k^δ corresponding to $\bar{c}_*^\delta(\lambda_n^*)$ should be added to the RMP if $\bar{c}_*^\delta(\lambda_n^*) < \bar{z}_n^\delta$, because the reduced costs of the newly added column are then lower than the reduced costs of the best column in the current pool. This implies that the newly added column would improve the solution of the Lagrangian subproblem. Moreover, $LB_n := \Theta_n^* - \sum_{\delta\in\bar{\Delta}}\left(\bar{z}_n^\delta(\lambda_n^*) - \bar{c}_*^\delta(\lambda_n^*)\right)$ is a lower bound on Θ^*.

Algorithm 4 Greedy procedure to construct feasible solutions

1: Order the original duties $\delta \in \bar{\Delta}$ randomly
2: **for all** $\delta \in \bar{\Delta}$ **do**
3: Choose $k_* \in \arg\min\{\bar{c}_k^\delta(\lambda) : k \in \bar{K}_n^\delta\}$ and set the corresponding $x_*^\delta = 1$
4: Set $\lambda_i = 0$ for all $i \in \bar{I}$ covered by x_*^δ
5: Set $\lambda_i = f_i$ for all $i \in \bar{I}$ not covered by x_*^δ for all $\delta \in \bar{\Delta}$
6: Make set of idle reserve duties $\bar{\Delta}_I := \{\delta \in \bar{\Delta}_R : x_*^\delta$ covers no tasks$\}$
7: **for all** $\delta \in \bar{\Delta}_I$ **do**
8: Set $x_*^\delta = 0$
9: Choose $k_* \in \arg\min\{\bar{c}_k^\delta(\lambda) : k \in \bar{K}_n^\delta\}$ and set the corresponding $x_*^\delta = 1$
10: Set $\lambda_i = 0$ for all $i \in \bar{I}$ covered by x_*^δ
11: Set $y_i = 1$ for all tasks not covered by x_*^δ for all δ

10.4.2.6 Column Fixing

At a certain point in the column generation process, we start fixing columns. This means that we set, for a fixed percentage of original duties, all feasible extensions to either 1 or 0. The feasible extension with the lowest reduced cost $\bar{c}_k^\delta(\lambda_n^*)$ is set to 1, all others to 0. Moreover, we set the Lagrangian multipliers of all covered tasks to 0.

After μ percent of the original duties is fixed, the pricing problem for the remaining original duties is solved with the updated multiplier vector. Good columns are added to the RMP, and we continue in the same way. At the moment that we cannot find any attractive columns or if the variables for more than $100 - \mu$ percent of the original duties have been fixed, we stop the fixing phase.

10.4.2.7 Pricing Problems

In the pricing problem, new feasible extensions are generated for every original duty. To solve the pricing problem, we construct a graph $\bar{G}^{\delta}$ with nodes for the departure and arrival of each task. In addition, there is a source and sink node. The source node represents the position of the original duty at the time of rescheduling. Note that there are three possibilities. (1) The duty has not started yet. (2) The driver is performing a task, or (3) is at a station. Furthermore, the sink node represents the end of an extension. There are arcs from the source and to the sink if the locations match, and there are arcs from an arrival node of task i to a departure node of task j, if a driver can execute task j immediately after task i. Finally, the Lagrangian multipliers are assigned to the arcs such that the cost of each path corresponds to the reduced cost of the feasible extension.

10.4.3 Some Numerical Results

The algorithm presented in this section has been tested in [8] on several scenarios for NS. In this chapter, we summarize some of the main computational results of the original paper [8].

In Table 10.4, we give a description and the results of the 10 cases (at 5 locations spread over the country, see Fig. 10.4). The different disruptions vary in size, some routes contain up to 4 trains per hour per direction, while other contain heavily used routes. Examples of the latter are disruptions around Abcoude, located between Amsterdam and Utrecht, and around 's-Hertogenbosch, in the South of the Netherlands. In these instances, up to 60 original duties are affected. The disruptions near Beilen disconnects the northern part of the railway network from the remainder of the country. Finally, the disruptions near Zoetermeer contain a partial blockage (one of the two tracks was closed) resulting in fewer trains. The duration of all disruptions is 3 hours, which is a typical duration for an accident.

The right part of Table 10.4 reports the considered number of original duties and tasks, and the relative gap between lower and upper bound, in the initial core problem. The reported gaps are less than 2.0%. Finally, the number of uncovered tasks is reported. After solving the initial core problem, at most one task was uncovered in all instances. One task was uncovered in the instances Ac A, Ht B, and Ztm B. For the instances Ac A and Ztm B, this task could be covered in the second iteration by extending the core problem. Only for instance Ht B, this could not be solved.

Computation times of at most a few minutes were reported in [8]. These results showed to be very promising from a practical perspective. Therefore, this algorithm has been fully implemented as part of a commercial crew planning and dispatching

Table 10.4 Summary of the different scenarios used in [8]. The columns AD and UT report the number of affected duties and the number of uncovered tasks, respectively

Location	ID	Time	Type	AD	$\|\bar{\Delta}\|$	$\|\bar{I}\|$	Gap (%)	UT
Abcoude	Ac A	11:00-14:00	Full	59	163	602	0.5	1
Abcoude	Ac B	16:30-19:30	Full	53	143	736	0.6	0
Beilen	Bl A	07:00-10:00	Full	15	80	234	0.0	0
Beilen	Bl B	16:00-19:00	Full	15	63	186	0.0	0
's-Hertogenbosch	Ht A	08:00-11:00	Full	55	134	617	1.8	0
's-Hertogenbosch	Ht B	15:30-18:30	Full	51	118	658	0.7	1
Lelystad	Lls A	04:00-07:00	Full	25	81	239	0.6	0
Lelystad	Lls B	13:00-16:00	Full	22	149	402	0.0	0
Zoetermeer	Ztm A	08:00-11:00	Partial	21	112	508	0.6	0
Zoetermeer	Ztm B	11:30-14:30	Partial	25	110	423	<0.1	1

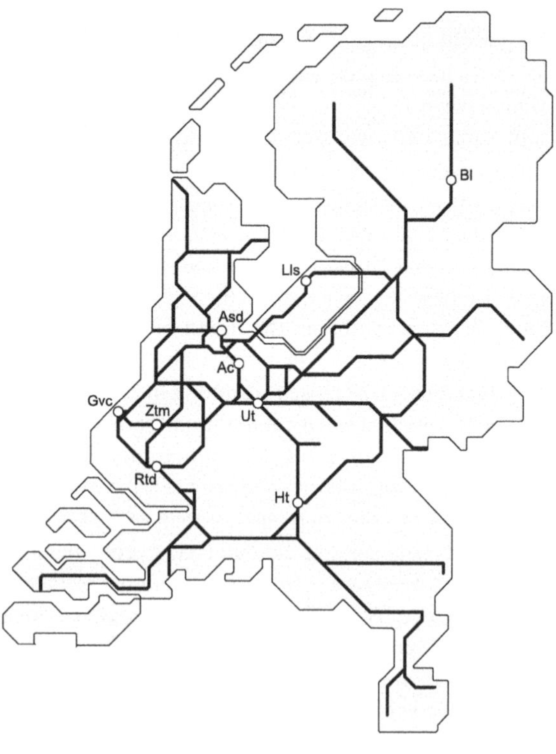

Fig. 10.4 Map of the NS network with the four major cities Amsterdam (Asd), Rotterdam (Rtd), The Hague (Gvc), and Utrecht (Ut) plus the 5 disruption locations

system, which is among other countries used in the Netherlands. In particular, it has been used in full operation at Netherlands Railways since 2013.

10.5 Concluding Remarks

We have considered large crew (re)scheduling problems arising in passenger railway transportation. We have formulated these problems as large-scale set covering problems with a huge number of variables. Therefore, column generation is required to solve these large integer programming problems. We combined column generation with Lagrangian relaxation, either in an exact or heuristic fashion. The benefits of Lagrangian relaxation over LP-relaxation are that the Lagrangian multipliers are more stable than dual variables, and that solutions to the Lagrangian subproblem are generally integer valued, making it easier to transform them into feasible solutions. In contrast, LP-relaxations can be embedded more easily in branch-and-price approaches, because fractional solutions provide a natural branching scheme.

Especially for large real-world instances, the problems need to be solved in a heuristic way. For example, at NS, these methods are used in day-to-day operation. Once a year, a nationwide crew scheduling problem of about 70,000 tasks is solved to schedule Monday to Sunday in one iteration. In a computation time of several days, a close to optimal solution can be found. Similar algorithms are used to reschedule the plan during the year. On the day of operations, drivers and guards are rescheduled during disruptions such as track blockages with the help of an algorithm as presented in Sect. 10.4. For large disruptions, a reasonable solution can be found in a few minutes.

The computational results that we have presented rely on several improvements to the column generation algorithm. Desaulniers et al. [4] propose a wide range of strategies to improve the performance of column generation algorithms in general. We will briefly describe a few examples now. First, multiple duties can be added to the restricted master problem in every column generation iteration. Adding more duties with negative reduced cost reduces the number of column generation iterations that are required. However, it also increases the size of the restricted master problem. Thus, an important design choice is how many duties to add in each iteration. Second, it is not necessary to solve the pricing problem for every depot or for every original duty in every column generation iteration. Instead, the pricing problem can be solved for a subset of the depots or original duties. If duties with negative reduced cost are found, these are added to the restricted master problem, and the next iteration starts. Otherwise, the pricing problem is solved for the remaining depots or original duties. By doing so, a trade-off can be made between the times spent on the restricted master problem and pricing problem. Note that the pricing problems for different depots or original duties can be solved in parallel. Third, a heuristic can be used to solve the pricing problem. If a duty with negative reduced cost is found, that duty is added. Otherwise, an exact method can be applied. An example is to first discard the break requirement and check whether the resulting duty contains a break or not.

Concluding, we have demonstrated how crew (re)scheduling problems can be solved and we hope to have convinced the reader that these solution methods can be really used in the real-world operations of a large railway operator. Interesting directions for further research include the application of Lagrangian relaxation and column generation to other problems. The method that we have presented here can be applied to all problems that can be modeled as a set covering or set partitioning problem with many columns. Note that such a formulation allows to incorporate all kinds of domain-specific requirements in the definition of the columns. Another interesting direction for future research is to further improve the performance of the approach we presented, for example by solving the pricing problem more efficiently.

10.6 Appendix A. Illustration of Subgradient Optimization Algorithm

In this appendix, we illustrate some iterations of the subgradient optimization algorithm of Sect. 10.3.3.4. We consider the set of tasks that are given in Table 10.1 and the set of duties that are specified in Table 10.2. We assume that Algorithm 2 is called with parameters $\lambda = 0$, $\alpha = 1$, $n_{\max} = 15$, $\gamma = 5$, and $\epsilon = 1$.

First Iteration

For $\lambda = 0$, it holds that $\bar{c}_k(\lambda) = c_k$ for all $k \in K$. Thus, $\bar{c}_k(\lambda) > 0$ for all $k \in K$. The optimal solution to the Lagrangian subproblem is to set $x_k = 0$ for all $k \in K$. We obtain a lower bound $\theta(\lambda^0) = 0$.

To obtain an upper bound, we apply Algorithm 1. Given that $\lambda = 0$, duties are selected in the order of their costs. We first select duty $k = 5$, then duties $k = 2$, $k = 3$, and $k = 4$, and finally duty $k = 1$. The upper bound UB^0 is equal to 875.

We then update the Lagrangian multipliers. Given that $x_k = 0$ for all $k \in K$, no tasks are covered in the solution to the Lagrangian subproblem. Thus, $d_i^0 = 1$ for all $i \in I$. It follows that $\|d^0\|^2 = 10$. We update the Lagrangian multipliers using (10.9) as follows:

$$\lambda_i^1 = \max\left\{\lambda_i^0 + \left(\alpha^0 \frac{UB - \theta(\lambda^0)}{\|d^0\|^2}\right) d_i^0, 0\right\} = 0 + 1\frac{875 - 0}{10}1 = 87.5$$

for all $i \in I$.

Second Iteration

Given that $\lambda_i^1 = 87.5$ for all $i \in I$, we obtain the following reduced costs: $\bar{c}_1(\lambda^1) = -165, \bar{c}_2(\lambda^1) = \bar{c}_3(\lambda^1) = \bar{c}_4(\lambda^1) = -82.5$, and $\bar{c}_5(\lambda^1) = -25$. It holds that $\bar{c}_k(\lambda) < 0$ for all $k \in K$. The optimal solution to the Lagrangian subproblem is to set $x_k = 1$ for all $k \in K$. We obtain a lower bound $\theta(\lambda^1) = 437.5$.

To obtain an upper bound, we apply Algorithm 1. Given that duty $k = 1$ has the lowest reduced cost, we first select that duty and then update $\hat{\lambda}_1 = \hat{\lambda}_4 = \hat{\lambda}_7 = \hat{\lambda}_9 = 0$. Using this updated vector of Lagrangian multipliers, we obtain $\bar{c}_2(\hat{\lambda}) = \bar{c}_3(\hat{\lambda}) = 5$, $\bar{c}_4(\hat{\lambda}) = -82.5$, and $\bar{c}_5(\hat{\lambda}) = 62.5$. We thus select duty $k = 4$ and update $\hat{\lambda}_3 = \hat{\lambda}_6 = \hat{\lambda}_{10} = 0$. The new reduced costs are $\bar{c}_2(\hat{\lambda}) = \bar{c}_3(\hat{\lambda}) = 5$, and $\bar{c}_5(\hat{\lambda}) = 150$. We select duty $k = 2$, set $\hat{\lambda}_2 = \hat{\lambda}_5 = \hat{\lambda}_9 = 0$ and obtain the new reduced costs $\bar{c}_4(\hat{\lambda}) = 92.5$ and $\bar{c}_5(\hat{\lambda}) = 150$. Finally, we select duty $k = 3$, and have then covered all tasks. The solution selects duties $\{1, 2, 3, 4\}$ and has cost equal to 725.

We then update the Lagrangian multipliers. Given that $x_k = 1$ for all $k \in K$, all tasks are covered in the solution to the Lagrangian subproblem at least once. Given that tasks 1, 5, 7, 9, and 10 are covered twice, $d_i^1 = -1$ for $i \in \{1, 5, 7, 9, 10\}$. It follows that $\|d^1\|^2 = 5$. We update the Lagrangian multipliers using (10.9) as follows:

$$\lambda_i^2 = \max\left\{\lambda_i^1 + \left(\alpha^1 \frac{UB - \theta(\lambda^1)}{\|d^1\|^2}\right) d_i^1, 0\right\} = 87.5 + 1\frac{725 - 437.5}{5}(-1) = 30$$

for all $i \in \{1, 5, 7, 9, 10\}$. For other values of i, $d_i^1 = 0$, and thus $\lambda_i^2 = \lambda_i^1 = 87.5$.

Third Iteration

Given λ^2, we obtain the following reduced costs: $\bar{c}_1(\lambda^2) = 7.5, \bar{c}_2(\lambda^2) = \bar{c}_3(\lambda^2) = 32.5, \bar{c}_4(\lambda^2) = -25$, and $\bar{c}_5(\lambda^2) = 90$. It holds that $\bar{c}_k(\lambda) < 0$ for $k = 4$. The optimal solution to the Lagrangian subproblem is to set $x_4 = 1$ and $x_k = 0$ for $k \in \{1, 2, 3, 5\}$. We obtain a lower bound $\theta(\lambda^2) = 562.5$.

To obtain an upper bound, we apply Algorithm 1. We again obtain a solution $x_k = 1$ for $k = \{1, 2, 3, 4\}$ and $x_5 = 0$. The cost of this solution equals 725.

We then update the Lagrangian multipliers. Given that tasks 3, 6, and 10 are covered in the solution of the Lagrangian subproblem, $d_i^2 = 0$ for $i \in \{3, 6, 10\}$. For other i, $d_i^2 = 1$. It follows that $\|d^2\|^2 = 7$. We obtain the following Lagrangian multipliers:

$$\lambda^3 \approx (53.2, 110.7, 87.5, 110.7, 53.2, 87.5, 53.2, 110.7, 53.2, 30).$$

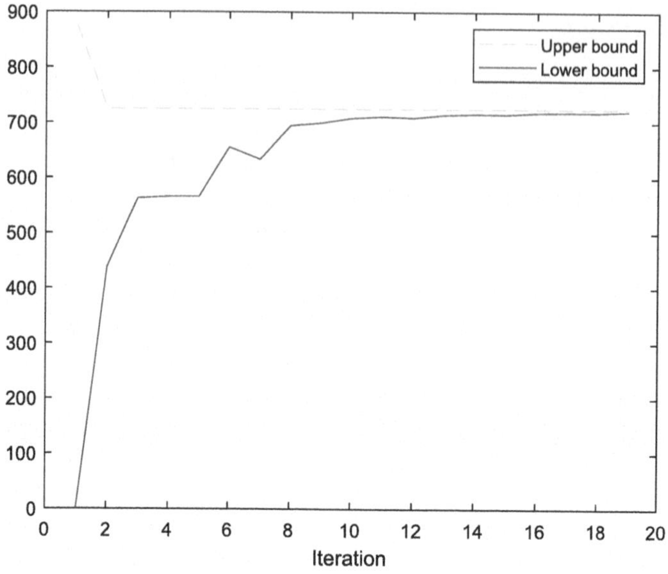

Fig. 10.5 Progress of the lower and upper bound

Remaining Iterations

To illustrate the convergence of the subgradient optimization method, Fig. 10.5 reports the progress of the lower bound and the upper bound over the iterations. The figure shows that both the lower bound and the upper bound converge to a value of 725. Note that in general, the lower bound and the upper bound might not converge to the same value.

References

1. Abbink E, Fischetti M, Kroon L, Timmer G, Vromans M (2005) Reinventing crew scheduling at Netherlands Railways. INFORMS J Appl Anal 35(5):393–401
2. Abbink EJ, Albino L, Dollevoet T, Huisman D, Roussado J, Saldanha RL (2011) Solving large scale crew scheduling problems in practice. Public Transp 3 (2):149–164
3. Caprara A, Fischetti M, Toth P (1999) A heuristic method for the set covering problem. Oper Res 47(5):730–743
4. Desaulniers G, Desrosiers J, Solomon MM (2002) Accelerating strategies in column generation methods for vehicle routing and crew scheduling problems. Springer US, Boston, MA, pp 309–324. ISBN 978-1-4615-1507-4
5. Heil J, Hoffmann K, Buscher U (2020) Railway crew scheduling: Models, methods and applications. Eur J Oper Res 283(2):405–425. ISSN 0377-2217
6. Huisman D (2007) A column generation approach for the rail crew re-scheduling problem. Eur J Oper Res 180(1):163–173. ISSN 0377-2217

7. Irnich S, Desaulniers G (2005) Shortest path problems with resource constraints. In: Desaulniers G, Desrosiers J, Solomon MM (eds) Column generation, Chapter 2. Springer, New York, pp 33–65
8. Potthoff D, Huisman D, Desaulniers G (2010) Column generation with dynamic duty selection for railway crew rescheduling. Transp Sci 44(4):493–505
9. Wolsey LA (2020) Lagrangian duality. In: Integer programming, Chapter 10. Wiley, New York, pp 195–212. ISBN 9781119606475

Chapter 11
Dantzig–Wolfe Reformulation and Column Generation: Optimization in the Chemical Tanker Industry

Anurag Ladage, Ashutosh Mahajan, Andreas Ernst, and Mohan Krishnamoorthy

Abstract Global chemical supply chains rely on chemical tankers for the movement of their raw materials and products. A chemical tanker is a sophisticated multi-compartment cargo ship that transports chemicals between ports. This chapter describes some of the decision problems arising while planning operations of such tankers. Assignment of chemicals to compartments of the tanker is considered in particular. Several domain-specific assignment decisions related to cargo-cargo and cargo-compartment compatibility rules, and tanker balancing safety restrictions make this assignment problem unique. A mixed-integer linear programming (MILP) model for the assignment problem is presented along with some insights on its structure and complexity. A different formulation based on the Dantzig–Wolfe reformulation is derived. The reformulation is solved using a customized column generation framework. This approach enables one to take advantage of multiple special structures in the problem. Some discussions around the implementation nuances are also presented.

A. Ladage
IITB-Monash Research Academy, IIT Bombay, Powai, Mumbai 400076, India
e-mail: ladageanurag@gmail.com

A. Mahajan (✉)
Department of Industrial Engineering and Operations Research, IIT Bombay, Powai, Mumbai 400076, India
e-mail: amahajan@iitb.ac.in

A. Ernst
School of Mathematics, Monash University, Melbourne, Victoria 3800, Australia
e-mail: Andreas.Ernst@monash.edu

M. Krishnamoorthy
Stem College, RMIT City campus, Melbourne, Victoria 3000, Australia
e-mail: mohan.krishnamoorthy@rmit.edu.au

F. Hamid (ed.), *Optimization Essentials*, International Series in Operations Research & Management Science 353, https://doi.org/10.1007/978-981-99-5491-9_11

11.1 Introduction

The maritime industry accounts for about 80% of the world merchandise trade (and more than 90% in developing countries) according to [21]. A minor change in the efficiency of these networks will result in savings amounting to billions of dollars. Based on the planning requirements and the type of cargo transported, cargo shipping is classified into liner and tramp shipping. Liner shipping (also known as container shipping) is analogous to a bus service. It transports large volumes of cargo (inside containers) at a relatively cheaper cost, but it has a pre-planned schedule. The tramp shipping segment is like a cab service. It is scheduled according to the needs of the customer. Generally, liner ships transport cargoes within containers, while tramp ships bulk cargoes. Special types of tramp ships that transport liquid bulk cargoes like crude oil, petroleum, and chemicals are called parcel tankers.

This chapter focuses on a problem often faced in the chemical tanker industry. Chemical tankers are a special type of parcel tankers with multiple cargo holds. Their sophisticated design enables simultaneous transport of multiple non-mixable chemicals between ports. Figure 11.1 shows the different types of parcel tankers. The figure also highlights the chemical tanker, and presents its structure.

The chapter introduces a *multi-period cargo-assignment problem* (mp-CAP) often observed while transporting liquid bulk cargoes (chemicals). For a fixed route, and a fixed set of cargoes, the mp-CAP tries to optimally assign cargoes to the compartments of the ship throughout its route. At most one cargo can be assigned to a compartment, but a cargo can be distributed into multiple compartments. The mp-CAP tries to minimize the total number of changeover or cargo-compartment swaps required during the ship's voyage. The problem is formulated initially as a mixed-integer problem (MIP), within which some *special structures* are observed. The model is rewritten as a Dantzig–Wolfe (DW) reformulation to leverage these special structures. A typical DW reformulation increases the number of decision variables exponentially while reducing the total number of constraints in the formulation. A customized column generation framework is used to solve the DW reformulation.

Column generation is an iterative technique that solves a linear program with an exponentially large number of decision variables. In the context of handling integer programs, the column generation framework is solved at every node of the Branch and Bound tree to solve an integer program. This method is often referred to as Branch and Price. The readers are referred to [2, 9], for an introduction to integer programming, the Dantzig–Wolfe reformulation, and the column generation algorithm. A realistic model for some tanker operations, christened *mp-CAP* is used to illustrate application of these advanced optimization techniques in this chapter.

The chapter is further organized as follows. Section 11.2 presents a brief background of ship scheduling problems, and the chemical tanker scheduling domain. Section 11.3 introduces the mp-CAP. Section 11.4 presents the MILP model for the mp-CAP. Subsequently, an application of the DW reformulation, decomposition, and the column generation framework to the[1] mp-CAP are discussed in Sect. 11.5.

[1] Descriptions in this section are derived from [17].

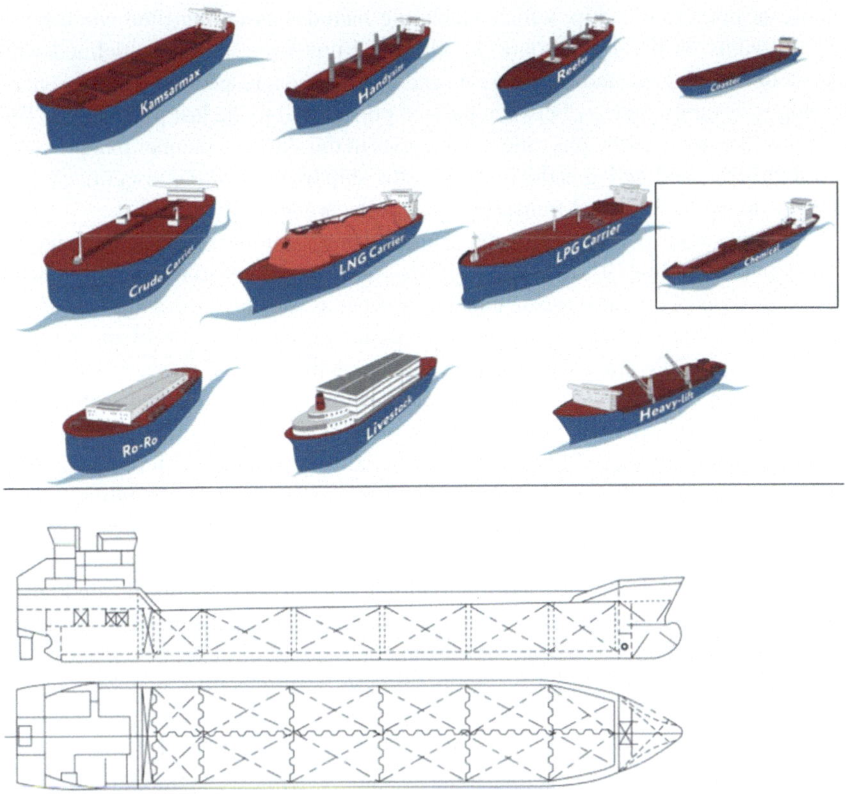

Fig. 11.1 Different types of tankers, and the structural layout of a typical chemical tanker

11.2 Cargo Ship Scheduling Overview

The cargo shipping industry can be broadly classified into two classes: liner shipping and tramp shipping. Liner shipping is cheaper compared to tramp shipping. However, liner ships have a fixed pre-published schedule, similar to a bus service. In contrast, Tramp ships are more expensive, but operate according to the needs of their customers. They transport different types of dry bulk cargoes (grain, coal, or metal) or liquid bulk cargoes (petroleum products, oil, and chemicals). Cargo commodities that are directly stored in the ship's compartments are called bulk cargoes. A fifty-year review of maritime transportation is presented in [20]. They also discuss various advances and challenges related to this domain. We briefly introduce some terminologies commonly used in the shipping industry, followed by a short discussion related to research in the chemical tanker industry.

Shipping generally refers to the act of transporting cargoes through ships, while shipper refers to the entity that provides the cargo. An operator generally refers to the entity responsible for scheduling the cargo ships. Routing relates to assigning a series of ports to the ship, while scheduling includes assigning time windows to various events on the ship's route. Loading and unloading ports are defined with respect to the cargoes, and their origins and destinations, respectively. A voyage of the ship is its entire journey between the first port visited to the last port visited. The term stowage refers to the placement of cargoes in the ship in a manner that provides ship stability. A sailing leg is the journey of the ship between two consecutive ports. Tankers are cargo ships that transport liquid bulk cargoes.

One of the first papers to discuss mathematical programming in the shipping industry was published by [19]. References [4–7, 16] also present excellent overview of the applications of operations research in both liner as well as tramp shipping. Christiansen et al. [6] specifically focus on the evolution of scheduling problems in tramp shipping. Further, [7, 22] extensively discuss the application of operations research to tramp shipping since 1983. Researchers have tackled the different scheduling problems by modeling them as an extension of the traveling salesman problem with pickups and deliveries (TSPPD) or the pickup delivery problem with time windows (PDPTW). The TSPPD-based formulations are used to model scheduling activities within a port. In contrast, the PDPTW extensions are used to model inter port scheduling problems.

Tramp ships capable of transporting multiple non-mixable chemicals are known as chemical tankers. Given a list of ports and potential cargoes, the problem is finding a schedule for a chemical tanker. A schedule consists of a subset of ports to visit, the planned arrival times at each port, cargoes to be serviced by the ship at each of these ports, and the cargo weights assigned to each compartment. According to [4], the maritime industry accounts for about 80–90% of the total world merchandise trade by volume and 60% of the total trade by value. Chemicals, including petroleum products, constitute an essential class of merchandise.

For chemical tanker scheduling problems, a sailing leg of the tanker is analogous to an arc on the network formed by ports, while a voyage refers to the route or path of the tanker. A chemical tanker has multiple cargo holds, enabling it to transport multiple chemicals simultaneously. Ladage et al. [17] present a chemical tanker scheduling problem modeled as an extension of the PDPTW. A summary of this scheduling problem is presented next in order to explain important decisions involved while transporting chemicals over a network of ports. The tanker must transport onboard chemicals to their corresponding unloading ports by determining the least expensive path or voyage. At the same time, the tanker can serve additional cargoes to increase the overall profit. Often, time windows are associated with cargoes, which restrict their pickup and drop-off times.

A schedule is generated to maximize the difference between the revenue and four different costs. These costs include the fuel cost incurred while traveling between ports, the time charter cost, a fixed port cost incurred while visiting a new port, and a changeover cost incurred if cargoes are swapped within compartments. Important decisions that need to be optimized include the cargoes to be picked up, the ports to be visited, the sequence in which these ports are visited, the arrival times at each port, and a feasible cargo-compartment allocation.

During each port visit, a constant administrative time incorporates berth allocation waiting time, bunkering activities, and customs and security clearances. Berths are different locations within a port where a ship needs dock in order to load or unload cargoes. A port typically has multiple berths. Bunkering refers to refueling the ship, and its associated time and cost. Time windows restrict cargo pickups. The pickup time windows are specified in units of days (fractional days are allowed). Any cargo picked up must be delivered within a maximum period, which is the number of days for which the schedule is generated. The temporal activities considered in this problem are loading/unloading times of the cargo, travel time between ports, and a fixed administrative time at each port. For simplicity, [17] assumes that all temporal activities are sequential.

Every chemical tanker has a loading capacity primarily dictated by draft restrictions. Draft limit is the maximum portion of the ship that can be submerged under water after loading it. Additional heel and trim movement restrictions are also considered by [17]. Heel is defined as the rotational movement of the ship along its longitudinal or length wise axis, while trim is rotational movement of the ship along its lateral or breath wise axis of the ship. Heel and trim movement limits ensure the safety of the tanker. These movements influence the distribution of cargoes in different compartments of the tanker.

These factors make the chemical tanker scheduling problem unique compared to other domains. Additionally, the presence of multiple compartments poses additional challenges in assigning cargoes to different compartments. We call this assignment problem the *multi-period cargo-assignment problem* (mp-CAP). Having understood the primary operations of a chemical tanker scheduling problem, the following section presents an overview of the mp-CAP.

11.3 The Multi-period Cargo-Assignment Problem (mp-CAP)

The presence of multiple compartments provides added flexibility to transport multiple chemicals. However, it also makes the problem more complicated. In the literature, this specific problem of assigning cargoes to compartments during a single sailing leg is called the cargo-assignment problem (CAP). CAP generates a feasible plan for a given set of cargoes and compartments. The cargo-compartment plan must adhere to compatibility, stability, and compartment capacity constraints. Some other modeling constraints like at most one cargo can be assigned to a compartment, cargo has to be assigned to at least one compartment, and the total weight of cargo distributed in all compartments should be equal to the total available weight of the cargo, need to be respected. Some recent publications that discuss the chemical tanker scheduling problem while addressing the difficulties originating due to the cargo-assignment aspect of the problem are [12, 15, 17, 18, 23].

Traditionally, the assignment plans are generated individually during every sailing leg. One major drawback of generating a cargo-compartment allocation plan at different sailing legs individually is that it might unnecessarily re-arrange chemicals between compartments to make room for new cargo that might be picked up. Even worse, re-arranging chemicals might not be allowed in many practical applications as they can prove quite time-consuming and expensive. However, if cargo re-arrangements are not allowed, cargo-compartment allocations during previous sailing legs might make a future profitable cargo un-serviceable due to the unavailability of compartments. Re-arrangements or not, a cargo-assignment problem that considers possible future cargo-compartment placements is needed from a practical point of view. As such, we propose a different variation of the CAP, which we refer to as the *multi-period cargo-assignment problem* (mp-CAP).

Here is a small example illustrating the mp-CAP. Assume that the chemical tanker (with four compartments) visits three ports on its route and transports four chemicals during its voyage. Cargo 1 remains on the tanker for all three ports. Cargo 2 is loaded at Port 1 and dropped off at Port 3, while Cargo 3 is loaded at Port 2 and remains on the tanker at Port 3. Cargo 4 is loaded at Port 2 and delivered at Port 3. Additionally, Cargo 1 and Cargo 4 are incompatible with each other, while Cargo 2 and Cargo 3 are incompatible. Assume that incompatible cargoes cannot be stored in neighboring (vertically or horizontally adjacent) compartments.

Figure 11.2 portrays a feasible solution generated by the mp-CAP. Cargo 1 remains assigned to Compartment 1 through the tanker's voyage. At the first port, Cargo 2 was stored in Compartment 4. Considering the incompatibilities among the cargoes, at Port 2, Cargo 2 had to be reassigned to Compartment 3, while Cargo 3 and 4 were placed in Compartments 2 and 4, respectively. At Port 3, to maintain the balance of the tanker, Cargo 3 was reassigned into Compartments 2, 3, and 4. The directed

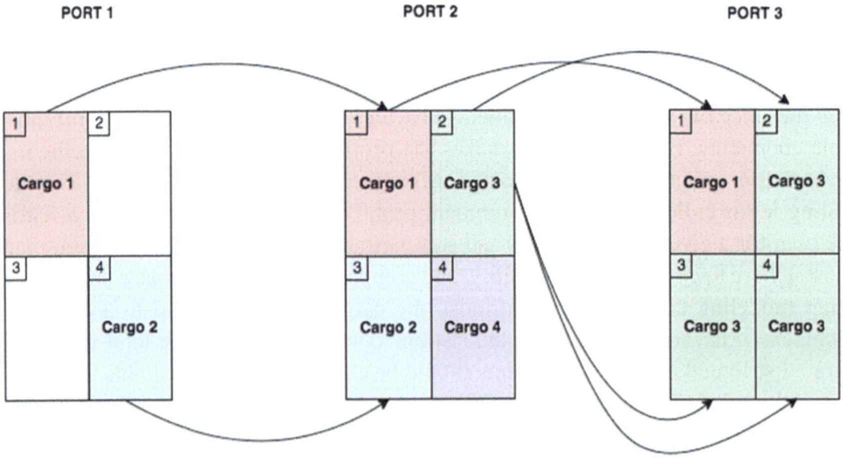

Fig. 11.2 toy size example presenting a solution instance of the multi-period cargo-assignment problem

arrows in Fig. 11.2 show the flow of cargoes from one port to another. The readers should be able to see that the solution instance presented in Fig. 11.2 results in seven cargo swaps, namely, two at Port 1, three at Port 2, and two at Port 3. Note that for completeness, for the loading leg of the cargo, the number of changeovers is equal to the number of cargo-compartment assignments.

The mp-CAP is defined to generate a cargo-compartment assignment plan at each port on the tanker's voyage (route) while minimizing the total number of changeovers or cargo swaps. The following section elaborates the mp-CAP further by presenting its mixed-integer linear programming (MILP) formulation.

11.4 MILP Formulation for the mp-CAP

The *multi-period cargo-assignment problem* (mp-CAP) minimizes the total number of cargo changeovers throughout the voyage of the chemical tanker. A changeover is expensive as it incurs tank cleaning costs, time, and labor costs throughout the voyage. The mp-CAP simultaneously generates a feasible cargo-compartment allocation plan for the chemical tanker's entire voyage. This would generate an assignment plan with the least number of changeovers, significantly reducing the associated costs while providing the necessary flexibility to the model for future cargo assignments. This section further presents a mathematical formulation[2] for the mp-CAP.

Sets :

K = Set of sailing legs,

N^G = Set of all cargoes/goods assigned to the chemical tanker during the voyage,

N_k^O = Set of cargoes/goods loaded on the chemical tanker at the end of leg $k \subset K$,

K_j = Set of sailing legs for cargo $j \in N^G$ is onboard the chemical tanker,

N_j^I = Set of cargoes incompatible with cargo $j \in N^G$, $N_j^I \subset N^G$,

N^H = Set of compartments (cargo holds) in the ship,

N_h^B = Set of neighboring/bordering compartments for compartment $h \in N^H$, $N_h^B \subset N^H$,

N_h^X = Set of cargoes that cannot be stored in compartment $h \in N^H$, $N_h^X \subset N^G$.

Indices :

k = Index for sailing leg (Index 0 indicates that the chemical tanker is at its starting port),

j = Index for cargo,

h = Index for compartment (cargo hold).

---

[2] The formulation in this section is heavily derived from [17].

Decision variables :

c_{kjh} = 1 if the chemical tanker at the end of leg $k \in K_j$ carries cargo $j \in N^G$ in compartment $h \in N^H$ (Binary),

w_{kjh} = Weight of cargo $j \in N^G$ assigned to compartment $h \in N^H$ of chemical tanker at end of leg $k \in K_j$ (Continuous),

b_{kjh} = 1 if the chemical tanker at the end of leg $k \in K_j$ replaces any cargo $j \in N^G$ (other than itself) with cargo $j \in N^G\backslash\{0\}$ in compartment $h \in N^H$ (Binary).

Parameters :

K_j^L = Loading leg for cargo $j \in N^G$,

K_j^D = Unloading leg for cargo $j \in N^G$,

C^S = Cost per changeover/swap including cleaning, labor, etc., related to swapping cargoes within compartments. Also represents the cost incurred if a cargo $j \in N^G\backslash\{0\}$ is filled in an empty compartment,

W_j = Weight of the cargo $j \in N^G$,

ρ_j = Density of the cargo $j \in N^G$,

V_h = Volume of compartment $h \in N^H$,

κ_h = Lateral distance from compartment $h \in N^H$ to the center of the chemical tanker,

ι_h = Longitudinal distance from compartment $h \in N^H$ to the center of the chemical tanker,

α = Maximum absolute permissible trim causing moment of the chemical tanker,

β = Maximum absolute permissible heel causing moment of the chemical tanker,

Objective Function:

$$\textbf{Minimize} \quad \mathbf{z} = C^S \sum_{k \in K} \sum_{j \in N^G\backslash\{0\}} \sum_{h \in N^H} b_{kjh} \tag{11.1}$$

The objective function (11.1) minimizes the total changeover cost (C^S) throughout the voyage of the chemical tanker.

Set of constraints:

$$\sum_{h \in N^H} c_{kjh} \geq 1 \qquad \forall j \in N^G, k \in K_j, \tag{11.2}$$

$$\sum_{j \in N_k^O} c_{kjh} \leq 1 \qquad \forall k \in K, h \in N^H, \tag{11.3}$$

$$b_{kjh} \geq c_{kjh} - c_{(k-1)jh} \qquad \forall j \in N^G, k \in K_j \backslash \{K_j^L\}, h \in N^H, \tag{11.4}$$

$$b_{K_j^L jh} = c_{K_j^L jh} \qquad \forall j \in N^G, h \in N^H, \tag{11.5}$$

$$c_{kjh} + \sum_{j' \in \{N_j^I \cap N_k^O\}} c_{kj'h'} \leq 1 \qquad \forall k \in K, j \in N_k^O, h \in N^H, h' \in N_h^B, \tag{11.6}$$

$$w_{kjh} \leq V_h^H \rho_j c_{kjh} \qquad \forall j \in N^G, k \in K_j, h \in N^H, \tag{11.7}$$

$$\sum_{h \in N^H} w_{kjh} = W_j \qquad \forall j \in N^G, k \in K_j, \tag{11.8}$$

$$-\alpha \leq \sum_{h \in N^H} \sum_{j \in N_k^O} w_{kjh} \iota_h \leq \alpha \qquad k \in K, \tag{11.9}$$

$$-\beta \leq \sum_{h \in N^H} \sum_{j \in N_k^O} w_{kjh} \kappa_h \leq \beta \qquad \forall k \in K. \tag{11.10}$$

Constraint (11.2) states that for every cargo and for every leg it is on board the ship, the summation of assignment variables (c_{kjh}) should at least be one. Essentially, it means that cargo $j \in N^G$ has to be assigned to at least one compartment between its pickup and discharge legs. K is the set of sailing legs, while N^H is the set of compartments. Thus, Constraint (11.3) ensures that during every sailing leg $k \in K$, for each compartment $h \in N^H$ at most one assignment variable c_{kjh} can be one. This ensures that a compartment can either be empty or hold at most one cargo.

Constraint (11.4) models changeovers. It tracks the swapping of cargoes within the chemical tanker's compartments. For two consecutive sailing legs k and $k-1$, for every cargo and for every compartment, Constraint (11.4) states that the changeover variable (b_{kjh}) is 1 if cargo j is assigned to compartment h in leg k ($c_{kjh} - 1$) and was not assigned ($c_{(k-1)jh} = 0$) to compartment h in leg $k-1$. Readers can observe that for any other combination of c_{kjh} and $c_{(k-1)jh}$, variable b_{kjh} becomes a free variable. However, the objective function forces it to zero. Constraint (11.5) defines the changeover variables for each cargo j and each compartment h. It is defined for the loading leg of each cargo j.

Constraint (11.7) ensures that the compartment capacity is not exceeded. Essentially, it makes sure that if a cargo j is assigned to a compartment h ($c_{kjh} = 1$), then the quantity of that cargo (w_{kjh}) assigned to the compartment cannot exceed the compartment capacity ($V_h^H \rho_j$). Constraint (11.8) ensures that the quantity of a cargo that is distributed in multiple compartments during a sailing leg is equal to the total weight of the cargo. Constraints (11.9) and (11.10) ensure that the maximum allowable trim and heel moments are not exceeded. Constraints (11.9) and

(11.10) are essential for the stability of the ship. These constraints ensure proper (stable) distribution of cargoes in different compartments. The cargo-compartments are enforced in pre-processing by fixing the bounds of the c_{kjh} variables. If a cargo j cannot be stored in compartment h, then for all sailing legs $k \in K$ the upper bounds on variables c_{kjh} are set to 0.

At this point, the readers are suggested to pause and study the MILP formulation carefully. Remember the earlier statement about the presence of *special structure* as a motivation toward employing DW decomposition and reformulation? Can the readers identify such structures in the MILP formulation of the mp-CAP? Before answering these questions, it is essential to understand the theory of both the Dantzig–Wolfe reformulation and the column generation algorithm. Section 11.5 discusses and motivates both these techniques.

11.5 Dantzig–Wolfe Reformulation and Column Generation

The concept of Dantzig–Wolfe (DW) decomposition and reformulation was first introduced by [10]. DW reformulation and decomposition are often employed when the set of constraints for a given MIP formulation can be separated into *easy* and *complicated* constraints. Ideally, the set of *easy* constraints would define a feasible region over which one can quickly evaluate the solution to a given objective function. From a practical perspective, the *niceness* of the feasible region defined by the *easy* constraints is due to some unique structure or the availability of good heuristics to evaluate a given objective over it.

$$\textbf{MILP Formulation}: \quad \begin{aligned} &\text{Min } c^T x \\ &A_1 x \geq b_1, \\ &A_2 x \geq b_2, \\ &x_j \in \mathbb{Z} \;\; \forall j = 1, \ldots, p, \\ &x_j \geq 0 \quad \forall j = 1, \ldots, n. \end{aligned} \tag{11.11}$$

Consider the MILP (11.11), where $p \leq n$, and all the data is assumed to be rational. Assume that $A_1 x \geq b_1$ are *complicating* constraints, while $A_2 x \geq b_2$ are *nice* constraints. DW reformulation technique uses Minkowski's Double Description Theorem, also known as the Minkowski-Weyl Theorem, to reformulate the feasible region represented by the complicating constraints. Formally, the Minkowski-Weyl theorem is stated as follows.

Theorem 11.1 (Minkowshi-Weyl Theorem) *Any polytope $P \in \mathbb{R}^n$ can be defined by either*

- $P = \{x \in \mathbb{R}^n : Ax \geq b$ *for some* $m \times n$, *matrix* A *and vector* $b \in \mathbb{R}^m\}$, *or*

- *For some set of vertices $f_1, \ldots, f_M \in \mathbb{R}^n$ and extreme rays $r_1, \ldots, r_N \in \mathbb{R}^n$ (where potentially M=0 or N=0): $P = \{\sum_{i=1}^{M} \lambda_i f_i + \sum_{j=1}^{N} \alpha_j r_j : \lambda \in \mathbb{R}^M, \alpha \in \mathbb{R}^N, \sum_{i=1}^{M} \lambda_i = 1, \lambda_i, \alpha_j \geq 0 \ \forall i, j\}$.*

Theorem 11.1 states that the solution to any linear program can be represented as a convex combination of the vertices (corner points) and extreme rays. It is assumed that the feasible region is always bounded for all further discussions, which is the case in most practical integer formulations. This means there are no extreme rays ($N = 0$). The feasible region of the *mp-CAP* is also bounded.

DW Relaxation:

$$\begin{aligned} \text{Min} \sum_{v \in V} c f^v \lambda^v, \\ \sum_{v \in V} A_1 f^v \lambda^v \geq b_1, \\ \sum_{v \in V} \lambda^v = 1, \\ \lambda \in \mathbb{R}^{V+}, \end{aligned} \tag{11.12}$$

$$\sum_{v \in V} f_j^v \lambda^v \in \mathbb{Z}^+ \quad \forall j = 1, \ldots, p. \tag{11.13}$$

Let V be a finite set of corner points of the feasible region defined by $Q := \{x \in \mathbb{R}_+^n : A_2 x \geq b_2 \ \& \ x \in \mathbb{Z}\}$. Let $f^v \in V$ represent a corner point. Assuming the absence of extreme rays, formulation (11.12) represents the DW relaxation of the LP relaxation of the MILP (11.11). Equation (11.13) is used to enforce the integrality condition.

This transformation is simply obtained by substituting $x_j = \sum_{v \in V} f_j^v \lambda^v$, where $\sum_{v \in V} \lambda = 1$ and f^v are vertices of the feasible region Q. Readers must observe that the number of constraints has reduced. In addition to reducing the number of constraints under favorable conditions, the DW reformulation also provides tighter bounds than the LP relaxation. However, the number of decision variables increased exponentially. A DW reformulation has at least as many decision variables as the number of corner points (and extreme rays, if present).

Consider the LP (11.12). Based on the knowledge of the simplex algorithm, readers can observe that the number of columns in the simplex table is equivalent to the number of corner points in set V. Generating all the columns beforehand is computationally quite expensive for practical problems. Nevertheless, readers must recall the following optimality condition defined by the simplex algorithm.

Theorem 11.2 *Consider a basic feasible solution x associated with a basis matrix B, and let $\bar{rc}$ be the vector of reduced costs. Then, the following statements are true for a minimization problem.*

- *If $\bar{rc} \geq 0$, then x is optimal.*
- *If x is optimal and non-degenerate, then $\bar{rc} \geq 0$.*

If the above optimality conditions are met, then the optimal solution $x^* = B^{-1}b$, where b is the R.H.S. of the constraints defining the feasible region. Observe that only a small subset of columns with non-negative reduced costs are required to prove optimality and calculate the optimal solution.

Column generation (also often referred to as delayed column generation) uses this fact to generate only those columns that improve the solution iteratively. Column generations algorithm was first discussed by [13] in the context of the cutting stock problem. Theoretically, this method terminates when there are no more columns that can enter the basis B, meaning no additional column with a non-negative reduced cost exists.

Restricted master problem (RMP):

$$\text{Min} \sum_{v \in V'} c f^v \lambda^v, \tag{11.14}$$

$$\sum_{v \in V'} A_1 f^v \lambda^v \geq b_1, \tag{11.15}$$

$$\sum_{v \in V'} \lambda^v = 1, \tag{11.16}$$

$$\lambda \in \mathbb{R}^{V'+}. \tag{11.17}$$

In the context of the DW reformulation (11.12), the relaxation (11.14)–(11.17), where $V' \subseteq V$ such that the problem is feasible. Formulation (11.14)–(11.17) is referred to as the restricted master problem (RMP). If the RMP is unbounded, the DW relaxation (11.12) is also unbounded.

Otherwise, for a given subset of column V', let $\bar{\lambda}$ be the optimal solution for the RMP (11.14). Moreover, let $\bar{\mu_1}$ and $\bar{\mu_2}$ be the optimal dual solution corresponding to the RMP Constraints (11.15) and (11.16), respectively. Observe that $\bar{\mu_1}$ is a vector, while $\bar{\mu_2}$ is a scalar. Populating V' with any new columns with non-negative reduced costs can only improve the solution. In other words, if the RMP has an optimal solution, then it is an upper bound on the full DW reformulation. New columns with negative reduced cost can be generated by solving the Formulation (11.18).

Sub-problem (SP):

$$Min_{f \in V} \quad f^T c - f^T {A_1}^T \bar{\mu_1} - \bar{\mu_2} \tag{11.18}$$

$$Min_{x \in Q} \quad (c - {A_1}^T \bar{\mu_1})^T x - \bar{\mu_2} \tag{11.19}$$

As f is just a corner point (or a basic solution) of Q, Formulation (11.18) can be rewritten in terms of x as Formulation (11.19). Thus, the sub-problem is nothing but our original problem (11.11) without the constraints $A_1 x \leq b_1$ and the objective function equivalent to the reduced cost in terms of the dual values $(\bar{\mu}_1, \bar{\mu}_2)$. Since the feasible region is assumed to be bounded, only two possible results can be obtained by solving the SP (11.19). If the objective is non-negative, then $(\bar{\lambda})$ is the optimal solution to the DW relaxation (11.12), where the values of variables λ_v, $v \in V \backslash V'$ are set to zero. Otherwise, one or more columns (variable) with negative reduced cost enter the restricted master problem.

It is imperative to understand the bounds generated by the reformulation. If the RMP has an optimal solution, then it is an upper bound on the full DW reformulation. In the absence of degeneracy, this upper bound keeps improving with each iteration of the column generation algorithm. However, how does the optimal solution to the full DW reformulation compare to the MILP formulation (11.11) and its LP relaxation? Let IP^*, LP^*, and DW^* be the optimal solutions of the MILP formulation (11.11), its relaxation, and the full DW reformulation, respectively. From the theory of integer programming, readers are already aware that for a minimization problem, $IP^* \geq LP^*$. The following theorem gives the relationship between the LP^* and DW^*.

Theorem 11.3 *For minimization problems defined as (11.11), $DW^* \geq LP^*$. Furthermore, $DW^* = LP^*$ if $conv(Q) = \{x \in \mathbb{R}^n : A_2 x \geq b_2\}$.*

The above theorem effectively means that tighter bounds from the DW reformulation are achieved if the sub-problems are harder to solve (harder in the sense that at least branching within the branch & bound tree is required to get an integer solution for the sub-problem). However, even this does not assure a better bound. If readers are aware of the Lagrangian relaxation, then they should derive and observe the fact that the Lagrangian dual is the LP dual of the DW relaxation (11.12). However, this proof is outside the scope of this chapter.

To solve the MILP (11.11), the DW relaxation is solved at each branch and bound tree node. This procedure is called Branch & Price (B&P). B&P can be considered an extension of the Branch & Bound (B&B) algorithm. Let Z_{ip}, Z_{lp}, and Z_{dw} denote the MILP formulation (11.11), its LP relaxation, and its DW relaxation (11.12), respectively. Recall that $x_j^* \in \mathbb{Z}$, $j = \{1, \ldots, p\}$ defined for Z_{ip} are integer variables. B&B solves Z_{lp} at each node of the tree while B&P solves Z_{dw} at each tree node.

An implementation of the B&P starts by solving the Z_{dw} at the root node. B&B employs the Simplex method to solve Z_{lp}, while B&P uses column generation to solve Z_{dw}. Theoretically, Z_{dw} is a form of linear relaxation and can be solved using the Simplex method. However, column generation or delayed column generation is very effective at handling formulations with an exponential number of variables, which is typically the case for Z_{dw}. At the root node, Z_{dw} is solved to optimality using column generation, which generates fractional λ's. The corresponding fractional x variable vector can be obtained from the Equation $x_j = \sum_{v \in V} f_j^v \lambda^v$, where f^v are the corner points defined in formulation (11.12).

If any $x_j^* \in \mathbb{Z}$, $j = \{1, \ldots, p' | p' \subseteq p\}$ take fractional values in the optimal solution of Z_{dw}, then the B&P algorithm creates $2p'$ branches to create as many additional DW relaxations. Every fractional integer variable x_j^* gives rise to two new DW relaxations, which have bounding constraints, $x_j^* \leq \lfloor x_j^* \rfloor$ and $x_j^* \geq \lceil x_j^* \rceil$. In terms of λ, these constraints can be written as $\sum_{v \in V} f_j^v \lambda_v^* \leq \lfloor x_j^* \rfloor$ and $\sum_{v \in V} f_j^v \lambda_v^* \geq \lceil x_j^* \rceil$. For example, if $x_1 = 2.5$ in the optimal solution of the Z_{dw}, then the bounding constraints are $x_1 <= 2$ and $x_1 >= 3$.

The B&P algorithm solves the new DW relaxations and the branches generated from them. All nodes resulting in optimal solutions with variables $x_j^* \in \mathbb{Z}$, $j = \{1, \ldots, p\}$ having integer values are known as the leaf node. The optimal solution at any leaf node is a feasible solution for Z_{ip}. Each leaf node generates a feasible solution or an upper bound, while each DW relaxation generates a lower bound for the Z_{ip}. Depending on the termination criteria, the B&P algorithm terminates after a certain time or relative gap (%). An outline of the B&P procedure is given below.

1. Solve the DW relaxation at a node to obtain the optimal solution λ^*, and calculate the corresponding optimal solution $x^*(\sum_{v \in V} f^v \lambda_v^*)$ for the LP relaxation of the MILP (11.11).
2. If x^* satisfied the integrality constraint $x_j^* \in \mathbb{Z}$, $j = 1, \ldots, p$, then a leaf node has been reached.
3. Otherwise, create new nodes by branching on every fractional x_j^* variables by introducing the following constraints for every new node j.

$$\begin{aligned} \sum_{v \in V} f_j^v \lambda_v^* &\leq \lfloor x_j^* \rfloor \\ \sum_{v \in V} f_j^v \lambda_v^* &\geq \lceil x_j^* \rceil \end{aligned} \tag{11.20}$$

Practical implementations of B&P have to be significantly improved using problem-specific cuts, heuristics to generate better upper bounds, and better formulations to reduce the computational effort required to solve DW relaxation at every node. Readers should also note that the sub-problem's structure changes every time a bounding constraint is added to the problem. This could destroy any special structure inherently present in the DW reformulation at the root node. These challenges must be addressed carefully while implementing column generation and the B&P algorithm.

To summarize the general procedure of the DW reformulation as applied to MILPs, the readers must start by identifying a problem for which the constraint set can be separated into *easy* and *complicated* constraints. The problem can be separated into a restricted master problem and one or many sub-problems. Subsequently, a column generation approach can be employed within a Branch and Price framework to obtain the integer solution for the MILP. Having understood the basics of DW reformulation and column generation, the readers should ponder upon the following questions.

- Can a decomposed problem have multiple sub-problems?
- Can a formulation be decomposed in multiple ways? Moreover, if multiple decompositions do exist, will they be equally efficient?
- Other than the decomposition into a master problem and a sub-problem, are there any additional special properties or structures worth exploiting?
- Is Branch and Bound algorithm always required to generate an MILP solution from a given set of master problem columns?

The discussion in the next section hopes to answer most of the above questions using the *mp-CAP*. The DW reformulation and column generation framework is applied to the *mp-CAP* to understand the practical aspects of implementing these techniques on a real-world problem.

11.6 Dantzig–Wolfe Reformulation and Column Generation Framework for the mp-CAP

This section discusses the DW decomposition and reformulation of the mixed-integer linear program (MILP) presented in Sect. 11.4. The unique structures in this problem provide two advantages. First, the constraint set can be divided into *nice* and *complicated* constraints. This is used to decompose the problem with a DW reformulation. This results in a formulation with fewer constraints but an exponentially large number of decision variables. The decomposed reformulation is solved within a column generation framework, also presented in this section. A unique secondary structure helps to generate heuristic MILP solutions at any iteration of the column generation procedure, without the Branch and Bound algorithm.

Observe the formulation presented in Sect. 11.4. The reader might observe that Constraint (11.4) is the only constraint linking decisions from subsequent sailing legs. Ignoring this *complicated* or linking constraint decomposes the problem into multiple sub-problems. Nonetheless, this is not the only decomposition possible for the mp-CAP MILP formulation.

Another decomposition can be obtained by retaining Constraints (11.3), (11.6), (11.9), and (11.10) in the master problem. Coincidentally, even this decomposition results in multiple sub-problems. Similarly, there exist multiple decompositions based on the MILP formulation presented in Sect. 11.4. However, readers will understand the multiple advantages of the problem decomposition presented in this section as the discussion progresses. Interestingly, there even exists a Bender's decomposition for the mp-CAP MILP formulation. However, that discussion is beyond the scope of this chapter. Interested readers should refer to Chap. 8 of [9], which provides an excellent overview of reformulations and relaxations in the context of integer programming.

In this section, we present the decomposition that retains Constraints (11.4) and (11.5) in master problem. The rest of the problem disintegrates into as many sub-problems as the number of ports visited by the tanker on its voyage. Each sub-problem deals with generating a cargoes-compartment assignment plan at each port. Such a special structure is known as a block diagonal structure. This decomposition allows us to leverage this structure inherently present in the MILP formulation. Any problem with a block diagonal structure can be decomposed into one master problem and multiple sub-problems. The DW reformulation is presented in Sects. 11.6.1 and 11.6.2.

11.6.1 Restricted Master Problem

This section presents the restricted master problem (RMP) of the DW reformulation for the mp-CAP MILP formulation (11.4). Constraints (11.4) and (11.5) are retained in the master problem. Let $\{V_k|k \in K\}$ be the set of all corner points of the feasible region (polytope) of the kth sub-problem. Let $V'_k \subseteq V_k$ be the set of corner points present in the restricted master problem at any iteration. $\{f^v_{jh}|k \in K, j \in N^O_k, h \in H, v \in V_k\}$ is used to denote a corner point. A corner point or a column in the RMP is a cargo-compartment allocation plan for the tanker's entire voyage (route). Every new column represents a different cargo-compartment allocation plan. As the readers can imagine, there is an exponential number of feasible allocations or columns, which necessitates the use of column generation to solve this reformulation.

λ^v are the multipliers dictating the convex combinations of the corner points. Every unique vector λ defines a unique cargo-compartment allocation plan. Based on Theorem 11.1, note the substitution $c_{kjh} = \sum_{v \in V'_k} f^v_{jh}\lambda^v$ while formulating the below master problem. The mathematical formulation of the RMP is as follows.

Objective Function:

$$\textbf{Minimize} \quad \mathbf{z_{mp}} = C^S \sum_{j \in N^G} \sum_{k \in K_j} \sum_{h \in N^H} b_{kjh} \tag{11.21}$$

Set of constraints:

$$\sum_{v' \in V'_{(k-1)}} f^{v'}_{jh}\lambda^{v'} - \sum_{v \in V'_k} f^v_{jh}\lambda^v \geq -b_{kjh} \qquad \forall j \in N^G, k \in K_j \backslash \{K^L_j\}, h \in N^H, \tag{11.22}$$

$$\sum_{v \in V'_{K^L_j}} f^v_{jh}\lambda^v = b_{K^L_j jh} \qquad \forall j \in N^G, h \in N^H, \tag{11.23}$$

$$\sum_{v \in V_k} \lambda^v = 1 \qquad \forall k \in K, \tag{11.24}$$

$$\lambda^v \geq 0 \qquad \forall k \in K, v \in V_k. \tag{11.25}$$

Objective function (11.21) is same as that of the original problem. It minimizes the total number of changeovers throughout the voyage of the ship. Constraints (11.22) and (11.23) are the reformulated constraints corresponding to Constraints (11.4) and (11.5), respectively. Constraint (11.22) defines the changeover activity throughout the voyage of the tanker. For the loading leg (K^L_j) of the cargo $j \in N^G$, Constraint (11.23) states that the number of changeovers ($b_{K^L_j jh}$) is equal to the convex combination of compartment assignments of cargo $j \in N^G$. Constraint (11.24) is the convexity constraint, which forces the sum of λ's for every $k \in K$ to be equal to 1.

At every iteration, new decision variables λ^v's along with its corresponding new columns/corner points $\{f^v|v \in V_k, k \in K\}$ are added to the RMP. If new columns were added, the RMP is re-optimized for decision variables b_{kjh} and λ^{v}. The updated λ vector is passed to the sub-problem. The formulation and implementation of these sub-problems are presented in the next section.

11.6.2 Dantzig–Wolfe Sub-problem

The multi-period cargo-assignment problem (mp-CAP) has a block diagonal structure while considering the proposed decomposition. This results in multiple sub-problems. Each sub-problem generates partial columns during every iteration of the column generation procedure. These partial columns are aggregated together to form columns entering the master problem.

Due to the exponential number of columns, the traditional DW relaxation and column generation often leads to slow convergence. As a result, some form of guidance toward generating good quality columns is often introduced to accelerate the algorithm's convergence. Some such customizations for the mp-CAP are also discussed further in this section.

Let $\bar{\lambda}^{v}$ the optimal solution to the master problem and $\bar{\mu}^{1}_{kjh}, \bar{\mu}^{2}_{K^L_j jh}, \bar{\mu}^{3}_{k}$ be the corresponding optimal dual values for Constraints (11.22), (11.23), and (11.24), respectively. Let μ^{LP}_{kjh} be the dual values corresponding to Constraints (11.4) and (11.5) obtained by solving the LP relaxation of the TAP formulation (Eqs. (11.1)–(11.10)). Similarly, let μ^{3LP}_{k} be the dual value associated with Eq. (11.2). In the below sub-problems N^{O} denotes the onboard cargoes on the ship during that specific leg. The sub-problem can be decomposed into the number of sailing legs. Thus, for a fixed $k \in K$ the sub-problem is as follows.

Objective Function:

$$\textbf{Minimize} \quad \mathbf{z_k} = \sum_{j \in N^G} \sum_{h \in H} \Delta_{kjh} c_{kjh} \tag{11.26}$$

$$\Delta_{kjh} = \alpha \bar{\mu}_{kjh} + (1-\alpha)\mu^{LP}_{kjh} + c^{heur}_{kjh} \qquad \forall k \in K, j \in N^G, h \in N^H, \alpha \in [0,1]$$

$$\mu^{3}_{k} = \alpha \bar{\mu}^{3}_{k} + (1-\alpha)\mu^{3LP}_{k}$$

$$\bar{\mu}_{kjh} = \begin{cases} \bar{\mu}^{2}_{K^L_j jh} - \bar{\mu}^{1}_{(k+1)jh}, & \text{if } |K_j| > 1 \text{ and } k = K^L_j \\ \bar{\mu}^{2}_{K^L_j jh}, & \text{if } |K_j| = 1 \text{ and } k = K^L_j \\ \bar{\mu}^{1}_{kjh}, & \text{if } |K_j| > 1 \text{ and } k = |K_j| \\ \bar{\mu}^{1}_{kjh} - \bar{\mu}^{1}_{(k+1)jh}, & \text{otherwise} \end{cases}$$

$$c^{heur}_{kjh} = \begin{cases} 0, & \text{if } k = K^L_j \text{ or } c_{(k-1)jh} = 1 \\ \text{Penalty Constant}, & \text{otherwise} \end{cases}$$

Note: $|K_j| = 1$ signifies that the cargo was picked up and immediately dropped off at the next port. Additionally, columns are accepted into the master problem only if $z_k < \mu_k^3$.

Set of constraints:

$$\sum_{h \in N^H} c_{kjh} \geq 1 \qquad \forall j \in N^G, \tag{11.27}$$

$$\sum_{j \in \{N_k^O \cup 0\}} c_{kjh} = 1 \qquad \forall h \in N^H, \tag{11.28}$$

$$c_{kjh} + \sum_{j' \in \{N_j^I \cap N_k^O\}} c_{kj'h'} \leq 1 \qquad \forall j \in N_k^O, h \in N^H, h' \in N_h^B, \tag{11.29}$$

$$w_{jh} \leq V_h^H \rho_j \bar{c}_{jh} \qquad \forall j \in N^G, h \in N^H, \tag{11.30}$$

$$\sum_{h \in N^H} w_{jh} = W_j \qquad \forall j \in N^G, \tag{11.31}$$

$$\sum_{h \in N^H} \sum_{j \in N_k^O} \iota_h w_{jh} \geq -\alpha, \tag{11.32}$$

$$\sum_{h \in N^H} \sum_{j \in N_k^O} \iota_h w_{jh} \leq \alpha, \tag{11.33}$$

$$\sum_{h \in N^H} \sum_{j \in N_k^O} \kappa_h w_{jh} \geq -\beta, \tag{11.34}$$

$$\sum_{h \in N^H} \sum_{j \in N_k^O} \kappa_h w_{jh} \leq \beta. \tag{11.35}$$

Traditionally, the column generation framework is used to solve linear programs, which means that both the master problem and the sub-problems are linear. However, in the proposed framework, the master problem is linear, while the sub-problems are MILPs. These MILP sub-problems generate only integer columns that enter into the master problem. As the procedure progresses, only columns fulfilling the reduced cost criteria are accepted into the master problem, ensuring improvement in the objective function and termination of the algorithm.

Column generation theory also dictates that the sub-problem objective should minimize the reduced cost equation to find columns with negative reduced cost. However, in the proposed framework, the sub-problem objective function uses a customized cost coefficient (Δ) that improves the convergence of the overall framework.

The cost coefficient vector (Δ) is generated as a linear combination of three sets of parameters. The first set of parameters includes the duals ($\bar{\mu}$) generated by solving the LP master problem at every iteration. These are the reduced cost coefficients. The second set of parameters contains the dual values (μ^{LP}) generated from the LP relaxation of the TAP formulation. Vector (μ^{LP}) corresponds to Constraints (11.5),

(11.4), and (11.3). μ^{LP} helps warm start the algorithm and guide the search during the initial iterations. These also ensure that the lower bound (minimization problems) generated by the DW relaxation is at least as good as the bound generated by the LP relaxation. However, as the column generation algorithm progresses, the effect of μ^{LP} is reduced to let $\bar{\mu}$ dictate the search direction.

The final set of parameters (c^{heur}) is generated using a simple repair heuristic. These are introduced to help guide the column generation process by transmitting information between sub-problems. As the readers can surmise, solving the sub-problems separately can sometimes lead to the generation of partial columns, which are feasible but too disjoint in terms of the overall objective function.

For example, consider a small example. Remember that the overall objective is to minimize the total number of changeovers. For simplicity, ignore the ship balancing criteria, and assume that only one cargo needs to be assigned to a tanker (with two compartments) during two consecutive sailing legs. Recall that a column generated by the sub-problems is nothing but different cargo-compartment allocations for the tanker's route. Consider two different columns that can be generated. The first column gives an allocation that assigns the cargo to compartment one during both the sailing legs. Alternatively, the second column defines an allocation that assigns the cargo to compartment one during the first sailing leg and compartment two during the second sailing leg.

The readers should be able to recognize that one of these columns is superior. Even though both columns can enter the master problem, the first column is superior as it results in zero changeovers. The cargo remains assigned to the same compartment during both legs. The c^{heur} parameters enable us to transmit this information between consecutive sub-problems. The repair heuristic is applied intermittently to guide the solution search. Additionally, the iterations for which the repair heuristic is applied need not fulfill the reduced cost criteria.

The DW relaxation (RMP and SP) is solved until a termination criterion is met. The solution obtained at any iteration or even the optimal solution to the DW relaxation is not feasible for the MILP formulation. This is because, for the original formulation, the decision variables b_{kjh} are still fractional. To generate integer solutions at the end of any iteration, one would need to solve the Branch and Bound (B&P) tree by branching on the b_{kjh} variables. This would generate a feasible solution for the MILP from the corresponding DW relaxation solution. However, a unique structure exists in the master problem, eliminating the need to solve the B&B tree. The following section explores this unique structure of the master problem.

11.6.3 Master Problem—Shortest Path Problem Representation

The master problem is solved as a linear program (LP) within the proposed solution framework. Consequently, solving the restricted master problem to optimality produces a fractional solution that is not feasible for the MILP formulation (Eqs. (11.1)–(11.10)) of the TAP. In this section, we show that the restricted master problem (Sect. 11.6.1) can be represented as the shortest path problem. This enables us to quickly generate a heuristic MILP solution for the TAP from master problem columns at any iteration without applying the Branch & Bound (B&B) procedure.

Let $\{V_k | k \in K\}$ and $\{f^{v_k} | v_k \in V_k\}$ denote the set of corner points and a corner point for the kth sub-problem, respectively. Let each corner point be represented by a vertex or node on the directed graph. Directed edges can only exist between corner points of two consecutive sub-problems. Moreover, let the edge weights represent the total changeovers ($B_{v_k, v_k^{'}}$) between any two vertices. Simple arithmetic gives us the total number of changeovers ($B_{v_k, v_k^{'}}$) for any pair of vertices (v_k,$v_k^{'}$). An artificial origin node and a destination node are introduced to complete the shortest path representation. The problem then becomes finding the shortest path (least number of changeovers) between the origin node and the destination node.

Figure 11.3 presents a small example. Assume, that three ports are visited giving rise to three sub-problems. At the present iteration, we have three corner points for each of the sub-problems. Figure 11.3, vertex $f^{1,1}$ represents corner point f_{1_1}. Vertices O and D are the artificial origin and destination nodes. The solution to the instance represented in Fig. 11.3 is as follows. Among the set of corner points, corner points (allocations) $f_{2_1} -> f_{1_2} -> f_{2_3}$ result in the least number of changeovers, i.e. two. Thus, at any iteration in the column generation framework, we can generate an MILP solution for the TAP from a set of master problem columns.

This chapter tries to explain the fundamentals of DW reformulation and the column generation algorithm with the help of a real-world example. The readers must also realize that the practical implementation of these advanced optimization techniques requires consideration of some additional parameters. These parameters change considerably from problem to problem. However, the next section summarizes some important points that will help the readers implement these techniques.

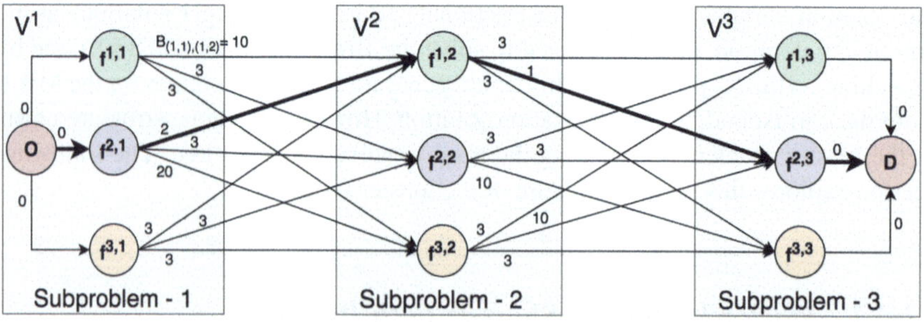

Fig. 11.3 Shortest path representation of the master problem presented in Sect. 11.6.1

11.7 Notes for Practical Implementation

In the previous section, a real-world problem was studied to understand the theory of DW decomposition and column generation. However, practical implementation of these methods would require the reader to consider the following points.

11.7.1 Termination Criteria

By now, the readers are aware that theoretically, the column generation technique terminates if there are no new columns that can enter the master problem. However, even if more columns are added, the improvement in objective function value can be minimal. As a result, generating all the columns can be pretty time-consuming. Instead, some widely used practical termination criteria are *total number of iterations*, *time limit*, and *gap (%)* between the upper and lower bounds. The total number of iterations states that the algorithm must terminate as soon as a certain number of iterations have elapsed. The time limit criterion terminates the algorithm after a fixed amount of time. Similarly, the gap (%) criterion terminates the algorithm when a particular gap (%) between the bounds is reached.

11.7.2 Generation of a Set of Initial Columns

The column generation algorithm requires at least one initial column to begin its execution. Initial columns can be introduced by generating a trivial solution to the sub-problems and constructing the corresponding column. However, generating columns from trivial solutions can lead to extremely slow convergence. It is helpful to have some heuristics that can guide the initial process of column generation. For example, in the proposed decomposition of the mp-CAP, duals from the LP relaxation of the mp-CAP are used during initial iterations to guide the generation of columns.

11.7.3 Adding Multiple Columns Per Iteration of Column Generation

Readers must observe that theory dictates the selection of only one entering column per iteration of the column generation algorithm by solving the sub-problems to optimality. However, many practical implementations of column generation apply a fast heuristic that generates multiple alternative columns from the original column. These alternative columns are still feasible for the sub-problems and adhere to the reduced cost criteria. Often for frameworks where MILP sub-problems are solved, traversing through the Branch and Bound tree leads to multiple integer feasible solutions. All the feasible integer solutions that satisfy the reduced cost criteria can be used to generate multiple columns during a single iteration. Adding multiple columns during every iteration of the column generation algorithm can significantly expedite its convergence.

11.7.4 Embedding Heuristics Within the Framework

The column generation algorithm helps us split the DW reformulation into a restricted master problem (RMP) and sub-problem(s) (SP). In a simple implementation, the RMP and SP are solved using a solver, which selects some linear programming or integer programming algorithm to solve the problem. However, some problems can have additional unique structures within the RMP, SP, or both. These structures might enable the readers to solve the RMP or SP using polynomial-time heuristics or simple arithmetic. For example, the shortest path structure in the master problem for the mp-CAP helped eliminate the need to solve the Branch and Bound algorithm in order to generate the MILP solution from a given set of master problem columns.

11.7.5 Symmetry Breaking

The DW reformulation and the block diagonal structure of the mp-CAP have been discussed earlier. However, imagine that the feasible regions defined by all the sub-problems were identical. For the sub-problems (11.26)–(11.35), this would be equivalent to saying that all the tanker compartments are identical, and for every sailing leg, the same set of cargoes has to be assigned to the compartments. In essence, the corner points $f_{k-1}^{v} = f_{k}^{v} = f_{k+1}^{v} \forall k \in K$, during each iteration. Additionally, the master problem simplifies as follows.

Restricted master problem for identical sub-problems:

$$\begin{aligned}
\textbf{Minimize}\quad & z_{mp} = C^S \sum_{j\in N^G}\sum_{k\in K_j}\sum_{h\in N^H} b_{kjh} \\
& \sum_{v\in V_1'} f_{jh}^{v}\lambda^{v} = b_{1jh} && \forall j \in N^G, h \in N^H, \\
& \sum_{v\in V_1} \lambda^{v} = |K|, \\
& \lambda^{v} \geq 0 && \forall k \in K, v \in V_k.
\end{aligned} \tag{11.36}$$

Constraint (11.22) disappears because the first and second terms of its L.H.S cancel out each other. Thus, the problem essentially becomes that of placing the cargoes in the least number of compartments during the first leg and maintaining the same placement till the end of the voyage. Readers should observe that the symmetry disappears in the master problem, which significantly simplifies the problem.

11.7.6 Parallel Computing

The readers may recall, the proposed reformulation helps decompose the problem into one master problem and multiple sub-problems. As the sub-problems are disjoint, they can be solved in parallel by using multiple CPUs. This would accelerate the time for convergence. For large-scale problems, such frameworks can also be designed to leverage distributed computing, where both memory and CPU speeds can be distributed across a network of computers. A recent study by [1] discusses the application of distributed computing with column generation.

11.8 Computational Results

Computational experiments based on the above formulations are now described. Since these problems require several inputs, this section starts by describing synthetically generated test instances of the mp-CAP and other related chemical scheduling problems. While some of the inputs are generated randomly, the range of random parameters is close to realistic values seen in practice.

11.8.1 Test Instances for Numerical Experimentation

The test instances and an instance generator presented by [17] are the starting points for computational work described in this section. They use benchmark test instances[3] to perform the various computations on general ship routing problems. These instances were created from different combinations of the inputs including 13 diverse chemical tankers. The test instances were divided into 44 instance sets depending on the different input characteristics. Inputs include the total number of cargoes, the port network number, the maximum number of sailing legs, the total planning time, and the cargo characteristics like total weight, density, loading port, unloading port, pickup time windows, and compatibility criteria. The total number of compartments in a tanker ranges from 16 to 52. For each instance, the total number of cargoes is 40, 80, or 120. Every instance is named according to the instance set to which it belongs and the chemical tanker that is used to generate the instance. For example, INST_1_1 indicates that the instance belongs to instance set 1 and is generated using chemical tanker 1.

Ladage et al. [17] also provide solution files for the benchmark instances generated by solving a chemical tanker scheduling problem using a commercial optimization solver. Since mp-CAP models a smaller set of decisions as compared to the general scheduling and routing problem, the instance set and the solution sets together are used for inputs to mp-CAP. The solution files provide a list of cargo loaded on the tanker at every port. The mp-CAP tries to assign these loaded cargoes to different compartments of the tanker. Additionally, the chemical tanker scheduling problem is defined in such a way that the list of cargoes used as an input for the mp-CAP has at least one feasible cargo-compartment allocation plan for the mp-CAP formulations described in this Sects. 11.3 and 11.6. The study presented in the following section discusses 15 instances selected from the above set.

11.8.2 Performance Analysis of the MILP Model and Column Generation Framework

This section discusses an empirical study conducted to test the mixed-integer linear programming (MILP) formulation and the DW reformulation solved using the customized column generation (CG) framework. Ideally, the CG framework would be embedded in a Branch-Price-and-Cut tree, with the CG framework being used at every node of the tree. Implementation of a user-defined Branch-Price-and-Cut tree requires knowledge of branching rules, cutting planes, and tree pruning strategies, which are out of the scope of this chapter. Hence, the CG framework presented here only solves the root node LP and does not solve the mp-CAP entirely.

[3] https://ladageanurag.shinyapps.io/s-PDP-TWTAC/.

The MILP and DW formulations are built using Julia 1.6 and JuMP. Julia was introduced by [3] and is a programming language suitable for scientific and engineering applications. JuMP is a modeling language designed by [11] for mathematical optimization in Julia. It enables easy construction of optimization models and allows one to call optimization solvers, like Gurobi 9.1 used here from a Julia program. Source files and the fifteen test instances have been provided[4] as a GitHub repository. Interested readers can implement these to further improve their understanding of the framework and recreate the results. The 15 test instances are solved in two ways:

- Using the Gurobi 9.1 MILP solver taking the mp-CAP formulation (Sect. 11.4) as an input. The formulation is modeled in JuMP. In this chapter, we refer to these runs as MILP runs.
- The root node of the CG framework (Sect. 11.6) implemented using JuMP. The master problem and sub-problems are solved using Gurobi 9.1 MILP solver. In this chapter, we refer to these runs as CG runs.

The MILP formulation and its LP relaxation are solved using the [14] commercial solver. Tables 11.1 and 11.2 record important problem statistics and performance parameters for the MILP runs, and CG runs. The total time limit for all instances is five minutes wall clock time.

Table 11.1 presents the problem sizes and solving times for the MILP and CG runs. Initial two columns give the *Instance set* and *Ship number* of the instances. Both these parameters aggregate specific problem characteristics as explained in Sect. 11.8.1. The following two columns present the total number of variables and constraints for the MILP runs. Columns *CG master vars* and *CG master constrs* show the number of variables and constraints for the master problem of the CG framework. Similarly, the following two columns tabulate the total number of variables and constraints for all sub-problems. Column *Sub-problems per iter* gives the number of sub-problems solved during each iteration of the CG runs, while the final two columns report the wall clock time (in seconds) for the MILP runs, and the CG runs. The last row of Table 11.1 gives the average value of every parameter for the 15 instances.

Table 11.1 shows that the average number of variables and constraints are significantly less for the CG runs compared to the MILP runs. Remember that the CG framework uses a delayed column generation procedure, which only generates columns as and when required. This results in a significant reduction in the number of variables. The reduction in the number of constraints is a property of the DW reformulation. Table 11.1 also shows that the average times taken by MILP runs and CG runs are 34 and 21 s, respectively.

Table 11.2 presents the upper and lower bounds generated during the MILP runs, and the CG runs. CG and MILP runs terminate optimally within the time limit for all 15 instances. Like Table 11.1, the first and second columns tabulate the instance set number and the ship number. Following two columns, tabulate the lower bounds generated during the CG and MILP runs. Column *LB diff* shows the difference between

[4] https://github.com/anuragladage/mp-CAP.git.

Table 11.1 Problem size and solving time comparison for MILP, and CG framework solved at root node

Instance set	Ship no.	MILP vars	MILP constrs	CG master vars	CG master constrs	Sub-problem vars	Sub-problem constrs	Sub-problem per iter	MILP time (s)	CG time (s)
2	9	2304	6004	106	208	294	1036	8	2	2
22	3	2304	5930	65	128	147	758	16	1	2
10	4	6336	15848	149	296	402	1616	16	2	3
14	17	4032	10492	177	352	510	1780	8	3	4
18	3	2640	7082	160	310	441	1534	11	91	6
6	1	12012	30618	376	750	1099	4274	11	4	8
13	18	21216	55310	1003	1992	2983	9834	8	81	22
2	1	11232	29194	478	952	1413	4894	8	54	22
15	9	3072	8498	573	496	735	2374	8	10	24
8	9	4224	11402	609	566	833	2742	11	5	32
23	18	24960	61956	537	1072	1570	5836	16	2	32
35	11	15360	38280	337	672	968	3896	16	22	35
33	10	24816	61492	535	1066	1562	5700	16	209	36
1	3	3072	8420	836	464	686	2204	8	19	37
3	7	12480	33516	784	128	147	758	16	5	51
Average		10004	25603	448	630	919	3282	12	34	21

Table 11.2 Comparison of upper and lower bounds generated by MILP runs and CG runs. Bound diff (CG) gives the difference between upper and lower bounds generated during CG runs

Instance set	Ship no.	Lower bounds			Upper bounds			Bound diff (CG)
		CG root	LP value	Diff	CG root	MILP obj	Diff	
2	9	450	350	100	450	450	0	0
22	3	200	200	0	200	200	0	0
10	4	350	350	0	350	350	0	0
14	17	300	300	0	350	300	50	50
18	3	300	250	50	350	350	0	50
6	1	400	350	50	400	400	0	0
13	18	950	850	100	950	950	0	0
2	1	450	450	0	450	450	0	0
15	9	650	400	250	850	650	200	200
8	9	650	400	250	850	650	200	200
23	18	500	500	0	500	500	0	0
35	11	400	400	0	400	400	0	0
33	10	550	550	0	550	550	0	0
1	3	400	200	200	650	400	250	250
3	7	600	500	100	650	650	0	50

these two lower bounds. A positive value (eight instances) in Column *LB diff* gives instances for which the CG runs found a tighter lower bound than the LP relaxation.

The upper bounds of the CG and MILP runs, and the difference between these upper bounds are tabulated in the following three columns. Zero value in the Column *UB diff* signifies that the CG framework could find the optimal solution for the mp-CAP MILP formulation for eleven instances. The CG framework discovered these optimal solutions at the root node.

Column *Bound diff (CG)* presents the difference between the upper and lower bounds generated during the CG runs. Recall that the CG upper bounds are generated from the shortest path representation of the master problem, while the CG lower bounds are the optimal master problem objective value. Observe the six instances with positive values in this column. For all of these six instances, to prove optimality, the CG framework has to be embedded into a Branch-Price-and-Cut tree.

These instances can be bifurcated into two sets. Four out of six of these instances also have a positive value in Column *UB diff*. For these four instances, optimality could not be proved because at the root node, the upper bound is higher than the optimal solution (MILP Obj). This is because the CG lower bounds are constructed from a fractional combination of optimal columns (fractional b_{kjh} variable values). In contrast, the CG upper bounds are constructed from optimal integer columns (binary b_{kjh} variable values).

On the other hand, for instances INST_18_3 and INST_3_7, the optimality could not be proved as the lower bounds are lower than the optimal solution (MILP Obj). The following section discusses some important conclusions for this chapter.

11.9 Concluding Remarks

This chapter discusses the theory of DW reformulation along with its implementation for the *multi-period cargo-assignment problem* (mp-CAP). This assignment problem is useful for scheduling operations related to chemical tankers. The presence of domain-specific constraints, multi-period approach, and flexibility to re-arrange cargoes in compartments makes this assignment problem unique.

There are a significant number of constraints that only apply locally, that is, relating to loading a ship for a single leg of the journey. These can be *hidden* in the sub-problem. Then the master problem only has to focus on determining how to select load plans for each leg so that every cargo is accommodated with a minimum number of changeovers in each hold. This is commonly the case in other situations. For example, in staff rostering, many conditions apply locally to a single line of work (roster for one person). Again, these are combined with a master problem selecting lines of work subject to only a small number of constraints going across all of the columns, in this case, counting shifts on each day or tasks completed. From a domain perspective, the primary idea is to observe whether a problem can be differentiated between global and local activities. Then, global activities-related constraints might be placed in the master problem, while the local activities-related constraints would constitute the sub-problem.

The presence of *easy* and *complicated* constraints along with the shortest path type unique structure motivates applying the DW reformulation to the mixed-integer linear programming (MILP) formulation. The DW reformulation represents the problem defined by the original objective function and *complicated* constraints in terms of the corner points or vertices of the feasible region defined by the *easy* set of constraints. Such a reformulation has multiple advantages.

First and foremost, it reduces the total number of constraints in the problem. The DW reformulation helps to break the symmetry in the problem. This symmetry occurs due to the presence of similar allocations with the same objective value. The DW reformulation also helps us decompose the problem into master and multiple sub-problems. These multiple sub-problems disintegrate into separate problems, thus allowing us to solve them in parallel. The master problem also helps us exploit a unique shortest path structure within the mp-CAP. This eliminates the need to solve the Branch and Bound algorithm to get an MILP solution from a given set of master problem columns.

However, the DW reformulation generally results in an exponential number of decision variables or columns. This necessitates using delayed column generation, which generates columns as and when required. The experiments showed that the CG framework could find good MILP solutions using fewer variables. Moreover, the

DW formulation solved using the CG framework found tighter lower bounds than the LP relaxation. For bigger instances, the CG framework converges extremely slowly. In practice, the CG framework needs to be embedded within a Branch-Price-and-Cut tree. This would require a very fast convergence of the CG framework.

A further study of the symmetric solutions could be a promising research avenue to discover stronger cuts. The problem's symmetry increases with the increase in the number of compartments. The right cuts would enable the CG framework to avoid symmetric solutions and significantly boost its convergence. Another promising research avenue could be to introduce stronger cover cuts for the mp-CAP MILP formulation at each node of the Branch-Price-and-Cut tree. A study related to the ship's compartment structure and cargo characteristics could help discover cover cuts that can limit the total number of cargoes assigned to the ship as a function of empty compartments. Other avenues include generating reasonable heuristic solutions for sub-problems and an improved sub-problem objective function that can improve the quality of columns entering the master problem.

Interested researchers could also study some extensions of the mp-CAP. A critical extension of the mp-CAP arises by including a more realistic version of the compatibility constraints. For example, the current definition of mp-CAP considers cargo-compartment incompatibility as a function of compartment material. However, cargo-compartment incompatibility also results from cargo stored in a compartment in the preceding sailing legs. Incorporating this into the model will change the structure of the problem and might impact the DW-CG framework. We encourage researchers interested in multi-compartment problems to refer to research presented by [8]. They present extensive findings related to multi-compartment problems in multiple domains.

Different problems have different decompositions and implementation hurdles than those discussed in this chapter. It is often not possible to predict whether one approach will be faster or yield better results than others. DW reformulation is one promising approach, but it must be used carefully to be effective.

References

1. Basso S, Ceselli A (2022) Distributed asynchronous column generation. Comput Oper Res 105894
2. Bertsimas D, Tsitsiklis JN (1997) Introduction to linear optimization, vol 6. Athena Scientific Belmont, MA
3. Bezanson J, Edelman A, Karpinski S, Shah VB (2017) Julia: A fresh approach to numerical computing. SIAM Rev 59(1):65–98. https://doi.org/10.1137/141000671
4. Brooks MR, Faust P (2018) 50 years of review of maritime transport, 1968–2018: Reflecting on the past, exploring the future. Technical report, UNCTAD
5. Christiansen M, Fagerholt K, Ronen D (2004) Ship routing and scheduling: Status and perspectives. Transp Sci 38(1):1–18. https://doi.org/10.1287/trsc.1030.0036

6. Christiansen M, Fagerholt K, Nygreen B, Ronen D (2007) Chapter 4 maritime transportation. In: Barnhart, C. & Laporte, G. (ed) Handbooks in operations research and management science: Transportation. Elsevier, pp 189–284. http://www.sciencedirect.com/science/article/pii/S0927050706140049
7. Christiansen M, Fagerholt K, Nygreen B, Ronen D (2013) Ship routing and scheduling in the new millennium. Eur J Oper Res 228(3):467–483. https://doi.org/10.1016/j.ejor.2012.12.002, http://www.sciencedirect.com/science/article/pii/S0377221712009125
8. Coelho LC, Laporte G (2015) Classification, models and exact algorithms for multi-compartment delivery problems. Eur J Oper Res 242(3):854–864
9. Conforti M, Cornuéjols G, Zambelli G et al (2014) Integer programming, vol 271. Springer, Berlin
10. Dantzig GB, Wolfe P (1960) Decomposition principle for linear programs. Oper Res 8(1):101–111
11. Dunning I, Huchette J, Lubin M (2017) JuMP: A modeling language for mathematical optimization. SIAM Rev 59(2):295–320. https://doi.org/10.1137/15M1020575
12. Giavarina dos Santos PT, Kretschmann E, Borenstein D, Guedes PC (2020) Cargo routing and scheduling problem in deep-sea transportation: Case study from a fertilizer company. Comput Oper Res 119:104934
13. Gilmore PC, Gomory RE (1961) A linear programming approach to the cutting-stock problem. Oper Res 9(6):849–859
14. Gurobi Optimization (2022) Gurobi Optimizer Reference Manual. https://www.gurobi.com
15. Hvattum LM, Fagerholt K, Armentano VA (2009) Tank allocation problems in maritime bulk shipping. Comput Oper Res 36(11):3051–3060. https://doi.org/10.1016/j.cor.2009.02.002, http://linkinghub.elsevier.com/retrieve/pii/S0305054809000422
16. Kjeldsen KH (2011) Classification of ship routing and scheduling problems in liner shipping. INFOR: Inf Syst Oper Res 49(2):139–152. http://www.utpjournals.press/doi/abs/10.3138/infor.49.2.139
17. Ladage A, Baatar D, Krishnamoorthy M, Mahajan A (2021) A revised formulation, library and heuristic for a chemical tanker scheduling problem. Comput Oper Res 133:105345
18. Neo KH, Oh HC, Karimi I (2006) Routing and cargo allocation planning of a parcel tanker. In: Computer aided chemical engineering, vol 21, Elsevier, pp 1985–1990
19. Ronen D (1983) Cargo ships routing and scheduling: Survey of models and problems. Eur J Oper Res 12(2):119–126
20. United Nations (2018) Review of maritime transport 2018. United Nations, UNCTAD 2018. https://unctad.org/en/PublicationsLibrary/rmt2018_en.pdf
21. United Nations (2019) Review of maritime transport 2019. United Nations, UNCTAD 2019. https://unctad.org/en/PublicationsLibrary/rmt2019_en.pdf
22. Vilhelmsen C, Larsen J, Lusby RM (2015) Tramp ship routing and scheduling-models, methods and opportunities. Techncal report, DTU Management Engineering. https://core.ac.uk/reader/43253760
23. Wang X, Arnesen MJ, Fagerholt K, Gjestvang M, Thun K (2018) A two-phase heuristic for an in-port ship routing problem with tank allocation. Comput Oper Res 91:37–47

Chapter 12
Volume Algorithm: Training Neural Networks and Solving Difficult Linear Programs

Mourad Baiou, Francisco Barahona, and Joao Goncalves

Abstract Subgradient methods have a long history of applications in Optimization. They are known for producing fast approximate solutions in cases where more exact methods are prohibitive to use. The Volume algorithm (VA) is a subgradient method, it was developed to deal with large-scale linear programs that were too difficult for traditional algorithms. Its main advantage is that it produces approximate primal solutions as well as dual solutions. It has also shown fast convergence in practice. Here we review this method and present experiments in three different areas, which demonstrate its good performance and wide applicability. First we deal with training neural networks. We show that replacing the widely used Stochastic Gradient Descent method with the VA leads to very good results. Then we study bus routing in a medium size city. Here we tackle a linear programming relaxation that is too large to be treated with traditional linear programming algorithms. Here the VA gives tight lower bounds, and the approximate primal solution is used as the starting point for several heuristics that produce the final routing. Finally we turn our attention to the p-median problem for restructuring semistructured data. We apply the VA to the linear programming relaxation, and we decompose it in a simple way so it can be treated in parallel. We use a parallel implementation combined with a heuristic. This heuristic starts with the approximate primal solution and derives an integer solution. We show experiments with large instances coming from semistructured databases. For the majority of these cases the gap from optimality was less than 1%.

M. Baiou
Université Clermont Auvergne, CNRS, Clermont Auvergne INP, Mines Saint-Etienne, LIMOS, 63000 Clermont-Ferrand, France
e-mail: baiou@isima.fr

F. Barahona (✉) · J. Goncalves
IBM T. J. Watson Research Center, P. O. Box 218, Yorktown Heights, NY 10598, USA
e-mail: barahon@us.ibm.com

J. Goncalves
e-mail: jpgoncal@us.ibm.com

F. Hamid (ed.), *Optimization Essentials*, International Series in Operations Research & Management Science 353, https://doi.org/10.1007/978-981-99-5491-9_12

Keywords Subgradient method · Volume algorithm · Training neural networks · Bus routing · p-median · Semistructured data

MSC Classification 68T07 · 90C25

12.1 Introduction

The Volume Algorithm [4] is a subgradient method designed for minimizing a convex piecewise linear function, later it was extended to convex non-differentiable functions in [3]. Since its publication in 2000, it has shown better convergence in practice than the traditional subgradient method [28]. It has been particularly useful for large-scale instances for which other algorithms have been impractical, see [19, 25, 45], for instance.

Here we report experiments with this algorithm in three different areas. We first use it for training neural networks and compare with other well-known algorithms. Then we use it for linear programs needed to design the bus routes in a medium size city, and we compare with CPLEX [16]. For a third type of experiments we tackle linear programs associated with the p-median problem, and we show computations in parallel. The instances that we treat here come from restructuring semistructured data [36].

12.2 The Volume Algorithm

We start with an example to motivate the choice of the direction in our method, and then we give a high level description of the algorithm.

12.2.1 An Example

Suppose that we want to maximize the piecewise linear function below

$$f(x, y) = \min\{2x - y + 2,\ -2/3x - y + 2,\ x + y + 10,\ -x + y + 10\}.$$

Consider the point $(\bar{x}, \bar{y}) = (0, 2)$. Here we have $f(\bar{x}, \bar{y}) = 0 = 2\bar{x} - \bar{y} + 2 = -2/3\bar{x} - \bar{y} + 2$. Thus at the point $(\bar{x}, \bar{y})$ the function has the subgradients $g_1 = (2, -1)$ and $g_2 = (-2/3, -1)$. Notice that the direction that gives the largest improvement per unit of movement is $d = (0, -1)$. It can be obtained as a convex combination of the subgradients above as $d = 1/4g_1 + 3/4g_2$. Now we discuss how to obtain the coefficients of this convex combination. Notice that the area of

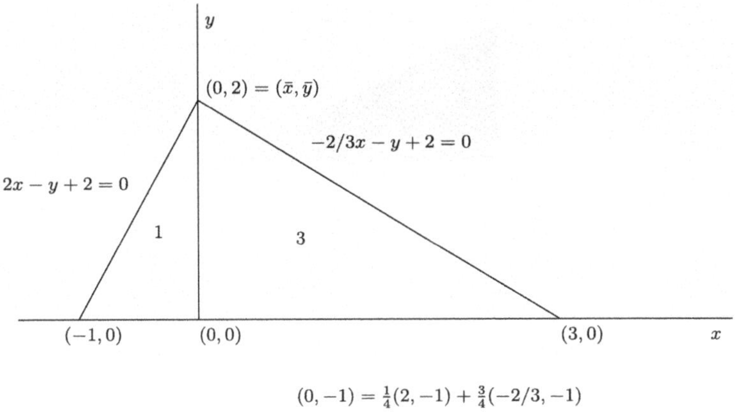

Fig. 12.1 Obtaining the direction of improvement as a convex combination of subgradients

the triangle with vertices $(0, 2)$, $(0, 0)$, $(-1, 0)$ is 1, and the area of the triangle with vertices $(0, 2)$, $(0, 0)$, $(3, 0)$ is 3. Thus if we use each of these areas divided by the sum of the areas we obtain the coefficients of the convex combination, see Fig. 12.1. In higher dimensions we should use volumes instead of areas. In the next subsection, we shall see that this procedure is not particular to this example, but it corresponds to a method to obtain the direction of improvement using volumes.

12.2.2 Volume and Duality

Now we give a theoretical justification for what we did in the example above. Recall that maximizing a concave piecewise linear function can be written as

$$\max w \tag{12.1}$$

$$w + a_i x \le d_i,\ i = 1, \dots, n. \tag{12.2}$$

The theorem below was published in [4].

Theorem 12.1 *Consider the linear program* (12.1)–(12.2)*. Suppose that* $(\hat{w}, \hat{x})$ *is an optimal solution, and that inequalities* $1, \dots, n'$, $n' \le n$, *hold with equation for this vector. Let* $\bar{w} < \hat{w}$ *and suppose that*

$$w + a_i x \le d_i,\ \textit{for}\ 1 \le i \le n',$$

$$w \ge \bar{w},$$

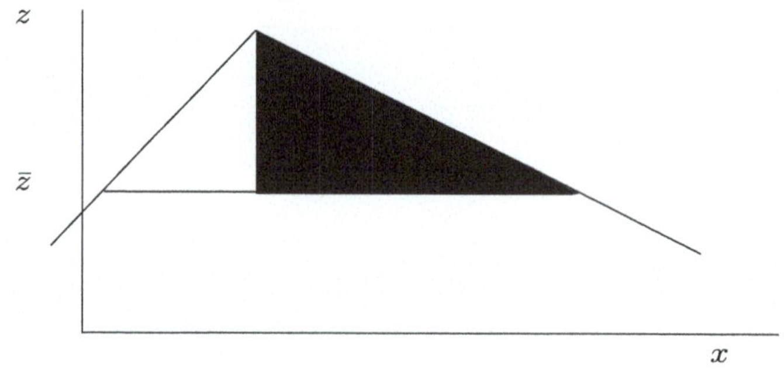

Fig. 12.2 Volume and duality

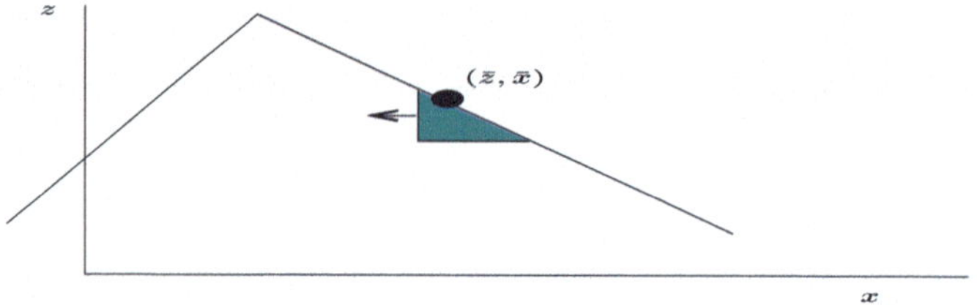

Fig. 12.3 Choice of a direction

describes a bounded polyhedron. Let δ_i, $i = 1, \ldots, n'$, *be the volume included between the face given by* $w + a_i x \leq d_i$ *and the hyperplane given by* $w = \bar{w}$. *The dark region in Fig. 12.2 depicts this. Then an optimal dual vector is defined by*

$$\alpha_i = \frac{\delta_i}{\sum_{k=1}^{n'} \delta_k}.$$

This suggests that for a vector $(\bar{w}, \bar{x})$, the active faces should be identified. Then the volume of their projection over the hyperplane $w = \bar{w} - \epsilon$, in a neighborhood of $\bar{w}$, and for some small value of ϵ, should be computed. If F_i is the face given by $w + a_i x \leq d_i$, and α_i is the ratio of the volume below the face F_i to the total, then we should look at the vector

$$(1, 0, \ldots, 0) - \sum_{i=1}^{n'} \lambda_i (1, a_i).$$

When it is a vector of all zeroes, this gives a proof of optimality. If not, this gives a direction of improvement, see Fig. 12.3. This is what we did in the example of Sect. 12.2.1. In the next subsection we describe a procedure that computes approximations of these volumes.

12.2.3 Basic Idea

As previously suggested, for a vector $(\bar{z}, \bar{x})$, we could try a set of random points in the hyperplane defined by $z = \bar{z} - \epsilon$. For each of these points we can identify the face above it. Theorem 12.1 suggests that we should look at the coefficients $\{\lambda_i\}$ as the probabilities that the different faces will appear. Thus we look at the coefficients $\{\lambda_i\}$ as the probabilities that the algorithm will find the different subgradients. Then to compute the direction of improvement we propose to keep a direction d and update it in every iteration as

$$d \leftarrow \alpha g + (1 - \alpha)d, \tag{12.3}$$

where g is the last subgradient produced, and α is a small value. Therefore if $g_0, \ldots, g_k$ is the sequence of subgradients produced, the direction is

$$d = (1 - \alpha)^k g_0 + \alpha(1 - \alpha)^{k-1} g_1 + \cdots + \alpha(1 - \alpha)g_{k-1} + \alpha g_k.$$

The coefficients of this convex combination are an approximation of the values $\{\lambda_i\}$ of Theorem 12.1, this is explained in [4]. This type of update scheme has also been used in Adam [32] and Momentum [39] (algorithms that are used in the experiments of Sect. 12.3). One difference with our algorithm is in the strategy to increase or decrease the step size. As we shall see in the next section, this makes a significant difference in terms of convergence.

Once the direction d has been chosen, ideally we would do a line search, but since this is prohibitively expensive, we use the following scheme. Suppose we use a step size l, and let $\bar{g}$ be the new subgradient obtained. If the inner product $d \cdot \bar{g}$ is positive, this means that the step size was too small, a larger step size would have resulted in a better solution. We call this a "green" iteration. If $d \cdot \bar{g}$ is negative, this means that a smaller step size would have given a better solution. We call this a "yellow" iteration. Then we keep a "green-yellow" indicator gy that is a number between 0 and 1. At each green iteration we update gy as $gy \leftarrow (1 - \alpha)gy + \alpha\, 1$. At each yellow iteration we update gy as $gy \leftarrow (1 - \alpha)gy + \alpha\, 0$. Each time that $gy > 0.9$ we increase the step size. When $gy < 0.8$ we decrease the step size. We see this as low cost approximation of a line search. Based on this idea, in the next three sections we present slightly different versions of this algorithm. The first one is for training neural networks, and the others are to tackle the Lagrangian dual of some network design problems.

12.3 Training Neural Networks

Here we present experiments for training different types of neural networks. Since in this case we deal with a non-linear non-differentiable function we made some modifications to the algorithm as follows.

Volume Algorithm

- Input: We assume that we have to maximize a function f of variables v. We also receive a value $\alpha \in [0, 1]$, and an initial step size s. We set an upper bound for the step size $U = 2s$, and a lower bound $L = 0.2s$. Initially we set $gy = 0.5$.
  At each iteration we keep a vector d that corresponds to a direction. We initialize it to be equal to the first subgradient. At each iteration we receive a new subgradient g that is used to update the direction. The different steps are below.
- **Step 1.** Compute $p = d \cdot g$. This is the inner product of the direction and the new subgradient.
  If $p > 0$ increase gy as $\quad gy \leftarrow (1-\alpha)gy + \alpha\, 1,$
  otherwise decrease gy as $\quad gy \leftarrow (1-\alpha)gy + \alpha\, 0.$
  If $gy > 0.9$ increase s as $s \leftarrow 1.01s$.
  If $gy < 0.8$ decrease s as $s \leftarrow 0.99s$.
  If $s > U$, set $s = U$. If $s < L$, set $s = L$.
  Update the direction as
  $$d \leftarrow (1-\alpha)d + \alpha\, g.$$
- **Step 2.** Update the variables v as
  $$v \leftarrow v + \frac{s}{\|d\|}d.$$
  Go to Step 1.

This is for maximization, if we want to minimize we update the variables as $v \leftarrow v - \frac{s}{\|d\|}d$. In our experiments we used $\alpha = 0.1$.

12.3.1 Experiments

Here we compare the Volume Algorithm with three of the most frequently used methods, namely Momentum [39], Adam [32], and RMSProp [54], we also compare with the more recent algorithm COCOB [38]. For the first three algorithms we used their implementation found in [49], for COCOB we used the implementation in [37]. The Volume Algorithm was implemented in TensorFlow. For the choice of the learning rate, for each algorithm we looked for the best value in $\{0.00001, 0.000025, 0.00005, 0.000075, 0.0001, 0.00025, 0.0005, 0.00075, 0.001, 0.0025, 0.005, 0.0075, 0.01, 0.025, 0.05, 0.075, 0.1\}$. For the Volume Algorithm the initial step size is set equal to the learning rate, and then the step size is increased or decreased as described in Sect. 12.2.3. The set of values above was used in [38] for their comparisons. Recall that COCOB has the interesting feature of not requiring an initial learning rate. In the figures below we show the best learning rate used for each algorithm. For Adam we used $\beta_1 = 0.9$ and $\beta_2 = 0.999$. For Momentum we used 0.9 for the momentum value.

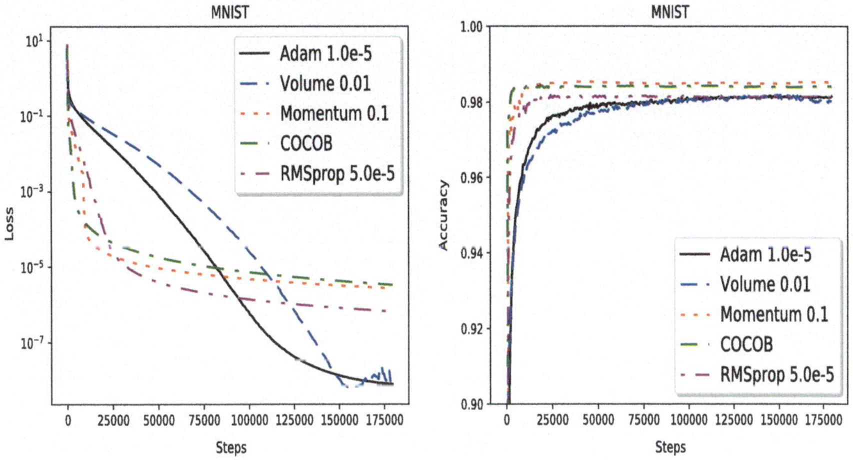

Fig. 12.4 MNIST. Loss and accuracy as functions of the number of steps. The best learning rates for each algorithm are shown in the legend

We used seven test sets available in the public domain, as follows. We have downloaded tensorflow implementations of different neural networks in the public domain. We used the same number of steps, epochs, batch-size, etc., as in these implementations. We make our code publicly available in [5], and thus these experiments are easy to reproduce.

12.3.1.1 MNIST

This is a well-known digits-recognition data-set, see [34]. We used the neural network implemented in [37]. In Fig. 12.4 we plot train loss and accuracy as functions of the number of steps.

12.3.1.2 CIFAR10

This is a data-set for classifying 32×32 RGB images across 10 object categories, see [12]. Here we used the code from the TensorFlow tutorial [50]. In Fig. 12.5 we plot train loss and accuracy as functions of the number of steps.

12.3.1.3 Recurrent Neural Networks

Here we use the Penn Tree Bank data-set [47] to train a Recurrent Neural Network. The goal is to predict next words in a text given a history of previous words. We use

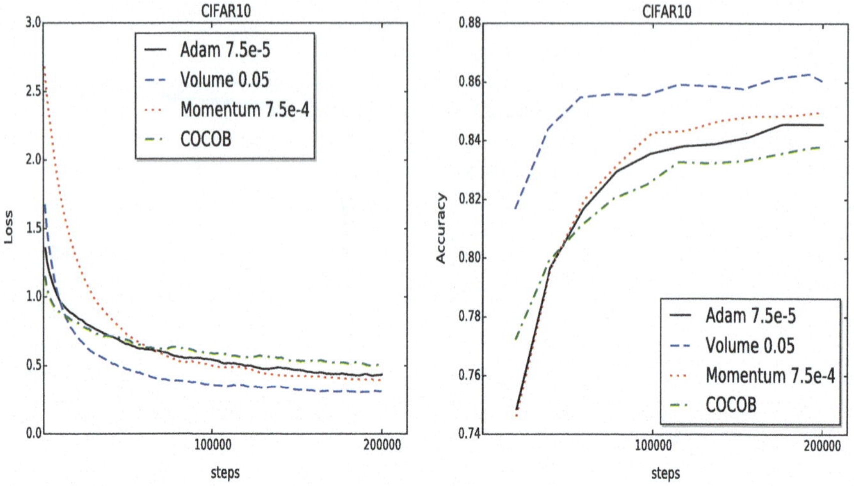

Fig. 12.5 CIFAR10. Loss and Accuracy as functions of the number of steps. The best learning rates for each algorithm are shown in the legend

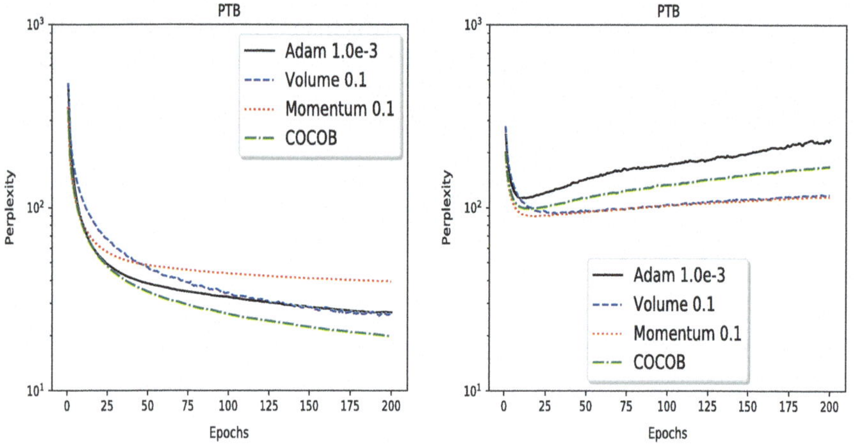

Fig. 12.6 PTB. Perplexity as a function of the number of epochs, for the train and test sets. The best learning rates for each algorithm are shown in the legend

the code in the TensorFlow tutorial [51]. In Fig. 12.6 we plot perplexity as a function of epochs for the train set and for the test set.

12.3.1.4 Retraining Inception V3

Here we use a model trained on the Imagenet data-set [30] to classify images of flowers. We use the code from TensorFlow for Poets [53] to download a pre-trained

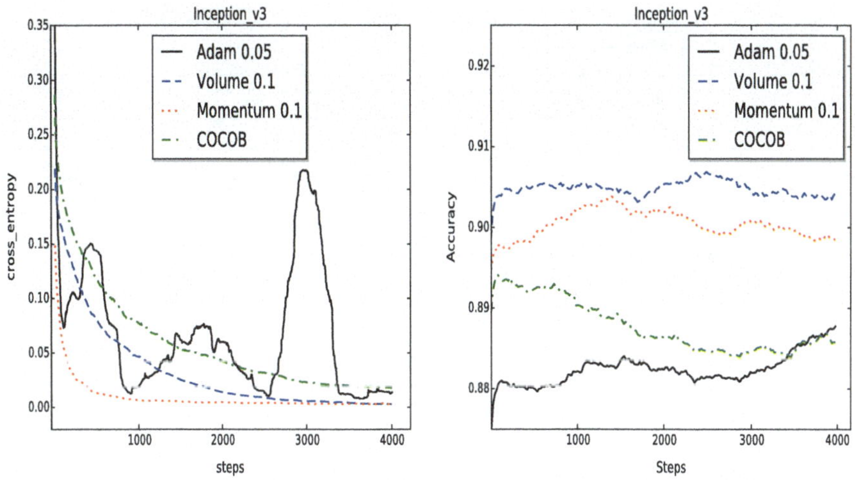

Fig. 12.7 Inception V3. Cross entropy for the training set and accuracy for the evaluation set, as functions of the number of steps. The best learning rates for each algorithm are shown in the legend

model, add a new final layer, and train that layer on a set of flower photos. Here we use an Inception V3 model, see [46]. In Fig. 12.7 we plot cross entropy for the training set, and accuracy for the evaluation set as a function of the number of steps.

12.3.1.5 Retraining Mobilenet_0.25_160

This time we retrain using a Mobilenet model [35]. The relative size of the model is 0.25 and the image size is 160. In Fig. 12.8 we plot cross entropy for the training set, and accuracy for the evaluation set as a function of the number of steps.

12.3.1.6 Retraining Mobilenet_0.25_128

Here we retrained using a Mobilenet model [35]. The relative size of the model is 0.25 and the image size is 128. In Fig. 12.9 we plot cross entropy for the training set, and accuracy for the evaluation set as a function of the number of steps.

12.3.1.7 Simple Audio Recognition

Here we use a basic speech recognition network that recognizes ten different words from the TensorFlow tutorial [52]. In Fig. 12.10 we plot cross entropy for the training set, and accuracy for the evaluation set as a function of the number of steps.

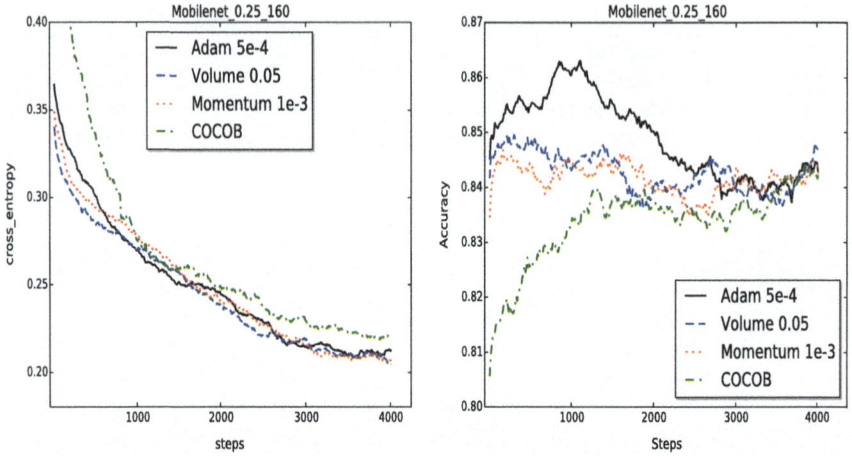

Fig. 12.8 Mobilenet_0.25_160. Cross entropy for the training set and accuracy for the evaluation set, as functions of the number of steps. The best learning rates for each algorithm are shown in the legend

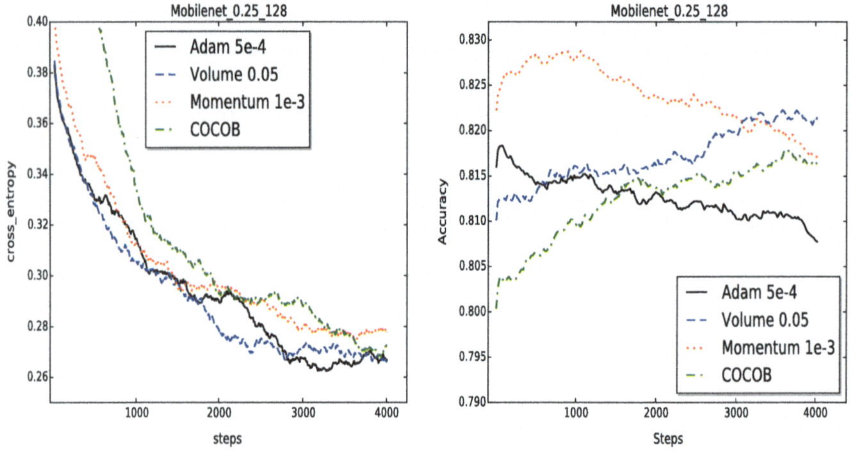

Fig. 12.9 Mobilenet_0.25_128. Cross entropy for the training set and accuracy for the evaluation set, as functions of the number of steps. The best learning rates for each algorithm are shown in the legend

12.3.2 Comparison

We have presented a comparison of several algorithms for training neural networks. In Table 12.1 we give labels from 1 to 4 to display the order from best to worst of the different algorithms on each data-set. If for one data-set several algorithms have been best, we give the label 1 to each of them. The Volume Algorithm received the label 1 five times and the label 2 two times.

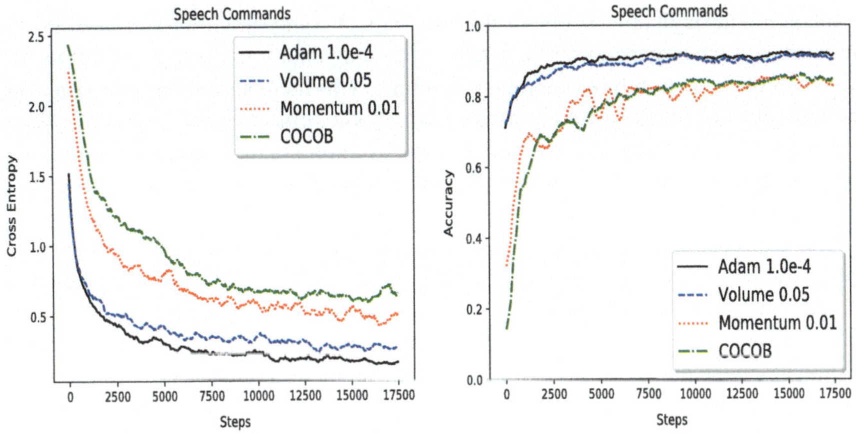

Fig. 12.10 Audio recognition. Cross entropy for the training set and accuracy for the evaluation set, as functions of the number of steps. The best learning rates for each algorithm are shown in the legend

Table 12.1 Comparison of training algorithms

	MNIST	CIFAR10	PTB	Inc V3	Mb_.25_160	Mb_.25_128	Audio
Adam	2	3	2	3	1	1	1
Momentum	3	2	4	1	1	4	3
COCOB	4	4	1	4	4	1	4
Volume	1	1	2	1	1	1	2

12.4 Bus Routing

Now we present experiments with optimization problems related to the design of the bus routes in a medium size city.

As stated by Ceder and Wilson [10] and many others, the bus planning process is usually divided into the following activities: bus network design, setting frequencies, timetable development, bus scheduling, and driver scheduling. The reason for this breakup is the complexity of each of these tasks. This makes it very difficult to treat all of them in one single model. Here we give an algorithm for the first task, namely, we develop a procedure that produces the routes for the buses that cover the transport demand.

In our case, we have a network whose nodes correspond to street corners, and whose arcs correspond to the streets. Some preprocessing is required to identify all forbidden crossings, forbidden turns, and unusable streets. Then for every arc we assume that we have the time that it takes for a bus to drive through, and for a person to walk through. We also have an *origin-destination (OD) matrix* that specifies the demand for transportation between a selected set of pairs of nodes. Also in our case there are three terminal nodes, so that each route corresponds to a cycle in this graph going through at least one of the terminals. A route should take at most thirty minutes, and when generating the routes we use a penalty to discourage left turns. Our objective is to minimize the total travel time. This includes the time walking and the time in a bus for the different users. Several other objective functions have been considered in recent and older literature. Some examples of these include minimizing the passenger waiting time, number of transfers, operating costs, fleet size, and many others.

Our approach consists of breaking down the original problem into two smaller subproblems that we call *route generation phase* and *route optimization phase*. In the former, we study and compare different heuristics for generating a large set of candidate routes. In the latter, we take the candidate routes as an input and we formulate an integer linear program to choose the final set of routes. For this model we first tried the commercial solver CPLEX [16], but due to the large computing time we had to switch to a Lagrangian relaxation approach that was much faster and produced solutions of similar quality.

There is a large number of articles written on Transit Network Design, we mention just a few of the different techniques used to attack this problem. Heuristics for route generation have been presented in Ceder and Wilson [10], Baaj and Mahmassani [2], and many others. Integer programming techniques have been used in Bussieck et al. [9], Claessens et al. [13], and Goossens et al. [24]. Column generation has been used by Schöbel and Scholl [44] and Borndörfer et al. [6]. Meta-heuristics have also been widely used, see Fan et al. [20, 21], and Zhao et al. [56], for instance. For a more complete set of references see the surveys by Desaulniers and Hickman [18], Kepaptsoglou and Karlaftis [31], and Schöbel [43].

12.4.1 General Approach

We break down the original problem into two smaller subproblems: the route generation phase and route optimization phase. In the former, we look for ways (or heuristics) to generate "good" candidate routes. In the latter, given a fixed (large) number of (previously generated) candidate routes, we use an integer programming model to choose the final routes out of the candidate ones. We next discuss each of these two phases in more detail.

12.4.2 Route Generation

We implement and compare different heuristics for generating candidate routes. Our route generation process breaks down into two stages: route skeletons creation and route skeletons expansion. This two-phase generation procedure is very similar to the one discussed in [2] by Baaj and Mahmassani. However, some of our route expansion heuristics are different from the ones discussed there. Moreover, [2] does not deal with the additional constraint that every route has to pass through one of the terminals.

In the first stage of the generation process we try to create three different skeletons for each OD pair. A skeleton is just a (directed) cycle that includes an OD pair, or one node from an OD pair and a terminal. We create three such skeletons as follows. Given an OD pair, we look at the different terminals and pick the one that leads to the shortest cycle including the OD pair and the terminal. If this cycle does not exceed the maximum allowed duration for a route, we add it to the set of route skeletons. In a similar way, we take the origin of the OD pair, look at the different terminals, and pick the one that leads to the shortest cycle containing the origin and the terminal. If this cycle does not exceed the maximum allowed duration we add it to the set of skeletons. Finally, we repeat the same procedure for the destination node. One of our data-sets had 160 OD pairs, thus if we can successfully add all the generated skeletons, we end up with a total of (160 + 320 =) 480 route skeletons.

The second stage of the generation process is the expansion stage. Here, we expand the initial skeletons to actual routes. The expansion process consists of two steps: node(s) selection and node(s) insertion. In the former, we use different heuristics to decide what is (are) the node(s) to be added next to the current route in expansion. In the latter, we have a procedure for adding a node (or nodes) to a route in current expansion. We discuss the node insertion process first.

Given an initial route skeleton or a route under current expansion, we say that a given node is *fixed* if it is an OD node (i.e. a node from an OD pair) or a terminal, and *nonfixed* otherwise. We use this terminology since the idea is that once a fixed node is added to a route under current expansion, it will never be removed from that route in posterior insertion steps. On the other hand, nonfixed nodes can potentially be removed from the route at later insertion steps.

The insertion step works as follows. Assume we have a route under current expansion and we want to add a node v to it (see Fig. 12.11). Then, we first find the node w in the current route that is closest to v (see Fig. 12.11a). We have two possible cases: w is either fixed or nonfixed. If w is a nonfixed node (i.e. w is neither a terminal nor an OD node) we look at the fixed node in the route preceding w (denote it by u_1) and proceeding w (denote it by u_2). That is, the segment $u_1a_1a_2...a_iwa_{i+1}...a_pu_2$ is part of the route, where all the nodes a_i are nonfixed and the nodes u_1, u_2 are fixed. Then, we perform the insertion by removing all the arcs in the route connecting u_1 with u_2 (i.e. the arcs $(u_1, a_1), (a_1, a_2), \ldots, (a_p, u_2)$), and adding the two new segments $P_1 := u_1b_1b_2...b_lv$ and $P_2 := vc_1c_2...c_{l'}u_2$, where P_1 and P_2 denote the shortest paths from u_1 to v and from v to u_2, respectively (see Fig. 12.11b). In the case that w is a fixed node, we repeat the same procedure as above and identify the fixed preceding and proceeding nodes u_1 and u_2, respectively. Also, let P_1 be the shortest path from u_1 to v, P_2 the shortest path from v to u_2, Q_1 the shortest path from v to w, Q_2 the shortest path from w to v, R_1 the shortest path from w to u_2, and R_2 the shortest path from u_1 to w. Notice that R_1 is just the segment in the route joining w and u_2, and R_2 the segment in the route joining u_1 and w. Then we compare the paths $P_1Q_1R_1$ and $R_2Q_2P_2$ and use the shortest of the two alternatives to expand the route. That is, if $P_1Q_1R_1$ is shorter than $R_2Q_2P_2$, we perform the insertion by

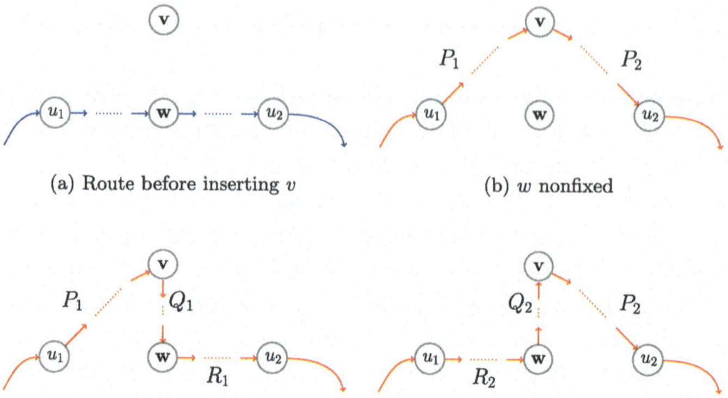

(a) Route before inserting v (b) w nonfixed (c) w fixed and $P_1Q_1R_1$ shorter (d) w fixed and $R_2Q_2P_2$ shorter

Fig. 12.11 Node insertion

removing all the arcs in the route connecting u_1 and w (i.e. R_2), and then adding P_1Q_1 (see Fig. 12.11c). On the contrary (i.e. $R_2Q_2P_2$ shorter than $P_1Q_1R_1$), we perform the insertion by removing all the arcs in the route connecting w and u_2 (i.e. R_1), and then adding Q_2P_2 (see Fig. 12.11d).

Finally, in the case we want to add two different nodes (e.g. an OD pair) v_1, v_2 to the route instead of only one, we repeat the above (adding a single node) procedure for both v_1 and v_2 separately. This completes the discussion about the node(s) insertion process.

We now discuss the node(s) selection stage. Recall that this step consists of deciding what is (are) the node(s) to be added next to the route in current expansion. We always start with a route skeleton and keep adding nodes until no more can be added. We consider several heuristics based on the idea of either adding the node or pair that is closest to the route (hence having longer routes that contain a larger number of OD nodes) or adding the node or pair with the highest demands (thus having routes that could be potentially shorter but that contain OD nodes and OD pairs with high demand). By a node's demand (or demand associated to a node) we mean the sum of the demands of all OD pairs that contain such node. Several of these ideas were motivated from the work in [2]. We consider the following five node(s) selection strategies:

1. **Random Shuffle** (RS): Repeat the following twice: Shuffle the OD pairs in a random order. Go over the sorted list of OD pairs and try to add first the OD pair. If this fails (i.e. inserting the OD pair violates the maximum allowed duration), try to add the OD origin node. If the latter insertion also fails, try to add the OD destination. Move to the next OD pair in the sorted list and repeat.
2. **Greedy on the Distance** (GD): Try to add the OD node that is closest to the current route in expansion. If it cannot be added stop expanding the route.
3. **Greedy on the Pairs Demand** (GPD): Presort the OD pairs by demand in decreasing order. Go over the whole sorted list trying to add the OD pairs. Then, go again

over the whole list and try to add the OD origin nodes in the cases where the corresponding OD pair could not be added before. Finally, go one more time over the list and try to add the OD destinations in the cases where neither the OD pair nor the OD origin could be previously added.

4. **Greedy on the Nodes Demand** (GND): Presort the OD nodes by demand in decreasing order. Go over the whole sorted list trying to add the nodes.
5. **Greedy on the nodes demand/distance ratio** (GNDDR): Try to add the OD node that maximizes the ratio: (node demand/node distance from the route). If it cannot be added stop expanding the route.

Finally, after completing the generation process, we do some postprocessing where we delete routes that are duplicated and routes that do not satisfy the minimum duration constraint. Notice that we never generate routes that exceed the maximum time allowance given the way the insertion step works (i.e. before performing the node(s) insertion we always check whether this would lead to exceeding the maximum time allowance).

12.4.3 Route Optimization

Recall that in the route optimization phase, given a set of candidates routes (previously generated in the route generation phase) the goal is to choose a subset of N routes that minimizes the total traveling time of the passengers. We do this by formulating an integer linear programming model, and solve it using both CPLEX and Lagrangian relaxation. Several comparisons of the performance of these two methods are presented in Sect. 12.4.8.2.

In this phase we have a *walking graph* $G_W = (V, A_W)$ and a *bus routes graph* $G_R = (V, A_R)$. The former comes from the original input data of the problem, where each arc $(u, v) = a \in A_W$ corresponds to going from u to v by walk. The bus routes graph is built after running the route generation phase by putting together all the arcs of the generated routes. That is, if we generated routes (where we see routes as subsets of subsequent arcs) $R_1, R_2, \dots, R_M$, we let A_R be the union of the R_i's, where if an arc a belongs to p different candidate routes we include p different copies of a in the set A_R, one for each candidate route. So we can think of A_R as containing M disjoint (or non-overlapping) routes. This fact will be important when formulating the integer program. We now introduce the rest of the notation for the variables and parameters.

x_a^k : number of people riding a bus on arc a for OD pair k
w_a^k : number of people walking on arc a for OD pair k
y_r : takes the value 1 if route r is used
u_r : bus capacity in route r
c_a : time it takes to travel arc a by bus
e_a : time it takes to travel arc a by walk
$\delta_R^-(u)$: set of arcs in A_R entering u

$\delta_W^-(u)$: set of arcs in A_W entering u
f_k: demand for OD pair k
o_k: origin node of OD pair k
d_k: destination node of OD pair k
$G_W = (V, A_W)$: walking graph
$G_R = (V, A_R)$: bus routes graph
A_r: set of arcs in route r
$\delta_R^+(u)$: set of arcs in A_R leaving u
$\delta_W^+(u)$: set of arcs in A_W leaving u.

Then the integer program can be written as follows:

$$\min \sum_k \sum_{a\in A_R} c_a x_a^k + \sum_k \sum_{a\in A_W} e_a w_a^k \tag{12.4}$$

subject to

$$\sum_{a\in\delta_R^-(v)} x_a^k + \sum_{a\in\delta_W^-(v)} w_a^k - \sum_{a\in\delta_R^+(v)} x_a^k - \sum_{a\in\delta_W^+(v)} w_a^k = \begin{cases} f_k, & \text{if } v = d_k \\ -f_k, & \text{if } v = o_k \\ 0, & \text{otherwise} \end{cases}, \quad \forall\, k, \forall\, v \in V \tag{12.5}$$

$$\sum_k x_a^k \leq u_r y_r\ , \qquad \forall r,\ \forall a \in A_r \tag{12.6}$$

$$x_a^k \leq f_k y_r\ , \qquad \forall k,\ \forall r,\ \forall a \in A_r \tag{12.7}$$

$$\sum_r y_r \leq N \tag{12.8}$$

$$0 \leq x_a^k \leq \min\{u_r, f_k\},\quad x_a^k \in \mathcal{Z},\quad \forall k, \forall r,\ \forall a \in A_r \tag{12.9}$$

$$0 \leq w_a^k \leq f_k,\quad w_a^k \in \mathcal{Z},\qquad \forall k,\ \forall a \in A_W \tag{12.10}$$

$$y_r \in \{0, 1\},\qquad \forall r \tag{12.11}$$

Indeed, notice that the objective is just the total traveling time of the passengers, i.e. total walking time plus total bus riding time. Regarding the constraints, (12.8) imposes that the number of chosen routes can be at most N. Constraint (12.6) makes sure that for each route r and each arc a belonging to route r, the bus capacity u_r for that route is not exceeded. Constraint (12.7) guarantees that if there is flow through arc a in route r then y_r should be greater than zero. Finally, (12.5) are the well-known flow conservation constraints. A simplification in this model is that it assumes that users can enter/leave a bus at any street corner. This is because the bus stops have not been defined. If the bus stops had been defined, then the nodes of the bus routes graph would be the bus stops.

12.4.4 Lagrangian Relaxation

As an alternative to using a commercial integer programming solver, we tried Lagrangian relaxation. We start by explaining the basics of this. Given a linear program

$$\min cx \tag{12.12}$$
$$\text{subject to}$$
$$Ax = b \tag{12.13}$$
$$Dx = r \tag{12.14}$$
$$x \geq 0, \tag{12.15}$$

we can multiply the constraints (12.13) by a set of *Lagrangian multipliers* π and solve

$$\theta(\pi) = \min cx - \pi Ax + \pi b$$
$$\text{subject to}$$
$$Dx = r$$
$$x \geq 0.$$

Then for any vector π, the value $\theta(\pi)$ is a lower bound for the minimum in (12.12)–(12.15). Moreover the maximum of $\theta(\pi)$ over all π, is exactly the value of the minimum in (12.12)–(12.15). This procedure is effective if removing constraints (12.13) makes the resulting linear program much easier to solve. Here we say that constraints (12.13) have been *dualized*.

In our case, removing constraints (12.5) makes the problem considerably easier, we explain this below. The flow conservation constraints (12.5) are dualized with multipliers π_v^k, for $v \in V$ and all k. Let m be the number of OD pairs. Let $\pi = (\pi_v^k) \in \mathbb{R}^{m|V|}$ be the resulting vector of Lagrangian multipliers. We have to solve:

$$\theta(\pi) = \min \sum_k \sum_{a \in A_R} g_a^k x_a^k + \sum_k \sum_{a \in A_W} h_a^k w_a^k + \sum_k f_k(\pi_{d_k}^k - \pi_{o_k}^k) \tag{12.16}$$
$$\text{subject to}$$
$$\sum_k x_a^k \leq u_r y_r \,, \qquad \forall r,\ \forall a \in A_r \tag{12.17}$$
$$x_a^k \leq f_k y_r \,, \qquad \forall k,\ \forall r,\ \forall a \in A_r \tag{12.18}$$
$$\sum_r y_r \leq N \tag{12.19}$$
$$0 \leq x_a^k \leq \min\{u_r, f_k\}, \quad x_a^k \in \mathcal{Z}, \quad \forall k, \forall r, \forall a \in A_r \tag{12.20}$$
$$0 \leq w_a^k \leq f_k, \quad w_a^k \in \mathcal{Z}, \qquad \forall k,\ \forall a \in A_W \tag{12.21}$$
$$y_r \in \{0, 1\}, \qquad \forall r. \tag{12.22}$$

Here for each arc $a = (u, v) \in A_R$, $g_a^k = c_a + \pi_u^k - \pi_v^k$. And for each $a = (u, v) \in A_W$, $h_a^k = e_a + \pi_u^k - \pi_v^k$. This is called the *Lagrangian subproblem*. In the next subsection we give a simple method to solve this.

12.4.5 Solving the Lagrangian Subproblem

For any fixed vector of Lagrangian multipliers π, we have to solve (12.16)–(12.22). First notice that variables w_a^k only appear in (12.21), thus if $h_a^k < 0$ we set $w_a^k = f_k$, and $w_a^k = 0$ otherwise. Also if $g_a^k \geq 0$ we should set $x_a^k = 0$, so in what follows we have to concentrate on the remaining variables.

Because of constraint (12.19) we have to find N variables y_r that take the value one. Consider a route r and suppose that $y_r = 1$. We have to study constraints (12.17) to find the values for the variables x. Consider an arc a in route r. We treat the variables x_a^k with the procedure below.

Step 0. Set $\alpha_a = 0$, $U = u_r$, $L = \{k \mid g_a^k < 0\}$, $x_a^k = 0$ for all k.
Step 1. If $L = \emptyset$ stop. Otherwise let $l = \arg\min\{g_a^k \mid k \in L\}$. Remove l from L.
Step 2. If $f_l \leq U$ set $x_a^l = f_l$, otherwise set $x_a^l = U$. Set $U \leftarrow U - x_a^l$, $\alpha_a \leftarrow \alpha_a + g_a^l x_a^l$. Go to Step 1.

This gives tentative values for the variables x_a^k and the number α_a is their possible contribution to the objective function. Then for the route r we define $\beta_r = \sum_{a \in A_r} \alpha_a$. This is the contribution to the objective function obtained by setting $y_r = 1$. Thus we should order the values $\{\beta_r\}$ and pick the N smallest values. This gives the set of routes whose variables should take the value one. For those routes, we use the tentative values defined above for the variables x. All other variables y and x should take the value zero. This procedure is easy to implement and computationally fast.

12.4.6 Solving the Lagrangian Dual Problem

For a given vector π we have seen how to compute the lower bound $\theta(\pi)$. Now we have to see how to improve this lower bound, this is called the *Lagrangian dual problem*. For that we use the *Volume algorithm* that can be seen as a fast way to approximate Dantzig-Wolfe decomposition [17]. Recall that the name "volume" is inspired by the fact that the primal values come from computing the volumes below the faces of the dual problem. A description of the algorithm when applied to the Lagrangian dual problem is below.

Volume algorithm

- **Step 0.**
  Let π^* be a vector of Lagrangian multipliers.
  Solve (12.16)–(12.22) with $\pi := \pi^*$.
  Let $(\hat{x}_0, \hat{w}_0, \hat{y}_0)$ be an optimal solution.
  Let $z^* = \theta(\pi^*)$. Set $t := 0$ and $(\bar{x}_0, \bar{w}_0, \bar{y}_0) := (\hat{x}_0, \hat{w}_0, \hat{y}_0)$.
- **Step 1a.**
  Let $\bar{v}_t = b - A^1\bar{x}_t - A^2\bar{w}_t$,
  where $A^1x + A^2w = b$ is the system (12.5).
  Perform a *subgradient displacement*:

$$\bar{\pi} := \pi^* + s_t\bar{v}_t \,,$$

  where the step size s_t is discussed below.
- **Step 1b.**
  Solve (12.16)–(12.22) with $\pi := \bar{\pi}$.
  Let $(\hat{x}_{t+1}, \hat{w}_{t+1}, \hat{y}_{t+1})$ be an optimal solution, and $z_{t+1} = \theta(\bar{\pi})$.
- **Step 2.**
  Compute

$$(\bar{x}_{t+1}, \bar{w}_{t+1}, \bar{y}_{t+1}) := \alpha\,(\hat{x}_{t+1}, \hat{w}_{t+1}, \hat{y}_{t+1}) \;+\; (1-\alpha)\,(\bar{x}_t, \bar{w}_t, \bar{y}_t). \qquad (12.23)$$

  The value of α is discussed below.
- **Step 3.**
  Perform the *Ascent-Test*:

$$\text{if } z_{t+1} > z^*, \text{ then update } z^* := z_{t+1},\ \pi^* := \bar{\pi}. \qquad (12.24)$$

- **Step 4.**
  Check stopping criteria. Do $t := t+1$ and go to STEP 1a.

  The step size in Step 1a is given by

$$s_t = \lambda \frac{T - z^*}{\|\bar{v}_t\|^2},$$

where $0 < \lambda < 2$, and T is a *target* value. A more complete discussion on how to choose λ appears in [4].

To choose the value of α, an initial value 0.1 is used. Then every 50 iterations, the progress in the objective value is monitored. If the increase is less than 1% and $\alpha > 10^{-5}$, then α is divided by 2.

The stopping criteria was based on a maximum number of iterations, a time limit, and a bound for the gap between the lower bound $\theta(\pi)$ and the value of an integer solution produced by the procedure below (whichever was attained first).

We used the implementation of the Volume algorithm that is in COIN-OR [14].

12.4.7 Producing a Heuristic Solution

Every ten iterations of the Volume algorithm we use the (fractional) vector $\bar{y}_t$ to produce a $0-1$ vector that defines the set of routes to be open. This heuristic is as follows. For each route r let ν_r be the component of $\bar{y}_t$ associated with route r, we add a random number $\mu_r \in [0, 1]$ to ν_r to produce the value γ_r. Then we pick the largest N values $\{\gamma_r\}$, this defines which routes should be open. Given a set of open routes, we treat the OD pairs sequentially. For each of them we find a shortest (time) path from origin to destination, using the walk arcs and the route arcs that are not full. We send flow through this path until an arc is saturated, or until the demand of the OD pair is satisfied. In the former case, we have to look for a new shortest path, and continue until the demand is satisfied. Once the demand of the OD pair is satisfied, we treat the next OD pair.

12.4.8 Computational Results

We now present and discuss the results of our computational experiments. We had to deal with a city of about 55000 inhabitants in the US. Our network has a total of 2487 nodes and 5192 arcs, where three of the nodes are terminals. Our OD-matrix contained travel demand among centroids of different zones. Statistics show that the demand is low, and buses run with only a small occupancy. Thus the OD-matrix had very low values, and we ignored all the ones below a certain threshold. We repeated this procedure for two data-sets, one for peak hours and the other for off peak. The first data-set has a total of 160 pairs while the second one has 157. We want to design a total of 16 routes (i.e. $N = 16$). Our bus routes have a minimum duration of 20 min and a maximum duration of 30 min. All buses had a capacity of forty passengers, and in our routings this capacity was never reached. All travel times are in minutes. All experiments were performed on a 2.33 GHz Intel Xeon E5410 machine with 32 GB of RAM, running Linux.

We are mainly interested in the following two questions:

1. What are good heuristics for generating candidate routes?
2. Is the Volume algorithm a better alternative than CPLEX when solving the integer program?

12.4.8.1 Heuristics Performance

We generate candidate routes following each of the five heuristics described in Sect. 12.4.2. We do this for the two different data-sets of customer demands (i.e. peak hours and off peak). We present the total number of generated routes for each case in Table 12.2. Recall that we delete duplicated routes and routes that do not meet the minimum duration criterion of 20 min.

Table 12.2 Number of generated routes

		RS	GD	GPD	GND	GNDDR
# Routes	Data-set 1	302	138	173	169	123
	Data-set 2	248	114	135	120	119

Table 12.3 Heuristics comparison (first data-set)

Volume Algorithm (parameters: 3 h, 40 000 iter, 3% gap)					
	RS	GD	GPD	GND	GNDDR
# Routes	302	138	173	169	123
Travel time	599 729	618 712	595 934	598 629	608 308
Lower bound	574 763	593 712	578 841	578 105	590 746
Relative gap	4.5%	4%	3%	3.5%	3%
Memory	0.5%	0.3%	0.3%	0.3%	0.3%
Time	3:00	2:30	2:50	2:55	2:00

In order to study which sets of candidate routes lead to a better solution, we use the Volume algorithm to solve the corresponding integer program (see Sect. 12.4.3) associated to each set of candidate routes. We justify this choice (i.e. using Volume instead of CPLEX) later on Sect. 12.4.8.2 where we show that the Volume algorithm runs much faster than CPLEX for instances of this size, and moreover, it also tends to give a better solution than CPLEX.

We run the Volume algorithm with the following stopping criteria (whichever it is reached first): a maximum of either 3 h, 40 000 iterations, or a 3% gap between the integer solution value and the lower bound. We describe in Table 12.3 the results obtained for the first data-set of customer demands, where for each set of candidate routes we show the total traveling time, the best lower bound found, the relative gap between these two, the share (in percentage) of total available memory used by the algorithm, and the total elapsed time (in hh:mm format) before the algorithm stopped.

We can observe from Table 12.3 that the best results are given by the heuristics GPD (Greedy on the Pairs Demand), GND (Greedy on the Nodes Demand), and RS (Random Shuffle), in this order.

Now, we create a new set of candidate routes by combining the N chosen routes (recall that in our case $N = 16$) from each of the five different heuristics discussed above. Certainly, we know that the objective value for this new set of candidate routes will not be worse than the best value attained for each of the heuristics separately. However, there is no obvious reason to think that the N optimal routes from each case will "complement" each other and give an improvement over each of the heuristics considered separately. Surprisingly, this seems to be the case with our data.

We can observe in Table 12.4 that by running the Volume algorithm for 3 h in this new candidate set, we obtain an objective value which is 2.5% better than the

Table 12.4 Combination of optimal routes (first data-set)

# Routes	80	80	80
Travel time	581 043	574 675	568 787
Lower bound	562 735	562 794	562 989
Relative gap	3%	2%	1%
Time	3:00	4:00	5:00

Table 12.5 Heuristics comparison (second data-set)

Volume Algorithm (parameters: 3 h, 40 000 iter, 3% gap)

	RS	GD	GPD	GND	GNDDR
# Routes	248	114	135	120	119
Travel time	472 146	480 593	463 629	482 336	472 907
Lower bound	450 000	466 596	451 051	472 200	460 332
Relative gap	5%	3%	3%	2%	3%
Memory	0.4%	0.3%	0.3%	0.3%	0.3%
Time	3:00	1:10	2:15	1:55	1:45

best solution we previously had. In addition, we also see how the lower bound in this case is better (i.e. lower) than the ones obtained for each of the heuristics separately, leading us to think that the objective value could be further improved. In this spirit, we run the Volume algorithm for an additional 1 and 2 h, respectively, and observe how in fact this is the case. That is, we see how the total traveling time decreases further, and how the relative gap with the lower bound is closed to 1%. Moreover, notice that there is an almost 5% improvement between the best solution found by combining the N optimal routes from each of the heuristics (with value 568 787) and the one found by optimizing for each heuristic separately (with value 595 934).

We perform the same computational experiments for our second data-set of customer demands. The network data (i.e. nodes, arcs, terminals, etc.) stays the same, but we now have a different set of OD pairs (and different demands) containing a total of 157 pairs. Table 12.5 shows the results obtained for this case.

Notice that for this new data-set of demands the number of generated routes for each heuristic is lower than for the previous data-set (see Table 12.2). Hence, the Volume algorithm tends to run faster than before, since its running time is proportional to the number of candidate routes that we are optimizing over. In addition, we can observe how the heuristics GPD, RS, and GNDDR give the best results in this order. We create again a new set of candidate routes by combining the N optimal routes chosen for each heuristic. Table 12.6 shows the results obtained after running Volume over this new set of candidate routes. We can observe how by combining the optimal routes we get again something strictly better, although in this case the improvement is not as significant as for the previous data-set.

Table 12.6 Combination of optimal routes (second data-set)

# Routes	80	80	80
Travel time	462 476	461 946	457 654
Lower bound	445 231	445 868	445 916
Relative gap	4%	3.5%	2.5%
Time	3:00	4:00	5:00

Table 12.7 CPLEX performance (first data-set)

CPLEX stopping parameters: 10 hr optimizer, 5% gap

	RS	GD	GPD	GND	GNDDR
# Routes	302	138	173	169	123
Travel time	829 851	888 908	930 000	656 578	598 194
Optimality gap	45%	50%	60%	14%	<1%
Memory	75%	39%	45%	45%	35%
Total time (LP time)	15:00 (5:00)	18:00 (8:00)	28:00 (18:00)	12:00 (2:00)	14:30 (7:00)
# Variables	4 900 000	2 700 000	3 200 000	3 100 000	2 300 000

We conclude this section by pointing out that our computational results seem to imply the following:

1. GPD gives the best results among the different heuristics for generating candidate routes.
2. Combining the optimal routes from the five different heuristics and optimizing again gives something strictly better than the best value found by optimizing over each candidate set separately.

12.4.8.2 CPLEX Versus Volume Algorithm

We now focus on the route optimization part of the problem and compare the performance of two different software approaches: CPLEX and the Volume algorithm. Recall that in the optimization part we are interested in finding the N optimal routes out of a larger number M (usually $N << M$) of candidate routes.

We present in Table 12.7 the results obtained after running CPLEX on the five different sets of candidate routes (see Table 12.2) generated from the first data-set of customer demands. We run CPLEX with a time limit of 10 h for the integer programming optimizer (notice that this does not include the time that it takes to solve the LP-relaxation) and a 5% optimality gap.

By comparing these results with the ones obtained by running the Volume algorithm on the same instances (see Table 12.3), we observe two interesting facts. First,

Table 12.8 Volume versus CPLEX comparison (first data-set)

	RS	GD	GPD	GND	GNDDR
Travel time	Vol 40%	Vol 43%	Vol 53%	Vol 10%	CPLEX 3%
Running time	Vol 500%	Vol 600%	Vol 1400%	Vol 400%	Vol 1000%

Table 12.9 CPLEX performance (second data-set)

CPLEX (parameters: 10 hr optimizer, 5% gap)

	RS	GD	GPD	GND	GNDDR
# Routes	248	114	135	120	119
Travel time	830 000	478 832	695 365	495 862	468 939
Optimality gap	85%	2%	53%	5%	1%
Memory	70%	38%	38%	38%	38%
Total Time (LP time)	19:00 (9:00)	13:00 (7:00)	18:30 (8:30)	15:00 (7:30)	17:00 (9:00)
# variables	4 200 000	2 300 000	2 600 000	2 400 000	2 300 000

it seems clear that Volume algorithm runs much faster than CPLEX for problems of this size. In particular, just solving the LP-relaxation in CPLEX seems to take a long time, and in most cases just solving the LP in CPLEX takes even longer than solving the whole problem using Volume algorithm. Second, it seems to be the case that Volume algorithm gives a much better solution than CPLEX, even though as we mentioned before, it runs for a much shorter amount of time. We should also notice that while Volume algorithm uses less than 1% of the total amount of available memory, CPLEX requires 35% or more of the available memory.

We present in Table 12.8 a case by case comparison between CPLEX and Volume algorithm for each of the five sets of candidate routes generated from the first data-set. We show the difference (in %) between the best solution found and between the running times. For instance, in the case of RS heuristic (second column of the table), we write in the first row "Vol 40%" to denote that the traveling time found by Volume algorithm is 40% better than the one found by CPLEX, and we write in the second row "Vol 500%" to denote that the running time of Volume algorithm is five times faster than the one of CPLEX.

We repeat the same computations for the second data-set of demands. That is, we run CPLEX on the five different sets of candidate routes generated from the second data-set and compare these results with the ones obtained by running Volume algorithm over the same instances (see Table 12.5). We present in Table 12.9 the results obtained with CPLEX and in Table 12.10 the comparison between CPLEX and Volume algorithm.

We can observe that the Volume algorithm is much faster than CPLEX for this problem. Also, Volume algorithm either gives a (usually much) better solution than CPLEX or in the cases where CPLEX outperforms Volume algorithm, this difference is very small (less than 1%).

Table 12.10 Volume versus CPLEX comparison (second data-set)

	RS	GD	GPD	GND	GNDDR
Travel time	Vol 76%	CPLEX <1%	Vol 50%	Vol 5%	CPLEX <1%
Running time	Vol 600%	Vol 1100%	Vol 800%	Vol 800%	Vol 1000%

12.4.9 Conclusions on Bus Routing

We considered the problem of designing the public bus transit network of a medium size city. The goal is to design a total of N routes so that the overall traveling time of the passengers is minimized. Our approach consisted of breaking down the original problem into two subproblems: route generation and route optimization. In the former, we are interested in finding heuristics to generate good candidate routes. In the latter, we formulate an integer program that chooses N optimal routes out of a larger number of candidates routes, and we study efficient ways to solve this integer program for our given data.

Our computational results seem to imply that the GPD (Greedy on the Pairs Demand) heuristic is a good way to generate candidate routes. This heuristic is very simple to implement and consists of initially presorting the OD pairs by demand in decreasing order, then trying to add first as many pairs as possible, and finally adding as many single nodes as possible. In addition, the best results were achieved by first optimizing separately over the candidate set generated by each heuristic, and then combining these five sets of optimal routes into a new set of candidate routes.

On the route optimization part we solve the integer program with both CPLEX and the Volume algorithm and compare their performances. Our computational results in this case suggest that Volume algorithm is a better choice than CPLEX for approaching this problem. In particular, Volume algorithm was much faster than CPLEX in all the instances that we considered. It also gave much better solutions than CPLEX in many of the instances, while in the few cases where CPLEX outperformed Volume algorithm this difference was very small (less than 1% in two cases and 3% in one case). Also the amount of memory that Volume algorithm requires is much less than the memory used by CPLEX.

12.5 The p-Median Problem: Application to Restructuring Semistructured Data

The main motivation of this section is the treatment of large instances of the p-median problem coming from semistructured data. In the *p-median problem* (pMP) we are given a directed graph $G = (V, A)$, and costs $c(u, v)$ for each arc $(u, v) \in A$, and $w(v)$ for each node $v \in V$. The problem consists of selecting p nodes, usually called *centers*, and then assign, through a single arc, each non-selected node to a selected one. The goal is thus to select exactly p nodes minimizing the total cost of the selected nodes plus the assignment cost. Thus, if we associate a variable $x(u, v)$ with each arc $(u, v) \in A$ and a variable $y(v)$ with each $v \in V$, an integer linear programming formulation of pMP is:

$$\text{minimize} \sum_{(u,v)\in A} c(u, v)x(u, v) + \sum_{u\in V} w(u)y(u), \tag{12.25}$$

$$\sum_{v\in V} y(v) = p, \tag{12.26}$$

$$\sum_{v:(u,v)\in A} x(u, v) + y(u) = 1 \quad \forall u \in V, \tag{12.27}$$

$$x(u, v) \le y(v) \qquad \forall (u, v) \in A, \tag{12.28}$$

$$1 \ge y(v) \ge 0 \qquad \forall v \in V, \tag{12.29}$$

$$x(u, v) \ge 0 \qquad \forall (u, v) \in A, \tag{12.30}$$

$$x \text{ and } y \text{ integer.} \tag{12.31}$$

This problem has been studied since the sixties, dynamic programming methods, branch and bound methods and heuristics have been proposed. See [26] for a survey on location theory. We deal with cases whose size cannot be treated with commercial integer programming solvers. Even the linear programming relaxation obtained by removing (12.31), cannot be handled with commercial solvers, so we use Lagrangian relaxation to compute a lower bound. When Eq. (12.27) is dualized, the remaining model is quite easy to solve. This type of relaxation has been used in [1, 7, 11, 15] and others. To produce good feasible solutions we use heuristics.

A new feature of our work is that the lower bound is computed in a multicore computer. Our implementation is based on the observation that when Eq. (12.27) is dualized, then the relaxed problem decomposes into simple independent pieces, one for each node of the graph. The Lagrangian problem is treated with the Volume algorithm [4]. This produces an approximate primal solution that we used to modify the costs before applying a greedy heuristic. This has lead to better upper bounds than the ones given by the original greedy heuristic. Finally the upper bound is improved with the interchange heuristic [15, 26, 27, 42, 48]. With this procedure we have obtained solutions that are on average within 0.1% from optimality. The worst gap from optimality that we obtained was 1.1%.

The p-median problem has been widely studied from an experimental point of view along different approaches, the most effective approaches that may solve some big instances may be found in [1, 22, 41]. Our motivation is to develop a method that exploits the LP-relaxation and solves huge instances that do not depend on the structural properties of the problem nor its cost function. For this end the method must be not memory consuming. In Sects. 12.5.1–12.5.3 we give a description of the Volume algorithm when applied to the p-median problem, its parallel implementation, and we describe the greedy and the local search heuristics to obtain a feasible solution from the approximate solution produced by the algorithm. In Sect. 12.5.4

we present an application in the field of databases. These databases consist of a huge number of objects in general and any method for restructuring this data must be not memory consuming, this consideration is our main motivation.

12.5.1 *The Volume Algorithm Applied to the p-Median Problem*

In an attempt to solve the linear program (12.25)–(12.30) we use Lagrangian relaxation. When we dualize Eq. (12.27), the Lagrangian problem becomes

$$\theta(\pi) = \min \sum_{(u,v)\in A} \big(c(u,v) - \pi(u)\big)x(u,v) + \sum_{u\in V} \big(w(u) - \pi(u)\big)y(u) + \sum_{u\in V} \pi(u) \tag{12.32}$$

$$\sum_{v\in V} y(v) = p \tag{12.33}$$

$$x(u,v) \le y(v) \quad \forall (u,v) \in A \tag{12.34}$$

$$1 \ge y(v) \ge 0 \quad \forall v \in V \tag{12.35}$$

$$x(u,v) \ge 0 \quad \forall (u,v) \in A. \tag{12.36}$$

Given a vector of multipliers π, to compute $\theta(\pi)$ one has to choose a set of p variables y that should take the value one. For that we compute for each node v the value

$$\mu(v) = \sum_{(u,v)\in A} \min\left\{0, \big(c(u,v) - \pi(u)\big)\right\} + \big(w(v) - \pi(v)\big).$$

Then we have to choose p smallest values μ. Here one should notice that the computation of the values μ takes the largest effort, but since they can be computed independently, this part can be done in parallel. This is the key in our implementation.

For any vector π, a lower bound of the optimal value of (12.25)–(12.30) is given by $\theta(\pi)$. So we should try to maximize $\theta(\cdot)$, the maximum value is exactly the optimal value of (12.25)–(12.30). For that we use the Volume algorithm.

The Volume algorithm to solve (12.25)–(12.30)

- **Step 0.**
  Let π^* be a vector of Lagrangian multipliers. Solve (12.32)–(12.36) with $\pi := \pi^*$. Let $(\hat{x}_0, \hat{y}_0)$ be the optimal solution.
  Set $z^* := \theta(\pi^*)$. Set $t := 0$ and $(\bar{x}_0, \bar{y}_0) := (\hat{x}_0, \hat{y}_0)$.
- **Step 1a.**

Compute $\bar{v}_t(u) = 1 - \bar{y}_t(u) - \sum_{(u,v)} \bar{x}_t(u, v)$, for each $u \in V$. Perform a *subgradient displacement*:

$$\bar{\pi} := \pi^* + s_t \bar{v}_t ,$$

where the step size s_t is as discussed in Sect. 12.4.6.

- **Step 1b.**
  Solve (12.32)–(12.36) with $\pi := \bar{\pi}$, let $(\hat{x}_{t+1}, \hat{y}_{t+1})$ be the solution thus obtained and $z_{t+1} = \theta(\bar{\pi})$.
- **Step 2.**
  Compute $(\bar{x}_{t+1}, \bar{y}_{t+1}) := \alpha(\hat{x}_{t+1}, \hat{y}_{t+1}) + (1 - \alpha)(\bar{x}_t, \bar{y}_t)$, where α is a number between 0 and 1 discussed in Sect. 12.4.6.
- **Step 3.**
  Perform the *Ascent-Test*: If $z_{t+1} > z^*$, then update $z^* := z_{t+1}, \pi^* := \bar{\pi}$.
- **Step 4.** Check stopping criteria. Do $t := t + 1$ and go to STEP 1a.

We stop if one of the following hold:

- the solution $(\hat{x}_{t+1}, \hat{y}_{t+1})$ satisfies (12.27),
- $T - \theta(\pi^*) < granularity$. Here granularity means the smallest difference between two different values of the cost function. It can be set to 1 when the costs are integers, the target value T is an upper bound obtained by any heuristic.
- $| \sum_{(u,v)} c(u, v)\bar{x}_{t+1}(u, v) + \sum_u w(u)\bar{y}_{t+1}(u) - z_{t+1} | < \epsilon | z_{t+1} |$, and $\sum | \bar{v}_t(u) | < \epsilon$. This means that the difference between the primal objective value and $\theta(\bar{\pi})$ is small, and the sum of the violations of Eq. (12.27) is small.
- There is not a significant change in the value of $\theta(\pi^t)$ after a certain number of iterations or that a limit in the number of iterations is reached.

12.5.2 Parallel Implementation

Our parallel implementation is based on a master/slave scheme on a multicore computer with 32 processors. In a sequential implementation we noticed that more than 90% of the computing time was taken by solving the Lagrangian problem (12.32)–(12.36). As seen above, problem (12.32)–(12.36) may be decomposed easily, where each value $\mu(v)$ may be computed independently. Hence most of the processors will be devoted to the computation of these values. The scheme of our implementation is given in Fig. 12.12.

We use 28 processors to compute μ. Each of them receives a node v, computes its value $\mu(v)$ and sends it with a list of arcs (u, v) such that $\bar{c}(u, v) < 0$ to the $(\hat{x}_t, \hat{y}_t)$ processor. This is described in the algorithm below.

The μ processor's algorithm

- **Step 0.**
  Receive a node v. Set $\rho := 0$. Let $u \in V$ such that $(u, v) \in A$.

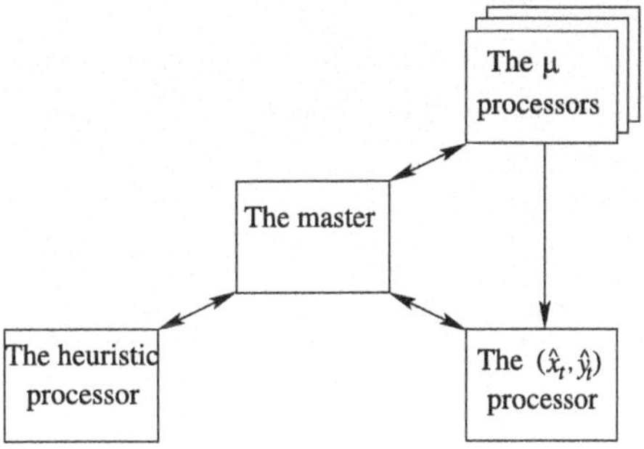

Fig. 12.12 The parallel implementation scheme for solving (12.25)–(12.30) with Lagrangean relaxation and the Volume algorithm

- **Step 1.** Set $\bar{c}(u, v) := c(u, v) - \pi(u)$.
  If $\bar{c}(u, v) < 0$: store($list$,$(u, \bar{c}(u, v))$) and set $\rho := \rho + \bar{c}(u, v)$. Set $V := V \setminus \{u\}$.
- **Step 2.**
  If $V \neq \emptyset$: choose a node $u \in V$ such that $(u, v) \in A$ and go to Step 1.
- **Step 3.**
  $\mu \leftarrow \rho - \pi(v)$. Send $(v,\mu,list)$.

To compute $(\hat{x}_t, \hat{y}_t)$ we need to choose p smallest values μ. For that we use a priority queue. This is described in the following algorithm.

The $(\hat{x}_t, \hat{y}_t)$ processor's algorithm

- **Step 0.**
  InitPriorityQueue(FP). Set $nbr\_received := 0$.
- **Step 1.**
  If $nbr\_received < n$: receive(u,$\mu(u)$,$list$), enqueue(FP,$\mu(u)$,u) and updateSolution($(\hat{x}_t, \hat{y}_t)$,u,$list$).
- **Step 2.**
  If size(FP) $> p$: set $v :=$ dequeue (FP) and updateSolution($(\hat{x}_t, \hat{y}_t)$,v,$list$).
- **Step 3.**
  Set $nbr\_received := nbr\_received + 1$ and go to Step 1.

Finally, the master processor receives the solution $(\hat{x}_t, \hat{y}_t)$ updates π and performs some elementary operations like checking the stopping criteria, if it is false, a new iteration is carried out. See the next algorithm.

The master processor's algorithm

- **Step 0.**
  InitParallel(). Let π^* be a vector Lagrangian multipliers.
- **Step 1.**
  Send each node $v \in V$ to the μ processor's algorithm with $\pi = \pi^*$.
- **Step 2.**
  Receive $z^* = \theta(\pi)$,$(\hat{x}_0, \hat{y}_0)$, $\bar{v}_0$, from the $(\hat{x}_t, \hat{y}_t)$ processor. Set $t := 0$, $(\bar{x}_0, \bar{y}_0) := (\hat{x}_0, \hat{y}_0)$.

- **Step 3.**
  Perform a *subgradient displacement*: $\bar{\pi} := \pi^* + s\bar{v}_t$. Send each node $v \in V$ to one μ processor with $\pi = \bar{\pi}$.
- **Step 4.**
  Receive $z_{t+1} = \theta(\pi)$,$(\hat{x}_{t+1}, \hat{y}_{t+1})$,$\bar{v}_{t+1}$, from the $(\hat{x}_t, \hat{y}_t)$ processor. Set $(\bar{x}_{t+1}, \bar{y}_{t+1}) := \alpha(\hat{x}_{t+1}, \hat{y}_{t+1}) + (1-\alpha)(\bar{x}_t, \bar{y}_t)$.
- **Step 5.** If $z_{t+1} > z^*$: set $\pi^* := \bar{\pi}$; $z^* := z_{t+1}$.
- **Step 6.** Check the stopping criteria. Do $t := t + 1$ and go to STEP 3.

12.5.3 Improving the Upper Bound

Our heuristic method that obtains feasible solutions starts with the primal solution found by the Volume algorithm and uses the greedy and local search methods.

Let $G = (V, A)$ be a directed graph, with cost functions c and w, associated with the arc set A and the node set V, respectively. With an integer constant p, we obtain an instance of the p-median problem defined by G, c, w and p. Given a subset of nodes $S \subseteq V$ with $|S| = p$, the cost of an optimal solution that uses the nodes in S as centers, is

$$cost(S) = \sum_{u \in V \setminus S} \min_{v \in S} c(u, v) + \sum_{v \in S} w(v).$$

For each pair of nodes u and v and a subset S of V, let $S_{uv} = (S \setminus \{u\}) \cup \{v\}$. The local search heuristic simply begins with a subset $S \subseteq V$, with $|S| = p$, it finds a pair of nodes $u^* \in S$ and $v^* \in V \setminus S$ that minimizes $cost(S_{uv})$ over all the pair of nodes $u \in S$ and $v \in V \setminus S$. If there is an improvement, that is $cost(S_{u^*v^*}) < cost(S)$, then S is replaced by $S_{u^*v^*}$. We repeat this procedure until no improvement is possible. Methods based on local search have been widely investigated for combinatorial optimization problems and especially for location problems, see [23, 26, 27, 29, 33, 41, 48].

The starting set S of the local search method is obtained by the greedy heuristic. The greedy heuristic simply begins with an empty set S and increases its size until $|S| = p$. At each step we add to S a node v^* that minimizes $cost(S \cup \{v\})$ over all $v \in V \setminus S$.

The naive implementation of the local search and the greedy methods is time consuming. We noticed that the clever implementation of Whitaker [55] reduces the running time of the naive method by approximately one hundred times, in almost all the instances. Better again, the implementation of Resende and Werneck [42] outperforms the previous implementation where we noticed an acceleration of about eight hundred times for some instances of TSPLIB [40]. However, we did not find any fast implementation for the greedy heuristic.

12.5.3.1 A Trick in Implementing the Greedy Algorithm

We will show how the greedy method reduces to a local search method and so one can use the implementation in [42] to considerably increase the speedup. Let $\bar{v}$ be an artificial node, with $w(\bar{v}) = M$ and $c(u, \bar{v}) = M$ and $c(\bar{v}, u) = 0$ for each $u \in V$, where M is a large positive constant. We begin a local search with $S = \emptyset$ and at each iteration we inject $\bar{v}$ in S and we make the best *virtual swap*. By definition this swap is made with $\bar{v}$ and v^* where v^* is the node that minimize $cost(S \cup \{v\})$ over all $v \in V \setminus S$. This is repeated until $|S| = p$. The solution obtained is exactly the one given by the greedy method. Since this implementation of the greedy heuristic is very fast, we have used it to produce an upper bound.

VGL method

- **Step 0.**
  Let (x_v, y_v) the LP-relaxation solution given by the volume algorithm.
- **Step 1.**
  For each arc $(u, v) \in A$, set $c'(u, v) := (1 - x_v(u, v))c(u, v)$.
- **Step 2.**
  Let $S \subseteq V$, $|S| = p$ the solution given by greedy heuristic when applied with respect to the costs c'.
- **Step 3.**
  Apply the local search heuristic with S as the initial solution.

12.5.4 Application to Restructuring Semistructured Data

Semistructured data is characterized by a structure which may be absent, irregular, implicit, or partial. Recent research work has led to the emergence of new data models making it possible to represent this type of data in a way "self-describing", i.e. without resorting to a rigid schema and predefined in advance (e.g. the XML language). However, in many applications, the data has a certain regularity. Ignoring the structure (even partially) of this data limits the possibilities of querying efficiently such data. In this context, we are interested in extracting structures in semistructured databases. This was reduced to a p-median problem in [36].

In [8] a semistructured data is represented by a directed graph $G = (V, A, L)$ where the set of vertices V represents the objects, the set of arcs A represents the relationships between the objects and L is a function that associates a label to each arc, see Fig. 12.13. The pending nodes (without outgoing arcs) are called *atomic objects*, the other nodes are called *complex objects*. Each atomic object has fixed value while each complex object has a type defined as below. Notice that the atomic objects are all represented by 0, since they have the same role in restructuring the database (they are fixed values).

For example, the database above may be interpreted as follows: object O_1 is a *student*, objects O_2, O_3, and O_4 are *courses*, and O_5, O_6 are the name of the courses

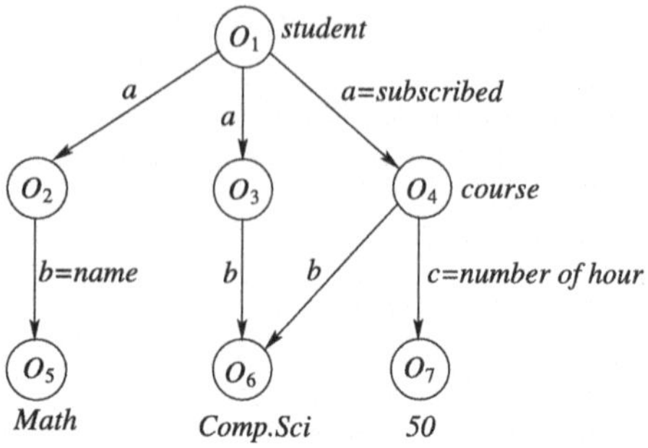

Fig. 12.13 A simple semistructured database

and O_7 number of hours (called atomic objects since they are assigned with a fixed value). Then the label a linking O_1 with O_2 indicates that O_1 is subscribed to course O_2, then the object O_1 contains the *typed link* $\overset{\rightarrow 2}{a}$ and vice-versa O_2 contains the typed link $\overset{\leftarrow 1}{a}$. In our example, the label b linking two objects specifies the name of the course and c specifies that the number of hours of a given course. So a type of an object is a set of typed links.

With these considerations the type t_1 expresses the fact the student O_1 is subscribed to O_2, O_3 and O_4 and the type of the object O_2, $t_2 = \overset{\leftarrow 1}{a}, \overset{\rightarrow 0}{b}$, expresses the fact that O_2 is followed by O_1 ($\overset{\leftarrow 1}{a}$) and its name is some fixed value ($\overset{\rightarrow 0}{b}$), these fixed values play no role in restructuring the data. These relationships are used to define the perfect type of an object. The perfect types in our example are τ_1, τ_2, and τ_3, given in the expressions below. Here objects O_2 and O_3 have the same type represented by τ_2, as we will see later this type τ_2 will be assigned a weight of 2 indicating the number of objects in our original graph having this type.

$$t_1 = \overset{\rightarrow 2}{a}, \overset{\rightarrow 3}{a}, \overset{\rightarrow 4}{a} = \tau_1$$

$$t_2 = \overset{\leftarrow 1}{a}, \overset{\rightarrow 0}{b} = \tau_2$$

$$t_3 = \overset{\leftarrow 1}{a}, \overset{\rightarrow 0}{b} = \tau_2$$

$$t_4 = \overset{\leftarrow 1}{a}, \overset{\rightarrow 0}{b}, \overset{\rightarrow 0}{c} = \tau_3.$$

In general, the number of objects is huge and one cannot create a type for each object. In order to have an effective way of storing and querying such data, the idea developed in [36] is to create a fixed number of types and then associate each object to one of these types by minimizing a total distance defined by the excess and the deficit of information between the original type and the fixed one. To reduce this problem to a p-median problem, we will create a complete directed graph $G = (T, A)$, each node in T will refer to a perfect type, and there are two arcs between each pair of nodes i and j. A distance $d(i, j) = d(j, i)$ is the Manhattan distance between the two types points τ_i and τ_j in the binary hypercube. In other words, this distance is

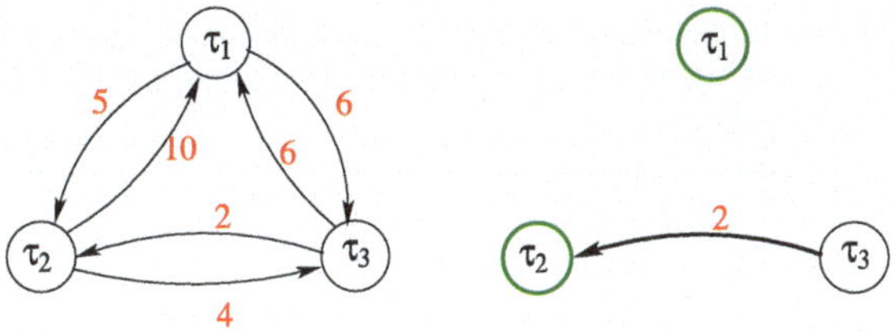

Fig. 12.14 The left figure represents the graph $G = (T, A)$, having three nodes τ_1, τ_2, and τ_3. The red number attached to each arc represents the cost. The optimal solution is reported in the right figure

the cardinality of the symmetric difference between these two types. For example $d(1, 2) = d(2, 1) = 5$, $d(2, 3) = d(3, 2) = 1$. Using this distance we define a cost $c(i, j)$ associated to each arc as $c(i, j) = w_i d(i, j)$, where w_i is the number of objects with a perfect type τ_i, in our example $w_1 = 1$, $w_2 = 2$, and $w_3 = 1$. Figure 12.14 shows the graph $G = (T, A)$ with the associated costs of the example in Fig. 12.13. Also the optimal solution is shown when the desired number of types to be fixed is $p = 2$.

12.5.5 Experimental Results

Table 12.11 gives some experimental results on instances from TSPLIB [40]. Table 12.12 concerns huge instances generated randomly and are structurally closed to semistructured database instances. In fact, a type of an object may be seen as an extreme point of a binary hypercube of dimension d, where d is the number of typed links. We generate randomly extreme points of a hypercube with a given dimension. A given extreme point may be generated several times, this number represents its weight in a perfect typing. Once these points with their weights are known, it is straightforward to construct the directed graph $G = (T, A)$ with the corresponding costs.[1]

Table 12.11 illustrates the solution of the linear relaxation (12.25)–(12.30) by CPLEX and the Volume algorithm with both implementations sequential and parallel. The first column of this table contains the reference of the instance with the number of nodes at the end. For each pair of nodes i and j, there are the arcs (i, j) and (j, i) and the costs are the Euclidean distance between these two nodes represented by points on the plane. The values L_{CPLEX} and L_{Volume} are respectively the values of linear relaxation solutions obtained by CPLEX and the Volume algorithm. The running times (in seconds) of CPLEX and of the sequential and parallel versions of the Volume algorithm are respectively reported in the last three columns of Table

[1] These experimental tests were conducted as part of JC Gay's Ph.D. thesis under the supervision of M. Baiou. Gay JC (2011) Résolution du problème du p-médian, application à la restructuration de bases de données semi-structurées, Université Blaise Pascal, Clermont-Ferrand, France.

Table 12.11 Comparing the running times of CPLEX with the running times of the Volume algorithm, for solving the linear relaxation (12.25)–(12.30). The values in bold are the values of the optimal solutions known in the literature

instances		Linear relaxation		Running time			Exact solution/Gap	
Name	p	L_{CPLEX}	L_{Volume}	T_{CPLEX}	$T_{VolumeSeq}$	$T_{VolumePar}$	UB_{VGL}	Gap (%)
rl1304	5	3 099 070	3 097 291.05	4582.28	88	295	**3 099 073**	0.00
rl1304	10	–	2 131 065.12	–	133	404	2 141 048	0.47
rl1304	20	1412 110	1 412 044.84	3 326.96	116	309	**1 412 108**	0.00
rl1304	50	795 012	794 941	1 232.08	114	134	**795 012**	0.00
fl1400	5	174 877	1 74 874.76	974.18	98	376	**174 877**	0.00
fl1400	10	100 601	100 585.38	710.58	145	405	**100 601**	0.00
fl1400	20	57 191	57 171.98	571.26	159	365	**57 191**	0.00
fl1400	50	28 486	28 482.53	530.68	174	249	**28 486**	0.00
fl1400	100	15 960.90	15 950.12	762.91	160	155	15 970	0.12
ul1432	5	1 210 126	1 210 016.66	2 321.94	105	303	**1 210 126**	0.00
ul1432	10	849 759	849 737.30	8 004.57	130	373	**849 759**	0.00
ul1432	20	588 720	588 649.29	9 319.99	154	325	588 875	0.03
d2103	50	–	301 141.53	–	397	466	303 228	0.7
d2103	300	90 423	90 208.26	2225.22	328	123	90 536	0.37
d2103	500	63 951	63 890.11	1446.77	312	221	64 394	0.79
pcb3038	5	–	1 777 293.71	–	598	2082	1 778 273	0.05
pcb3038	10	–	1 211 703.82	–	593	1227	**1 211 704**	0.00
pcb3038	20	–	837 783.72	–	747	1047	840 027	0.26
pcb3038	50	–	505 990.03	–	731	634	506 509	0.10
pcb3038	100	–	351 161.57	–	744	574	351 681	0.14
pcb3038	150	–	279 959.77	–	722	414	280 209	0.08
pcb3038	200	–	237 023.77	–	596	326	237 531	0.21
pcb3038	300	–	186 698.90	–	685	300	186 896	0.10
pcb3038	400	–	156 173.22	–	637	265	156 335	0.10
pcb3038	500	–	134 733.44	–	580	282	134 822	0.06
fl3795	5	–	1 052 626.93	–	911	3 083	**1 052 627**	0.00
fl3795	10	–	520 934.25	–	1158	1902	**520 940**	0.00
fl3795	20	–	319 704.37	–	1287	2089	**319 722**	0.00
fl3795	50	–	150 699.13	–	1386	1241	150 942	0.16
fl3795	100	–	88 115.78	–	1364	1025	88 319	0.23
fl3795	150	–	65 708.13	–	1340	835	65 891	0.27
fl3795	300	–	39 544.08	–	1464	622	39 616	0.18
fl3795	500	–	25 968.28	–	1409	443	26 003	0.13

Table 12.12 Exact solutions for big random instances following the semistructured database structures

n	p	L_{Volume}	UB_{VGL}	T_{VGL}	Gap (%)
14960	1000	49 582.85	50 131	7 218	1.10
14960	1500	45 276.35	45 684	6 572	0.89
14960	2000	41 937.05	42 186	6 731	0.59
14960	2500	438 996.67	39 147	8 242	0.38
14997	1000	71 695.78	72 421	7 134	1.00
14997	1500	66 268.27	66 718	6 956	0.67
14997	2000	61 833.46	62 146	7 099	0.50
14997	2500	57 903.97	58 145	7 183	0.41
15000	1000	261 521.15	262 378	6 097	0.32
15000	1500	248 829.83	249 226	5 768	0.16
15000	2000	236 702.47	237 270	5 929	0.24
15994	1000	70 798.70	71 560	7 886	1.07
15994	2000	61 159.78	61 570	8 004	0.67
15994	2500	57 377.51	57 573	9 011	0.34
20000	1500	164 199.21	165 902	10 895	1.03
20000	2000	155 329.32	156 465	12 482	0.73
20000	2500	147 469.12	148 589	13 430	0.75

12.11. We remark that CPLEX is unable to solve the linear relaxation whenever the size reaches the 3000 nodes, while the sequential implementation of the Volume algorithm produces a solution in a couple of minutes. This time is divided by two when using the parallel implementation. Notice that the parallel implementation is not effective for small values of p, which is expected since most of the running time of the Volume algorithm is spent in computing the p smallest values of μ. However, the effect of the parallelization is noticeable when p increases. The column UB_{VGL} contains the value of an integer feasible solution. This value is obtained as described above by combining the solution of the linear relaxation provided by the Volume algorithm with a greedy heuristic that produces an initial solution. Then the interchange heuristic improves the upper bound. The last column contains the value $Gap = \frac{UB_{VGL}-L_{Volume}}{L_{Volume}}$. This expresses the fact that the value of our exact solution is at most $Gap \times z_{Opt}$ away from the value of the optimal solution z_{Opt}.

The first column of Table 12.12 contains the number of nodes. We observe that despite the huge size of the instances the running time, T_{VGL} in seconds, is between 1 and 3 h and the gap is always around 1%. This does not depend on the size of the instances which shows that our approach is stable with respect to this parameter.

12.6 Final Remarks

The Volume algorithm is a modification of the subgradient method. We have shown three applications of it to different optimization problems. Their common characteristic is their large scale and the need for fast approximate solutions. We hope that our results encourage the readers to use it in many other new areas.

References

1. Avella P, Sassano A, Vasil'ev I (2007) Computational study of large-scale p-median problems. Math Program 109:89–114
2. Baaj MH, Mahmassani HS (1995) Hybrid route generation heuristic algorithm for the design of transit networks. Transp Res Part C: Emerg Technol 3:31–50
3. Bahiense L, Maculan N, Sagastizábal C (2002) The volume algorithm revisited: Relation with bundle methods. Math Program 94:41–69
4. Barahona F, Anbil R (2000) The volume algorithm: Producing primal solutions with a subgradient method. Math Program 87:385–399
5. Barahona F, Goncalves J (2021) Volume optimizer for tensorflow. https://github.com/IBM/volume-optimizer-for-tensorflow
6. Borndörfer R, Grötschel M, Pfetsch ME (2007) A column-generation approach to line planning in public transport. Transp Sci 41:123–132
7. Briant O, Naddef D (2004) The optimal diversity management problem. Oper Res 52:515–526
8. Buneman P, Davidson S, Hillebrand G, Suciu D (1996) A query language and optimization techniques for unstructured data. In: Proceedings of the 1996 ACM SIGMOD international conference on Management of data, pp 505–516
9. Bussieck MR, Kreuzer P, Zimmermann UT (1997) Optimal lines for railway systems. Eur J Oper Res 96:54–63
10. Ceder A, Wilson NH (1986) Bus network design. Transp Res Part B: Methodol 20:331–344
11. Christofides N, Beasley JE (1982) A tree search algorithm for the p-median problem. Eur J Oper Res 10:196–204
12. CIFAR-10 The CIFAR-10 dataset (2009). https://www.cs.toronto.edu/~kriz/cifar.html
13. Claessens M, van Dijk NM, Zwaneveld PJ (1998) Cost optimal allocation of rail passenger lines. Eur J Oper Res 110:474–489
14. COIN-OR. https://www.coin-or.org/
15. Cornuejols G, Fisher ML, Nemhauser GL (1977) Exceptional paper-location of bank accounts to optimize float: An analytic study of exact and approximate algorithms. Manag Sci 23:789–810
16. CPLEX II (2015) V12. 6: User's manual for CPLEX. In: International Business Machines Corporation
17. Dantzig GB, Wolfe P (1960) Decomposition principle for linear programs. Oper Res 8:101–111
18. Desaulniers G, Hickman MD (2007) Public transit. Handbooks Oper Res Manag Sci 14:69–127
19. Escudero LF, Landete M, Rodriguez-Chia AM (2011) Stochastic set packing problem. Eur J Oper Res 211:232–240
20. Fan W, Machemehl RB (2006) Optimal transit route network design problem with variable transit demand: genetic algorithm approach. J Transp Eng 132:40–51
21. Fan W, Machemehl RB (2006) Using a simulated annealing algorithm to solve the transit route network design problem. J Transp Eng 132:122–132
22. García S, Labbé M, Marín A (2011) Solving large p-median problems with a radius formulation. INFORMS J Comput 23:546–556

23. García-López F, Melián-Batista B, Moreno-Pérez JA, Moreno-Vega JM (2002) The parallel variable neighborhood search for the p-median problem. J Heuristics 8:375–388
24. Goossens J-W, Van Hoesel S, Kroon L (2004) A branch-and-cut approach for solving railway line-planning problems. Transp Sci 38:379–393
25. Günlük O, Kimbrel T, Ladanyi L, Schieber B, Sorkin GB (2006) Vehicle routing and staffing for sedan service. Transp Sci 40:313–326
26. Hansen P, Mladenović N (1997) Variable neighborhood search for the p-median. Location Sci 5:207–226
27. Hansen P, Mladenović N, Perez-Britos D (2001) Variable neighborhood decomposition search. J Heuristics 7:335–350
28. Held M, Wolfe P, Crowder HP (1974) Validation of subgradient optimization. Math Program 6:62–88
29. Hodgson MJ (1978) Toward more realistic allocation in location-allocation models: An interaction approach. Environ Plan A 10:1273–1285
30. IMAGENET (2016) http://image-net.org/
31. Kepaptsoglou K, Karlaftis M (2009) Transit route network design problem: Review. J Transp Eng 135:491–505
32. Kingma DP, Ba J (2014) Adam: A method for stochastic optimization. arXiv:1412.6980
33. Manne AS (1964) Plant location under economies-of-scale-decentralization and computation. Manag Sci 11:213–235
34. MNIST (2010) The MNIST database of handwritten digits. http://yann.lecun.com/exdb/mnist/
35. Mobilenets (2017) MobileNets: open-source models for efficient on-device vision. https://ai.googleblog.com/2017/06/mobilenets-open-source-models-for.html
36. Nestorov S, Abiteboul S, Motwani R (1998) Extracting schema from semistructured data. In: Proceedings of the 1998 ACM SIGMOD international conference on Management of data, pp 295–306
37. Orabona F, Tommasi T (2017) COCOB. https://github.com/bremen79/cocob
38. Orabona F, Tommasi T (2017) Training deep networks without learning rates through coin betting. In: Advances in neural information processing systems, pp 2157–2167
39. Qian N (1999) On the momentum term in gradient descent learning algorithms. Neural Netw 12:145–151
40. Reinelt G (1991) TSPLIB-a traveling salesman problem library. ORSA J Comput 3:376–384
41. Resende MG, Werneck RF (2004) A hybrid heuristic for the p-median problem. J Heuristics 10:59–88
42. Resende MG, Werneck RF (2007) A fast swap-based local search procedure for location problems. Ann Oper Res 150:205–230
43. Schöbel A (2012) Line planning in public transportation: Models and methods. OR Spectr 34:491–510
44. Schöbel A, Scholl S (2006) Line planning with minimal traveling time. In: ATMOS 2005–5th workshop on algorithmic methods and models for optimization of railways. Schloss Dagstuhl, Internationales Begegnungs-und Forschungszentrum für Informatik (IBFI)
45. Shibasaki RS, Baïou M, Barahona F, Mahey P, de Souza MC (2021) Lagrangian bounds for large-scale multicommodity network design: A comparison between volume and bundle methods. Int Trans Oper Res 28:296–326
46. Szegedy C, Vanhoucke V, Ioffe S, Shlens J, Wojna Z (2015) Rethinking the inception architecture for computer vision. CoRR. arxiv:abs/1512.00567
47. Taylor A, Marcus M, Santorini B (2003) The Penn treebank: An overview
48. Teitz MB, Bart P (1968) Heuristic methods for estimating the generalized vertex median of a weighted graph. Oper Res 16:955–961
49. TensorFlow (2015) https://www.tensorflow.org/
50. Tensorflow (2015) Convolutional neural networks. https://www.tensorflow.org/tutorials/images/cnn
51. TensorFlow (2015) Recurrent neural networks. https://github.com/IBM/Recurrent-Neural-Networks-using-TensorFlow-Keras/blob/main/examples/RecurrentNeuralNetworkUsingTensorFlow_with_output.ipynb

52. TensorFlow (2015) Simple audio recognition. https://www.tensorflow.org/tutorials/audio/simple_audio
53. TensorFlow For Poets (2016) https://kiosk-dot-codelabs-site.appspot.com/codelabs/tensorflow-for-poets/#0
54. Tieleman T, Hinton G (2012) Lecture 6.5-rmsprop, coursera: Neural networks for machine learning, University of Toronto, Technical Report
55. Whitaker R (1983) A fast algorithm for the greedy interchange for large-scale clustering and median location problems. INFOR: Inf Syst Oper Res 21:95–108
56. Zhao F, Zeng X (2006) Optimization of transit network layout and headway with a combined genetic algorithm and simulated annealing method. Eng Optim 38:701–722

Chapter 13
Exact and Approximation Methods for Mixed-Integer Nonlinear Programming Problem

Amit Kumar Vatsa and Saurabh Chandra

Abstract Mixed-integer nonlinear programming problems (MINLPs) are frequently observed in the real world. These problems generally require more computational effort than their counterpart mixed-integer linear programming problems (MILPs). The solvers and solution technique improvements in the last few decades have enabled us to solve MINLPs of practical sizes. This chapter presents various techniques that can be used to solve MINLPs. We divide this chapter into approximation algorithms and exact algorithms. We discuss two approximation algorithms in detail: piece-wise linearization and linearization based on the McCormick envelope. These methods can be used to obtain feasible solutions with a guarantee on bound on a wide variety of problems, including non-convex problems, which are notoriously difficult to solve with an exact method. Next, we present the exact solution methods, which are guaranteed to give an optimal solution if the algorithms are allowed to run indefinitely. These methods, such as outer approximation and generalized Benders decomposition can be applied to problems where relaxation of integrality constraint makes the problem a convex optimization problem. Some non-convex problems can be convexified using the transformation of non-convex functions. These transformations might be computationally expensive but ensure an exact solution as a trade-off. With toy examples, we demonstrate the use of exact and approximation methods on MINLPs. We also provide a real-world case study for the location-allocation of shared wastewater treatment plants and show how different methods can be used to solve that problem.

13.1 Introduction

In many real-world problems (e.g., engineering, logistics, finance, public policy), the decision variables are continuous as well as discrete. Furthermore, there might exist a nonlinear relationship between the variables. Consequently, many real-world

A. Kumar Vatsa (✉) · S. Chandra
Indian Institute of Management, Indore 453556, MP, India
e-mail: amitv@iimidr.ac.in

S. Chandra
e-mail: saurabh@iimidr.ac.in

F. Hamid (ed.), *Optimization Essentials*, International Series in Operations Research & Management Science 353, https://doi.org/10.1007/978-981-99-5491-9_13

problems can be formulated as mixed-integer nonlinear programming problems (MINLPs). Some examples of MINLP applications include:

- Logistics operations in presence of economies/dis-economies of scale
- Production planning and scheduling
- Design of robotic and automation systems
- Optimizing power systems
- Process optimization in chemical and oil industries
- Locating cold storage and warehouses.

Even though a wide range of problems can be formulated as MINLP and it is of interest to various industries, MINLPs were hardly used till a decade back. There are three major reasons that make MINLPs challenging:

1. Presence of discrete variables
2. Numerical difficulties of handling nonlinear functions
3. The nonlinear function can be non-convex.

Consequently, state-of-the-art optimization solvers still struggle to handle MINLPs, limiting their use in the industry. However, the two essential MINLP sub-systems, mixed-integer linear programming (MILP) and nonlinear programming (NLP) have made great strides in the last decade. In this chapter, we will elaborate on some methods for solving MINLPs relying on the fact that modern commercial solvers are quite good at handling many practical-size MILPs even though such problems are NP-Hard.

MINLPs that we will discuss in this chapter can be represented by the following generalized formulation.

[MINLP:]

$$\min_{x} f(x) \tag{13.1}$$

$$s.t. \quad g_i(x) \leq 0 \qquad \forall i \in I \tag{13.2}$$

$$x \in X \tag{13.3}$$

$$x_j \in \mathbb{Z} \qquad \forall j \in J \tag{13.4}$$

where X is a bounded polyhedral subset of $\mathbb{R}^n$ (e.g., $X = \{x \mid x \in \mathbb{R}^n_+, Ax \leq b\}$. The functions $f : \mathbb{R}^n \to \mathbb{R}$ and $g_i : \mathbb{R}^n \to \mathbb{R}$ are twice continuously differentiable functions. The set I is the index set of nonlinear constraints and the set J is the index set of discrete variables.

The presence of discrete variables makes all the MINLPs non-convex. However, the term non-convex MINLP is typically used when any of the functions f or g_i is non-convex. In the remainder of this chapter, we refer to the MINLP to be convex if all the functions f and g_i are convex, i.e., if the problem becomes a convex optimization problem after relaxing discrete variables as continuous. Further, note that the MINLP formulation given above is quite general and can be easily modified for handling maximization objectives, "=" or "≥" constraints. However, such modifications will determine if the problem is a convex or non-convex MINLP.

In this chapter, we discuss techniques that can be used to solve MINLP problems, classifying them into approximation algorithms and exact algorithms. Approximation algorithms are designed to find a solution that is close to the optimal solution and is typically based on heuristics or metaheuristics. Two specific approximation algorithms that we elaborate on are piece-wise linearization and linearization based on the McCormick envelope. These methods can be used to obtain feasible solutions with a guarantee on the bound for a wide range of problems, including non-convex MINLPs which are difficult to solve with exact methods.

Exact algorithms, on the other hand, are designed to find the optimal solution if the algorithms are allowed to run indefinitely. Two specific exact algorithms that are discussed are an outer approximation and generalized Benders decomposition. These methods can be applied to problems where the relaxation of the integrality constraint makes the MINLP a convex optimization problem.

To illustrate the use of both exact and approximation methods on MINLP problems, toy examples are provided. Additionally, a real-world case study for the location-allocation of shared wastewater treatment plants is discussed to demonstrate how different methods can be used to solve that problem. Some non-convex problems can be convexified using the transformation of non-convex functions, which can be computationally expensive but ensure an exact solution as a trade-off. A real-world example of this convexification strategy is discussed.

The outline of the chapter is as follows. Section 13.2 presents approximate algorithms for solving MINLPs. In Sect. 13.3 exact algorithms are discussed. Section 13.4 presents the methods discussed in earlier sections using a simple example. We conclude and provide future research directions in Sect. 13.5.

13.2 Approximation Algorithms

13.2.1 Linearization Based on McCormick Envelope

McCormick envelope is a useful approximation method for solving MINLPs. It is based on relaxing nonlinear functions (specifically the product of two variables) using linear bounds or envelops. The nonlinear function in the MINLP formulation is replaced by this envelope. Further, in many problems, it can be reasoned which of the constraints in the envelope will be binding at optimality. Consequently, the non-binding constraints of the envelope can be omitted.

Solving an MINLP with McCormick envelope involves the following steps:

1. Identify the nonlinear (or non-convex) functions in the MINLP problem. In the classic implementation of the McCormick envelope, a product of two variables (a non-convex function) is taken.
2. Replace each such function with continuous variables called McCormick envelope variables.

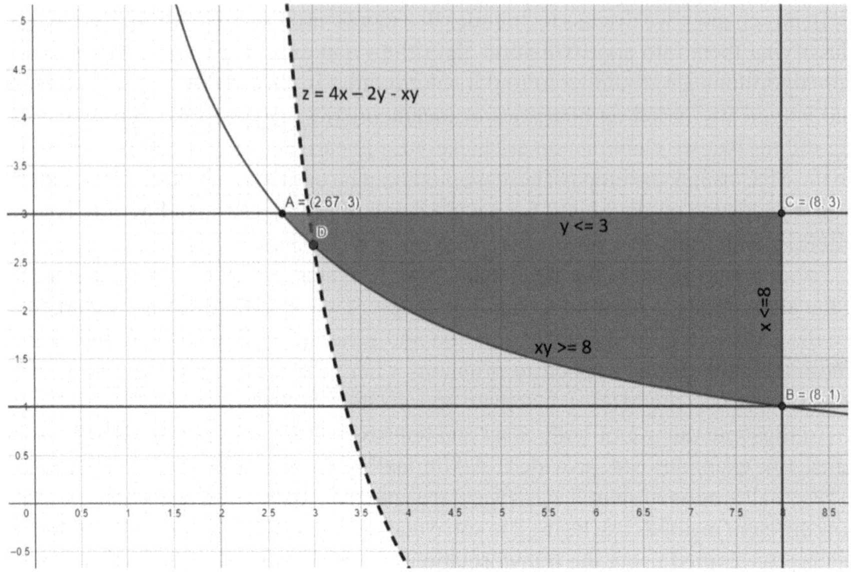

Fig. 13.1 Graphical representation of $\mathcal{P}$

3. Add linear bounds for the McCormick envelope variables.
4. Solve the resulting MILP.

Consider the following example where McCormick envelopes can be used.
[$\mathcal{P}$:]

$$\min_{x,y} \; z = 4x - 2y + xy$$
$$s.t. \; xy \geq 8$$
$$2 \leq x \leq 8$$
$$1 \leq y \leq 3$$
$$x \in \mathbb{Z}, \; y \in \mathbb{R}$$

The function xy is non-convex. Figure 13.1 presents the graph of $\mathcal{P}$. The dark-shaded region ABC represents the feasible region and the stripped curve shows the objective function. The optimal solution is shown at point $D(3, 2.67)$ with a value $z = 14.67$.

For making a linear approximation, we would like to replace xy with α, and create linear bounds for α using the following four relations:

$$(x_u - x)(y_u - y) \geq 0 \implies (8 - x)(3 - y) \geq 0 \implies \alpha \geq 3x + 8y - 24 \tag{13.5}$$

$$(x_u - x)(y - y_l) \geq 0 \implies (8 - x)(y - 1) \geq 0 \implies \alpha \leq x + 8y - 8 \tag{13.6}$$

$$(x - x_l)(y_u - y) \geq 0 \implies (x - 2)(3 - y) \geq 0 \implies \alpha \leq 3x + 2y - 6 \quad (13.7)$$

$$(x - x_l)(y - y_l) \geq 0 \implies (x - 2)(y - 1) \geq 0 \implies \alpha \geq x + 2y - 2 \quad (13.8)$$

However, given the objective function would ensure α is minimized, only the constraint for lower bound of α (i.e., (13.5) and (13.8)) will be binding at optimality. The formulation for linear relaxation of this problem using McCormick envelopes would be

[$\mathcal{P} - env$:]

$$\begin{aligned}
\min_{x,y,\alpha} \ & z_{LB} = 4x - 2y + \alpha \\
s.t. \ & \alpha \geq 8 \\
& 2 \leq x \leq 8 \\
& 1 \leq y \leq 3 \\
& \alpha \geq 3x + 8y - 24 \\
& \alpha \geq x + 2y - 2 \\
& x \in \mathbb{Z}, y \in \mathbb{R}, \alpha \in \mathbb{R}
\end{aligned}$$

The problem $\mathcal{P} - env$ is a MILP that can be solved by any MILP solver. Though this approach is quite straightforward, it suffers from a drawback in that it gives a solution to a relaxed problem. Hence, the solution to $\mathcal{P} - env$ will be a lower bound to the solution of $\mathcal{P}$. The optimal solution to the two problems are

- $\mathcal{P} : z^* = 14.67, x^* = 3, y^* = 2.67$
- $\mathcal{P} - env : z^*_{LB} = 14.25, x^* = 3, y^* = 2.875, \alpha^* = 8$

From the lower bound solution (i.e., $x = 3$, $y = 2.875$), an upper bound can be obtained, as shown in Fig. 13.2. Hence, $z_{UB} = 4 \times 3 - 2 \times 2.875 + 3 \times 2.875 = 14.875$. Therefore, McCormick envelope even though will not be able to provide the exact solution, provides an approximate solution that might be of good quality for many problem instances.

13.2.2 Piece-Wise Linearization

We provide the piece-wise linear approximation (PWL) based on the triangular method of piece-wise linear approximation [3]. To explain the workings, we continue with the same example of problem $\mathcal{P}$. To implement the triangular method of PWL approximation, we estimate the breakpoints for the values of the variables x and y and calculate the value of the nonlinear term xy at these breakpoints. As illustrated in Fig. 13.3, any section of the nonlinear surface $f(x, y)$ bounded by (x_i, x_{i+1}) and (y_i, y_{i+1}) is approximated by a linear surface $ABDC$, where the value of the actual

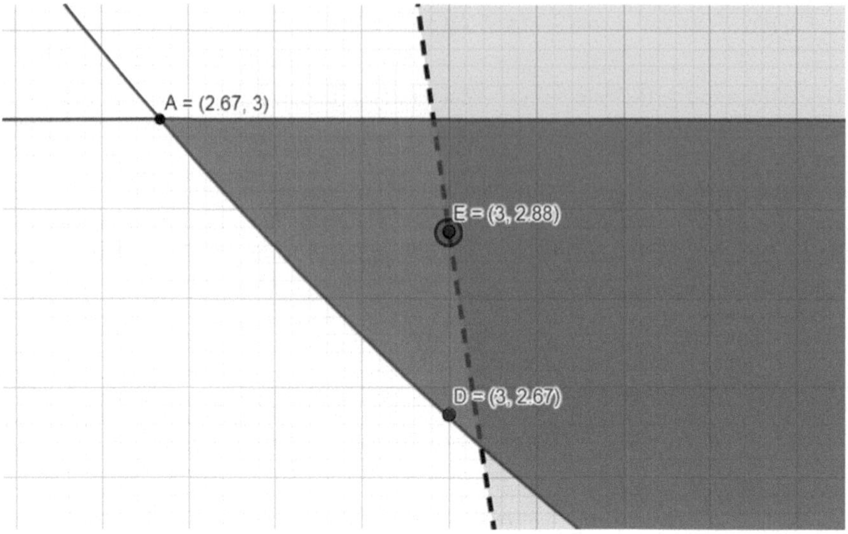

Fig. 13.2 McCormick upper bound solution

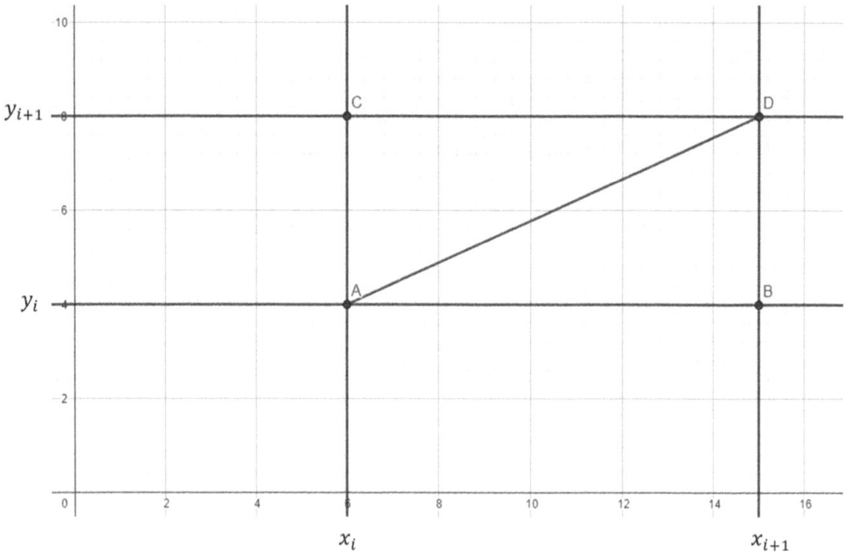

Fig. 13.3 Piece-wise linear approximation in two variables [3]

function for any point (x, y) is replaced by a convex combination of the corner points on the triangle ACD (or ABD).

Let us define the notations used in the piece-wise linear formulation of the model.

Notations used in the PWL formulation

$\mathcal{N}_x$ set of breakpoints considered for variable x, indexed by a
$\mathcal{N}_y$ set of breakpoints considered for variable y, indexed by b
x_a value of the variable x at the break-point a
y_b value of the variable y at the break-point b
xy the nonlinear function value at the combination of breakpoints a and b

Additional decision variables introduced in the PWL formulation

λ_{ab} continuous variable [0,1] for a break-point (a, b) to establish convexity constraints in three dimensions
h^u_{ab} 1, if the upper triangle is selected in the break-point (a, b), 0 otherwise
h^l_{ab} 1, if the lower triangle is selected in the break-point (a, b), 0 otherwise

The values of x, y and the nonlinear function (here the product xy) at different breakpoints are computed as shown in Eqs. (13.9)–(13.11).

$$x_a = x_{\min} + \frac{(x_{\max} - x_{\min})}{(|\mathcal{N}_x| - 1)} a \qquad a \in \mathcal{N}_x \tag{13.9}$$

$$y_b = y_{\min} + \frac{(y_{\max} - y_{\min})}{(|\mathcal{N}_y| - 1)} b \qquad b \in \mathcal{N}_y \tag{13.10}$$

$$\alpha_{ab} = (x_a)(y_b) \qquad a \in \mathcal{N}_x, b \in \mathcal{N}_y \tag{13.11}$$

In this problem since $x \in \mathbb{Z}$ and $y \in \mathbb{R}$, we take $\mathcal{N}_x = \{2, 3, 4, 5, 6, 7, 8\}$ and $\mathcal{N}_y = \{1, 1.5, 2, \ldots, 2.5, 3\}$. Consequently $|\mathcal{N}_x| = 7$, and $|\mathcal{N}_y| = 5$.

Thus, the MINLP model $\mathcal{P}$ is approximated as:

$[\mathcal{P} - PWL:]$

$$\min\ 4x - 2y + \alpha \tag{13.12}$$

$$s.t.\ \alpha \geq 8 \tag{13.13}$$

$$2 \leq x \leq 8 \tag{13.14}$$

$$1 \leq y \leq 3 \tag{13.15}$$

$$\sum_{a \in \mathcal{N}_x} \sum_{b \in \mathcal{N}_y} \alpha_{ab} = 1 \tag{13.16}$$

$$x = \sum_{a \in \mathcal{N}_x} \sum_{b \in \mathcal{N}_y} \alpha_{ab} x_a \tag{13.17}$$

$$y = \sum_{a \in \mathcal{N}_x} \sum_{b \in \mathcal{N}_y} \alpha_{ab} y_b \tag{13.18}$$

$$\alpha = \sum_{a \in \mathcal{N}_x} \sum_{b \in \mathcal{N}_y} \lambda_{ab} \alpha_{ab} \tag{13.19}$$

$$\sum_{a \in \mathcal{N}_x} \sum_{b \in \mathcal{N}_y} h^u_{ab} + \sum_{a \in \mathcal{N}_x} \sum_{b \in \mathcal{N}_y} h^l_{ab} = 1 \tag{13.20}$$

$$\lambda_{ab} \le h^u_{ab} + h^l_{ab} + h^u_{a,b-1} + h^l_{a-1,b-1} + h^u_{a-1,b-1} + h^l_{a-1,b}, \quad \forall a \in \mathcal{N}_x, b \in \mathcal{N}_y \tag{13.21}$$

$$\lambda_{ab} \in [0, 1] \quad \forall a \in \mathcal{N}_x, b \in \mathcal{N}_p \tag{13.22}$$

$$h^u_{ab}, h^l_{ab} \in \{0, 1\} \quad \forall a \in \mathcal{N}_x, b \in \mathcal{N}_y \tag{13.23}$$

$$x \in \mathbb{Z}, y \in \mathbb{R}, \alpha \in \mathbb{R} \tag{13.24}$$

13.3 Exact Algorithms

13.3.1 Outer Approximation

One of the exact methods of solving convex MINLPs is outer approximation, which was proposed by [2]. This approach is based on the concept of relaxation and outer approximation of the feasible region. A nonlinear constraint at the boundary of the feasible region is approximated using a set of tangent hyperplanes. These hyperplanes are added iteratively while solving the MILPs.

Consider a general convex MINLP of the form:

$$\begin{aligned} z = \min_x\ & f(x) \\ s.t.\ \ & g_i(x) \le 0 && \forall i \in I \\ & x \in X \\ & x_j \in \mathbb{Z} && \forall j \in J \end{aligned}$$

For the above MINLP, an outer approximating MINLP can be specified as

$$\begin{aligned} z_{OA} = \ & \min_x \eta \\ s.t.\ \ & g_i(\bar{x}) + \nabla g_i(\bar{x})^T(x - \bar{x}) \le 0 && \forall i \in I, \bar{x} \in K \\ & f(\bar{x}) + \nabla f(\bar{x})^T(x - \bar{x}) \le \eta && \forall \bar{x} \in K \\ & x \in X \\ & x_j \in \mathbb{Z} && \forall j \in J \end{aligned}$$

The original nonlinear functions have been relaxed as a linear function. Further the linear functions (tangent hyperplane) are lower bound to the respective convex function. Consequently, z_{OA} is a lower bound to z. We illustrate the outer approximation along with a related concept of Kelley's cutting plane, by solving two convex MINLPs "$\mathcal{P}1$" and "$\mathcal{P}2$".

13.3.1.1 Example 1

$$\mathcal{P}1 : \max_{x,y} \; 5x + 2y \tag{13.25}$$

$$s.t. \; g(x, y) = 4x^2 - 2xy + y^2 \leq 40 \tag{13.26}$$

$$-10 \leq x \leq 10 \tag{13.27}$$

$$-10 \leq y \leq 10 \tag{13.28}$$

$$x \in \mathbb{R}, \; y \in \mathbb{Z} \tag{13.29}$$

It can be seen from this problem that it is a convex MINLP. The objective function is linear (which is concave as well as convex) and the nonlinear function $g(x, y)$ is convex. To check if a function is convex, we can find its Hessian matrix, which for $g(x, y)$ is as follows:

$$\mathbf{H} = \begin{bmatrix} 8 & -2 \\ -2 & 1 \end{bmatrix}$$

$$\det(\mathbf{H}) = 8 - 4 = 4 > 0$$

As $g(x, y)$ is a bivariate function, it is convex if and only if the following conditions are satisfied:

1. Diagonal elements of the matrix $\mathbf{H}$ (here 8 and 1) are non-negative:
2. $\det(\mathbf{H})$ is non-negative.

Since, $g(x, y)$ satisfies these conditions, it is a convex function. Further, for any convex function it is a known property that the tangent hyperplane drawn at any point will be below the function, i.e., will be a lower bound to the function. Mathematically,

$$g(x_t, y_t) + \begin{bmatrix} x - x_t & y - y_t \end{bmatrix} \nabla g(x_t, y_t) \leq 4x^2 - 2xy + y^2 \leq 40 \quad \forall (x_t, y_t) \tag{13.30}$$

Since, at point (x_t, y_t), the gradient $\nabla_x g(x, y) = 8x_t - 2y_t$, $\nabla_y g(x, y) = -2x_t + 2y_t$. Thus,

$$\nabla g(x_t, y_t) = \begin{bmatrix} \frac{\partial g(x,y)}{\partial x} \\ \frac{\partial g(x,y)}{\partial y} \end{bmatrix}_{(x_t, y_t)} = \begin{bmatrix} 8x_t - 2y_t \\ -2x_t + 2y_t \end{bmatrix}$$

Therefore, the inequality (13.30) can be written as:

$$\begin{aligned} & g(x_t, y_t) + (x - x_t)(8x_t - 2y_t) + (y - y_t)(-2x_t + 2y_t) \leq 4x^2 - 2xy + y^2 \leq 40 \quad \forall (x_t, y_t) \\ \implies & (8x_t - 2y_t)x + (-2x_t + 2y_t)y - 4x_t^2 + 2x_t y_t - y_t^2 \leq 40 \quad \forall (x_t, y_t) \end{aligned} \tag{13.31}$$

Hence, (13.26) can be replaced with constraint set (13.31). Note that (13.31) acts as a cutting plane that cuts off the incumbent solution till the stopping criteria is not met. The outer approximating problem is

$$\mathcal{P}1_{OA} : \max_{x,y} \ 5x + 2y$$
$$s.t. \ \ (27) - (29), (31)$$

Solution Method 1: Based on Kelley's cutting plane
The following is a valid method (though inefficient because of the reasons listed subsequently) for solving $\mathcal{P}1$.

1. Solve $\mathcal{P}1_{OA}$. The solution obtained will be an Upper Bound (UB) to an optimal objective function value.
2. Based on the solution x_t, y_t obtained in iteration t, obtain and add the cut (13.31) to the problem.
3. Repeat Step 1 and 2 till $g(x_t, y_t) - 40$ is within tolerance. Alternatively, a Lower Bound (LB) can be calculated at every iteration by using y_t and obtaining x that satisfies (13.26). In that case, the iterations will stop when $UB - LB$ is within tolerance.

The following iterations would be performed for solving $\mathcal{P}1$ by Method 1:

- iteration 1: $UB = 70, x = 10, y = 10, g(x_t, y_t) = 300$. Cut added based on (13.31):
$$60x + 0y \leq 340$$
- iteration 2: $UB = 48.33, x = 5.67, y = 10, g(x_t, y_t) = 115.1$. Cut added based on (13.31):
$$25.33x + 8.67y \leq 155.11$$
- iteration 3: $UB = 33.51, x = 2.70, y = 10, g(x_t, y_t) = 75.2$. Cut added based on (13.31):
$$1.61x + 14.6y \leq 115.16$$
- iteration 4: $UB = 32.61, x = 3.73, y = 7, g(x_t, y_t) = 52.4$. Cut added based on (13.31):
$$15.82x + 6.54y \leq 92.4$$
- iteration 5: $UB = 29.13, x = 5.43, y = 1, g(x_t, y_t) = 107.9$. Cut added based on (13.31):
$$41.4x - 8.85y \leq 147.9$$
- iteration 6: $UB = 28.92, x = 4.18, y = 4, g(x_t, y_t) = 52.6$. Cut added based on (13.31):
$$25.48x - 0.37y \leq 92.58$$

- iteration 7: $UB = 28.79, x = 3.36, y = 6, g(x_t, y_t) = 40.8$. Cut added based on (13.31):
$$14.86x + 5.28y \leq 80.81$$
- iteration 8: $UB = 28.72, x = 2.94, y = 7, g(x_t, y_t) = 40.5$. Cut added based on (13.31):
$$9.56x + 8.11y \leq 82.46$$
- iteration 9: $UB = 28.518, x = 3.3036, y = 6, g(x_t, y_t) = 40.01$. Cut added based on (13.31):
$$14.43x + 5.39y \leq 80.01$$
- iteration 10: $UB = 28.514, x = 3.303, y = 6$.

$$\text{STOP as} \quad g(x_t, y_t) - 40 = 2.7 \times 10^{-}6 < tolerance$$

It can be observed that by adding tangent hyperplanes as cuts in place of $g(x, y)$, MILP converges to an optimal solution. Further in every iteration the formulation gets tighter and the UB improves. However, note that in the few iterations, tangent hyperplanes were added at the same integer solution. For example, in the first three iterations (i.e., $y = 10$), three tangent hyperplanes were added. As we had to solve three MILPs at $y = 10$, Method 1 is inferior. The Outer Approximation described next is an improvement over Method 1.

Solution Method 2: Outer Approximation (OA)

We would like to use the fact that convex NLPs can be solved very efficiently. For a given integer solution $\bar{y}$, we want to add a single tangent hyperplane. In other words, we would like to avoid getting the same integer solution in different iterations as solving MILPs is expensive. Hence, for a given $\bar{y}$, the following problem can be solved to optimality.

$$\begin{aligned}
\mathcal{P}1_y : &\max_x \; 5x + 2\bar{y} \\
s.t. \;\; g(x, \bar{y}) = \; & 4x^2 - 2x\bar{y} + \bar{y}^2 \leq 40 \\
& -10 \leq x \leq 10 \\
& x \in \mathbb{R}
\end{aligned}$$

The tangent hyperplanes based on (13.31) will be added at the point which is a solution to $\mathcal{P}1_y$. Furthermore, if $\mathcal{P}1_y$ is infeasible for any $\bar{y}$, any modern solver reports the point where infeasiblity is minimized, and cut is added at that point. The following iterations would be performed for solving $\mathcal{P}1$ by OA:

- iteration 1: Solve $\mathcal{P}1_{OA}$ to obtain $UB = 70, x = 10, y = 10$. From this solution we retain the y and UB. With $\bar{y} = 10$, solve $\mathcal{P}1_y$ and find that $\mathcal{P}1_y$ is infeasible. While reporting infeasibility, the solver gives a point $x = 2.5$, where infeasibility is minimized. Cut added (for this solution) based on (13.31):

$$15y \leq 115$$

- iteration 2: Solve $\mathcal{P}1_{OA}$ to obtain $UB = 64, x = 10, y = 7$. From this solution we retain the y and UB. With $\bar{y} = 7$, solve $\mathcal{P}1_y$ to obtain $x = 2.6514, LB = 27.257$. Cut added based on (13.31):

$$7.2112x + 8.6972y \leq 80$$

- iteration 3: Solve $\mathcal{P}1_{OA}$ to obtain $UB = 51.439, x = 9.888, y = 1$. From this solution we retain the y and UB. With $\bar{y} = 1$, solve $\mathcal{P}1_y$ to obtain $x = 3.3825, LB = 27.257$. Cut added based on (13.31):

$$25.06x - 4.765y \leq 80$$

- iteration 4: Solve $\mathcal{P}1_{OA}$ to obtain $UB = 31.287, x = 3.86, y = 6$. From this solution we retain the y and UB. With $\bar{y} = 6$, solve $\mathcal{P}1_y$ to obtain $x = 3.3028, LB = 28.514$. Cut added based on (13.31):

$$14.422x + 5.394y \leq 80$$

- iteration 5: Solve $\mathcal{P}1_{OA}$ to obtain $UB = 28.514, x = 3.3028, y = 6$. STOP as LB and UB have converged.

Clearly, the solution method 2 is a superior method as a lot fewer MILPs need to be solved and convergence is a lot faster. Let us take another convex MINLP where outer approximation can be performed.

13.3.1.2 Example 2

$$\mathcal{P}2 : \min_{x,y} \; x^2 + y \tag{13.32}$$

$$s.t. \;\; 3x + 2y \geq 8 \tag{13.33}$$

$$x + 2y \geq 6 \tag{13.34}$$

$$x, y \geq 0, x \in \mathbb{R}, y \in \mathbb{Z} \tag{13.35}$$

$\mathcal{P}2$ is represented graphically in Fig. 13.4 with the optimal solution as $D(0.67, 3)$. Here again, $f(x) = x^2$ is a convex function and the tangent hyperplane gives a lower bound to the function value. Hence,

$$f(x_t) + (x - x_t)\nabla f(x_t) \leq x^2 \leq \theta \quad \forall x_t$$

At point (x_t), $\nabla f(x) = 2x_t$. Thus,

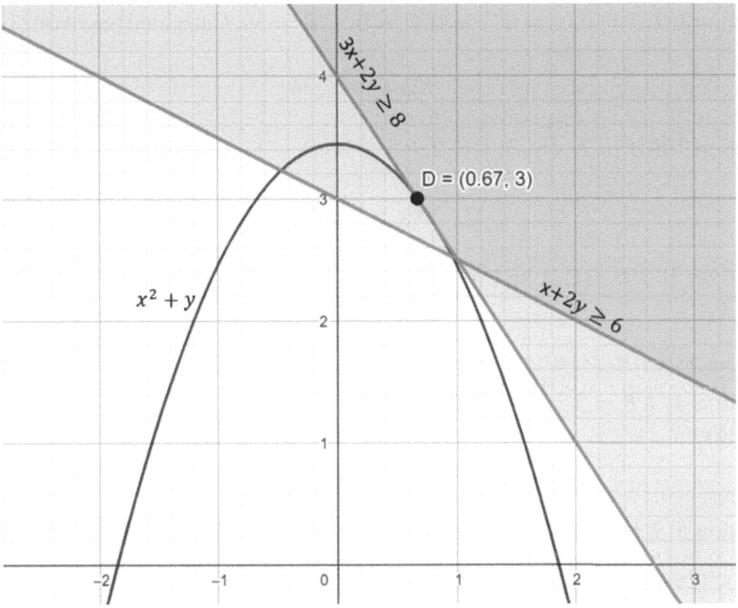

Fig. 13.4 Graphical representation of $\mathcal{P}2$

$$\begin{aligned} & x_t^2 + (x - x_t)2x_t \le \theta \quad \forall x_t \\ \implies & (2x_t)x - x_t^2 \le \theta \quad \forall x_t \end{aligned} \tag{13.36}$$

For solving this problem with outer approximation, we replace the nonlinear term from the objective function and instead add it as a constraint. The resulting problem is as follows:

$$\begin{aligned} \mathcal{P}2_{OA}: \quad & \min_{x,y,\theta} \; \theta + y \\ s.t. \quad & (33) - (35), (36) \\ & x, y, \theta \ge 0, x \in \mathbb{R}, y \in \mathbb{Z} \end{aligned}$$

Solution Method 1: Based on Kelley's cutting plane
The following is a valid method (though inefficient) for solving $\mathcal{P}2$

1. Solve $\mathcal{P}2_{OA}$. Notice now the solution obtained will be a Lower Bound (LB) to an optimal objective function value.
2. Based on the solution x_t obtained in iteration t, obtain and add the cut (13.36) to the problem.
3. Repeat Step 1 and 2 till $f(x_t) - \theta$ is within tolerance. Alternatively, obtain UB as $x_t^2 + y_t$. Stop when $UB - LB$ is within tolerance.

Solving $\mathcal{P}2$ by outer approximation leads to following iterations:

- iteration 1: $LB = 0, x = 6, y = 0, \theta = 0, UB = 36$. Cut added based on (13.36):

$$12x - \theta \leq 36$$

- iteration 2: $LB = 2, x = 3, y = 2, \theta = 0, UB = 11$. Cut added based on (13.36):

$$6x - \theta \leq 9$$

- iteration 3: $LB = 3, x = 1.5, y = 3, \theta = 0, UB = 5.25$. Cut added based on (13.36):

$$3x - \theta \leq 2.25$$

- iteration 4: $LB = 3, x = 0.667, y = 3, \theta = 0, UB = 3.44$. Cut added based on (13.36):

$$1.33x - \theta \leq 0.444$$

- iteration 5: $LB = 3.444, x = 0.667, y = 3, \theta = 0.444, UB = 3.444$. STOP.

It can be observed that with the addition of tangent hyperplanes as cuts in place of $f(x) = x^2$, MILP converges to an optimal solution. However, again as first method of example 1, multiple iterations are happening at the same integer solution ($y = 3$), which slows the convergence.

Solution Method 2: Outer Approximation (OA)

For a given integer solution $\bar{y}$, we want to add a single tangent hyperplane. Hence, for a given $\bar{y}$, the following problem can be solved to optimality.

$$\begin{aligned} \mathcal{P}2_y: \quad & \bar{y} + \min_x \ x^2 \\ s.t. \quad & 3x \geq 8 - 2\bar{y} \\ & x \geq 6 - 2\bar{y} \\ & x \geq 0, x \in \mathbb{R} \end{aligned}$$

The tangent hyperplanes based on (13.36) will be added at the point which is a solution to $\mathcal{P}2_y$. The following iterations would be performed for solving $\mathcal{P}2$ by OA:

- iteration 1: Solve $\mathcal{P}2_{OA}$ to obtain $LB = 0, x = 6, y = 0, \theta = 0$. From this solution we retain the y and LB. With $\bar{y} = 0$, solve $\mathcal{P}2_y$ to obtain $x = 6, UB = 36$. Cut added based on (13.36):

$$12x - \theta \leq 36$$

- iteration 2: Solve $\mathcal{P}2_{OA}$ to obtain $LB = 2, x = 3, y = 2, \theta = 0$. From this solution we retain the y and LB. With $\bar{y} = 2$, solve $\mathcal{P}2_y$ to obtain $x = 2, UB = 6$. Cut added based on (13.36):

$$4x - \theta \leq 4$$

- iteration 3: Solve $\mathcal{P}2_{OA}$ to obtain $LB = 3, x = \frac{2}{3}, y = 3, \theta = 0$. From this solution we retain the y and LB. With $\bar{y} = 3$, solve $\mathcal{P}2_y$ to obtain $x = \frac{2}{3}, UB = 3\frac{4}{9}$. Cut added based on (13.36):

$$\frac{4}{3}x - \theta \leq \frac{4}{9}$$

- iteration 4: $LB = 3\frac{4}{9}, x = \frac{2}{3}, y = 3, \theta = \frac{4}{9}$. STOP as LB and UB have converged.

Here again as expected, the solution method 2 is a superior method as a lot fewer MILPs need to be solved and convergence is a lot faster.

13.3.2 *Generalized Benders Decomposition*

Generalized benders decomposition(GBD) is a partitioning-based technique in which the problem is partitioned into a master problem and sub-problem. The master problem contains integer variables while the sub-problem is a problem with continuous variables. In other words, it can be said that the entire problem is projected onto the solution space of the master problem variables. GBD is widely used to solve convex MINLPs. In this chapter we consider separable MINLPs to keep things simple. Consider the following convex MINLP which has discrete as well as continuous variables.

13.3.2.1 Example : $\mathcal{P}2$

$$\begin{aligned}
\min_{x,y}\ & x^2 + y \\
s.t.\ & 3x + 2y \geq 8 \\
& x + 2y \geq 6 \\
& x, y \geq 0, x \in \mathbb{R}, y \in \mathbb{Z}
\end{aligned}$$

Since this is a very small problem let us understand the solution with exhaustive enumerations (Table 13.1).

Projecting this problem on the master problem variable y, we get the following equivalent problem

$$\min_{y,\theta}\ \theta + y \tag{13.37}$$

$$s.t.\ \theta = \min_{x\geq 0} x^2 : 3x + 2y \geq 8, x + 2y \geq 6 \tag{13.38}$$

$$y, \theta \geq 0,\ y \in \mathbb{Z} \tag{13.39}$$

The sub-problem in constraint (13.38) for a given master problem solution y_t can be written as:

Table 13.1 Solution for $\mathcal{P}2$

y	Constraints	x	$x^2 + y$
0	$x \geq 8/3, x \geq 6$	6	36
1	$x \geq 2, x \geq 4$	4	17
2	$x \geq 4/3, x \geq 2$	2	6
3	$x \geq 2/3, x \geq 0$	2/3	3.44
4	$x \geq 0, x \geq -2$	0	4

Opt:
$x = 2/3,\ y = 3,\ obj = 3.44$

[sub-problem]

$$\begin{aligned} SP_t &= \min_x \ x^2 \\ s.t.\ & 3x \geq 8 - 2y_t \\ & x \geq 6 - 2y_t \\ & x \geq 0 \end{aligned}$$

The sub-problem is a convex NLP, consequently, strong duality holds. The optimal Lagrangian dual can be written as

$$SP_t = \max_{\lambda_1 \geq 0, \lambda_2 \geq 0} \min_{x \geq 0} x^2 + \lambda_1(8 - 3x - 2y_t) + \lambda_2(6 - x - 2y_t)$$

Because of strong duality, the Benders cut (13.38) can be written in terms of Lagrangian dual as follows:

$$\theta = \max_{\lambda_1 \geq 0, \lambda_2 \geq 0} \min_{x \geq 0} x^2 + \lambda_1(8 - 3x - 2y) + \lambda_2(6 - x - 2y)$$

$$\Longrightarrow \quad \theta \geq \min_{x \geq 0} x^2 + \lambda_{1t}(8 - 3x - 2y) + \lambda_{2t}(6 - x - 2y) \quad \forall \lambda_{1t} \geq 0, \lambda_{2t} \geq 0$$

Let $x_t, \lambda_{1t}, \lambda_{1t}$ be the sub-problem solution in iteration t and SP_t be the optimal objective function value of the sub-problem. The Benders cut expression can be simplified as follows. Benders cut:

$$\theta \geq SP_t + 2\lambda_{1t}(y_t - y) + 2\lambda_{2t}(y_t - y) \quad \forall \lambda_{1t} \geq 0, \lambda_{2t} \geq 0$$

Hence, the Benders Master problem can be written as:
[Master Problem]

$$\min_{y,\theta} \ \theta + y \tag{13.40}$$

$$s.t. \ \theta \geq SP_t + 2\lambda_{1t}(y_t - y) + 2\lambda_{2t}(y_t - y) \quad \forall \lambda_{1t} \geq 0, \lambda_{2t} \geq 0 \tag{13.41}$$

$$y, \theta \geq 0, \ y \in \mathbb{Z} \tag{13.42}$$

Solving $\mathcal{P}2$ by Generalized Benders Decomposition (GBD), involves the following steps:

1. Solve the Master problem (13.40)–(13.42) and obtain θ, y and the Lower Bound (LB) to the optimal objective function value.
2. Based on the master problem solution y_t obtained in iteration t, solve the sub-problem using KKT conditions.

   a. Solve for x and identify which of the two sub-problem constraints are non-binding. λ for that constraint will have a value of zero due to complementary slackness.
   b. Further, one of the KKT conditions (by differentiating the Lagrangian function w.r.t. x) is $2x - 3\lambda_1 - \lambda_2 = 0$. Hence, obtain SP_t, λ_{1t}, and λ_{2t}, and add the Benders cut (13.41) corresponding to this sub-problem solution to the master problem.
   c. Obtain Upper Bound (UB) as $SP_t + y_t$.

3. Repeat Steps 1 and 2 till $UB - LB$ is within tolerance.

Solving the above problem by GBD and adding Benders cuts to the Master problem leads to the following iterations:

- iteration 1: Master Problem: $y = 0, \theta = 0, LB = 0$. Sub-problem constraints: $3x \geq 8, x \geq 6$. Using KKT conditions $x = 6, \lambda_1 = 0, \lambda_2 = 2x = 12, SP_t = 36, UB = SP_t + y_t = 36$. Cut added to MP: $\theta \geq 36 - 24y$.
- iteration 2: Master Problem: $y = 2, \theta = 0, LB = 2$. Sub-problem constraints: $3x \geq 4, x \geq 2$. Using KKT conditions $x = 2, \lambda_1 = 0, \lambda_2 = 2x = 4, SP_t = 4, UB = SP_t + y_t = 6$. Cut added to MP: $\theta \geq 20 - 8y$.
- iteration 3: Master Problem: $y = 3, \theta = 0, LB = 3$. Sub-problem constraints: $3x \geq 2, x \geq 0$. Using KKT conditions $x = 2/3, \lambda_1 = 2x/3 = 4/9, \lambda_2 = 0, SP_t = 4/9, UB = SP_t + y_t = 3\frac{4}{9}$. Cut added to MP: $\theta \geq 3\frac{1}{9} - \frac{8}{9}y$.
- iteration 4: Master Problem: $y = 3, \theta = 4/9, LB = 3\frac{4}{9}$. Since, $UB = LB = 3\frac{4}{9}$, STOP.

13.4 Illustrative Example

We present an application of non-convex MINLP in addressing a significant problem. Processing units use water as an essential input like leather processing, garment dying, food processing, etc., and consequently discharge contaminated wastewater as a byproduct. Government regulations worldwide mandate the treatment of this effluent wastewater to reduce contamination to prescribed levels before final discharge into the open. For small and medium enterprises installing an effluent treatment plant may be cost-prohibitive. To this end, firms operating similar processing units in a cluster, install common effluent treatment plants (CETP) to avail the benefits of economies of scale in handling larger effluent volumes from member units. Installing a network of CETPs for an industrial cluster is a facility location problem with the following decisions:

- selecting locations to install CETPs,
- capacity of each installed CETP, and
- which processing units are to be connected to each installed facility.

The costs involved in the CETP facility location problem are as follows:

- fixed cost of installing CETPs,
- transportation cost of connecting processing units to CETPs, and
- effluent treatment cost.

We give a brief overview of the CETP facility location-allocation problem. For a detailed description, readers are referred to [1]. An industrial cluster comprising multiple processing units engaged in the production of product category. All these processing units discharge effluent wastewater of varying volumes and concentrations. These units lack the resources to set up their own effluent treatment plants, so they collaborate with each other to install a network of CETPs serving all the member units. Next, we present the mathematical formulation for the problem.

13.4.1 *Mathematical Model*

$\mathcal{P}$ is a cluster of processing units co-located in a cluster producing a similar product, p representing a single processing unit. A number of potential locations have been identified for the installation of CETPs, this set being represented as $\mathcal{L}$, l denoting a single location. A discrete set $\mathcal{D}$ of CETP types, varying in size, indexed as d is available for installation across the potential locations. F_{dl} is the fixed cost of installing a CETP type d at a location l. The transportation cost when connecting processing unit p to a CETP installed at a location l is T_{pl}. The CETP of type t has the capacity to process Q_t units (in volume) of effluent. A processing unit p generates F_p^v in volume and F_p^c in concentration (weight/unit volume) of pollutants present. The treatment cost term comprises a constant multiplier K and exponents a and b for volume and concentration, respectively. Let us define decision variables. h_{dl} is a binary variable, which equals 1 when a CETP type d is installed at location l. g_{pl} is a binary variables, which equals 1 when a processing unit p is allocated to the CETP at location l. m_l is the volume of wastewater processed by a CETP installed at location l. n_l is the concentration (weight/volume) of pollutants processed by CETP at l. w_l is the weight of pollutant processed at CETP at l. Now, we can present the mathematical model for the CETP location-allocation problem.

$$\text{P-CETP: Min} \sum_{l\in\mathcal{L}}\sum_{d\in\mathcal{D}} F_{dl}h_{dl} + \sum_{l\in\mathcal{L}}\sum_{p\in\mathcal{P}} T_{pl}g_{pl} + \sum_{l\in\mathcal{L}} K m_l^a n_l^b \tag{13.43}$$

$$\text{s.t.} \sum_{l\in\mathcal{L}} g_{pl} = 1 \qquad \forall p \in \mathcal{P} \tag{13.44}$$

$$\sum_{d\in\mathcal{D}} h_{dl} \le 1 \qquad \forall l \in \mathcal{L} \tag{13.45}$$

$$g_{pl} \le \sum_{d\in\mathcal{D}} h_{dl} \qquad \forall l \in \mathcal{L}, p \in \mathcal{P} \tag{13.46}$$

$$m_l = \sum_{p\in\mathcal{P}} F_p^v g_{pl} \qquad \forall l \in \mathcal{L} \tag{13.47}$$

$$\sum_{p\in\mathcal{P}} F_p^v g_{pl} \le \sum_{d\in\mathcal{D}} Q_t h_{dl} \qquad \forall l \in \mathcal{L} \tag{13.48}$$

$$n_l \sum_{p\in\mathcal{P}} F_p^v g_{pl} = \sum_{p\in\mathcal{P}} F_p^v F_p^c g_{pl} \qquad \forall l \in \mathcal{L} \tag{13.49}$$

$$m_l, n_l \ge 0 \qquad \forall l \in \mathcal{L} \tag{13.50}$$

$$g_{pl} \in \{0, 1\} \qquad \forall l \in \mathcal{L}, p \in \mathcal{P} \tag{13.51}$$

$$h_{dl} \in \{0, 1\} \qquad \forall l \in \mathcal{L}, d \subset \mathcal{D} \tag{13.52}$$

The objective function (13.43) minimizes the sum of installation, transportation, and treatment costs respectively. Constraints (13.44) ensure that each processing unit is connected to exactly one location. Constraints (13.45) ensure that a maximum of one type of CETP can be installed at a location. Constraints (13.46) ensure that a processing unit can be allocated to a location only when a CETP has been installed there. The volume processed by a location is calculated by the constraints (13.47). Constraints (13.48) ensure that the total volume of effluent processed by a CETP is bound by the capacity installed at that location. The concentration of effluent treated at a location is calculated by the constraints (13.49). Finally, constraints (13.50)–(13.52) put domain restrictions on the decision variables.

In the following subsections, we present the characteristics of the above MINLP and the solution techniques applied as solution strategies in brief.

13.4.2 Characteristics of the MINLP Model

The mathematical model $P - CETP$ presented above has the following characteristics.

- The last term of the objective function (13.43), the treatment cost is a nonlinear term derived as a Cobb-Douglas function for wastewater treatment plants [5].
- Treatment cost if a function of wastewater volume and concentration processed by a CETP.
- There exist economies of scale in treatment cost w.r.t. wastewater volume.
- For all values of the exponent a such that the treatment cost shows economies of scale w.r.t. volume, i.e., $a < 1$, the treatment cost is a non-convex function. For detailed proof, please refer to [1].
- The non-convex treatment cost term makes $P - CETP$ a non-convex MINLP.

To solve non-convex MINLP, we give a brief description of the exact and approximate methods applicable to this problem next. We present an alternative formula for the treatment cost function, which we use in the subsequent sections. In equation (13.53), treatment cost TC is expressed as a function of wastewater flow volume (m_l) and wastewater pollutant weight (r_l). We introduce a new exponent c, such that $c = a - b$.

$$TC = \sum_{l \in \mathcal{L}} K m_l^a \frac{r_l^b}{m_l^b} = \sum_{l \in \mathcal{L}} K m_l^{(a-b)} r_l^b = \sum_{l \in \mathcal{L}} K m_l^c r_l^b \tag{13.53}$$

13.4.3 Exact Solution Methods for $P - CETP$

To implement exact methods to the $P - CETP$, we must convert the non-convex treatment cost function to a convex function. While this is not possible for all values of the exponents a and b, under special conditions it is possible.
Steps to convexify the nonlinear treatment cost:

- Replace the variable m_l by another variable q_l, such that

$$q_l = (m_l)^2 \tag{13.54}$$

- Replace all terms involving m_l with q_l in the model $P - CETP$. The new treatment cost function can be represented as follows.

$$TC = \sum_{l \in \mathcal{L}} K (q_l)^{\frac{c}{2}} r_l^b \tag{13.55}$$

- The treatment cost becomes convex under the following condition shown in Eq. (13.56). For detailed proof please refer to [1].

$$\frac{c}{2} + b \geq 1 \quad i.e., \quad a + b \geq 2 \tag{13.56}$$

- Incorporating the treatment cost term as shown in Eq. (13.4.3) to $P - CETP$, TC becomes a convex function under the condition outlined in the previous bullet point.
- Now, the exact solution procedures explained in the remaining section can be applied to the problem.

13.4.3.1 Generalized Benders Decomposition (GBD)

1. The nonlinear cost term TC along with the constraints involving continuous variables, (13.47)–(13.50) are modeled as a sub-problem ($P - CETP_{SP}$).
2. The master problem ($P - CETP_{MP}$) comprises of the objective terms and constraints in the binary variable only.
3. A continuous variable μ is added to the model.
4. The resulting master problem is solved and the optimal solution to the binary decision variables is derived as $\bar{g}_{pl}$ & $\bar{h}_{dl}$ for all index pairs. Also, the values of dual variables corresponding to each of the sub-problem constraints are extracted.
5. A GBD cut is generated from the current binary solutions and the dual values derived. For the formulation of the GBD cut, please refer to [4].
6. The master problem is solved after adding the above cut and the lower bound (LB) and the upper bound (UB) are derived as the MP objective and MP+SP objective, respectively.
7. Steps 2–5 are repeated till the $LB = UB$.

13.4.3.2 Outer Approximation (OA)

1. The nonlinear term is removed from the main model and a new continuous variable μ is introduced. This is the master problem (MP).
2. An incumbent solution to the MP is derived by solving the resulting MILP. The values of the decision variables from the MP are stored from the current solution.
3. Estimate the values of the continuous decision variables from the binary decision variables derived above.
4. Generate the OA cut from the incumbent solution. For details on how to generate the OA cut for this problem, please refer to [1].
5. Add the cut to the MP and solve the resulting MILP again till the lower bound (LB) is equal to the upper bound (UB). LB is derived as the MP solution and the UB is derived as the LB plus the resulting treatment cost in each iteration.

13.4.4 Approximate Solution Methods for $P - CETP$

The approximate solution methods can be applied regardless of the nature of the nonlinearity in the MINLP.

13.4.4.1 Piece-Wise Linear (PWL) Approximation

- Generate multiple breakpoints as pairs of flow (m_l) and concentration (n_l) variables and resulting treatment cost values.
- Add linear relaxations of the areas under each successive break-point pair using the two-variable PWL approach suggested by [3].
- Solve the resulting MILP after replacing the nonlinear terms with PWL relaxations. This would give an upper bound to the original MINLP model.

13.4.4.2 McCormick Envelope

- Replace the nonlinear term $m_l^a n_l^b$ by a new variable α_l.
- Derive the bounds of the continuous decision variables, namely, flow (m_l) and concentration (n_l) as $m_{\max,l}$ & $m_{\min,l}$ and $n_{\max,l}$ & $n_{\min,l}$, respectively.
- Add the McCormick envelopes by using the following inequalities and deriving linear bounds to the resulting expressions.

$$(m_{\max,l}^a - m_l^a)(n_{\max,l}^b - n_l^b) \geq 0 \tag{13.57}$$

$$(m_l^a - m_{\min,l}^a)(n_l^b - n_{\min,l}^b) \geq 0 \tag{13.58}$$

For details refer to [1].
- Run the resulting MILP after adding the McCormick envelopes derived from the constraints (13.57) and (13.58). This would give an upper bound to the original model.

13.5 Conclusions and Research Directions

Many real-life problems can be modeled using MINLP formulations. The presence of nonlinear terms involving decision variables makes the problem difficult to solve, esp. when some of the nonlinear functions are non-convex in nature. MINLP models come in great variety and there is no single technique suitable to address this range. To this end, we present some common approximate and exact solution approaches which can be implemented to solve MINLPs. The approximate methods discussed are the two-variable piece-wise linear approximation method and McCormick envelope methods. These methods can be implemented easily in MINLPs although do not guarantee optimal solutions or even good lower bounds to the problems. Exact methods, GBD and OA can only be applied if the nonlinear functions are convex. We present a convexification strategy applied to a specific problem.

Future research direction in this field includes identifying interesting application areas for MINLP formulation and solution. Tighter bounds need to be found for implementing the linearization techniques common to all solution methods in MINLP. While implementing branch-and-cut-based exact methods like GBD and OA, heuristic methods guaranteeing optimal solutions to sub-problems need to be found to improve computational times.

References

1. Chandra S, Sarkhel M, Vatsa AK (2021) Capacitated facility location-allocation problem for wastewater treatment in an industrial cluster. Comput Oper Res 132:105338
2. Duran MA, Grossmann IE (1986) An outer-approximation algorithm for a class of mixed-integer nonlinear programs. Math Program 36(3):307–339
3. D'Ambrosio C, Lodi A, Martello S (2010) Piecewise linear approximation of functions of two variables in MILP models. Oper Res Lett 38(1):39–46
4. Geoffrion AM (1972) Generalized Benders decomposition. J Optim Theory Appl 10:237–260
5. Mundle S, Shankar U, Mehta S (1995) Incentives and regulation for pollution abatement with an application to waste water treatment. Technical report, Asian Development Bank

Chapter 14
Lagrangian Methods and Dynamic Programming-Based MIP Formulations for the Unit Commitment Problem

Tiziano Bacci, Antonio Frangioni, and Claudio Gentile

Abstract Exactly solving hard optimization problems, in particular those mixing both nonlinear components and combinatorial ones, crucially requires the ability to derive tight bounds on the optimal value, which are obtained via relaxations. A fundamental trade-off exists between the tightness of the bound, which is important to avoid the explosion of the number of nodes in branch-and-bound (B&B) approaches and better drive the heuristics and the branching decisions, and the cost of the solution of the continuous relaxation. This is in particular true because "strong" relaxations (providing tight bounds) are also very often "large" ones, i.e., with a very large—often exponential—number of variables and/or constraints. Navigating this trade-off is crucial to develop overall efficient solution algorithms, especially since apparently minor aspects may have a disproportionate large impacts depending on the fine details of the computational environment in which the algorithm is implemented. This chapter provides an extensive illustration of this process using as test bed the Unit Commitment problem in electrical power production. In particular we will review several Mixed-Integer NonLinear formulations of (some of the most common variants of) the problem, with different trade-offs between size and "tightness", as well as algorithms for the efficient solutions of single-unit versions of the problem that motivate Lagrangian approaches. We will then compare the different variants, mostly in terms of their running time versus the quality of the provided bound, in the context of a state-of-the-art software implementation, highlighting the several nontrivial choices that have to be navigated in order to choose the best approach for each given application.

T. Bacci · C. Gentile
Istituto di Analisi dei Sistemi ed Informatica "A. Ruberti", C.N.R, Via dei Taurini 19, 00185 Rome, Italy
e-mail: tiziano.bacci@iasi.cnr.it

C. Gentile
e-mail: gentile@iasi.cnr.it

A. Frangioni (✉)
Dipartimento di Informatica, Università di Pisa, Largo B. Pontecorvo 3, 56127 Pisa, Italy
e-mail: frangio@di.unipi.it

F. Hamid (ed.), *Optimization Essentials*, International Series in Operations Research & Management Science 353, https://doi.org/10.1007/978-981-99-5491-9_14

14.1 Introduction

The Unit Commitment problem (UC) in electrical power production is a large family of models related with optimally managing a set of units producing energy at minimum cost, usually over a short time horizon (one day, one week), while satisfying a given energy demand. Several complex constraints regulate the energy production of each unit, such as minimum and maximum power output, minimum up- and down-time, ramp-up and ramp-down limits, and start-up and shut-down limits. The deterministic version of the problem is already hard to solve with high accuracy in the "unreasonably short" time allowed by actual operational procedures, and this only gets worse when the many sources of uncertainty are taken into account, which justifies the continuous interest in solution approaches to the many variants of the problem [46]. In this chapter we will focus on deterministic problems, but the lessons learned from this case can be applied to the versions incorporating information about the uncertainty in the data.

Traditionally, the most effective techniques for solving UC large-scale deterministic (and stochastic) instances have been Lagrangian relaxation algorithms [26, 43], which exploit the (geographical, temporal, scenario, …) structure of the problem by relaxing appropriately chosen "linking constraints" in a Lagrangian fashion (cf. Sect. 14.7), thereby yielding a set of independent problems that can be efficiently solved. In the deterministic case, the most widely used approach is to relax the energy balance constraints, yielding single-unit commitment problems that can be efficiently solved by dynamic programming (DP) techniques (cf. Sect. 14.4, [22]). However, the development of more effective formulations better exploiting the different forms of structure of the problem, among which the presence of semi-continuous variables with nonlinear cost that can be exploited by means of the Perspective Reformulation (PR) technique (cf. Sect. 14.5, [21]), has gradually made the use of general-purpose Mixed-Integer [Non]Linear Programming (MI[N]LP) solvers more attractive (cf. Sect. 14.3, [27]). The trade-off between the cost of computing the relaxation and the corresponding bound quality is nontrivial to navigate. This is even more true after the recent proposal of the first formulation that describes the convex hull of the single-unit commitment with convex power generation costs that has a polynomial (but rather large) number of constraints and variables, based on the combination of the above DP approach and the PR technique (cf. Sect. 14.6, [3]). This prompted the development of several formulations that have a wide range of trade-offs between size and bound quality [2].

In this chapter we review and compare different techniques to yield tight lower bounds for UC. Our aim is to provide a didactic, yet complete and up-to-date with the current state of the art, illustration of some of the different ways that can be explored to devise "tight" formulations of hard optimization problems, and of the highly nontrivial trade-offs that need to be successfully navigated to choose the ones that are more suited to a particular application. The structure of the chapter is the following. In Sect. 14.2 we provide a general introduction to UC. In Sect. 14.3 we review the history of MI[N]LP formulations for the problem, focusing on the aspects related to thermal units, which usually have the most complex constraints. For these units, efficient algorithms have been developed for the so-called *self-scheduling* (or single-unit) versions of the problem where each unit can be optimized separately, as discussed in Sect. 14.4. Then, Sect. 14.5 reviews one important concept for strengthening MINLP formulations with semi-continuous variables that have proven useful to improve the performances of approaches based on general-purpose solvers. The last two concepts are brought together in Sect. 14.6, where new formulations of (thermal units in) the problem are proposed. Yet, efficient approaches for single-unit optimization can be employed in a different way, using Lagrangian techniques, to derive tight bounds, as recalled in Sect. 14.7. All the different approaches are computationally compared in Sect. 14.8, after which the learnt lessons are summarized in Sect. 14.9.

14.2 The Unit Commitment Problem

A fundamental problem in electrical energy production and distribution systems is to compute the production plan of the available generating units, accordingly with their technical and operational constraints, to satisfy global equilibrium between energy demand and energy production (and possibly other constraints related to network security and robustness). The corresponding family of mathematical models is known as the *Unit Commitment* problem (UC), and its practical impact is clearly proven by the huge number of scientific publications devoted to its solution. Yet, UC is a large-scale complex MINLP that cannot be considered "well-solved", in particular, because the requirements of actual operational environments dictate that it must be solved in short time. Furthermore, the increasing relevance of intermittent (renewable) production sources such as solar and wind power is continuously increasing the variability in the electrical system, pushing toward incorporation of information about uncertainty (in future production and demand) in the models. However, the deterministic versions of UC are already difficult enough, which is why in this chapter we will restrict to these; the reader interested in a review of methodologies for stochastic or robust versions of the problem can consult, e.g., [46].

One of the characteristics of UC is that it is not a single problem, but rather a sprawling family of problems that often share many aspects, but that in fact contains models that are almost entirely different. This is due to the fact that the electrical (and, more in general, energy) system is organized differently in different countries, and is continuously evolving to meet the always shifting needs of the society, the economical, ecological, and socio-political conditions, and the evolution of the underlying technologies. Hence, the mathematical models, and their corresponding solution processes, need to be continuously revised over time. For instance, the initial UC models were formulated in the setting of a Monopolistic Producer (MP) owning the production, transmission, and distribution assets in some large region/nation. Later on, many nations moved to liberalized markets where the companies owning generation assets (GENCOs) have to bid their capacity on the energy market(s) managed by a Market Operator (MO), while a Transmission System Operator (TSO) manages the transmission infrastructure to ensure safe and non-discriminatory delivery of the energy to the end-users. As a consequence, GENCOs do not need to consider energy balance and network considerations while scheduling their own production units, but rather have to provide bids to the market anticipating the varying energy prices along the day, which leads to *Self Scheduling* problems where each unit is in fact scheduled independently, thereby highly reducing the overall complexity. This is unless the producer has market power and wants to try to exercise it to increase its own profits via forms of *strategic bidding* [7], a practice that again requires coordination between the units. Furthermore, MOs and TSOs have to solve variants of the problem that coordinate the units' production, albeit possibly with a drastically simplified view of the units' technical constraints.

All these aspects are implemented in many different ways in power systems across the world, giving rise to a large number of different variants of UC. In the following we will try to provide a condensed introduction to the different forms of the problem that is sufficient to provide context for the technical discussion, referring the interested reader to the extensive literature on the subject (e.g., [35, 46] and the many references therein) for more details.

In broad terms, UC requires to maximize the benefit obtained by the production plan for the considered generating units over a time horizon, denoted by $\mathcal{T}$ and composed by a finite number of *time instants* $t \in \mathcal{T}$. These are usually hours or fractions over one or two days; thus, the typical value of $n = |\mathcal{T}|$ ranges from 24 to a few hundred, but longer-term of the problem have been considered where n can be as large as the number of hours in a year (8760).

Along the time horizon, two main processes are happening: generation of energy by the units, and distribution of energy via the interconnection network. These can have very different relevance depending on the version of the problem at hand. For GENCOs, for instance, the technical constraints of generation are paramount and the network may not even be a concern, while for a TSO the converse holds. Also, different units can have very different technical constraints. In the following we will provide a quick rundown of the most relevant aspects differentiating the many variants of UC.

The objective function of UC reflects the different organizations of the energy system: when the demand needs to be satisfied (MP or TSO cases) the objective is to minimize energy production costs, while in the liberalized market regime GENCOs aim at maximizing profits. Yet, profits are revenues minus costs, so a large part of the objective still carries over. However, estimating the revenues requires nontrivial forecast models even when the producer is a *price taker*, that is, its production is small enough not to materially impact the realized prices; even more so for a producer that can exercise market power.

Thermal power units are typically the ones with the most complex technical constraints. These units are either offline or online (giving rise to the standard so-called *commitment binary variables* distinguishing the two cases): in the former case the power output is zero (or even negative but fixed, due to the power used by auxiliary installations, e.g., for nuclear plants), in the former case it must be between some nonzero Minimal Stable Generation (MSG) value and the maximal power output.

That is, power production variables are *semi-continuous ones*. Besides power, however, an important consideration is the so-called *primary and secondary spinning reserve*, a measure of the extra generating capacity that is available by increasing the power production of generators that are already active and linked to the power system. A proper amount of such a reserve must always be kept in the system to be able to face the inevitable discrepancy between the energy demand forecasted when UC is solved and the actual one when the system is in operation. This is because thermal units typically have *Operating Ramp Rate* (ORR, also known as ramp-up and ramp-down rate) constraints limiting the increment and decrement of the generated energy from a time step to another. Usually, larger units have cheaper costs but more stringent ORR constraints (nuclear ones being an extreme case), leading to nontrivial trade-offs between cost and flexibility that are one of the main driver of the savings that can be obtained by optimally solving UC problems. Besides these, generating units (especially, again, larger ones) have constraints on the maximum numbers of starts along the time horizon, imposed not so much for technical reasons as for reducing the wear and-tear involved in these operations and therefore the long-term maintenance costs of the units; these are typically implemented by means of *Minimum Up/Down Time* constraints, imposing that the unit has to remain online/offline for at least a specific amount of time. Again for the largest units, in particular nuclear ones, constraints may be imposed that require the unit to remain *stable*, i.e., keeping the same generation level, for a minimum number of time periods before changing it, reducing the possibility for the unit to be *in modulation*, i.e., actively changing its output level. Specific units may have even more peculiar constraints, such as feasible state transition for combined cycle gas turbine units or starting/stopping curves; all these make the models of the units more complex, but this is typically "on top" of a set of "standard" constraints, which are those upon which we will concentrate in the next section.

Energy production costs for fuel-burning units are usually well represented by convex piecewise linear or quadratic functions separable over each $t \in \mathcal{T}$, but technical characteristics of generating systems may introduce nonconvex elements. These comprise a *fixed cost* to be paid at all $t \in \mathcal{T}$ in which the unit is producing, irrespective of the generated power, and *start-up* (sometimes, also shut-down) costs that are incurred when a unit is brought on(off)line after a period of (in)activity. The simplest form of start-up cost is a constant, but frequently it rather depends on the time the unit has been off before restarting, which makes it not separable across time instants. Other types of costs may need to be considered for specific generation technologies: for instance, nuclear units incur significant *modulation costs* whenever changes of power output are required (again creating a non-separability).

Hydro generation units comprise reservoirs and downstream turbines that generate energy when water is released; sometimes pumps are also present that do the opposite. In general whole hydro valleys need to be modeled, as flow from upstream reservoirs eventually gets to further downstream ones. Typical constraints in hydro units require the level of water in each reservoir to be between a lower and an upper bound (most often less tied to the physics of the system than to its long-term economics, as discussed below), and bounds on the water flowing through each turbine/pump, some of which might have a minimal production level akin to the MSG of thermal units (introducing nonconvexities), as well as similar ORR constraints. Characteristic of hydro units are the equations that describe how water moves across the various basins (and is, ultimately, released in the downstream rivers) corresponding to the turbining and pumping (and spilling, whenever this is allowed) decisions, also taking into account the delay—that can possibly be of several hours—for the water reaching the next reservoir; most often these equations are simple linear flow ones. However, depending on the accuracy of the modeling several complex constraints may be added. These comprise turbining/pumping incompatibility (where turbines and pumps actually use the same equipment in opposite directions and therefore cannot be performed simultaneously), smooth turbining (where turbines output should not increase and immediately afterward decrease during some period), and forbidden zones modeling effects like mechanical vibrations and cavitation that strongly discourage using certain intervals of turbined water. Furthermore, accurate modeling of the physics of energy production requires to take into account the *water head* effect, where the amount of energy produced by a given amount of water is not constant but depends on the level of the receiving reservoir; this gives rise to complex nonlinear constraints that are often discretized at the cost of the introduction of many new binary (and continuous) variables.

Hydro units generally have no "real" energy production cost. Therefore UC, being a short-term model, suffers from the so-called "border effect" whereby what happens after the end of its time horizon $\mathcal{T}$ is ignored; this would lead to the systematic depletion of all available water. To avoid this, a *future value of water* function is added to the objective to represent the expected value of reserves remaining in the reservoirs at the end of $\mathcal{T}$; this is often a piecewise-linear (polyhedral) convex function. Alternatively, very tight bounds are set on the maximum amount of expended water across the whole of $\mathcal{T}$, which will almost surely be entirely used. The value function/bounds are chosen via mid-term optimization models, which are highly complex optimization problems of themselves but with a rather different structure than UC (save of, in fact, comprising "many" UC subproblems) and therefore are out of scope for our discussion.

Last, but not least, Renewable Energy Sources (RES) in UC can be wind farms, solar generation, run-of-river hydro units (not downstream to any controllable reservoir), and geothermal units. RES typically cannot be *modulated*: the amount of produced energy—and even if energy is produced at all—is decided by external factors. Furthermore, most RES generation (most notably solar and wind) is difficult to accurately forecast, even for short time horizons (say, day-ahead forecasts). In the end, the estimated contribution from these sources is directly deducted from the demand; hence, they do not directly increase the complexity of the model in terms of variables or constraints. However, the increases in the uncertainty of net load have a very significant impact on the practical behavior of the energy system, forcing *programmable* generation units (thermal and hydro ones) to serve primarily—or at least much more often than they were designed to—as backup production in case of fluctuations, rather than as primary production systems. This underlines the need of properly integrating information about uncertainty in the UC problem, a very significant trend in the recent literature that we will not discuss here.

The increase in uncertainty also brings the need for an increase of flexibility in the system. Besides from existing programmable units, this can come from *demand response* programs and (distributed) *energy storage*. The first ones are contracts with users (typically energy-intensive industries or large groups of customers co-ordinated by an *aggregator* in an *energy community* [11]) who accept, typically in exchange for economic incentives, to shift or shed a certain amount of load for a limited set of times over the time horizon. In a similar way, energy storage makes it possible to adapt electricity consumption (and generation), satisfying the need for flexibility. Different energy storage technologies are being developed and deployed, among which mechanical storage (compressed air and flywheels), electrochemical (including flow batteries), chemical (hydrogen and synthetic natural gas), electrical (double-layer capacitors), and thermal. An issue of all these storage forms is that their overall capacity is still limited w.r.t. the scale of the energy system, where large hydro units still represent the lion's share of storage capacity; yet, even limited amount of capacity can be precious. Furthermore, one of the most promising forms of large-scale energy capacity is the one related to the electric vehicles, provided they are equipped with vehicle-to-grid mechanisms, that are foreseen to rapidly increase in number in the near future. In terms of UC, demand response and energy storage are usually represented by mathematical models not unlike those of "simple" hydro units.

Besides production, the other fundamental aspect of UC (except in the self-scheduling version, where each unit reacts independently to price signals) concerns energy distribution from the generating units to the demand points, in order to satisfy the *load constraints*, i.e., that global energy demand is met for each $t \in \mathcal{T}$.

Load constraints are sufficient in UC only if the transmission network has large enough capacity to allow energy to freely flow from any two points of the system; this is referred to as the *bus model*, and it is a common choice in UC formulations. However, in many applications the network capacity cannot be disregarded, which leads to the need to explicitly modeling its functioning. This can be done at different levels of sophistication, starting with the *DC model* where the network topology and the capacity of the transmission links are modeled via Kirchhoff's laws, which yield linear constraints. Since such a model may overlook some dynamical aspects, the *AC model* may have to be used instead where both Kirchhoff's and Ohm's laws are used, leading to highly nonlinear and nonconvex constraints; indeed, even if the generation decisions are fixed, the corresponding *optimal power flow* problem is *NP*-hard. A recent interesting stream of research concerns the fact that the nonconvex AC constraints can be written as quadratic relations, which opens the way for convex relaxations using semidefinite programming approaches [5]. Incorporating such constraints into UC clearly makes it way more complex in practice.

A different source of complexity in transmission constraints is related to the fact that electrical networks, unlike most other networks, are "not routable": the current is determined by the nodal power injection and the network topology. While the first is partly fixed, at least as far as demand is concerned, the second can actually be changed by deciding to opening (interrupting) an existing line. Indeed, a somewhat counterintuitive consequence of Kirchhoff laws is that doing that, maybe even for a congested one, causes a global re-routing of electrical energy and may reduce the overall cost. This requires adding binary variables modeling the corresponding decisions, and highly nontrivial constraints to strengthen the corresponding (typically, weak) formulations, making the corresponding versions of UC even more challenging.

Other common forms of system-wide constraints are spinning reserve ones, requiring that the committed units must be able to provide a fraction of the total load to satisfy unexpected peaks in demand or failures of generating units and/or transmission equipment. Besides these, many other special constraints have been considered in the literature, such as global limits on the generation of pollutants (CO_2, SO_x, NO_x, particles, ...) by fossil-fuel burning units within the time horizon, limits on availability of fuel (typically gas) for groups of geographically close units feeding from the same reservoirs, and many others.

Even this streamlined discussion reveals how UC is a huge family of problems of high practical importance that are most often beyond the capabilities of current solution systems given the size and the limited amount of time available to solve them; this would be especially true for models taking into account all facets of the real systems, which leads to painful and nontrivial decisions about which aspects need be ignored or simplified in order to keep solution times in check. Any progress in dealing with common aspects of these mathematical models, such as formulations of the typical constraints of thermal units (still the backbone of most energy systems), can therefore have an impact on the efficiency of electrical systems around the globe, an issue that has perhaps never been pressing like in these days and years. In the rest of this chapter, we will illustrate some recent progresses on this front.

14.3 Traditional MIP Formulations

In this section we describe the main traditional MIP formulations for the Unit Commitment problem. Section 14.2 has shown that UC has two types of constraints: system-wide constraints that enforce limits for the global system, such as demand constraints, reserve constraints, and network constraints, and power generation unit constraints that describe their functional limits and strongly depend on their type and technology. In the following we concentrate on the formulations for the thermal generation units (Sect. 14.3.1), whose constraints have the most complex and interesting combinatorial structure, before quickly discussing the formulation for the other types of units (Sect. 14.3.2) and for the system-wide constraints (Sect. 14.3.3).

14.3.1 Thermal Units

Here we summarize the most popular MIP formulations of the Unit Commitment problem for thermal units, that are usually called *1-bin* and *3-bin* according to the number of vectors of binary variables that are required.

Let I be the set of thermal generators, with $m = |I|$, and $\mathcal{T} = \{1, \dots, n\}$ be the set of time periods in the planning horizon. Given two time periods t' and t'', we denote by $\mathcal{T}(t', t'')$ the set of all the time periods between t' and t'', extremes included (obviously, $\mathcal{T}(t', t'') = \emptyset$ if $t' > t''$). For each $i \in I$ and $t \in \mathcal{T}$, let p_{it} be the power level of unit i at time period t (the *power variables*), and x_{it} be the binary variable denoting the ON/OFF state of unit i at time period t (the *commitment variables*).

We first describe the *1-bin* formulation by listing its constraints and objective function. The *minimum and maximum power output constraints* are simply

$$l_i x_{it} \le p_{it} \le u_i x_{it} \qquad t \in \mathcal{T} \tag{14.1}$$

where l_i and u_i are the minimum and the maximum power output for unit $i \in I$, respectively. As seen in Sect. 14.2, thermal units are subject to *minimum up- and down-time constraints* in order to reduce the wear and frequence of maintenance operations. With τ_i^+ and τ_i^- being the minimum number of consecutive time periods that unit i has to be in ON and OFF state, respectively, these can be expressed as

$$x_{it} \ge x_{ir} - x_{i,r-1} \qquad t \in \mathcal{T}(\tau_i^+ + 1, n), \quad r \in \mathcal{T}(t - \tau_i^+ + 1, t - 1), \tag{14.2}$$

$$x_{it} \le 1 - x_{i,r-1} + x_{ir} \qquad t \in \mathcal{T}(\tau_i^- + 1, n), \quad r \in \mathcal{T}(t - \tau_i^- + 1, t - 1). \tag{14.3}$$

Other constraints are necessary to specify the initial conditions of the unit. Let τ_i^0 denote the initial state of unit i as follows: if $\tau_i^0 > 0$ then unit i has been in ON state for τ_i^0 time periods before time 0, thus one needs to impose the condition

$$x_{it} = 1 \qquad t \in \mathcal{T}(1, \tau_i^+ - \tau_i^0). \tag{14.4}$$

Of course, this is only required if, besides $\tau_i^0 > 0$, one also has $\tau_i^0 < \tau_i^+$. Similarly, $\tau_i^0 \le 0$ means that unit i has been in OFF state for $-\tau_i^0$ time periods, and one needs to impose the condition

$$x_{it} = 0 \qquad t \in \mathcal{T}(1, \tau_i^- + \tau_i^0) \tag{14.5}$$

(again, this is only significant if $-\tau_i^0 < \tau_i^-$). The above inequalities are not sufficient to describe the convex hull $P_n(\tau_i^+, \tau_i^-)$ of the integer points belonging to the unit hypercube and satisfying (14.2)–(14.3); this was observed in [38] where two classes of valid inequalities were proposed.

Definition 14.1 For $k \in \mathcal{T} = \{1, \dots, n\}$, let $\phi(j)$ for $j = 1, \dots, k+1$ and $\psi(j)$ for $j = 1, \dots, k$ be two sequences of indices in T such that

$$1 \leq \phi(1) < \psi(1) < \phi(2) < \psi(2) < \ldots < \phi(k) < \psi(k) < \phi(k+1) \leq n$$

and $\phi(k+1) - \phi(1) \leq \tau_i^+$, then the following *alternating up inequality* is valid:

$$-\sum_{j=1}^{k+1} x_{\phi(j)} + \sum_{j=1}^{k} x_{\psi(j)} \leq 0 \tag{14.6}$$

Definition 14.2 For $k \in \mathcal{T} = \{1, \ldots, n\}$. Let $\phi(j)$ for $j = 1, \ldots, k+1$ and $\psi(j)$ for $j = 1, \ldots, k$ be two sequences of indices in $\mathcal{T}$ such that

$$1 \leq \phi(1) < \psi(1) < \phi(2) < \psi(2) < \ldots < \phi(k) < \psi(k) < \phi(k+1) \leq n$$

and $\phi(k+1) - \phi(1) \leq \tau_i^-$, then the following *alternating down inequality* is valid:

$$\sum_{j=1}^{k+1} x_{\phi(j)} - \sum_{j=1}^{k} x_{\psi(j)} \leq 1 \tag{14.7}$$

In [38] it was proved that the alternating up and down inequalities (14.6) and (14.7) define a linear description of $P_n(\tau_i^+, \tau_i^-)$. Furthermore, the separation problem to find an alternating up (down) inequality violated by a point $\tilde{x}_i \in [0, 1]^n$, if one exists, can be solved in $O(n)$ time.

In Sect. 14.2 we also introduced Operating Ramp Rate constraints. Let Δ_i^+ and Δ_i^- be the ramp-up and ramp-down rates for unit i, respectively, i.e., the maximum increase/decrease of power output w.r.t. the previous time period. As start-up and shut-down periods must satisfy specific rules, it is common to introduce two particular values $\bar{l}_i$ and $\bar{u}_i$, named start-up and shut-down limits, for each unit i. These limits are the maximum power that the unit can produce in a start-up and shut-down period, respectively, and they can be different from l_i and u_i; to be consistent with the other constraints, the conditions $l_i \leq \bar{l}_i \leq u_i$ and $l_i \leq \bar{u}_i \leq u_i$ must hold. Then, the *ramp constraints* can be formulated as follows:

$$p_{it} - p_{i,t-1} \leq \Delta_i^+ x_{i,t-1} + \bar{l}_i(1 - x_{i,t-1}) \qquad t \in \mathcal{T}, \tag{14.8}$$

$$p_{i,t-1} - p_{it} \leq \Delta_i^- x_{it} + \bar{u}_i(1 - x_{it}) \qquad t \in \mathcal{T}. \tag{14.9}$$

Indeed, if we consider the ramp-up constraint (14.8) and we suppose that $x_{i,t-1} = 1$, then (14.8) becomes $p_{it} - p_{i,t-1} \leq \Delta_i^+$ and expresses the ramp up constraint if $p_{it} \geq p_{i,t-1}$ and it is trivially satisfied otherwise (included when $p_{it} = 0$); if $x_{i,t-1} = 0$, then (14.8) boils down to $p_{it} \leq \bar{l}_i$ expressing the start-up limit. Similarly, (14.9) expresses either the ramp-down constraint or the shut-down limit. We assume that $0 \leq \Delta_i^+ \leq u_i - l_i$ and $0 \leq \Delta_i^- \leq u_i - l_i$, otherwise constraints (14.8)–(14.9) are either redundant or not feasible. For $t = 1$ the constraints (14.8)–(14.9) refer to values p_{i0} and x_{i0}, which are parameters defined by initial conditions (cf. τ_i^0 above).

The objective function usually considers the minimization of the production costs, that are composed of two main contributions: the generation costs and the start-up costs. For each unit i and time period t the generation costs are usually represented by a convex quadratic function

$$f_i(p_{it}) = a_i p_{it}^2 + b_i p_{it}, \tag{14.10}$$

possibly plus a fixed cost $c_i x_{it}$.

The start-up costs are in general expressed as a function $s_i(x_i)$ of the complete state vector x_i, because they depend on the time intervals that unit i has been off. An accurate discussion about start-up costs can be found, e.g., in [3]. For our purposes, this complex function only needs to be known at the discrete set of values $\sigma_{i\tau}$ for $\tau \in \mathcal{T}(\tau_i^-, \overline{\tau}_i)$, where τ represents the length of the off period and $\overline{\tau}_i$ is the time such that the unit has reached ambient temperature and the start-up cost is maximum. The values $\sigma_{i\tau}$ are supposed to be non-decreasing with respect to τ. Using this property, [40] proposed to formulate the start-up costs by means of a single extra new variable s_{it} and $\overline{\tau}_i - \tau_i^- + 1$ additional constraints (for each unit and time instant), as follows:

$$s_i(x_i) = \sum_{t=1}^{n} s_{it} \tag{14.11}$$

$$s_{it} \geq \sigma_{i\tau}\left(x_{it} - \sum_{j=1}^{\tau} x_{i,t-j}\right) \qquad t \in \mathcal{T}, \quad \tau \in \mathcal{T}(\tau_i^-, \overline{\tau}_i) \tag{14.12}$$

$$s_{it} \geq 0 \qquad t \in \mathcal{T} \tag{14.13}$$

The detailed representation of the start-up cost (14.11)–(14.13) can have a consistent impact on the performances; this is why, the start-up costs are often approximated with a fixed value (e.g., the maximum cost). We refer to the parameters $\sigma_{i\tau}$ as *history-dependent start-up costs* if $\overline{\tau}_i > \tau_i^-$, while we refer to *fixed start-up costs* when $\overline{\tau}_i = \tau_i^-$.

The *3-bin* formulation was introduced as early as [32], but only in [42], and independently in [39], some interesting theoretical results about it were proven. It is based on an exact formulation of the minimum up- and down-time constraints only, that is obtained by considering 3 vectors of binary variables: the commitment variables x_{it}, already considered in the *1-bin* formulation, plus the *start-up variables* v_{it} denoting if unit i has been started up at time period t (i.e., $x_{it} = 1$ and $x_{i,t-1} = 0$), and the *shut-down variables* w_{it} denoting if i has been shut-down at t (i.e., $x_{it} = 0$ and $x_{i,t-1} = 1$). Using the above defined variables, the *minimum up- and down-time constraints* can be redefined by

$$\sum_{s \in \mathcal{T}(t-\tau_i^+ + 1, t)} v_{is} \leq x_{it} \qquad t \in \mathcal{T}(\tau_i^+, n) \tag{14.14}$$

$$\sum_{s \in \mathcal{T}(t-\tau_i^- + 1, t)} w_{is} \leq 1 - x_{it} \qquad t \in \mathcal{T}(\tau_i^-, n) \tag{14.15}$$

$$x_{it} - x_{i,t-1} = v_{it} - w_{it} \qquad t \in \mathcal{T}(1, n) \tag{14.16}$$

Initial conditions expressed by constraints (14.4) and (14.5) as well as minimum and maximum power output (14.1), the power generation cost (14.10) and fixed costs $c_i x_{it}$ in the objective function must be considered also in the *3-bin* formulation. The *ramp constraints* can be strengthened as follows [41]:

$$p_{it} - p_{i,t-1} \leq \Delta_i^+ x_{i,t-1} + \bar{l}_i v_{it} \qquad t \in \mathcal{T} \qquad (14.17)$$

$$p_{i,t-1} - p_{it} \leq \Delta_i^- x_{it} + \bar{u}_i w_{it} \qquad t \in \mathcal{T} \qquad (14.18)$$

Similarly, also the *start-up cost formulation* can be strengthened as follows [41]:

$$s_i(x_i) = \sum_{t=1}^{n} s_{it} \qquad (14.19)$$

$$s_{it} \geq \sigma_{i\tau}(v_{it} - \sum_{j=2}^{\tau} w_{i,t-j+1}) \qquad t \in \mathcal{T}, \quad \tau \in \mathcal{T}(\tau_i^-, \overline{\tau}_i) \qquad (14.20)$$

$$s_{it} \geq 0 \qquad t \in \mathcal{T} \qquad (14.21)$$

If we consider fixed start-up costs s_i, the *3-bin* formulation can be significantly simplified, as the start-up cost is represented by adding the simple term $\sum_{t \in \mathcal{T}} s_i v_{it}$ to the objective function, while the extra variables s_{it} and the constraints (14.19)–(14.21) are no longer needed. The interested reader can find tighter formulations for history-dependent start-up costs in [8, 44]. Because the *3-bin* formulation is generally accepted as a better starting formulation, we take it as a benchmark for the new formulations in the computational results. The UC on thermal unit has a very challenging combinatorial structure that motivated several authors to study its linear description: many of these results were summarized in [37], and some additional references can be found in [3].

14.3.2 Other Generation Units

Hydro generation units are another traditional type of generation unit that share with thermal units the feature to be programmable. Hydro units may belong to a cascade of reservoirs whose water capacity can be controlled by turbining for power generation or by water spillage.

A basic version of the local constraints for hydro cascade units is described in the following. Each cascade $h \in H$ consists of a set $H(h)$ of individual hydro units, possibly containing only one element; for each $j \in H(h)$, variables q_{jt}, ν_{jt}, and ξ_{jt} represent discharged water, the volume of the reservoir, and the spilled water at time period $t \in \mathcal{T}$, respectively. Parameters $\bar{\nu}_j^{\min}$ and $\bar{\nu}_j^{\max}$ denote the minimum and maximum volume for the reservoir, respectively, $\bar{q}_j^{\max}$ denotes the technical maximum of discharged water (the technical minimum is assumed to be zero in order to avoid the addition of new binary variables in the model), while $\bar{\xi}_{jt}$ denotes the natural inflows at time period $t \in \mathcal{T}$. Let $S(j)$ be the (possibly empty) set of the immediate predecessors of unit j—those whose discharge and spillage reaches j without passing through other reservoirs—and t_{kj} be the water time delay from plant $k \in S(j)$ to the basin feeding plant j. Finally, let α_j be the coefficient of transformation from discharged water to generated power, denoted by $\tilde{p}_{jt}$. Then, the local constraints corresponding to each unit $j \in H(h)$ are:

$$0 \leq q_{jt} \leq \bar{q}_j^{\max} \qquad t \in \mathcal{T} \tag{14.22}$$

$$\bar{\nu}_j^{\min} \leq \nu_{jt} \leq \bar{\nu}_j^{\max} \qquad t \in \mathcal{T} \tag{14.23}$$

$$\nu_{jt} - \nu_{j,t-1} = \bar{\xi}_{jt} - \xi_{jt} - q_{jt} + \sum_{k \in S(j)} \left(q_{k,t-t_{kj}} + \xi_{k,t-t_{kj}}\right) \qquad t \in \mathcal{T} \tag{14.24}$$

$$\tilde{p}_{jt} = \alpha_j q_{jt} \qquad t \in \mathcal{T} \tag{14.25}$$

Initial conditions at time 0 should be properly considered for balance equations (14.24). More complicated constraints for hydro units are described in [45] considering, e.g., pumping operations to reload upstream reservoirs in low consumption time periods and more precise relationship between discharged water and generated power whose relationship is in general nonlinear.

Other types of units that are gaining increasing relevance are Renewable Energy Sources (RES); mainly, *wind farms* (collections of wind generation units) and *photovoltaic units*. These units present basically no constraints and no variables, as the generated power is only subject to weather conditions. In general it would be possible to curtail or even completely discard the power output of these units, thereby introducing some decisions, but this is typically avoided and it is even prohibited in some systems. Hence, in the deterministic UC formulations we consider here the RES production is taken as part of the input and generically cumulated in the parameter RES_t for each time period $t \in \mathcal{T}$, that is just subtracted to the demand (possibly at the corresponding node of the transmission network if these are defined, as discussed next). However, since this production is highly uncertain and can change in a short time, the correct handling of these factors would require optimization models under uncertainty such as robust or stochastic optimization methods, for which we refer the interested reader to [46].

14.3.3 System-Wide Constraints

System-wide constraints link and coordinate the decisions of the different generating units. The simplest and most common form of those constraints is that of the demand constraints

$$\sum_{i \in I} p_{it} + \sum_{h \in H} \sum_{j \in H(j)} \tilde{p}_{jt} = d_t - RES_t \qquad t \in \mathcal{T} \tag{14.26}$$

where d_t is the (forecasted) total energy demand at time period t. The network constraints form another class of system-wide constraints. They can be formulated in different ways. The simplest ones (Direct Current, DC model) reduce to a set of linear constraints, while the most accurate ones (Alternate Current, AC model) involve highly nonlinear terms. This has recently motivated a rather active research line where formulations or tight relaxations of AC constraints are proposed using Second-Order Cone or SemiDefinite constraints (e.g., [4, 10] and the references therein). As an example of network constraints we present the case of the DC model that is usually used as a rough approximation of the real system. The network will be characterized by a set of nodes V called buses which can be production and/or consumptions sites and a set L of links. Each node of the network will require a power demand d_{it} for each time period t, i.e., $d_t = \sum_{i \in V} d_{it}$, a current f_{ijt} will flow in each link $ij \in L$ of the network, that is also characterized by a line *susceptance* B_{ij}; this yields the constraints

$$p_{it} - d_{it} = \sum_{j \in \delta^+(i)} f_{ijt} - \sum_{j \in \delta^-(i)} f_{jit} \qquad i \in V,\ t \in \mathcal{T} \tag{14.27}$$

$$B_{ij}(\theta_{it} - \theta_{jt}) = f_{ijt} \qquad ij \in L,\ t \in \mathcal{T} \tag{14.28}$$

$$-\bar{f}_{ij} \le f_{ijt} \le \bar{f}_{ij} \qquad ij \in L,\ t \in \mathcal{T} \tag{14.29}$$

where (14.27) represent power flow constraints on the lines, (14.28) represent Kirchoff and Ohm laws establishing that the current is not freely distributed in network, but it depends on voltage angle variables θ_{it} that can be controlled at each bus and line susceptances B_{ij}, while (14.29) model the maximum capacity of each line.

Finally, other system-wide constraints concern "reserve" (primary, secondary, or inertia) that are needed to guarantee that the system remains operational even if the real conditions deviate from the expected ones. An example is the primary spinning reserve that is related to the maximum power r_{it} that can be promptly produced without changing the schedule of the thermal units

$$r_{it} \le u_i x_{it} \qquad i \in I,\ t \in \mathcal{T} \tag{14.30}$$

$$r_{it} \le p_{i,t-1} + \Delta_i^+ x_{i,t-1} + \bar{l}_i v_{it} \qquad i \in I,\ t \in \mathcal{T} \tag{14.31}$$

$$\sum_{i \in I} r_{it} + \sum_{h \in H} \sum_{j \in H(j)} \tilde{p}_{jt} \ge (1 + \rho) d_t - RES_t \qquad t \in \mathcal{T} \tag{14.32}$$

where ρ is the percentage of reserve with respect to the expected demand d_t.

14.4 Self-UC and Dynamic Programming

An important variant of UC is Self-UC, that is, UC without system-wide constraints and with the addition of revenues in the objective function. If π_t is the price of energy at time period t, then Self-UC aims at determining the schedule of the units that maximize the net profit, and it can be formulated as

$$\min \sum_{i \in I} \sum_{t \in \mathcal{T}} a_i (p_{it})^2 + b_i p_{it} + c_i x_{it} + s_{it} - \pi_t p_{it} \tag{14.33}$$

$$\left.\begin{array}{l}(14.1),\ \ (14.4)\ \text{or}\ (14.5),\ \ (14.14) - (14.16)\\ (14.17) - (14.18),\ \ (14.19) - (14.21)\end{array}\right\} \quad i \in I \tag{14.34}$$

$$(14.22) - (14.25) \qquad j \in H(h),\ h \in H \tag{14.35}$$

Clearly, Self-UC can be disaggregated at each single generating unit, that is, by solving an optimization problem for each thermal unit and for each hydro unit. We call UC restricted to a single unit *single-unit commitment problem* (1UC).

1UC for hydro units with the features here considered is a simple network flow problem. Instead, 1UC for thermal units is nontrivial to describe in the variable space proposed in the previous sections, even if it is actually relatively easy to solve. Indeed, several variants of 1UC have been studied in the literature and different Dynamic Programming (DP) algorithms have been proposed for their solution. All 1UC variants share the following basic features:

- minimum and maximum power output;
- minimum up- and down-time constraints;
- quadratic power generation costs.

However, as discussed in Sect. 14.3.1, UC may also consider other features associated with the single-unit functioning, such as ramp constraints and history-dependent start-up costs, that can be added to the definition of 1UC. The complexity of the solution algorithms for 1UC depends on these additional features. In particular, we can list the following three variants with increasing difficulty:

1. no ramp constraints and fixed start-up costs;
2. no ramp constraints and history-dependent start-up costs;
3. ramp constraints and history-dependent start-up costs.

All of them can be solved in polynomial time and in the following we will recall the solution algorithms proposed in the literature.

14.4.1 1UC with No Ramp Constraints and Fixed Start-Up Costs

The 1UC problem in this case considers only the following features:

- minimum and maximum power output;
- minimum up- and down-time constraints;
- quadratic power generation costs and fixed start-up costs.

and can be formulated as follows:

$$\begin{array}{ll} \min \sum_{t\in T} s_i v_{it} + \sum_{t\in T} f_i(p_{it}) + c_i x_{it} - \pi_t p_{it} & \\ (14.1),\quad (14.4) \text{ or } (14.5),\quad (14.14)-(14.16) & \\ x_{it},\, v_{it},\, w_{it} \in \{0,1\} & t\in T \\ p_{it} \in \mathbb{R}_+ & t \in T \end{array} \quad (14.36)$$

This problem can be solved by DP, defining a state-space graph G_i^F (see Fig. 14.1) with the following nodes:

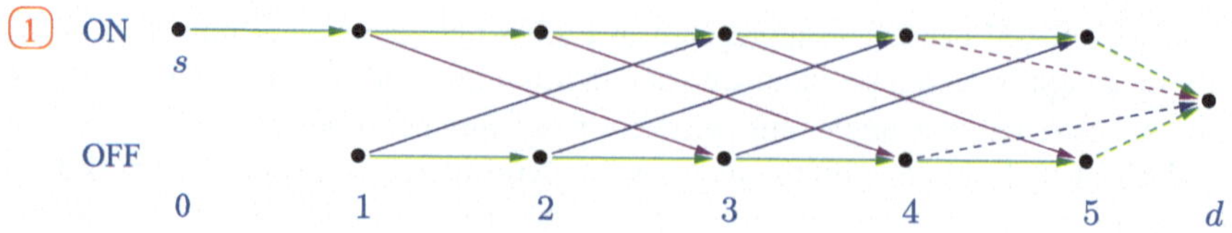

Fig. 14.1 An example of graph G_i^F with $\tau_i^+ = \tau_i^- = 2$ and $\tau_i^0 = 1$

- *ON nodes:* ON_t for each $t \in \mathcal{T}$ denoting that the unit is ON at time instant t;
- *OFF nodes:* OFF_t for each $t \in \mathcal{T}$ denoting that the unit is OFF at time instant t;
- two additional nodes s and d denoting the initial and the final state.

The arcs of the graph describe all the possible valid changes of the state, where the tail of the node represents the starting state and the head of the node represents the new state of the unit:

- $(OFF_t, ON_{t+\tau_i^+})$ arcs are associated with the event that the unit is started up at time instant $t+1$ and that remains ON until (at least) $t+\tau_i^+$ to satisfy the minimum up-time constraint;
- $(ON_t, OFF_{t+\tau_i^-})$ arcs are associated with the event that the unit is shut down at time instant $t+1$ and that the unit remains OFF until (at least) $t+\tau_i^-$ to satisfy the minimum down-time constraint;
- (ON_t, ON_{t+1}) arcs are associated with the condition that the unit is ON at time t and remains ON also at time instant $t+1$;
- (OFF_t, OFF_{t+1}) arcs are associated with the condition that the unit is OFF at time t and remains OFF also at time instant $t+1$.

In order to describe the costs associated with the arcs, we define the following 1-variable optimization problem (whose solution is trivial because f is convex):

$$Z_{i,\tau} = c_i + \min\{f_i(p_{i\tau}) - \pi_\tau p_{i\tau} \; : \; l_i \leq p_{i\tau} \leq u_i\}. \tag{14.37}$$

Then the costs of the arcs can be described by the following different cases. The arcs of type (OFF_t, OFF_{t+1}) and $(ON_t, OFF_{t+\tau_i^-})$ have zero cost, because the new associated state does not imply additional costs. The arcs of type (ON_t, ON_{t+1}) involve only the cost of keeping the unit ON for an additional time instant $t+1$. As there are no ramp constraints this cost is separable on the time instants and corresponds to $Z_{i,t+1}$. The solution $p^*_{i,t+1}$ of (14.37) for $\tau = t+1$ will represent the optimal power level if the arc is traversed in an optimal solution of 1UC. The arcs of the type $(OFF_t, ON_{t+\tau_i^+})$ are assigned two costs: the fixed start-up cost s_i plus $\sum_{\tau=t+1}^{t+\tau_i^+} Z_{i,\tau}$. The power levels from $t+1$ to $t+\tau_i^+$ are defined by the optimal solutions of the subproblems (14.37).

In addition, there are arcs representing initial conditions and final ones. For the former we have to distinguish if the unit i is ON or OFF at time 0. If it is OFF for $-\tau_i^0 \geq 0$ time periods two cases may apply: (i) $-\tau_i^0 < \tau_i^-$, then there is an arc $(s, OFF_{\tau_i^- + \tau_i^0})$ with zero cost representing that the unit remains OFF until period

$\tau_i^- + \tau_i^0$; (ii) $-\tau_i^0 \geq \tau_i^-$ then there is an arc (s, OFF_1) with zero cost, representing that the unit remains OFF an additional time unit, and an arc $(s, ON_{\tau_i^+})$ representing that the unit is started up with cost $s_i + \sum_{\tau=1}^{\tau_i^+} Z_{i\tau}$. Symmetrically, if the unit is ON for $\tau_i^+ > 0$ time periods there are also two cases: (i) $\tau_i^0 < \tau_i^+$, then there is an arc $(s, ON_{\tau_i^+ - \tau_i^0})$ representing that the unit remains ON until $\tau_i^+ - \tau_i^0$ with cost $\sum_{\tau=1}^{\tau_i^+ - \tau_i^0} Z_{i\tau}$; (ii) $\tau_i^0 \geq \tau_i^+$, there is an arc (s, ON_1) with cost Z_{i1} representing that the unit remains ON one additional time unit and, if $p_{i0} \leq \bar{u}$, an arc $(s, OFF_{\tau_i^-})$ with zero cost representing that the unit is shut-down at time 1 and remains OFF until τ_i^-. Finally, there are arcs associated with final conditions. As in the previous cases these can be of four types: (i) arcs (ON_n, d) and (ii) (OFF_n, d) represent that the unit has completed its scheduling period; (iii) arcs (ON_t, d) with $n - \tau_i^- < t < n$ represent that the unit is switched OFF at time $t + 1$ and must remain OFF until $t + \tau_i^- > n$ (beyond the planning period); (iv) arcs (OFF_t, d) with $n - \tau_i^+ < t < n$ represent that the unit is switched ON at time $t + 1$ and must remain ON until $t + \tau_i^+ > n$; in the last case there is also a cost on the arc that is defined by $s_i + \sum_{\tau=t+1}^{n} Z_{i\tau}$.

Computing the shortest path from s to d in G_i^F corresponds to solve 1UC as defined by the MINLP (14.36). As the graph G_i^F is acyclic and has $2n + 2$ nodes and at most $4n + 2$ arcs, computing the shortest path has $O(n)$ computational complexity.

14.4.2 1UC with No Ramp Constraints and History-Dependent Start-Up Costs

A 1UC that considers history-dependent start-up costs, but still no ramp constraints, can be solved with a shortest path computation on a state-space graph G_i^H (see Fig. 14.2) obtained from slight modifications of G_i^F that only involve the OFF nodes and related arcs. Let $\sigma_{i,\tau_i^-}, \sigma_{i,\tau_i^- +1}, \ldots, \sigma_{\tau_i^- + \bar{\tau}_i}$ be all the possible different start-up costs, associated with the cases where the unit remains OFF for the corresponding number of time periods; that is, the start-up costs remain constant after $\tau_i^- + \bar{\tau}_i$ periods. Then, for each t, G_i^H contains the nodes $OFF_{t,r}$ with $r \in \{\tau_i^-, \ldots, \bar{\tau}_i\}$, instead of the single OFF_t node in G_i^F, associated with the state "the unit i is OFF at time t and has been OFF for r time periods". The change in the set of arcs is straightforward, involving the arcs defining that the unit remains in OFF states, that is, arcs $(OFF_{t,r}, OFF_{t+1,r+1})$ and arcs $(OFF_{t,\bar{\tau}_i}, OFF_{t+1,\bar{\tau}_i})$, and the arcs defining that the unit is started up, that is, arcs $(OFF_{t,r}, ON_{t+\tau_i^+})$ with cost $\sigma_{ir} + \sum_{\tau=t+1}^{t+\tau_i^+} Z_{i\tau}$, as well as those exiting from s and entering in d.

Computing the shortest path from s to d in G_i^H corresponds to solve 1UC with no ramp constraints and history-dependent start-up costs. As the graph G_i^H is acyclic and has $n \times (\bar{\tau}_i - \tau_i^- + 2) + 2$ nodes and at most 2 arcs exiting for each node different from d, computing the shortest path has $O(n \times (\bar{\tau}_i - \tau_i^-))$ computational complexity, that is bounded by $O(n^2)$.

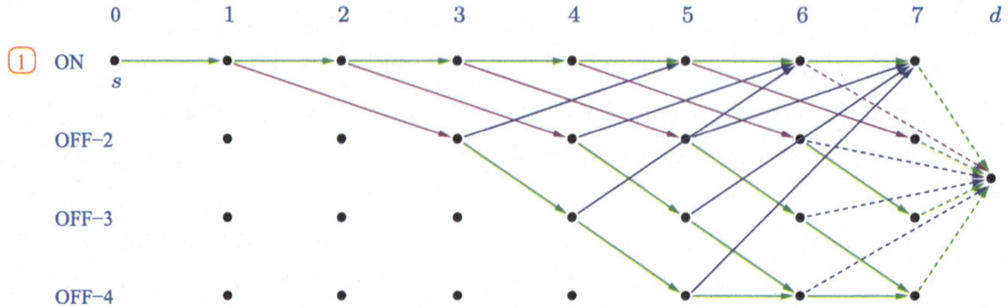

Fig. 14.2 An example of graph G_i^H with $\tau_i^+ = \tau_i^- = 2$, $\tau_i^0 = 1$, and $\overline{\tau}_i = 4$

14.4.3 1UC with Ramp Constraints and History-Dependent Start-Up Costs

In [22] a DP algorithm was proposed to solve 1UC with all the constraints—minimum up- and down-time, ramp and generation limits—in $O(n^3)$ with the standard quadratic separable cost function (14.10), and that can be generalized to more complex objectives. Here we present the improved version of [24] (see also [3]), based on the state-space graph $G_i = (N_i, A_i)$ (see Fig. 14.3) defined in the following. The nodes in N_i are of two types: ON_t and OFF_t for each $t \in \mathcal{T}$, plus the source s and the sink d. The arcs in A_i are of two types: (i) *ON arcs* (OFF_h, ON_k) denote that the unit is turned ON at the beginning of time period h and remains ON until the end of time period k, (ii) *OFF arcs* (ON_k, OFF_r) denote that the unit is OFF from time periods $k+1$ to time period $r-1$. The OFF arcs satisfy minimum down-time constraints, that is $(ON_k, OFF_r) \in A_i$ if and only if $r \geq k + \tau_i^- + 1$, and are labeled with the start-up cost of the off period. The most general time-dependent start-up costs $\sigma_{i\tau}$ are easily handled within this framework, since the computation is done entirely offline. Indeed, the cost on the OFF arc (ON_k, OFF_r) will be $\sigma_{i,r-k-1}$, if $r-k-1 < \overline{\tau}_i$, or $\sigma_{i,\overline{\tau}_i}$, if $r-k-1 \geq \overline{\tau}_i$. The ON arcs satisfy minimum up-time constraints, that is $(OFF_h, ON_k) \in A_i$ if and only if $k \geq h + \tau_i^+ - 1$, and are labeled with the cost of the optimal dispatch in the associated period. This is composed of two parts: fixed cost and variable cost. The first is just $\sum_{i=h}^{k} c_i$ since the unit will be committed in the

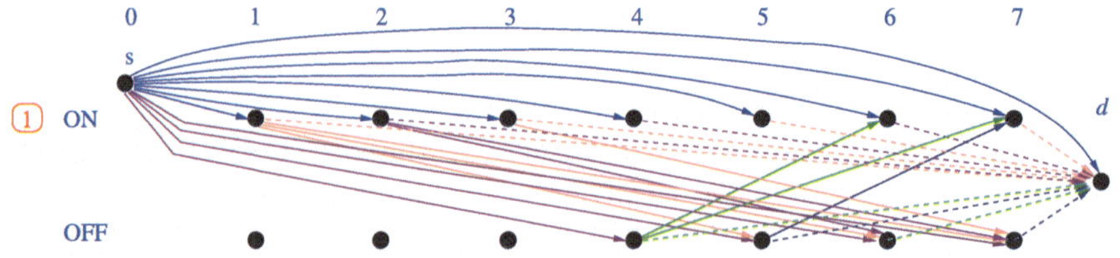

Fig. 14.3 An example of graph G_i with $\tau_i^+ = \tau_i^- = 3$ and $\tau_i^0 \geq 3$

interval $\mathcal{T}(h, k)$. The variable cost, that depends on the p_{it} variables, is the optimal value of the following *Economic Dispatch problem with Ramping Constraints*

$$
(ED^{hk}) \quad
\begin{array}{lll}
z^{hk} = \min & \sum_{t \in \mathcal{T}(h,k)} f(p_{it}) - \pi_t p_{it} & \\
& l_i \le p_{it} \le u_i & t \in \mathcal{T}(h, k) \\
& p_{ih} \le \bar{l}_i & \\
& p_{i,t+1} \le p_{it} + \Delta_i^+ & t \in \mathcal{T}(h, k-1) \\
& p_{it} \le p_{i,t+1} + \Delta_i^- & t \in \mathcal{T}(h, k-1) \\
& p_{ik} \le \bar{u}_i &
\end{array}
$$

Since all the binary variables have been fixed, (ED^{hk}) is an optimization problem with convex objective function and linear constraints. Hence, its optimal objective function value z^{hk} can be computed in polynomial time.

Moreover, there are the connections between the source node s and the ON and OFF nodes defined according to the initial state of the unit; all the missing details for the construction of G_i are quite easy and can be found in [3].

In conclusion, the state-space (acyclic) graph G_i has $2n + 2$ nodes and $O(n^2)$ arcs and every s–d path on G_i represents a feasible production plan for the unit: hence, 1UC is reduced to a shortest path problem on G_i, that has complexity $O(n^2)$ once that all the data has been computed. In [22] it is shown how all $O(n^2)$ Economic Dispatch problems with Ramping Constraints (ED^{hk}) can be solved in $O(n^3)$ by means of another DP algorithm, which is therefore the cost of the overall procedure. In Appendix A we report a simple example of execution of the above DP algorithm.

14.5 Perspective Reformulation

We now recall a reformulation technique for MINLPs with convex objective functions and variables $p_i \in \mathbb{R}$ for $i \in N = \{1, \ldots, n\}$ assuming either the value 0, or a value in some given compact nonempty interval $\mathcal{P}_i = [\, l_i\,,\; u_i\,]$ $(-\infty < l_i < u_i < \infty)$. This class of variables is named *semi-continuous* and requires the addition of corresponding binary variables x_i, as in

$$\min \; g(y) + \textstyle\sum_{i \in N} f_i(p_i) + c_i x_i \tag{14.38}$$

$$l_i x_i \le p_i \le u_i x_i \qquad i \in N \tag{14.39}$$

$$(p, x, y) \in \mathcal{O} \tag{14.40}$$

$$x \in \{0, 1\}^n, \quad p \in \mathbb{R}^n \tag{14.41}$$

We assume that the functions f_i are closed convex; this is the case in UC due to the power cost function of the thermal units, cf. (14.10) and (14.1). However, with UC, the binary variables x_i are not only useful to prescribe the semi-continuous status of the corresponding p_i, but, as they are identified with the commitment variables, for representing many other constraints of the model. The function g in the "other vari-

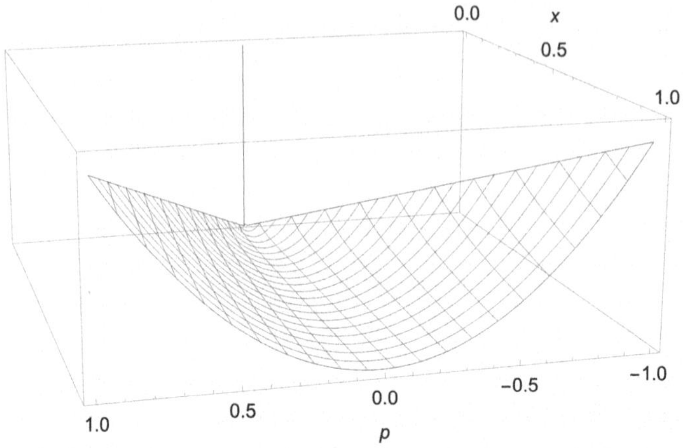

Fig. 14.4 The perspective function $h(p, x) = p^2/x$ of $f(p) = p^2$

ables y" and the "other constraints (14.40)" do not play any role in our development, but the *Perspective Reformulation* technique, proposed in [22], is particularly suitable for the case where the objective function and all the constraints in (14.38)–(14.40) are convex, so that the corresponding continuous relaxation is a convex program. Besides UC, problems (14.38)–(14.41) can be applied to many real-world problems such as distribution and production planning problems, financial trading and planning problems, and many others; see [15] and the references therein.

The *Perspective Reformulation* of (14.38)–(14.41) is

$$\text{(PR)} \qquad \min\{\, g(y) + \textstyle\sum_{i \in N} h_i(p_i, x_i) + c_i x_i \;:\; (14.39)-(14.41) \,\} \qquad (14.42)$$

where $h_i(p_i, x_i) = x_i f_i(p_i/x_i)$ is the *perspective function* of $f_i(p_i)$. Figure 14.4 shows the perspective function $h(p, x) = p^2/x$ corresponding to $f(p) = p^2$, the same class of functions of (14.10); the figure justifies the "perspective" moniker, as the graph is a cone pointed in $(p, u) = (0, 0)$ and whose "lower shape" is the graph of f put at some finite distance (say, $x = 1$), as if making a perspective of it.

This choice is motivated by the fact that h_i is the *convex envelope*, i.e., the best convex underestimator, of f_i over (14.39) and (14.41); hence, PR is equivalent to the original problem when integrality of the variables is enforced, but its continuous relaxation $\underline{\text{PR}}$, known as the *Perspective Relaxation*, provides better lower bounds than that of (14.38)–(14.41). However, $\underline{\text{PR}}$ is also "significantly more nonlinear" than the original problems, yielding yet another nontrivial trade-off between bound tightness and computational cost. In order to solve $\underline{\text{PR}}$ with present MIP solvers one typically needs to reformulate it in a suitable way. A first chance is to write (14.42) as a Mixed-Integer Second-Order Cone Program (MI-SOCP) [23]: assuming that the original objective function is SOCP-representable, its perspective function also is. For instance, if f is convex quadratic as in (14.10) then its $\underline{\text{PR}}$

$$\begin{aligned} &\min g(y) + \textstyle\sum_{i \in N} z_i + b_i p_i + c_i x_i \\ &(14.39)–(14.41) \\ &z_i x_i \geq a_i p_i^2 \qquad\qquad i \in N \end{aligned} \tag{14.43}$$

involves *rotated SOCP constraints* that current MIP solvers can tackle. An alternative reformulation of (14.42) is the Semi-Infinite MINLP (SI-MINLP)

$$\min g(y) + \textstyle\sum_{i \in N} z_i + c_i x_i \tag{14.44}$$

$$(14.39)–(14.41) \tag{14.45}$$

$$z_i \geq f_i'(\bar{p}_i) p_i + [f_i(\bar{p}_i) - f'(\bar{p}_i)\bar{p}_i\,] x_i \qquad 0 \leq \bar{p}_i \leq u_i, \quad i \in N \tag{14.46}$$

The infinitely many inequalities (14.46), indexed over the continuous quantities p_i, are called *Perspective Cuts* (P/C). Note that, by using the derivative f_i', the formula would seem to imply the need for f to be continuously differentiable, but in fact a more general formula using subdifferentials is available for the nondifferentiable case [21]. By requiring only the computation of values of f_i and f_i', P/C are trivial to separate given the value of $\bar{p}_i$: hence, they can be used in the standard branch-and-cut MIP solvers, where they are separated with callback functions (for both lazy constraints and user cuts). The process typically starts with only the two cuts corresponding to $\bar{p}_i = 0$ and $\bar{p}_i = u_i$ in the formulation besides (14.44)–(14.45). For each solution (be it integral or not), P/Cs are separated for each $i \in N$ and $\bar{p}_i = p_i \,/\, x_i$, until the desired optimality tolerance is reached. Notably, unlike the case of MISOCP, this procedure can be used with any convex function f (not only SOCP-representable ones), and even with a MILP solver if the only nonlinearities to the formulation are in f_i, as it frequently happens in applications (among which, UC).

Other approaches for solving Perspective Reformulations have been proposed in [25] for special cases and in [15, 16], where the *Approximated Projected PR* (AP^2R) method is proposed as an alternative to SOCP and P/C that is competitive for some applications. In all cases, the use of PR-based approaches strengthens the formulation considerably, dramatically improving the efficiency of the solution process.

14.6 New DP-Based MIP Formulations

Bringing together the concepts developed in the two previous sections we now recall two "strong" UC formulations proposed in [3]. The first one, presented in Sect. 14.6.1, is obtained by writing the convex hull description for the thermal 1UC polytope using the flow formulation of the corresponding state-space graph G_i from Sect. 14.4.3. Since this formulation has a large number of variables and constraints, in Sect. 14.6.2 we propose a simplified version that trades a reduction of the bound for a smaller number of variables.

14.6.1 A Convex Hull Description for the Thermal 1UC Polytope

Based on the DP algorithm presented in Sect. 14.4.3, it is now easy to describe the convex hull of the feasible solutions for the thermal single-unit 1UC polytope. Two types of variables are considered:

- binary variables y associated with the arcs of the state-space graph G_i of the DP algorithm; in particular, variables y_{i+}^{hk} defined on ON arcs (OFF_h, ON_k) and y_{i-}^{hk} defined on OFF arcs (ON_h, OFF_k), including arcs starting from the source node and ending at the sink node;
- continuous variable p_{it}^{hk} associated with each ON arc (OFF_h, ON_k) and each time period $t \in \mathcal{T}(h, k)$ and representing the power level for each time instant.

The constraints in this new formulation are of two types:

- *Shortest Path* constraints defined on the state-space graph G_i of the DP algorithm, that are succinctly expressed as follows:

$$E^i y_i = \delta_i, \quad y_i \geq 0 \tag{14.47}$$

  where E^i is the node-arc incidence matrix of G_i and $\delta \in \mathbb{R}^{2n+2}$ is the vector with all zero entries except $\delta_s = -1$ and $\delta_d = 1$ for the source node s and the sink node d, respectively.
- *Extended Economic Dispatch* constraints defined on each ON arc (h, k) of the graph G_i as follows:

$$(EED^{hk}) \quad \left.\begin{array}{ll} l_i y_{i+}^{hk} \leq p_{ih}^{hk} \leq \bar{l}_i y_{i+}^{hk} & \\ l_i y_{i+}^{hk} \leq p_{it}^{hk} \leq u_i y_{i+}^{hk} & t \in \mathcal{T}(h+1, k-1) \\ l_i y_{i+}^{hk} \leq p_{ik}^{hk} \leq \bar{u}_i y_{i+}^{hk} & \\ p_{i,t+1}^{hk} \leq p_{it}^{hk} + y_{i+}^{hk} \Delta_i^+ \; t \in \mathcal{T}(h, k-1) & \\ p_{it}^{hk} \leq p_{i,t+1}^{hk} + y_{i+}^{hk} \Delta_i^- \; t \in \mathcal{T}(h, k-1) & \end{array}\right\} \tag{14.48}$$

Finally, the objective function is

$$\sum_{(h,k)\in A_+^i} c_{i+}^{hk} y_{i+}^{hk} + \sum_{(h,k)\in A_-^i} c_{i-}^{hk} y_{i-}^{hk} + \sum_{(h,k)\in A_+^i} \sum_{t\in \mathcal{T}(h,k)} f(p_{it}^{hk}) \tag{14.49}$$

where A_+^i is the set of ON arcs in G_i, A_-^i is the set of OFF arcs in G_i, $c_{i+}^{hk} = (h-k+1)c_i$ is equal to the sum of the fixed costs for the interval $\mathcal{T}(h, k)$, c_{i-}^{hk} is equal to the start-up cost depending on the length of the shut-down period $\mathcal{T}(h+1, k-1)$, and $f(p_{it}^{hk})$ is the production cost for power level p_{it}^{hk}. In [3] it was proved that, under normal constraint qualifying conditions, by replacing the objective function (14.49) with its *perspective reformulation*

$$\sum_{(h,k)\in A^i_+} c^{hk}_{i+} y^{hk}_{i+} + \sum_{(h,k)\in A^i_-} c^{hk}_{i-} y^{hk}_{i-} + \sum_{(h,k)\in A^i_+} \sum_{t\in \mathcal{T}(h,k)} y^{hk}_{i+} f(p^{hk}_{it} / y^{hk}_{i+}) \tag{14.50}$$

then the model that includes (14.50), (14.47), and (14.48) for each $(h, k) \in A^i_+$ describes the convex envelope of the 1UC problem defined by (14.49), (14.47), and (14.48) for each $(h, k) \in A_+$. Hence, a DP-based UC formulation based on the exact 1UC formulation is

$$\begin{array}{lll} \min (14.50) & & \\ \sum_{i\in I} \sum_{(h,k)\in A^i_+ : t\in \mathcal{T}(h,k)} p^{hk}_{it} = d_t & t \in \mathcal{T} & \\ (14.47),\ \ (14.48) & i \subset I & (14.51) \\ y^{hk}_{i+} \in \{0, 1\} & i \in I,\ (h, k) \in A^i_+ & \\ y^{hk}_{i-} \in \{0, 1\} & i \in I,\ (h, k) \in A^i_- & \end{array}$$

In the rest of the chapter we will refer to (14.51) as the *DP formulation*. The number of variables is $O(n^2|I|)$ for the network flow system (14.47) and $O(n)$ for each of the $O(n^2|I|)$ subproblems (14.48) associated with each pair (h, k); hence, the total number of variables in (14.51) is $O(n^3|I|)$. The specific form of PR used in (14.51), using auxiliary variables z^{hk}_{it}, underlines that both the MI-SOCP approach (14.43) and the SI-MINLP one (14.46) can be used to actually implement the DP formulation, of course at the cost of further conic or (in principle, infinitely many) linear constraints. Other PR approaches like AP^{2}R may provide more compact formulations (in particular, doing away with the auxiliary variables z^{hk}_{it} entirely) at the possible cost of a weaker bound at non-root nodes in the branch-and-cut; yet another trade-off that we will leave the interested reader to autonomously explore.

14.6.2 Additional DP-Based Formulations

In [2, 3], four formulations were introduced, based on the above DP formulation, that attain different trade-offs between size and quality of the continuous relaxation lower bound. When restricted to 1UC, all of them are less tight than the exact formulation given by (14.50), (14.47), and (14.48), but they have considerably fewer variables and constraints. Indeed, they all keep the network constraints (14.47), but each one has a different set of continuous variables to express the generated power. The first one uses the power variables p_{it}, as in the *3-bin* model, while the others present new types of variables whose cardinality is intermediate between *3-bin* and *DP* formulations. For space and clarity reasons, here we limit the discussion only to the first one of these new formulations, called the p_t *model*, and we refer the interested reader to [2, 3] for the other formulations and their comparison.

Given a unit $i \in I$, consider the commitment variable x_{it}, the start-up/shut-down variables v_{it}/w_{it} and the set of variables y_{i+}^{hk} associated with ON arcs (OFF_h, ON_k) such that $t \in \mathcal{T}(h,k)$. It is easy to see that, by definition, these variables are related by the equations

$$x_{it} = \sum_{(h,k):t\in\mathcal{T}(h,k)} y_{i+}^{hk}, \quad v_{it} = \sum_{k\geq t} y_{i+}^{tk}, \quad w_{it+1} = \sum_{h\leq t} y_{i+}^{ht}. \tag{14.52}$$

For each $t \in \mathcal{T}(2,n)$, the ramp-up/ramp-down constraints assume, respectively, the form

$$p_{it} - p_{it-1} \leq \Delta_i^+ \sum_{(h,k):t-1\in\mathcal{T}(h,k-1)} y_{i+}^{hk} + \bar{l}_i \sum_{k:k\geq t} y_{i+}^{tk} \tag{14.53}$$

$$p_{it-1} - p_{it} \leq \Delta_i^- \sum_{(h,k):t-1\in\mathcal{T}(h,k-1)} y_{i+}^{hk} + \bar{u}_i \sum_{h:h\leq t-1} y_{i+}^{ht-1} \tag{14.54}$$

Note that, in case the unit is on at the beginning of time horizon ($\tau_i^0 > 0$), the initial ramp-up/ramp-down conditions have to be set by

$$p_{i1} \leq (\Delta_i^+ + p_{i0}) \sum_{k:1\leq k} y_{i+}^{0k} \tag{14.55}$$

$$-p_{i1} \leq (\Delta_i^- - p_{i0}) \sum_{k:1\leq k} y_{i+}^{0k} \tag{14.56}$$

Then, the minimum and maximum power output constraints for each $t \in \mathcal{T}$ can be rewritten as

$$l_i \sum_{(h,k):t\in\mathcal{T}(h,k)} y_{i+}^{hk} \leq p_{it} \leq u_i \sum_{(h,k):t\in\mathcal{T}(h,k)} y_{i+}^{hk} \tag{14.57}$$

The right-hand side of (14.57) can be reinforced as follows. Assuming that $\tau_i^+ \geq 2$, if a unit i is switched on at time t then $\sum_{k:k\geq t} y_{i+}^{tk} = 1$ and the power p_{it} is bounded by $\bar{l}_i$. If the unit is switched off at time t then $\sum_{h:h\leq t} y_{i+}^{ht} = 1$ and the power p_{it} does not exceed $\bar{u}_i$. In case the unit does not turn on or off but it is committed at time t, then $\sum_{(h,k):h<t<k} y_{i+}^{hk} = 1$ holds. It follows that there exists (h,k) such that $h < t < k$ and $y_{i+}^{hk} = 1$. Thus, because of the maximum power output and the ramp-up/ramp-down constraints, the power p_{it} is bounded by $\psi_{it}^{hk} = \min\{u_i, \bar{l}_i + \Delta_i^+(t-h), \bar{u}_i + \Delta_i^-(k-t)\}$. Furthermore, if the unit is initially committed ($\tau_i^0 > 0$) then $\sum_{k:1\leq k} y_{i+}^{0k} = 1$ and we have to set $\psi_{it}^{0k} = \min\{u_i, p_{i0} + \Delta_i^+ \cdot t, \bar{u}_i + \Delta_i^-(k-t)\}$. Note that, if $\tau_i^+ \geq 2$, the variable y_{i+}^{tt} is not defined. Then, if $\tau_i^+ = 1$ and $y_{i+}^{tt} = 1$, the power p_{it} is bounded by the minimum between $\bar{l}_i$ and $\bar{u}_i$. Hence, if $\tau_i^+ \geq 2$, the right-hand side of constraints (14.57) can be reinforced as

$$p_{it} \leq \bar{l}_i \sum_{k:k\geq t} y_{i+}^{tk} + \bar{u}_i \sum_{h:h\leq t} y_{i+}^{ht} + \sum_{(h,k):h<t<k} \psi_{it}^{hk} y_{i+}^{hk} \tag{14.58}$$

while if $\tau_i^+ = 1$ this becomes

$$p_{it} \leq \bar{l}_i \sum_{k:k>t} y_{i+}^{tk} + \bar{u}_i \sum_{h:h<t} y_{i+}^{ht} + \sum_{(h,k):h<t<k} \psi_{it}^{hk} y_{i+}^{hk} + \min\{\bar{l}_i, \bar{u}_i\} y_{i+}^{tt} \tag{14.59}$$

Finally, the objective function of the p_t *model* is

$$\min \ \sum_{i \in I} \sum_{(h,k) \in A_+^i} c_{i+}^{hk} y_{i+}^{hk} + \sum_{(h,k) \in A_-^i} c_{i-}^{hk} y_{i-}^{hk} + \sum_{t \in \mathcal{T}} z_{it}, \tag{14.60}$$

with the constraints (for each $i \in I$, $t \in \mathcal{T}$)

$$z_{it} \geq \sum_{t \in \mathcal{T}} \Big(\sum_{(h,k):t \in \mathcal{T}(h,k)} y_{i+}^{hk} \Big) f^i \Big(p_{it} \,/\, \Big(\sum_{(h,k):t \in \mathcal{T}(h,k)} y_{i+}^{hk} \Big) \Big) \tag{14.61}$$

that can be implemented, e.g., in the MI-SOCP or SI-MILP way. To summarize, the p_t model for pure thermal units is

$$\begin{array}{lll} \min & (14.60) & \\ & (14.26) & \\ & (14.47),\ (14.55) - (14.56) & i \in I \\ & (14.53) - (14.54) & t \in \mathcal{T}(2,n),\ i \in I \\ & (14.57) - (14.59),\ (14.61) & t \in \mathcal{T},\ i \in I \\ & y_{i+}^{hk} \in \{0,1\} & i \in I,\ (h,k) \in A_+^i \\ & y_{i-}^{hk} \in \{0,1\} & i \in I,\ (h,k) \in A_-^i \end{array} \tag{14.62}$$

This is the DP-based model that experimentally has shown to have the best trade-off between tightness of the bound and computational cost—much better, in fact, than the original DP formulation, as we shall see–and therefore we leave the interested reader to consult [2, 3] for the other variants.

14.7 Lagrangian Relaxation

As the previous sections have shown, UC is indeed a large-scale model, especially as the number of units or the time horizon (or both) increases. In fact, for a long time UC has been considered to be too large to be addressable by off-the-shelf, general-purpose MILP (or MIQP) tools. However, UC has a *block structure* that makes it amenable to alternative approaches, in particular, the *Lagrangian Relaxation* (LR)/*Dantzig–Wolfe Decomposition* (DW) one that we briefly recall in this section.

To make the presentation easier we consider the schematic version

$$(P) \qquad \min\Big\{ \sum_{i \in I} c^i(x^i) \ :\ \sum_{i \in I} A^i x^i = b,\ \ x^i \in X^i \ \ i \in I \Big\} \tag{14.63}$$

The notation conveniently hides in X^i all the details of the representation of the units, comprised the integrality requirements on some of the different types of variables inside each unit, and in A^i these of the system-wide constraints. In fact, the technique is fairly general and it works for many other applications (e.g., [9, 30]) under relatively weak conditions; in particular, linearity of the *linking constraints* $\sum_{i \in I} A^i x^i = b$ is not required (but separability between different units is), and the approach is easily extended to the case of inequality constraints. The fundamental observation is that the linking constraints can be considered the *complicating constraints*, i.e., (14.63) would be "much easier" if they could be removed. This is clearly true in general: without the complicating constraints, (14.63) decomposes into a (large) number of (much) smaller independent subproblems. Since the complexity of basically any problem grows (highly) superlinearly, and often exponentially, with its size, the reduction in computational effort can clearly be dramatic. Furthermore, this is compounded with the fact that each of the individual sets X^i can be "easy", in the sense that the corresponding problem allows specialized solution methods.

Of course, removing the constraints outright is not really an option: however, a feasible alternative is to form the *Lagrangian relaxation* of (P) with respect to the complicating constraints, i.e., the (m independent sub)problem(s)

$$(P_\lambda) \qquad \min\{ \lambda b + \textstyle\sum_{i \in I} c^i(x^i) - (\lambda A^i)x^i \ : \ x^i \in X^i \}$$

for a fixed vector λ of *Lagrangian multipliers*, whose solution can be performed with m parallel processes since there are no longer constraints linking them. Indeed, the single-unit subproblems

$$(P^i_\lambda) \qquad \min\{ c^i(x^i) - (\lambda A^i)x^i \ : \ x^i \in X^i \}$$

can be solved as the 1UC problems discussed in Sect. 14.4, where the term $\pi = \lambda A^i$ plays the role of the revenue; yet, in (P_λ) it acts as a *penalty term* to discourage the violation of the relaxed complicating constraints. Hence, in this case (P_λ) is significantly easier than (P) to solve not only because of decomposability, but also because it allows to exploit structure in the (sub)problem(s). Note that this requires that the extra *Lagrangian term* $-(\lambda A^i)x^i$ in the objective does not significantly change the complexity of the (sub)problem(s), which is most often true (and linearity of the linking constraints often does help here, in that it makes the term linear in x^i), but not always. For instance, in Multicommodity flows applications [17] the subproblems should be elementary shortest paths, a problem that is polynomial if there are no negative-cost cycles (e.g., all arc costs are non-negative), but becomes $\mathcal{NP}$-hard if there are: hence, changing the costs in the subproblem(s) may change the complexity. This is not the case in UC, though.

The number of complicating constraints (equivalently, of Lagrangian multipliers) for UC is at least as many as the energy balance constraints (14.26), and therefore as the n instants in the time horizon; in this section we will therefore denote it with n, with the provision that it could be considerably larger than $|\mathcal{T}|$ if more complicated system-wide constraints need be considered (cf., e.g., (14.27)–(14.28)). The crucial issue is clearly how to properly choose the value of $\lambda \in \mathbb{R}^n$. Denoting by $v(\cdot)$ the (possibly infinite) optimal value of a given optimization problem, it is easy to check that $v(P_\lambda) \leq v(P)$ for each choice of $\lambda \in \mathbb{R}^n$, i.e., that every (P_λ) is a *relaxation* of (P). The case of complicating constraints being inequalities just requires the additional condition that $\lambda \geq 0$. Since (P_λ) can be efficiently solved, one can consider finding the *best possible* Lagrangian relaxation, i.e., the solution of the *Lagrangian dual*

$$(D) \qquad \max\{\, v(P_\lambda) \;:\; \lambda \in \mathbb{R}^n \,\}$$

of (P) with respect to the complicating constraints. Note that $v(P_\lambda) \leq v(P) \;\; \forall \lambda \in \mathbb{R}^n$ implies $v(D) \leq v(P)$, i.e., (D) is still a relaxation of (P). Solving the Lagrangian dual of (P), possibly approximately, yields information that can be used to drive solution methods to (P), as described in the following. This is what is generally referred to as a Lagrangian Relaxation (LR) approach to (P). Since one of the main reasons for the efficiency of the approach is the fact that the subproblem decomposes, this could also reasonably be called a Lagrangian *Decomposition* approach, but the moniker is historically reserved to a special form of the approach whereby the decomposable structure is not originally there and it is created by an appropriate reformulation before applying the LR (cf. [34]), hence we will stick with the generally approved nomenclature.

The LR technique has been successfully used to both compute tight lower bounds and to help implementing effective heuristics for many difficult optimization problems, among which UC; see, e.g., [13] and the references therein for a partial list. Yet, successful LR approaches have to overcome two nontrivial obstacles:

1. Finding the best possible Lagrangian multipliers λ^* requires solving (D), a—possibly, large-scale—nondifferentiable optimization problem in that $v(P_\lambda)$ is concave but typically lacks continuous derivatives.
2. Even with the optimal Lagrangian multipliers λ^*, the penalty term is typically not capable of enforcing feasibility of the optimal solution of (P_{λ^*}) w.r.t. the relaxed constraints. If this happened one would in fact have found an optimal solution to (P), but this is most often not the case because (D) is usually (and in particular in the UC case) *not* equivalent to (P), but rather a strict relaxation.

Solving both points requires a better understanding of the mathematical properties of (D) and its relationships with (P), which we now briefly recall.

14.7.1 Lagrangian Relaxation: Theory and Algorithms

To simplify the discussion we further simplify the notation to the case where there is only one subproblem, i.e., either $m = 1$ or, more properly, we only consider the set $X = X^1 \times X^2 \times \cdots \times X^m$ (the Cartesian product of all the feasible regions of the units' subproblems) and its elements $x = [\, x^1 \,,\, x^2 \,,\, \ldots \, x^m \,]$ disregarding their structure. Doing so in not (necessarily) a good approach computationally but it allows to more succinctly describe the problems as

$$(P) \qquad \min\{\, c(x) \;:\; Ax = b, \;\; x \in X \,\}$$

$$(D) \qquad \max\Big\{\ \phi(\lambda) = \min\{\ c(x) + \lambda(b - Ax)\ :\ x \in X\ \}\ \Big\}$$

highlighting the role of the Lagrangian function $\phi(\lambda) = v(P_\lambda)$. As already mentioned, (D) is a concave maximization problem and therefore "easy"; this is easily seen by considering X rewritten as the (possibly, infinite) collection $X = \{\bar{x}_1\ ,\ \bar{x}_2\ ,\ \dots\}$ of its elements, and therefore (D), with a standard reformulation trick, as

$$(D) \qquad \max\{\ v\ :\ v \leq c(\bar{x}) + \lambda(b - A\bar{x}),\quad \bar{x} \in X\ \}. \tag{14.64}$$

This reveals that (D) is "just a linear program", albeit possibly a semi-infinite one. In many cases, X can be restricted to a finite set, say because of the integrality requirements on the variables and/or the fact that it only has a finite number of extreme points; yet, (14.64) is still a "very large" (typically, exponential-size) LP since it has as many constraints as the elements of X. To even further simplifying the discussion we will however assume that X is at least *compact*; dropped this only requires some complication in the notation (and the algorithms), cf., e.g., [13], but since it holds true for UC we do not need to consider the issue here. We can also assume that X is nonempty; this not being the case is found out the first time (P_λ) is solved, and immediately implies that (P) has no feasible solution. Therefore, ϕ is finite everywhere and, being the pointwise minimum of a set of linear functions, is concave. In particular, if X is a discrete set then ϕ is *polyhedral*, i.e., its *ipograph*

$$Ipo(\phi) = \{\ (\ v\ ,\ \lambda\ )\ :\ v \leq \phi(\lambda)\ \}$$

is a polyhedron, hence convex. Each optimal solution $\bar{x}$ to (P_λ) gives the *supporting hyperplane* $b - A\bar{x}$ to $Ipo(\phi)$, known as a *supergradient* of ϕ in λ. This immediately provides an algorithmic venue for attacking (D) under the form of *subgradient-type* methods, that surely have the advantage of an extreme simplicity: once $g = b - A\bar{x}$ is computed, the vector of Lagrangian multipliers is updated as $\lambda \leftarrow \lambda + \alpha g$ for a properly chosen stepsize $\alpha > 0$. With simple choices of α (even fixed a-priori) the process can be proven to converge toward an optimal solution λ^* of (D), if any exists. More sophisticated choices of α may result in faster convergence, making the approach possibly competitive for finding solutions of (D) with low-to-mid accuracy (say, with a relative gap of `1e-3` to `1e-4`) [20]. In fact, the subgradient algorithm has for a long time been viewed as the only viable approach for solving (D), up to the point that "LR approach" used to be a synonym of "Lagrangian dual solved by a subgradient algorithm". However, such an approach is basically incapable of producing substantially more accurate solutions in reasonable times. Furthermore, it does not "naturally" produce any kind of feasible solution, as the $\bar{x}$ produced at each iteration would typically remain highly infeasible; although the approaches can be endowed with techniques to produce some, in the sense discussed right below, these are not really practical.

A different approach is simply to go back to the formulation (14.64) as a "very large scale LP". While of course the full formulation cannot even be written in practice, it is well-known that LPs with "many" constraints can still be efficiently solved by *row generation* techniques, provided that an efficient *separation procedure* is available. Indeed, typically only a very small fraction of the constraints in (14.64) are actually necessary, as readily shown by realizing that only the $n + 1$ constraints associated with an optimal base would suffice. While the optimal base is unknown, one can resort to a *constraints generation* approach where a "small" $\mathcal{B} \subset X$ defines the (Dual) *Master Problem*

$$(D_{\mathcal{B}}) \qquad \max\{\ v\ :\ v \leq c(x) + \lambda(b - Ax),\quad x \in \mathcal{B}\ \}, \tag{14.65}$$

which corresponds to

$$\max\big\{\ \phi_{\mathcal{B}}(\lambda) = \min\{\ c(x) + \lambda(b - Ax)\ :\ x \in \mathcal{B}\ \}\ \big\},$$

i.e., to maximizing the *cutting plane model* $\phi_{\mathcal{B}}$, a concave polyhedral upper approximation of the true Lagrangian function ϕ. Given an optimal solution $(\tilde{v}, \tilde{\lambda})$ of $(D_{\mathcal{B}})$, it is immediate to realize that $\tilde{v} \geq v(D)$. It is then possible to efficiently check whether or not it is also optimal for (D) by just solving $(P_{\tilde{\lambda}})$, i.e., computing $\phi(\tilde{\lambda})$. If $\tilde{v} \leq \phi(\tilde{\lambda})$, then clearly $\tilde{v} = \phi(\tilde{\lambda}) = v(D)$ and the problem is solved. Otherwise, any $\bar{x}$ optimal for $(P_{\tilde{\lambda}})$ yields a violated constraint of (14.64) that can be added to $\mathcal{B}$, and the process iterated. It can be shown that the corresponding *Cutting Plane algorithm* (CP) is globally convergent; this is readily apparent when X is finite, as it will necessarily terminate in a finite number of solutions, although possibly after having collected an excessively large number of cuts. Indeed, the by-the-book CP algorithm is often not very efficient in practice. This starts with the fact that there are in principle no guarantees that (14.65) is bounded, especially at the beginning of the process when $\mathcal{B}$ is "small"; but in fact even in general, since the whole of (D) may be unbounded below if (P) has no feasible solution (but the individual (P_λ) have). In fact, if (P) is feasible and a feasible solution $\hat{x}$ for (P) is known, then (14.65) is bounded provided just that $\hat{x} \subset \mathcal{B}$, since this corresponds to inserting the constraint $v \leq c(\hat{x})$ into $(D_{\mathcal{B}})$. Alternatively, some "phase 0" approach a-la simplex method can be used: clearly, once (14.65) is bounded it will remain so if constraints are only added, as the base version of the approach postulates. Yet, the CP approach can still suffer from unduly slow convergence, and improvements are needed. We leave this discussion for later; at least, the CP algorithm determines at each step a lower and an upper bound on $v(D)$, thereby having a fully implementable stopping rule for (approximate) termination, the lack of which is perhaps the most significant drawback of subgradient-type methods.

The latter property is in fact deeply routed in a fundamental trait of the Lagrangian dual: since it is a (semi-infinite) LP it has a dual, which provides compact optimality certificates. This is by no means a theoretical construct: $(D_{\mathcal{B}})$ is a simple finite LP, and therefore it is equivalent to its own linear dual

$$(P_{\mathcal{B}}) \qquad \min\big\{\ \textstyle\sum_{x\in\mathcal{B}} c(x)\theta_x\ :\ \sum_{x\in\mathcal{B}} x\theta_x(b - Ax) = 0,\quad \theta \in \Theta\ \big\} \tag{14.66}$$

where $\Theta = \{ \theta \geq 0 : \sum_{x \in \mathcal{B}} \theta_x = 1 \}$. The *Primal Master Problem* $(P_{\mathcal{B}})$ has a variable for each row of $(D_{\mathcal{B}})$, i.e., for each $x \in \mathcal{B}$. Furthermore, Θ clearly indicates that a convex hull operation is taking place. Indeed, the result is way more intuitive if discussed for the even simpler case where the objective $c(x)$ is linear, although it holds in general. Using linearity of c and the constraints allows to rewrite (14.66) as

$$(P_{\mathcal{B}}) \qquad \min \left\{ c \left(\sum_{x \in \mathcal{B}} x\theta_x \right) : A \left(\sum_{x \in \mathcal{B}} x\theta_x \right) = b, \ \ \theta \in \Theta \right\} \tag{14.67}$$

which immediately reveals itself as the "explicit form" of $(P_{\mathcal{B}})$ written as

$$(P_{\mathcal{B}}) \qquad \min\{ \, cx \, : \, Ax = b, \ \ x \in X_{\mathcal{B}} = conv(\mathcal{B}) \, \}, \tag{14.68}$$

where $conv(\cdot)$ denotes the convex hull. In plain words, $(P_{\mathcal{B}})$ optimizes the "natural" objective on the intersection between the (feasible region defined by the) complicating constraints and the convex hull $X_{\mathcal{B}}$ generated by the known set of points in $\mathcal{B}$. Note that $X_{\mathcal{B}}$ is an inner approximation of X, i.e., $X_{\mathcal{B}} \subset X$, if X itself is convex, which is *not* the case of UC. Yet, since the result holds for any $\mathcal{B}$ it also does for $\mathcal{B} = X$, which yields a celebrated theoretical result: the Lagrangian dual (D) is equivalent to the *convexified relaxation* of (P)

$$(\tilde{P}) \qquad \min\{ \, cx \, : \, Ax = b, \ \ x \in conv(X) \, \}. \tag{14.69}$$

The result can be proven by elementary means when X is discrete and $c()$ linear (e.g., [13]) but requires some further *constraint qualification* in the general case; we will not delve into this as the constraint qualification typically holds in practice, and it does in the case of UC since all constraints are linear and only the objective function is not [3]. Similar, but more notationally complex, results hold if $c()$ is nonconvex or the complicating constraints are nonlinear. In cases such as UC, (D) is a *partial convexification* operation that is applied to the "easy part" X of the problem only, leaving the complicating constraints $Ax = b$ unchanged. This has the following useful corollaries:

- If $X = \{ x \in \mathbb{R}^k \, : \, Ex \leq d \}$ is a polyhedron, or more generally a convex set (subject to constraint qualification), (P) and $(\tilde{P})$ coincide; in particular, the Lagrangian dual of an LP is the classical linear dual, with λ being the dual variables of the relaxed constraints $Ax = b$.
- If X is a polyhedron (convex set) with further integrality restrictions imposed, say $X = \{ x \in \mathbb{Z}^k \, : \, Ex \leq d \}$, then the Lagrangian dual provides a bound that is not worse than that of the continuous relaxation $(\bar{P})$ where the integrality constraints are relaxed. In fact, $(\tilde{P})$ has no more solutions than $(\bar{P})$, and the two problems have the same objective function. It is immediate to construct examples where the dominance is strict, as well as examples where $v(D) < v(P)$: that is, the Lagrangian dual is better (not worse) than the continuous relaxation, but a relaxation of the original problem nonetheless, unless of course (P) is convex (which is not the case for UC).

- However, if in the previous case the constraints $Ex \leq d$ have the *integrality property*, i.e., the optimal value of $v(P_\lambda)$ does not change if the integrality restrictions are relaxed (equivalently, all vertices of X are integral), then $v(D) = v(\tilde{P}) = v(\bar{P})$. This is a sort of "no free lunch" principle: (D) can only provide a better bound than $(\bar{P})$ if (P_λ) is "more difficult" than an LP, i.e., an "exact" description of $conv(X)$ must not be already provided by just the $Ex \leq d$ constraints.
- From the arguments above it immediately follows that using any formulation of the Thermal Units except the "exact" DP one will result in the lower bound of the continuous relaxation being weaker (not stronger) than that of the Lagrangian dual. Conversely, the continuous relaxation bound of the formulation using the "exact" DP formulation of the Thermal Units is poised to be identical to that of the Lagrangian dual.

All the above observations only hinge on the $v(D) = v(\tilde{P})$ equality. However, the result runs much deeper, as it says that (D) and $(\tilde{P})$ are actually *the linear dual* of each other. Hence, *proving optimality* of some λ^* to (D) necessarily entails constructing an *optimal solution* $\tilde{x}$ of $(\tilde{P})$. Not surprisingly, whenever the CP algorithm finally terminates it has in fact constructed such an optimal primal solution.

Indeed, the CP algorithm is the "dual view" of the well-known *Dantzig–Wolfe decomposition method*. This is an approach for solving convex problems of the form (14.69) where X is explicitly given in terms of its (extreme) solutions rather than, as more customary, as a set of defining constraints. The issue with (14.69) written in its "explicit form" (14.67) (with $\mathcal{B} = X$) is that the problem typically has (exponentially, or even infinitely) many columns: however, it is well-known that problems like these can still be efficiently solved by *column generation* approaches provided that an efficient *pricer* is available to generate the columns. In fact, the CP algorithm is nothing but the *Dantzig–Wolfe decomposition approach*, i.e., a column generation technique, applied to solving the "explicit form" (14.67) of $(\tilde{P})$. At each step, the (primal) Master Problem $(\tilde{P}_\mathcal{B})$ yields a primal optimal solution θ^*, and thus $\tilde{x} = \sum_{x \in \mathcal{B}} x\theta^*_x$ feasible for $(\tilde{P})$, and a dual solution $\tilde{\lambda}$ corresponding to constraints $Ax = b$. The *pricing problem* $(P_{\tilde{\lambda}})$ determines the column $\bar{x} \in X$ of least *reduced cost* $(c - \tilde{\lambda}A)\bar{x}$: if $(c - \tilde{\lambda}A)\bar{x} \geq 0$ then $\tilde{x} = x^*$ is optimal for $(\tilde{P})$, otherwise $\bar{x}$ is added to $\mathcal{B}$ in order to construct a better primal solution at the next iteration. It has to be stressed that this is not a different algorithm, as these steps are *exactly* those of the CP, just seen from the primal viewpoint. Indeed, the case where (14.65) is unbounded below exactly corresponds to $\mathcal{B}$ being "too small", so that $X_\mathcal{B}$ has empty intersection with $Ax = b$. Thus, the Cutting Plane/Dantzig–Wolfe solves (D) together with $(\tilde{P})$ via the *inner approximation* $X_\mathcal{B}$ of $conv(X)$, that is iteratively enlarged with the solutions of the pricing problem until it contains one optimal solution of $(\tilde{P})$. Dually, the Master Problem optimizes $\phi_\mathcal{B}$, an *outer approximation* of ϕ, that is iteratively "thinned" by the subgradients $b - A\bar{x}$ generated by the separation problem until—roughly speaking—its optima coincide with those of ϕ. Yet, the two processes are just the two faces of the same indivisible coin.

To summarize, LR is an approach to define and solve continuous relaxations of a problem corresponding to "strong" formulations, providing "tight" bounds together with the same primal and dual information that solving an ordinary continuous relaxation with standard tools would provide. Since it exploits the structure of (P) by repeatedly generating (extreme) points of $conv(X)$, the approach is particularly attractive whenever optimizing over X is more convenient than obtaining an explicit representation of $conv(X)$. However, the development so far has only illustrated the basic ideas: turning these into a competitive implementation requires overcoming several hurdles, which will be described in the following.

14.7.2 Lagrangian Relaxation: Practical Aspects

The practical behavior of the standard CP/DW approach, even disregarding the possible need for a "phase 0", is often unsatisfactory: convergence is, overall, rather slow. This is mostly related to the fact that the approach is *unstable*: the produced iterates λ^1, λ^2, ...usually behave "wildly", with subsequent iterates possibly "very far" from each other dashing almost randomly across the search space. This means that often the new element $\bar{x}$, when added to $\mathcal{B}$, fails to convey information capable of effectively driving the iterates toward λ^*. This is clearly illustrated by the fact that, as shown, e.g., in [14], starting the approach with $\lambda^1 \approx \lambda^*$ does not typically result in significantly reduced iterations, in stark contrast with the best smooth optimization algorithms that, owing to the use of second-order models having very good approximations of the curvature information, would converge extremely fast in the same setting. The CP algorithm, instead, has to rely on the cutting plane model $\phi_{\mathcal{B}}$, a piecewise-linear (polyhedral) function that can only approximate the curvature information by painstakingly accruing it piecemeal via first-order information. This phenomenon is well understood and grounded in theory: while optimization of (strongly) convex smooth functions can find ε-optimal solutions in $O(\log(1/\varepsilon))$ iterations for just first-order approaches (even less for second-order ones), nonsmooth optimization is proven to be $\Omega(1/\varepsilon^2)$ in general.

This is not to say that the (often, abysmal) practical behavior of the CP algorithm cannot be improved upon. Perhaps the most popular approach is to try to add "poorman's" second-order information to $\phi_{\mathcal{B}}$ by choosing a *stability center* $\bar{\lambda}$, a *stabilization parameter* t, and solve the *Stabilized (Dual) Master Problem*

$$(D_{\mathcal{B},\bar{\lambda},t}) \qquad \max\{\ \phi_{\mathcal{B}}(\lambda) - \tfrac{1}{2t} \|\ \lambda - \bar{\lambda}\ \|^2\ \} \tag{14.70}$$

instead of (14.65). The added *stabilizing term* provided (arguably, very crude) curvature information to the model, governed by the (arbitrary, and not easy to tune) parameter t. More sophisticated approaches using, say, quasi-Newton-type techniques have been proposed, but they require considerably more sophisticated theory and implementation; furthermore, they rely on the assumption that some second-order information exists, which is not the case in our application since ϕ (and not only its model) is inherently polyhedral. Even if it mostly fails to capture "true" curvature information, the stabilizing term at least ensures that the iterates do not stray "too far" from $\bar{\lambda}$. If the latter is a good approximation of λ^* this lends to the iterates sequence the locality properties necessary to enforce faster convergence (in practice, if not in theory). Otherwise, $\bar{\lambda}$ has to be revised, typically when an iterate λ^i is found such that $\phi(\lambda^i) \ll \phi(\bar{\lambda})$. In all this t could even be kept fixed, but choosing it dynamically is important for practical performances. The above *proximal stabilization* of the CP algorithm is just one of several different variants of "bundle methods" (such as those employing trust-region or level stabilization instead) proposed to improve its performances, together with center-based methods and other forms of stabilization [14]. Each of these has a "primal interpretation" as a variant of the DW approach: for instance, the proximal stabilization leads to a (generalized) Augmented Lagrangian of (14.67). We will not delve on the fine details here, that can be found in the literature [12, 14]; the relevant remark is that in all cases the approaches will necessarily have to (asymptotically) produce an optimal solution x^* of (14.68), typically via producing an optimal θ^* for its "explicit form" (14.67). Indeed, stabilization is only an algorithmic device to improve the performances, but it does not change the nature of the problem.

Besides stabilization, other aspects of the implementation of the CP algorithm can have a very significant impact on its performances. A very important one is to *not* disregard the Cartesian product nature of X whenever it is present, such as in our application. This means that the Lagrangian function is actually the sum of $m + 1$ terms, i.e.,

$$\phi(\lambda) = \lambda b + \textstyle\sum_{i \in I} \left(\phi^i(\lambda) = \min\{ \, (c^i - \lambda A^i) x^i \; : \; x^i \in X^i \, \} \right).$$

Besides the fact that the individual components ϕ^i of the Lagrangian function can be computed in parallel, each of them is a polyhedral function for which individual subgradients $-A^i \bar{x}^i$ are produced. Rather than aggregating them in a unique subgradient of ϕ, *individual models* $\phi^i_{\mathcal{B}}$ can be separately constructed for each component. The corresponding *disaggregated Dual Master Problem* is roughly a factor of m larger—after the same number of iterations, and save *selection* and *compression* techniques that can be used to reduce the size of $\mathcal{B}$ and that we will not discuss here—and therefore significantly more costly to solve. However, it typically also results in a much faster convergence. The reason is easily understood by looking at its dual, the *disaggregated Primal Master Problem* (here in its simple non-stabilized form, but stabilization against does not change the fundamental nature of the problem)

$$(P_{\mathcal{B}}) \qquad \min \left\{ \begin{array}{l} \sum_{i \in I} c^i \sum_{x^i \in \mathcal{B}} x^i \theta_{x^i} \\ \sum_{i \in I} A^i \sum_{x^i \in \mathcal{B}} x^i \theta_{x^i} = b \\ \sum_{x^i \in \mathcal{B}} \theta_{x^i} = 1 \qquad i \in I \qquad \theta \geq 0 \end{array} \right. \tag{14.71}$$

In (14.71), each of the solutions generated during the process is split into its components x^i, which are separately kept and have an independent dual variable θ_{x^i}. It is plain to see that, for the same $\mathcal{B} \subseteq X$, (14.71) has more solutions than (14.67); in fact, (14.67) is the restriction of (14.71) where all the components $\bar{x}^i$ corresponding to the solution $\bar{x}$ obtained at some iteration must have the same convex multiplier.

Due to X being a Cartesian product, forcing each component of $\bar{x}$ to have the same multiplier does not conceptually make sense: in other words, the Cartesian product of the convex hulls of the individual $\bar{x}^i$ is a better approximation of $conv(X)$ than $conv(\mathcal{B})$. There actually is a potential reason for wanting to aggregate, in that the larger disaggregated Master Problems are more costly to solve than the aggregated ones. However, they use more efficiently the generated information about (14.69), which typically results in much faster convergence, and ultimately in better overall performances. Indeed, while the Master Problems grow in size much faster in the disaggregated case, a very significant reduction in the number of iterations often leads to the algorithm terminating before they reach very large sizes, unlike in the aggregated case. Yet, *partial aggregation* may be considered to fully exploit the trade-off between convergence speed and master problem cost, especially for applications where the cost of the Lagrangian problem is low and therefore the master problem is the computational bottleneck. As there are no clear guidelines for choosing the right level of aggregation, the fully disaggregated case is typically preferred.

A somehow related, although different, improvement can be obtained by exploiting the fact that the different Lagrangian subproblems need not be all of the same type. This is for instance the case in Hydro-Thermal UC problems with "easy" hydro units such as (14.22)–(14.25). We can abstract the situation by considering a simplified version of (P) like

$$(P) \qquad \min \begin{cases} c^1x^1 + c^2x^2 \\ A^1x^1 + A^2x^2 = b \\ x^1 \in X, \quad Ex^2 \leq d \end{cases} \tag{14.72}$$

Since the problem has two separate blocks of variables once the complicating constraints are removed, it would seem to be natural to apply the disaggregated approach (14.71). Of course, the corresponding set X^2 could be considered as the convex hull of its extreme points, and optimizing over X^2 is clearly "easy". However, this is likely not the right idea: the representation of X^2 via its vertices likely has an exponential number of variables, while a compact linear formulation is available. Thus, an arguably better solution is to form the analogous of (14.71) in which only the first group of variables is reformulated, while the second is just kept as it is

$$(P_{\mathcal{B}}) \qquad \min \begin{cases} c^1\left(\sum_{x^1 \in \mathcal{B}} x^1\theta_{x^1}\right) + c^2x^2 \\ A^1\left(\sum_{x^1 \in \mathcal{B}} x^1\theta_{x^1}\right) + A^2x^2 = b \\ \sum_{x^1 \in \mathcal{B}} \theta_{x^1} = 1, \quad \theta \geq 0, \quad Ex^2 \leq d \end{cases} \tag{14.73}$$

The idea clearly carries over to any number of "easy" and "difficult" components of the problem. Again, a trade-off exists between the size of the Master Problem, that initially can be larger, and the fact that having "perfect" information on the feasible set of some variables—as opposed to having to painstakingly accrue it iteratively—usually decreases the total number of iterations required to solve the problem. This is easily seen by considering the "extreme" case in which all the components are "easy", and therefore the first Master Problem is already equivalent to the whole of $(\tilde{P})$ and only one iteration is performed. Again, the trade-off is usually in favor of using this "easy components" idea, in particular, for the UC case [43]. Using stabilization techniques in conjunction with this approach is not necessarily trivial, but it is nonetheless possible: while the master problem can become even more complex, and hence costly, the resulting performance improvements can be nothing short of dramatic [29].

It should be remarked that all the above developments point toward the fact that it is usually beneficial to invest in having "large and complex Master Problems". In fact, a generalization of the Dantzig–Wolfe decomposition approach has been defined for the case where the reformulation of X need not necessarily be in terms of the convex hull of its extreme points, but it can be "any" large-scale (e.g., pseudo-polynomial) formulation. Provided that the Lagrangian relaxation can be efficiently solved (which does not in any way depend on how the Master Problem is formulated), a more structured Master Problem can be defined whose variables and constraints can be dynamically added in a way that exactly parallels the standard CP/DW approach [18]. The resulting algorithm can also be stabilized [19] and it has shown to be able to further significantly improve the performances of LR approaches. Interestingly, the DP formulation of UC precisely has all the required characteristics, and therefore it is conceptually possible to apply the *Stabilized Structured Dantzig–Wolfe approach* to the problem. Unfortunately, the implementation of such an approach is highly nontrivial and there currently are few (if any) software environments that properly support it, which is why we will not further describe this option here.

In fact, the issue about LR approaches being usually far less trivial to implement than those based on standard MILP/MIQP formulations is true for the more "standard" versions as well. However, recently the new software framework `SMS++` has been released [1] that makes it easier to develop LR approaches. All the computational results discussed in the next section have been obtained using `SMS++` components. In particular, in `SMS++` optimization problems are represented by derived classes of the base `Block` class, meant to represent one specific (family of) optimization models that may comprise any number of other `Block` (recursively) as components. For our results we have mainly used the `UCBlock` class, that represents several different versions of UC problems. In particular, in `UCBlock` the individual units in the problem are represented by instances of classes derived by `UnitBlock`, and abstract base class (deriving from `Block`) that defines a common interface for "any production unit". The UC instances that we used for our tests are mainly composed of `ThermalUnitBlock` and `HydroUnitBlock`, but the project already provides classes for several other types of generation units as well as for different kinds of network constraints. Also, the flexibility of `Block` allows to produce different formulations for the same problem: this is used in `ThermalUnitBlock`, which can produce all the formulations discussed in the previous sections (and more) by just changing a configuration parameter.

Once a `UCBlock` is loaded, it can be solved with standard MILP approaches by classes derived from `MILPSolver`, that implement interfaces with general-purpose MILP solvers like `CPLEX` and `SCIP`. Alternatively, the Lagrangian dual w.r.t. the constraints linking the different units and the network (energy, primary and secondary reserve, pollutants emissions, ...) can be automatically constructed by just registering the LagrangianDualSolver to the UCBlock. In turn, LagrangianDualSolver employs a state-of-the-art implementation of "bundle method", BundleSolver, to solve the Lagrangian dual and provide the optimal dual and primal solution λ^* and x^* of the convexified relaxation of the problem. BundleSolver implements all the above mentioned computational improvements (except the Stabilized Structured Dantzig–Wolfe approach), i.e., stabilization, disaggregate Master Problem, and "easy components". A further derived class ParallelBundleSolver can be used instead which also implements parallel computation of the Lagrangian subproblems. Also, LagrangianDualSolver allows to use specialized algorithms to solve the Lagrangian subproblems; in particular, for ThermalUnitBlock we can use the DPThermalUnitSolver, a state-of-the-art implementation of the DP algorithms described in Sect. 14.4.3.

Remarkably, LagrangianDualSolver and [Parallel]BundleSolver are not specific for UCBlock but can be used with any Block with the same structure, finally providing convenient support for easily and efficiently solving Lagrangian duals of structured problems. In fact they have already been used to solve other classes of structured optimization problems like multicommodity flows, multiple knapsack assignment problems, and capacitated facility location problems. Hopefully, SMS++ will revive interest in LR approaches by providing a much more convenient way of using them w.r.t. the relative dearth of workable software solutions so far.

This is not to say that SMS++ already provides all the necessary components to implement a full-fledged LR approach. Indeed, so far is only possible to easily and efficiently compute the primal and dual optimal solution to the convexified relaxation, but this alone does not solve the problem. These elements could of course be in principle used to implement an enumerative (Branch&Bound) approach and/or heuristics for UC. In fact, LR-based heuristics for UC have been implemented with good success [6, 43]. These can exploit the fact that an LR approach actually provides two different forms of primal information: the integer but unfeasible (w.r.t. the complicating constraints) solutions $\bar{x}$ of the Lagrangian subproblem(s), and the continuous but feasible solution x^* of the convexified relaxation. The latter is actually not really feasible for most of the iterations (stabilization in the dual corresponds to relaxation in the primal, as clearly shown by the fact that the stabilized Primal Master Problem is always feasible), but it typically becomes "almost feasible" rapidly and therefore it usually is a good "guide" to construct feasible solutions. Combining all these aspects to yield an efficient and effective heuristic is possible but nontrivial, and it has not been ported yet to SMS++ so that it will not be discussed in the computational section. Similarly, a generic B&B approach based on LR is currently under development using SMS++, but it is not yet in a shape that allows a fair comparison with the highly optimized branch-and-cut (B&C) in general-purpose MILP solver. Hopefully these will be available in the near future, not only for UC but for all other problems with analogous structure. For now, the computational section will concentrate on the comparison between the efficiency and effectiveness of the bounding procedures based on the use of MILP/MIQP solvers for the different formulation and on LR. Anyway, computing tight lower bounds on the optimal value is well-known to be the decisive (although not the only one) aspect of exact approaches for the solution of hard problems such as UC.

14.8 Computational Results

We now try to paint a picture, although at wide strokes, of the complexity of choosing the right combination of formulation and algorithmic parameters to compute bounds for a hard optimization problem such as UC. We concentrate all the necessary information in just Table 14.1. In the table we report results about the lower bound obtained by applying standard MILP approaches, in particular `CPLEX` version 12.10, to the different formulations presented in Sects. 14.3 and 14.6. In particular, for realistic UC instances with $m = 10$, 20, and 50 that have been a staple of computational experiments for many years [2, 3, 6, 7, 15, 22, 26–28, 33] and for the three representative formulations *3-bin* (as defined in Sect. 14.3), *DP* (14.51), and p_t (14.62), we report the relative gap (in percentage) between the lower bound provided by the corresponding formulation and the optimal value (which is otherwise known), as well as the time to compute the corresponding bound.

However, the "basic" formulation is not the only factor at play. First of all, each formulation may or may not employ the Perspective Reformulation technique for the quadratic objective function (cf., e.g., (14.49) versus (14.50)); in our case we have chosen to implement that via using *Perspective Cuts* (P/C) (14.46), as these are easy to include in any modern B&C solver like `CPLEX` via the appropriate `callback` routines and are most often the best performing option when compared, e.g., to the (slightly simpler to implement) SOCP formulation (14.43) [23]. However, MILP solvers also automatically improve the formulation via two techniques: *preprocessing* and *general-purpose valid inequalities* (a.k.a. "cuts"). Both these operations have a (usually slight in the preprocessing case) time impact, but can (significantly, especially in the cuts case) improve the produced lower bound. Although in general one would just leave these options activated, in Table 14.1 we provide results for all the eight combinations of P/C, cuts and preprocessing being active or not. Notice that the "no cuts, no pre" option corresponds to the "natural" bound provided by the corresponding formulation(s, two ones corresponding to using or not P/C), with the others providing improvements over that which depend on the fine details of the inner workings of the specific solver, and that therefore could be different if using anything else than `CPLEX` (or even different versions of it). In fact, solvers typically provide *algorithmic parameters* to control both preprocessing and especially cuts; for the latter, different families of *separators* can be individually activated with different levels of aggressiveness. Thus, a more detailed computational study should also try to find the combination of the algorithmic parameters that provides the best gap/time trade-off, notoriously a time-consuming and nontrivial task that may require sophisticated approaches to be performed instance-wise [36]. For the sake of conciseness and clarity we will refrain from that here, presenting the results corresponding to the corresponding feature being either completely off, or active at the "default" value;

this is what most users would do anyway, and efficient solvers typically have highly tuned choices for their "default" options that provide competitive results in most cases.

Together with the results from the MILP solver we also report in Table 14.1 those for the bounds obtained by solving the Lagrangian Dual of UC w.r.t. the linking constraints (14.26), using the state-of-the-art bundle-type algorithm implemented in `BundleSolver` and solving the subproblems via an efficient implementation of the DP algorithm of 1UC described in Sect. 14.4. This is visually located at the same height as one of the eight MILP options for which the *DP* formulation is guaranteed to provide the same bound, i.e., "P/C, no cuts, no pre". Notice that in fact applying both preprocessing and general-purpose valid inequalities to the Lagrangian formulation is in principle possible, and could result in analogous improvements in bound quality (at a computational cost). However, software-wise this is more complex to obtain and the corresponding options are not yet available in `SMS++`, so we only report the "base" case that can be easily obtained. All in all, Table 14.1 paints a varied picture that requires a detailed analysis.

Preprocessing is almost invariably the right choice: it usually has a positive effect on the bound, sometimes surprisingly pronounced (e.g., $m = 20$ for p_t). It often has a positive impact on the running time, also, especially on larger formulations like *DP*; in fact, preprocessing is more focused on reducing the size of the problem for efficiency reasons than on improving the bound, although techniques specifically designed for the latter purpose are also used, as the results imply. Furthermore, the impact on time may not always be positive; cf., e.g., p_t in particular in the "no P/C" half. However, the real impact should be measured across the whole branch-and-cut process, as opposed to only at the root node; the final effect may then still be positive even in those cases. Yet, the effect of preprocessing on the gap is more evident when no other techniques are used to improve it (cuts, be them general-purpose ones or P/C), but it clearly reduces—up to vanishing—when they are, especially for very strong formulations like *DP*. Finally, the effect also depends on size, being (most of the time, but with exceptions) somewhat larger for the smallest instances but tapering off as the size increases. All in all, however, preprocessing is typically best left active.

The picture for cuts is similarly clear, but with a more evident trade-off with time. Pulling all the stops with cuts (using them all, be them P/C or not) has the potential to very significantly reduce the gap, even by a whole order of magnitude for the formulations that "start weak" like *3-bin*. In fact, it is almost surprising that *3-bin* with all three options on provides gaps that are significantly smaller than *DP* (with the same option), which is "naturally" a much stronger formulation. In fact, the effect of cuts on the *DP* formulation is overall limited, especially for general-purpose ones (but also, albeit to a lesser extent, for P/C as well); clearly, the stronger formulation already "naturally contains" many of them, as opposed to the weaker ones, and yet cut separation has a rather sizable impact on its (already, rather large) running times. Indeed, the "intermediate" p_t formulation seems to capture the sweet spot here: cuts have a significant but not dramatic impact on running times while very significantly improving the gap. The relative improvement of P/C versus general-purpose cuts is varied across the board, with either family dominating in some cases and losing out

Table 14.1 Computational results on different bounding procedures for UC

				3-bin		DP		p_t		LR	
			m	Time	Gap	Time	Gap	Time	Gap	Time	Gap
P/C	Cuts	Pre	10	1.73	0.11	92.09	0.58	2.42	0.06		
			20	4.64	0.14	451.63	0.48	6.68	0.08		
			50	15.40	0.12	4407.44	0.08	18.78	0.04		
		No pre	10	2.65	0.13	124.12	0.54	2.12	0.06		
			20	4.42	0.14	696.72	0.46	6.10	0.24		
			50	19.63	0.12	5331.71	0.08	17.13	0.05		
	No cuts	Pre	10	0.21	1.03	78.56	0.67	1.00	0.53		
			20	0.90	0.93	480.02	0.51	2.58	0.27		
			50	4.18	0.81	3836.78	0.08	9.92	0.09		
		No pre	10	0.14	1.16	101.31	0.67	0.63	0.68	0.46	0.67
			20	0.85	0.95	652.31	0.51	2.16	0.52	0.83	0.51
			50	4.10	0.82	4365.31	0.08	8.65	0.09	1.19	0.07
No P/C	Cuts	Pre	10	1.08	0.49	43.25	0.84	2.07	0.35		
			20	3.81	0.46	146.50	0.68	5.27	0.33		
			50	10.00	0.34	1644.14	0.18	14.04	0.18		
		No pre	10	1.09	0.52	65.18	0.83	1.84	0.37		
			20	4.40	0.51	226.73	0.69	2.52	0.61		
			50	15.42	0.44	1740.14	0.25	7.26	0.25		
	No cuts	Pre	10	0.12	1.41	32.51	0.88	0.84	0.71		
			20	0.31	1.29	189.79	0.68	2.69	0.43		
			50	1.06	1.16	1406.02	0.18	11.67	0.18		
		No pre	10	0.09	1.66	43.31	0.88	0.61	0.99		
			20	0.32	1.43	201.20	0.69	1.38	0.69		
			50	0.86	1.32	1698.47	0.25	4.51	0.25		

in others; however, in general, using all available cuts looks to be the best option, especially if starting from a proper middle ground like p_t, which is strong enough to yield the best final gaps but not too large as to require too much time for its solution. It should be noted, however, that again the returns of using cuts seems to be somewhat diminishing with the size.

Finally, the LR approach has a very specific set of strengths and weaknesses. By not being (currently) capable of (easily) exploiting preprocessing and general-purpose cuts, it does not yield the overall better gaps; note, however, that by doing convexification it automatically achieves the same bound as that provided by the Perspective Reformulation without any need of the corresponding P/C. Yet, it is very competitive in running times, save against the *3-bin* formulation without cuts, which however produces way worse bounds. Furthermore, as expected from a decomposition algorithm, the computational cost tends to scale at most linearly with the number of units m, as opposed to the MILP approaches whose growth is clearly superlinear (and, of course, way worse for *DP*). This is due to the fact that the size of the dual space, n, does not depend on the number of units; therefore, the number of iterations remain similar as m grows. Since one fundamental driver for iterations cost—having to solve more subproblems—is clearly linear in m, this justifies the observed behavior. It should be noted that the cost of the "master problem" in the approach may instead grow superlinearly, as its size in principle grows linearly in m, but this seems not to be relevant in our UC instances. Furthermore, the subproblems may actually be solved in parallel, thereby further reducing the wall-clock time of the solution: this is in fact easily possible in our computational environment by just using `ParallelBundleSolver` instead of `BundleSolver` as the core algorithm in `LagrangianDualSolver`, but for the sake of conciseness we do not further delve on the matter. Anyway, the (sequential) LR approach as it is already provides competitive gaps, especially as the size grows, in a much reduced running time than MILP-based ones, and therefore it is promising for the solution of large-scale UC instances. This would only be improved by the ability of applying general-purpose cuts and preprocessing techniques to it, which is not a methodological issue but "only" a software-related one that `SMS++` will hopefully tackle in the future. Yet, it should be remarked that these figures only relate to one-shot solution of the corresponding relaxations: it is well-known that standard LP techniques (in particular, simplex-based methods) have excellent reoptimization capabilities that allow them to re-solve the problems after a branching or cutting operation in much shorter time than the one required to solve the problem from scratch. This is less obvious for LR approaches, but it does not mean is not doable; in fact, `BundleSolver` implements extensive reoptimization capabilities that may allow it to be competitive in this respect. Further experiments will be necessary to verify the degree to which this is actually true, that are clearly out of the scope of the current presentation.

14.9 Conclusions

This chapter has hopefully illustrated, with the use of an important practical application like the Unit Commitment problem, at least a part of the complexity involved in the construction of efficient approaches to hard optimization problems. In particular for exact approaches (which are naturally turned into heuristic ones by introducing time limits), this typically requires first and foremost to choose "the right formulation" to the model. Yet, this is a highly nontrivial process, as there are in general many possible formulations, possibly very different ones (cf. the Dantzig–Wolfe reformulation underlying the LR approach), some of which may provide distinctly different points in the trade-off curve of bound accuracy (gap) versus bound solution time. Which one is "the right" one is then influenced by many factors pertaining to the fine details of the algorithmic, and more properly software, environment: which solver one is using, which algorithmic parameters are available and how they must be set, … Some approaches, like the LR one, may present very favorable trade-offs in principle (like in the UC case, especially as m grows while n does not) but may still remain at a fundamental disadvantage due to the lack of proper software tools capable of fully exploiting their strengths in combination with other necessary methodologies, like general-purpose cuts, preprocessing, heuristics and so on. Indeed, solution algorithms for complex programs are inherently (and likely unavoidably) complex, and no single component in isolation is sufficient; rather, the harmonious integration of different parts is required. Current general-purpose solvers offer many of these aspects pre-integrated and finely optimized for an effective collaboration; yet, they may not be able to exploit all relevant algorithmic approaches, such as decomposition ones.

All in all, finding the right combination of formulation, solving environment and algorithmic parameters that provide the most efficient solution process for (each given instance of) a hard optimization problem is a demanding process that requires specific, highly nontrivial skills. Making this process simpler would therefore be useful to allow more complex real-world optimization problems to be efficiently solved than is currently possible due to the shortage (and cost) of properly skilled personnel. Several different aspects would need to be significantly improved for a substantial breakthrough in this respect to be achieved: allowing easier combination of diverse specialized solution approaches (what `SMS++` aims at providing), making it possible to automatically choose "good" settings of the very many algorithmic parameters of the corresponding complex approaches [36], and especially being able to efficiently traverse the (huge) space of possible formulations to identify "the best" one, a task that is in principle possible [31] but for which there is nowadays, unfortunately, hardly any support. More work is needed to provide domain experts requiring the use of optimization techniques—without being optimization experts—with dependable tools for constructing efficient solution approaches for their problem; so far, the intervention of optimization experts capable of disentangling all the complexities of the solution process is still required, and this may remain true for a long time. We hope that this chapter provides a useful reference for the current and next generations of those called to perform such a challenging but highly necessary task.

Acknowledgements This study received funding from the European Union - Next-GenerationEU - National Recovery and Resilience Plan (NRRP) - MISSION 4 COMPONENT 2, INVESTIMENT N. 1.1, CALL PRIN 2022 D.D. 104 02-02-2022, project title "Large-scale optimization for sustainable and resilient energy systems", CUP I53D23002310006.

A Appendix: An Example of the DP Algorithm

We now provide a complete example of the DP algorithm on the DP graph in Fig. 14.3, i.e., with $n = |\mathcal{T}| = 7$, $\tau^+ = \tau^- = 3$, and $\tau^0 = 3$; that is, the unit is ON at time 0. We are also assuming that the initial power is $p_0 = 40$, i.e., equal to the shut down limit: together with the fact that the unit has already been ON for the minimum up-time, this implies that it can go OFF immediately if desired, as well as remain ON.

A.1 A Complete Description of the DP Graph

The DP graph, as described in Sect. 14.4.3 (and as depicted in Fig. 14.3), has the following $2n + 2 = 16$ nodes:

- the 7 ON_t nodes for $t = 1, \ldots, 7$;
- the 7 OFF_t nodes for $t = 1, \ldots, 7$;
- the single source node s and sink node d.

The arcs of the graph are logically subdivided into 4 groups:

- The 8 *ON arcs* outgoing from s, i.e., (s, ON_t) for $t = 1, \ldots, 7$ and (s, d). Indeed, since the unit is ON at time 0 and $\tau^0 \geq \tau^+$, it is possible to stay ON and switch off at any time from 1 to 7, as well as to remain ON without ever switching OFF (arc $ON(s, d)$). Recall that the unit is ON at time periods associated to both extremes of an ON arc.
- The 5 *OFF arcs* outgoing from s, i.e., (s, OFF_t) for $t = 4, \ldots, 7$ plus the OFF arc $OFF(s, d)$, denoting that the unit is immediately switched OFF and not switched ON again in the time horizon. Indeed, since the unit is ON at time 0 and $\tau^0 \geq \tau^+$, it is possible to immediately switch OFF the unit. In this case, the unit must be OFF for at least $\tau^- = 3$, that is, at least in time instants 1, 2, and 3. Recall that the unit is OFF on the time instants that are *strictly* contained in the extremes of the OFF arc; say, arc (s, OFF_6) means that the unit is OFF at time instants from 1 to 5 (which is allowed since these are 5 instants, i.e., more than $\tau^- = 3$).
- The remaining 7 *ON arcs* (OFF_4, ON_6), (OFF_4, ON_7), (OFF_4, d), (OFF_5, ON_7), (OFF_5, d), (OFF_6, d), (OFF_7, d) outgoing by the OFF nodes. Note that none of these arcs starts from OFF_1, OFF_2, OFF_3, because the unit is initially ON, and even if it is immediately switched OFF (which is possible), it must remain OFF at least at the time instants 1, 2, and 3, hence OFF_1, OFF_2, OFF_3 cannot be nodes where the unit is again switched ON.
- The remaining 13 *OFF arcs* (ON_1, OFF_5), (ON_1, OFF_6), (ON_1, OFF_7), (ON_1, d), (ON_2, OFF_6), (ON_2, OFF_7), (ON_2, d), (ON_3, OFF_7), (ON_3, d), (ON_4, d), (ON_5, d), (ON_6, d), (ON_7, d) outgoing from the ON nodes. Recall that, e.g., arc (ON_1, OFF_5) denotes that the unit—being ON at time 1—is OFF at time instants 2, 3, and 4 and is again switched ON at time instant 5. Note also that there is an arc from each node ON to d, representing that the unit is switched OFF and it will not be switched ON again until (possibly) in the next planning horizon.

In our example, we assume a simple version of time-dependent start-up cost whereby the cost is 100 if the unit is OFF for exactly 3 time instants (the minimum), and it is 150 if the unit is OFF for more than 3 time instants. As a consequence, the OFF arcs:

- (s, OFF_4), (ON_1, OFF_5), (ON_2, OFF_6), and (ON_3, OFF_7) will be assigned start-up cost 100;
- (s, OFF_5), (s, OFF_6), (s, OFF_7), (ON_1, OFF_6), (ON_1, OFF_7), (ON_2, OFF_7), will be assigned start-up cost 150;

- (ON_1, d), (ON_2, d), (ON_3, d), (ON_4, d), (ON_5, d), (ON_6, d), (ON_7, d), and the OFF arc $OFF(s, d)$ will be assigned start-up cost 0.

In order to define the generation costs, we take:

- the quadratic function as $f_t(p_t) = 0.0023p_t^2 + 4.1271p_t$;
- the selling prices of energy π_t at time t (also corresponding to the dual multipliers associated with demand constraints at time period t) as, respectively, $\pi_1 = 4.8883, \pi_2 = 4.9004, \pi_3 = 2.335, \pi_4 = 8.5049, \pi_5 = 8.6616, \pi_6 = 9.0161$, and $\pi_7 = 8.982$;
- the fixed cost as $c = 134.432$.

These are realistic values taken from instances from the literature. Finally, the other necessary data is:

- Minimum and maximum power: $l = 40$, $u = 107$;
- Maximum ramp-up and ramp-down rate: $\Delta^+ = 42.8$, $\Delta^- = 34.8$;
- Maximum start-up and shut-down limit: $\bar{l} = 48$, $\bar{u} = 40$.

A.2 Computation of the Costs of ON Arcs

Now that all the data is specified, it is possible to assign the costs to the ON arcs. For compactness we only report the first four digits (at most) after the point of the data and the results.

- (s, ON_1): besides the fixed cost for one instant ($c = 134.432$), the cost requires the "solution" of the following "problem" (ED^{s1})

$$\begin{aligned} \min\ & 0.0023p_1^2 + 4.1271p_1 - 4.8883p_1 = 0.0023p_1^2 - 0.7612p_1 \\ \text{s.t. } & 40 \leq p_1 \leq 40 \end{aligned}$$

  i.e., $p_1 = 40$, yielding a total generation cost of $c + z^{s1} = 134.432 + 0.0023 \times 40^2 - 0.7612 \times 40 = 107.664$.
- (s, ON_2): besides the fixed cost for two instants ($2 \times c = 268.864$), the cost requires the solution of the following problem (ED^{s2})

$$\begin{aligned} \min\ & 0.0023p_1^2 + 4.1271p_1 + 0.0023p_2^2 + 4.1271p_2 - 4.8883p_1 - 4.9004p_2 \\ \text{s.t. } & 40 \leq p_1 \leq 107\ ,\ 40 \leq p_2 \leq \min\{107, 40\} \\ & p_1 - p_0 \leq 42.8 \iff p_1 \leq 42.8 + p_0 = 82.8 \\ & p_0 - p_1 \leq 34.8 \iff p_1 \geq p_0 - 34.8 = 5.2 \\ & p_2 - p_1 \leq 42.8\ ,\ p_1 - p_2 \leq 34.8 \end{aligned}$$

  that after easy simplifications becomes:

$$\begin{aligned}&\min 0.0023p_1^2 - 0.7612p_1 + 0.0023p_2^2 - 0.7733p_2\\&\text{s.t. } 40 \le p_1 \le 82.8\,,\ p_2 = 40\end{aligned}$$

It is then easy to verify that the optimal solution is $p_1 = 74.8$, $p_2 = 40$ with cost $z^{s2} = -71.3212$, yielding a total generation cost of 197.5428.

- (s, ON_3): besides the fixed cost for three instants ($3 \times c = 403.296$), the cost requires the solution of the following problem (ED^{s3})

$$\begin{aligned}\min\ & 0.0023p_1^2 + 4.1271p_1 + 0.0023p_2^2 + 4.1271p_2 + 0.0023p_3^2 + 4.1271p_3\\&- 4.8883p_1 - 4.9004p_2 - 2.335p_3\\\text{s.t. }& 40 \le p_1 \le 107\,,\ 40 \le p_2 \le 107\,,\ 40 \le p_3 \le \min\{107\,,\ 40\}\\&p_1 - p_0 \le 42.8\,,\ p_2 - p_1 \le 42.8\,,\ p_3 - p_2 \le 42.8\\&p_0 - p_1 \le 34.8\,,\ p_1 - p_2 \le 34.8\,,\ p_2 - p_3 \le 34.8\end{aligned}$$

that after easy simplifications becomes:

$$\begin{aligned}\min\ & 0.0023p_1^2 - 0.7612p_1 + 0.0023p_2^2 - 0.7733p_2 + 0.0023p_3^2 + 1.792p_3\\\text{s.t. }& 40 \le p_1 \le 82.8\,,\ 40 \le p_2 \le 74.8\,,\ p_3 = 40\\&p_2 - p_1 \le 42.8\,,\ p_1 - p_2 \le 34.8\end{aligned}$$

This problem has (after reduction) only 2 variables, and it can be easily verified that its solution is $p_1 = 82.8$, $p_2 = 74.8$, $p_3 = 40$, whose cost $z^{s3} = -16.8692$ yields a total generation cost of 386.4268.

- (s, ON_4): besides the fixed cost for four instants ($4 \times c = 537.728$), the cost requires the solution of the following problem (ED^{s4})

$$\begin{aligned}\min\ & \textstyle\sum_{t=1}^{4} 0.0023p_t^2 + 4.1271p_t - 4.8883p_1 - 4.9004p_2 \quad -2.335p_3 - 8.5049p_4\\\text{s.t. }& 40 \le p_t \le 107 && t = 1, \ldots, 4\\&p_t - p_{t-1} \le 42.8 && t = 1, \ldots, 4\\&p_{t-1} - p_t \le 34.8 && t = 1, \ldots, 4\\&p_4 \le 40\end{aligned}$$

We leave as an exercise to the reader the possible reductions, however the above problem can be fed to a QP solver as it is, obtaining optimal solution $p_1 = 82.8$, $p_2 = 74.8$, $p_3 = p_4 = 40$ with cost $z^{s4} = -188, 3012$, thus yielding a total generation cost of 349.4268.

- (s, ON_5): besides the fixed cost for five instants ($5 \times c = 672.16$), the cost requires the solution of the following problem (ED^{s5})

$$\begin{aligned} \min\ & \sum_{t=1}^{5} 0.0023p_t^2 + 4.1271p_t \\ & - 4.8883p_1 - 4.9004p_2 - 2.335p_3 - 8.5049p_4 - 8.6616p_5 \\ \text{s.t. } & 40 \le p_t \le 107 && t = 1, \dots, 5 \\ & p_t - p_{t-1} \le 42.8 && t = 1, \dots, 5 \\ & p_{t-1} - p_t \le 34.8 && t = 1, \dots, 5 \\ & p_5 \le 40 \end{aligned}$$

whose optimal solution $p_1 = 82.8$, $p_2 = 74.8$, $p_3 = 40$, $p_4 = 74.8$, $p_5 = 40$ has cost $z^{s5} = -509.16$, thus yielding a total generation cost of 163.

- (s, ON_6): besides the fixed cost for six instants ($6 \times c = 806.592$), the cost requires the solution of the following problem (ED^{s6})

$$\begin{aligned} \min\ & \sum_{t=1}^{6} 0.0023p_t^2 + 4.1271p_t \\ & - 4.8883p_1 - 4.9004p_2 - 2.335p_3 - 8.5049p_4 \quad -8.6616p_5 - 9.0161p_6 \\ \text{s.t. } & 40 \le p_t \le 107 && t = 1, \dots, 6 \\ & p_t - p_{t-1} \le 42.8 && t = 1, \dots, 6 \\ & p_{t-1} - p_t \le 34.8 && t = 1, \dots, 6 \\ & p_6 \le 40 \end{aligned}$$

whose optimal solution $p_1 = 82.8, p_2 = 99, p_3 = 64.2, p_4 = 107, p_5 = 74.8, p_6 = 40$ has cost $z^{s6} = -937.0246$, thus yielding a total generation cost of -130.4326.

- (s, ON_7): besides the fixed cost for seven instants ($7 \times c = 941.024$), the cost requires the solution of the following problem (ED^{s7})

$$\begin{aligned} \min\ & \sum_{t=1}^{7} 0.0023p_t^2 + 4.1271p_t - 4.8883p_1 \\ & - 4.9004p_2 - 2.335p_3 - 8.5049p_4 - 8.6616p_5 \quad -9.0161p_6 - 8.982p_7 \\ \text{s.t. } & 40 \le p_t \le 107 && t = 1, \dots, 7 \\ & p_t - p_{t-1} \le 42.8 && t = 1, \dots, 7 \\ & p_{t-1} - p_t \le 34.8 && t = 1, \dots, 7 \\ & p_7 \le 40 \end{aligned}$$

whose optimal solution $p_1 = 82.8, p_2 = 99, p_3 = 64.2, p_4 = p_5 = 107, p_6 = 74.8$, $p_7 = 40$ has cost $z^{s7} = -1421.0396$, thus yielding a total generation cost of -480.0156.

- ON arc $ON(s, d)$: the cost on ON arc $ON(s, d)$ differs from the one of arc (s, ON_7) only because the shut-down limit constraint $p_7 \le 40$ is not present in (ED^{sd}); this changes the optimal solution in $p_1 = 82.8$, $p_2 = 99$, $p_3 = 64.2$, $p_4 = p_5 = p_6 = p_7 = 107$, with cost $z^{sd} = -1867.6233$, and therefore the total generation cost in -926.5993.
- (OFF_4, ON_6): the arc covers $6 - 4 + 1 = 3$ instants, and therefore it has a fixed cost of $3 \times c = 403.296$ plus the optimal value of (ED^{46})

$$
\begin{aligned}
\min\ & \textstyle\sum_{t=4}^{6} 0.0023p_t^2 + 4.1271p_t - 8.5049p_4 - 8.6616p_5 & -9.0161p_6 \\
\text{s.t. } & 40 \le p_t \le 107 & t = 4, 5, 6 \\
& p_t - p_{t-1} \le 42.8 & t = 5, 6 \\
& p_{t-1} - p_t \le 34.8 & t = 5, 6 \\
& p_4 \le 48\,,\ p_6 \le 40
\end{aligned}
$$

whose solution is $p_4 = 48$, $p_5 = 74.8$, $p_6 = 40$, with cost $z^{46} = -723.0272$, thus yielding a total generation cost of -319.7312.

- (OFF_4, ON_7): the arc covers $7 - 4 + 1 = 4$ instants, and therefore it has a fixed cost of $4 \times c = 537.728$ plus the optimal value of (ED^{47})

$$
\begin{aligned}
\min\ & \textstyle\sum_{t=4}^{7} 0.0023p_t^2 + 4.1271p_t \\
& - 8.5049p_4 - 8.6616p_5 - 9.0161p_6 - 8.982p_7 \\
\text{s.t. } & 40 \le p_t \le 107 & t = 4, \ldots, 7 \\
& p_t - p_{t-1} \le 42.8 & t = 5, 6, 7 \\
& p_{t-1} - p_t \le 34.8 & t = 5, 6, 7 \\
& p_4 \le 48\,,\ p_7 \le 40
\end{aligned}
$$

whose solution is $p_4 = 48, p_5 = 90.8, p_6 = 74.8, p_7 = 40$, with cost $z^{47} = -1140.9497$, thus yielding a total generation cost of -603.2217.

- (OFF_4, d): the cost on arc (OFF_4, d) differs from the one of arc (OFF_4, ON_7) only because the shut-down limit constraint $p_7 \le 40$ is not present in (ED^{4d}); this changes the optimal solution in $p_4 = 48, p_5 = 90.8, p_6 = p_7 = 107$, with cost $z^{4d} = -1587.5370$, and therefore the total generation cost in -1049.8090.
- (OFF_5, ON_7): the arc covers $7 - 5 + 1 = 3$ instants, and therefore it has a fixed cost of $3 \times c = 403.296$ plus the optimal value of (ED^{57})

$$
\begin{aligned}
\min\ & \textstyle\sum_{t=5}^{7} 0.0023p_t^2 + 4.1271p_t - 8.6616p_5 - 9.0161p_6 - 8.982p_7 \\
\text{s.t. } & 40 \le p_t \le 107 & t = 5, 6, 7 \\
& p_t - p_{t-1} \le 42.8 & t = 6, 7 \\
& p_{t-1} - p_t \le 34.8 & t = 6, 7 \\
& p_5 \le 48\,,\ p_7 \le 40
\end{aligned}
$$

whose solution is $p_5 = 48$, $p_6 = 74.8$, $p_7 = 40$, with cost $z^{57} = -755.7014$, thus yielding a total generation cost of -352.4054.

- (OFF_5, d): the cost on arc (OFF_5, d) differs from the one of arc (OFF_5, ON_7) only because the shut-down limit constraint $p_7 \le 40$ is not present in (ED^{5d}); this changes the optimal solution in $p_5 = 48$, $p_6 = 90.8$, $p_7 = 107$, with cost $z^{5d} = -1130.4569$, and therefore the total generation cost in -727.1609.
- (OFF_6, d): the arc covers 2 instants, and therefore it has a fixed cost of $2 \times c = 268.864$ plus the optimal value of (ED^{6d})

$$\begin{aligned}&\min\ 0.0023p_6^2+4.1271p_6+0.0023p_7^2+4.1271p_7-9.0161p_6-8.982p_7\\&\text{s.t. } 40\le p_6\le 48\,,\ 40\le p_7\le 107\\&\qquad p_7-p_6\le 42.8\,,\ p_6-p_7\le 34.8\end{aligned}$$

whose solution is $p_6 = 48$, $p_7 = 90.8$, with cost $z^{6d} = -651.235$, thus yielding a total generation cost of -382.3710.

- (OFF_7, d): besides the fixed cost $c = 134.432$, we need to compute the solution to the (very simple) (ED^{7d})

$$\min\{\,0.0023p_7^2+4.1271p_7-8.982p_7\ :\ 40\le p_7\le 48\,\}$$

which is $p_7 = 48$, with cost $z^{7d} = -227.736$, thus yielding a total generation cost of -93.3040.

We summarize the computed arc costs in Table 14.2.

Table 14.2 Arc costs

ON arcs		OFF arcs	
Arc	Cost	Arc	Cost
(s, ON_1)	107.6640	(s, OFF_4)	100
(s, ON_2)	197.5428	(s, OFF_5)	150
(s, ON_3)	386.4268	(s, OFF_6)	150
(s, ON_4)	349.4268	(s, OFF_7)	150
(s, ON_5)	163.0000	$OFF(s, d)$	0
(s, ON_6)	–130.4326	(ON_1, OFF_5)	100
(s, ON_7)	-480.0156	(ON_1, OFF_6)	150
$ON(s, d)$	-926.5993	(ON_1, OFF_7)	150
(OFF_4, ON_6)	-319.7312	(ON_1, d)	0
(OFF_4, ON_7)	-603.2217	(ON_2, OFF_6)	100
(OFF_4, d)	–1049.8090	(ON_2, OFF_7)	150
(OFF_5, ON_7)	-352.4054	(ON_2, d)	0
(OFF_5, d)	-727.1609	(ON_3, OFF_7)	100
(OFF_6, d)	-382.3710	(ON_3, d)	0
(OFF_7, d)	-93.3040	(ON_4, d)	0
		(ON_5, d)	0
		(ON_6, d)	0
		(ON_7, d)	0

A.3 Final Solution of the DP

We can now compute the optimal solution to our problem by finding the shortest s-d path on the above described acyclic graph with the computed weight. This is obtained by visiting the graph in any topological order, i.e., a node is only visited after all the nodes from which it is reachable. In the case of a graph depending on a temporal index like our own a topological order is immediately established: s is the first node, then nodes ON_t and OFF_t (in any order between them) for $t = 1, \ldots, 7$, and finally node d. All nodes are initially given a label $+\infty$ (save s that is given label 0) and no predecessor node (save s that is given itself as predecessor). When a node i with label l_i is visited, for all nodes j such that the arc (i, j) exists in the graph the tentative value $l_i + c_{ij}$ is computed: if this is lower than the current label l_j then l_j is changed to $l_i + c_{ij}$ and the predecessor of j is set to i, otherwise nothing is done. At the end of the process, the optimal path can be reconstructed starting from d and following backward the established predecessor function until s is reached.

We now succinctly describe the iterations of the algorithm:

- We start by visiting node s, which changes the labels of all nodes that are directly reached from it via an arc (i.e., all but OFF_1, OFF_2 and OFF_3) since their current labels (∞) are surely larger than the tentative values, producing the labeling of the nodes as in Table 14.3.
- We then visit node ON_1, from which there are OFF arcs leading to nodes OFF_5, OFF_6, OFF_7 and d. The corresponding tentative values are $107.664 + 100 = 207.664$ for node OFF_5, $107.664 + 150 = 257.664$ for node OFF_6 and node

Table 14.3 Labels after visiting node s

Node	Label	Predecessor
ON_1	107.6640	s
OFF_1	∞	None
ON_2	197.5428	s
OFF_2	∞	None
ON_3	386.4268	s
OFF_3	∞	None
ON_4	349.4268	s
OFF_4	100	s
ON_5	163.0000	s
OFF_5	150	s
ON_6	–130.4326	s
OFF_6	150	s
ON_7	–480.0156	s
OFF_7	150	s
d	–926.5993	s

Table 14.4 Labels after visiting node OFF_4

Node	Label	Predecessor
ON_1	107.6640	s
OFF_1	∞	None
ON_2	197.5428	s
OFF_2	∞	None
ON_3	386.4268	s
OFF_3	∞	None
ON_4	349.4268	s
OFF_4	100	s
ON_5	163.0000	s
OFF_5	150	s
ON_6	–219.7312	OFF_4
OFF_6	150	s
ON_7	–503.2217	OFF_4
OFF_7	150	s
d	–949.8090	OFF_4

OFF_7, and $107.664 + 0 = 107.664$ for node d. None of the tentative values are less than the existing node (set by s), and therefore none of the labels will be updated.

- We then visit node OFF_1, that has a label of ∞ (not being reachable from s) and that therefore can never change any label.
- Similar situations occur when visiting nodes ON_2, OFF_2, ON_3, OFF_3, and ON_4.
- The next node to visit is OFF_4, from which there are three arcs leading to nodes ON_6, ON_7, and d. The corresponding tentative values are $100 - 319.7312 = -219.7312$ for node ON_6, $100 - 603.2217 = -503.2217$ for node ON_7, and $100 - 1049.8090 = -949.8090$ for node d. These values produce the update of labels and of the predecessor for all three nodes ON_6, ON_7, and d, yielding the labeling of the nodes reported in Table 14.4.
- The next node to visit is ON_5, from which there is only an arc to node d with weight 100: the corresponding tentative value is $163 + 100 = 263$, so no update is performed.
- The next node to visit is OFF_5, from which there are two arcs leading to nodes ON_7 and d. The corresponding tentative values are $150 - 352.4054$ for node ON_7 and $150 - 727.1609$ for node d, so no update is performed.
- It is easy to verify that, similarly, when visiting ON_6, OFF_6, ON_7 and OFF_7 no update is performed. There is no need of visiting d since it has no outgoing arcs (and anyway it is the destination node).

Hence, the shortest path from s to d is $s \rightarrow OFF_4 \rightarrow d$, with cost -949.8090 (the label of d at termination). This is associated with the solution of the problem $p_1 =$

$p_2 = p_3 = 0$, $p_4 = 48$, $p_5 = 90.8$, $p_6 = p_7 = 107$. Indeed, the first three values are fixed due to the use of the OFF arc (s, OFF_4), meaning that the unit shuts down immediately at $t = 1$ and remains OFF until $t = 4$ (excluded). The remaining four values have been found when (ED^{4d}) has been solved to compute the cost of the ON arc (OFF_4, d), as discussed above.

Note that this would seem to imply that one has to store all the solutions of all the $O(n^2)$ (ED^{hk}) problems, which would require $O(n^3)$ overall. However, the computation of the solutions need only be done after the shortest path has been computed, and therefore can be done efficiently only for the ON arcs in the optimal path in $O(n)$ (memory) overall; the interested reader can consult [22] for details, or study the `ThermalUnitDPSolver` in the `UCBlock` repository of the `SMS++` project [1].

References

1. SMS++: A Structured Modeling System for mathematical models. https://gitlab.com/smspp/smspp-project
2. Bacci T, Frangioni A, Gentile C (2021) Start-up/shut-down MINLP formulations for the unit commitment with ramp constraints. In: Gentile C, Stecca G, Ventura P (eds) Graphs and combinatorial optimization: From theory to applications—CTW2020 proceedings. AIRO-Springer series, vol 5, pp 335–347
3. Bacci T, Frangioni A, Gentile C, Tavlaridis-Gyparakis K (2024, to appear) New MINLP formulations for the unit commitment problem with ramping constraints. Oper Res
4. Bienstock D (2013) Progress on solving power flow problems. Optima 93:1–8
5. Bienstock D, Escobar M, Gentile C, Liberti L (2022) Mathematical programming formulations for the alternating current optimal power flow problem. Ann Oper Res 314:277–315
6. Borghetti A, Frangioni A, Lacalandra F, Nucci CA (2003) Lagrangian heuristics based on disaggregated bundle methods for hydrothermal unit commitment. IEEE T Power Syst 18:313–323
7. Borghetti A, Frangioni A, Lacalandra F, Nucci CA, Pelacchi P (2003) Using of a cost-based unit commitment algorithm to assist bidding strategy decisions. In: Borghetti A, Nucci CA, Paolone M (eds) Proceedings IEEE 2003 Powerteck Bologna conference, p 547
8. Brandenberg R, Huber M, Silbernagl M (2017) The summed start-up costs in a unit commitment problem. EURO J Comput Optim 5(1–2):203–238
9. Crainic TG, Frangioni A, Gendron B (2001) Bundle-based relaxation methods for multicommodity capacitated fixed charge network design problems. Discret Appl Math 112:73–99
10. Fattahi S, Ashraphijuo M, Lavaei J, Atamtürk A (2017) Conic relaxations of the unit commitment problem. Energy 134:1079–1095
11. Fioriti D, Frangioni A, Poli D (2021) Optimal sizing of energy communities with fair revenue sharing and exit clauses: Value, role and business model of aggregators and users. Appl Energy 299:117328
12. Frangioni A (2002) Generalized bundle methods. SIAM J Optim 13(1):117–156
13. Frangioni A (2005) About Lagrangian methods in integer optimization. Ann Oper Res 139(1):163–193
14. Frangioni A (2020) Standard bundle methods: Untrusted models and duality. In: Bagirov A M, Gaudioso M, Karmitsa N, Mäkelä M, Taheri S (eds) Numerical nonsmooth optimization: State of the art algorithms, Chap. 2. Springer, Berlin, pp 61–116
15. Frangioni A, Furini F, Gentile C (2016) Approximated perspective relaxations: A project and lift approach. Comput Optim Appl 63(3):705–735

16. Frangioni A, Furini F, Gentile C (2017) Improving the approximated projected perspective reformulation by dual information. Oper Res Lett 45:519–524
17. Frangioni A, Gallo G (1999) A bundle type dual-ascent approach to linear multicommodity min cost flow problems. INFORMS J Comput 11(4):370–393
18. Frangioni A, Gendron B (2009) 0–1 reformulations of the multicommodity capacitated network design problem. Discret Appl Math 157(6):1229–1241
19. Frangioni A, Gendron B (2013) A Stabilized Structured Dantzig-Wolfe Decomposition Method. Math Program 140:45–76
20. Frangioni A, Gendron B, Gorgone E (2017) On the computational efficiency of subgradient methods: A case study with Lagrangian bounds. Math Program Comput 9(4):573–604
21. Frangioni A, Gentile C (2006) Perspective cuts for a class of convex 0–1 mixed integer programs. Math Program 106(2):225–236
22. Frangioni A, Gentile C (2006) Solving Nonlinear Single-Unit Commitment Problems with Ramping Constraints. Oper Res 54(4):767–775
23. Frangioni A, Gentile C (2009) A computational comparison of reformulations of the perspective relaxation: SOCP versus cutting planes. Oper Res Lett 37(3):206–210
24. Frangioni A, Gentile C (2015) New MIP formulations for the single-unit commitment problems with ramping constraints. Research Report 15-06, IASI–CNR
25. Frangioni A, Gentile C, Grande E, Pacifici A (2011) Projected perspective reformulations with applications in design problems. Oper Res 59(5):1225–1232
26. Frangioni A, Gentile C, Lacalandra F (2008) Solving unit commitment problems with general ramp contraints. Int J Electri Power Energy Syst 30:316–326
27. Frangioni A, Gentile C, Lacalandra F (2009) Tighter approximated MILP formulations for unit commitment problems. IEEE T Power Syst 24(1):105–113
28. Frangioni A, Gentile C, Lacalandra F (2011) Sequential Lagrangian-MILP approaches for unit commitment problems. Int J Electri Power Energy Syst 33:585–593
29. Frangioni A, Gorgone E (2014) Generalized bundle methods for sum-functions with "easy" components: Applications to multicommodity network design. Math Program 145(1):133–161
30. Frangioni A, Lodi A, Rinaldi G (2005) New approaches for optimizing over the semimetric polytope. Math Program 104(2–3):375–388
31. Frangioni A, Perez Sanchez L (2010) Searching the best (formulation, solver, configuration) for structured problems. In: Aiguier M, Bretaudeau F, Krob D (eds) Complex systems design and management: proceedings of the first international conference on complex systems design and management CSDM 2010. Springer, Berlin, pp 85–97
32. Garver LL (1962) Power generation scheduling by integer programming—development and theory. Trans Am Inst Electr Eng Part III: Power Appl Syst 81(3):730–734
33. Gentile C, Morales-España G, Ramos A (2017) A tight MIP formulation of the unit commitment problem with start-up and shut-down constraints. EURO J Comput Optim 5(1):117–201
34. Guignard M, Kim S (1987) Lagrangian decomposition: A model yelding stronger Lagrangian bounds. Math Program 39:215–228
35. Hadjidimitriou NS, Frangioni A, Koch T, Lodi A (2020) Mathematical optimization for efficient and robust energy networks, volume 4 of AIRO springer series. Springer International Publishing
36. Iommazzo G, d'Ambrosio C, Frangioni A, Liberti L (2020) Learning to configure mathematical programming solvers by mathematical programming. In: Kotsireas IS, Pardalos PM (eds) Learning and intelligent optimization—LION 2020, volume 12096 of Lecture notes in computer science. Springer, Berlin, pp 377–389
37. Knueven B, Ostrowski J, Watson J (2020) On mixed-integer programming formulations for the unit commitment problem. INFORMS J Comput 32(4):857–876
38. Lee J, Leung J, Margot F (2004) Min-up/min-down polytopes. Discret Optim 1(1):77–85
39. Malkin P, Wolsey LA (2004) Minimum runtime and stoptime polyhedra. Report, CORE, Université catholique de Louvain
40. Nowak MP, Römisch W (2000) Stochastic Lagrangian relaxation applied to power scheduling in a hydro-thermal system under uncertainty. Ann Oper Res 100:251–272

41. Ostrowski J, Anjos MF, Vannelli A (2012) Tight mixed integer linear programming formulations for the unit commitment problem. IEEE T Power Syst 27(1):39–46
42. Rajan D, Takriti S (2005) Minimum Up/Down polytopes of the unit commitment problem with start-up costs. Research Report RC23628, IBM
43. Scuzziato MR, Finardi EC, Frangioni A (2018) Comparing spatial and scenario decomposition for stochastic hydrothermal unit commitment problems. IEEE Trans Sustain Energy 9(3):1307–1317
44. Silbernagl M, Huber M, Brandenberg R (2016) Improving accuracy and efficiency of start-up cost formulations in MIP unit commitment by modeling power plant temperatures. IEEE T Power Syst 31(4)
45. Taktak R, d'Ambrosio C (2017) An overview on mathematical programming approaches for the deterministic unit commitment problem in hydro valleys. Energy Syst 8:57–79
46. van Ackooij W, Danti Lopez I, Frangioni A, Lacalandra F, Tahanan M (2018) Large-scale unit commitment under uncertainty: An updated literature survey. Ann Oper Res 271(1):11–85

Chapter 15
Bilevel Optimization: Applications, Models and Solution Approaches

Sachin Jayaswal and Ankur Sinha

Abstract Bilevel optimization is a difficult class of optimization problems, which contains an inner optimization problem as a constraint in an outer optimization problem. Such optimization problems are commonly referred to as Stackelberg games in the area of game theory, where a hierarchical interaction between a leader and a follower is modeled. This chapter presents several examples of bilevel optimization problems arising in various contexts, e.g., the product line selection problem and the shortest-path interdiction problem. Depending on the context of the problem, the leader and the follower may have the same objective function but with conflicting objectives (max-min in the shortest-path interdiction), or may have different objective functions (as in the product line selection problem). Under this hierarchical setting, the leader tries to optimize their own decision by taking into account the rational response of the follower. Bilevel optimization problems are NP-hard even in the simplest case in which the problems of the leader and the follower are both linear programs. This chapter discusses classical solution approaches that are based on the reformulation of the bilevel problem into a single-level problem. It also discusses several alternate single-level reformulations for the application problems considered in this chapter.

Keywords Bilevel optimization · Shortest-path interdiction · Product line design

15.1 Introduction

The interest in bilevel programming has been rising over the past few decades, primarily because of new applications that are being identified in various domains. For instance, one of the applications of bilevel programming that is receiving attention is in the area of machine learning, where bilevel optimization provides a way to

S. Jayaswal (✉) · A. Sinha
Operations & Decision Sciences, Indian Institute of Management, Ahmedabad 380015, India
e-mail: sachin@iima.ac.in

A. Sinha
e-mail: asinha@iima.ac.in

F. Hamid (ed.), *Optimization Essentials*, International Series in Operations Research & Management Science 353, https://doi.org/10.1007/978-981-99-5491-9_15

optimize hyperparameters that need to be appropriately chosen before the training process starts for the model parameters. The hyperparameters can appear in various forms, such as network architecture, regularization parameters, or learning rate, among others. Some of the studies that have attempted a bilevel approach to handling hyperparameters are Bennett et al. [16], Sinha et al. [63]. The hyperparameter optimization problem is also important for researchers in other domains. For example, in the area of evolutionary computation, the methods contain hyperparameters, a good choice of which impacts the algorithm performance [70]. Another example is from the area of transportation, in the context of a toll-setting problem, which is inherently a bilevel program [19, 45, 52, 56, 65]. Consider a large network of highways operated by the government or a private entity that desires to optimize the tolls for maximizing its revenues. A high toll would deter users from using the network, thereby reducing revenues due to the low volume, while a low amount of the toll would increase the volume but may not be optimal for maximizing revenues. Under such a situation, the government cannot state its own optimization problem by ignoring the highway users; rather a two-level optimization problem that is hierarchical in nature needs to be stated. In such a problem, the upper level is the government, which has the objective to maximize revenues under a set of constraints, with an additional constraint represented by the highway users' optimization problem. The highway users' optimization problem consists of an objective function that minimizes a generalized cost, which may include their travel distances, travel times, and toll costs. Solving the highway users' optimization problem allows the government to figure out their response for any value of toll that is set, which in turn allows the government to compute its revenues. In this problem, at the upper level is the government, which is often referred to as the leader, and the lower level is the set of highway users, which are referred to as the followers. Such scenarios with a leader-follower interaction within an optimization problem are studied in the area of game theory and economics as Stackelberg games [76, 80].

Applications of bilevel optimization exist in almost all domains. For example, in engineering, they are common in structural optimization [22, 75], where the design variables in an integrated system are decomposed into upper-level (system level) and lower-level (sub-system level) design variables. Similarly, a common problem in chemical engineering is deciding the optimal state variables and quantities of reactants, in which the lower-level optimization problem appears as an equilibrium condition requiring entropy functional minimization [24, 35, 59, 74]. In defense, bilevel problems commonly appear as attacker-defender or defender-attacker problems [17, 20, 57, 60, 61, 81]. In the area of business and management, bilevel problems are common in designing optimal tax policies [45, 68], investigation of strategic behavior in deregulated markets [37], agri-business management [18, 84], and supply chain and location problems [44, 77]. The readers may refer to review papers and books for additional information about the bilevel methods and applications [7, 26, 41, 67].

General bilevel problems are NP-hard [36, 79]. Even the linear bilevel problems with continuous variables are not easy, which are often reduced to mixed-integer linear programs (MILP) using Big$M$ constraints. One of the most straightforward ways for handling bilevel programs has been the Karush-Kuhn-Tucker (KKT) approach, where the lower-level optimization problem is replaced by the set of its KKT constraints to reduce the bilevel program to a single-level problem that can be handled by standard solvers. However, the KKT approach can be used only in special cases, where the lower-level problem adheres to certain regularity conditions [12]. Linear bilevel programs [8, 15, 33, 78, 82] and quadratic bilevel problems [3, 9, 30] have been widely solved using this approach. Other approaches to handle bilevel problems include gradient descent [62, 79], trust-region [25, 49, 53], and penalty [1, 2, 38, 43, 72, 83] methods.

Given the difficulty in handling bilevel programs, researchers and practitioners have often resorted to solving bilevel problems using computationally intensive approaches, such as nested search. In such an approach the lower-level optimization problem is solved corresponding to a large number of upper-level decision vectors and a bilevel solution is *searched*. Evolutionary algorithms have been the most widely used approach for implementing nested search [4, 39, 47, 69]. Recently, the algorithmic development on bilevel optimization focused on exploiting two important mappings; namely, the lower-level reaction set mapping, and the lower-level optimal value function mapping [5, 64, 66, 71]. The details of these two mappings are discussed later in this chapter.

The focus of the chapter is on two classes of bilevel optimization problems, namely the product line selection problem, and the shortest-path interdiction problem. However, we begin the chapter by discussing the basics of bilevel optimization that will be useful in solving these two classes of problems. The structure of the chapter is as follows. In Sect. 15.2, a general bilevel problem is discussed, along with its various formulations that can be written with the help of different mappings. This is followed by Sects. 15.3 and 15.4, where the product line selection problem and the shortest-path interdiction problem are discussed, along with their respective solution methodologies. Finally, we conclude in Sect. 15.5. The chapter also includes the AMPL codes[1] and C++ codes[2] to handle the two classes of optimization problems. The codes and the ideas in this chapter can be adapted to solve various classes of bilevel optimization problems.

15.2 An Overview of Bilevel Optimization

A general bilevel program with the upper-level objective function $F : \mathbb{R}^n \times \mathbb{R}^m \to \mathbb{R}$ and lower-level objective function $f : \mathbb{R}^n \times \mathbb{R}^m \to \mathbb{R}$ is given as follows:

[1] Refer to Appendices A and B. Alternatively, refer to https://github.com/sachinjayaswal/Product-Line-Selection-AMPL; https://github.com/sachinjayaswal/Shortest-Path-Interdiction-AMPL.

[2] https://github.com/sachinjayaswal/Product-Line-Selection.

$$\text{“}\min_{x,y}\text{”} \quad F(x, y) \tag{15.1}$$

subject to

$$G_k(x, y) \leq 0, \qquad k = 1, \ldots, K \tag{15.2}$$

$$H_l(x, y) = 0, \qquad l = 1, \ldots, L \tag{15.3}$$

$$y \in \operatorname*{argmin}_y \{f(x, y) : g_i(x, y) \leq 0, i = 1, \ldots, I, h_j(x, y) = 0, j = 1, \ldots, J\} \tag{15.4}$$

where $g_i : \mathbb{R}^n \times \mathbb{R}^m \to \mathbb{R}$ represents the lower-level inequality constraints, $h_j : \mathbb{R}^n \times \mathbb{R}^m \to \mathbb{R}$ represents the lower-level equality constraints, $G_k : \mathbb{R}^n \times \mathbb{R}^m \to \mathbb{R}$ represents the upper-level inequality constraints, and $H_l : \mathbb{R}^n \times \mathbb{R}^m \to \mathbb{R}$ represents the upper-level equality constraints. The upper and lower-level variables, x and y, in the above definition may be continuous, integer, or mixed-integer.

The reason for using double quotes ($\text{“}\min_{x,y}\text{”}$) in the above definition is to highlight that the formulation is ill-defined. In the context of multiple lower-level optimal solutions corresponding to a given upper-level decision vector, the above formulation does not clearly define which solution from the lower-level optimal set should be considered in the upper level. To handle this ambiguity, let us first define the lower-level optimality set. Let $\Psi : \mathbb{R}^n \rightrightarrows \mathbb{R}^m$ be a set-valued mapping that defines the solution set of the lower-level decision-maker for every upper-level decision vector. Mathematically, it is defined as:

$$\Psi(x) = \operatorname*{argmin}_y \{f(x, y) : g(x, y) \leq 0, h(x, y) = 0\}$$

where $g : \mathbb{R}^n \times \mathbb{R}^m \to \mathbb{R}^I$ and $h : \mathbb{R}^n \times \mathbb{R}^m \to \mathbb{R}^J$ represent vector constraints. The ambiguity on the choice of the solution from $\Psi(x)$ for any given x can be handled by specifying the position assumed by the upper-level decision-maker. Below we discuss the two common positions that the upper-level decision-maker assumes in the presence of multiple lower-level optimal solutions for one or more upper-level vectors.

15.2.1 Optimistic Position

If the upper level is optimistic that the lower level will cooperate and choose that solution from its optimal set that is the most favorable to the upper level, then the formulation is said to model its optimistic position. In an optimistic position, the choice function of the follower is given as follows:

$$\psi_o(x) = \operatorname*{argmin}_y \{F(x, y) : y \in \Psi(x)\}$$

This leads to the following bilevel formulation:

$$\begin{aligned}
\min_{x,y} \quad & F(x, y) \\
\text{subject to} \quad & \\
& G(x, y) \leq 0 \\
& H(x, y) \leq 0 \\
& y = \psi_{\mathrm{o}}(x)
\end{aligned}$$

15.2.2 *Pessimistic Position*

If the upper level is pessimistic that the lower level may choose that solution from its optimal set that is the least favorable to the upper level, then the formulation is said to model its pessimistic position [27, 50, 51]. In a pessimistic position, the choice function of the follower is given as follows:

$$\psi_{\mathrm{p}}(x) = \operatorname*{argmax}_{y}\{F(x, y) : y \in \Psi(x)\}$$

This leads to the following bilevel formulation:

$$\begin{aligned}
\min_{x,y} \quad & F(x, y) \\
\text{subject to} \quad & \\
& G(x, y) \leq 0 \\
& H(x, y) \leq 0 \\
& y = \psi_{\mathrm{p}}(x)
\end{aligned}$$

Since the optimistic position is more tractable as compared to the pessimistic position, it is more commonly used in the literature. Therefore, unless specifically mentioned in the context of bilevel programs, an optimistic position is commonly assumed. There are intermediate formulations possible between the optimistic and the pessimistic positions. However, if the lower-level optimization problem is known to have only a single optimal solution for all possible upper-level decision vectors, then these alternative formulations do not arise.

15.2.3 *Single-Level Reformulations*

In this section, we discuss various single-level formulations of a bilevel program. In all these discussions, we assume an optimistic position for the upper-level decision-maker.

15.2.3.1 Using Set-Valued Mapping

Using $\Psi(x)$ that defines the solution set of the lower-level decision-maker for every upper-level decision vector, the bilevel optimization problem can be expressed as a constrained optimization problem as follows:

$$\begin{aligned} &\min_{x,y} \quad F(x, y) \\ &\text{subject to} \\ &\quad G(x, y) \leq 0 \\ &\quad H(x, y) = 0 \\ &\quad y \in \Psi(x) \end{aligned}$$

where Ψ is a parameterized range-constraint for the lower-level decision vector y. The Ψ mapping is unknown a priori and some algorithms rely on the estimation of this mapping to solve the bilevel optimization problem.

15.2.3.2 Using Lower-Level Karush-Kuhn-Tucker Conditions

When the lower-level optimization problem adheres to certain regularity conditions (like Slater's condition, linear independence constraint qualification, Mangasarian-Fromovitz constraint qualification, etc.) and is a convex optimization problem, it can be replaced by its Karush-Kuhn-Tucker (KKT) conditions, thereby reducing the bilevel optimization problem to a single-level constrained optimization problem. In that case, (15.1)–(15.4) can be equivalently rewritten as

$$\min_{x,y,\lambda,\mu} \quad F(x, y) \tag{15.5}$$

subject to

$$G_k(x, y) \leq 0, \qquad k = 1, \ldots, K \tag{15.6}$$

$$H_l(x, y) = 0, \qquad l = 1, \ldots, L \tag{15.7}$$

$$g_i(x, y) \leq 0, \qquad i = 1, \ldots, I \tag{15.8}$$

$$h_j(x, y) = 0, \qquad j = 1, \ldots, J \tag{15.9}$$

$$\nabla_y L(x, y, \lambda, \mu) = 0 \tag{15.10}$$

$$\lambda_i g_i(x, y) = 0, \qquad i = 1, \ldots, I \tag{15.11}$$

$$\lambda_i \geq 0, \qquad i = 1, \ldots, I \tag{15.12}$$

where

$$L(x, y, \lambda, \mu) = f(x, y) + \sum_{i=1}^{I} \lambda_i g_i(x, y) + \sum_{j=1}^{J} \mu_j h_j(x, y) \tag{15.13}$$

The KKT conditions appear as stationary conditions (15.10), complementary slackness conditions (15.11), primal feasibility constraints (15.8)–(15.9), and dual feasibility constraints (15.12). Each of the complementary slackness conditions represents a product of two terms containing different variables, and therefore, induces non-convexity. Similarly, the set of stationary conditions corresponding to a non-linear lower-level problem also introduces non-linear equality constraints. The combination of the two may render the single-level optimization problem into a highly non-linear mixed-integer program. The complementary slackness conditions can be linearized with the help of combinatorial variables (z_i) and large numbers (M_1 and M_2), popularly referred to as Big$M$, as follows:

$$\begin{aligned} &\lambda_i \leq M_1 z_i && i = 1, \ldots, I \\ &g_i(x, y) \geq -M_2(1 - z_i) && i = 1, \ldots, I \end{aligned}$$

A right choice of M_1 and M_2 can make the single-level formulation computationally faster; however, it has been recently shown that choosing tight values for Big$M$ is not a straightforward task even in the context of linear bilevel optimization [42]. The above two constraint sets replace (15.11) since they, together with (15.8) and (15.12), ensure that either λ_i or $g_i(x, y)$ is set to zero. The complementary slackness conditions can also be represented as Special Ordered Sets of type 1 (SOS1), which are products of two variables, at most one of which can be non-zero [13]. Thus, (15.11) can be represented as:

$$s_i = g_i(x, y); \quad \{\lambda_i, s_i\} \text{ is SOS1} \qquad i = 1, \ldots, I$$

Modern solvers handle SOS1-type constraints by automatically reformulating them as mixed-integer programs using the Big$M$ approach discussed above, if appropriate values of Big$M$ can be found, or by branching on the variables in the SOS1 constraints.

15.2.3.3 Using Lower-Level Optimal Value Function

The lower-level optimal value function (φ) is commonly used to state an alternate formulation for the general bilevel optimization problem. The optimal value function $\varphi : \mathbb{R}^n \to \mathbb{R}$ is given as follows:

$$\varphi(x) = \min_{y \in Y}\{f(x, y) : g(x, y) \leq 0, h(x, y) = 0\}$$

This mapping provides the optimal lower-level objective function value for any given upper-level decision vector. Using the φ-mapping, the bilevel problem can be written as follows [87]:

$$\min_{x,y} \quad F(x, y)$$

subject to

$$\begin{aligned}
g_i(x, y) &\leq 0, & i &= 1, \ldots, I \\
h_j(x, y) &= 0, & j &= 1, \ldots, J \\
G_k(x, y) &\leq 0, & k &= 1, \ldots, K \\
H_l(x, y) &= 0, & l &= 1, \ldots, L \\
f(x, y) &\leq \varphi(x)
\end{aligned}$$

Since the value function is almost never known a priori, most of the algorithms that rely on using this mapping estimate it during the iterations of the algorithm. The estimation of this mapping, which is actually a function, is often easier than estimating the $\Psi(x)$ mapping, which returns a set of vectors for any given x.

15.2.3.4 Using Duality-Based Approach

For a convex lower-level program that satisfies strong duality, the bilevel program can be reformulated into a single-level program using the duality of the lower-level optimization problem. In this section, we demonstrate this approach on a general linear bilevel program, but the ideas can be extended to other classes of bilevel problems that satisfy the convexity and strong-duality conditions. Consider a general linear bilevel program with continuous variables. For consistency with the literature on linear programs (LP), we represent the inequality constraints with the "≥" sign for minimization problems.

$$\min_{x,y} \; c_x^\top x + c_y^\top y \tag{15.14}$$

subject to

$$A_x x + A_y y \geq a \tag{15.15}$$

$$y \in \underset{y}{\operatorname{argmin}} \left\{ d^\top y : B_x x + B_y y \geq b \right\} \tag{15.16}$$

with $c_x \in \mathbb{R}^n$, $c_y, d \in \mathbb{R}^m$, $A_x \in \mathbb{R}^{K\times n}$, $A_y \in \mathbb{R}^{K\times m}$, and $a \in \mathbb{R}^K$, $B_x \in \mathbb{R}^{I\times n}$, $B_y \in \mathbb{R}^{I\times m}$, and $b \in \mathbb{R}^I$. We have omitted the linear term corresponding to the upper-level variables in the lower-level objective function without any loss of generality. First, let us write the KKT-based single-level reduction of the above problem.

$$\min_{x,y,\lambda} \; c_x^\top x + c_y^\top y \tag{15.17}$$

$$\text{s.t. } A_x x + A_y y \geq a \tag{15.18}$$

$$B_x x + B_y y \geq b \tag{15.19}$$

$$B_y^\top \lambda = d \tag{15.20}$$

$$\lambda_i \left(B_x + B_y y - b\right)_i = 0 \quad \forall\, i = 1, \ldots, I \tag{15.21}$$

$$\lambda \geq 0 \tag{15.22}$$

The bilinear term in the complementary slackness conditions can be handled as discussed in Sect. 15.2.3.2. Next, let us write the single-level reformulation of the linear bilevel program in (15.14)–(15.16) using the duality-based approach. The dual of the lower-level LP is given as follows:

$$\max_{\lambda} \; (b - B_x x)^\top \lambda \quad \text{s.t.} \quad B_y^\top \lambda = d, \lambda \geq 0$$

Using the principle of weak duality, the following is valid for every feasible primal-dual pair (y, λ).

$$d^\top y \geq (b - B_x x)^\top \lambda$$

Using strong duality, any feasible pair that satisfies the following condition is an optimal solution to the LP.

$$d^\top y \leq (b - B_x x)^\top \lambda$$

Using the above, the linear bilevel program (15.14)–(15.16) can be reformulated as a single-level problem as follows:

$$\min_{x,y,\lambda} \; c_x^\top x + c_y^\top y \tag{15.23}$$

$$\text{s.t. } A_x x + A_y y \geq a, \; B_x x + B_y y \geq b \tag{15.24}$$

$$B_y^\top \lambda = d, \lambda \geq 0 \tag{15.25}$$

$$d^\top y \leq (b - B_x x)^\top \lambda \tag{15.26}$$

The difference between (15.17)–(15.22) and (15.23)–(15.26) is that the set of complementary conditions (15.21) has been replaced with the strong-duality constraint (15.26). The constraint set (15.26) in the duality-based single-level reformulation is a product of variables x and λ, and is therefore bilinear in nature. Such bilinear constraints are not easy to linearize if both x and λ are continuous. One often resorts to McCormick envelopes [55] to handle such bilinear constraints.

Linear bilevel programs in which the upper-level variables are mixed-integers can also be handled using the duality-based approach discussed above. In fact, when x is combinatorial, linearization of the bilinear constraint can be achieved easily using Big$M$ constraints (refer to Sect. 15.3 for an illustration). The duality-based method is often a preferred approach for a number of classes of bilevel problems, for instance, the min-max (or max-min) optimization problems (refer to Sect. 15.4 for details).

15.3 Product Line Selection

New products are commonly introduced in several configurations so as to appeal to different market segments [54]. For example, a refrigerator may be introduced in various capacities (95, 420, 570 L, etc.), number of doors (single or double), and colors (white, gray, red, etc.). Capacity, number of doors, and color are three of the several attributes based on which a refrigerator can be differentiated for the different market segments. Thus, each configuration of a refrigerator can be viewed as a bundle of attributes (e.g., capacity) fixed at some levels (e.g., 570 L). Using just these three attributes, and just two possible levels for each, gives rise to 8 possible configurations. Developing each of these configurations entails some fixed cost. Moreover, developing too many variants can cannibalize each other's sales with the result that some of them may fail to recoup through sales the investments in their development. Firms, therefore, face the problem of selecting a subset from the set of alternative product configurations. This problem is commonly referred to as the product line selection (PLS) problem.

Let us understand the problem using an illustrative example. Assume that a firm can introduce a product in 10 different configurations (bundles of attributes), the development costs for which are given in Table 15.1. The table also provides the profit per unit sale of each of these configurations. The firm has to decide a subset of these configurations to develop so as to maximize its profit from their sales. The sales of each configuration will depend on the customers' preferences among the set of configurations available to them. For this, firms need to predict which one among a set of available configurations will a given customer buy. Conjoint analysis[3] makes this possible by estimating the potential customers' "value system", i.e., their utility from every level of each attribute of a product. For this, the potential customers are asked to make judgments about a small subset of the firm's envisaged product configurations, based on which it recovers the customers' value systems. These value systems are, in turn, used to estimate the customers' utility for any product configuration that can be composed of the basic attribute levels used to represent their value systems. Further, different customer segments can be identified by performing clustering[4] on the customers' value systems.

Suppose the conjoint analysis and the subsequent cluster analysis produce the utility matrix as shown in Table 15.1 for the 5 customer segments identified. Given this utility matrix, the firm can predict which product configuration will be purchased by a given customer segment. There are two alternate rules to translate the utility

[3] Readers interested in the details of conjoint analysis can refer to Green and Krieger [34], Dolan [29].

[4] Readers interested in the details of clustering can refer to Provost and Fawcett [58].

Table 15.1 Product line selection parameters

Firms' parameters			Customer segments and their utilities				
Configurations	Fixed cost	Unit profit	1	2	3	4	5
1	15,000	50	6	9	8	4	10
2	12,000	60	4	3	9	5	4
3	9,000	40	−2	10	7	1	9
4	7,000	55	4	5	7	8	9
5	6,000	35	−4	9	4	2	5
6	5,000	45	6	9	9	6	10
7	6,000	55	7	9	3	7	1
8	10,000	60	9	2	4	7	3
9	8,000	55	8	8	10	−1	−4
10	9,500	45	6	5	10	6	10
Customer segment sizes			6,000	8,500	9,500	7,000	9,000
Customer segment reservation utilities			4	4	4	4	4

matrix into the customers' choice predictions: (i) first (deterministic) choice rule; (ii) probabilistic choice rule [29]. In the first choice rule, a customer is assumed to always buy the product configuration that gives her the highest utility among all the configurations available, provided its utility is no less than her reservation utility.[5] For example, if all the 10 product configurations in Table 15.1 are developed by the firm, then given the utility matrix in Table 15.1, customer segment 1 will choose product 8, while customer segment 2 will choose product 3. If we assume the customers' buying decisions are governed by the first choice rule, and the size of each customer segment is as given in the last row of Table 15.1, then which of these 10 product configurations should the firm develop to maximize its total profit?[6] This is the question the firm's problem of PLS tries to answer. PLS can also be studied using the probabilistic choice rule wherein the probability of buying a particular configuration is proportional to its utility. In the rest of the chapter, the discussion is restricted to the first choice rule.

Clearly, PLS is a Stackelberg game between the firm (leader), who decides the subset of the given product configurations to develop to maximize its profit, and the customers (followers), who decide which among the developed configurations to purchase to maximize their utility. As such, PLS can be mathematically stated as a bilevel optimization problem. To state it formally, the following notation is first introduced.

[5] A customer's reservation utility may represent her utility from some other competing product/brand that she will buy in the absence of any product configuration offered by this firm.

[6] Here, we assume that each customer will buy only one unit of the product configuration in the planning horizon for which the firm is developing its PLS. This is not an unrealistic assumption for consumer durable goods like refrigerators.

Indices and Parameters:

P	:	Set of product configurations
p, q	:	Indices for product configurations; $p, q \in P$
S	:	Set of customer segments
s	:	Index for customer segment; $s \in S$
f_p	:	Fixed cost of developing product configuration p
π_p	:	Unit profit from product configuration p
k_s	:	Size of market segment s
u_{sp}	:	Utility of customer segment s from product configuration p
u_{s0}	:	Reservation utility of customer segment s

Decision Variables:

x_p	:	1 if product configuration p is developed, 0 otherwise
y_{sp}	:	1 if customer segment s buys product configuration p, 0 otherwise

Using the above notation, the bilevel optimization model of PLS can be stated as follows:

$$\max_x \sum_{s \in S} k_s \sum_{p \in P} \pi_p y_{sp} - \sum_{p \in P} f_p x_p \tag{15.27}$$

$$s.t.\ x_p \in \{0, 1\} \qquad \forall p \in P \tag{15.28}$$

$$\max_y \sum_{s \in S} \sum_{p \in P} u_{sp} y_{sp} \tag{15.29}$$

$$s.t.\ \ y_{sp} \leq x_p \qquad \forall s \in S, p \in P \tag{15.30}$$

$$\sum_p y_{sp} \leq 1 \qquad \forall s \in S \tag{15.31}$$

$$y_{sp} \leq 0 \qquad \forall s \in S, p \in P : u_{sp} < u_{s0} \tag{15.32}$$

$$y_{sp} \in \{0, 1\} \qquad \forall s \in S, p \in P \tag{15.33}$$

where (15.27) represents the firm's (leader's) objective of profit maximization, which clearly depends on the customers' buying decision (given by the y_{sp} variables), besides its own product development decision (given by the x_p variables). The inner optimization problem (15.29)–(15.33) represents the product choice problem of all the customer segments. Inequality (15.29) represents the customers' utility maximization objective. Inequality (15.30) ensures that customers can buy only the product configurations that are developed by the firm, while (15.31) ensures that each customer segment buys at most one configuration. Inequality (15.32) ensures that a customer does not buy a product configuration if the utility derived from it is less than her reservation utility. In other words, she considers only those configurations that give her a utility at least as large as her reservation utility. Note that (15.32) should ideally be an equality constraint, but may also be written as an inequality constraint.

It is easy to verify that if the constraints of the customers' problem are summarized as $Ax \leq b$, then A (called the constraint matrix of the customers' problem) is a Totally Unimodular Matrix (TUM), i.e., its every square submatrix has a determinant 0, +1, or −1 [85]. Therefore, the customers' problem exhibits the integrality property, i.e., its LP relaxation results in an all-integer optimal solution,[7] due to which the binary constraints in (15.33) can be replaced by its continuous relaxation ($0 \leq y_{sp} \leq 1$). This transforms the customers' problem into a continuous LP, which allows us to reformulate (15.27)–(15.33) into a single-level program using the KKT conditions, as described in Sect. 15.2.3.2, or using the duality-based approach, as described in Sect. 15.2.3.4. We describe the applications of these two approaches to PLS in Sects. 15.3.1 and 15.3.2, respectively, followed by other single-level formulations in Sect. 15.3.3.

15.3.1 Single-Level Reformulation of PLS Using KKT Conditions

To reduce (15.27)–(15.33) to a single level, first note that the upper bounds on y_{sp} are redundant in the presence of (15.31). Hence, (15.33) can be simply replaced by the non-negative constraints of the form ($y_{sp} \geq 0$). Associating the dual variables α_{sp}, β_s, γ_{sp} and δ_{sp} with the constraint sets (15.30), (15.31), (15.32), and $y_{sp} \geq 0$, respectively, the single-level reformulation of (15.27)–(15.33) can be written as

$$\max_{x,y,\alpha,\beta,\gamma} \sum_{s\in S} k_s \sum_{p\in P} \pi_p y_{sp} - \sum_{p\in P} f_p x_p \tag{15.34}$$

$$\text{s.t. } (15.28), (15.30) - (15.32)$$

$$y_{sp} \geq 0 \qquad \forall s \in S, p \in P \tag{15.35}$$

$$\alpha_{sp} + \beta_s + \gamma_{sp} - \delta_{sp} = u_{sp} \qquad \forall s \in S, p \in P : u_{sp} < u_{s0} \tag{15.36}$$

$$\alpha_{sp} + \beta_s - \delta_{sp} = u_{sp} \qquad \forall s \in S, p \in P : u_{sp} \geq u_{s0} \tag{15.37}$$

$$\alpha_{sp} > 0 \qquad \forall s \in S, p \in P \tag{15.38}$$

$$\beta_s \geq 0 \qquad \forall s \in S \tag{15.39}$$

$$\gamma_{sp} \geq 0 \qquad \forall s \in S, p \in P \tag{15.40}$$

[7] Readers unfamiliar with TUM can see the integrality property using the following argument. Let us assume that the integrality property does not hold. Specifically, assume that a customer segment s buys a fraction f unit of one product configuration p and a fraction $1 - f$ of another configuration q, i.e., $y_{sp} = f$, $y_{sq} = 1 - f$ (note that constraint (15.31) requires $y_{sp} + y_{sq} \leq 1$ but the objective function (15.29) forces this constraint to be binding at optimality). If $u_{sp} > u_{sq}$, then the above solution cannot be optimal (since $1 - f$ should be 0 in the optimal solution). If, on the other hand, $u_{sp} < u_{sq}$, then the above solution cannot be optimal once again (since now f should be 0 in the optimal solution). So, the only case in which the optimal solution can be fractional is when $u_{sp} = u_{sq}$. But, in such a case, $f = 1, 1 - f = 0$ and $f = 0, 1 - f = 1$ are also alternate optimal solutions. Since these two are extreme point solutions, either of them will get selected as an optimal solution by any simplex-based linear program.

$$\delta_{sp} \geq 0 \qquad \forall s \in S, p \in P \tag{15.41}$$

$$\left(y_{sp} - x_p\right)\alpha_{sp} = 0 \qquad \forall s \in S, p \in P \tag{15.42}$$

$$\left(\sum_{p \in P} y_{sp} - 1\right)\beta_s = 0 \qquad \forall s \in S \tag{15.43}$$

$$y_{sp}\gamma_{sp} = 0 \qquad \forall s \in S, p \in P : u_{sp} < u_{s0} \tag{15.44}$$

$$y_{sp}\delta_{sp} = 0 \qquad \forall s \in S, p \in P \tag{15.45}$$

Here, (15.30)–(15.32), (15.35) are primal feasibility constraints, (15.36)–(15.37) are stationary conditions, (15.38)–(15.41) are dual feasibility constraints, and (15.42)–(15.45) are the complementary slackness conditions for the lower-level problem. (15.34) represents the objective function of the upper-level problem, while (15.28) represents its constraints. For LPs, the stationary conditions and the dual feasibility constraints together represent the set of constraints in the dual problem.

Note that the complementary slackness conditions are non-linear, due to the presence of bilinear terms, which can be handled using the Big$M$ constraints, as described in Sect. 15.2.3.2. However, choosing appropriate values of Big$M$ is not trivial, and we defer its discussion to Sect. 15.3.2, where they appear again. The bilinear terms in (15.42)–(15.45) can alternatively be handled as SOS1 constraints, which the modern solvers (CPLEX, Gurobi, etc.) are designed to handle, as discussed in Sect. 15.2.3.2.

15.3.2 Single-Level Reformulation of PLS Using Duality-Based Approach

In the context of LPs, the duality-based approach, like the KKT-based approach, also uses the primal and dual constraints of the lower-level problem. The difference arises only in the usage of the strong-duality condition, as given below, in place of the complementary slackness conditions (15.42)–(15.45).

$$\sum_{s \in S}\sum_{p \in P} u_{sp}y_{sp} \geq \sum_{s \in S}\sum_{p \in P} \alpha_{sp}x_p + \sum_{s \in S} \beta_s \tag{15.46}$$

Again, note that like the complementary slackness conditions in the KKT-based approach, (15.46) is non-linear, due to the bilinear term $\alpha_{sp}x_p$, which can be replaced by the following set of linear constraints when the upper-level decision variables (x_{sp}) are binary.

$$\sum_{s\in S}\sum_{p\in P} u_{sp}y_{sp} \geq \sum_{s\in S}\sum_{p\in P} z_{sp} + \sum_{s\in S}\beta_s \tag{15.47}$$

$$z_{sp} \leq \alpha_{sp} \qquad \forall s \in S, p \in P \tag{15.48}$$

$$z_{sp} \leq M_{sp}x_p \qquad \forall p \in P \tag{15.49}$$

$$z_{sp} \geq \alpha_{sp} - M_{sp}(1 - x_p) \qquad \forall s \in S, p \in P \tag{15.50}$$

$$z_{sp} \geq 0 \qquad \forall s \in S, p \in P \tag{15.51}$$

where M_{sp} appearing in (15.49) and (15.50) is a large number (Big$M$). The tightness of its LP relaxation, and hence its computational efficiency, depends on the choice of M_{sp}: the lower the value of M_{sp}, the better it is [21]. Therefore, M_{sp} can be replaced by its lower bound, given by the following proposition.

Proposition 15.1 *A valid lower bound on M_{sp} appearing in* (15.49)–(15.50) *is given by $\tilde{M}_{sp}$, where:*

$$\tilde{M}_{sp} = \begin{cases} u_{sp}, & if\ u_{sp} \geq u_{s0} \\ 0, & if\ u_{sp} < u_{s0} \end{cases} \tag{15.52}$$

Proof From (15.48) and (15.49), it is clear that a valid M_{sp} should satisfy $M_{sp} \geq \alpha_{sp}$. Since α_{sp} is the dual variable corresponding to (15.30), the largest value it can take is given by the maximum change in the objective function (15.29) when x_p in the RHS of (15.30) changes from 0 to 1. Clearly, the maximum change in (15.29) when x_p changes from 0 to 1 is given by $\tilde{M}_{sp}$ in (15.52). Hence, $M_{sp} \geq \alpha_{sp} \geq \tilde{M}_{sp}$. □

15.3.3 *Other Single-Level Reformulations of PLS*

For some bilevel problems, it is possible to replace the follower's problem with constraints not necessarily obtained from the KKT conditions or the duality-based approach. For example, Ramamoorthy et al. [60, 61] reduce the bilevel hub-interdiction problem[8] to a single level by replacing the follower's problem by the so-called "closest assignment constraints". Similarly, the lower-level customers' product choice problem in PLS can be replaced with the following set of constraints [54].

$$u_{sp} \geq u_{sq}x_q - M_{spq}(1 - y_{sp}) \qquad \forall s \in S, p \in P, q \in P : p \neq q \tag{15.53}$$

where M_{spq} is a Big$M$. Inequality (15.53) models the customers' first choice rule (described earlier) by imposing the constraint that if the customer segment s selects the product configuration p (i.e., $y_{sp} = 1$), then its utility from that configuration should be at least as high as that from any other configuration q that is also developed (i.e., $u_{sp} \geq u_{sq}$ if $x_q = 1$). Using (15.53), PLS can be restated as the following single-

[8] An example of an interdiction problem is discussed in Sect. 15.4.

level integer program, which can be solved directly using any standard off-the-shelf integer program solver.

$$(15.27), (15.28), (15.30) - (15.33), (15.53)$$

Proposition 15.2 *A valid lower bound on M_{spq} appearing in* (15.53) *is given by $\tilde{M}_{spq}$, where:*

$$\tilde{M}_{spq} = |u_{sq}| + |u_{sp}|$$

Proof For $y_{sp} = 1$, (15.53) is always valid, independent of the value of M_{spq}. For $y_{sp} = 0$, (15.53) reduces to $u_{sp} \geq u_{sq}x_q - M_{spq}$, which is valid for any $M_{spq} \geq |u_{sp}| + |u_{sp}|$. □

Below, we provide alternate ways to model the customers' first choice rule, where one of the following constraint sets that do not require Big$M$ can be used instead of (15.53).

$$y_{sp} \geq y_{sq} - (1 - x_p) \qquad \forall s \in S, p \in P, q \in P : u_{sp} > u_{sq} \tag{15.53a}$$

$$\sum_{q:u_{sq}<u_{sp}} y_{sq} \leq 1 - x_p \qquad \forall s \in S, p \in P \tag{15.53b}$$

$$\sum_{q:u_{sq}\geq u_{sp}} y_{sq} \geq x_p \qquad \forall s \in S, p \in P \tag{15.53c}$$

$$\sum_{q} u_{sq}y_{sq} \geq u_{sp}x_p \qquad \forall s \in S, p \in P \tag{15.53d}$$

(15.53a) makes any solution infeasible in which product configuration p is developed by the firm (i.e., $x_p = 1$) and any customer segment s does not buy it but buys some other product q with a lower utility (i.e., $y_{sp} = 0$, $y_{sq} = 1$, where $u_{sp} > u_{sq}$). Inequality (15.53b) ensures that if product configuration p is developed by the firm (i.e., $x_p = 1$), then any customer segment s cannot buy any other product configuration q that gives it strictly less utility than configuration p (i.e., $y_{sq} = 0\ \forall q : u_{sq} < u_{sp}$). Inequalities (15.53c) and (15.53d) ensure, although in slightly different ways, that if product configuration p is developed by the firm (i.e., $x_p = 1$), then any customer segment s can only buy either configuration p or any other configuration q with utility at least as high as configuration p (i.e., $y_{sq} = 1$ only for one of those q for which $u_{sq} \geq u_{sp}$).

The optimal solution to the problem, obtained using any of the above three reformulations, is summarized in Table 15.2, and its AMPL code for the single-level

Table 15.2 Optimal product line selection

Configurations developed	Customer segments and their configuration choices				
	1	2	3	4	5
2	0	0	1	0	1
7	0	1	0	0	0
8	1	0	0	1	0

reformulation using (15.53) is provided in Appendix A.[9] It is worth noting that in the optimal solution, customer segment 4 chooses product configuration 8, and not 7, even though both give it the same utility (see Table 15.1). Further, note that product configuration 8 is more profitable to the firm than configuration 7: it gives the firm a profit of 60 per unit, compared to only 55 for configuration 7. Hence, the above optimal solution represents an optimistic solution to the problem, and all the alternate formulations of the problem discussed above are optimistic formulations, as discussed in Sect. 15.2. On the other hand, if the firm is pessimistic that customer segment 4 may select product configuration 7, instead of 8, then the above formulations need to be modified to take into account the firm's pessimistic position.

One natural question that arises at this stage is about the difference among the different reformulations obtained using the three approaches in solving PLS. All 7 reformulations (one using KKT-based, one using duality-based, and 5 other single-level reformulations obtained using (15.53), (15.53a), (15.53b), (15.53c), and (15.53d), respectively) of course, produce the same optimal solution; the difference generally lies in the efficiency with which they solve the problem, as reflected in their CPU times. To highlight this difference, we summarize in Table 15.3 their computational performances on two different instances. In general, the efficiency of a formulation depends on its size, measured in terms of the number of variables and constraints, and the sharpness of its LP bound. Based on these criteria, for the first instance ($|P| = 10$, $|S| = 5$), single-level reformulations using (15.53b) and (15.53c) are expected to be the most efficient, which is indeed reflected in their CPU times. However, this instance turned out to be easy for all 7 formulations, and the difference in the CPU times is not significant. For the larger problem instance ($|P| = 500$, $|S| = 25$), the LP relaxations of all formulations produce the integer optimal solution. However, they differ in CPU times due to the time required to solve the corresponding LP relaxation models, which vary drastically in size. Since the reformulations using (15.53) and (15.53a) are the largest in size, they take longer to solve compared to the other reformulations. Interestingly, the KKT-based and duality-based approaches, which are not reported in the literature on PLS, outperform the rest. More challenging instances may require advanced methods (e.g., Lagrangian

[9] C++ codes for all single-level reformulations are publicly available at https://github.com/sachinjayaswal/Product-Line-Selection. AMPL codes are also publicly available at https://github.com/sachinjayaswal/Product-Line-Selection-AMPL.

Table 15.3 Comparison among computational performances of different reformulations of PLS

Prob. size		Reform-ulation	#Vars	#Bin vars	#Cons	#SOS1	Opt. Obj. ($\times 10^6$)	LP gap%	CPU (s)	#B&B nodes	#Cplex cuts
$\|P\|$	$\|S\|$										
10	5	KKT+SOS1	231	10	171	116	2.33	–[a]	0.04	16	0
		Duality	176	10	267	0	2.33	0.42	0.03	0	11
		Other (a)	60	10	516	0	2.33	0.41	0.03	0	5
		Other (b)	60	10	270	0	2.33	0.17	0.06	0	2
		Other (c)	60	10	116	0	2.33	0.00	0.01	0	0
		Other (d)	60	10	116	0	2.33	0.00	0.01	0	0
		Other (e)	60	10	116	0	2.33	0.41	0.02	0	22
500	25	KKT+SOS1	50,703	500	37,703	25,178	13.91	–	1.93	0	0
		Duality	38,178	500	62,679	0	13.91	0.00	0.63	0	0
		Other (a)	13,000	500	62,50,178	0	13.91	0.00	62.16	0	0
		Other (b)	13,000	500	31,25,296	0	13.91	0.00	49.84	0	0
		Other (c)	13,000	500	25,178	0	13.91	0.00	3.14	0	0
		Other (d)	13,000	500	25,178	0	13.91	0.00	4.01	0	0
		Other (e)	13,000	500	25,178	0	13.91	0.00	7.61	0	0

Other (a): single-level reformulation using (15.53); Other (b): single-level reformulation using (15.53a); Other (c): single-level reformulation using (15.53b); Other (d): single-level reformulation using (15.53c); other (e): single-level reformulation using (15.53d)

[a] The continuous relaxation of the reformulation is a non-linear program; hence, no LP Gap reported

relaxation, valid inequalities, heuristics, etc.) to solve them efficiently. Belloni et al. [14] presents a comparison among several such methods.

15.4 Shortest-Path Interdiction

The shortest-path interdiction problem entails destroying (interdicting) a subset of arcs, using a given set of resources/budget, in a given network so as to make the shortest path between a source-destination pair as long as possible [40]. The problem belongs to the general class of network interdiction problems, which has applications in infectious disease control [6], counter-terrorism [31], cyber-physical systems, illicit supply chains, etc. [73]. Let us understand the problem using a simple 5-node directed network given in Fig. 15.1, in which node s is the source and node t is the destination. Each arc is labeled with a number, and the corresponding triplet represents its normal length (or time), the delay on it caused by its interdiction, and the amount of resource required to interdict it, respectively. Assume the interdictor has a limited budget sufficient to interdict a maximum of two arcs. With this budget, the possible subset of arcs that can be interdicted and the resulting lengths of the different paths between the origin-destination pair are given in Table 15.4. Clearly, there are five alternate shortest paths, each with a length of 16, that are the longest possible. The five shortest paths and the corresponding subsets of arcs to interdict are summarized in Table 15.5.

The problem in the above illustrative example is small enough to be solved by complete enumeration. However, for a network of a reasonable size, the number of solutions will be too huge to be enumerated, thus requiring a more practical approach to solve the problem. Therefore, we next present a mathematical model of the problem, which turns out to be a bilevel program. Subsequently, we show how to reduce the bilevel program to a single level, which can be solved using an off-the-shelf mixed-integer programming solver. For the mathematical model, consider the following notation:

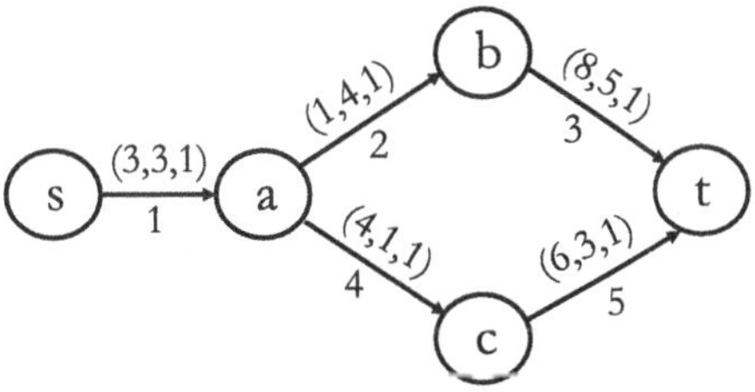

Fig. 15.1 A 5-node illustrative network

Table 15.4 Alternate O-D paths

Interdicted	Possible paths		Shortest path	
Arcs	Path	Path length	Path	Path length
{1, 2}	1–2–3	(3+3)+(1+4)+(8+0) = 19	1–4–5	16
	1–4–5	(3+3)+(4+0)+(6+0) = 16		
{1, 3}	1–2–3	(3+3)+(1+0)+(8+5) = 20	1–4–5	16
	1–4–5	(3+3)+(4+0)+(6+0) =16		
{1, 4}	1–2–3	(3+3)+(1+0)+(8+0) = 15	1–2–3	15
	1–4–5	(3+3)+(4+1)+(6+0) = 17		
{1, 5}	1–2–3	(3+3)+(1+0)+(8+0) = 15	1–2–3	15
	1–4–5	(3+3)+(4+0)+(6+3) = 19		
{2, 3}	1–2–3	(3+0)+(1+4)+(8+5) =21	1–4–5	13
	1–4–5	(3+0)+(4+0)+(6+0) = 13		
{2, 4}	1–2–3	(3+0)+(1+4)+(8+0) = 16	1–4–5	14
	1–4–5	(3+0)+(4+1)+(6+0) = 14		
{2, 5}	1–2–3	(3+0)+(1+4)+(8+0) = 16	1–2–3	16
	1–4–5	(3+0)+(4+0)+(6+3) = 16	1–4–5	16
{3, 4}	1–2–3	(3+0)+(1+0)+(8+5) = 17	1–4–5	14
	1–4–5	(3+0)+(4+1)+(6+0) = 14		
{3, 5}	1–2–3	(3+0)+(1+0)+(8+5) = 17	1–4–5	16
	1–4–5	(3+0)+(4+0)+(6+3) = 16		
{4, 5}	1–2–3	(3+0)+(1+0)+(8+0) = 12	1–2–3	12
	1–4–5	(3+0)+(4+1)+(6+3) = 17		

Table 15.5 Alternate optimal solutions

Optimal solution no.	Arcs interdicted (Leader's strategy)	Shortest path (Follower's best response)
1	{1, 2}	1–4–5
2	{1, 3}	1–4–5
3	{2, 5}	1–2–3
		1–4–5
4	{3, 5}	1–4–5

Indices and Parameters:

N : Set of nodes in the network
i, j : Indices for nodes; $i, j \in N$
s : Index for the source node
t : Index for the destination node
A : Set of arcs in the network
k : Index for arcs; $k \in A$
$I(i)$: Subset of arcs directed into node i
$O(i)$: Subset of arcs directed out of node i
c_k : Normal length of arc k

d_k : Delay on arc k caused by its interdiction
r_k : Resource required to interdict arc k
B : Budget (interdiction resource) available

Decision Variables:

x_k : 1 if arc k is interdicted, 0 otherwise
y_k : 1 if arc k lies on a shortest path

Using the above notation, the mathematical model for the shortest-path interdiction problem can be stated as follows:

$$\max_{x} \; Z \tag{15.54}$$

$$s.t. \sum_{k \in A} r_k x_k \leq B \tag{15.55}$$

$$x_k \in \{0, 1\} \qquad \forall k \in A \tag{15.56}$$

$$Z = \min_{y} \sum_{k \in A} (c_k + d_k x_k) y_k \tag{15.57}$$

$$s.t. \sum_{k \in O(i)} y_k - \sum_{k \in I(i)} y_k = \begin{cases} 1, & \text{if } i = s \\ -1, & \text{if } i = t \\ 0, & \text{if } i \in N \setminus \{s, t\} \end{cases} \qquad \forall i \in N \tag{15.58}$$

$$y_k \in \{0, 1\} \qquad \forall k \in A \tag{15.59}$$

(15.54)–(15.59) is a bilevel integer program. (15.57)–(15.59) represent the inner (follower's) optimization problem. The follower's problem (15.57)–(15.59) is a shortest-path problem, in which the objective function (15.57) captures the normal length (or time) of arc k, and the added delay if arc k is interdicted. Constraint set (15.58) represents the flow balance constraints typically used in a shortest-path problem. The leader's objective in (15.54) is to maximize the length of the follower's shortest path, using his limited set of resources given by the budget constraint (15.55). The above problem can be solved using the KKT-based approach or the duality-based approach; however, the duality-based approach is known to outperform the KKT-based approach. While using the duality-based approach, we will notice that for a max-min problem, the strong-duality condition involving non-linearity is not required. Also, the primal variables and constraints of the lower-level problem can be avoided in the equivalent single-level reformulation. This provides significant computational advantages compared to the KKT-based approach.

Note that the follower's problem (15.57)–(15.59), which is a shortest-path problem, is known to exhibit integrality property, i.e., its LP relaxation results in an all-integer optimal solution [11]. This means that the binary constraints in (15.59) can be replaced by their continuous relaxations ($0 \leq y_k \leq 1$). This makes the follower's optimization problem a continuous LP, which can be replaced by its dual using the strong law of duality. Further, the upper bounds on y_k are redundant due to (15.58). Hence, (15.59) can be simply replaced by the non-negativity constraints of the form ($y_k \geq 0$). Let us illustrate this for the 5-node, 5-arc network in Fig. 15.1. For the case that none of the arcs is interdicted (i.e., $x_k = 0 \; \forall k \in A$), the follower's shortest-path problem can be written as follows:

$$
\begin{aligned}
Z = \min \;\; & 3y_1 + y_2 + 8y_3 + 4y_4 + 6y_5 && \text{(Dual Variable)} \\
\text{s.t. } y_1 &= 1 && (\pi_s) \\
-y_1 + y_2 + y_4 &= 0 && (\pi_a) \\
-y_2 + y_3 &= 0 && (\pi_b) \\
-y_4 + y_5 &= 0 && (\pi_c) \\
-y_3 - y_5 &= -1 && (\pi_t) \\
y_k &\geq 0 && \forall k \in \{1, 2, 3, 4, 5\}
\end{aligned}
$$

The dual of the above model can be written as follows:

$$
\begin{aligned}
Z = \max \;\; & \pi_s - \pi_t \\
\text{s.t. } & \pi_s - \pi_a \leq 3 \\
& \pi_a - \pi_b \leq 1 \\
& \pi_b - \pi_t \leq 8 \\
& \pi_a - \pi_c \leq 4 \\
& \pi_c - \pi_t \leq 6 \\
& \pi_i \text{ free} && \forall i \in \{s, a, b, c, t\}
\end{aligned}
$$

The above dual problem has multiple optimal solutions: for instance, consider $Z = 5$, then ($\pi_s = 0$, $\pi_t = -5$) and ($\pi_s = 1$, $\pi_t = -6$) represent two possible optimal solutions. In general, if $\{\pi_i^*\}_{i \in N}$ is an optimal solution, then so is $\{\pi_i^* - n\}_{i \in N}$, for every $n > 0$. Therefore, we can arbitrarily set $\pi_s = 0$. In that case, for any node i that lies on the shortest path, the value of π_i in the optimal solution gives the negative of the length of the shortest path from the origin s to node i. If we, however, replace π_i by -π_i' in the above dual, then π_i' in the optimal solution has a much nicer interpretation: it represents the length of the shortest path from the origin s to node i if node i lies on the shortest path [11].

Replacing the lower-level problem by its dual, the bilevel problem (15.54)–(15.59) can, in general, be written as follows:

$$\max_{x} \ Z \tag{15.60}$$

$$s.t. \sum_{k \in A} r_k x_k \le B \tag{15.61}$$

$$x_k \in \{0, 1\} \qquad \forall k \in A \tag{15.62}$$

$$Z = \max_{\pi_i'} \ \pi_t' \tag{15.63}$$

$$s.t. \ \pi_j' - \pi_i' \le c_k + d_k x_k \qquad \forall k = (i, j) \in A \tag{15.64}$$

$$\pi_s' = 0 \tag{15.65}$$

$$\pi_i' \text{ free} \qquad \forall i \in N \tag{15.66}$$

This simple trick allows us to reduce the bilevel program (15.60)–(15.66) to the following single-level mixed-integer program:

$$\max_{x, \pi_i'} \ \pi_t' \tag{15.67}$$

$$s.t. \ (15.61), (15.62), (15.64) - (15.66)$$

Solving the above single-level reformulation of the shortest-path interdiction problem for the 5-node network in Fig. 15.1 results in the following optimal solution: $x_3 = x_5 = 1$, $y_1 = y_4 = y_5 = 1$ (the values of the y variables are recovered using the primal-dual relationship between the y and π variables). This, in the parlance of game theory, implies that if the leader (interdictor) is a rational person and knows that the follower is also rational, then the interdictor will interdict arcs 3 and 5. The follower, if rational, will choose 1-4-5 as the shortest path in the interdicted network. This solution is an equilibrium solution under a hierarchical setting, commonly referred to as the Stackelberg equilibrium.

The resulting single-level program for the 5-node illustrative example can be solved by any standard off-the-shelf mixed-integer programming solver. Its AMPL code and the output are provided in Appendix B.[10] However, the model becomes difficult to solve for practical-size problems and the interested reader is referred to Israeli and Wood [40] for a discussion on some advanced methods (Benders decomposition and supervalid inequalities) to solve it more efficiently. Interdiction problems have also been studied in the context of network flow [86], multicommodity network flow [48], facility location [23, 57], hub location [17, 46, 60, 61], packing [28], matching [88], knapsack [32], etc. A natural extension of these interdiction problems arises when one of the agents (leader or follower) has incomplete information related to some aspect of the problem. Such an extension in which the follower has incomplete information about the arc lengths, in the context of shortest-path interdiction, is studied by Bayrak and Bailey [10].

[10] AMPL codes are also publicly available at https://github.com/sachinjayaswal/Shortest-Path-Interdiction-AMPL.

15.5 Conclusions

This chapter provided a discussion on the basics of bilevel optimization and the various ways to reformulate a bilevel optimization problem into a single-level optimization problem. It considered two classes of bilevel problems, namely, the product line selection problem and the shortest-path interdiction problem, and showed how to solve them exactly. In the context of the product line selection problem, it also showed how the bilevel structure can be bypassed in special cases using constraints similar to the so-called “closest assignment constraints”. Using such constraints, it provided some new single-level reformulations for the product line selection problem not reported earlier in the literature. For the ease of implementation of the discussed ideas in the context of the larger instances of the product line selection and the shortest-path interdiction problems, and also other application problems, it provided the AMPL codes in the appendix.

Acknowledgements The authors acknowledge the support provided by the Research & Publications Cell and Brij Disa Centre for Data Science and Artificial Intelligence, Indian Institute of Management Ahmedabad.

Appendix A. AMPL Codes for Product Line Selection

```
#----------DATA File-------------
#Name of Data file: PLS_10_5.dat

param N_Product:= 10;
param N_Segment:= 5;

param:
dev_cost    profit:=
    1   15000   50
    2   12000   60
    3   9000    40
    4   7000    55
    5   6000    35
    6   5000    45
    7   6000    55
    8   10000   60
    9   8000    55
    10  9500    45;

param:
seg_size     reserv_util:=
    1   6000    4
    2   8500    4
    3   9500    4
    4   7000    4
    5   9000    4;

param utility(tr):
    1   2   3   4   5:=
1   6   9   8   4   10
2   4   3   9   5   4
3   -2  10  7   1   9
4   4   5   7   8   9
```

```
5  -4  9   4   2   5
6   6  9   9   6   10
7   7  9   3   7   1
8   9  2   4   7   3
9   8  8   10  -1  -4
10  6  5   10  6   10;
#-----------END OF DATA File---------

#---------MODEL FILE----------------
#Name of Model file: model PLS.mod (Other Single-level Reformulation (a); refer to Table 3
in Section 3.3)
param N_Product;
param N_Segment;
set Product:= 1..N_Product;
set Segment:= 1..N_Segment;

param utility {Segment, Product};        #Utility of a product to a given customer segment
param seg_size {Segment} >=0;            #Size of each customer segment
param reserv_util{Segment} >=0;              #UReservation tility of a given customer segment
param dev_cost {Product} >=0;                #Product Development Cost
param profit {Product} >=0;              #Unit profit on each product
param M{s in Segment, p in product, q in product: q!=p}:= abs(utility[s, q])+ abs(utility
[s, p]);#Large number

var Develop {Product} binary;            #1 if the particular product is selected, 0 otherwise
var Buy {Segment, Product} binary;   #1 if the product is chosen by the customer segment,
0 otherwise

maximize
    Profit: sum {s in Segment, p in Product}profit[p]*seg_size[s]*Buy[s, p] - sum {p in
Product}dev_cost[p]*Develop[p];
subject to
    Buy_only_if_developed{s in Segment, p in Product}: Buy[s, p] <= Develop[p];
    Buy_max_one_product{s in Segment}: sum {p in Product}Buy[s, p] <= 1;
    Dont_buy_if_below_reserve_util{s in Segment, p in Product: utility[s, p]
< reserv_util[s]}: Buy[s, p] <= 0;
    Max_Util_Choice{s in Segment, p in Product, q in Product: q != p}:
            utility[s, p] >= utility[s, q]*Develop[q] - M[s,p,q]*(1-Buy[s, p]);
#---------END OF MODEL FILE----------------

#---------RUN FILE----------------
reset;
model PLS.mod;
data PLS_10_5.dat;
option solver cplex;
option show_stats 1;
option cplex_options 'mipdisplay=4 mipinterval=2';#Extent to which display B&B Search
info

solve;

option omit_zero_rows 1;#Do not display variables that have 0 values;
option omit_zero_cols 1;#Do not display variables that have 0 values;

display _total_solve_time >Output_PLS.txt;
display Profit>Output_PLS.txt;
display Develop>Output_PLS.txt;
display Buy>Output_PLS.txt;
expand>Output_PLS.txt;
close Output_PLS.txt;
#---------END OF RUN FILE----------------

#---------OUTPUT FILE----------------
#Output_PLS.txt
_total_solve_time = 0.0

Profit = 2329500.0
```

```
Develop [*] :=
 2  1.0
 7  1.0
 8  1.0;

Buy :=
1 8   1.0
2 7   1.0
3 2   1.0
4 8   1.0
5 2   1.0;
#---------END OF OUTPUT FILE----------------
```

Appendix B. AMPL Codes for Shortest-Path Interdiction

```
#----------DATA File-------------
#Name of Data file: SPI_5_5.dat
param N_Nodes:= 5;
param N_Arcs:= 5;
param Budget:= 2;#Total resource available to the Interdictor
param Source_Node:= 1;
param Dest_Node:= 5;
param Node_Arc:
    1   2   3   4   5:=
1   -1  1   0   0   0
2   0   -1  1   0   0
3   0   0   -1  0   1
4   0   -1  0   1   0
5   0   0   0   -1  1; #Each Row is an Arc Number
param: Arc_Length   Arc_Delay   Arc_Resource:=
1   3   3   1
2   1   4   1
3   8   5   1
4   4   1   1
5   6   3   1;
#-----------END OF DATA File---------

#---------MODEL FILE----------------
#Name of Model file: model SP_Interdict.mod
param N_Nodes;
param N_Arcs;
param Budget;#Total resource available to the Interdictor
param Source_Node;
param Dest_Node;

set NODES:= 1..N_Nodes;
set ARCS:= 1..N_Arcs;

param Node_Arc{ARCS, NODES};
param Arc_Length{ARCS};
param Arc_Delay{ARCS};
param Arc_Resource{ARCS}; #amount of resource required to interdict the arc

var Interdict{ARCS} binary; #1 if Arc a is interdicted, 0 otherwise
var Dual{NODES}; #Dual variable for Node i. It represents the length of the shortest path
to node i from source s

maximize Longest_shortest_path: Dual[Dest_Node] - Dual[Source_Node];
subject to
Dual_Constraint{a in ARCS}: sum{n in NODES} Node_Arc[a,n]*Dual[n] -
Arc_Delay[a]*Interdict[a] <= Arc_Length[a];
```

```
Dual_node_source: Dual[Source_Node] = 0;
Interdiction_budget: sum{a in ARCS} Arc_Resource[a]*Interdict[a] <= Budget;
#---------END OF MODEL FILE----------------

#---------RUN FILE----------------
reset;
model SP_Interdict.mod;
data SP_5_5.dat;
option solver cplex;
option show_stats 1;
option cplex_options 'mipdisplay=4 mipinterval=2';#Extent to which display B&B Search
info

solve;

option omit_zero_rows 1;# Do not display variables that have 0 values
option omit_zero_cols 1;# Do not display variables that have 0 values

display _total_solve_time>SP_Interdict.txt;#Display the total CPU time
display Longest_shortest_path>SP_Interdict.txt;
display Interdict>SP_Interdict.txt;
display Dual>SP_Interdict.txt;
close >SP_Interdict.txt;
#---------END OF RUN FILE----------------

#---------OUTPUT FILE----------------
#SP_Interdict.txt
_total_solve_time = 0.1
Longest_shortest_path = 16.0
Interdict [*] :=
3  1.0
5  1.0;

Dual [*] :=
2   3.0
3   3.0
4   7.0
5  16.0;
#---------END OF OUTPUT FILE----------------
```

References

1. Aiyoshi E, Shimizu K (1981) Hierarchical decentralized systems and its new solution by a barrier method. IEEE Trans Syst Man Cybern 6:444–449
2. Aiyoshi E, Shimizu K (1984) A solution method for the static constrained Stackelberg problem via penalty method. IEEE Trans Automat Control 29:1111–1114
3. Al-Khayyal F, Horst R, Pardalos P (1992) Global optimization of concave functions subject to quadratic constraints: An application in nonlinear bilevel programming. Ann Oper Res 34:125–147
4. Angelo J, Krempser E, Barbosa H (2013) Differential evolution for bilevel programming. In: Proceedings of the 2013 congress on evolutionary computation (CEC-2013). IEEE Press
5. Angelo JS, Krempser E, Barbosa HJC (2014) Differential evolution assisted by a surrogate model for bilevel programming problems. In: IEEE congress on evolutionary computation (CEC). IEEE, pp 1784–1791
6. Assimakopoulos N (1987) A network interdiction model for hospital infection control. Comput Biol Med 17(6):413–422
7. Bard J (1998) Practical bilevel optimization: Algorithms and applications. Kluwer, The Netherlands

8. Bard J, Falk J (1982) An explicit solution to the multi-level programming problem. Comput Oper Res 9:77–100
9. Bard J, Moore J (1990) A branch and bound algorithm for the bilevel programming problem. SIAM J Sci Stat Comput 11:281–292
10. Bayrak H, Bailey MD (2008) Shortest path network interdiction with asymmetric information. Netw Int J 52(3):133–140
11. Bazaraa MS, Jarvis JJ, Sherali HD (2008) Linear programming and network flows. Wiley, New York
12. Bazaraa MS, Sherali HD, Shetty CM (2013) Nonlinear programming: Theory and algorithms. Wiley, New York
13. Beale EML, Tomlin JA (1970) Special facilities in a general mathematical programming system for non-convex problems using ordered sets of variables. OR 69(447–454):99
14. Belloni A, Freund R, Selove M, Simester D (2008) Optimizing product line designs: Efficient methods and comparisons. Manag Sci 54(9):1544–1552
15. Ben-Ayed O (1993) Bilevel linear programming. Comput Oper Res 20:485–501
16. Bennett KP, Kunapuli G, Hu J, Pang J-S (2008) Bilevel optimization and machine learning. In: Computational intelligence: research frontiers. Springer, Berlin, pp 25–47
17. Bhatt S, Sinha A, Jayaswal S (2021) Hub interdiction problem under congestion: Formulations and solution methods. Working Paper
18. Bostian M, Whittaker G, Sinha A, Barnhart B (2015) Incorporating data envelopment analysis solution methods into bilevel multi-objective optimization. In: 2015 IEEE congress on evolutionary computation (CEC). IEEE, pp 1667–1674
19. Brotcorne L, Labbe M, Marcotte P, Savard G (2001) A bilevel model for toll optimization on a multicommodity transportation network. Transp Sci 35(4):345–358
20. Brown G, Carlyle M, Diehl D, Kline J, Wood K (2005) A two-sided optimization for theater ballistic missile defense. Oper Res 53(5):745–763
21. Camm JD, Raturi AS, Tsubakitani S (1990) Cutting big M down to size. Interfaces 20(5):61–66
22. Christiansen S, Patriksson M, Wynter L (2001) Stochastic bilevel programming in structural optimization. Structural Multi Optim 21(5):361–371
23. Church RL, Scaparra MP, Middleton RS (2004) Identifying critical infrastructure: The median and covering facility interdiction problems. Ann Assoc Am Geogr 94(3):491–502
24. Clark PA, Westerberg AW (1990) Bilevel programming for steady-state chemical process design-I. Fundamentals and algorithms. Comput Chem Eng 14(1):87–97
25. Colson B, Marcotte P, Savard G (2005) A trust-region method for nonlinear bilevel programming: Algorithm and computational experience. Comput Optim Appl 30(3):211–227
26. Dempe S (2002) Foundations of bilevel programming. Kluwer Academic Publishers, Secaucus, NJ, USA
27. Dempe S, Mordukhovich BS, Zemkoho AB (2014) Necessary optimality conditions in pessimistic bilevel programming. Optimization 63(4):505–533
28. Dinitz M, Gupta A (2013) Packing interdiction and partial covering problems. In: International conference on integer programming and combinatorial optimization. Springer, Berlin, pp 157–168
29. Dolan RJ (2001) Analyzing consumer preferences. Harvard Business School Publications
30. Edmunds T, Bard J (1991) Algorithms for nonlinear bilevel mathematical programming. IEEE Trans Syst Man Cybern 21:83–89
31. Farley JD (2003) Breaking Al Qaeda cells: A mathematical analysis of counterterrorism operations (A guide for risk assessment and decision making). Stud Conflict Terrorism 26(6):399–411
32. Fischetti M, Ljubić I, Monaci M, Sinnl M (2019) Interdiction games and monotonicity, with application to knapsack problems. INFORMS J Comput 31(2):390–410
33. Fortuny-Amat J, McCarl B (1981) A representation and economic interpretation of a two-level programming problem. J Oper Res Soc 32:783–792
34. Green PE, Krieger AM (1985) Models and heuristics for product line selection. Mark Sci 4(1):1–19

35. Halter W, Mostaghim S (2006) Bilevel optimization of multi-component chemical systems using particle swarm optimization. In: IEEE congress on evolutionary computation (CEC). IEEE, pp 1240–1247
36. Hansen P, Jaumard B, Savard G (1992) New branch-and-bound rules for linear bilevel programming. SIAM J Sci Stat Comput 13(5):1194–1217
37. Hu X, Ralph D (2007) Using EPECs to model bilevel games in restructured electricity markets with locational prices. Oper Res 55(5):809–827
38. Ishizuka Y, Aiyoshi E (1992) Double penalty method for bilevel optimization problems. Ann Oper Res 34:73–88
39. Islam MM, Singh HK, Ray T (2017) A surrogate assisted approach for single-objective bilevel optimization. IEEE Trans Evol Comput 21(5):681–696
40. Israeli E, Wood RK (2002) Shortest-path network interdiction. Netw: Int J 40(2):97–111
41. Kleinert T, Labbé M, Ljubić I, Schmidt M (2021) A survey on mixed-integer programming techniques in bilevel optimization. EURO J Comput Optim 9:100007
42. Kleinert T, Labbé M, Plein FA, Schmidt M (2020) There's no free lunch: On the hardness of choosing a correct big-M in bilevel optimization. Oper Res 68(6):1716–1721
43. Kleinert T, Schmidt M (2021) Computing feasible points of bilevel problems with a penalty alternating direction method. INFORMS J Comput 33(1):198–215
44. Küçükaydin H, Aras N, Altınel IK (2011) Competitive facility location problem with attractiveness adjustment of the follower: A bilevel programming model and its solution. Eur J Oper Res 208(3):206–220
45. Labbé M, Marcotte P, Savard G (1998) A bilevel model of taxation and its application to optimal highway pricing. Manag Sci 44(12):1608–1622
46. Lei TL (2013) Identifying critical facilities in hub-and-spoke networks: A hub interdiction median problem. Geogr Anal 45(2):105–122
47. Li X, Tian P, Min X (2006) A hierarchical particle swarm optimization for solving bilevel programming problems. In: Rutkowski L, Tadeusiewicz R, Zadeh LA, Zurada JM (eds) Artificial intelligence and soft computing—ICAISC 2006, vol 4029. Lecture notes in computer science. Springer, Berlin, pp 1169–1178
48. Lim C, Smith JC (2007) Algorithms for discrete and continuous multicommodity flow network interdiction problems. IIE Trans 39(1):15–26
49. Liu G, Han J, Wang S (1998) A trust region algorithm for bilevel programing problems. Chinese Scie Bull 43(10):820–824
50. Loridan P, Morgan J (1996) Weak via strong Stackelberg problems. J Global Optim 8:263–287
51. Lucchetti R, Mignanego F, Pieri G (1987) Existence theorem of equilibrium points in Stackelberg games with constraints. Optimization 18:857–866
52. Marcotte P, Savard G, Semet F (2004) A bilevel programming approach to the travelling salesman problem. Oper Res Lett 32(3):240–248
53. Marcotte P, Savard G, Zhu DL (2001) A trust region algorithm for nonlinear bilevel programming. Oper Res Lett 29(4):171–179
54. McBride RD, Zufryden FS (1988) An integer programming approach to the optimal product line selection problem. Mark Sci 7(2):126–140
55. McCormick GP (1976) Computability of global solutions to factorable nonconvex programs: Part I-convex underestimating problems. Math Program 10(1):147–175
56. Migdalas A (1995) Bilevel programming in traffic planning: Models, methods and challenge. J Global Optim 7(4):381–405
57. Nigudkar S, Jayaswal S, Sinha A, Nair A (2021) Design for failure of food banking networks. Working Paper
58. Provost F, Fawcett T (2013) Data science for business: What you need to know about data mining and data-analytic thinking. O'Reilly Media, Inc
59. Raghunathan AU, Biegler LT (2003) Mathematical programs with equilibrium constraints (MPECs) in process engineering. Comput Chem Eng 27(10):1381–1392
60. Ramamoorthy P, Jayaswal S, Sinha A, Vidyarthi N (2017) Hub-and-spoke network design under the risk of interdiction. (No. WP 2017-05-01), Technical report, Indian Institute of Management Ahmedabad

61. Ramamoorthy P, Jayaswal S, Sinha A, Vidyarthi N (2018) Multiple allocation hub interdiction and protection problems: Model formulations and solution approaches. Eur J Oper Res 270(1):230–245
62. Savard G, Gauvin J (1994) The steepest descent direction for the nonlinear bilevel programming problem. Oper Res Lett 15:275–282
63. Sinha A, Khandait T, Mohanty R (2020) A gradient-based bilevel optimization approach for tuning hyperparameters in machine learning. CoRR. arxiv:abs/2007.11022
64. Sinha A, Lu Z, Deb K, Malo P (2020) Bilevel optimization based on iterative approximation of multiple mappings. J Heuristics 26(2):151–185
65. Sinha A, Malo P, Deb K (2015) Transportation policy formulation as a multi-objective bilevel optimization problem. In: 2015 IEEE congress on evolutionary computation (CEC-2015). IEEE Press
66. Sinha A, Malo P, Deb K (2017) Evolutionary algorithm for bilevel optimization using approximations of the lower level optimal solution mapping. Eur J Oper Res 257(2):395–411
67. Sinha A, Malo P, Deb K (2018) A review on bilevel optimization: From classical to evolutionary approaches and applications. IEEE Trans Evol Comput 22(2):276–295
68. Sinha A, Malo P, Frantsev A, Deb K (2013) Multi-objective stackelberg game between a regulating authority and a mining company: A case study in environmental economics. In: 2013 IEEE congress on evolutionary computation (CEC-2013). IEEE Press
69. Sinha A, Malo P, Frantsev A, Deb K (2014) Finding optimal strategies in a multi-period multi-leader-follower stackelberg game using an evolutionary algorithm. Comput Oper Res 41:374–385
70. Sinha A, Malo P, Xu P, Deb K (2014) A bilevel optimization approach to automated parameter tuning. In: Proceedings of the 16th annual genetic and evolutionary computation conference (GECCO 2014). ACM Press, New York
71. Sinha A, Shaikh V (2021) Solving bilevel optimization problems using Kriging approximations. IEEE Trans Cybern
72. Sinha A, Soun T, Deb K (2019) Using Karush-Kuhn-Tucker proximity measure for solving bilevel optimization problems. Swarm Evol Comput 44:496–510
73. Smith JC, Song Y (2020) A survey of network interdiction models and algorithms. Eur J Oper Res 283(3):797–811
74. Smith WR, Missen RW (1982) Chemical reaction equilibrium analysis: Theory and algorithms. Wiley, New York
75. Sobieszczanski-Sobieski J, Agte JS, Sandusky RR Jr (2000) Bilevel integrated system synthesis. AIAA J 38(1):164–172
76. Stackelberg H (1952) The theory of the market economy. Oxford University Press, New York, Oxford
77. Sun H, Gao Z, Wu J (2008) A bi-level programming model and solution algorithm for the location of logistics distribution centers. Appl Math Model 32(4):610–616
78. Tuy H, Migdalas A, Värbrand P (1993) A global optimization approach for the linear two-level program. J Global Optim 3:1–23
79. Vicente L, Savard G, Júdice J (1994) Descent approaches for quadratic bilevel programming. J Optim Theory Appl 81:379–399
80. Von Stackelberg H (1934) Marktform und gleichgewicht. J. springer
81. Wein L (2009) Homeland security: From mathematical models to policy implementation: The 2008 philip mccord morse lecture. Oper Res 57(4):801–811
82. Wen U, Hsu S (1991) Linear bi-level programming problems—a review. J Oper Res Soc 42:125–133
83. White D, Anandalingam G (1993) A penalty function approach for solving bi-level linear programs. J Global Optim 3:397–419
84. Whittaker G, Färe R, Grosskopf S, Barnhart B, Bostian M, Mueller-Warrant G, Griffith S (2017) Spatial targeting of agri-environmental policy using bilevel evolutionary optimization. Omega 66:15–27
85. Wolsey LA (2020) Integer programming. Wiley, New York

86. Wood RK (1993) Deterministic network interdiction. Math Comput Model 17(2):1–18
87. Ye JJ, Zhu D (2010) New necessary optimality conditions for bilevel programs by combining the MPEC and value function approaches. SIAM J Optim 20(4):1885–1905
88. Zenklusen R (2010) Matching interdiction. Discret Appl Math 158(15):1676–1690

Chapter 16
Modeling and Solving Robust Chance-Constrained Binary Programs Using Sample Average Approximations

Shanshan Wang, Mohsen Mohammadi, and Sanjay Mehrotra

Abstract This chapter describes the modeling and solution approaches for binary programs involving chance constraints with ambiguously described probability distributions followed by the parameters. Here the chance constraints are satisfied over all possible probability distributions that belong to an ambiguity set. We use a stochastic bin packing model as an illustrative example. A bin packing problem allocates items of a certain size to a bin of a fixed capacity while minimizing the total allocation cost. The study of distributional robustness in chance-constrained is a topic of recent development. In a chance-constrained packing problem, we allocate items to a bin so that the bin capacity exceeds only with the desired probability. This chapter also describes a probability cut-based algorithm for solving such problems and presents computational results for their effectiveness. The computational results are presented using an ambiguity set of distributions described using the Wasserstein metric. The value of using an ambiguity set in improving the out-of-sample performance for satisfying the chance constraint is also illustrated.

Keywords Distributionally robust chance-constrained program · Bin packing · Branch-and-cut

---

This chapter is based on the results presented in [41, 42].

S. Wang
School of Management and Economics, Beijing Institute of Technology, Beijing 100081, China
e-mail: shshwang_bit@163.com

M. Mohammadi · S. Mehrotra (✉)
Department of Industrial Engineering and Management Sciences, Northwestern University, Evanston, IL 60208, USA
e-mail: mehrotra@iems.northwestern.edu

M. Mohammadi
e-mail: mohsen.mohammadi@northwestern.edu

F. Hamid (ed.), *Optimization Essentials*, International Series in Operations Research & Management Science 353, https://doi.org/10.1007/978-981-99-5491-9_16

16.1 Introduction

A general deterministic binary linear optimization model is given as:

$$\underset{x,y}{\text{minimize}} \sum_{j\in\mathcal{J}} c_j x_j \tag{16.1a}$$

$$\text{subject to} \sum_{i=1}^{n} a_{ij} x_j \leq b_i \qquad \forall i \in \mathcal{I}, \tag{16.1b}$$

$$x_j \in \{0, 1\} \qquad \forall j \in \mathcal{J}. \tag{16.1c}$$

If the parameters a_{ij} specifying the constraint in (16.1) are random, an approach to develop a well-posed model is to specify chance constraints replacing (16.1b) as follows:

$$\mathbb{P}_i \left\{ \sum_{j\in\mathcal{J}} \tilde{a}_{ij} x_j \leq b_j \right\} \geq 1 - \varepsilon \qquad \forall i \in \mathcal{I}, \tag{16.2}$$

where the random parameters $\tilde{a}_{ij}$ follow a probability distribution $\mathbb{P}_i$. It is possible that the probability distribution $\mathbb{P}_i$ is not known exactly. However, partial information on $\mathbb{P}_i$ is known. This partial information provides a set of probability distributions denoted by $\mathcal{P}_i$. If this is the case, we can model the chance constraint (16.2) as follows:

$$\inf_{\mathbb{P}_i \in \mathcal{P}_i} \left(\mathbb{P}_i \left\{ \sum_{j\in\mathcal{J}} \tilde{a}_{ij} x_j \leq b_j \right\} \right) \geq 1 - \varepsilon \qquad \forall i \in \mathcal{I}. \tag{16.3}$$

The model in (16.3) is called a *distributionally robust chance constraint model.* The set $\mathcal{P}_i$ is called an *ambiguity set.* This modeling framework ensures that the chance constraints are satisfied for every probability distribution belonging to this set; thus resulting in a worst-case satisfaction for the solution obtained from the model. Because of the worst-case specification in (16.3), there is a possibility that the solution obtained from the model may also satisfy the true unknown distribution. It is also possible to have elements in $\tilde{a}_{ij}$ follow a joint probability distribution over all elements, not just elements in $\tilde{a}_{ij}$ for a fixed i as in (16.1b). We will provide a discussion on this later in the chapter in the context of a model, referred to as the *distributionally robust chance-constrained bin packing model.*

16.1.1 Deterministic Bin Packing

Consider assigning a set of items to a set of bins of certain sizes in the best possible way. This is called the optimal bin packing. Here in order to achieve efficient packing, we minimize the total cost (i.e., the cost of assigning items and utilizing the bins), while additional constraints are met. When the item sizes are given, the bin capacity is fixed, and assignment costs are known, the deterministic bin packing problem has the following formulation:

$$\underset{x,y}{\text{minimize}} \sum_{j\in\mathcal{J}} c_j^a x_j + \sum_{i\in\mathcal{I}}\sum_{j\in\mathcal{J}} c_{ij}^b y_{ij} \tag{16.4a}$$

$$\text{subject to } y_{ij} \le x_j \quad \forall i \in \mathcal{I} j \in \mathcal{J}, \tag{16.4b}$$

$$\sum_{j\in\mathcal{J}} y_{ij} = 1 \quad \forall i \in \mathcal{I}, \tag{16.4c}$$

$$\sum_{i\in\mathcal{I}} \xi_i y_{ij} \le t_j \quad \forall j \in \mathcal{J}, \tag{16.4d}$$

$$x_j \in \{0,1\}, y_{ij} \in \{0,1\} \quad \forall i \in \mathcal{I}, j \in \mathcal{J}, \tag{16.4e}$$

where $\mathcal{I} := \{1, \ldots, |\mathcal{I}|\}$ denotes the set of items and $\mathcal{J} := \{1, \ldots, |\mathcal{J}|\}$ denotes the set of bins. We use $|\cdot|$ to denote the cardinality of a set. The cost for opening bin j is denoted by $c_j^a \ge 0$, and cost of assigning item i to bin j is denoted by $c_{ij}^b \ge 0$. Vector $\boldsymbol{\xi} = (\xi_1, \ldots, \xi_{|\mathcal{I}|})^\top$ is a vector of non-negative item sizes. The binary variable x_j for all $j \in \mathcal{J}$ is equal to 1 if bin j is open, and 0 otherwise. The binary variable y_{ij} for all $i \in \mathcal{I}$ and $j \in \mathcal{J}$ is equal to 1 if we assign item i to bin j, 0 otherwise. Let $\boldsymbol{x} = (x_1, \ldots, x_{|\mathcal{J}|})^\top$, $\boldsymbol{y}_j = (y_{1j}, \ldots, y_{|\mathcal{I}|j})^\top$, $\boldsymbol{y} = (\boldsymbol{y}_1, \ldots, \boldsymbol{y}_{|\mathcal{J}|})^\top$. The objective (16.4a) minimizes the total cost that includes the cost of opening a bin and subsequently assigning items to an opened bin. Constraints (16.4b) ensure that we can assign item i to bin j only if bin j is open. Constraints (16.4c) ensure that each item i is assigned to only one bin. Constraints (16.4d) ensure that the total item sizes assigned to a bin cannot exceed the capacity of the bin. Constraints (16.4e) define binary variables x_j and y_{ij}. If all c_j^a are 1, and $c_{ij}^b = 0$, $\forall i \in \mathcal{I}, j \in \mathcal{J}$; then the formulation minimizes the number of opened bins to pack the items. An example of the deterministic bin packing problem, which will be further extended throughout the chapter, is discussed below.

Example 16.1 Consider a problem instance of (16.4) with 5 items and 3 bins. Suppose the cost of assigning each item to each bin, the cost of opening each bin, capacity of each bin, and item sizes are as follows:

$$\mathbf{c^b} = \begin{pmatrix} 12 & 23 & 26 \\ 24 & 13 & 17 \\ 12 & 13 & 21 \\ 20 & 17 & 13 \\ 15 & 21 & 13 \end{pmatrix}, \quad \mathbf{c^a} = \begin{pmatrix} 50 \\ 85 \\ 70 \end{pmatrix}, \quad \mathbf{t} = \begin{pmatrix} 13 \\ 19 \\ 14 \end{pmatrix}, \quad \boldsymbol{\xi} = \begin{pmatrix} 8 \\ 12 \\ 7 \\ 5 \\ 12 \end{pmatrix}.$$

The problem formulation is given as

$$\left\{\begin{array}{ll}\text{minimize } 50x_1 + \cdots + 70x_3 + 12y_{11} + \cdots + 13y_{53} & \\ \text{subject to} \qquad\qquad\qquad y_{11} \leq x_1 & \\ \qquad\qquad \vdots\ \vdots\ \vdots & \\ \qquad y_{11} + y_{12} + y_{13} = 1 & \\ \qquad\qquad \vdots\ \vdots\ \vdots & \\ \quad 8y_{11} + 12y_{12} + \cdots + 12y_{51} \leq 13 & \\ \qquad\qquad \vdots\ \vdots\ \vdots & \\ x_j \in \{0, 1\}, y_{ij} \in \{0, 1\}, \quad \forall i \in \mathcal{I}, j \in \mathcal{J} & \end{array}\right\}, \qquad (16.5)$$

which admits an optimal solution of $y^*_{11} = y^*_{22} = y^*_{32} = y^*_{41} = y^*_{53} = 1$ and $x^*_1 = x^*_2 = x^*_3 = 1$, with the optimal value of 276.

The deterministic bin packing problem has been extensively studied both from theoretical and algorithmic perspectives. It is known to be strongly NP-Complete [e.g., 9]. Algorithms for solving this problem exactly and approximately are also well developed [e.g., see 37]. Algorithms such as branch and price have been developed [e.g., 44]. Even genetic algorithms have been used for solving these problems [e.g., 27]. Several variations and generalizations of the model (16.4) have also been studied. Well known variations and generalizations of the classical bin packing problem include multidimensional bin packing problem [e.g., 29, 31], variable size bin packing problem [e.g., 12, 25], and the dynamic bin packing problem [e.g., 3, 10]. More details on this topic can be found in the review papers by [9, 28] and the references therein.

The bin packing problem has many applications in operations research and management sciences. These examples arise in healthcare operating room surgery assignments [e.g., 16], scheduling of vehicle crash testing [e.g., 36], machine availability problems [e.g., 39], capacity planning in logistics [e.g., 13], and those in supply chain [e.g., 34]. For example, in the healthcare operating room surgery assignments problem, the decision-maker assigns a set of surgeries (item set $\mathcal{I}$) with service durations ($\boldsymbol{\xi}$) to a set of operating rooms (bin set $\mathcal{J}$) with time limits ($\boldsymbol{t}$), while minimizing the total assignment and opening cost. For scheduling of vehicle crash testing, a set of crash tests are assigned to the prototype vehicles. The capacity constraints restrict the number of tests assigned to a vehicle. Machine availability problems assign a set of jobs with processing time to a set of machines having a deadline. The constraints in this case ensure that the total work on a machine cannot exceed the deadline of the machine. The capacity planning in logistics addresses the capacity of a company needed to satisfy the demand. In all these application examples, it is assumed that parameters are known exactly. However, in practice, it is quite common to confront imprecise input data, which might result in unusable solutions. The following example further elaborates on this.

Example 16.2 In Example 16.1, the constraint $8y_{11} + 12y_{21} + 7y_{31} + 5y_{41} + 12y_{51} \leq 13$ is binding at the optimal solution. Suppose the size of item four is six instead of five. It is easy to see that the current optimal solution would be infeasible to the problem, making it unusable.

16.1.2 Chance-Constrained Bin Packing with Distributional Robustness

If the space required by each item in a knapsack is random, the classical deterministic knapsack constraint generalizes to a stochastic knapsack constraint. In the case of multiple bins, with random item sizes, the model becomes a stochastic multiple bin packing problem. Recent research has focused on such problems, as it allows to model examples such as operating room planning/scheduling, where the surgery duration is typically random [41]. In non-routine surgeries, the actual duration is estimated by not knowing exactly in advance. By not incorporating the surgical duration uncertainty one may create significant backlogs when multiple surgeries are scheduled in a room. Hence, one needs to consider the randomness of the likelihood of being able to perform all the tasks within the scheduled shift as desired. This naturally involves the use of a probability-based constraint, called a chance constraint that pre-specifies the probability of finishing the task on time.

A recent work [42] considered a joint chance-constrained decision framework, for which big-M strengthening results were obtained in the general context, and a probability cut algorithm was developed. In order to illustrate the ideas developed in [42] in this chapter we consider the following distributionally robust chance-constrained bin packing problem referred to as (DR-CBP):

$$\text{(DR-CBP)} \qquad \underset{x,y}{\text{minimize}} \sum_{j \in \mathcal{J}} c_j^a x_j + \sum_{i \in \mathcal{I}} \sum_{j \in \mathcal{J}} c_{ij}^b y_{ij} \tag{16.6a}$$

$$\text{subject to } (16.4b), (16.4c), (16.4e),$$

$$\inf_{\mathbb{P} \in \mathcal{P}} \mathbb{P} \left\{ \sum_{i \in \mathcal{I}} \xi_i y_{ij} \leq t_j \right\} \geq 1 - \varepsilon \qquad \forall j \in \mathcal{J}, \tag{16.6b}$$

where $\boldsymbol{\xi} = (\xi_1, \ldots, \xi_{|\mathcal{I}|})^\top$ is a vector of random sizes and follows probability distribution $\mathbb{P} \in \mathcal{P}$. We assume that the random size $\boldsymbol{\xi}$ has finite support set with N scenarios $\{\boldsymbol{\xi}^\omega\}_{\omega \in \Omega}$, where $\Omega = \{1, \ldots, N\}$. Hence, the distribution $\mathbb{P}$ can be characterized using a probability vector $(p_1, \ldots, p_N)^\top$ such that $p_\omega \geq 0$ and $\sum_{\omega \in \Omega} p_\omega = 1$. Let ξ_i^ω be the size of item i under the scenario $\omega \in \Omega$, let $\varepsilon \in [0, 1]$ be the level of chance satisfaction. Without loss of generality, we assume that $\xi_i^{\omega} < t_j$, for all $i \in \mathcal{I}, j \in \mathcal{J}, \omega \in \Omega$. Constraints (16.6b) are the chance constraints that the bin capacity constraints should be satisfied for probability distributions in the ambiguity set $\mathscr{P}$ with probability $1 - \varepsilon$.

If the items follow a normal distribution $\Phi(\mu_i, \sigma_i)$, for $\epsilon \leq 0.5$, constraint (16.6b) can be reformulated as a second-order cone constraint as follows:

$$\boldsymbol{\mu}^{\top} \boldsymbol{y}_j + \Phi^{-1}(1-\varepsilon)\sqrt{\sum_{i \in \mathcal{I}} \sigma_i^2 y_{ij}^2} \leq t_j \qquad \forall j \in \mathcal{J}.$$

Example 16.3 Consider again Example 16.1. Let item sizes be independent and normally distributed with mean $\boldsymbol{\mu} = (8, 12, 7, 5, 12)$ and variance $\boldsymbol{\sigma}^2 = (4, 3, 5, 1, 2)$, and $\varepsilon = 0.2$. One can then reformulate (16.6b) as

$$8y_{11} + \cdots + 12y_{51} + 0.84\sqrt{4y_{11}^2 + \cdots + 2y_{51}^2} \leq 13,$$
$$8y_{12} + \cdots + 12y_{52} + 0.84\sqrt{4y_{12}^2 + \cdots + 2y_{52}^2} \leq 19,$$
$$8y_{13} + \cdots + 12y_{53} + 0.84\sqrt{4y_{13}^2 + \cdots + 2y_{53}^2} \leq 14,$$

which reduces (DR-CBP) to a binary conic program.

However, when the items follow an arbitrary distribution, such reformulations may not be possible, and the chance-constrained problems can be much more difficult to solve than a typical binary linear program. In the chance-constrained programming literature, a common assumption is that the probability distributions have a finite support set, i.e., the probability distribution specifying the items is discrete. In Sects. 16.1.3 and 16.1.4, we review the literature on the chance-constrained bin packing problem and distributionally robust optimization (DRO) problem.

16.1.3 Literature Review on Chance-Constrained Bin Packing Problem

In terms of the chance-constrained bin packing problem, [26] developed an approximation algorithm to solve stochastic chance-constrained bin packing for a bandwidth allocation problem. Shylo et al. [38] studied the stochastic operating room scheduling problem where surgery durations follow a multivariate normal distribution. Deng and Shen [15] investigated an appointment scheduling problem with chance constraints to determine operating room allocation and surgery scheduling and proposed some strengthening strategies within a decomposition algorithm to solve the reformulation of this problem. Zhang et al. [52] investigated two-stage stochastic and distributionally robust chance-constrained bin packing problems with moment information. They solved the two model formulations by using a column generation-based branch-and-price method. Zhang et al. [51] investigated a distributionally robust chance-constrained bin packing problem where only moment information is available, and solved a second-order cone reformulation by developing a branch-and-cut algorithm with valid inequalities. Song et al. [40] studied a chance-constrained single bin problem. They provided a probabilistic cover reformulation and used the big-M and a lifting technique proposed by [50] to derive the probabilistic cover inequalities. Hoogervorst et al. [21] investigated the privacy and solution quality trade-off in a k-anonymized database used as input to the bin packing problem and proposed a stochastic model and a robust model with chance constraints. Cohen et al. [11] studied a model that quantifies the value of overcommitment by modeling the resource allocation problem (e.g., cloud computing) as a CCBP and developed a family of online algorithms that provide a constant factor guarantee from optimal. Wang et al. [41] studied a chance-constrained bin packing problem. They formulated the problem as an integer program with the help of the big-M technique and a bilinear program. By exploiting the structure of a bilinear knapsack set, they obtained three new classes of inequalities. A computational study was carried out with real data from a large public hospital to show the effectiveness of the proposed inequalities. More recently, [32] studied the problem of assigning stochastic jobs to a minimum number of servers, which is formulated as a stochastic CCBP with conflicts and finally derived a MILP. Wang and Mehrotra [43] have recently studied a robust expected utility bin packing problem with a chance constraint where the information on utility function is incomplete. Using a discretization approach, they reformulated the problem as a mixed-integer second-order cone program, which can be efficiently solved by a row generation algorithm.

In recent work, [42] study the assignment problems with a distributionally robust chance constraint. By taking advantage of the ambiguity set, a novel technique to estimate big-M in a semi-infinite reformulation is presented. A numerical study is performed with the idea on the surgery planning problem. This chapter employs the techniques developed by [42] to investigate the problem (DR-CBP). Different from [42], this chapter studies a distributionally robust chance-constrained bin packing problem where we also determine the number of opening bins to satisfy the chance constraint on the bin capacity. However, the method used to solve this problem is similar to the one developed in [42]. While presenting computational results for a general bin packing problem, we also provide out-of-sample performance results of the integer solutions obtained from the instances generated by using different underlying probability distributions. The results show that the solutions of (DR-CBP) are robust to the misestimation of the probability distribution.

16.1.4 Literature Review on Distributionally Robust Optimization

The DRO problems under the moment-based ambiguity sets and the statistical distance-based sets have been widely studied (see, e.g., [1, 2, 14, 17, 24, 30, 33, 42, 45]). For the distributionally robust chance-constrained programs, [5] studied a linear program with a probability constraint where distributions belong to an ambiguity set. They provided some tractable conditions that can be used to ensure that the chance satisfaction belongs to some sets. Xu et al. [48] studied a probabilistic envelope constraint to control the violation of the chance constraint. For the stochastic problem, they showed that the problem is equivalent to a semi-infinite optimization problem. For the comprehensive robust optimization problem, tight probabilistic bounds are obtained. References [6, 53] studied a moment-based ambiguity set and approximated distributionally robust chance constraints as tractable reformulations. Cheng et al. [8] considered a quadratic knapsack problem with distributionally robust chance constraints where the uncertainty set is characterized by first, second moments, and the joint support. They formulated the model as a semi-definite programming (SDP) problem. Yang and Xu [49] studied a distributionally robust chance constraint and probabilistic envelope constraints for non-linear uncertainty where only mean and variance information is available. For the previous problem, they showed it is tractable when the constraint function is quasi-convex in the random parameters and is concave in the variables. For the latter problem, they gave some conditions under which the problem is tractable. [19] presented a convex reformulation of a joint distributionally robust chance-constrained problem with the ambiguity set including the mean and an upper bound on the dispersion information. They showed that the joint chance constraints are conic representable. Jiang and Guan [23] studied a distributionally robust chance-constrained problem where the ambiguity set of the probability distribution is a ϕ-divergence measure-based set. They showed that the problem can be reformulated as a traditional chance-constrained problem by defining a risk level. Xie et al. [47] studied a distributionally robust joint chance constraint, and obtained a Bonferroni approximation of the problem by using the union bound. The approximation involves a set of single chance constraints, which can be reformulated as a deterministic convex problem. They also provided a condition that when the uncertain parameters are separable, then the optimized Bonferroni approximation is exact. When the uncertain set is characterized by marginal or by moments distributions, they also showed that the optimized Bonferroni approximation is tractable under various conditions. Ghosal and Wiesemann [18] studied a distributionally robust capacitated vehicle routing problem with probability constraints. They showed that if the ambiguity set satisfies a subadditivity condition, then the problem can be solved by using a branch-and-cut algorithm.

For the Wasserstein ambiguity set, as stated by [4, 17], comparing other popular notions of dissimilarity, such as Kullback-Leibler (KL) or other f-divergences with absolute continuous distributions, Wasserstein distance is well-defined between any pair of distributions over a sample space. Thus, Wasserstein distances between empirical and continuous distributions are asymptotic, finite, and informative (e.g., become the same if decay to 0). Moreover, the Wasserstein metric captures the closeness between support points while other metrics only consider the probabilities on them, and is well-defined for distributions with nonoverlapping supports. Moreover, distributions within a KL divergence or total variation distance, do not allow for probability mass outside the support of the empirical distribution. Recently, [7, 20, 22, 46] have investigated the distributionally robust chance-constrained problems where the ambiguity set is defined by using the Wasserstein ball. More specifically, [46] first provided a conditional value-at-risk (CVaR) interpretation, they rewrite the problem as a mixed-integer program. Then they used a CVaR reformulation to obtain inner and outer approximations. Moreover, they exploited the problem structure and proposed mixed-integer linear programs for the problem with only 0-1 variables. Chen et al. [7] investigated individual and joint chance-constrained problems with right-hand side uncertainty where the support set is continuous. They obtained a mixed-integer conic reformulation for the problem. In the special case where the Wasserstein ambiguity set is defined by 1-norm or ∞-norm, they provided a mixed-integer linear program reformulation. Ji and Lejeune [22] considered finite support and a continuum of realizations. When the support set is finite, they reformulated the problem as a mixed-integer linear program by using the dual theorem. This reformulation involves a big-M obtained by the bounds of the decision variables. Ho-Nguyen et al. [20] obtained a strengthened reformulation of a chance-constrained problem with right-hand side uncertainty. They also proposed two families of valid inequalities for the problem by exploring the connections between the problem and the sample average approximation, mixing sets, and robust binary program. Finally, more details on DRO can be found in a recent survey by [35].

16.1.5 Organization

The remainder of this chapter is organized as follows. In Sect. 16.2, we provide a semi-infinite programming reformulation of (DR-CBP). This section also presents a big-M coefficient strengthening procedure. From the solution methodology perspective, a branch-and-cut algorithm with probability cuts is proposed to solve (DR-CBP) in Sect. 16.3, which is adapted from [42]. For the sake of conciseness, convergence proof is omitted. Section 16.4 describes the bin packing problem with joint distributionally robust chance constraints. In this section, we extend the single chance constraints in (DR-CBP) to a joint one. Sect. 16.5 reports computational results on (DR-CBP) formulation to illustrate the importance of the distributional robustness approach to ensure the out-of-sample performance of a chance constraint model. It is important to note that with a smaller sample size the chance constraint models are solvable, and they become increasingly difficult to solve as the sample size increases. Distributional robustness ensures a significant improvement in the out-of-sample performance.

16.2 A Semi-infinite Programming Reformulation

Examples of ambiguity sets. While the results presented in this and subsequent sections are for a general ambiguity set of probability distributions, in the following, we give two examples for the ambiguity sets described on a finite support. The first example is the l_1-Wasserstein ambiguity set:

$$\mathscr{P}_W = \{\boldsymbol{p} \in \mathbb{R}^N_+ \mid \sum_{\omega\in\Omega}\sum_{k\in\Omega} \|\boldsymbol{\xi}^\omega - \boldsymbol{\xi}^k\| v_{\omega k} \le \eta,\ \sum_{k\in\Omega} v_{\omega k} = p_\omega,\ \forall\omega\in\Omega,$$
$$\sum_{\omega\in\Omega} v_{\omega k} = p^*_k,\ \forall k\in\Omega,\ \sum_{\omega\in\Omega} p_\omega = 1,\ \ v_{\omega k}\ge 0,\ \forall\omega,k\in\Omega\}, \tag{16.7}$$

where $\eta \ge 0$ is radius of Wasserstein ball and $\{p^*_k\}_{k\in\Omega}$ is an empirical probability distribution of $\boldsymbol{\xi}$. It is known that when $\eta = 0$, then p_ω reduces to p^*_ω for all $\omega\in\Omega$ and (DR-CBP) is a traditional chance-constrained bin packing problem. Under the definition, constraint (16.6b) using the Wasserstein ambiguity set can be rewritten as follows:

$$\inf\left\{\sum_{\omega\in\Omega} p_\omega \mathbf{1}\left(\sum_{i\in\mathcal{I}} \xi^\omega_i y_{ij} \le t_j\right) \,\middle|\, \boldsymbol{p}\in\mathscr{P}_W\right\} \ge 1-\varepsilon \qquad \forall j\in\mathcal{J},$$

where $\mathbf{1}(\cdot)$ is an indicator function.

Another example is called the moment-based ambiguity set which is given as follows:

$$\mathscr{P}_M = \left\{\boldsymbol{p}\in\mathbb{R}^N_+ \,\middle|\, \sum_{\omega\in\Omega} p_\omega = 1, l_{ik} \le \sum_{\omega\in\Omega} p_\omega(\xi^\omega_i)^k \le u_{ik}, \forall i\in\mathcal{I}, k=\{1,\cdots,K\}\right\},$$

where u_{ik} and l_{ik} are upper and lower bounds on kth moment, respectively, for $i\in\mathcal{I}, k=\{1,\cdots,K\}$.

Reformulation with finite samples. For $j\in\mathcal{J}$, let $z_j = (z^1_j,\ldots,z^N_j)^\top$, $z = (z_1,\ldots,z_{|\mathcal{J}|})^\top$. Now, let us introduce a binary variable z^ω_j to indicate if bin j satisfies the bin capacity constraint for each scenario $\omega\in\Omega$, namely,

$$z^\omega_j = \begin{cases} 1 & \text{if } \sum_{i\in\mathcal{I}} \xi^\omega_i y_{ij} \le t_j, \\ 0 & \text{otherwise.} \end{cases}$$

Theorem 16.1 shows that (DR-CBP) with the general ambiguity set $\mathscr{P}$ can be reformulated as a semi-infinite program with constraints (16.8b). The result stated below holds for any ambiguity set having a finite support that provides a polyhedral structure.

Theorem 16.1 (Theorem 1 [42]) *The model (DR-CBP) is equivalent to the following semi-infinite program:*

$$\text{(SIP)} \quad \underset{\boldsymbol{x},\boldsymbol{y},\boldsymbol{z}}{\text{minimize}} \sum_{j\in\mathcal{J}} c^a_j x_j + \sum_{i\in\mathcal{I}}\sum_{j\in\mathcal{J}} c^b_{ij} y_{ij} \tag{16.8a}$$

subject to (16.4b), (16.4c), (16.4e),

$$\inf_{p \in \mathscr{P}} \sum_{\omega \in \Omega} p_\omega z_j^\omega \geq 1 - \varepsilon \qquad \forall j \in \mathcal{J}, \tag{16.8b}$$

$$\sum_{i \in \mathcal{I}} \xi_i^\omega y_{ij} + \left(m_j^\omega(k_{\bar{q}}) - m_j^\omega(\omega)\right) z_j^\omega \leq m_j^\omega(k_{\bar{q}}) \qquad \forall j \in \mathcal{J}, \omega \in \Omega, \tag{16.8c}$$

where

$$m_j^\omega(k) := \underset{y_j \in \{0,1\}^{|\mathcal{I}|}}{\text{maximize}} \left\{ \sum_{i \in \mathcal{I}} \xi_i^\omega y_{ij} \middle| \sum_{i \in \mathcal{I}} \xi_i^k y_{ij} \leq t_j \right\}, \tag{16.9}$$

for $j \in \mathcal{J}$, *and* $\omega, k \in \Omega$. *Parameters* $m_j^\omega(\cdot)$ *are sorted in a non-decreasing order such that* $m_j^\omega(k_1) \leq \cdots \leq m_j^\omega(k_N)$, *and* $\bar{q} := \min\{l \mid \sup_{p \in \mathscr{P}} \sum_{j=1}^{l} p_{k_j} > \varepsilon\}$.

Note that the Wasserstein ambiguity set (16.7) is polyhedral which can be represented by its extreme points by the representation theorem. Thus, by considering the Wasserstein ambiguity set, constraints (16.8b) can be represented by a finite number of chance constraints where each is specified for a vertex solution of the polyhedron describing the Wasserstein set, and the semi-infinite reformulation reduces to an integer program. This result holds for any ambiguity set that has a polyhedral structure [1].

Now note also that the computation of $\bar{q}$ in the case $\mathscr{P} := \mathscr{P}_W$, in Theorem 16.1 requires to solve a series of linear programs. For a given z_j^ω, the left-hand side of (16.8b) is a linear programming-based formulation. However, for all $j \in \mathcal{J}$ and $\omega \in \Omega$, solving the linear programs for $\bar{q}$ can be computationally expensive.

The following corollary from [42] shows that we can use any probability distribution in $\mathscr{P}$ to estimate the big-M coefficients in the semi-infinite program. We can derive such probability distributions by taking the probability cuts described in Sect. 16.3. For example, the nominal distribution used to define the Wasserstein provides a valid probability distribution that can be used.

Corollary 16.1 ([42, Corollary 1]) *Let* $\{p'_\omega\}_{\omega \in \Omega} \in \mathscr{P}$, *and* $\hat{q} = \min\{l \mid \sum_{j=1}^{l} p'_{k_j} > \varepsilon\}$. *Then,* $\bar{q} \leq \hat{q}$ *and* $m_j^\omega(k_{\bar{q}}) \leq m_j^\omega(k_{\hat{q}})$.

The following example describes some numerical results on big-M coefficients yielded by $\bar{q}$ and $\hat{q}$.

Example 16.4 Suppose that in Example 16.1, we have ten different scenarios (Table 16.1) for item sizes ($N = 10$) and $\mathscr{P} = \mathscr{P}_W$, with $\eta = 1$. Let $\varepsilon = 0.5$ and $p_k^* = 0.1$ for all $k \in \Omega$. For each scenario $\omega \in \Omega$ and $j \in \mathcal{J}$, one would need to solve (16.9) for all $k \in \Omega$ (i.e., solving 300 problems in total). Here, we discuss the process for the first scenario and the capacity constraint of the first bin (i.e., $j = 1$, $\omega = 1$, and $k \in \{1, \ldots, 10\}$); other scenarios and constraints can be handled likewise. For the first scenario, by (16.9), we have

Table 16.1 Ten different scenarios for all $i \in \mathcal{I}$

Item (i)	Scenarios (ω) 1	2	3	4	5	6	7	8	9	10
1	5	5	4	5	10	12	10	8	6	9
2	8	8	13	11	15	15	12	9	12	11
3	6	3	9	3	4	8	7	4	7	3
4	3	5	7	4	6	3	6	9	3	5
5	16	13	8	10	16	16	8	9	15	16

Table 16.2 big-M coefficients for all $j \in \mathcal{J}$ and $\omega \in \Omega$

	j	ω 1	2	3	4	5	6	7	8	9	10
$\bar{q}$	1	11	10	16	11	16	20	17	13	13	14
	2	19	18	22	16	25	27	22	18	22	21
	3	14	11	16	14	19	20	17	17	18	14
$\hat{q}$	1	13	13	16	11	16	20	17	13	15	16
	2	22	18	22	18	26	28	23	21	22	25
	3	14	13	16	14	20	23	18	17	19	19

$$m_1^1(1) = 13,\ m_1^1(2) = 16,\ m_1^1(3) = 21,\ m_1^1(4) = 22,\ m_1^1(5) = 9,$$
$$m_1^1(6) = 9,\ m_1^1(7) = 16,\ m_1^1(8) = 22,\ m_1^1(9) = 11,\ m_1^1(10) = 11.$$

After sorting them in non-decreasing order, by Theorem 16.1, one gets

$$\begin{aligned}
&\texttt{iteration1}: l = 1, \quad p_5 = 0.23\\
&\texttt{iteration2}: l = 2, \quad p_5 + p_6 = 0.33\\
&\texttt{iteration3}: l = 3, \quad p_5 + p_6 + p_9 = 0.44\\
&\texttt{iteration4}: l = 4, \quad p_5 + p_6 + p_9 + p_{10} = 0.53;
\end{aligned}$$

hence, $\bar{q} = 4$ and $m_1^1(k_{\bar{q}}) = 11$. In Corollary 16.1, let $\{p'_\omega\}_{\omega\in\Omega}$ be equal to the empirical probability distribution (i.e., p_k^*). We have $\hat{q} = 5$, and thus $m_1^1(k_{\hat{q}}) = 13$, which is larger than $m_1^1(k_{\bar{q}})$. Table 16.2 reports all the big-M coefficients yielded by $\bar{q}$ and $\hat{q}$. One can see that in 17 cases, $\bar{q}$ returns smaller big-M coefficients, which also required solving 121 more linear programs.

16.3 A Probability Cut Algorithm

The algorithm presented below is based on taking an outer approximation of (DR-CBP). This algorithm was first proposed in [42]. The algorithm strengthens the outer approximation using a sequence of probability distributions. The use of a probability distribution generates a non-linear cut. This cut strengthens the outer approximation of the problem. The finite convergence of such an algorithm is ensured because of the polyhedral properties (finite number of vertices) of the Wasserstein ambiguity set. This result is formally stated in Theorem 16.2. We point out that the problem of generating the probability distribution specifying the cut is a linear program (referred to as (SP_j)). Consequently, (SP_j) is also solved to optimality and generates a vertex specifying the cut. We also point out that instead of using the probability cut approach, we can dualize the problem in (16.8), resulting in a more explicit model with dual variables added to the model. We experimented with the problem obtained from the dual formulation, but it turned out to be less efficient to solve.

Let us consider the model

$$\text{(MP)} \qquad \underset{(\boldsymbol{x},\boldsymbol{y},\boldsymbol{z})\in\left\{\{0,1\}^{|\mathcal{J}|}\times\{0,1\}^{|\mathcal{I}||\mathcal{J}|}\times\{0,1\}^{|\mathcal{J}|N}\right\}\cap\mathcal{X}}{\text{minimize}} \quad \sum_{j\in\mathcal{J}} c^a_j x_j + \sum_{i\in\mathcal{I}}\sum_{j\in\mathcal{J}} c^b_{ij} y_{ij}$$

$$\text{subject to (16.4b), (16.4c), (16.7c),}$$

where the set $\mathcal{X}$ is defined by any feasibility cuts added to the model based on the previously identified probability distributions and any subsequent strengthening of the problem. Given a feasible solution $(\hat{\boldsymbol{y}}, \hat{\boldsymbol{z}})$ of (MP) for each $j \in \mathcal{J}$, we define the following distribution separation problem:

$$(\text{SP}_j) \qquad \mathcal{S}_j(\hat{\boldsymbol{z}}) := \inf_{\boldsymbol{p}\in\mathscr{P}} \sum_{\omega\in\Omega} p_\omega \hat{z}_{j\omega}.$$

The separation problem defined in (SP_j) verifies the feasibility of $(\hat{\boldsymbol{y}}_j, \hat{\boldsymbol{z}}_j)$ to (DR-CBP). If $\mathcal{S}_j(\hat{\boldsymbol{z}}) \geq 1 - \varepsilon$, $(\hat{\boldsymbol{y}}_j, \hat{\boldsymbol{z}}_j)$ is feasible to (DR-CBP). Otherwise, we add the following probability and feasibility cuts to (MP).

Let $\{\hat{p}_\omega\}_{\omega\in\Omega} \in \mathscr{P}$ be an optimal solution of (SP_j) corresponding to $\hat{\boldsymbol{z}}$, then a probability cut is given as follows:

$$\sum_{\omega\in\Omega} \hat{p}_\omega z_{j\omega} \geq 1 - \varepsilon. \tag{16.11}$$

Let $\mathcal{I}^1_j = \{i \in \mathcal{I} | \hat{y}_{ij} = 1\}$. The following cut, referred to as a feasibility cut is also added to (MP):

$$\sum_{i\in\mathcal{I}^1_j} y_{ij} \leq |\mathcal{I}^1_j| - 1. \tag{16.12}$$

Example 16.5 Consider Example 16.4. Let us formulate the (MP) with big-M coefficients yielded by $\bar{q}$ (see Table 16.2). It is easy to verify that the following is a feasible solution to (MP):

$$y_{11} = y_{23} = y_{31} = y_{42} = y_{52} = 1,$$
$$z_{11} = z_{12} = z_{14} = z_{18} = z_{19} = z_{21} = z_{22} = z_{24} = z_{28} = z_{31} = z_{32} = z_{34} = z_{310} = 1,$$

while other y_{ij} and $z_{j\omega}$ not mentioned above are equal to zero. Given this feasible solution, one gets $\mathrm{SP}_1 = 0.40$, $\mathrm{SP}_2 = 0.27$, and $\mathrm{SP}_3 = 0.26$. Since $\mathrm{SP}_j < 0.5$ for all the three bins, one can add the following three cuts to (MP):

$$0.1z_{11} + 0z_{12} + 0.1z_{13} + 0.1z_{14} + 0.1z_{15} + 0.1z_{16} + 0.1z_{17} + 0.1z_{18} + 0.1z_{19} + 0.2z_{110} \geq 0.5,$$
$$0z_{21} + 0.07z_{22} + 0.1z_{23} + 0.1z_{24} + 0.1z_{25} + 0.1z_{26} + 0.1z_{27} + 0.1z_{28} + 0.2z_{29} + 0.13z_{210} \geq 0.5,$$
$$0.06z_{31} + 0.1z_{32} + 0.1z_{33} + 0.1z_{34} + 0.2z_{35} + 0.1z_{36} + 0.1z_{37} + 0.1z_{38} + 0.14z_{39} + 0z_{310} \geq 0.5.$$

Similarly, the following feasibility cuts can be added to (MP):

$$\{y_{11} + y_{31} \leq 1,\ \ y_{42} + y_{52} \leq 1,\ \ y_{23} \leq 0\}.$$

The upper and lower bounds are respectively denoted by UB and LB in Algorithm 1. Algorithm 1 is initialized by setting the iteration counter k to 0, UB to positive infinity, and LB to negative infinity. Nodes generated during the branching process are added to the list of nodes $\mathcal{N}$. We let (LMP) denote the linear program relaxation of (MP), and o denote a selected node. In Algorithm 1, lines 4–5 derive an optimal solution $(\boldsymbol{x}^k, \boldsymbol{y}^k, \boldsymbol{z}^k)$ by solving (LMP). The corresponding optimal value is denoted by obj^k. If obj^k is smaller than the upper bound (UB), then line 7 checks if the binary requirement is satisfied by $(\boldsymbol{x}^k, \boldsymbol{y}^k, \boldsymbol{z}^k)$. If $(\boldsymbol{x}^k, \boldsymbol{y}^k, \boldsymbol{z}^k)$ satisfies the binary requirement, the distribution separation problem (SP_j) is solved for all $j \in \mathcal{J}$ with $\mathscr{P}$, and we obtain the optimal solution $\{p_\omega^k\}_{\omega \in \Omega}$; and the corresponding optimal value $uobj^k$. In lines 10-13, if $uobj^k$ is smaller than $1 - \varepsilon$, the probability and feasibility cuts are added to (LMP). We return to line 4, if probability and feasibility cuts are found and perform the next iteration. Otherwise, we have a feasible solution $(\boldsymbol{x}^k, \boldsymbol{y}^k, \boldsymbol{z}^k)$ to (DR-CBP). In this case, line 15 updates the upper bound and save $(\boldsymbol{x}^k, \boldsymbol{y}^k, \boldsymbol{z}^k)$. On the other hand if $(\boldsymbol{x}^k, \boldsymbol{y}^k, \boldsymbol{z}^k)$ is fractional, violated inequalities are added or branching is performed as indicated in lines 17–19. As in the case of general branch-and-cut algorithms, termination takes place if $\mathcal{N}$ is empty. Upon termination the value UB is optimal, and the corresponding solution $(\boldsymbol{x}^*, \boldsymbol{y}^*, \boldsymbol{z}^*)$ is returned as an optimal solution in line 22.

Theorem 16.2 ([42]) *If there exists an oracle that can solve the distribution separation problem (SP_j) optimally, then Algorithm 1 terminates after a finite number of iterations. If UB $< +\infty$, then Algorithm 1 returns an optimal solution $(\boldsymbol{x}^*, \boldsymbol{y}^*, \boldsymbol{z}^*)$ of (DR-CBP) corresponding to the optimal value UB at termination.*

Algorithm 1: Branch-and-Cut Algorithm with Probability and feasibility Cuts

1 ***Initialize*** upper bound UB $= +\infty$, and lower bound LB $= -\infty$.
2 ***Initialize*** the number of iteration $k = 0$, node list $\mathcal{N} = \{o\}$, o has no branching constraints.
3 **while** *($\mathcal{N}$ is nonempty)* **do**
4   Solve the LP relaxation of (MP) at a node $o \in \mathcal{N}$, $\mathcal{N} \leftarrow \mathcal{N}/\{o\}$ and $k = k + 1$.
5   Derive an optimal objective value obj^k and solution $(\boldsymbol{x}^k, \boldsymbol{y}^k, \boldsymbol{z}^k)$.
6   **if** $obj^k < UB$ **then**
7   **if** $(\boldsymbol{x}^k, \boldsymbol{y}^k, \boldsymbol{z}^k)$ *is integer* **then**
8   **for** $j \in \mathcal{J}$ **do**
9   Derive an optimal objective value $uobj^k$ and solution $\boldsymbol{p}^k$ by solving (SP$_j$).
10   **if** $uobj^k < 1 - \varepsilon$ **then**
11   Add the probability cut (16.10) and feasibility cut (16.11) to the LP relaxation of (MP).
12   Go to step 4.
13   **end**
14   **end**
15   UB $= uobj^k$, $(\boldsymbol{x}^*, \boldsymbol{y}^*, \boldsymbol{z}^*) = (\boldsymbol{x}^k, \boldsymbol{y}^k, \boldsymbol{z}^k)$.
16   **end**
17   **if** $(\boldsymbol{x}^k, \boldsymbol{y}^k, \boldsymbol{z}^k)$ *is fractional* **then**
18   Branch, resulting in nodes o^* and o^{**}, $\mathcal{N} \leftarrow \mathcal{N} \cup \{o^*, o^{**}\}$.
19   **end**
20   **end**
21 **end**
22 **return** UB and its corresponding optimal solution $(\boldsymbol{x}^*, \boldsymbol{y}^*, \boldsymbol{z}^*)$.

16.4 Generalization to Joint Chance-Constrained Problems

In [42], we extended a distributionally robust chance-constrained problem to one involving multiple constraints over which the probability is specified. This problem is called a joint chance-constrained problem. In this section, we refer to the problem as (DR-JCBP). Specifically, the jointly distributionally robust chance constraint is given as follows:

$$\inf_{\mathbb{P} \in \mathscr{P}} \mathbb{P}\left\{\sum_{i \in \mathcal{I}} \xi_i y_{ij} \le t_j, \quad \forall j \in \mathcal{J}\right\} \ge 1 - \varepsilon. \tag{16.13}$$

Note that each scenario in the joint constraint corresponds to a matrix of items and bin-capacities. This allows us to model the possible dependence between the space used by the items and the bins in which they are assigned. It is easy to show that, under the finite support assumption, as in the previous case, (16.13) can be rewritten as

$$\inf_{p \in \mathscr{P}} \sum_{\omega \in \Omega} p_\omega z'_\omega \geq 1 - \varepsilon \qquad \forall j \in \mathcal{J}, \tag{16.14a}$$

$$\sum_{i \in \mathcal{I}} \xi_i^\omega y_{ij} + \left(m_j^\omega(k_{\bar{q}}) - m_j^\omega(\omega)\right) z'_\omega \leq m_j^\omega(k_{\bar{q}}) \qquad \forall j \in \mathcal{J}, \omega \in \Omega, \tag{16.14b}$$

where $m_j^\omega(\cdot)$ and $\bar{q}$ is defined similarly as in Sect. 16.2, with the following definition of the binary variable z'_ω used to write the chance constraint as a mixed-binary program:

$$z'_\omega = \begin{cases} 1 & \text{if } \sum_{i \in \mathcal{I}} \xi_i^\omega y_{ij} \leq t_j, \ \forall j \in \mathcal{J}, \\ 0 & \text{otherwise.} \end{cases}$$

Now the distribution separation problem is defined as

$$\text{(SP)} \qquad \mathcal{S}(\hat{z}') := \inf_{p \in \mathscr{P}} \sum_{\omega \in \Omega} p_\omega \hat{z}'_\omega,$$

and as before, its optimal solution $\{\hat{p}_\omega\}_{\omega \in \Omega}$ is used to generate the probability cut

$$\sum_{\omega \in \Omega} \hat{p}_\omega z'_\omega \geq 1 - \varepsilon. \tag{16.15}$$

An algorithmic analog of Algorithm 1 can also be developed. This algorithm for the joint constraint problem is outlined as Algorithm 2. Theorem 16.3 provides a formal statement about the finite convergence for Algorithm 2 in solving (DR-JCBP).

Theorem 16.3 ([42]) *If there exists an oracle that can solve the distribution separation problem (SP) optimally, then Algorithm 1 terminates after a finite number of iterations. If UB* $< +\infty$, *then Algorithm 2 returns an optimal solution* $(\boldsymbol{x}^*, \boldsymbol{y}^*, z'^*)$ *of (DR-JCBP) corresponding to the optimal value UB at termination.*

16.5 Computational Experiments

In this section, we present computational experiments whose setting is set similar to the operating room planning problem. Briefly, the problem seeks to assign a set of surgeries ($\mathcal{I}$) of certain types (e.g., Gynecology, Galactophore, and Lymphatic) with random duration ξ_i, $\forall i \in \mathcal{I}$, to a set of operating rooms $\mathcal{J}$, while minimizing the number of utilized operating rooms. In this application example, each operating room j has capacity t_j, $\forall j \in \mathcal{J}$, which needs to be taken into account. Hence, the overcapacity constraints for bins are modeled by chance constraints, where one restricts the probability of overtime to be less than pre-defined ε.

In what follows, in Sect. 16.5.1 we provide additional implementation details. Section 16.5.2 provides the performance of the proposed algorithm (i.e., Algorithm

Algorithm 2: Branch-and-Cut Algorithm with Probability and feasibility Cuts for (DR-JCBP)

```
1  Initialize upper bound UB = +∞, and lower bound LB = −∞.
2  Initialize the number of iteration k = 0, node list 𝒩 = {o}, o has no branching constraints.
3  while (𝒩 is nonempty) do
4      Solve the LP relaxation of (JMP) at a node o ∈ 𝒩, 𝒩 ← 𝒩/{o} and k = k + 1.
5      (JMP)  [see display below]
6      Derive an optimal objective value obj^k and solution (x^k, y^k, z'^k).
7      if obj^k < UB then
8          if (x^k, y^k, z'^k) is integer then
9              Derive an optimal objective value uobj^k and solution p^k by solving (SP).
10             if uobj^k < 1 − ε then
11                 Add the probability cut (16.14) and Σ_{j∈𝒥} Σ_{i∈ℐ_j^1} y_ij ≤ |ℐ| − 1 to the LP
                   relaxation of (JMP).
12                 Go to step 4.
13             end
14             UB = uobj^k, (x*, y*, z'*) = (x^k, y^k, z'^k).
15         end
16         if (x^k, y^k, z'^k) is fractional then
17             Branch, resulting in nodes o* and o**, 𝒩 ← 𝒩 ∪ {o*, o**}.
18         end
19     end
20 end
21 return UB and its corresponding optimal solution (x*, y*, z'*).
```

Line 5:

$$\text{(JMP)} \qquad \underset{(x,y,z')\in\{\{0,1\}^{|\mathcal{J}|}\times\{0,1\}^{|\mathcal{I}||\mathcal{J}|}\times\{0,1\}^{N}\}\cap\mathcal{X}}{\text{minimize}} \quad \sum_{j\in\mathcal{J}} c_j^a x_j + \sum_{i\in\mathcal{I}}\sum_{j\in\mathcal{J}} c_{ij}^b y_{ij}$$

$$\text{subject to (16.4b), (16.4c), (16.13b),}$$

1 described in Sect. 16.3) to solve (DR-CBP) with the Wasserstein ambiguity set. An out-of-sample analysis is presented in Sect. 16.5.3.

16.5.1 Implementation Details

Mimicking the setting of the operating room problem, in our test examples, we consider 14 items ($|\mathcal{I}|$) of nine types and six bins ($|\mathcal{J}|$). We consider the bin capacity to be ten, that is, $t_j := 10$ for $j \in \mathcal{J}$. Table 16.3 presents the mean and standard deviation of the item size for each type and the proportion for each item type. The proportion in Table 16.3 is used to calculate the number of items for each type in our problem instance generator. Also, the item sizes are rounded to the nearest integer. We choose log-normal distribution and Gamma distribution for generating problem instances. (The distributions are parameterized to ensure that the mean and standard

Table 16.3 The mean (Mean), standard deviation (Std), and proportion (Proportion) for each item type are reported

Item type	Mean	Std	Proportion
Type 1	1.0	1.2	0.29
Type 2	1.4	0.8	0.15
Type 3	3.1	1.0	0.14
Type 4	2.6	1.5	0.13
Type 5	2.2	1.6	0.07
Type 6	2.4	1.3	0.07
Type 7	1.4	0.4	0.06
Type 8	2.6	1.1	0.06
Type 9	3.1	1.7	0.03

deviation are equal to that given in Table 16.3.) Note also that the item sizes are non-negative. We generated five instances for each sample size and each distribution type. Note that for the problem that minimizes the number of opened bins $c^a_j := 1$, and $c^b_{ij} := 0, \forall i \in \mathcal{I}, j \in \mathcal{J}$.

All algorithms are coded in the programming language C and optimization problems were solved using IBM CPLEX version 12.10 callable libraries. All experiments are conducted on the cedar cluster of Compute Canada with a single CPU core and 64 GB memory. Only one thread is used for all computations. We turned off the CPLEX pre-solve procedure when implementing the branch-and-cut algorithm because we needed to use the CPLEX callback function to work on the original problem in our testing. A prioritized node selection strategy is used in the branch-and-cut algorithm. Specifically, branching on x is preferred over y, and y is preferred over z. For all instances, we use the time limit of two hours. We report the solution time (in seconds) for the instances solved to optimality within the time limit.

16.5.2 Computational Results

In this section, we show the performance of Algorithm 1 for solving the reformulation (16.8) of (DR-CBP). Based on Corollary 16.1, we let $\bar{q}$ defined in Theorem 16.1 be equal to $\hat{q}$ by using the empirical probability distribution. The sample average distribution with independent and identically distributed samples is used as the empirical probability distribution. The problems were generated using $N = 200$ scenarios. Different from the above discussion, we can also obtain the following reformulation by using linear programming duality to solve (DR-CBP):

$$\text{minimize} \sum_{j \in \mathcal{J}} c_j^a x_j + \sum_{i \in \mathcal{I}} \sum_{j \in \mathcal{J}} c_{ij}^b y_{ij}$$

subject to (16.4b), (16.4c), (16.4e), (16.7c),

$$\rho_j^1 - \eta \rho_j^2 + \sum_{k \in \Omega} p_k^* \rho_{jk}^4 \geq 1 - \varepsilon \qquad \forall j \in \mathcal{J},$$

$$-\rho_j^1 + \rho_{j\omega}^3 + z_{j\omega} \geq 0 \qquad \forall j \in \mathcal{J}, \omega \in \Omega,$$

$$||\boldsymbol{\xi}^\omega - \boldsymbol{\xi}^k|| \rho_j^2 - \rho_{j\omega}^3 - \rho_{jk}^4 \geq 0 \qquad \forall j \in \mathcal{J}, \omega, k \in \Omega,$$

$$\rho_j^2 \geq 0 \qquad \forall j \in \mathcal{J},$$

where $\boldsymbol{\rho}^1$, $\boldsymbol{\rho}^2$, $\boldsymbol{\rho}^3$, and $\boldsymbol{\rho}^4$ are dual variables of the constraints in the Wasserstein ambiguity set in (16.7). When solving the reformulation obtained from taking the dual, we found that it cannot solve any generated instances to optimality within the run-time limit. In particular, for most of the instances, the average optimality gap is larger than 70%, and for the rest of the instances, it cannot even return a feasible solution. We conjecture that this is because, despite providing an explicit formulation, taking the dual introduces new continuous variables, which makes the models much harder to solve for an off-the-shelf solver. This is evidenced by a significant increase in the size of the branch-and-cut tree.

Table 16.4 presents the average CPU time (in seconds) for Algorithm 1 and the separation problem, the average number of nodes, the number of probability and feasibility cuts, the number of instances that can be solved optimally within the two-hour run-time limit over the total five instances, and the opened bins for the instances generated by the log-normal distribution.

From Table 16.4 we observe that as the level of chance satisfaction ε increases, the average number of nodes also increases significantly. This results in a substantial increase in the average solution time. In particular, when $\varepsilon = 0.15$ the average solution time for (DR-CBP) is about two times higher than the one when $\varepsilon = 0.1$. We obtain similar results for $\varepsilon = 0.2$ compared to $\varepsilon = 0.15$. This can be explained by

Table 16.4 The computational results of (DR-CBP) for the log-normal distribution of the item size

ε	η	AvT-B&CP	AvT-SP	# of nodes	# of p&f-cuts	solved	# of bins
0.1	0.1	665.6	2.8	20,206	2	5/5	(4, 4, 4, 4, 4)
	0.5	687.0	5.2	18,862	8	5/5	(4, 4, 4, 4, 4)
	1	979.6	8.5	29,064	22	5/5	(4, 4, 4, 4, 4)
0.15	0.1	1,411.8	2.6	68,251	1	5/5	(4, 4, 4, 4, 4)
	0.5	1,514.2	2.7	74,474	4	5/5	(4, 4, 4, 4, 4)
	1	1,614.5	3.4	76,711	6	5/5	(4, 4, 4, 4, 4)
0.2	0.1	3,273.7	2.5	305,332	0.4	5/5	(4, 4, 4, 4, 4)
	0.5	2,811.5	2.4	239,960	1	5/5	(4, 4, 4, 4, 4)
	1	3,045.4	2.9	262,355	2	5/5	(4, 4, 4, 4, 4)

Table 16.5 The computational results of (DR-CBP) for the Gamma distribution of the item size

ε	η	AvT-B&CP	AvT-SP	# of nodes	# of p&f-cuts	solved	# of bins
0.1	0.1	392.7	1.9	18,029	2	5/5	(4, 4, 4, 4, 4)
	0.5	411.3	6.1	17,892	26	5/5	(4, 4, 4, 4, 4)
	1	566.7	18.6	25,763	132	5/5	(4, 4, 4, 4, 4)
0.15	0.1	543.9	1.4	32,169	0	5/5	(4, 4, 4, 4, 4)
	0.5	637.3	1.6	35,817	1	5/5	(4, 4, 4, 4, 4)
	1	523.7	2.4	32,111	5	5/5	(4, 4, 4, 4, 4)
0.2	0.1	2,138.7	2.6	135,230	1	5/5	(4, 4, 4, 4, 4)
	0.5	2,233.2	2.8	140,934	3	5/5	(4, 4, 4, 4, 4)
	1	2,169.9	3.5	133,828	6	5/5	(4, 4, 4, 4, 4)

the fact that when ε increases, it makes the problem hard to decide how many bins are required to open while satisfying the chance constraint, and minimizing the total cost. We can also observe that no significant changes occur in average solution times when the Wasserstein radius η varies from 0.1 to 1. Furthermore, we found that the average number of probability and feasibility cuts for (DR-CBP) is usually under 30, and this number decreases with the level of chance satisfaction and grows with the Wasserstein radius. This is expected as the size of the Wasserstein ambiguity set is increasing with a larger radius, the algorithm generates more solutions that are infeasible based on the ambiguity sets. Note that after adding a probability cut, we can use the branch-and-cut tree from the incumbent problem as a warm-start for the new problem.

Table 16.5 provides the computational results for the instances generated by the Gamma distribution. Table 16.5 further confirms the effectiveness of Algorithm 1. Similar to the previous table, as the level of chance satisfaction ε increases, the solution time increases dramatically, while it doesn't change significantly when η varies from 0.1 to 1. In addition, compared with Table 16.4, the solution time for the instances generated by the Gamma distribution is slightly smaller than the one of the log-normal distribution, even though they have the same number of opened bins.

16.5.3 Out-of-Sample Performance of (DR-CBP) Solutions

In Sect. 16.5.2, we used a finite support set drawn from a probability distribution to solve the distributionally robust chance constraints in (DR-CBP). This section evaluates the solutions obtained from the discrete model via an out-of-sample performance analysis. Thus, we generate a large number of scenarios (i.e., 1,500,000) from the log-normal distribution. Table 16.6 reports the average and the worst-case overcapacity probability, the average overcapacity (in minutes), and 85, 95, 99%

Table 16.6 The out-of-sample performance for the log-normal distribution of the item size

ε	η	Mean-p	Worst-p	Mean-over	85%	95%	99%
0.1	0.1	0.066	0.113	5.9	0.0	23.5	90.5
	0.5	0.057	0.073	4.7	0.0	14.0	90.5
	1	0.057	0.081	4.8	0.0	13.0	90.5
0.15	0.1	0.080	0.136	7.4	1.0	33.0	96.5
	0.5	0.066	0.130	5.9	0.5	23.0	90.0
	1	0.065	0.110	5.7	0.0	23.0	90.5
0.2	0.1	0.128	0.165	12.3	4.0	49.0	112.0
	0.5	0.111	0.162	10.5	3.0	43.5	105.5
	1	0.089	0.156	8.1	2.0	38.5	100.5

quantiles (in minutes) of the overcapacity for the integer solution obtained from the instances generated by the log-normal distribution.

We observe from results in Table 16.6 that the average out-of-sample probability for the overcapacity is typically under the ε value for all the generated instances, and the worst-case out-of-sample probability for the overcapacity is also around the ε value. Table 16.6 also shows that as the size of the Wasserstein ball η increases, the average and worst-case probability for the overcapacity decreases in (DR-CBP). We have similar observations for the average overcapacity and 85, 95, and 99% quantiles for the overcapacity. Considering the level of chance constraint satisfaction into account, we can use a larger η value, i.e., let η be equal to 1, without increasing the computational burden as observed in Table 16.4. From Table 16.6, we observe that selecting the Wasserstein radius η is important in (DR-CBP) and will affect the optimal opening and assignment decisions. We can use a data-driven bisection method to estimate the optimal η value. In this method, we can use $\eta \in \{0, \eta_{\max}\}$ at the first step, where $\eta_{\max}$ is the maximal Wasserstein radius given the discrete support set, and this number can be obtained from the available finite samples. By assuming that the out-of-sample performance is monotonous, we can iteratively let $\eta_{\max}^{k+1} := \eta_{\max}^{k}/2$, where k is the index of iterations. Then we evaluate the out-of-sample performance by using the optimal integer solutions derived from (DR-CBP) with $\eta_{\max}^{k+1}$ and stop when the out-of-sample performance cannot achieve the level of chance satisfaction.

Table 16.7 presents the out-of-sample performance for the solution obtained from the instances generated by the Gamma distribution. The general observations are similar to those for the log-normal case. For example, the average out-of-sample probability for the overcapacity is smaller than the ε value for all the generated instances, and the worst-case probability is around the ε value. As η increases, the probability and overcapacity is decreasing for most of the instances. However, compared with Table 16.6, the out-of-sample performance of the optimal solution obtained by the instances generated by the Gamma distribution is slightly worse than the one by the log-normal distribution. More specifically, the average and worst-case over-

Table 16.7 The out-of-sample performance for the Gamma distribution of the item size

ε	η	Avg-prob	Worst-prob	Avg-over	85%	95%	99%
0.1	0.1	0.076	0.114	6.8	0.0	28.5	108.0
	0.5	0.073	0.102	6.5	0.0	27.0	108.5
	1	0.073	0.108	6.6	0.0	26.5	109.0
0.15	0.1	0.092	0.162	8.6	3.5	38.5	112.5
	0.5	0.084	0.152	7.8	2.0	33.0	110.0
	1	0.084	0.132	7.7	0.5	34.5	109.5
0.2	0.1	0.124	0.195	12.5	8.5	54.5	118.0
	0.5	0.098	0.173	9.5	3.0	46.5	112.5
	1	0.099	0.155	9.4	2.5	48.0	111.5

capacity probability are larger for most of the instances when compared with the log-normal distribution. Similar observations hold for the 85, 95, and 99% quantiles of the overcapacity. Although we see from Table 16.5 that the average solution time for the Gamma distribution is lower compared to the log-normal distribution, this is achieved at the cost of the out-of-sample performance. In summary, from Tables 16.6 and 16.7 we conclude that the optimal solutions obtained by (DR-CBP) are robust to misestimation of the item sizes' probability distribution.

16.6 Concluding Remarks

In this chapter, we illustrate the use of the big-M technique to reformulate (DR-CBP) as a semi-infinite program. We also describe the use of a branch-and-cut framework with probability and feasibility cuts to solve the problem. The computational results show that the probability cut algorithm can solve the instances with $N = 200$ scenarios within two hours time limit. As ε increases, the average solution time for the problem increases significantly. However, the average solution time is less sensitive to the Wasserstein radius η in the range from 0.1 to 1. By using an appropriate Wasserstein radius with the number of samples $N = 200$, the integer solutions obtained from the instances generated by log-normal and Gamma distributions achieve the desirable out-of-sample performance of the chance satisfaction and overcapacity. Generally, the solutions are robust to the misestimation of the probability distribution even though they are obtained using a small number of samples. The chapter illustrates the value of modeling chance constraints within the framework of distributional robustness; specifically, for the problem that can only be solved with small sample sizes. Although we focused on the bin packing problem, we also remark that the algorithmic framework discussed in this chapter has wider applicability when the problem structure provides constraints of the form appearing in this chapter.

References

1. Bansal M, Huang KL, Mehrotra S (2018) Decomposition algorithms for two-stage distributionally robust mixed binary programs. SIAM J Optim 28:2360–2383
2. Ben-Tal A, Den Hertog D, De Waegenaere A, Melenberg B, Rennen G (2013) Robust solutions of optimization problems affected by uncertain probabilities. Manag Sci 59:341–357
3. Berndt S, Jansen K, Klein KM (2020) Fully dynamic bin packing revisited. Math Program 179:109–155
4. Bernton E, Jacob PE, Gerber M, Robert CP (2019) On parameter estimation with the Wasserstein distance. Inf Inference: J IMA 8:657–676
5. Calafiore GC, El Ghaoui L (2006) On distributionally robust chance-constrained linear programs. J Optim Theory Appl 130:1–22
6. Chen W, Sim M, Sun J, Teo CP (2010) From CVaR to uncertainty set: Implications in joint chance-constrained optimization. Oper Res 58:470–485
7. Chen Z, Kuhn D, Wiesemann W (2018) Data-driven chance constrained programs over Wasserstein balls. arXiv:1809.00210
8. Cheng J, Delage E, Lisser A (2014) Distributionally robust stochastic knapsack problem. SIAM J Optim 24:1485–1506
9. Coffman EG, Csirik J, Galambos G, Martello S, Vigo D (2013) Bin packing approximation algorithms: Survey and classification. In: Handbook of combinatorial optimization. Springer, New York, pp 455–531
10. Coffman EG Jr, Garey MR, Johnson DS (1983) Dynamic bin packing. SIAM J Comput 12:227–258
11. Cohen MC, Keller PW, Mirrokni V, Zadimoghaddam M (2019) Overcommitment in cloud services: Bin packing with chance constraints. Manag Sci 65:3255–3271
12. Correia I, Gouveia L, Saldanha-da Gama F (2008) Solving the variable size bin packing problem with discretized formulations. Comput Oper Res 35:2103–2113
13. Crainic TG, Gobbato L, Perboli G, Rei W (2016) Logistics capacity planning: A stochastic bin packing formulation and a progressive hedging meta-heuristic. Eur J Oper Res 253:404–417
14. Delage E, Ye Y (2010) Distributionally robust optimization under moment uncertainty with application to data-driven problems. Oper Res 58:595–612
15. Deng Y, Shen S (2016) Decomposition algorithms for optimizing multi-server appointment scheduling with chance constraints. Math Program 157:245–276
16. Denton BT, Miller AJ, Balasubramanian HJ, Huschka TR (2010) Optimal allocation of surgery blocks to operating rooms under uncertainty. Oper Res 58:802–816
17. Esfahani PM, Kuhn D (2018) Data-driven distributionally robust optimization using the Wasserstein metric: Performance guarantees and tractable reformulations. Math Program 171:115–166
18. Ghosal S, Wiesemann W (2020) The distributionally robust chance-constrained vehicle routing problem. Oper Res 68:716–732
19. Hanasusanto GA, Roitch V, Kuhn D, Wiesemann W (2017) Ambiguous joint chance constraints under mean and dispersion information. Oper Res 65:751–767
20. Ho-Nguyen N, Kılınç-Karzan F, Küçükyavuz S, Lee D (2021) Distributionally robust chance-constrained programs with right-hand side uncertainty under Wasserstein ambiguity. Math Program 1–32
21. Hoogervorst R, Zhang Y, Tillem G, Erkin Z, Verwer S (2019) Solving bin-packing problems under privacy preservation: possibilities and trade-offs. Inf Sci 500:203–216
22. Ji R, Lejeune MA (2020) Data-driven distributionally robust chance-constrained optimization with Wasserstein metric. J Global Optim 1–33
23. Jiang R, Guan Y (2016) Data-driven chance constrained stochastic program. Math Program 158:291–327
24. Jiang R, Guan Y (2018) Risk-averse two-stage stochastic program with distributional ambiguity. Oper Res 66:1390–1405
25. Kang J, Park S (2003) Algorithms for the variable sized bin packing problem. Eur J Oper Res 147:365–372

26. Kleinberg J, Rabani Y, Tardos É (2000) Allocating bandwidth for bursty connections. SIAM J Comput 30:191–217
27. Kucukyilmaz T, Kiziloz HE (2018) Cooperative parallel grouping genetic algorithm for the one-dimensional bin packing problem. Comput Ind Eng 125:157–170
28. Lodi A, Martello S, Monaci M (2002) Two-dimensional packing problems: A survey. Eur J Oper Res 141:241–252
29. Lodi A, Martello S, Vigo D (1999) Approximation algorithms for the oriented two-dimensional bin packing problem. Eur J Oper Res 112:158–166
30. Luo F, Mehrotra S (2019) Decomposition algorithm for distributionally robust optimization using Wasserstein metric with an application to a class of regression models. Eur J Oper Res 278:20–35
31. Martello S, Pisinger D, Vigo D (2000) The three-dimensional bin packing problem. Oper Res 48:256–267
32. Martinovic J, Hähnel M, Scheithauer G, Dargie W (2021) An introduction to stochastic bin packing-based server consolidation with conflicts. TOP 1–36
33. Mehrotra S, Papp D (2014) A cutting surface algorithm for semi-infinite convex programming with an application to moment robust optimization. SIAM J Optim 24:1670–1697
34. Perboli G, Gobbato L, Perfetti F (2014) Packing problems in transportation and supply chain: New problems and trends. Procedia-Soc Behav Sci 111:672–681
35. Rahimian H, Mehrotra S (2019) Distributionally robust optimization: A review. arXiv:1908.05659
36. Reich D, Shi Y, Epelman M, Cohn A, Barnes E, Arthurs K, Klampfl E (2016) Scheduling crash tests at Ford motor company. Interfaces 46:409–423
37. Scheithauer G (2017) Introduction to cutting and packing optimization: Problems, modeling approaches, solution methods, vol 263. Springer, Berlin
38. Shylo OV, Prokopyev OA, Schaefer AJ (2012) Stochastic operating room scheduling for high-volume specialties under block booking. INFORMS J Comput 25:682–692
39. Song G, Kowalczyk D, Leus R (2018) The robust machine availability problem-bin packing under uncertainty. IISE Trans 50:997–1012
40. Song Y, Luedtke JR, Küçükyavuz S (2014) Chance-constrained binary packing problems. INFORMS J Comput 26:735–747
41. Wang S, Li J, Mehrotra S (2021) Chance-constrained multiple bin packing problem with an application to operating room planning. INFORMS J Comput 33(4):1661–1677
42. Wang S, Li J, Mehrotra S (2022) A solution approach to distributionally robust chance-constrained assignment problems. INFORMS J Optim 4:125–247
43. Wang S, Mehrotra S (2021) Robust concave utility maximization over a chance-constraint. Working paper. http://www.optimization-online.org/DB_FILE/2021/12/8716.pdf
44. Wei L, Luo Z, Baldacci R, Lim A (2020) A new branch-and-price-and-cut algorithm for one-dimensional bin-packing problems. INFORMS J Comput 32:428–443
45. Wiesemann W, Kuhn D, Sim M (2014) Distributionally robust convex optimization. Oper Res 62:1358–1376
46. Xie W (2019) On distributionally robust chance constrained programs with Wasserstein distance. Math Program 1–41
47. Xie W, Ahmed S, Jiang R (2022) Optimized Bonferroni approximations of distributionally robust joint chance constraints. Math Program 191:79–112
48. Xu H, Caramanis C, Mannor S (2012) Optimization under probabilistic envelope constraints. Oper Res 60:682–699
49. Yang W, Xu H (2016) Distributionally robust chance constraints for non-linear uncertainties. Math Program 155:231–265
50. Zemel E (1989) Easily computable facets of the knapsack polytope. Math Oper Res 14:760–764
51. Zhang Y, Jiang R, Shen S (2018) Ambiguous chance-constrained binary programs under mean-covariance information. SIAM J Optim 28:2922–2944
52. Zhang Z, Denton BT, Xie X (2021) Branch and price for chance-constrained bin packing. INFORMS J Comput 32(3):547–564
53. Zymler S, Kuhn D, Rustem B (2013) Distributionally robust joint chance constraints with second-order moment information. Math Program 137:167–198

Part II
Applications

Chapter 17
Production Planning and Scheduling for Parallel Machines with Sequence-Dependent Setup Times

Nadjib Brahimi, Abrar Khalaf, Rim Larbi, and Hebah Al-Hammadi

Abstract The authors present a production planning and scheduling problem inspired by a collaboration project with an aluminum production company. The objective is to meet customers' demands for some aluminum products with different characteristics while minimizing the total processing time. In the first part of the chapter, the authors present a literature review of production planning and scheduling problems, together with their different solution approaches. The second part focuses on aluminum production processes. It presents research work on production planning and scheduling in the aluminum industry. Then, it describes the problem and its characteristics together with the solution approach. The chapter finishes with numerical results, conclusions, and recommendations for future research work. The numerical results are validated by the planning service of the company. The approach proposed saves several days of work for the planners as the planning problem is solved within a few seconds. We discuss the reasons for such a very small CPU time and provide recommendations on how to solve harder real-life problems.

Keywords Sequence dependent · Mixed integer linear programming · Aluminum · Lot sizing · Scheduling

17.1 Introduction

Production planning and scheduling are challenging optimization problems to solve. They are even more complex when it comes to real industrial applications, especially the scheduling of continuous or batch production in chemical and manufacturing

N. Brahimi (✉)
Rennes School of Business, Rennes, France
e-mail: nadjib.brahimi@rennes-sb.com

A. Khalaf
Emirates Global Aluminium, Dubai, United Arab Emirates

R. Larbi
Ecole de Technologie Supérieure, Montreal, Canada

H. Al-Hammadi
Dubai Aviation Engineering Projects, Dubai, United Arab Emirates

F. Hamid (ed.), *Optimization Essentials*, International Series in Operations Research & Management Science 353, https://doi.org/10.1007/978-981-99-5491-9_17

industries, for example. This complexity is due to the highly combinatorial nature of these problems and to their very large size in practice.

With the advanced technology and the spread of Enterprise Resource Planning systems (ERP), many companies are now equipped with Advanced Planning Software (APS) which provides a solid decision support system to rely on. Such systems enable the automation and optimization of production planning and scheduling tasks. However, there are some special requirements and constraints that the planning software tools are not able to accommodate. The complexity of the processes in some manufacturing sectors can often lead to unsuccessful implementations of those systems. In Aluminum industry, the production process is very complex with its different stages such as casting, rolling, homogenizing, and cutting.

This project was done as a part of the collaboration of the authors with an aluminum company in the United Arab Emirates. We will call this company DXB company in the rest of the chapter. The project mainly concerns the billet scheduling problem at the casting stage. This stage is where the molten metal is poured into the molds to solidify and form the required shape based on the order's specifications. The casting stage is considered as the bottleneck in the production process and it determines the throughput of the company. The authors formulate the problem as a mixed integer linear programming formulation (MILP) which is solved using a commercial solver. The results are presented at the end of this chapter.

17.2 Production Planning and Scheduling

Manufacturing companies transform raw materials into finished products (sold to external customers) through their production processes. To achieve this, the utilization of resources such as budget, machinery, and labor needs to be properly planned. This section presents some of the concepts related to decisions in production planning, scheduling, and their integration.

17.2.1 Production Planning and Lot Sizing

Good production planning results in better utilization of resources (avoiding waste) and higher customer satisfaction. Production planning decisions are usually considered as mid-term tactical decisions spanning over several days to a few months. Production plans, as stated by [24], are inherently optimization problems where the aim is to maximize the productivity by minimizing the whole costs associated. In production planning, many decisions should be considered. This part will expose some of those considerations briefly. The most important thing to start with is what planning decisions to include [24] such as the production and inventory levels to attain. Also, the hierarchical level to include in the planning should be reflected especially if the product is assembled or fabricated in parallel. It is significant to integrate and coordinate the processes of those kinds of manufacturing environments (see for example [10]). Equivalently, plans are always described in terms of time. Production plans are usually built on a rolling horizon basis. For example, a company plans production over a horizon of 4 weeks. The plans are frozen and executed for the first week at the end of which a new plan is built for the next 4 weeks, etc. Another important concept in production planning is the aggregation of resources, products, and time in order to reduce the complexity of their optimization problems. The number of products can be reduced from several hundred to a few families, which results in much easier optimization problems to solve.

The single-item lot sizing problem The dynamic lot sizing problem represents a family of production planning problems where demand is time-dependent and setups are required at each production run. The most basic problem in this category is the single-item dynamic lot sizing problem. Wagner and Whitin [50] solved this problem using a $O(T^2)$ time dynamic programming algorithm (T is the length of the planning horizon). Faster algorithms were developed 30 years later by different researchers to solve it in $O(T \log T)$ (e.g., [49]). Besides the algorithmic contributions, researchers have studied many extensions of the single-item lot sizing problem. These include the consideration of capacity constraints, product perishability, consideration of returns, in addition to consideration of multi-item and multi-level cases including assembly and disassembly operations. For an overview of these extensions, the reader is invited to read [13, 27]).

The multi-item lot sizing problem with limited production capacity As a first step toward the formulation of the real aluminum production planning and scheduling problem on which the authors have worked, the mixed integer linear programming formulation of a multi-item lot sizing problem with setup times and limited production capacity (CLSP-ST) is presented below (e.g., [45]). The objective is to determine the production plan of N items over a planning horizon of T time periods while satisfying customer demands D_{it} for each item i and each time period t without any shortages or backorders. The variable production cost per unit produced of item i in time period t is (p_{it}). Each unit consumes σ_i units of the capacity of a single resource. The available capacity per period t is C_t. The inventory holding cost and the setup cost are h_{it} and s_{it}, respectively. A fixed setup time of τ_i unit is required to start a new production lot of item i.

The decision variables for each item i and each time period t are X_{it}, I_{it}, and Y_{it} which represent the produced quantity, the ending inventory level, and the binary setup variable, respectively. Variable Y_{it} takes the value 1 if there is a setup of i in t and 0 otherwise ($X_{it} > 0$ only if $Y_{it} = 1$).

Model CLSP-ST

$$\text{Minimize} \sum_{i=1}^{N}\sum_{t=1}^{T}(s_{it}Y_{it} + p_{it}X_{it} + h_{it}I_{it}) \tag{17.1}$$

subject to:

$$I_{i,t-1} + X_{it} = D_{it} + I_{it} \qquad \forall i,t \tag{17.2}$$

$$\sum_{i=1}^{N}(\sigma_i X_{it} + \tau_i Y_{it}) \le C_t \qquad \forall t \tag{17.3}$$

$$X_{it} \le \left(\sum_{l=t}^{T} D_{il}\right) Y_{it} \qquad \forall i,t \tag{17.4}$$

$$Y_{it} \in \{0, 1\} \qquad \forall i,t \tag{17.5}$$

$$X_{it}, I_{it} \ge 0 \qquad \forall i,t \tag{17.6}$$

This formulation is a direct extension of the aggregate formulation of the capacitated multi-item lot sizing problem (see, e.g., [9, 34]). The objective (17.1) is to minimize the total cost composed of the fixed and variable production costs in addition to the inventory holding costs. Inventory balance equations are represented by constraints (17.2). Constraints (17.4) are used to link setup variables to the continuous production variable. Integrality and non-negativity restrictions are represented by (17.5) and (17.6).

Capacity constraints (17.3) are the only ones that differ from lot sizing problem without setup times (CLSP). When setup times are neglected in the CLSP, this constraint is expressed as $\sum_{i=1}^{N}(\sigma_i X_{it}) \le C_t$ or equivalently as $\sum_{i=1}^{N} X'_{it} \le C_t$ (by applying the right variable change). In Constraints (17.3), the left-hand side of the constraints takes into consideration the time required to set up the resource.

The CLSP-ST is NP-hard in the strong sense since it is an extension of CLSP for which this property was proven by [17] through a transformation from the three partition problem. Notice that the single-item lot sizing problem with capacity constraints is also NP-hard but in the weak sense. Researchers, such as [22], have solved it to optimality using pseudo-polynomial time dynamic programming algorithms. In terms of feasibility, consider the following conditions:

$$\sum_{k=1}^{t} C_k \ge \sum_{i=1}^{N}\sigma_i \sum_{k=1}^{t} D_{ik}$$

This condition is necessary and sufficient to find a feasible solution in the case of the CLSP. In the case of the CLSP-SP, this condition is not sufficient. Actually, the feasibility check of the CLSP-ST is a NP-complete problem as it was proved by [33].

Several heuristics were proposed to solve the CLSP-ST. The most successful among these is the Lagrangian relaxation-based heuristic developed by [45]. It starts by relaxing the capacity constraints (linking constraints) which are dualized in the objective function with corresponding Lagrangian multipliers π_t.

The resulting relaxed problem can be decomposed into N independent single-item lot sizing problems without any capacity restrictions. The aggregate formulation for each product i is:

$$\text{Minimize} \sum_{t=1}^{T} \left(s_{it}^{L} Y_{it} + p_{it}^{L} X_{it} + h_{it} I_{it} \right) + \text{Constant} \tag{17.7}$$

Subject to:

$$I_{i,t-1} + X_{it} = D_{it} + I_{it} \qquad \forall t \tag{17.8}$$

$$X_{it} \le \left(\sum_{l=t}^{T} D_{il} \right) Y_{it} \qquad \forall t \tag{17.9}$$

$$Y_{it} \in \{0, 1\} \qquad \forall t \tag{17.10}$$

$$X_{it}, I_{it} \ge 0 \qquad \forall t \tag{17.11}$$

where

$$s_{it}^{L} = s_{it} + \tau_i \pi_t$$

and

$$p_{it}^{L} = p_{it} + \sigma_i \pi_t$$

Each of these sub-problems can be solved in $O(T^2)$ using algorithm of [50] or in $O(T \log T)$ using any one among the three algorithms of [1, 21], or [49]. The resulting solution does not necessarily respect all capacity constraints but it provides a lower bound on the optimal solution which can be used to evaluate any heuristic solution. Furthermore, the Lagrangian procedure takes the a priori infeasible solution and fixes it by shifting production and setups to different periods. At each iteration of the heuristic, Lagrangian multipliers are updated using the subgradient method. Consequently, the gap between the lower bound and the upper bound (feasible solution values) decreases over time and reaches zero in several cases, which means that the optimal solution is found. Recent efforts to improve the performance of the heuristic of [45] included mainly attempts to strengthen the lower bounds (e.g., [47, 48]) or to relax other constraints (e.g., [26]). For some recent applications of Lagrangian relaxation to lot sizing problems, see [11, 12, 16].

The dynamic lot sizing with sequence-dependent setups A particular lot sizing model that integrates sequencing decisions into lot sizing is the multi-item lot sizing problem with sequence-dependent setup times and/or costs [5, 25]. In this case, and contrary to the classical multi-item lot sizing problem, the sequence in which products are to be processed on machines during a time period is important and can impact both feasibility and optimality of the problem. Besides the aluminum industry and the project presented in this chapter, there are several process industries where sequence-dependent setups take place. These include glass containers, metallurgy, and some chemical industries.

The formulation below (slightly modified from [5]) corresponds to a single-stage single resource (machine) problem. Besides the notation introduced in the beginning of this section, the following extensions and modifications are needed. It is assumed for each product that the inventory holding costs are time independent and are denoted by h_i. Since the setup time is sequence-dependent, the setup time becomes τ_{ij} which represents the time needed to set up the machine from product i to product j. The setup cost is now time-independent. It is represented by s_{ij} and indicates the cost to switch production from product i to product j. An important assumption in this situation is the triangular inequality of both the setup costs and times. It suggests that $s_{ik} \leq s_{ij} + s_{jk}$ and $\tau_{ik} \leq \tau_{ij} + \tau_{jk}$. The decision variables are the production quantity of item i in time period t (X_{it}) and the inventory level at the end of each period (I_{it}). Variable M_{ijt} is equal to 1 if a setup occurs on the machine from product i to product j ($i \neq j$) in time period t, otherwise zero. It is also assumed that setup carryover is allowed. That is, if the machine is set up to produce product i at the end of time period t, it does not need to be set up again at the beginning of period $t+1$ if it starts production with the same item. Hence, decision variable B_{it} is introduced. It is equal to 1 if the machine is set up to produce item i at the beginning of period t, otherwise zero. Finally, decision variable U_{it} is used for the sub-tour elimination constraints which will be explained later.

The formulation is as follows:

$$\text{Minimize} \sum_{i=1}^{N} \sum_{j=1 \backslash j \neq i}^{N} \sum_{t=1}^{T} s_{ij} M_{ijt} + \sum_{i=1}^{N} \sum_{t=1}^{T} (p_i X_{it} + h_i I_{it}) \tag{17.12}$$

Subject to:

$$I_{i,t-1} + X_{it} = D_{it} + I_{it} \qquad \forall\, i, t \tag{17.13}$$

$$\sum_{i=1}^{N} \sigma_i X_{it} + \sum_{i=1}^{N} \sum_{j=1 \backslash j \neq i}^{N} \tau_{ij} M_{ijt} \leq C_t \qquad \forall\, t \tag{17.14}$$

$$X_{it} \leq \sum_{l=t}^{T} D_{il} \left(\sum_{j=1 \backslash j \neq i}^{N} M_{ijt} + B_{it} \right) \qquad \forall\, i, t \qquad (17.15)$$

$$\sum_{i=1}^{N} B_{it} = 1 \qquad \forall\, t \qquad (17.16)$$

$$B_{it} + \sum_{j=1}^{N} M_{jit} = B_{i,t+1} + \sum_{j=1}^{N} M_{ijt} \qquad \forall\, i, t \qquad (17.17)$$

$$U_{it} + NT_{ijt} - (N-1) - NB_{it} \leq U_{jt} \qquad \forall\, , i, j \neq i, t \qquad (17.18)$$

$$X_{it}, I_{it}, B_{it}, U_{it} \geq 0,\ M_{ijt} \in \{0, 1\} \qquad \forall\, i, j, t \qquad (17.19)$$

The objective function (17.12) minimizes the total costs modified from the objective in (17.1) by considering the sequence-dependent setup costs. The inventory balance equations are constraints (17.13). Capacity constraints represented by (17.14) take into consideration processing times and the sequence-dependent setup times. Constraints (17.15) link the continuous production variables X_{it} with binary variables M_{ijt} and B_{it}. Constraints (17.16) indicate that the machine is set up for one and only one product at the beginning of each period t. The setup carryover and conservation from one period to another is represented by Constraints (17.17). Finally, Constraints (17.18) are the sub-tour elimination constraints. They are used to avoid disconnected cycles just as it is used in the traveling salesman problem.

17.2.2 Scheduling

The output of the tactical production planning decision becomes an input to the scheduling process which allows a set of jobs or tasks to be executed effectively. Scheduling decisions are short-term operational decisions that span over several hours or a few days. A particular decision in scheduling consists of allocating available production resources to perform certain tasks in a specific sequence; this is called the sequencing problem. General scheduling problems provide sequencing and timing of execution of tasks on some resources while minimizing or maximizing a certain objective.

Scheduling problems have attracted a lot of attention from researchers since the 1950s. Several journal papers, books, and textbooks have been published on the subject, including [15, 37], among others. References [2, 4] are among the most comprehensive surveys on scheduling problems, especially those with setup times/costs and parallel machines.

Examples of scheduling problems include flow shop and job shop problems, single or multi-stage problems with or without parallel processors, and open shop problems (see, e.g., [30, 37, 41]).

Because of the large variety of scheduling problems, it is preferable to refer to them using the three-field notation proposed by [3, 23]. Each problem is denoted by a combination of three fields $\alpha|\beta|\gamma$. The three fields represent the scheduling type, the information and conditions of scheduling, and the performance criteria, respectively. For example, $\alpha = 1$ indicates a single machine problem, while $\alpha = R_m$ indicates a scheduling problem with unrelated parallel machines. Example values of β can be "pmtn" to indicate preemption or p_{ij} which indicates a particular processing time value (e.g., $p_{ij} = 1$ unit of time). Finally, the performance criterion can be the makespan ($\gamma = C_{max}$) or the maximum lateness (($\gamma = L_{max}$), for example.

In general, the notation f_{max} is used to represent C_{max} or L_{max} and the notation $\sum f_j$ all the other criteria with a summation such as the total weighted completion time.

In the following, each condition is explained briefly. The preemptive (pmtn) condition is where the process is allowed to be interrupted in order to process another job. The precedence constraint (prec) represents the situation in which the job can be produced simultaneously on multiple machines. Otherwise, a job should be processed on several machines in a certain order and/or a precedence condition should be respected between different jobs.

The next conditions are based on the nature of setup times. The problem can be with sequence-dependent setup times (ST_{sd}) or with sequence-independent setups (ST_{si}). The problem addressed in this project is considered as unrelated parallel machines where setups are sequence dependent. The machines are classified as unrelated due to the fact that they differ in terms of size, capacity, and capability.

In parallel machine scheduling problems, performance measures usually include the minimization of the maximum lateness, the weighted sum of tardiness, or the makespan. To solve the problem with parallel and unrelated machines, several authors have proposed heuristic and exact algorithms. Because most of these problems are NP-hard, the exact algorithms are only able to solve instances with few machines and jobs [19]. Examples of such exact algorithms include the Branch and Bound (B&B) algorithm of [35] with the objective of minimizing the makespan. They use Lagrangian relaxation to determine lower bounds. Another B&B procedure was proposed by [31]. In this case, the objective was the minimization of the weighted sum of tardiness. None of these publications considered the setup times between jobs.

The problem in this chapter can be considered as an extension of the unrelated parallel machine scheduling problem with sequence-dependent setup times.

As an example, an integer linear programming formulation is presented here. The two problems to which this formulation applies are $Q|p_j = 1|\sum f_i$ and $Q|p_j = 1|f_{\max}$. The formulation is a transportation network model with n sources and mn sinks. That is, the positioning of each job J_j ($j = 1, ..., n$), is indicated with a machine index $i = 1, ..., m$ and a position index $k = 1, ..., n$. In the network model, we set the cost of arc $(j, (i, k))$ equal to $c_{ijk} = f_j(kq_i)$, where q_i is a speed factor of machine M_i.

The only set of decision variables is the binary variables x_{ijk} such that $x_{ijk} = 1$ if J_j is executed on machine M_i in the kth position and $x_{ijk} = 0$ otherwise.

$$\text{Minimize} \sum_{i=1}^{n}\sum_{j=1}^{m}\sum_{k=1}^{n} c_{ijk}x_{ijk} \quad \text{or} \qquad \text{Minimize} \max_{i,j,k}\{c_{ijk}x_{ijk}\}$$

Subject to: (17.20)

$$\sum_{i=1}^{n}\sum_{k=1}^{n} x_{ijk} = 1 \qquad \forall j \qquad (17.21)$$

$$\sum_{i=1}^{n} x_{ijk} \leq 1 \qquad \forall i, k \qquad (17.22)$$

$$x_{ijk} \geq 0 \qquad \forall i, j, k \qquad (17.23)$$

The optimal solution for this problem can be obtained in $O(n^3)$. For some particular objectives, there exists a faster algorithm that can run in $O(n \log n)$ [23]. It is well known in the scheduling literature that mathematical programming solvers applied to corresponding IP formulation are not a good choice. The brief literature review below shows that most combinatorial optimization algorithms (mainly heuristics) very often do not rely on mathematical programming formulations to solve such problems. However, the impressive increase in computer performance in addition to improvements in the modern mathematical and especially (mixed integer) linear solvers is making this approach more and more attractive. The results that were obtained in this project are an illustration of such a success.

Considering setup time in scheduling problems makes considerable cost reductions. In most industries, it is not reasonable to ignore them. This is the case in the aluminum industry. Furthermore, in most foundries, there are several furnaces or casting machines working in parallel and one of the decisions that has to be taken by the planner is the assignment of orders/jobs to the different parallel resources. Scheduling problems with parallel machines have also been extensively studied in the scheduling literature (See [37] for a detailed presentation of these problems).

A Branch and Bound algorithm was proposed by [38] to solve a problem with setup times with the objective of minimizing the weighted sum of tardiness. Tran and Beck [44] considered the minimization of the makespan and proposed Benders decomposition to solve their problem. In the last 20 years, there was a considerable increase in the number of publications on metaheuristics for scheduling problems with unrelated parallel machines ([4]. These include Genetic algorithms (e.g., [46]), simulated annealing (e.g., [51]), Ant Colony Optimization (e.g., [7]), and clonal selection (e.g., [19]). All of these studies had as an objective the minimization of the makespan. A comparison between several methods was made by [19] and showed that the clonal selection algorithm is efficient. Another stream of research on solution methods includes the use of mixed integer linear programming formulations combined with heuristics and metaheuristics. This is the case of [8], who proposed several mixed integer linear programming formulations for an unrelated parallel machine scheduling problem with sequence and machine-dependent setup times and makespan minimization.

A hybrid flow shop (HFS), as described in [32], is a series of production stages. Each stage consists of several parallel machines. It is possible that some stages will consist of a single machine, but at least one stage must have more than one machine in order to be considered as a hybrid flow shop. A job must go through all the stages, while it must be processed only on one machine at each stage. The measures of performance of the parallel machine scheduling problem include the minimization of the maximum makespan along with other performance measures such as minimizing the mean flow time, maximum lateness, total tardiness, and number of tardy jobs. This problem is presented here because it can very well model several situations in the aluminum industry. For example [29] presented an original formulation of a sequence-dependent hybrid flow shop of an aluminum foundry production situation. The first stage consists of parallel machines that share a common casting machine, which represents the second stage. The objective in their model is to minimize the sum of sequence-dependent setup costs and the holding cost as well as minimizing the sum of sequence-dependent product penalties. The developed mixed integer linear programming model was able to get a very good solution in a reasonable amount of time, however, the optimal solution was not found as it requires a lot of computing time by the used software.

The hybrid flow shop problem is known to be NP-hard even for very restrictive assumptions, such as allowing preemption. Besides its importance for aluminum production, it can be found in several other types of industries such as electronics, paper, and textile. It has been considerably studied in the literature and several exact and heuristic approaches were proposed to solve it and its extensions. For a comprehensive literature review, the reader is invited to see the surveys of [36, 39].

Real-world production scheduling problems often result in even more intractable models; only realistic modeling of the problem features can help managers in their decisions. Thus, developing specific algorithms to solve the problem in such a way that these algorithms are flexible enough to accommodate frequent changes in the problem description by the companies is very challenging. It is for this reason that the authors have chosen to build a mixed integer linear programming formulation (MILP) that can be solved using a MILP solver and that can be modified easily to accommodate very specific constraints. The mathematical programming formulation is efficient in the sense that it needs very few minutes to find the optimal solution. Once the problem is well defined and all constraints are identified, it will be possible to think about other extensions of this study by developing fast and dedicated algorithms, for example.

17.2.3 Integrated Lot Sizing and Scheduling

As explained earlier, the scope of production planning and scheduling differ from each other. However, they are complementary to each other. For that reason, considering the integration of two decisions in a single model is relevant and can result in several benefits. The ultimate goal of integration is to guarantee the feasibility of all the processes related to planning and scheduling. Still, it is not easy to do this job as many challenges could appear. A very interesting reference is [18], where the importance of integration is presented; giving more details about this topic.

17.3 Production Process at DXB Company

The company DXB was created in 1975 and has grown up to become the largest single-site primary aluminum producer in the world. In 2009, the company covered a 1.3 million tons per annum smelter, a 3,100 MW power station, and other facilities. The company produces high-quality aluminum products in four main forms: billets, re-melt ingots, slab ingots, and liquid metal. In this chapter, we are interested in the billet production. The billets are plain cylindrical shape bars. They are used for extrusion and forging processes in industry. In DXB company, it is considered as an end product since it is a direct input in the customers' processes. For that reason, the company takes intensive care in its production process. The company classifies the billet production process into two stages: casting stage and finishing stage.

17.3.1 Casting Stage

In the casting stage, billets are formed through direct casting (DC) machines (Fig. 17.1). Each casting unit is connected to three furnaces. The latter are classified into melters and holders. The melter furnace has the ability to melt a high amount of scrap while the holders are used to accommodate the direct hot metal poured from the crucibles. Each machine is connected to one melter furnace and two holder furnaces. Once the metal is ready in the furnace, it is poured into the launders where it passes through a degaser unit and a Ceramic Foam Filter unit (CFF). The purpose of these units is to remove the hydrogen and other inclusions from the metal before it is poured into the casting table, where it will be solidified in the required billet diameter. Each casting table can be used to produce billets of different diameters. Some diameters share the same table base while the molds are changed based on the size. The site contains a total of five direct casting (DC) machines that are different in size, capacity, and capability of producing the different diameters (See Fig. 17.1). During the furnace preparation, a unique recipe of elements is added to the pure aluminum such as silicon, magnesium, copper, and many others. This unique recipe

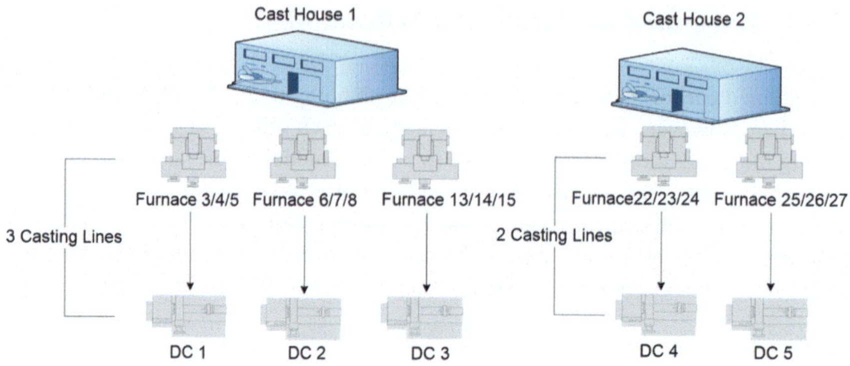

Fig. 17.1 The five available casting lines distribution in DXB company

is referred to as *alloy*. The company produces hundreds of different alloys based on customer specifications.

The process of preparing casting tables and fixing the required molds on the tables requires a period of at least three days. Preparing the casting table is a very critical job that requires a lot of manual work and costly consumables. So, it is not cost-effective to set up a diameter on a machine to produce a small quantity of billets with a given diameter. On the other hand, there is a control set by the process control department to limit the maximum number of drops (furnace batches) allowed on a casting table, after that, the table must be removed and sent for re-preparation. The constraints related to the casting table and molds are very critical to the schedule and must be considered wisely.

17.3.2 The Finishing Stage

The finishing stage starts after the solidification of the billets. At this point, billets will be taken from the table to the lay-down area where they will be processed in ultra-sonic devices to capture any defects (internal damages). After the inspection process, the billet logs enter the homogenizers. Homogenization is an important process in billets production. It strengthens the mechanical properties of the billets. These transformations will help the customers in extruding their billets which affects the quality of their final products. The company uses 2 types of homogenizers: batch and continuous. The type of alloy in a billet will decide which homogenizer to use. Some alloys can be processed in both homogenizers. In such scenario, the homogenizing path will be based on the customers' request or the availability of the homogenizers. The diameter size is also a key factor in determining the homogenizing path. The processing time depends on the diameter and some homogenizers can process only certain diameters. The two types of homogenizers differ in their process. In continuous homogenization, billet logs are fed into the homogenization machine log by log passing through the homogenizer. On the other hand, in batch homogenization, billet logs will be stacked on each other to build a load which is then charged to the homogenizer as one batch. After homogenization, the billets are moved to the sawing area where they will be cut to a certain length specified by the customers. Pieces of the billets of the same order are bundled and sent to the storage area for further inspection (mainly for external damages) by quality control team.

17.4 Objectives of the Study

The aim of this project is to find an optimal schedule for the billet production at the casting stage by minimizing the total processing time and total setup times. This can be obtained by achieving the following objectives:

- Develop mathematical model for the billet aluminum scheduling problem with the objective of minimizing total processing and setup times.
- Analyze the performance of the model on real data provided by the company and perform a sensitivity analysis to observe the impact of different parameters on the performance and on the structure of the solution.

From the literature of production planning and scheduling, this problem is closely related to the unrelated parallel machine scheduling problem with sequence-dependent setups. Several special constraints related to the aluminum billet production were integrated into the developed model. The fact that setups are sequence dependent makes the problem directly related to the traveling salesman problem and to vehicle routing problems.

To formulate the problem and solve it, the authors have followed the following methodology. The first step was to conduct a literature review about planning and scheduling and explore some of the previous work that was carried out on similar problems or in the same field. The outcome of this step allowed the authors to clearly identify the problem. A data collection was conducted in order to prepare all the necessary data to implement and test the solution. Next, using the knowledge gained from the literature review, a mathematical model is developed based on the real case constraints. Finally, the implementation of the MILP was carried out and the analysis of the generated results was presented.

17.5 Related Work

Very few studies on the aluminum industry address the scheduling problem. In [28], the scheduling problem on a single casting machine was addressed. In their model, the jobs are already assigned to the machines, so the objective was to find the sequence in which to cast aluminum logs such that setup times are minimized. The problem was formulated as a traveling salesman problem and solved using a Genetic Algorithm. In

[20], a scheduling problem for continuous aluminum casting lines was studied. Their problem is very close to the problem addressed here in the sense that they are solving a continuous aluminum casting scheduling problem with parallel machines. However, [20] just consider work balance as an objective instead of makespan minimization and do not consider the new case-specific constraints imposed by the company in our case.

17.6 Problem Formulation

In order to state the scheduling model minimizing the total processing time and total setup times on the casting centers, the following notations are used:

Indices and sets:
K: Set of casting machines available
J: Set of jobs to be scheduled where $J = 0, \ldots, |J|$
Δ: Set of diameters that can be manufactured
i, j: Job index ($i, j = 0, \ldots, |J|$). The first job ($j = 0$) is dummy
δ: Job diameter index ($\delta = 1, \ldots, |\Delta|$)
k: Machine index ($k = 1, \ldots, |K|$)
J_k: Set of jobs that can be processed on machine k

Parameters:
q_i: Size of job i in number of billets
δ_i: Diameter of job i
AV_k : Availability of machine k in hours based on shutdown plan
p_{ik} : Processing time of job i on machine k
S_{ijk}: Setup time required if job i immediately precedes job j on machine k
MRH_δ: Minimum run hours of diameter δ. It corresponds to the minimum time a diameter should be used on the casting table.
$Bs_{\delta k}$: Average batch size of diameter δ on machine k, i.e., the number of billets per batch.
$MaxC$: Maximum number of batches allowed in one diameter before changing to another

Decision Variables:
$y_{ik} = 1$, if job i is assigned to machine k, 0 otherwise.
$x_{ijk} = 1$, if job i is the immediate predecessor of job j on machine k, 0 otherwise.
$z_{\delta,k} = 1$, if diameter δ is assigned to casting machine k, 0 otherwise
U_{ik}: Continuous variable used in the sub-tour elimination constraints.

Using the above notation, a mixed integer linear programming model can be formulated as follows:

$$\text{Minimize} \sum_{k\in K}\sum_{i\in J_k} p_{ik}y_{ik} + \sum_{k=1}^{K}\sum_{i\in J_k}\sum_{j\in J_k} S_{ijk}x_{ijk} \tag{17.24}$$

subject to

$$y_{ik} = 0 \qquad \forall k, \forall i \notin J_k \tag{17.25}$$

$$\sum_{k\in K} y_{ik} = 1, \qquad \forall i \neq 0 \tag{17.26}$$

$$\sum_{i\in J_k} P_{ik}y_{ik} + \sum_{i\in J_k}\sum_{j\in J_k} S_{ijk}x_{ijk} \leq AV_k \qquad \forall k \tag{17.27}$$

$$y_{0k} = 1, \qquad \forall k \tag{17.28}$$

$$\sum_{i\in J_k} x_{0ik} = 1 \qquad \forall k \tag{17.29}$$

$$\sum_{j\in J_k} x_{ijk} = y_{ik} \qquad \forall k, \forall i \in J_k \tag{17.30}$$

$$\sum_{j\in J_k} x_{jik} = y_{ik} \qquad \forall k, \forall i \in J_k \tag{17.31}$$

$$\sum_{i\in J_k\backslash d_i=d} y_{ik} \leq |J| \times z_{dk} \qquad \forall d, k \tag{17.32}$$

$$|J| \times \sum_{i\in J_k\backslash d_i=d} y_{ik} \geq z_{dk} \qquad \forall d, k \tag{17.33}$$

$$\sum_{i\in J_k\backslash d_i=d} y_{ik}p_{ik} \geq MRH_d z_{dk} \qquad \forall d, k \tag{17.34}$$

$$\sum_{i\in J_k\backslash d_i=d} \frac{q_i}{Bs_{dk}} y_{ik} \leq MaxC \qquad \forall d, k \tag{17.35}$$

$$U_{ik} \geq p_{ik}y_{ik} \qquad \forall i, i \neq 0, \forall k \tag{17.36}$$

$$U_{ik} \leq AV_k \qquad \forall i, i \neq 0 \forall k \tag{17.37}$$

$$U_{ik} - U_{jk} + AV_k x_{ijk} \leq AV_k - p_{jk}y_{jk} \qquad \forall i\backslash i \neq 0, j, i \neq j, k \tag{17.38}$$

$$x_{ijk} \in \{0, 1\} \qquad \forall i, j, k \tag{17.39}$$

$$y_{ik} \in \{0, 1\} \qquad \forall i, k \tag{17.40}$$

$$z_{dk} \in \{0, 1\} \qquad \forall d, k \tag{17.41}$$

$$U_{ik} \geq 0 \qquad \forall i, k \tag{17.42}$$

The objective function (17.24) of the proposed model is to minimize the total processing time and setup time of all jobs on all machines. Constraints (17.25) impose that job i will not be assigned to machine k if it is not applicable on it ($i \notin J_k$). Constraints (17.26) ensure that all the jobs are assigned to a casting center. Constraints (17.27) state that the workload assigned to any line cannot exceed the number of hours during which the machine is available for production. This also helps in considering the planned shutdowns for the casting centers. By applying constraints (17.28) and (17.29), we ensure that the dummy job "0" is assigned to all machines to define the status of the machines while assigning the new jobs. Constraints (17.30) and (17.31) control the inflow and outflow of the jobs and ensure that a job succeeds/precedes only one job. Constraints (17.32) and (17.33) are added to define the value of z_{dk} as a function of y_{ik}. If any job of diameter d is assigned to machine k, then z_{dk} must be equal to 1, otherwise it is equal to 0. Constraints (17.34) restrict the minimum hours for the diameter molds to stay in production before another diameter molds are installed. This is imposed to avoid misusing the molds, as each time the molds are removed, they are sent to the mold shop to be prepared for the next run. It is also to provide the mold shop team enough time to prepare the molds required for the next Diameter. Constraints (17.35) guarantee that the number of drops casted continuously using the same diameter molds does not exceed MaxC drops. This limit is set by process control to avoid any quality issues in the product. For example, in the test case, MaxC=80 drops. After the 80th drop, the diameter molds should be changed. Constraints (17.36), (17.37), and (17.38) represent the Miller–Tucker–Zemlin (MTZ) constraints. These constraints are added for sub-tour elimination (See also [14] for a similar usage of these constraints). Constraints (17.39)–(17.42) are integrality and non-negativity constraints.

17.7 Computational Analysis

The model was implemented on FICO Xpress solver, with optimizer version 24 on a PC with CPU Intel i7 and 16GB RAM. The proposed model was applied to a real-world case. It was also tested on randomly generated problems. Sensitivity studies on a number of jobs, number of diameters, availability of machines, and density of the distribution of jobs on machines are also conducted.

17.7.1 Data

Data was collected from different departments in the cast house, mainly from procedures related to the casting process and from the production history which are available in the Manufacturing Execution Systems (MES) and company's records. For confidentiality reasons, some data will not be shared in detail.

The following data were collected:

- Billet diameters: The company produces billets in $|\Delta| = 13$ types of diameters. Their sizes are which are: 152, 155, 178, 203, 216, 228, 254, 273, 279, 305, 355, 381, and 406 millimeters. However, in the base case data, only 9 diameters are considered.

Table 17.1 Test case 1 data

Job	Required diameter index	Processing time (hrs)	MRH (hrs)	MaxC (hrs)
1	1	40.7	48	80
2	1	12.21	48	80
3	2	81.4	48	80
4	2	28.49	48	80
5	3	46.33	48	80
6	3	28.3	48	80
7	4	56.98	48	80
8	4	36.63	48	80
9	5	56.43	48	80
10	5	48.84	48	80
11	6	40.7	48	80
12	6	44.77	48	80
13	7	116.8	48	80
14	8	71.5	48	80
15	9	38.8	24	80

- Machines: The plant consists of $|K| = 5$ machines which differ in size and capability of producing all diameter sizes. The machines are referred to as: DC01, DC02, DC03, DC04, and DC05.
- Jobs and demand data: On average the aluminum plant receives between 1000 and 1500 orders each month. Each order is characterized by its diameter, alloy and quantity (number of billets). Through an order consolidation, the number of jobs was reduced to 50–60 per month.
- Planning horizon: The production control department issues the schedule to operations on a weekly basis. The scheduling takes place in the beginning of the week for all set of jobs to be processed during that week.
- Processing times of jobs (p_{ik}): The casting speed taken from the process control procedure was used to calculate the average processing time per metric ton for each product (Diameter/Homogenizer Path) on each machine. Table 17.1 shows the first scenario of processing times.
- Average Batch Size (Bs_{dk}): It varies between 128 for the smallest diameter and 16 for the largest diameter.
- Setup times (S_{ijk}) was set to 2.5 hrs.
- Minimum run hours (MRH_d) was set to 48 hrs for all diameters except for diameter $\delta = 9$ where it was set to 24 hrs.
- Maximum number of batches per diameter ($MaxC$) was set to 80 hrs.

The optimal solution is found in 1.5 s with the objective value of 761.49 hrs. The generated schedule is shown in Fig. 17.2.

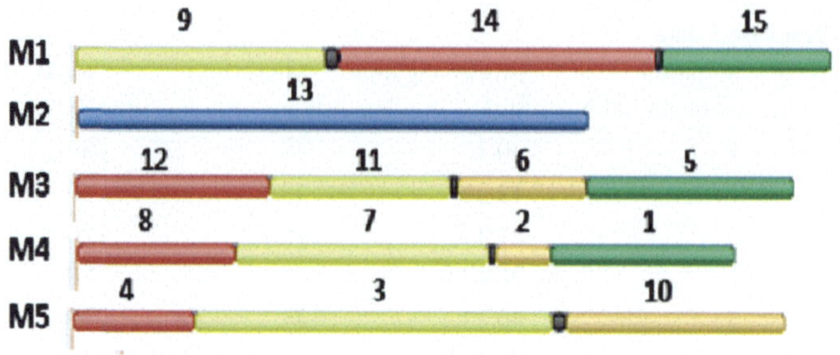

Fig. 17.2 Gantt chart for the company schedule weekly data

Table 17.2 CPU time in seconds for the different testing samples with different availability of machines

		Without shutdown	With shutdown
Jobs	20	0.534	0.381
	30	2.457	1.781
	40	63.700	34.798
	80	218.140	151.308
Diameters	7	72.000	102.883
	15	22.116	39.532
	Average	63.158	55.114

17.7.2 Experimentation on Randomly Generated Data

The authors have run simulations on instances with different sizes in terms of the number of jobs and the number of diameters. For each problem size, the effect of the availability of the machines is analyzed. The authors generated samples where all machines are functional (no planned shutdown) and samples where a shutdown occurs on one machine reducing its availability by 30%. The CPU times required by the solver to find the optimal solutions are recorded in Table 17.2. Table 17.3 presents the effect of the distribution of jobs on machines as the machines are not identical and each one has its set of jobs to process. In the generated data sets, the percentage of machines that can process a given job is defined as the problem *density*. Tests are carried out on instances with 70% density (on average jobs can be processed by 70% of the machines) or 50% density (jobs can be processed by 50% of the machines). The results are summarized in Table 17.3. For each case, five instances are generated and the average is calculated.

For all the samples tested, it can be noticed that the CPU time is relatively small; the maximum is approximately 3 min 30 s for 80 jobs. With the growth of problem size, one can observe that CPU time increases while on average CPU time decreases when the number of diameters increases. This can be explained by the limitation of feasible solutions when the number of diameters increases. Table 17.2 presents CPU

Table 17.3 CPU for the different testing samples with different density of distribution

		Density 70%	Density 50%
Jobs	20	0.7168	0.170
	30	63.0146	2.048
	40	6.2158	63.994
	80	144.4946	135.483
Diameters	7	95.887	63.678
	15	11.334	37.169
	Average	53.610	50.424

times when varying machine availability. These results indicate that instances with shutdowns are rather easier to solve. The CPU time remains small. From Table 17.3, it can be noticed that the CPU time is very sensitive to the density (percentage of machines that can process a given job).

17.8 Discussion

In this work, a mathematical model for scheduling aluminum billets on casting machines was developed. The proposed model takes into consideration all requirements and constraints that are related to the casting stage to provide the best allocation and sequencing of jobs on the casting centers. The model was implemented and solved using a state-of-the-art commercial solver where the optimal solution was found within seconds for real problem size. The provided model will be used by the company as an initial step toward automating the whole planning and scheduling process.

These models can also find their application in other industries than aluminum production. In food industry, for example, [42] discussed a similar problem with constraints specific to the case company in addition to some perishability constraints. Another interesting application is the pharmaceutical industry. Ruiz-Torres et al. [40] studied such a similar parallel machine production planning problem with complex setups. Compared to these studies, aluminum production process does not consider any perishability constraints. On the other hand, aluminum production planning can face challenges related to storage space constraints and a very large number of alloys required by the customers.

As was presented in the numerical results, though the problem is NP-hard, the commercial solver finds optimal solutions in very few minutes and in a few seconds in many cases. This can be explained by the size of the problem on the one hand and by the fact that it is a single-period problem. Planning over several time periods would results in much more challenging optimization problems. Further complexity would happen if the setup carryover is considered.

Despite the accomplishment, there are areas of improvement for the model and room for future work. The model presented here considers the casting stage only. It is recommended to include the homogenization stage and see what impact it will have on the schedule and total production output. It is also suggested to solve the model for longer planning horizons. For more complex extensions of this problem, it should be envisaged to design and implement efficient heuristics or metaheuristics to obtain good solutions within a very short CPU time. Examples can include variable neighborhood search heuristics (e.g., [6]) or matheuristics such as the relax-and-fix approach (e.g., [43]).

References

1. Aggarwal A, Park J (1993) Improved algorithms for economic lot-size problems. Oper Res 41:549–571
2. Allahverdi A (2015) The third comprehensive survey on scheduling problems with setup times/costs. Eur J Oper Res 246(2):345–378
3. Allahverdi A, Gupta JN, Aldowaisan T (1999) A review of scheduling research involving setup considerations. Omega 27(2):219–239
4. Allahverdi A, Ng CT, Cheng TE, Kovalyov MY (2008) A survey of scheduling problems with setup times or costs. Eur J Oper Res 187(3):985–1032
5. Almada-Lobo B, Klabjan D, Antónia Carravilla M, Oliveira JF (2007) Single machine multi-product capacitated lot sizing with sequence-dependent setups. Int J Prod Res 45(20):4873–4894
6. Almada-Lobo B, Oliveira JF, Carravilla MA (2008) Production planning and scheduling in the glass container industry: A VNS approach. Int J Prod Econ 114(1):363–375
7. Arnaout J-P, Musa R, Rabadi G (2014) A two-stage ant colony optimization algorithm to minimize the makespan on unrelated parallel machines–part II: Enhancements and experimentations. J Intell Manuf 25(1):43–53
8. Avalos-Rosales O, Angel-Bello F, Alvarez A (2015) Efficient metaheuristic algorithm and re-formulations for the unrelated parallel machine scheduling problem with sequence and machine-dependent setup times. Int J Adv Manuf Technol 76(9–12):1705–1718
9. Barany I, Van Roy T, Wolsey L (1984) Strong formulations for multi- item capacitated lot-sizing. Manag Sci 30:1255–1261
10. Bitran GR, Haas EA, Hax AC (1982) Hierarchical production planning: A two-stage system. Oper Res 30(2):232–251
11. Brahimi N, Dauzère-Pérès S (2015) A Lagrangian heuristic for capacitated single item lot sizing problems. 4OR 13(2):173–198
12. Brahimi N, Dauzère-Pérès S, Najid N (2006) Capacitated multi-multi item lot sizing problems with time windows. Oper Res 54(5):951–967
13. Brahimi N, Absi N, Dauzère-Pérès S, Nordli A (2017) Single-item dynamic lot-sizing problems: An updated survey. Eur J Oper Res 263(3):838–863
14. Brahimi N, Aouam T (2016) Multi-item production routing problem with backordering: A MILP approach. Int J Prod Res 54(4):1076–1093
15. Brucker P (2007) Scheduling algorithms. Springer, Berlin
16. Carvalho DM, Nascimento MC (2016) Lagrangian heuristics for the capacitated multi-plant lot sizing problem with multiple periods and items. Comput Oper Res 71:137–148
17. Chen W, Thizy J (1990) Analysis of relaxations for the multi-item capacitated lot-sizing problem. Ann Oper Res 26:29–72
18. Dauzère-Pérès S, Brahimi NNM, Nordli A (2002) Uncapacitated lot-sizing problems with time windows. Res Rep 02/4/AUTO, Ecole des Mines de Nantes, France

19. Diana ROM, de França Filho MF, de Souza SR, de Almeida Vitor JF (2015) An immune-inspired algorithm for an unrelated parallel machines' scheduling problem with sequence and machine dependent setup-times for makespan minimisation. Neurocomputing 163:94–105
20. Duman E, Yildirim MB, Alkaya AF (2008) Scheduling continuous aluminium casting lines. Int J Prod Res 46(20):5701–5718
21. Federgruen A, Tzur M (1991) Simple forward algorithm to solve general dynamic lot sizing models with n periods in O(n log n) or O(n) time. Manag Sci 37(8):909–925
22. Florian M, Lenstra JK, Rinnooy Kan AHG (1980) Deterministic production planning: Algorithms and complexity. Manag Sci 26(7):669–679
23. Graham RL, Lawler EL, Lenstra JK, Kan AR (1979) Optimization and approximation in deterministic sequencing and scheduling: A survey. Ann Discret Math 5:287–326. Elsevier
24. Graves SC (1999) Manufacturing planning and control. In: Massachusetts institute of technology, pp 1–26
25. Gupta D, Magnusson T (2015) The capacitated lot-sizing and scheduling problem with sequence-dependent setup costs and setup times. Comput Oper Res 32(4):727–747
26. Jans R, Degraeve Z (2004) Improved lower bounds for the capacitated lot sizing problem with setup times. Oper Res Lett 32(2):185–195
27. Jans R, Degraeve Z (2008) Modeling industrial lot sizing problems: A review. Int J Prod Res 46(6):1619–1643
28. Jensson P, Kristinsdottir BP, Gunnarsson HP (2005) Optimal sequencing of tasks in an aluminium smelter casthouse. Comput Ind 56(2):221–234
29. Khalaf A, Alhammadi H (2011) Optimizing planning and scheduling in aluminium production using mixed integer linear programming. MSc Thesis number Co-op/2011/Fall/002. University of Sharjah
30. Li K, Yang S-I (2009) Non-identical parallel-machine scheduling research with minimizing total weighted completion times: Models, relaxations and algorithms. Appl Math Model 33(40):2145–2158
31. Liaw C-F, Lin Y-K, Cheng C-Y, Chen M (2003) Scheduling unrelated parallel machines to minimize total weighted tardiness. Comput Oper Res 30(12):1777–1789
32. Low C, Hsu C-J, Su C-T (2008) A two-stage hybrid owshop scheduling problem with a function constraint and unrelated alternative machines. Comput Oper Res 35(3):845–853
33. Maes J, McClain J, Wassenhove L (1991)Multilevel capacitated lotsizing complexity and LP-based heuristics. Eur J Oper Res 53(2):131–148
34. Maes J, Van Wassenhove L (1986) A simple heuristic for the multi item single level capacitated lotsizing problem. Oper Res Lett 4(6):265–273
35. Martello S, Soumis F, Toth P (1997) Exact and approximation algorithms for makespan minimization on unrelated parallel machines. Discret Appl Math 75(2):169–188
36. Meng L, Zhang C, Shao X, Zhang B, Ren Y, Lin W (2020) More MILP models for hybrid flow shop scheduling problem and its extended problems. Int J Prod Res 58(13):3905–3930
37. Pinedo ML (2018) Scheduling: Theory, algorithms, and systems. Springer, Berlin
38. Rocha PL, Ravetti MG, Mateus GR, Pardalos PM (2008) Exact algorithms for a scheduling problem with unrelated parallel machines and sequence and machine-dependent setup times. Comput Oper Res 35(4):1250–1264
39. Ruiz R, Vázquez-Rodríguez JA (2010) The hybrid flow shop scheduling problem. Eur J Oper Res 205(1):1–18
40. Ruiz-Torres AJ, Ablanedo-Rosas JH, Ho JC (2012) The parallel machine scheduling problem with variable demand and a pre-defined lot size. Int J Oper Res 14(1):1–30
41. Senthilkumar P, Narayanan S (2010) Literature review of single machine scheduling problem with uniform parallel machines. Intell Inf Manag 2(8):1–18
42. Tempelmeier H, Copil K (2016) Capacitated lot sizing with parallel machines, sequence-dependent setups, and a common setup operator. OR Spectr 38(4):819–847
43. Toso E, Morabito R, Clark A (2009) Lot sizing and sequencing optimisation at an animal-feed plant. Comput Ind Eng 57(3):813–821

44. Tran TT, Beck JC (2012) Logic-based Benders decomposition for alternative resource scheduling with sequence dependent setups. In: ECAI, pp 774–779
45. Trigeiro W, Thomas L, McClain J (1989) Capacitated lot sizing with setup times. Manag Sci 35:353–366
46. Vallada E, Ruiz R (2011) A genetic algorithm for the unrelated parallel machine scheduling problem with sequence dependent setup times. Eur J Oper Res 211(3):612–622
47. Van Vyve M (2003) A solution approach of production planning problems based on compact formulations for single-item lot-sizing models. PhD thesis. UCL
48. Van Vyve M, Wolsey L (2006) Approximate extended formulations. Math Program B 105:501–522
49. Wagelmans A, Hoesel CV, Kolen A (1992) Economic lot sizing: An O(n log n) algorithm that runs in linear time in the Wagner-Whitin case. Oper Res 40:S145–S156
50. Wagner HM, Whitin TM (1958) Dynamic version of the economic lot size model. Manag Sci 5(1):89–96
51. Ying K-C, Lee Z-J, Lin S-W (2012) Makespan minimization for scheduling unrelated parallel machines with setup times. J Intell Manuf 23(5):1795–1803

Chapter 18
Facility Location Problem: A Case Study of School Consolidation

Abhishek Bhatnagar and Nomesh B. Bolia

Abstract This chapter studies the facility location problem and demonstrates its application through a real-life case of school consolidation. India has a large number of schools primarily established under the *Sarva Shiksha Abhiyan*, a flagship programme of the Government of India to promote accessibility. A large number of schools (including many small size schools) have been established to enhance accessibility, but it is difficult to provide good teaching and infrastructural resources in all schools. In recent years, the focus has shifted toward improving the quality of education delivery. Hence, there is a need to merge small schools to have better facilities at the expense of accessibility. However, accessibility cannot be compromised as it may create issues related to safe commutes to school and eventually lead to dropouts in some cases. Dropouts would violate the government's aim of universalization of education. The process of merging small schools and assigning students to new (alternate) schools is known as school consolidation. The school consolidation problem is formulated as an integer linear program to minimize the number of schools in operation and the total student disruption considering the accessibility and capacity requirements. Thus, the problem has multiple objectives and is solved using lexicographic ordering in two stages along with the ε-constraint method. The models have been demonstrated in a case study of Belgaum district in Karnataka, India. The number of schools closed and the average additional travel distance varies with the spatial distribution of schools and their enrollment. The results indicate an average additional travel between 200 and 700 m in the first stage of the model and 150 m to 350 m in the second stage of the model for different blocks in the district.

Keywords Integer programming · School consolidation · Student assignment · Lexicographic ordering

---

A. Bhatnagar · N. B. Bolia (✉)
Department of Mechanical Engineering, Indian Institute of Technology Delhi, Hauz Khas, New Delhi 110016, India
e-mail: nomesh@mech.iitd.ac.in

A. Bhatnagar
e-mail: abhishekjmi11@gmail.com

F. Hamid (ed.), *Optimization Essentials*, International Series in Operations Research & Management Science 353, https://doi.org/10.1007/978-981-99-5491-9_18

18.1 Introduction

Optimization refers to the act of making the best decisions given a set of available resources. A decision maker is often encountered with situations where it is cumbersome to take decisions manually, either due to problem complexity or computational complexity because of large scale of operations. In such cases, a decision problem is modeled as a mathematical program that is solved to find an optimal solution or in certain cases an acceptable solution. The aim of this chapter is to understand, model, and solve a real-life problem using integer linear programming. The real-life problem is related to the optimal location of educational facilities, that is, schools. Facility location problem refers to the optimal placement of a facility or a network of facilities from a set of potential candidates [1]. The aim is to select a set of locations that minimize and/or maximize certain objective(s). Facility location problem can be divided on the basis of the type of decisions that need to be taken:

(i) Opening of new facilities: here, the decision maker needs to find optimal locations and capacities of new facilities,
(ii) Simultaneous opening and closing of facilities: in this, the decision maker needs to find the optimal location of new facilities with their capacities, closure of existing facilities and resultant shift in demand, and expansion and reduction in the existing facility capacity, and
(iii) Pure closure of facilities: this refers to the closure of facilities without opening new facilities. This generally happens if the demand reduces over a period of time which restricts the opening of new facilities. This requires the closure of existing facilities and assignment of the existing demand preferably to the nearest available alternative. These problems need to be modeled for a network of facilities and not consider each facility individually [3].

Facility location is an important problem both in the manufacturing and the service domains and can be solved using mathematical programs. However, different considerations apply to both depending on the problem context and the industry-specific needs. The key criteria for the location of facilities include the fixed cost of land, cost of installations and fixtures, capacity of facility and scope of expansion, availability of raw material, availability of labor (cheap/skilled), service levels, safety and security considerations of the stakeholders (customers, manufacturers, and service providers), distance from market, availability and cost of different modes of transportation, and government policies. We limit our discussion to service facilities, which is the main idea of this chapter. The chapter is divided into seven sections: Sect. 18.2 details the problem context and the need for modeling this problem as a pure facility closure problem. Section 18.3 describes the history of integer programming, classification of optimization problems, and solution methods for multi-objective problems. Section 18.4 explains the facility location problem. Section 18.5 presents the nature of the school consolidation problem and its integer programming formulation followed by the solutions and results in Sect. 18.6. Section 18.7 concludes the chapter and provides a practice problem for the readers.

18.2 Problem Context and Description

The chapter discusses and addresses a pure facility closure problem (school consolidation) through mathematical programming in the context of school education in India. India has a large school system that consists of 15 lakh schools and a total enrollment of 26.5 crore as per Unified District Information System for Education 2020. The school system is subdivided into four sections where the students from the age group 6 to 18 are enrollled in different grades. These sections are: (i) Primary: grades I to V; (ii) Upper-primary: grades VI to VIII; (iii) Secondary: grades IX and X; and (iv) Senior secondary: grades XI and XII. Some schools provide all four sections, while others provide some of these sections.

The aim of an education system is to provide good quality education. For this purpose, a large number of schools have opened under the *Sarva Shiksha Abhiyan*, a flagship programme of the Government of India [11]. The government of India and state governments run a majority of schools, some of which may be as small as catering to only one student. It is not possible to provide quality educational services in all small schools due to budget constraints. For example, it is tough to have different subject teachers of desired qualifications and experience levels for very small schools. Further, shortage/improper distribution of teachers leads to multi-grade teaching in many cases, where students of different grades are taught in one classroom by the same teacher. Hence, the teaching and movable infrastructural resources need to be combined through the merger of schools to improve education quality. However, access to educational services should not be compromised as it would defeat the aim of universalization of education. This means that a remote school would always be functional irrespective of number of students enrolled in the school. Further, capacity considerations in the functional schools should be ensured to avoid any infrastructural and space issues. In practice, most of the low enrollment schools are located in a rural setting and are underutilized. These schools are often merged with big schools. However, two or more small schools can also be merged together in case they satisfy the accessibility and capacity criteria. Thus, school consolidation problem aims to find the best possible set of schools that can be closed and the resultant student assignment.

Mathematical models are often utilized to take key policy decisions. Models provide data-driven support and justification for these decisions since they affect a large number of people. Also, the models can incorporate the perspectives of stakeholders to reach a solution that maximizes/minimizes certain objective(s) while ensuring all the considerations are met. The chapter develops mathematical models for the school consolidation problem. The models are demonstrated through a case study of the Belgaum district in the state of Karnataka in India. The district is administratively divided into six blocks each having its own characteristics of schools in terms of enrollment, number of grades offered by schools, and the distance between schools. Since optimization is the core theme, the next section presents a contextual summary of optimization problems and follows it up with other sections that situate the school consolidation problems therein.

18.3 Optimization Problems

As mentioned in Sect. 18.1, optimization is the act of making the best decisions given the available resources. An optimization problem comprises the following key components [12] as given in Fig. 18.1.

Sections 3.1 and 3.2 describe the history of integer programming and the classification of optimization problems. These are followed by solution techniques for multi-objective optimization problems in Sect. 3.3.

18.3.1 History of Integer Programming

This subsection on the history draws from [12]. Integer programming has been used in a variety of applications with its usage increased since 1980s due to progress made in its solution methodology. The branch-and-bound approach has been refined since the late 1960s and early 1970s. The initial methodology was efficient for about 100 variables. There was an exponential increase in the solution time for a modest increase in the problem size. The practical problems were still out of scope to be solved. In 1980s and early 1990s, a few important and breakthrough studies published were [6, 13, 14]. Problems with up to 2756 variables were solved using an algorithmic approach. It is to be noted that these problems did not have any special structures.

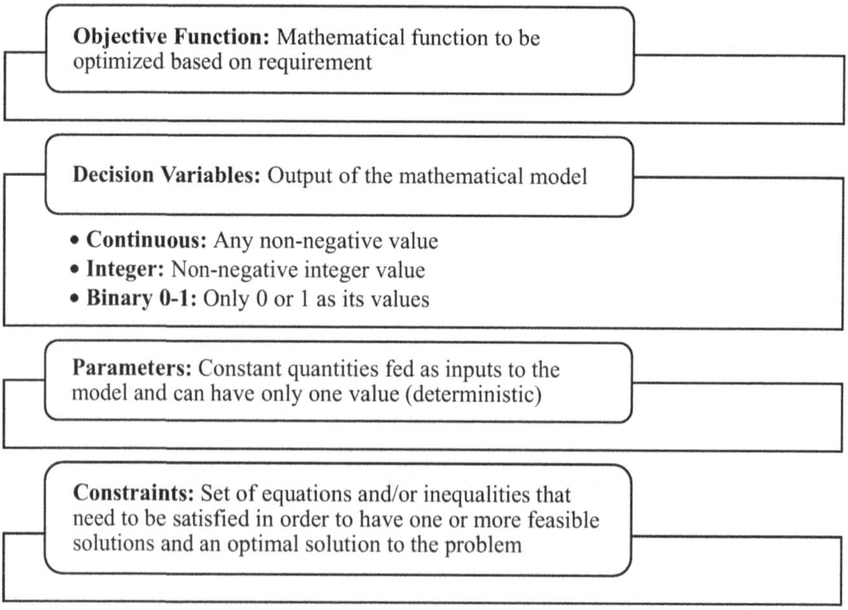

Fig. 18.1 Components of an Optimization Problem

The refinements to these algorithmic approaches continued for some years till the early 1990s. A key limitation till that point was the lack of ability to solve the mixed integer problems. By 1987, problems of about 1000 binary variables and a large number of continuous variables had been solved. The pure binary integer programming problems used special problem structures such as sparsity.

Another method, known as the cutting plane method was introduced in the 1950s by Gomory. However, this method was inefficient as large amounts of cuts were needed to reach the solution, making it impractical. However, it became useful when used with a branch-and-bound approach in the mid-1990s. Most available solvers nowadays use Gomory cuts in one form or the other to solve problems.

A widely used solver, CPLEX has been in use since 1980s and was acquired by ILOG in 1997, and then by IBM in 2009. The latest version of CPLEX available is 20.1, however, CPLEX 12.10 has been commonly used since 2019.

18.3.2 Classification of Optimization Problems

Optimization problems can be divided into several classes as given [20]:

- Continuous versus Discrete: optimization problems can be divided into continuous and discrete based on solution space. Continuous optimization problems have an infinitely large feasible solution set. The feasible solution set of discrete optimization problems is a subset of continuous optimization problems. However, discrete problems are tougher to solve than continuous problems due to the smoothness of functions in the continuous domain. Often, the discrete problems are relaxed and solved as continuous problems in an iterative fashion.
- No objective versus Single objective versus Multi-objective: an optimization problem can have none, single or multiple objectives. Problems without objectives are just to check feasibility. Single-objective problems are those with one type of objective, that is, either maximization or minimization. Multi-objective problems are those that have multiple conflicting objectives and a trade-off is determined between the objectives.
- Constrained versus Unconstrained: this classification is based on constraints that need to be specified in the model explicitly in terms of a set of equations and inequalities. Unconstrained optimization assumes any real value of the variables without any restrictions on them.
- Deterministic versus Stochastic: this classification is based on the properties of the data. Deterministic optimization assumes that the data is accurately known with some level of certainty. However, this may not always be the case due to measurement inaccuracies or uncertainty with the future. Stochastic optimization is used in such cases by determining the probability distribution of the data.

In the school consolidation problem illustrated through the case study, the problem is a multi-objective, discrete, constrained and deterministic optimization problem.

18.3.3 Solution Approaches for Multi-Objective Optimization Problems

This section discusses the solution approaches for integer programming models that are multi-objective since the case discussed in the chapter corresponds to an integer program with multiple objectives. A multi-objective problem generally has multiple optimal solutions based on the objectives. However, a decision maker can choose the best based on requirements. It is important to know the concept of pareto-optimality to understand the solution approaches (Refer [4] for detailed reading on Pareto-optimality). Let us understand this concept better. Suppose there are two objectives and five solutions obtained for an optimization problem, some of these solutions may be better with respect to only one of the objectives and the decision maker can address the trade-off as per the need. In some cases, let us say solution 1 is inferior to solution 2 with respect to both objectives. Then, solution 1 is *dominated* by solution 2. This implies that the decision maker should not choose solution 1 in any scenario as solution 2 is better with respect to both objectives. The aim is to check for dominated solutions and find a set of non-dominated solutions that constitute a pareto-optimal frontier [19]. The solutions need to be as close to this frontier as possible.

The solution methods for multi-objective problems are broadly classified into *apriori* and *posteriori* methods. In *apriori* methods, decision makers make the preference for the objectives before solving the optimization problem. Hence, the information about the objectives and their relative importance is incorporated, and a single solution is obtained. Alternatively, *posteriori* methods solve the problem to obtain a set of solutions and then select them based on the requirements [7].

There are several approaches to solving these problems. Some of them are as follows:

- ε-constraint method [16]: in this method, a multi-objective problem with N objectives is transformed into a single objective problem by choosing one of the objectives to be optimized, and the remaining $N - 1$ objectives are translated into constraints by incorporating a threshold value on each of these constraints. This ensures that the solution is optimal with respect to one objective and requirements of all other objectives are satisfied. It is, however, important to decide the objective that is to be optimized, and the threshold values on the constraints need to be varied (i) to avoid infeasible solutions, and (ii) to get a number of solutions to choose from since it is often difficult to find the threshold values.
- Weighted sum method (scalarization method) [8]: in this method, multiple objectives are transformed into a single objective function by assigning weights to the objectives. These weights are varied to obtain multiple solutions. Here, no additional constraint is added to the model since the objective function itself incorporates all the objectives. However, weights are subjective to individual judgments.

- Lexicographic ordering method (hierarchical approach): in this approach, the objectives are solved in the order of importance iteratively such that the solution obtained in any stage is not compromised in the next stage [4].

18.4 Facility Location Problem

A facility location problem refers to the location of facilities such that a certain objective function is optimized while considering the relevant constraints that may vary from one context to another. Although facility location has received significant research interest, each problem is unique in terms of types of facility and geographic location and requires different solution approaches. Any location problem is characterized by four main considerations [17], namely: (i) customer, whose location is assumed to be fixed and known, (ii) types of facilities to be located, (iii) a physical space where the customers are located and facilities need to be located, and (iv) distance or time between the customer and possible facility locations. There are a few important reviews in the area of facility location (Refer [9, 17, 18]). Recent reviews such as [5, 10, 15] can be helpful in understanding the facility location problems and the current state. The facility location problems are classified in Sect. 4.1 and distance calculations are given in Sect. 4.2.

18.4.1 Classification of Facility Location Problems

The location problems can be classified based on the following criteria:

- Endpoints and physical path [17]: this involves the segregation of location problems into two types: (a) planar: given an n-dimensional plane where the distance between any two points is the straight line connecting the two points. Hence, the distance depends on the coordinates of the endpoints; (b) network: where the path between two points is needed and not just the endpoints. This is calculated based on the shortest path in terms of distance or time taken.
- Location decisions: it refers to the physical space where the facilities are to be located. A facility can be located at any of the points in the space known as continuous facility location or at only certain specific and designated points, known as discrete facility location.
- Objectives: the objectives in location problems are classified into three broad areas: median, coverage, and fixed charge. In median problems, facilities are located such that the total distance between the facilities and customers is minimized. Coverage problems are broadly classified into two parts, namely, (i) set covering problems: all the customers are covered while the aim is to minimize the number of facilities and (ii) maximum covering problems: a fixed number of facilities are located such that maximum population gets covered. Fixed charge

problems minimize the total fixed costs as well as the variable costs between customers and facilities [18].

- Flow pattern: this classification is based on the flow in case of multiple levels. A pattern is a single flow if the product/customer goes from the lowest level to the highest level in the hierarchy, and covers all intermediate levels or vice versa. For example, a school education system where the student enters the lowest grade and exits from the highest grade after passing through all the grades. Further, if the product/customer goes between any two levels, it follows multiple flow patterns. For example, the healthcare delivery systems in which patients can be referred to a given level from any other level based on diagnosis and the criticality [18].
- Service varieties: a facility at a higher level of hierarchy provides all the same services provided by its immediate previous level along with at least one additional service. This is called nested service variety. For example, a patient going to a hospital for surgery may be given all the services that an outpatient receives (such as diagnosis and basic treatment) in addition to the surgery. In case of non-nested service variety, all hierarchical levels provide different services such as those provided in the case of school education system.

18.4.2 Distance Calculations

Facility location decisions rely heavily on the distances between the facility and customer points. Distances between the spatial entities, that is, the facilities (proposed locations for opening of new facilities or current locations for closures), and the customer points need to be calculated. The first task is to collect and validate the data on the latitudes and longitudes of these entities. These validated latitudes and longitudes can then be used to calculate the distances between these points. Broadly, distances between customer locations and facilities can be calculated using formulae based on Minkowski distance [1, 17]. Minkowski distance is a generalized form for distance calculation in an n-dimensional space (norm of a vector). Equation (18.1) gives the Minkowski distance for a norm of order p between points a and b with coordinates $(a_1, a_2, \ldots..a_n)$ and $(b_1, b_2, \ldots..b_n)$, respectively.

$$d_{a,b} = \left[\sum_{i=1}^{n} |a_i - b_i|^p\right]^{1/p} \tag{18.1}$$

Special cases of Minkowski distance are listed below:

- Manhattan distance (L1 norm): in this case, the value of p is 1, that is, the sum of absolute differences between the data points in all n-dimensions. Hence, Eq. (18.1) reduces to Eq. (18.2) as stated.

$$d_{a,b} = \sum_{i=1}^{n} |a_i - b_i| \tag{18.2}$$

- Euclidean distance (L2 norm): here, the value of p is two. Hence, this is the shortest distance between any two points in an n-dimensional plane and is a widely used distance calculation as given in Eq. (18.3).

$$d_{a,b} = \left[\sum_{i=1}^{n} |a_i - b_i|^2\right]^{1/2} \quad (18.3)$$

- Chebyshev distance (L∞ norm): in this distance calculation, the value of p is ∞. It is the maximum of the absolute difference between the points in an n-dimensional plane as given in Eq. (18.4).

$$d_{a,b} = \max_{i} |a_i - b_i| \quad (18.4)$$

The distance between any two points in the network when the actual path is considered is calculated by dividing the path into an infinite number of points close to each other, and the distance between two consecutive points is summed over the entire path to obtain the distance between the initial and final points.

18.5 School Consolidation Problem

In the school consolidation problem, some schools need to be closed and their operations shift to another school. The newly allocated school should be at an accessible location without a drastic increase in student travel. Based on the concepts explained earlier, it is needed to understand school consolidation in terms of the type of optimization and nature of the facility location problem. This section is divided into five subsections: Sect. 18.5.1 presents the nature of the school consolidation problem; Sect. 18.5.2 describes the constraints in terms of their physical significance; Sect. 18.5.3 discusses the mathematical model formulation including decision variables and the notations; Sect. 18.5.4 presents the solution procedure; Sect. 18.5.5 describes the performance indicators of school consolidation.

18.5.1 Nature of School Consolidation Problem

This section aims to provide an understanding of the nature of the school consolidation problem. Firstly, the problem can be modeled as an integer program as all the decision variables are restricted to integers. Binary decision variables of school closure and student assignment need to be integers 0 and 1; decision variables corresponding to the number of students transferred must be integers as the divisibility assumption for students cannot hold true. The decision variables are elaborated in Sect. 18.5.3.

In terms of solution space for the school consolidation problem, the problem is discrete as the schools are located at specific locations and transfers can take place between these schools based on constraints.

The objectives in the school consolidation problem can be divided in terms of the needs of the stakeholders. The government want to run a smaller number of schools without compromising on accessibility. Students want the school to be at the nearest possible location to the school. Hence, the two objectives would be to minimize the number of schools operational and to minimize the additional student travel.

Further, the problem is a constrained optimization problem since the constraints are explicitly defined. The available data does not have uncertainty and modeling is done for one period based on the current enrollment, hence the problem is deterministic in nature.

In terms of endpoints and the physical path, the school consolidation problem is modeled as a network problem, since the distances between schools need not be a straight line (as defined for planar problems). The main reason is that straight line distances are the shortest distances; however, students need walking accessibility, hence the paths should incorporate walking as a mode. For instance, consider two schools that are located on opposite sides of the road without a zebra crossing, which makes it infeasible to cross the road and reach the other school in a straight line. Students would need to walk a long distance due to the unavailability of walking path. In such a case, accessibility would be violated while considering planar distances. Hence, this problem needs to be modeled as a network, and distances calculated pertain to the shortest travel path or travel time as per requirement. Hence, the solution to the school consolidation problem would also consider the path to reach the new school from the old school.

In terms of flow pattern, a school system can be modeled as a single flow since all students start from the lowest grade, move through all intermediate grades, and then culminate school at the highest grade. Since all grades offer different curricula and are distinct, this is a case of a non-nested service variety.

18.5.2 Constraints

This subsection describes the considerations and corresponding constraints in the school consolidation problem:

- Accessibility constraints: it refers to students being transferred to a new (alternate) school within a stipulated distance from the current school. Accessibility measure varies depending upon factors such as availability of transportation facilities, demographics and terrain of an area, safe commute to school from residence, etc. Schools are located such that they are accessible from student habitations. However, in the current context, the exact student habitations are unknown. In such a case, there is a need to define accessibility in a slightly different way through a reasonable assumption. Due to the large number of schools established in all habitations, it is safe to assume that the students currently attend the nearest school. Hence, the accessibility measure in this case is the additional travel distance, that is, the distance between the current and the new (alternate) allocated school. Also, the distance between the residence and the new school would always be less than or equal to the sum of the distance between the residence and the current school, and the current school to the new school (triangle inequality).

- Assignment constraints: these constraints ensure the correct assignment of students to schools. This includes transfer to the same grade in the new school. Further, the transfers should be done to the schools that remain open after consolidation. In order to avoid unnecessary disruption of students, students of schools that remain open after consolidation should not be transferred.
- School-based constraints: this set of constraints is used to decide the candidates for closure. Schools below or equal to a threshold level of enrollment are possible candidates for closure. The threshold level of enrollment can be decided by the decision-makers.
- Capacity constraints: these impose an upper limit on the capacity of each grade of a school. These ensure each school has the required physical infrastructure to accommodate a given number of students.
- Flow balance constraints: these maintain the student flow balance so that all the students are allocated a new school in case their current school is closed.

18.5.3 Decision Variables, Notations, and Problem Formulation

The school consolidation problem is modeled as a multi-objective optimization problem. The objectives are minimization of number of schools operational and minimization of total student disruption. Total student disruption is defined as the sum of additional distances travelled by all students from their current to the new (alternate) school. Model M1 minimizes the number of schools operational and Model M2 minimizes the total student disruption using the output of Model M1. A lexicographic ordering approach along with the ε-constraint method has been used to solve the school consolidation problem in two stages, that is, the result of Model M1 is used as an input in Model M2. Model M1 keeps the accessibility distance within a threshold, that is, uses the ε-constraint approach followed by optimization of the distance itself in Model M2. The notations used in the models are given as follows:

Indices	i, j	Indices of schools
	g	Index of grades
Sets	S	Set of schools
	G	Set of grades

(continued)

(continued)

Parameters	N_{max}	Enrollment below or equal to which the schools are candidates for closure (also known as small schools)
	D_{max}	Accessibility distance within which transfers should take place
	$d_{i,j}$	Distance between schools i and j
	$C_{j,g}$	Capacity of grade g of school j
	$N_{j,g}$	Number of students in grade g of school j
	q	Number of schools to remain open as obtained from Model M1 and used in Model M2
Decision variables	y_j	Equals 1, if school j remains open after consolidation and 0, otherwise
	$x_{i,g,j}$	Equals 1, if students of grade g of school i are transferred to school j and 0, otherwise
	$z_{j,g}$	Number of students in grade g of school j after consolidation
	$t_{i,g,j}$	Number of students of grade g of school i transferred to school j

The number of decision variables can be calculated as follows:

Total number of binary variables:
$x_{i,g,j} + y_j \forall i \in S, g \in G, j \in S$
Hence, the total binary variables would be $(|S|^2 \times |G|) + |S|$.
Total number of integer variables:

$$t_{i,g,j} + z_{j,g} \forall i \in S, g \in G, j \in S$$

Hence, the total integer variables would be $(|S|^2 \times |G|) + (|S| \times |G|)$.

Model formulation draws heavily from [2], and extends that work. Bhatnagar and Bolia [2] allow for a single assignment, that is, all students of a grade of a school will move to the same grade of the new (alternate) school. Bhatnagar and Bolia [2] do not enforce strict capacity requirements since minor breaches in capacity are acceptable in case of rural areas, and a key requirement is to transfer students together. In this chapter, strict capacity constraints have been enforced, hence all students of a grade of a school may not move to the same new (alternate) school and maybe split based on the available capacity of the recipient school. Therefore, the closest assignment is not guaranteed for all students, but the overall student disruption is minimized in the second stage.

Model M1 minimizes the number of schools in operation and is stated as follows:

$$\text{Min} \sum_{j \in S} y_j \tag{18.5}$$

subject to

$$d_{i,j}x_{i,g,j} \le D_{max} \quad \forall i \in S, g \in G, j \in S, j \ne i \tag{18.6}$$

$$x_{i,g,j} \le y_j \quad \forall i \in S, g \in G, j \in S, j \ne i \tag{18.7}$$

$$x_{i,g,j} \le 1 - y_i \quad i \in S, g \in G, j \in S, j \ne i \tag{18.8}$$

$$t_{i,g,j} \le C_{j,g}x_{i,g,j} \quad \forall i \in S, g \in G, j \in S, j \ne i \tag{18.9}$$

$$x_{i,g,j} \le t_{i,g,j} \quad \forall i \in S, g \in G, j \in S, j \ne i \tag{18.10}$$

$$\sum_{j \in S} t_{i,g,j} = N_{i,g}(1 - y_i) \quad \forall i \in S, g \in G \tag{18.11}$$

$$\left(\sum_{g \in G} N_{j,g} - N_{max}\right)(1 - y_j) \le 0 \quad \forall j \in S \tag{18.12}$$

$$z_{j,g} = \sum_{i \in S} t_{i,g,j} + N_{j,g}y_j \quad \forall g \in G, j \in S \tag{18.13}$$

$$\sum_{j \in S} z_{j,g} = \sum\nolimits_{i \in S} N_{i,g} \quad \forall g \in G \tag{18.14}$$

$$z_{j,g} \le C_{j,g}y_j \quad \forall g \in G, j \in S \tag{18.15}$$

$$y_j, x_{i,g,j} \in [0, 1] \quad \forall i \in S, g \in G, j \in S$$

$$z_{j,g}, t_{i,g,j} \ge 0 \quad \forall i \in S, g \in G, j \in S$$

Model M2 stated below, minimizes additional student travel and uses the output of Model M1, that is, the number of schools to be kept open.

$$\text{Min} \sum_{i \in S} \sum_{g \in G} \sum_{j \in S} d_{i,j}t_{i,g,j} \tag{18.16}$$

subject to

$$\sum_{j \in S} y_j = q \tag{18.17}$$

and constraints (18.6)–(18.15) and

$$y_j, x_{i,g,j} \in [0, 1] \quad \forall i \in S, g \in G, j \in S$$

$$z_{j,g}, t_{i,g,j} \ge 0 \quad \forall i \in S, g \in G, j \in S$$

Now, the aim is to understand the logic behind these constraints and how they relate to the physical problem at hand. The objective function (18.5) is the summation of all the schools on S, that exist in the school network. It seeks to minimize the number of schools operational after the school consolidation process, subject to constraints given as equations and inequalities (18.6)–(18.15). Decision variable y_j is binary and indicates whether the school j remains open or closed after consolidation. Hence, minimizing its sum would minimize the number of schools in the network.

Inequalities (18.6) incorporate the accessibility constraint. The right-hand side (RHS) of these inequalities indicates the threshold additional distance. Refer to left hand side (LHS) of inequalities (18.6), it multiplies the assignment variable $x_{i,g,j}$ with the distance between a pair of schools, $d_{i,j}$. If the distance between any pair of schools is more than the threshold limit D_{max}, the assignment of students between that pair cannot happen, that is, $x_{i,g,j}$ will assume a value of 0 for inequalities (18.6) to be feasible. Note that these need to be repeated for all schools and grades.

Inequalities (18.7) specify transfer to schools that remain open after the consolidation process. If the RHS is 0, that is, school j is closed after consolidation, and LHS must assume the value 0. It means students cannot be transferred to school j if it is not open. Similarly, if RHS is 1, then the transfer may or may not take place depending upon other considerations. In that case, LHS can take either value, 0 or 1.

Inequalities (18.8) imply that no transfer should take place from a school that remains open after consolidation. To understand this better, assume that school i is open, and the corresponding value of y_i is 1, hence RHS reduces to 0 which forces LHS to be 0, therefore, students cannot be transferred from schools that remain open. This is required to avoid unnecessary disruption. Note the difference in the indices of y in inequalities (18.7) and (18.8). In (18.7), RHS pertains to the receiving school, hence index j, which should receive students only if it is open. In (18.8), RHS pertains to the school that may get closed after consolidation, which implies that students should not move out of this school if it is open.

Inequalities (18.9) and (18.10) link the assignment variable $x_{i,g,j}$ with number of students transferred $t_{i,g,j}$. In this problem, it is assumed that all students of a grade of a school may not be transferred to the same new (alternate) school and are split based on available capacity in the schools that remain open after consolidation. In inequalities (18.9), if the value of RHS is 0, then LHS needs to be 0. Otherwise, $t_{i,g,j}$ can assume a non-negative value depending on other constraints. Inequalities (18.10) imply that if a transfer happens, then at least one student will be transferred. Inequalities (18.9) and (18.10) together imply that if $x_{i,g,j}$ is 0, then $t_{i,g,j}$ has to be 0, and, if $x_{i,g,j} = 1$, then $t_{i,g,j}$ can be any value greater than or equal to 1.

Equalities (18.11) pertain to the transfer of all students of each closed school. RHS of (18.11) would be 0 if y_i is 1 and no students would be transferred, hence, LHS would be 0. If y_i is 0, then all students of that grade of the school, $N_{i,g}$ must be transferred to any of the open schools indicated by LHS.

Inequalities (18.12) consider an individual school to decide whether it is a potential candidate for closure or not. The term $\sum_{g \in G} N_{j,g}$ calculates the total enrollment in school j by summing up enrollments of each grade in the school. The difference between the total enrollment and threshold enrollment N_{max} determines whether the school is a potential candidate for closure (subject to other constraints). If $(\sum_{g \in G} N_{j,g} - N_{max})$ is negative or 0, then y_j can be either 0 or 1 and hence, a possible candidate for closure.

Equalities (18.13) ensure that the number of students in grade g of school j after consolidation is the sum of its enrollment and the transfers from all other schools.

Equalities (18.14) make sure that all students remain in the same grade after consolidation. That is, for all the grades, the sum of initial and final enrollment over all schools remains the same.

Inequalities (18.15) ensure the adherence to capacity. They ensure that the number of enrollments in a grade of an open school cannot exceed the capacity of a grade of the school. RHS would be zero in case the school closes and it implies that students in that grade- school combination would be zero (given by LHS).

Model M2 uses the number of schools operational as obtained from Model M1 as a constraint given by Eq. (18.17). The objective function (18.16) minimizes the total student disruption, measured as the sum of the additional distance travelled by all students. Model M2 has all the constraints (18.6)–(18.15) from Model M1 and constraint (18.17). Constraint (18.17) ensures that q number of schools remain functional; value of q being obtained as the output of Model M1. Overall, Model M1 optimizes the first objective (minimize the number of schools) only. Given the optimal solution of Model M1, Model M2 finds a solution where the additional distance travelled by all students is minimized.

The total number of constraints in models M1 and M2 can be determined by summing up all the constraints in inequalities and equations given in (18.6)–(18.15) for Model M1; and (18.6)–(18.15) and (18.17) for Model M2. These are detailed in Table 18.1.

Table 18.1 Number of constraints

Inequality/Equation no	Number of constraints	
	Model M1	Model M2
6	$\|S\| \times \|S-1\| \times \|G\|$	$\|S\| \times \|S-1\| \times \|G\|$
7	$\|S\| \times \|S-1\| \times \|G\|$	$\|S\| \times \|S-1\| \times \|G\|$
8	$\|S\| \times \|S-1\| \times \|G\|$	$\|S\| \times \|S-1\| \times \|G\|$
9	$\|S\| \times \|S-1\| \times \|G\|$	$\|S\| \times \|S-1\| \times \|G\|$
10	$\|S\| \times \|S-1\| \times \|G\|$	$\|S\| \times \|S-1\| \times \|G\|$
11	$\|S\| \times \|G\|$	$\|S\| \times \|G\|$
12	$\|S\|$	$\|S\|$
13	$\|S\| \times \|G\|$	$\|S\| \times \|G\|$
14	$\|G\|$	$\|G\|$
15	$\|S\| \times \|G\|$	$\|S\| \times \|G\|$
17	–	1
Total	$(5 \times \|S\| \times \|S-1\| \times \|G\|) + (3 \times \|S\| \times \|G\|) + \|S\| + \|G\|$	$(5 \times \|S\| \times \|S-1\| \times \|G\|) + (3 \times \|S\| \times \|G\|) + \|S\| + \|G\| + 1$

18.5.4 Solution Procedure

The problem has been formulated as an integer program. The next step is to optimally solve the problem. Given the large size of the problem, we pre-process the variables and then solve the problem using exact methods. The detailed solution procedure is given in the following steps:

- Determination of policy parameters: it is required that the parameters needed in the school consolidation process are finalized. Threshold enrollment and accessibility parameters need to be decided based on discussions with experts or a survey of stakeholders to reach a consensus. For accessibility parameter D_{max}, its threshold value is taken to be 1,000 m based on the requirement of the government. Enrollment and spatial distribution of each school play an important role in deciding the threshold enrollment and are subjective depending on the need of the decision-making body. In this chapter, the value of threshold enrollment starts from 10 to 50 in steps of 10. The capacity of each grade of a school is 30.
- Determination of distances: this involves the calculation of distance between all the pairs of schools. For a school network of n schools, a total of n^2 distances need to be computed. It is computationally expensive to determine paths for each of the n^2 combinations. Also, the origin–destination matrix would be asymmetric to account for the walking accessibility on roads. To reduce computational effort, the straight line distances (Euclidean distances) for all n^2 combinations have been calculated. Since these distances are the shortest distances between the points, the actual path would always be more than these distances (or equal in very rare cases). The actual path is computed only for those combinations where the straight line distances are below the threshold distance. All the unfeasible student assignments to new schools (Euclidean distance above threshold distance) have been removed and a distance value of ∞ is assigned to those combinations.
- Determination of schools to participate in the school consolidation process: in this step, all schools that can participate in the school consolidation process are identified, that is, they are either candidates for closures or potential recipients of students of closed schools. The schools that do not have another school within the accessibility distance cannot take part in the school consolidation process and hence, are removed from the process.
- Model solution: in the previous steps, policy parameters have been finalized. The mathematical model has two objectives: (i) minimize the number of schools operational and (ii) minimize the total student disruption. The approach that uses the ε-constraint method and lexicographic ordering is adopted. The objectives are ordered in relative need/importance. The first objective is to minimize the number of schools operational. In this model, the threshold travel distance is put as a constraint and the output of the model gives the minimum number of schools needed to ensure accessibility. The second objective is to determine the set of assignments that minimizes the total disruption. In this stage, the number of schools obtained in the first stage is used as a constraint. The model is then run using the set of policy parameters.

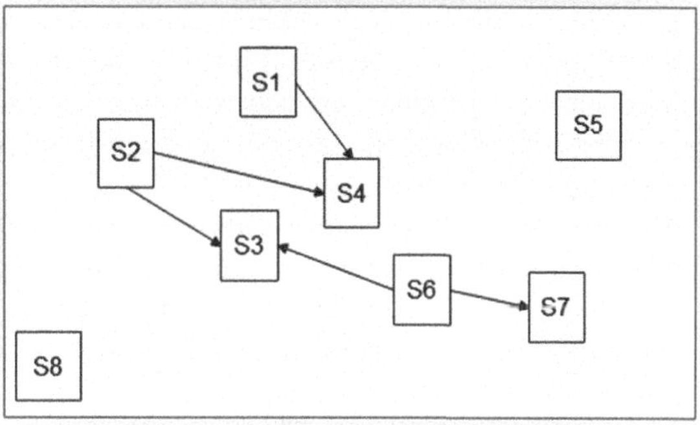

Fig. 18.2 Demonstration of school consolidation

Let us understand the solution procedure through Fig. 18.2. The school network has eight schools, namely, S1–S8. Refer to the solution steps given earlier in this section. In the first step, the policy parameter D_{max} is assigned a value 1,000 m upon discussion with stakeholders. In the second step, the origin–destination distance matrix between the eight schools is calculated. The matrix would have 64 elements, with diagonal elements being 0. The third step involves the elimination of those assignments whose distance lies above the threshold distance. Let us say S5 and S8 do not have any school within 1,000 m, and hence they cannot be closed nor they can receive any students from any of the closed schools. They are eliminated in this pre-processing stage and not fed into the optimization model. The next step involves running the model in CPLEX to yield the results. The student assignments from the schools closed are shown in Fig. 18.2. Further, the students of open schools cannot be transferred to any other school, hence the values $x_{3,g,j}$, $x_{4,g,j}$, $x_{7,g,j}$ and $x_{8,g,j}$ obtained from the model are 0 for all grades g and schools j. From Fig. 18.2, S1, S2, and S6 eventually get closed and resultant assignments can be seen along the direction of arrows. It is also clear that the demand from the closed school is served by more than one school. Further, open schools can receive students from more than one school.

18.5.5 Performance Indicators

School consolidation aims to close schools and transfer the students from the closed to new (alternate) schools. The results should be analyzed in terms of school-related variables. The performance indicators for the school consolidation process are given as follows:

- Average additional distance: This measures the average additional distance travelled by the students from the current school to the new (alternate) school. While the accessibility conditions put an upper limit on the additional student travel, it is necessary that the average distance within the accessibility limit is as low as possible. Distribution of the additional travel of students gives insights into the new travel profile and can be useful for practitioners. A high value of average travel may not be good since students may need to walk more or would use motorized vehicles which may discourage them from going to school due to limited transportation infrastructure in rural areas.

- Maximum additional distance: This refers to the maximum additional travel of any student. If strict accessibility conditions are imposed, then this should not be more than the threshold accessibility distance. However, in the absence of accessibility conditions, this can assume high value and may conflict with the government's aim of universalization of education.
- Number of schools closed: This is an important indicator of the school consolidation process. It depends on the spatial distribution of schools, their enrollment, and the choice of the threshold school enrollment (N_{max}) above which all schools should remain open. A higher value of N_{max} would lead to the closer of more schools. Hence, a proper value of N_{max} should be chosen in order to draft appropriate policies that consider all stakeholders.
- Number of students relocated: Students are important stakeholders in schools, hence the number of students transferred gives a measure of disruption. There may be two cases that lead to high disruption: (i) more students are transferred to a smaller distance or (ii) less students are transferred to a larger distance. Hence, these need to be balanced.
- School utilization: Utilization of any facility is an important measure. For a dense school network with low enrollment, most schools are underutilized. In such cases, school consolidation can lead to enhanced utilization.
- Pupil–teacher ratio (PTR): Generally, the preferred pupil–teacher ratio is 30:1 at the primary level and can be slightly increased for higher grades. Rural schools are characterized by typically low PTRs; however, lower number of teachers lead to multi-grade teaching, which is undesirable. A high PTR can lead to large class size and hamper individual attention to students.

All the mentioned performance indicators play a role in formulating the school consolidation policy. Some of these are discussed later in the results section. These are interlinked and final policy parameters should be decided based on these indicators. The results need to be validated for feasibility and actual implementation to consider the requirements beyond the scope of modeling. However, the actual school closures would be a subset of those suggested by the models.

18.6 Results

This section provides the results of the case study of Belgaum district in Karnataka, India. The district has six blocks, namely, Blocks 1–6 and the number of schools in each block is 263, 144, 306, 341, 216, and 244, respectively. The number of students in each block is 39,823; 20,843; 44,820; 25,022; 40,792, and 31,077, respectively.

Value of accessibility requirement, D_{max} is taken as 1,000 m, while the value of N_{max} is varied from 10 to 50 in steps of 10. Each school has 12 grades. The results of models M1 and M2 are summarized in Tables 18.2 and 18.3, respectively. The lower bounds have been obtained by LP relaxation of all integer decision variables. As evident from the tables, the average additional distance for the same model parameters decreases in Model M2 as compared with M1. All distances in Model M1 are also within the threshold accessibility, however, Model M2 further optimizes it. In most cases, the number of students transferred is equal to or less in Model M2 than in Model M1. In only one case for Block 3 and $N_{max} = 40$, number of students transferred are more in Model M2 than in Model M1, however, average additional distance in lower for Model M2. This indicates that more students are transferred but to a lower distance. Further, it may be noted that for the same set of parameters, there are differences in the number of schools closed and average additional distance in each block, as also suggested for the case mentioned in [2] where grades are not split and strict capacity is not enforced. Hence, the same policy parameters should not be used for all blocks and be done based on the enrollment and spatial distribution.

Table 18.2 Results of Model M1

Block (# schools)	N_{max}	Number of schools with enrollments below/equal N_{max}	# Schools operational	Lower bound	% Gap (Schools operational and lower bound)	# Students transferred	Average additional distance (m)	CPLEX time (s)	# Constraints	# Binary Variables	# Integer Variables
Block 1 (263)	10	6	258	257	0.39	32	235.69	3.47	4,144,103	830,291	833,184
	20	27	250	236	5.60	152	377.70	3.65			
	30	38	248	225	9.27	211	433.68	3.93			
	40	53	243	210	13.58	398	496.43	3.58			
	50	61	240	202	15.83	536	494.63	2.82			
Block 2 (144)	10	11	134	133	0.75	85	684.86	1.07	1,240,860	248,976	250,560
	20	20	125	124	0.80	231	666.52	0.98			
	30	39	110	105	4.55	597	653.50	1.02			
	40	47	104	97	6.73	813	625.59	1.20			
	50	53	99	91	8.08	1057	577.82	1.16			
Block 3 (306)	10	10	298	296	0.67	46	385.50	4.07	5,611,134	1,123,938	1,127,304
	20	33	278	273	1.80	344	311.73	5.43			
	30	50	272	256	5.88	502	326.85	5.21			
	40	74	265	232	12.45	734	395.47	4.58			
	50	90	260	216	16.92	879	339.08	4.98			
Block 4 (341)	10	42	329	299	9.12	73	290.49	6.74	6,969,029	1,395,713	1,399,464
	20	95	314	246	21.66	315	388.20	7.46			
	30	139	300	202	32.67	676	402.31	7.37			
	40	168	297	173	41.75	809	380.81	7.01			
	50	187	295	154	47.80	928	387.97	7.71			

(continued)

Table 18.2 (continued)

Block (# schools)	N_{max}	Number of schools with enrollments below/equal N_{max}	# Schools operational	Lower bound	% Gap (Schools operational and lower bound)	# Students transferred	Average additional distance (m)	CPLEX time (s)	# Constraints	# Binary Variables	# Integer Variables
Block 5 (216)	10	3	213	213	0.00	17	181.29	2.06	2,794,404	560,088	562,464
	20	16	210	200	4.76	59	499.36	3.80			
	30	26	207	190	8.21	139	563.48	4.83			
	40	38	206	178	13.59	171	569.14	5.50			
	50	58	202	158	21.78	388	566.66	3.81			
Block 6 (244)	10	2	243	242	0.41	7	426.86	4.55	3,566,560	714,676	717,360
	20	19	235	225	4.26	131	324.98	3.97			
	30	27	232	217	6.47	206	519.96	6.40			
	40	41	226	203	10.18	420	525.26	4.08			
	50	55	226	189	16.37	429	526.30	3.05			

Table 18.3 Results of Model M2

Block (# schools)	N_{max}	# Students transferred	# Schools operational	Total additional travel distance (m)	Lower Bound	% Gap (Total additional travel distance and lower bound)	# Average travel distance (m)	% Reduction in average travel distance from M1	CPLEX time (s)	# Constraints	# Binary Variables	# Integer Variables
Block 1 (263)	10	32	258	6278	6278	0.00	196.19	16.76	3.47	4,144,104	830,291	833,184
	20	152	250	42,861	30,164	29.62	281.98	25.34	4.02			
	30	198	248	54,123	40,930	24.38	273.35	36.97	3.13			
	40	380	243	130,978	71,991	45.04	344.68	30.57	2.98			
	50	518	240	181,792	102,496	43.62	350.95	29.05	3.34			
Block 2 (144)	10	85	134	20,944	20,944	0.00	246.40	64.02	0.84	1,240,861	248,976	250,560
	20	231	125	52,801	52,801	0.00	228.58	65.71	0.91			
	30	597	110	199,070	186,000	6.57	333.45	48.97	1.05			
	40	813	104	242,116	228,268	5.72	297.81	52.40	1.16			
	50	1048	99	320,380	282,588	11.80	305.71	47.09	1.28			
Block 3 (306)	10	46	298	5780	5780	0.00	125.65	67.41	5.67	5,611,135	1,123,938	1,127,304
	20	344	278	85,931	63,464	26.15	249.80	19.87	4.88			
	30	495	272	125,371	75,321	39.92	253.27	22.51	4.22			
	40	737	265	205,407	123,681	39.79	278.71	29.52	4.61			
	50	864	262	205,489	114,254	44.40	237.83	29.86	6.23			
Block 4 (341)	10	73	329	19,646	17,148	12.72	269.12	7.36	7.11	6,969,030	1,395,713	1,399,464
	20	315	314	113,436	107,209	5.49	360.11	7.24	5.72			

(continued)

Table 18.3 (continued)

Block (# schools)	N_{max}	# Students transferred	# Schools operational	Total additional travel distance (m)	Lower Bound	% Gap (Total additional travel distance and lower bound)	# Average travel distance (m)	% Reduction in average travel distance from M1	CPLEX time (s)	# Constraints	# Binary Variables	# Integer Variables
	30	664	300	227,108	186,453	17.90	342.03	14.98	6.86			
	40	786	297	265,163	206,820	22.00	337.36	12.41	8.56			
	50	877	295	292,180	221,121	24.32	333.16	14.13	8.45			
Block 5 (216)	10	17	213	1541	1541	0.00	90.65	50.00	3.73	2,794,405	560,088	562,464
	20	59	210	15,212	15,212	0.00	257.83	48.37	2.36			
	30	139	207	43,788	43,788	0.00	315.02	44.09	2.41			
	40	171	206	54,812	49,346	9.97	320.54	43.68	2.36			
	50	361	202	178,335	150,975	15.34	494.00	12.82	2.27			
Block 6 (244)	10	7	243	0	0	0	0.00	100.00	4.28	3,566,561	714,676	717,360
	20	131	235	34,998	31,882	8.90	267.16	17.79	4.05			
	30	206	232	81,338	62,012	23.76	394.84	24.06	5.86			
	40	420	226	204,865	161,867	20.99	487.77	7.14	3.25			
	50	420	226	204,865	161,639	21.10	487.77	7.32	6.14			

18.7 Conclusion

The models presented in the chapter can be used for a wide variety of applications that involve the pure closure of facilities with minor alterations based on the requirements and context. The models have been applied to one district in this context. These can be replicated in other locations by varying the parameters. The selection of parameters needs to be executed carefully based on the location and enrollment distribution as evident from our results. The future research in this context would include consolidation of schools based on the addresses of the students which are currently not available. Such a model would use principles of median and set covering approach to optimally close schools. There can be simultaneous opening and closing of schools where a number of schools to be opened can be decided by the policy makers as per requirement. In addition to consolidation, small schools can be attached to big schools to facilitate sharing of resources.

Practice Problem

The readers may use the following data to practice the model. Let us consider five schools with three grades each. The enrollments are given as follows:

School/Grade	Grade 1	Grade 2	Grade 3
School 1	4	9	7
School 2	10	22	4
School 3	12	23	11
School 4	5	2	9
School 5	6	10	8

The distance matrix (distance in meters) is given as follows:

	School 1	School 2	School 3	School 4	School 5
School 1	0	100	1200	500	700
School 2	100	0	700	1500	800
School 3	1200	700	0	400	500
School 4	500	1500	400	0	1100
School 5	700	800	500	1100	0

Find the optimal consolidation plan using the models for different values of D_{max} and $C_{j,g}$.

References

1. Adeleke OJ, Olukanni DO (2020) Facility location problems: Models, techniques, and applications in waste management 1–20
2. Bhatnagar A, Bolia NB (2019) Improved governance of Indian school system through school consolidation. J Policy Model 41(6):1160–1178. https://doi.org/10.1016/j.jpolmod.2019.05.001
3. Bhaumik PK (2010) Optimal shrinking of the distribution chain: The facilities delocation decision. Int J Syst Sci 41(3):271–280. https://doi.org/10.1080/00207720903326860
4. Castro-Gutierrez JP, Landa-Silva D, Moreno-Perez J (2009) Dynamic lexicographic approach for heuristic multi-objective optimization. In: Proceedings of the workshop on intelligent metaheuristics for logistic planning (CAEPIA-TTIA), pp 153–163
5. Church RL, Drezner Z (2022) Review of obnoxious facilities location problems. Comput Oper Res 138(July):105468. https://doi.org/10.1016/j.cor.2021.105468
6. Crowder H, Johnson EL, Padberg M (1983) Solving large-scale zero-one linear programming problems. Oper Res 31(5):803–834
7. Elaraby R (2019) Multi-objective optimization—no preference methods (ex: VEGA and NSGA). https://medium.com/@radwa.elaraby/multi-objective-optimization-no-preference-methods-ex-vega-and-nsga-9110d94e0c29
8. Elaraby R (2019). Multi-objective optimization—scalarization. https://medium.com/@radwa.elaraby/multi-objective-optimization-via-scalarization-approach-1e1e054506b6
9. Farahani RZ, Asgari N, Heidari N, Hosseininia M, Goh M (2012) Covering problems in facility location: A review. Comput Ind Eng 62(1):368–407. https://doi.org/10.1016/j.cie.2011.08.020
10. Farahani RZ, Fallah S, Ruiz R, Hosseini S, Asgari N (2019) OR models in urban service facility location: A critical review of applications and future developments. Eur J Oper Res 276(1):1–27. https://doi.org/10.1016/j.ejor.2018.07.036
11. GoI (2011) Sarva Shiksha Abhiyan: Framework for implementation
12. Hillier FS, Lieberman GJ (2001) Introduction to operations research, 7th. Tata Mcgraw-Hill Education
13. Hoffman KL, Padberg M (1991) Improving LP-representations of zero-one linear programs for branch-and-cut. ORSA J Comput 3(2):121–134
14. Johnson EL, Kostreva MM, Suhl UH (1985) Solving 0–1 integer programming problems arising from large scale planning models. Oper Res 33(4):803–819
15. Kchaou-Boujelben M (2021) Charging station location problem: A comprehensive review on models and solution approaches. Transp Res Part C: Emerg Technol 132:103376
16. Mavrotas G (2009) Effective implementation of the ε-constraint method in multi-objective mathematical programming problems. Appl Math Comput 213(2):455–465. https://doi.org/10.1016/j.amc.2009.03.037
17. ReVelle CS, Eiselt HA (2005) Location analysis: A synthesis and survey. Eur J Oper Res. https://doi.org/10.1016/j.ejor.2003.11.032
18. Şahin G, Süral H (2007) A review of hierarchical facility location models. Comput Oper Res 34(8):2310–2331. https://doi.org/10.1016/j.cor.2005.09.005
19. Taplin JHE, Sun Y (2020) Optimizing bus stop locations for walking access: Stops-first design of a feeder route to enhance a residential plan. Environ Plan B: Urban Anal City Sci 47(7):1237–1259. https://doi.org/10.1177/2399808318824108
20. Types of Optimization Problems (n.d.) https://neos-guide.org/optimization-tree. Accessed 29 Oct 2021

Chapter 19
Quantitative Models in Railway Operations Management

Narayan Rangaraj, Swapnesh Subramanian, and Shripad Salsingikar

Abstract Railway operations management involves planning and scheduling various resources to operate the railway services efficiently. The recent advancements in computing, communication, storage hardware, and algorithms help to automate the planning process and decision-making involved in railway operations management with the help of tools and techniques from Operations Research. Studies indicate that the use of such techniques can result in significant cost savings for railways. The techniques used to solve the decision problems include Integer Programming, Simulation, (Meta) Heuristics, and Reinforcement Learning. In this chapter, we cover the basics of railway operations management, and introduce various problems involved at different levels of planning where quantitative methods are applied. We explore a few problems in detail, including the problem statement, modeling, solution techniques, and validation of models. The specific problems taken up for illustrative analysis are (a) Large-scale timetabling, (b) Rolling stock circulation problem for long-distance passenger operations, and (c) Station level planning. We illustrate the representation of the timetable as a mathematical object and describe an application of simulation and optimization tools to a large-scale timetabling exercise. We provide an example of optimizing the rolling stock circulation in Indian Railways based on an integer programming model applied to a network graph. An approach for modeling the track infrastructure of a station for planning the train movement is also illustrated.

Keywords Operations research · Railway operations management

N. Rangaraj (✉)
Industrial Engineering and Operations Research, Indian Institute of Technology Bombay, Mumbai, Maharashtra, India
e-mail: narayan.rangaraj@iitb.ac

S. Subramanian
Rajiv Gandhi Institute of Technology, Kottayam, Kerala, India
e-mail: swapnesh.s@rit.ac.in

S. Salsingikar
TCS Research, Tata Consultancy Services, Thane, Maharashtra, India
e-mail: shripad.salsingikar@tcs.com

F. Hamid (ed.), *Optimization Essentials*, International Series in Operations Research & Management Science 353, https://doi.org/10.1007/978-981-99-5491-9_19

19.1 Introduction

In many countries, railways provide an efficient and reliable means of transportation. Railway infrastructure is capital-intensive, and the optimal use of these resources is essential. The railway industry offers scope for using Operations Research (OR) techniques to solve many practical problems and improve performance. Many of the planning or operational management problems in railways have a combinatorial nature, requiring the application of OR techniques. Huisman et al. [1] have described the different optimization problems in the context of passenger rail transportation and an overview of operations research models used. The increasing need for rail transport, capacity or resource constraints, and rapid increase in computational power have encouraged researchers to use optimization and simulation models. An example is a complete revamp of the Netherlands railway timetable through many years of cooperative work between railway personnel, academics, and software professionals resulting in significant cost savings [2].

The railway planning process varies from country to country, but in general, there are sequential stages in railway planning ranging from strategic planning to day-to-day operations, as shown in Fig. 19.1.

Depending on the planning horizon, the problems can be classified as strategic, tactical, and operational. Strategic decisions have a long-term impact and usually involve investments in resources or facilities. Tactical problems typically involve resource allocation over existing infrastructure. In operational issues, day-to-day

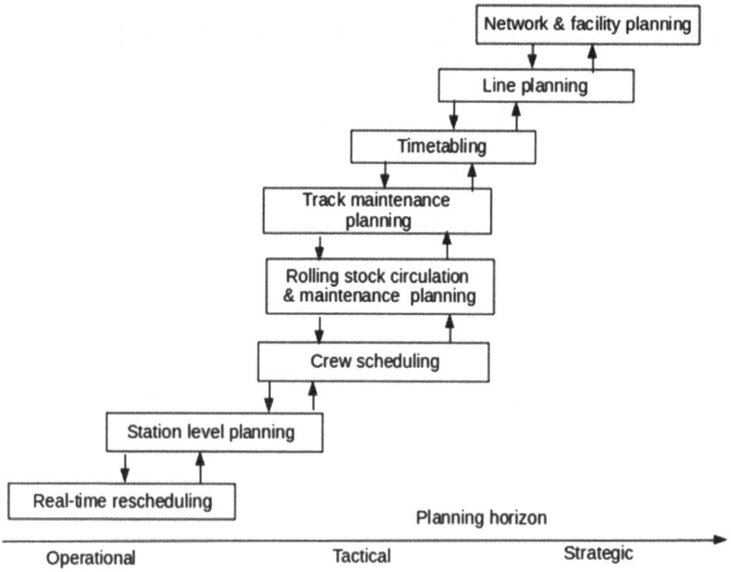

Fig. 19.1 Railway planning problems

Table 19.1 Comparison of decision-making at various planning horizon level [3]

Planning level	Horizon	Decision time	Decisions
Strategic	2–5 years or more	3–6 months	Infrastructure planning, capacity planning
Tactical scheduled services	6 months to 1 year	1–2 months	Service planning (line planning and routing), yearly timetabling, rolling stock and crew planning
Tactical non-scheduled services	1 week to 3 months	1 day to 2 weeks	Non-scheduled service planning, freight train planning, train routing in stations
Real time	1 h to 2 to 3 days	Few seconds to few minutes	Real-time rescheduling, delay management

activities are to be managed and often involve the readjustment of tactical-level decisions due to unforeseen circumstances. For example, creating or expanding a railway network and related facilities is a strategic decision, and the creation of a timetable is a tactical problem. In contrast, train unit shunting or real-time rescheduling in a railway terminal is an operational issue. Table 19.1 gives a brief of the typical planning horizon, decision time, and typical decisions from a planning level perspective.

This chapter gives an overview of some relevant railway planning problems, OR techniques used to solve these problems, and illustrates the modeling aspects using practical problems in Indian Railways. Indian Railways form the backbone of the transport system in India, and it is the fourth largest in the world in terms of network length. Indian Railways has a three-tier organizational structure with an apex body of the Railway Board, geographically distinct regional jurisdictions termed zones, and sub-regional units called divisions.

In the next section, we provide an overview of the entire railway planning problem, from the planning horizon point of view, and explain problems solved at each planning level.

19.1.1 Infrastructure Planning

Railway infrastructure planning involves decisions related to the desired service level to be provided to the customers (in terms of direct connections, frequencies, and reliability), and the capacities of the resources (infrastructure, rolling stock, etc.) that are required to accomplish these services [4]. A strategic decision in this connection involves creating the network and related infrastructure and deciding the quantity and type of rolling stock [1]. The above decisions have long-term implications compared to tactical problems like rolling stock circulation and maintenance planning.

Network and facility planning—This problem involves selecting nodes in space, the capacity of related facilities, and deciding on the network layout, where a rail network is represented by nodes corresponding to stations and links corresponding to tracks connecting these stations. In most cases, the scaling up of the railway network is incrementally done rather than creating a fully evolved network from scratch. Network planning also includes track infrastructure-related decisions like new routes to be laid out, conversion from a single line to double or multiple lines, electrification of lines, and conversion of gauges. Indian Institute of Technology Bombay developed a line capacity simulator for decision support for track infrastructure planning. This tool helps by providing insights into the system's behavior under different scenarios [5].

Rolling stock planning—Rolling stock in rail transport refers to wheeled vehicles like locomotives, freight, and passenger cars (coaches). Rolling stock is a highly capital-intensive component of a railway system, and the operational life can be in decades. Rolling stock planning decisions include the selection of the types of rolling stock to be used to cover passenger demand in the future using long-term forecasting models and life cycle analysis; acquisition; upgrading, selling, and destructing rolling stock. According to Huisman et al. [1] strategic or long-term rolling stock management has not received much attention from researchers.

19.1.2 Service Planning

Service Planning is about designing the service offerings and allocation of the resources for its execution. Service Planning includes the following modules.

Line planning—A fundamental structure in a rail planning process is a line connecting two terminals with interconnecting stations in between. Line planning and timetabling are two critical issues at the strategic and tactical levels influencing the service quality of railway operations [6]. Given a railway network and passenger demand between nodes, the problem is to decide the service lines to be operated along with the types and frequencies of the trains to be operated on each line. Subsequently, the departure and arrival timings in the line structure are decided in timetabling. The line system is to be designed to meet the travel demands and objectives like maximizing the service for passengers or minimizing the operational costs [7]. Goossens et al. [8] introduced models for finding solutions to line planning problems wherein lines have different halt patterns. The scientific software toolbox LinTim includes algorithms and libraries to optimize public transport [9]. This tool aids in solving different planning problems in railways, including line planning and timetabling.

Timetabling—Timetabling refers to deciding the arrival and departure time for each train at each station/section on the route for that train. It also fixes the order in which trains would utilize sections of the railway network (blocks, tracks, and platforms). The railway timetable is obtained through a detailed planning process.

The problem often involves finding a feasible (often periodic) schedule for trains that also satisfies the operational constraints. The timetable also makes it convenient for passengers to plan their journey. OR methods have been used in the planning and implementation of timetables which lead to effective utilization of rolling stock and crew [2].

Rolling stock circulation—The rolling stock circulation problem includes the allocation of vehicles to trips and deciding the vehicle composition based on demand. Rolling stock circulation is assigning the train units (locomotives and train rakes) to the train services according to the timetable. It also includes planning for empty wagons or deadheading locomotive movements. The coordination of the movement of rolling stock units in a network is one of the important planning problems for railways. Cordeau et al. [10] described the simultaneous assignment of locomotives and rakes to scheduled passenger trips based on multi-commodity flow in a network.

Maintenance Planning—Regular maintenance activities are essential for maintaining assets in good condition to ensure the reliability and safety of the railway system. There are two broad planning problems in this connection: (1) track maintenance planning along with timetabling, and (2) rolling stock maintenance planning integrated with circulation planning. Budai et al. [11] discussed track maintenance planning and propose clustering of maintenance activities in the same period in the network to minimize disturbance. This clustered maintenance plan can reduce track possession time.

Crew Scheduling Crew scheduling in a transport system is a well-known and challenging problem involving multiple constraints. Crew allocation involves human resources assignment for running the services as per timetable. The two standard formulations used in crew planning problems are: network flow formulation and set covering formulation [12]. Heil et al. [13] provided a detailed survey on crew scheduling models in railways. According to them, crew scheduling problems are typically formulated as mixed-integer programming formulations.

Station level planning—Timetabling for a railway network usually uses higher-level details and often does not include station-level planning problems. Each railway terminus has multiple facilities wherein train units can be parked. Platforming refers to assigning the platforms to the train in the stations. When not in use, the trains are parked in the stabling lines. Shunting yards are used for assembling and disassembling trains through tracks and railroad switches. Boysen et al. [14] presented a review of papers on shunting operations and models for predicting shunting yard performance.

Freight Train Planning problem—In many railway systems, resources are shared between passenger and freight trains. In India, passenger trains run as per a fixed schedule and get priority over freight trains. Freight train schedules are dynamically created based on demand on the available network. Ghorpade and Rangaraj [15] propose exact models and heuristics for getting optimal freight movement cycles. Upadhyay et al. [16] presented an optimization model for optimizing the loading plan of a double-stack container for a major Container Train Operator (CTO) in India. The CTO needs to utilize rolling stock efficiently, select containers, and assign containers to wagons optimally, considering conflicting requirements. Implementing a mathematical programming-based approach for train-load planning can save rail haulage costs [17]. Haehn et al. [18] present an algorithm based on constraint time graphs for finding feasible freight schedules.

19.1.3 Service Execution

Service execution planning is the planning during the execution of railway services. It deals with real-time railway traffic control (re-timetabling), rolling stock, and crew duty rescheduling. During the service execution, disturbances can result from external or internal factors such as maintenance activities, new service/train inclusion, weather, technical failures, track incidents, and run and dwell time delays. These disturbances impact the operations of all trains in the system, which makes the current timetable infeasible. By delay-induced disturbances, the existing train schedule and corresponding resource allocation become operationally nonviable or unsafe, thus warranting rescheduling.

Real-time rescheduling—The operational rescheduling includes live dispatching of established timetables and network recovery from delays or incidents. Recovery can take the form of attempting to return trains to their original timetables, or generating new temporary timetables for the remainder of their journeys and is called railway rescheduling [19]. Railway rescheduling involves modifying an existing timetable/schedule to ensure operationally viable and safe train movements while minimizing the impact of disturbances that occur during operations. The impact of disruption can be measured in a combination of objectives. Typical objectives for railway rescheduling are minimizing delays, (average, maximum, total, and only secondary). In many cases, no explicit objective function is used, and the only goal is to obtain a feasible solution [20]. The railway rescheduling problem involves the following decisions: (a) Re-timing deals with changing time instants of departures and arrival of trains via (i) changing the halt time of trains at station loops (Re-dwelling) and/or (ii) changing the speed profile of trains at open line section (Re-speeding); (b) Re-ordering is changing the order of trains on their common track segments; (c) Re-routing involves changing the routing of trains, via (i) changing platforms at a station or line section between stations (Local routing) or (ii) a completely different rerouting in the network, skipping stations and passing through different stations (Global routing); (d) Changing service involves canceling or truncating trains, canceling or adding stops.

When a decision regarding changing train routes or canceling scheduled train journeys or giving priority to delayed train is taken, it may impact possible connections or other adverse effects. Delay management focuses on minimizing the adverse effects of disturbances and disruptions for passengers or freight. In the case of passenger trains, it is deciding whether a connecting train should wait for a delayed feeder train or should depart on time.

The outbreak of the COVID-19 pandemic has resulted in travel restrictions, resulting in an adjustment of train schedules throughout different railway systems across the globe [21]. Haque and Hamid [22] proposed a seat assignment policy to mitigate virus diffusion risk among passengers and implemented a mixed-integer linear programming model based on real-life data from Indian Railways. Seat assignment using OR techniques can result in a reduction of risk and improvement in seat utilization compared to the current approach by railway operations [23].

19.2 Common OR Techniques in Railway Operations Management

A variety of techniques are used to solve railway planning problems including Mixed-Integer Programming, simulation, network flow models, and heuristics, and the implementation of these methods has resulted in significant benefits. The implementation of these techniques was aided by an exponential increase in computational power over the past few decades. Below are some examples of techniques used to solve problems in railway planning.

- **Exact approaches**—Branch and Bound, Dynamic Programming, Lagrangian Relaxation, Mixed-Integer Linear Programming (MILP)
- **Simulation**—Discrete Event Simulation, Train travel simulation
- **Heuristics**—Greedy algorithm, Local search
- **Evolutionary algorithms**—Genetic Algorithm, Simulated Annealing, Ant Colony Optimization, Tabu Search
- **Search-based techniques**—Constraint Programming, Artificial Intelligence, Fuzzy Logic, Knowledge-based systems
- **Hybrid**—MILP-based local search, MILP + Constraint Programming, MILP + Heuristics
- **Scheduling strategies**—Greedy heuristics, Shifting bottleneck, Right shift strategy
- **Dispatching rules**—Highest Priority, First-Come-First-Serve (FCFS), Brute force (BF), Avoid Most Critical Completion Time (AMCC)

Mixed-Integer Linear Programming models are used in solving different railway planning problems. Some examples are line capacity evaluation [24], vehicle scheduling [25], locomotive scheduling [26], freight train travel time estimation [27], maintenance scheduling [28], and rescheduling of disrupted train movements [29]. Gopalakrishnan and Rangaraj [30] discuss train capacity management among multiple travel segments in Indian Railways based on user and type of travel. A linear programming model is implemented for an optimal split of seat capacity in an end-to-end travel segment. Harrod [19] provides an overview of MILP models used in the railway domain.

The use of network flow models is common in railway planning problems. Ahuja et al. [26] provided an overview of the railroad planning and scheduling problems and a description of network models used in solving those problems. Maroti and Kroon [31] have used a multi-commodity flow model in a network graph for maintenance routing. Vaidyanathan et al. [32] have used a network flow-based crew scheduling model in the North American railroad scenario.

Both optimization and simulation are widely used in railway operations planning to aid decision-making. Optimization tends to be used when the planning problem is not too complex and an exact solution is preferred. Simulation is useful when the system is difficult to model mathematically, the randomness of the system cannot be ignored, and the impact of alternate scenarios needs to be studied. There exist many simulation tools such as RailSys [33], OpenTrack [34], FRISO [35], TOPSim [36], RAILS [37] and RTC (Rail Traffic Controller) Simulator [38]. Most of these tools are targeted toward long-distance and middle-distance railways [39].

Constraint Programming (CP) is used for finding feasible solutions from large candidate sets. The constraints put restrictions on the values that can be assigned to the variables simultaneously. Cheung et al. [40] have used CP for assigning railway tracks for scheduling maintenance tracks subject to a given set of constraints. Rodriguez [41] used CP along with simulation for routing of trains through a junction.

19.3 Applications to Large Scale Timetabling

19.3.1 Timetable as a Mathematical Object

A timetable is the combined determination of a path in the time-space of all scheduled trains in the network. In sequential terms, it involves determining the route of a train (including the origin-destination and halt pattern), start times, running times, and planned overtake for higher priority trains. Timetables need to incorporate allowances in running times and slack in turnaround to achieve robustness.

The key operational and safety requirements for timetabling are (a) There can be only one train at a time in a block section. A block section is that portion of the track between two stations on which no running train may enter until a 'line clear' signal has been received from the corresponding station. (b) Two consecutive trains in the same direction on a railway line should maintain a minimum separation time window (headway) with each other; (c) Arrival time of the train at station = Departure time of the train from the previous station + time required to travel between stations; (d) Contiguity of allocation between open line sections and station sections allocated to each train must be maintained; (e) The smooth operation of trains also requires additional operational time lag in the schedule, such as minimum time required to change interlocking and signals at a station once a train has passed from a loop or a line section, and safety margin which is an additional time lag between a pair of trains depending on their class/running speed.

Convenient representation of timetables facilitates operational planning, crew planning, and also other commercial purposes. Due to the large size of railway networks and operational complexity, the creation of a robust railway timetable is still a computationally challenging problem. In addition, rake links and maintenance

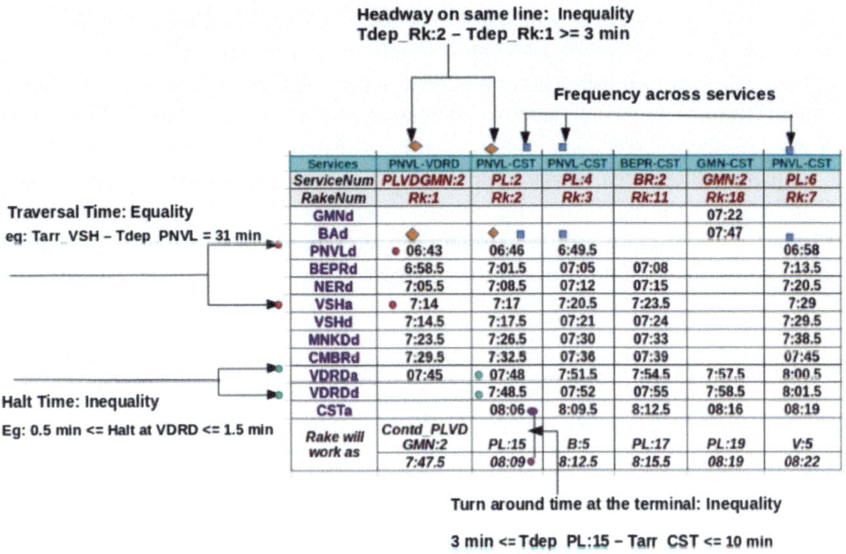

Services	PNVL-VDRD	PNVL-CST	PNVL-CST	BEPR-CST	GMN-CST	PNVL-CST
ServiceNum	*PLVDGMN:2*	*PL:2*	*PL:4*	*BR:2*	*GMN:2*	*PL:6*
RakeNum	*Rk:1*	*Rk:2*	*Rk:3*	*Rk:11*	*Rk:18*	*Rk:7*
GMNd					07:22	
BAd					07:47	
PNVLd	06:43	06:46	6:49.5			06:58
BEPRd	6:58.5	7:01.5	07:05	07:08		7:13.5
NERd	7:05.5	7:08.5	07:12	07:15		7:20.5
VSHa	7:14	7:17	7:20.5	7:23.5		7:29
VSHd	7:14.5	7:17.5	07:21	07:24		7:29.5
MNKDd	7:23.5	7:26.5	07:30	07:33		7:38.5
CMBRd	7:29.5	7:32.5	07:36	07:39		07:45
VDRDa	07:45	07:48	7:51.5	7:54.5	7:57.5	8:00.5
VDRDd		7:48.5	07:52	07:55	7:58.5	8:01.5
CSTa		08:06	8:09.5	8:12.5	08:16	08:19
Rake will work as	*Contd_PLVD GMN:2*	*PL:15*	*B:5*	*PL:17*	*PL:19*	*V:5*
	7:47.5	*08:09*	*8:12.5*	*8:15.5*	*08:19*	*08:22*

Fig. 19.2 A part of Mumbai suburban Harbour line timetable

need to be integrated for the feasibility of a working timetable. The timetable should find a fair balance between the customer's and the railway operator's goals.

Viewing the timetable as a mathematical object helps in generating new timetables and also in adding new services incrementally. As an example see a part of the Mumbai suburban Harbour line timetable in Fig. 19.2. In this suburban section, the traversal time is standardized and hence represented as an equality constraint between certain time variables. The halt times are represented as equality if it is standardized and inequality constraints with lower and upper bounds if there is some flexibility in operation. Similarly, the headway, frequency, and turnaround constraints are also represented. Generation of a timetable involves filling up values (arrival and departure times) that satisfy the various operational constraints. Constructive, search-based, optimization, and simulation approaches can be used in the creation of timetables [2, 42–44].

19.3.2 Suburban Timetabling

In India, suburban rail plays an important role in commuter transport in Mumbai, Kolkata, and Chennai and to a lesser extent in many other cities. Construction of timetables for suburban services has the additional considerations that (a) timetables have to be constructed in very close synchronization with rolling stock planning, (as the services are very frequent and the turnaround at terminals has to be planned together with timetables), and (b) commuter requirements often require some approximate periodicity of services which has to be integrated into the timetabling decision. The techniques used here are also applicable to planning metro rail operations. One way to handle these is through the application of the PESP (Periodic Event Scheduling Problem) framework, which leads to interesting and challenging optimization applications. The aperiodic version is also possible, with appropriate constraints related to the frequency of services. This is appropriate when there are well-differentiated peak periods of travel during the day, once in the morning and once in the evening. There could be an added constraint to maintain a fixed number of slots for long-distance train paths as the track infrastructure is shared.

19.3.3 Timetables for the Golden Quadrilateral and Diagonals on the Indian Railway Network

In this section, an application of simulation and optimization tools to a large-scale timetabling exercise on the Indian Railways network, done between March and December 2020 is discussed [45]. This was an exercise to generate *de novo* timetables of passenger trains on the Golden Quadrilateral (New Delhi—Howrah—Chennai—Mumbai—New Delhi) and the Diagonals (New Delhi—Chennai and Howrah—Mumbai), referred to in short as GQD (Golden Quadrilateral plus Diagonals). This is a significant high-density traffic part of the Indian Railways network and it was felt that streamlining the operations on this part of the network would generate significant benefits. The exercise was carried out by a team from the Railway Board, the Centre for Railway Information Systems, twelve Zonal Railways of Indian Railways, and a team from IIT Bombay.

The Indian Railway network is complex, and the simultaneous determination of train timings is a challenging task. Support using software tools allows for priority-wise scheduling and resource allocation. The project was done to create a new passenger timetable with many objectives, which included faster run times (possible due to the introduction of modern rolling stock), rationalized halt patterns, sufficient windows for maintenance activities, and a freight corridor for efficient freight operations.

The core technique used was a rail traffic simulator (in Java) developed over many years at IIT Bombay, with support from Indian Railways. The simulator helps in capacity building and identifying bottlenecks in a railway network. The simulator is based on a travel-advanced heuristic with asynchronous train movement. It can effectively handle global conflicts across the network and process inputs and outputs for various goals. There are several auxiliary tools for analytics and visualization.

Initially, high-priority trains are scheduled (simulated) followed by lower-priority trains by adding each train one by one into the system. The timetable generated with the support of the simulator is guaranteed to be feasible and conflict-free.

There were a number of pre-processing and post-processing steps for the massive amount of data to be handled and usable outcomes to be generated. The GQD network involves about 9000 route kilometres and about 19000 track kilometres (most of the route is a double line and there are a few sections with three and four lines) and 1382 stations. A total of 3521 trains per week were timetabled, which included daily and non-daily trains. The details include 3804 loop lines at stations, in addition to the main line (where trains can be safely handled), more than 1500 speed restriction areas, and the desired pattern of 99835 halts that had to be planned.

The main inputs were:

- Network information—Station list, inter-station distances, lines in each section, running lines at the station, block working times, and maximum permissible speed at the sections.
- Train details—proposed start time, route, halt duration, maximum speed, acceleration, deceleration, priority, and allowances.

The major outputs were traversal details for each train for each section, end-to-end travel time, diagnostic charts, reports, and network-level timetable.

The simulation also permits the identification of problematic inter-station sections and pinpoint causes for longer train running times. In addition, the simulation analysis helps to identify corridors for freight train traversal that reduce the en-route detention of freight trains due to higher priority passenger trains. A number of numerical and modeling issues had to be handled in this model.

- Rounding of times to an appropriate subdivision of a minute had to be decided. One interesting observation was that rounding up inter-station running times in minutes will cause an average 45 min increase in the Mumbai-Delhi route for all trains.
- Choice of up and down directions had to be made on each route because every track resource is potentially directional, i.e. usable in one or other direction only.
- Handling of paths of non-daily trains, which often share essentially the same time path on different days of the week, but which could be in different groups on different parts of the network.

The entire exercise was extended to include planning maintenance time windows and also windows of time when freight trains could be effectively run in the presence of passenger traffic, given that passenger traffic is usually of higher priority.

19.4 Rolling Stock Circulation

19.4.1 Introduction

Rolling stock is one of the most important cost components in a railway system, and the effective management of these assets is important. The number of trips that a railway enterprise can offer is dependent on the number of rolling stock units. Rolling stock planning aims at finding an efficient allocation of rolling stock units for a given set of trips as per the railway timetable [46]. Rolling stock circulation problems come after network planning, line planning, and timetabling. The services as per the timetable need to be allotted physical rolling stock units [47]. The units move along a pre-decided cyclic path in a network satisfying operational or infrastructural constraints.

Since rolling stock units need to undergo periodic maintenance activities regularly many studies focused on an approach in which rolling stock service and maintenance schedules are integrated. Periodic maintenance ensures that the assets are maintained in good shape thereby increasing the reliability and quality of service.

According to Borndörfer et al., [25] integration of different technical aspects like vehicle composition, maintenance constraints, infrastructure capacities, and regulatory requirements is a challenging problem. There is a need for an integrated approach by accommodating different requirements. They have used a hypergraph-based model and integer programming-based formulations for rolling stock allocation and rotation. Canha and Barrena [47] have used a sequential approach for rolling stock circulation in rapid transit systems. In the first stage, a minimum number of vehicles needed to perform the services as per the schedule is determined. In the second stage, a set of daily routes is obtained, and in the final stage, weekly circulations and depot locations are identified.

Rolling stock circulation in Indian Railways There are 67,597 passenger coaches used by Indian Railways, which are divided into 86 types. In India, a group of connected coaches, excluding the locomotive, that makes a train is called a rake. Rake linking is assigning physical rakes to the scheduled services as per the timetable. To enhance the utilization of rakes, Indian Railways is using complex rake linking patterns in many cases. Experienced railway planners iteratively generate feasible assignments. The current practice is to make modifications in the rolling stock schedules depending on the change in timetable.

In Fig. 19.3, the weekly cycle of a rake under the ownership of Central Railway, one of the zones of Indian Railways, is shown. The rake is used to run long-distance

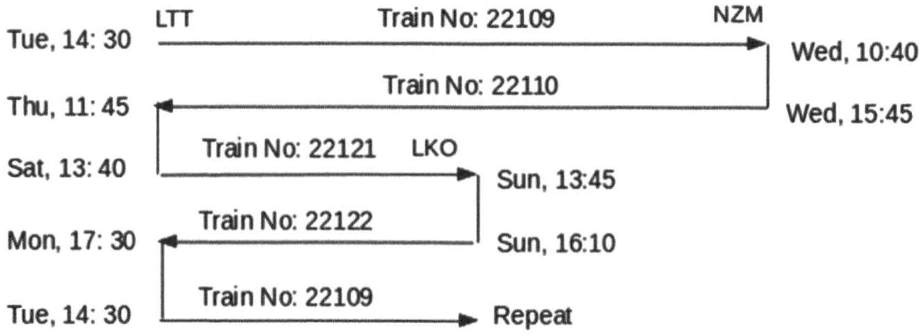

Fig. 19.3 Rake linking example (*Source* Rake Link, Central Railway, w.e.f, July 2019)

trains from Lokmanya Tilak Terminus to two different locations. In the figure, the abbreviations written on top refer to location IDs: LTT-Lokmanya Tilak Terminus, NZM-Hazrat Nizamuddin, LKO-Lucknow, and the numbers above the lines are train identification numbers. Each horizontal line in the diagram represents a trip, and the small vertical lines in between represent the time between the trips, which includes the maintenance and the waiting time.

The rakes have to be assigned to trips along with a periodic maintenance schedule so that rake movements are in periodic cycles. The current practice of rolling stock rostering and maintenance scheduling is done by making incremental changes as per requirements necessitated by changes in the timetable or other operational constraints. Rakes must be maintained in periodic intervals so that time and distance upper bounds of the maintenance are not violated. Maintenance upper bounds on distance and time vary with the type of rolling stock used. Two major types of periodic maintenance activities done on rakes used by Indian Railways are:

- Primary maintenance—Indian Railway schedules the primary maintenance if the journey length exceeds a distance threshold (e.g., 3000 KM for conventional, Integrated Coach Factory (ICF) coaches, or 4000 KM for Linke Hofmann Busch (LHB) coaches). The duration for the primary maintenance is six hours.
- Secondary maintenance—Secondary maintenance involves inspection, watering, cleaning, etc. The duration of secondary maintenance is two hours.

Due to the size of the network and the large scale of operations finding efficient cyclic rake links is a challenging task. Currently, the preparation of maintenance schedules and coaching rake links is largely manual in Indian Railways. The current practice of rolling stock rostering and maintenance scheduling is done by making incremental changes as per requirements necessitated by changes in the timetable or other operational constraints.

A model for creating optimal rake links that also satisfy the maintenance constraints is useful. The manual process can be time consuming and may not provide an optimal assignment that maximizes the rolling stock utilization.

19.4.2 Modeling for Generating Rake Links

A simple schematic sketch on modeling for creating rake links is given in Fig. 19.4. The problem's input includes demand requirements, rake information, timetable, maintenance location, and capacity information. In the present work, identification of the rolling stock composition based on passenger, demand is not included.

Creation of space-time network graph Initially, a space-time network graph is created and then a mixed-integer programming model is used to find cycles in the graph that minimizes the cost of flow in the network. The circulations generated by the model have to satisfy connection, composition, and maintenance requirements. It is assumed that the timetable is cyclic with a periodicity of one week.

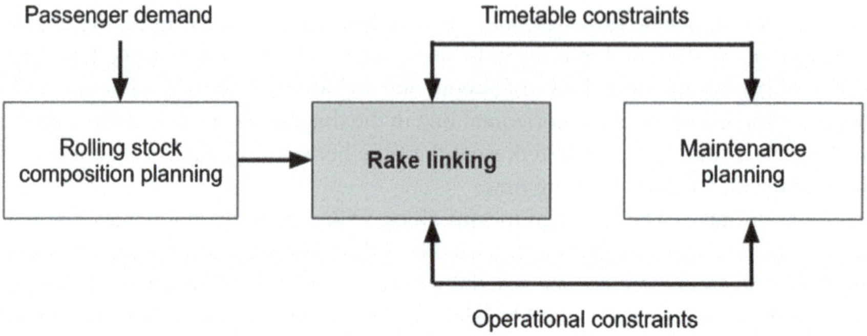

Fig. 19.4 Inputs and constraints in rake linking

The space-time network graph for the model is created as follows: Trip nodes are created corresponding to every departure and arrival as per the given timetable. Each node represents a time and location in the network graph. Maintenance activity nodes are created if a sufficient time gap is available between the arrival and departure. A pit line is required in the terminal to facilitate maintenance activities. Each directed arc indicates either a trip or a waiting time between the trips. Waiting arcs are created based on time precedence and rake compatibility between two corresponding trips. For maintenance activity, two arcs are created, one representing movement for the maintenance and the other representing movement after maintenance.

A time-space network graph is shown in Fig. 19.5. In this figure, arc 1–2 represents a trip between terminal-1 and terminal-2 which starts at t1 and ends at t2. Arc 2–3 is

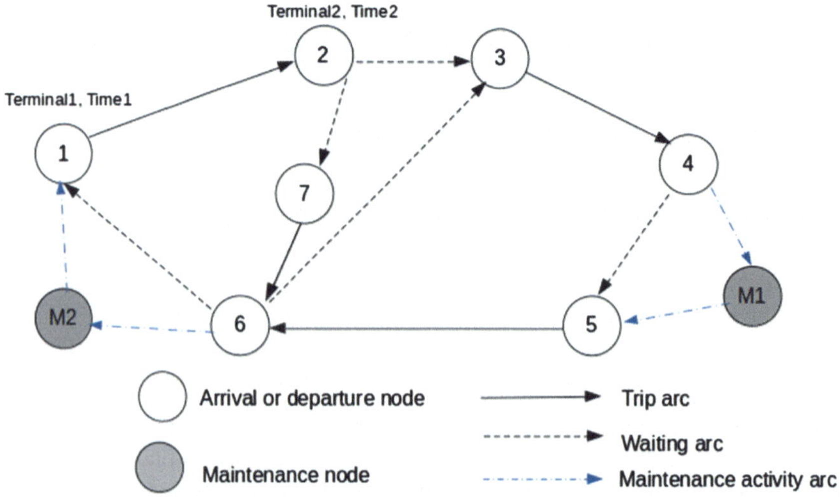

Fig. 19.5 Time-space network graph used in rake linking

Table 19.2 Sets, parameters, and variables

Notation	Description
N	Set of all trip nodes
L	Set of all maintenance activity nodes
M	Set of all maintenance constraints
A	Set of all arcs including trip and waiting arcs
S	Set of all service/trip nodes
c_a	Cost of arc $a \in A$
$r_m(n, o)$	Resource consumption of $m \in M$ as arc (n, o) is traversed
x_a	Binary variable indicating whether arc $a \in A$ is included in the cyclic linkage
y_a^m	Continuous variable to track the resource usage since arc, a has been covered
U_m	Resource usage upper bound corresponding to maintenance constraint $m \in M$

a waiting arc representing a possible linkage between trips 1–2 and 3–4. The model will select suitable linkage among different available linkages to minimize aggregate waiting time and satisfy all the constraints. The capacity constraints of maintenance facilities, parking lines, and platforms are not considered in modeling.

The costs of arcs are chosen so that the objective of minimizing the waiting time between the trips to enhance rolling stock utilization is attained. The trip arcs can be assigned any cost as all the trips will be included in the final solution. The waiting arcs are given a cost proportional to the waiting time. The maintenance arc costs are higher than the ordinary waiting arcs so that unnecessary visit to the maintenance facilities is avoided.

Mixed-Integer Programming Formulation We propose a Mixed-Integer Programming model for rake linking of long-distance passenger trains. The model generates rake cycles that will cover all trips as per the timetable. The notations are shown in Table 19.2.

For each node $n \in N \cup L$ the set of incoming and outgoing arcs are:

- $A(n)^{in} := \{(l, n) \in A\}$,
- $A(n)^{out} := \{(n, o) \in A\}$,

$$min \sum_{a \in A} c_a x_a, \tag{19.1}$$

subject to

$$x_a = 1, \qquad \forall a \in S, \tag{19.2}$$

$$\sum_{a \in A(n)_{in}} x_a = \sum_{a \in A(n)_{out}} x_a, \qquad \forall n \in N \cup L, \qquad (19.3)$$

$$y_a^m \leq U_m x_a, \qquad \forall a \in A, m \in M, \qquad (19.4)$$

$$\sum_{a \in A(n)^{out}} y_a^m = \sum_{a \in A(n)^{in}} y_a^m + \sum_{a \in A(n)^{out}} x_a r_m(n, o), \qquad \forall n \in N, m \in M, \qquad (19.5)$$

$$\sum_{a \in A(n)^{out}} y_a^m = \sum_{a \in A(n)^{out}} x_a r_m(n, o), \qquad \forall n \in N, m \in M, \qquad (19.6)$$

$$x_a \in \{0, 1\}, \qquad \forall a \in A, \qquad (19.7)$$

$$y_a^m \in [0, U_m] \subset Q+, \qquad \forall a \in A, m \in M. \qquad (19.8)$$

The objective minimizes the total cost of rake linkages. Constraints (19.2) ensures that all trip arcs are selected indicating that rakes will be allocated to all trip arcs in the network graph. Flow conservation constraints (19.3) dictate that, at every node, the inflow and outflow of rakes must be equal. Constraints (19.4) state that upper bound restrictions for maintenance (time/distance) should not be violated. According to constraints (19.5), the sum of outgoing resource consumption (time/distance) is equal to the incoming resource usage added to the resource consumption caused by the outbound arc. As per the constraints (19.6), the resource consumption is set back to zero at the maintenance node. Constraints (19.7) state that variables x_a's are binary and constraints (19.8) apply bounds on the values of the flow variables.

The model is implemented on different subsets of services originating from a busy terminal in Indian Railways and found that with the rake links generated and by standardization of rakes, there could be a significant improvement in the utilization of rakes [48].

19.5 Station Level Planning

A railway station is a place where embarking and disembarking of passengers, loading and unloading of goods, and meeting and passing of trains take place. A railway station consists of interconnected track resources such as loops (multiple parallel tracks), interchanges, switches, and cross-overs along with platforms. A station handles frequent train movement activities (arrival, halt, and departure) and infrequent shunting activities (train or loco reversal, train shunting, and traction change) as a part of regular train operations. Often, in a railway network, trains get delayed at or near large railway stations as they do not get the required traversal path. A station has interconnected, overlapping and crossing track resources, which cause hindrance in multiple train movements, which in turn leads to extended train travel times. Safe train movements in a railway station involve planning and executing various activities and decisions like finding traversal paths for trains through the station network, assigning platforms to trains, and planning local shunting activities. Planning these activities at large and complex railway stations is challenging, especially in situations of congestion, due to the number of decisions involved, details of infrastructure to be considered, and operational rules constraining the passage of trains through track sections.

19.5.1 Problem Definition

Train movement planning at a railway station is assigning track resources to a set of train movement activities such that operational and safety requirements are satisfied, planned train movements are carried out smoothly and overall delays are minimized. The problem is to find a conflict-free path for each train passing through the station. The outcome is represented as the occupation start time and end time for each train for each allocated track element in the station. Key decisions involve determining a route for each train (i.e. a sequence of track elements visited by each train), time intervals for which track elements will be occupied by each train [49].

A set of trains and related activities (traversal, reversal) at the station are assumed to be known in advance. Each train is defined in terms of an entry point and an exit point at the station, scheduled arrival time at the entry point, scheduled departure time from the platform at the station, and minimum halt time at the station. The trains are assumed to be running at a constant speed, but different speed for different paths and routes of trains is modeled. For safety reasons, trains cannot move faster than the maximum permissible speed of the track elements especially, at crossings/switches. No infrastructure element can be allowed to be allotted to two different trains at the same time. The traffic approaching already occupied tracks from other connected tracks is blocked. Minimum time between two trains must be maintained when they occupy the same resource consecutively.

The interlocking setting available at the station impacts train traversal through the station. The interlocking setting decides how are track elements reserved ahead and released later during the traversal of the train through the station. In *Route-lock route-release setting*, all path elements for a train are reserved simultaneously and all path elements other than the last element (platform or halt edge) are released simultaneously when the train reaches the last element. In route-lock route-release settings, all track elements in the chosen path are reserved simultaneously, and except the last element (loop or platform or halt edge) all other elements are released simultaneously when the train reaches the last element. When an element is released, it can be occupied by the next train only after some safety time.

19.5.2 Modeling Framework

Figure 19.6 shows the schematic layout of a typical railway station. The figure shows various elements of a station network like signals, crossings, and track segments.

The railway station network is represented as a directed graph where each signal or crossing is modeled as a node and track segment between two nodes as an arc. The underlying railway station network is modeled at a microscopic level.

A node is characterized by a node type (signal/crossing) and the direction of traffic allowed through that node. The signal node is classified as entry, exit, halt, intermediate, and terminal signal. An edge is characterized by from-node, to-node, length of the arc/track, speed limit while traversing through the edge, edge type (normal or crossing), and direction of traffic allowed through the edge. An edge is a directed edge representing the allowed direction for the train traffic. The edge 10-9 with signal G & F shown in Fig. 19.6 is a bidirectional arc.

All possible routes between all pairs of entry and exit signals (in a directed graph shown in Fig. 19.6 are obtained using a depth-first graph traversal algorithm. There may be multiple routes possible between a given entry signal and a given exit signal. Each route is broken into two paths, namely the inbound path and the outbound path. The inbound path is between an entry signal and a halt signal; the outbound path is between a halt signal and an exit signal. Please note that a train may occupy more than one track element at a time due to small length track elements at the station and is modeled explicitly.

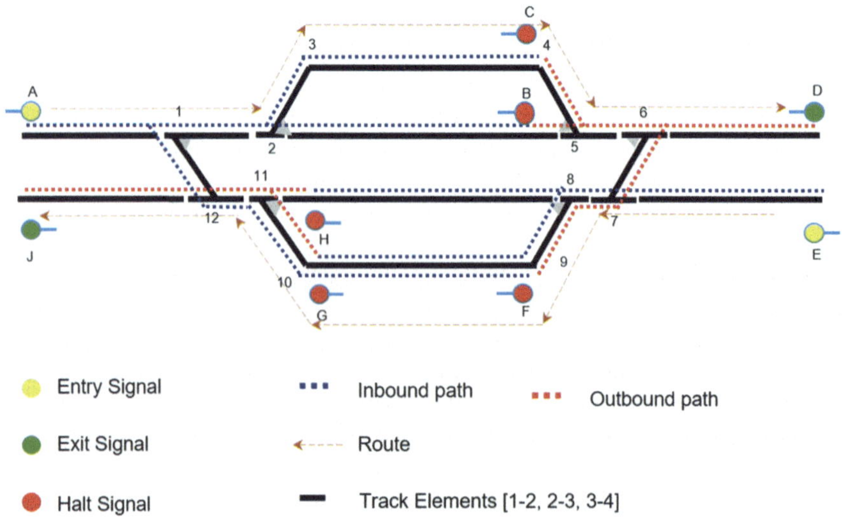

Fig. 19.6 Station schematic layout [50]

The train traversal through the station network is modeled under the route-lock route-release interlocking setting. The concept of lock time and release time facilitating the safety margin between two consecutive trains occupying a given rail infrastructure is used while modeling train traversal

19.5.3 Mathematical Model

This section presents an exact optimization-based approach for developing an operational plan for a station. It presents a Mixed-Integer Linear Programming (MILP) model, which considers micro-level infrastructure, and speed dynamics of trains (indirectly), covers associated local activities like train reversal, and implements route-lock route-release operating policy. The output of the model is conflict-free paths for each train through the station (including reversal) and produces an implementable schedule of trains at the station.

Assumptions made during modeling and solving train movement planning at a station are listed here. In case any assumption is relaxed, it is mentioned in the respective section. (a) Each train has only one entry signal and one exit signal in the network; (b) The model is a fixed speed model, i.e. speed variation is not modeled, but it can accommodate different speeds for each path the train may occupy; (c) All trains and their entry time at the station in known upfront before running the model (d) The planning horizon for the model is defined by the input data, i.e. trains and their scheduled arrival times (or inter-arrival time between trains) and exit time; (e) The model does not consider initial conditions, e.g. current occupation of track resources; (f) Exit and entry signals will have only one edge leading to them; (g) Train length is modeled using the clearance times concept; (h) No reversal movements are not modeled; and (i) The value of run time for trains is lower bound and planned run time may be higher than the input run time. It means a train can get delayed (j) All trains

Table 19.3 Sets, parameters and variables

Notation	Description
S	Set of signals/switches in the station network, indexed by i or j. It can be construed as nodes of a graph
E	Set of track elements in the station network connecting two signals/switches, indexed by e. It can be construed as the edges of a graph. $E = \{e = (i, j) : i, j$ are connected, $i \in S, j \in S\}$
$G(S, E)$	Track network at the railway station consisting of signals/switches as nodes (S) and track elements connecting nodes as arcs (E)

Table 19.3 (continued)

Notation	Description
$NS \subset S$	Set of entry signals at the station
$XS \subset S$	Set of exit signals at the station
$HS \subset S$	Set of halt signals at the station
T	Set of trains passing through the station, indexed by t
$NS_t \in NS$	Entry signal for train t at the station
$XS_t \in XS$	Exit signal for train t at the station
TR_t	Indicator, if train t needs to be reversed at the station
AT_t	Scheduled arrival time for train t at NS_t at the station
DT_t	Scheduled departure time for train t fro a halt signal at the station
HT_t	Minimum halt time for train t at the station
p_{ij}	A path is defined by a sequence of connected track elements between two signals. Thus p_{ij} is a path in $G(S, E)$ from $i \in S$ to $j \in S$.
	For this paper, a path is always defined between entry signals (NS) and halt signals (HS) or between halt signals (HS) and exit signals (XS).
P	A set of paths possible at the station, indexed by p. $P = PI \cup PO$
$PI \subset P$	A set of inbound paths, i.e. all paths between entry signals NS and halt signals HS at the station. $PI = \{p_{ij} : i \in NS, j \in HS\}$
$PO \subset P$	A set of outbound paths, i.e. all paths between halt signals HS and exit signals XS at the station. $PO = \{p_{ij} : i \in HS, j \in XS\}$
$P_t \subset P$	Set of paths train t can access. $P_t = PI_t \cup PO_t$
$PI_t \subset P_t$	Set of inbound paths train t can access. $PI_t = \{p_{ij} : i \in NS_t, j \in HS\}$
$PO_t \subset P_t$	Set of outbound paths train t can access. $PO_t = \{p_{ij} : i \in HS, j \in XS_t\}$
$E_p \in E$	Set of track elements that belongs to path p.
R	Set of routes possible at the station, indexed by r. A route is defined by an ordered sequence of an inbound path and an outbound path.
	$R = \{(p, p') : p \in PI \text{ is } p_{ih},\ p' \in PO \text{ is } p_{hj},\ i \in NS,\ j \in XS,\ h \in HS\}$
$R_t \in R$	Set of routes which train t is allowed to access. $R_t = \{(p, p') \in R : p \in PI_t,\ p' \in PO_t\}$

Table 19.3 (continued)

Notation	Description
$RT_{t,e,p}$	Minimum run time to cross given track element, (time between train head entering *e* and train head leaving *e*, i.e. head reaching the end of *e*)
	for train t, path p and track element e, where $t \in T,\ p \in P_t,\ e \in E_p$
$CT_{t,e,p}$	Minimum clearance time (time between head leaving e and tail leaving e) for train t, path p, and track element e where $t \in T,\ p \in P_t,\ e \in E_p$
HW_e	Minimum time gap between two trains occupying track element e consecutively, where $e \in E$
Variables	
$y_{t,p}$	Binary variable = 1 if train t is allocated to path p; = 0 otherwise
$x_{t,e}$	Binary variable = 1 if train t is allocated to track e; = 0 otherwise
$F_{t,t',e}$	Binary variable = 1 if train t occupy edge e after train t' ; = 0 otherwise, where $t \neq t'$
$a_{t,e}$	Continuous variable, Arrival time for train t at track element e, when head of train t start occupying e
$d_{t,e}$	Continuous variable, Departure time for train t at track element e, when tail of train t leaves e
$u_{t,e}$	Continuous variable, Locking start time for train t at track element e
$v_{t,e}$	Continuous variable, Locking end time for train t at track element e

which enter the station would eventually exit from the station to another station or to the yard. The text below describes the objective and constraint of the mathematical model for developing a train movement plan for a station. Table 19.3 states notations and their descriptions used to describe the model.

Objective: Minimize the sum of (a) total time spent by all trains in the station, (b) total track occupancy time by trains, and (c) total track locking time by trains

$$\begin{aligned} \text{minimize} \quad & \sum_{t\in T,\ e=(i,j):j=XS_t} (d_{t,e} - AT_t) + \\ & \sum_{t\in T,\ e\in E} ((d_{t,e} - a_{t,e}) + (v_{t,e} - u_{t,e}) \end{aligned} \tag{19.9}$$

The last two elements in the objective functions are added to make those respective variables bounded, else those variables get the default values.
Subject to following constraints:
(A train has to be allocated to exactly one inbound path)

$$\sum_{p \in PI_t} y_{t,p} = 1 \quad \forall\ t \in T \tag{19.10}$$

(A train has to be allocated to exactly one outbound path)

$$\sum_{p' \in PO_t} y_{t,p'} = 1 \quad \forall\ t \in T \tag{19.11}$$

(Both inbound and outbound paths a train is allocated to must be elements of the same route)

$$\sum_{p' \in PO_t : (p,p') \in R_t} y_{t,p'} \geq y_{t,p} \quad \forall\ t \in T,\ p \in PI_t \tag{19.12}$$

Constraints (19.10), (19.11), and (19.12) ensure that a train has exactly one inbound path and one outbound path and both these paths are part of same route.
(Each element in the path should be allocated when a path is chosen)

$$x_{t,e} \geq y_{t,p} \qquad \forall\ t \in T,\ p \in P_t,\ e \in E_p \tag{19.13}$$

$$x_{t,e} \leq \sum_{p : e \in E_p} y_{t,p} \qquad \forall\ t \in T,\ e \in E_p \tag{19.14}$$

Constraints (19.13) and (19.14) ensure that if a path is selected for a train then each element in that path should be allocated to the same train. Please note that a track element may also be part of two same direction paths for example in Fig. 19.6 track element A–1 is part of path A–B as well as path A–C, thus we can not have Constraint (19.13) as equating constraint.
(A train cannot arrive earlier than scheduled arrival time at the station)

$$\begin{aligned} a_{t,e} \geq AT_t \cdot x_{t,e} \quad & \forall\ t \in T,\ e \in E_p \\ \text{where } e = (i,j) : i = NS_t \text{ and } & p \in PI_t \end{aligned} \tag{19.15}$$

(A train must not depart earlier than the scheduled departure time at the station)

$$\begin{aligned} d_{t,e} \geq DT_t \cdot x_{t,e} \quad & \forall\ t \in T,\ e \in E_p \\ \text{where } e = (i,j) : i \in HS \text{ and } & p \in PO_t \end{aligned} \tag{19.16}$$

Constraints (19.15) and (19.16) enforce that a train cannot arrive earlier than the expected arrival time at the station and a train must not depart earlier than scheduled departure time at the station, respectively.
(Minimum headway and sequencing constraint between two trains for each track element)

$$\begin{aligned} &a_{t,e} - a_{t',e} \geq (HW_e \cdot F_{t,t',e}) - M\,(3 - F_{t,t',e} - x_{t,e} - x_{t',e}) \\ &\forall\ t \in T,\ t' \in T,\ e \in E_p \cap E_{p'} \text{ where } t \neq t',\ p \in P_t,\ p' \in P_{t'} \end{aligned} \tag{19.17}$$

$$\begin{aligned} &u_{t,e} - v_{t',e} \geq -M\,(3 - F_{t,t',e} - x_{t,e} - x_{t',e}) \\ &\forall\ t \in T,\ t' \in T,\ e \in E_p \cap E_{p'} \text{ where } t \neq t',\ p \in P_t,\ p' \in P_{t'} \end{aligned} \tag{19.18}$$

$$\begin{aligned} &F_{t,t',e} + F_{t',t,e} = 1 \\ &\forall\ \ t \in T,\ t' \in T,\ e \in E_p \cap E_{p'} \text{ where } t \neq t',\ p \in P_t,\ p' \in P_{t'} \end{aligned} \tag{19.19}$$

Constraints (19.17), (19.18), and (19.19) are for introducing minimum safety time and sequencing constraint between two trains for each track element that both trains can occupy.
(Track element occupation time-related constraints)

$$\begin{aligned} &\text{(departure } >= \text{ arrival + run time + clearance + halt)} \\ &d_{t,e} \geq a_{t,e} + (RT_{t,e,p} + CT_{t,e,p} + HT_{et}) \cdot y_{t,p} - M \cdot (1 - y_{t,p}) \\ &\text{where } HT_{et} = HT_t \text{ if } e = (i,j) : j \in HS \text{ else } 0 \\ &\forall\ t \in T,\ p \in P_t,\ e \in E_p \end{aligned} \tag{19.20}$$

$$\begin{aligned} &\text{(arrival next edge } >= \text{ arrival prev edge + run time)} \\ &\text{(departure prev edge } >= \text{ arrival next edge + clearance)} \\ &a_{t,e'} \geq a_{t,e} + RT_{t,e,p} \cdot y_{t,p} - M \cdot (1 - y_{t,p}) \\ &d_{t,e} \geq a_{t,e'} + CT_{t,e,p} \cdot x_{t,e} - M \cdot (1 - y_{t,p}) \\ &\forall\ t \in T,\ p \in P_t,\ e \in E_p,\ e' \in E_p \\ &e' \text{ follows } e \text{ in } p : e = (i,k),\ e' = (k,j),\ j > i \text{ in } p, (i,j,k) \in S \end{aligned} \tag{19.21}$$

$$\begin{aligned}
&\text{(arrival next edge } >= \text{ arrival prev edge + run time + halt)}\\
&\text{(arrival next edge } >= \text{ departure prev edge - clearance)}\\
&\text{for all trains, paths and edges ending at halt signals}\\
&a_{t,e'} \geq a_{t,e} + (RT_{t,e,p} + HT_t) \cdot y_{t,p} - M \cdot (2 - y_{t,p} - y_{t,p'})\\
&a_{t,e'} \geq d_{t,e} - CT_{t,e,p} \cdot y_{t,p} - M \cdot (2 - y_{t,p} - y_{t,p'})\\
&\forall\, t \in T,\ p \in PI_t,\ p' \in PO_t,\ e \in E_p,\ e' \in E_{p'}\ (p,' p) \in R_t\\
&e' \text{ follows } e \text{ in } R_t : e = (i,k),\ e' = (k,j)\ k \in HS,\ (i,j) \in S
\end{aligned} \tag{19.22}$$

Constraints (19.20)–(19.22) are track element occupation time-related constraints. Constraint (19.20) ensures that the departure time of a train at each track element in a path chosen should be greater than or equal to the sum of arrival time, run time over that track element, clearance time for that train for that track element, and halt time for that train at that station. The halt time is added only if the track element has halt signal as end node. Similarly, constraints (19.21) ensures the correct arrival time and departure time of a train at each track element. Constraint (19.22) ensures occupancy and time continuity among the tracks with halt signals which are common elements between two connected paths.

Interlock setting: Route-lock route-release

$$\begin{aligned}
&\text{lock start } \leq \text{ arrival for all trains, all edges in paths}\\
&u_{t,e'} \leq a_{t,e} - M \cdot (1 - y_{t,p})\\
&\forall\, t \in T, p \in P, e \in E_p, e' \in E_p\\
&p \in PI_t \implies e = (i,k) : i = NS_t\\
&p \in PO_t \implies e = (i,k) : i \in HS
\end{aligned} \tag{19.23}$$

$$\begin{aligned}
&\text{lock end } \geq \text{ departure for all trains, all edges in paths}\\
&v_{t,e'} \geq d_{t,e} - M \cdot (1 - y_{t,p})\\
&\forall\, t \in T, p \in P, e \in E_p, e' \in E_p\\
&p \in PI_t \implies e = (i,k) : i \in HS\\
&p \in PO_t \implies e = (i,k) : i = XS_t
\end{aligned} \tag{19.24}$$

$$\begin{aligned}
&\text{lock end time } \geq \text{ arrival time prev edge + run time}\\
&v_{t,e'} \geq (a_{t,e} + RT_{t,e,p}) \cdot y_{t,p} - M \cdot (1 - y_{t,p})\\
&\forall\, t \in T, p \in PI_t, e' \in E_p, e \in E_p : e = (i,k) : k = HS, e \neq e'
\end{aligned} \tag{19.25}$$

$$\begin{aligned}
&\text{relation between lock start/lock end of last edge of inbound path}\\
&\text{and that of first edge of outbound path}\\
&u_{t,e'} \leq u_{t,e} + M \cdot (2 - y_{t,p} - y_{t,p'})\\
&v_{t,e} \geq v_{t,e'} - M \cdot (2 - y_{t,p} - y_{t,p'}) \qquad (19.26)\\
&\forall\ t \in T,\ p \in PI_t,\ p' \in PO_t,\ e \in E_p,\ e' \in E_{p'}\\
&\text{where } e = (i,k),\ e' = (k,i),\ k \in HS,\ i \in S\\
&\text{i.e. } p \text{ and } p' \text{ are in same route and } e' \text{ follows } e \text{ in that route}
\end{aligned}$$

Constraints (19.23)–(19.26) implement route-lock route-release setting. Constraint (19.23) ensures that all elements of chosen inbound and outbound path for a train are locked as soon as the train arrives at the first element of that path. Constraint (19.24) ensures that all elements of the chosen path for a train are released or freed only after the train reaches the end node of the path. In case of the inbound path, the end node would be the halt signal, whereas for outbound path, it is the exit signal. Constraint (19.25) ensures that the lock end time of each track element should be greater than the sum of arrival time at the previous edge and the run time over the track element. Constraint (19.26) depicts the relation between the lock end of the last edge of the inbound path and that of first edge of the outbound path. It ensures that the lock start time of the first edge of the outbound path should be less than that of last edge of the inbound path. It also ensures that the lock end time of the last edge of the inbound path should be greater than that of the first edge of the outbound path.

19.5.4 Case Study

This section presents a case study and experiments performed on a real-life terminal station extracted from The Indian Railways network. The experiment demonstrates the impact of various values of inter-arrival time (i.e. time between two consecutive trains entering the station) on the overall delay, which in turn determines the capacity of the station. Delay encompasses both the delay acquired by trains while traversing within the network and the delay experienced by trains at the entry of the network. We demonstrate how this approach can be used for determining the timetable for the trains running in the network. Train movements in the station are modeled using the mathematical model presented in the section above.

Due to the network layout of the station, the traversal time of the train depends on which route the train takes in the station. Thus the minimum time when the next train can enter the terminal depends on what route the previous train has taken.

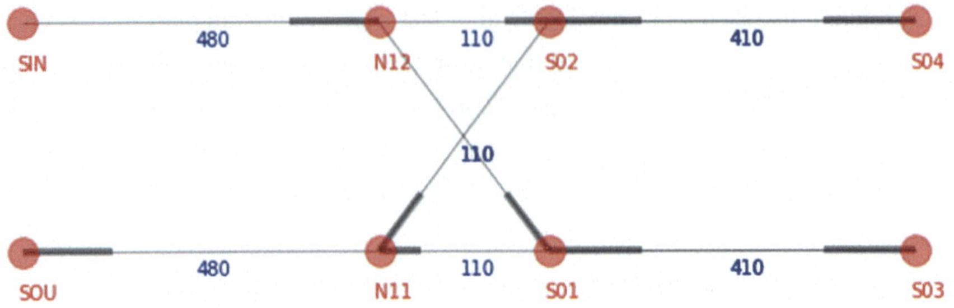

Fig. 19.7 CSMT Harbour terminal scissor layout

Station layout

For this case study, the Chhatrapati Shivaji Maharaj Terminal (CSMT)—Harbour station is used as an illustrative example. It is the starting/terminal station of the Harbour commuting local line in Mumbai, operated by Central Railway of Indian Railways, and is a conventional double line. CSMT Harbour terminal has only two terminating lines/platforms. These two lines are interconnected with each other at the start of the terminal to facilitate train reversal movement. A train enters the terminal using then UP line and leaves the terminal using the DOWN line thus crossing is involved either during the entry or during the exit of the train depending on the line/platform selected for the train for halting and reversing.

Figure 19.7 shows the layout of a CSMT Harbour station. The edge length (in metres) is shown just above the edge in blue color. The thick end of an edge represents the direction of traffic allowed on that arc. Nodes starting with 'S' are signals, whereas that with 'N' are crossings. The CSMT Harbour terminal has only two loops; thus, there are only four paths and two routes in this terminal. Table 19.4 gives a list of all paths, their respective routes, arcs in each path, and their runtime. Refer Sect. 19.5.2 and Fig. 19.6 for definition of routes and paths.

The edge N12-S01 connects Entry-Signal SIN and Platform S01-S03 in the up direction and is in path P1. The edge S02-N11 connects Platform S02-S04 and Exist-Signal SOU in the DOWN direction and is in path P4. These two arcs cross each other physically. Route R1 consists of paths P1, P2, platform S01-S03, and edge N12-S01. Route R2 consists of path P3, path P4, platform S02-S04, and edge S02-N11. The traversal time for route R1 is less than that of route R2. Please note that the shaded rectangle at the end of the edge represents the direction. For example, edge SIN-N12 has a direction SIN to N12, whereas edge SOU-N11 has a direction from N11 to SOU. Arcs S02-S04 and S01-S03 are bidirectional arcs, the rest are unidirectional.

Table 19.4 Runtime data for scissor layout

Route-Path	Edges	Travel time (s)	Total (s)
R1-P1	SIN-N12-S01-S03	40–25–095	160
R1-P2	S03-S01-N11-SOU	25–10–045	80
R2-P3	SIN-N12-S02-S04	35–10–035	80
R2-P4	S04-S02-N11-SOU	55–25–100	180

Experiment setup

This section demonstrates how proposed approaches can be used for the tactical purpose of determining the capacity of a terminal. The capacity of a terminal can be represented in terms of headway (inter-arrival time of two consecutive trains entering the terminal) at which delay is lowest; which in turn would give us the number of trains that can be handled by the station in the given planning horizon.

The experiment demonstrates the impact of various headway values on the overall delay. Delay includes delay acquired by trains in the network (i.e. traversal delay or addiction halt given to train due to unavailability of track resources) and delay experienced by trains at the entry of the network (i.e. train is ready to enter but can not enter the network due to the unavailability of track resources). Train movements in the terminal station are modeled using the mathematical model presented in the section above.

Test Instances: The data instance consists of 25 homogeneous trains where each train has some network entry time. The headway is referred to as the difference between the network entry time of two consecutive trains in the network. A headway of 300 s means that a train can enter the network only after 300 s from the entry of the previous train. A total of 21 data instances are generated using different headway values from 0 to 600 with a step size of 30 s. The unique input headway values used are 0, 30, 60, 90, 120, 150, 180, 210, 240, 270, and 300 seconds. Please note that the lower the headway value, the more congestion as more trains need to be scheduled in an hour. The number of trains in an hour for corresponding headway values are ∞, 120, 60, 40, 30, 24, 17, 15, 13, and 12. In each of these 21 instances, the scheduled arrival of consecutive trains differs exactly by the headway value for that instance. We assume that all trains are homogeneous and halt at the station for 180 s. The traversal time for each train is pre-calculated using standard motion equations taking entry speed, the speed limit at tracks, acceleration, and deceleration of the train.

Experiments: Experiments include generating a train movement plan for all 21 data instances using the mixed-integer linear programming model described in Sect. 19.5.3, which in turn implements route-lock route-release interlocking related constraints, analyzing the impact of headway on a given network and presenting the results. The model given in Sect. 19.5.3 is implemented in C++ and is solved using

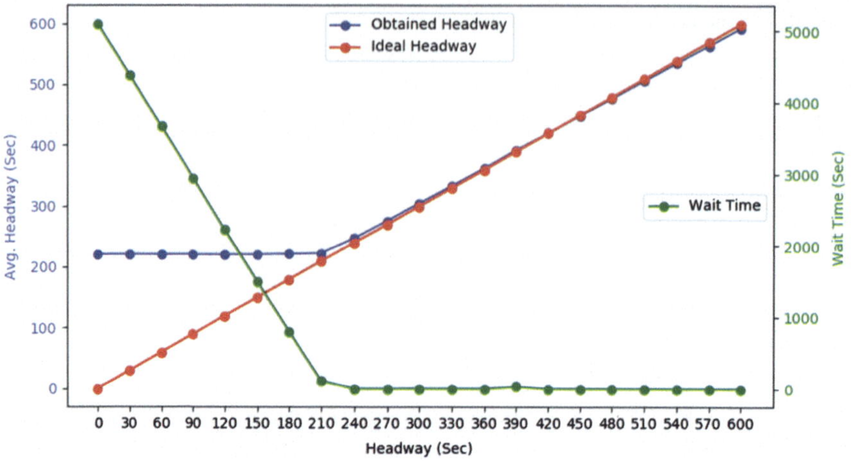

Fig. 19.8 Headway and entry delay

ILOG CPLEX 12.6. All experiments were conducted on a computer with Intel(R) Core(TM) i5-2520M processor with 2.5 GHz clock speed and 4.0 GB RAM running on Ubuntu 16.04 LTS. The solver time limit for each run was set to be 45000 s.

Results

This subsection gives four graphs, each depicting two or more performance parameters against the headway value used. Each graph depicts two or more performance parameters against the input headway value used. The details of the graphs are as below.

Headway and delay graph (Fig. 19.8) shows the average headway observed, input headway, and delay acquired by trains in the system trains. The obtained headway is the time between two consecutive trains arriving at the station as per the output of the model. The average headway is calculated by dividing the obtained headway by the number of trains in the network. The input (ideal) headway value is also shown for comparison.

The ideal headway to operate the CSMT network is 210 s. It is the point where average headway is almost equal to input headway and train delay is zero. Which in turn represents the capacity of the CSMT network. Figure 19.8 depicts that the CSMT network can handle 1 train every 210 s or 17 (3600/210) trains every hour without any significant delay. Running more than 17 trains will lead to an increase in entry delay for trains, which is represented by lower input headway.

Track occupancy graph (Fig. 19.9) shows the average time a train spends in the system. The graph shows the average track occupation (utilization) time by all trains

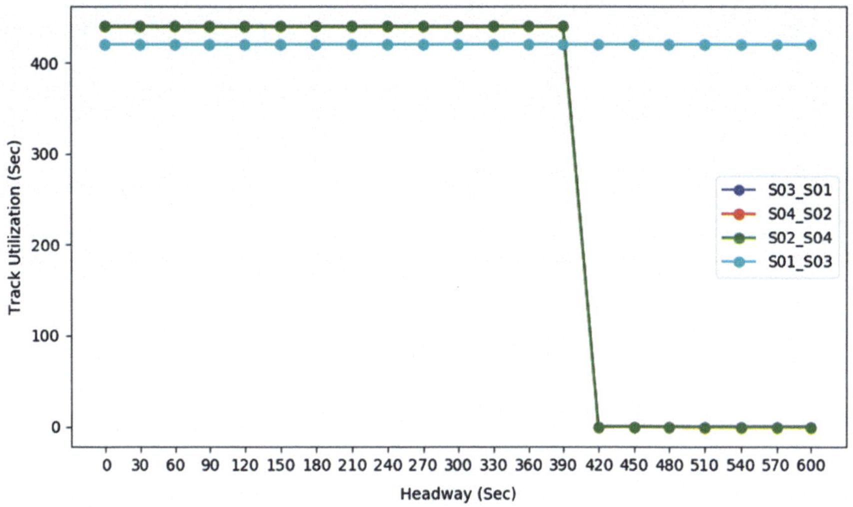

Fig. 19.9 Average track occupancy

for each route. S01-S03 shows the average time spent by trains occupying Route1, whereas S02-S04 is the average time spend by trains occupying Route2.

Occupation of edge S03-S01 (Platform1) is more as Route1 is the shortest of two available routes. All trains occupying respective routes have the same average occupation time (see Fig. 19.9). These values are the same as the travel time for the respective route and it indicates that there is no acquired delay by train during traversal. All delays are due to the entry delay, i.e. due to the unavailability of the track resources trains are not allowed to enter the network even if they are ready/available for entry.

Platform allocation graph (Fig. 19.10) depicts the number of trains allocated or occupying Platform S01-S03 and Platform S02-S04 for each input headway value. It indicates the pattern of platform allocation as headway between trains changes.

If the input headway is 420 s or above, all trains are allocated to 'Route1' (S03-S01) and if it is below 320 s, then trains are equally distributed among both routes/platforms. The headway between 340 and 420 s is the transition period. All trains are allocated to the shortest Route, i.e. Route1 after 420 headway value because a train needs a minimum of 420 s to enter the station traverse to Platform1 S01-S03 taking 160 s, halt at Platform1 for 180 s, and exit the network using Path2 for 80 s. Any trains in the data instances with input headway 420 or more will always take the shortest available route (Route1), and thus all trains are allocated to Platform1 (which is part of Route1).

Solver run time graph (Fig. 19.11) shows the time required to get the solution for each data instance.

In this section, we provided a mathematical optimization method for planning and analyzing train movement through the station. We modeled associated activi-

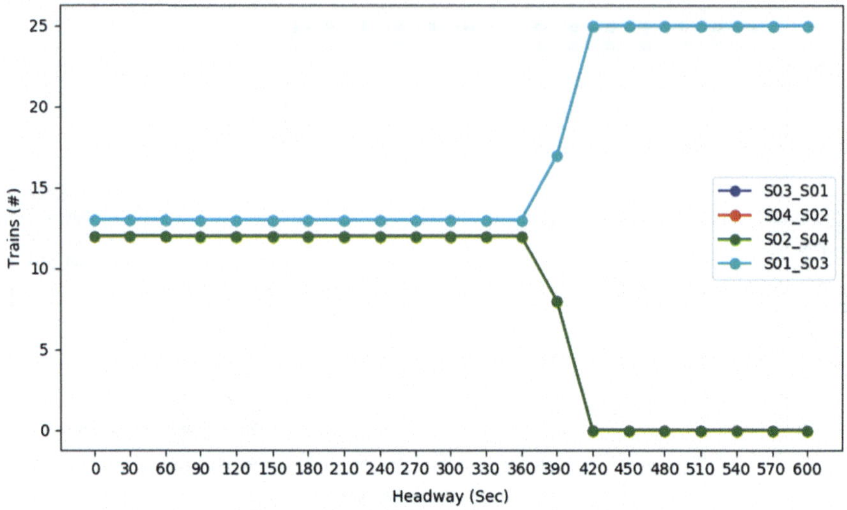

Fig. 19.10 Platform allocation

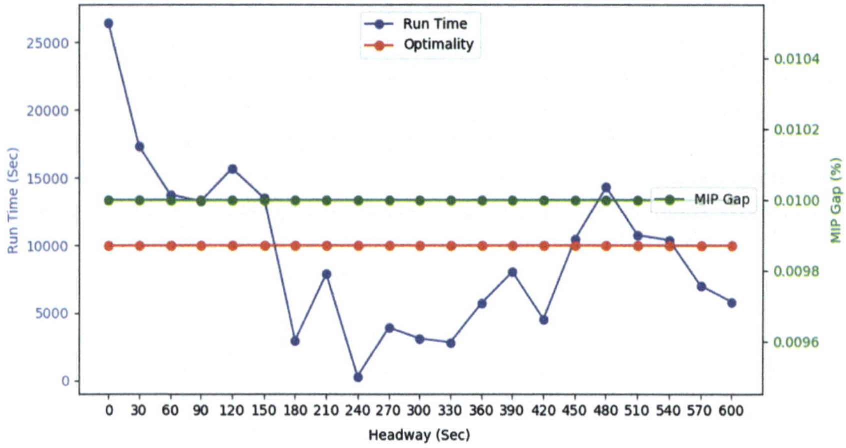

Fig. 19.11 Solver run time and optimality gap

ties such as train reversal and interlocking settings of route-lock route-release. The section demonstrated that the proposed MILP approach can be used for developing an operational plan for a station. It also demonstrated that the proposed approach could be used to determine the capacity of a railway terminal. Further, Salsingikar [3] shows that the proposed approach can also be used to compare different layout and their impact on train traversal through a terminal.

19.6 Conclusion and Future Scope

It is clear from the discussion that the use of quantitative models can aid in different areas like timetabling, rake linking, routing, crew scheduling, and maintenance management. As railway operations are becoming complex and technology-dependent there is a need for data-based input and evidence-based decisions in different aspects of railway operations management. Railway systems are converging toward a data-rich environment with many data streams and databases [51, 52].

Due to the size of the problem and the large number of practical constraints involved, planning in railways is often done relatively independently and in sequence [53]. The advantage of this approach is the decomposition of the whole problem into sub-problems of manageable size [54]. The difficulty in the sequential problem-solving approach is that in the final stages, the solution may be infeasible. With the evolution of faster computers and better algorithms, there has been a shift from sequential planning to integrated planning wherein two or more optimization problems are solved together [25, 47, 55, 56]. The integration of different planning problems would improve the quality of solutions.

In recent research, more attention has been given to robustness considerations in both research and application. Robustness is the capacity of a system to absorb or resist changes [57, 58]. There is scope for research in integrated planning problems while considering robustness. As planners often need to consider multiple conflicting objectives, such as costs and safety, future research should also aim to develop models for multi-objective optimization [59]. Railway operations are becoming more data-driven, and many researchers focus on using passenger behavior analysis to enhance service quality [60, 61].

Worldwide, railway organizations use tools ranging from optimization and operations research to machine learning to improve operations. This is true for mainline operations, including freight, suburban rail, and metro services. Optimization models integrated into solvers and software interfaces can be helpful and result in significant cost savings.

References

1. Huisman D, Kroon L, Lentink R, Vromans M (2005) Operations research in passenger railway transportation. Statistica Neerlandica 59(4):467–497
2. Kroon L, Huisman D, Abbink E, Fioole P-J, Fischetti M, Maróti G, Schrijver A, Steenbeek A, Ybema R (2009) The new Dutch timetable: the OR revolution. Interfaces INFORMS J Appl Anal 39(1):6–17
3. Salsingikar S (2021) Routing and scheduling of trains in a railway network. Phd thesis, Indian Institute of Technology Bombay, Mumbai, India
4. Narayanaswami S, Rangaraj N (2011) Scheduling and rescheduling of railway operations: A review and expository analysis. Technol Oper Manag 2(2):102–122
5. Rangaraj N, Ranade A, Moudgalya K, Konda C, Johri M, Naik R (2003) Simulator for railway line capacity planning. Oper Res Appl 516

6. Yan F, Goverde RMP (2019) Combined line planning and train timetabling for strongly heterogeneous railway lines with direct connections. Transp Res Part B: Methodol 127:20–46
7. Caprara A, Kroon L, Monaci M, Peeters M, Toth P (2007) Passenger railway optimization. Handbooks Oper Res Manag Sci 14:129–187
8. Goossens J-W, van Hoesel S, Kroon L (2006) On solving multi-type railway line planning problems. Eur J Oper Res 168(2):403–424
9. Schiewe A, Albert S, Pätzold J, Schiewe P, Schöbel A, Schulz J (2018) LinTim: An integrated environment for mathematical public transport optimization. Technical report, TU Kaiserslautern, pp 2018–08
10. Cordeau J-F, Soumis F, Desrosiers J (2001) Simultaneous assignment of locomotives and cars to passenger trains. Oper Res 49(4):531–548
11. Budai G, Huisman D, Dekker R (2006) Scheduling preventive railway maintenance activities. J Oper Res Soc 57:1035–1044
12. Şahin G, Yüceoğlu B (2011) Tactical crew planning in railways. Transp Res Part E: Logistics Transp Rev 47(6):1221–1243
13. Heil J, Hoffmann K, Buscher U (2020) Railway crew scheduling: Models, methods and applications. Eur J Oper Res 283(2):405–425
14. Boysen N, Fliedner M, Jaehn F, Pesch E (2012) Shunting yard operations: Theoretical aspects and applications. Eur J Oper Res 220(1):1–14
15. Ghorpade T, Rangaraj N (2019) Order first split second heuristic for alternative routing strategy for freight railways. Transp Policy
16. Upadhyay A, Weihua G, Bolia N (2017) Optimal loading of double-stack container trains. Transp Res Part E: Logistics Transp Rev 107:1–22
17. Upadhyay A (2020) Improving intermodal train operations in Indian Railways. INFORMS J Appl Anal 50(4):213–224
18. Haehn R, Ábrahám E, Nießen N (2020) Freight train scheduling in railway systems. International conference on measurement, modelling and evaluation of computing systems. Springer, Berlin, pp 225–241
19. Harrod S (2012) A tutorial on fundamental model structures for railway timetable optimization. Surv Oper Res Manag Sci 17(2):85–96
20. D'Ariano A (2008) Improving real-time train dispatching: Models, algorithms and applications. Phd thesis, Delft University of Technology, Delft, Netherlands
21. Kang L, Xiao Y, Sun H, Wu J, Luo S, Buhigiro N (2022) Decisions on train rescheduling and locomotive assignment during the COVID-19 outbreak: A case of the Beijing-Tianjin intercity railway. Decision Support Systems, 161, 113600
22. Haque MT, Hamid F (2022) An optimization model to assign seats in long distance trains to minimize SARS-CoV-2 diffusion. Transp Res Part A: Policy Pract 162:104–120
23. Haque MT, Hamid F (2022) Social distancing and revenue management—A post-pandemic adaptation for railways. Omega 102737
24. Yaghini M, Nikoo N, Ahadi HR (2014) An integer programming model for analysing impacts of different train types on railway line capacity. Transport 29(1):28–35
25. Borndörfer R, Reuther M, Schlechte T, Waas K, Weider S (2016) Integrated optimization of rolling stock rotations for intercity railways. Transp Sci 50(3):863–877
26. Ahuja R, Cunha C, ç Şahin G (2005) Network models in railroad planning and scheduling. INFORMS Tutorials Oper Res 54–101
27. Taslimi B, Sarijaloo FB, Liu H, Pardalos PM (2022) A novel mixed integer programming model for freight train travel time estimation. European J Oper Res 300(2):676–688
28. Mira L, Andrade AR, Gomes MC (2020) Maintenance scheduling within rolling stock planning in railway operations under uncertain maintenance durations. J Rail Transp Plan Manag 14:100177
29. Narayanaswami S, Rangaraj N (2013) Modelling disruptions and resolving conflicts optimally in a railway schedule. Comput Ind Eng 64(1):469–481
30. Gopalakrishnan R, Rangaraj N (2010) Capacity management on long-distance passenger trains of Indian Railways. Interfaces 40(4):291–302

31. Maróti G, Kroon L (2005) Maintenance routing for train units: The transition model. Transp Sci 39(4):518–525
32. Vaidyanathan B, Jha K, Ahuja R (2007) Multicommodity network flow approach to the railroad crew-scheduling problem. IBM J Res Dev 51(3.4):325–344
33. Bendfeldt J-P, Mohr U, Muller L (2000) RailSys, a system to plan future railway needs. Comput Railways VII 50
34. Nash A, Huerlimann D (2004) Railroad simulation using OpenTrack. Comput Railways IX 1:45–54
35. Middelkoop A, Loeve L (2006) Simulation of traffic management with FRISO. Comput Railways X 88:9
36. Hellström P, Kauppi A, Wikström J, Arne A, Sandblad B, Kvist T, Gideon G (2003) Experimental evaluation of decision support tools for train traffic control. In: The world congress on railway research, pp 670–677
37. Leilich R (1998) Application of simulation models in capacity constrained rail corridors. In: Proceedings of the 30th conference on winter simulation. IEEE Computer Society Press, Washington, DC, USA, pp 1125–1134
38. Pouryousef H, Lautala P (2015) Hybrid simulation approach for improving railway capacity and train schedules. J Rail Transp Plann Manag 5(4):211–224
39. Espinosa-Aranda J, García-Ródenas R (2012) A discrete event-based simulation model for real-time traffic management in railways. J Intell Transp Syst Technol Plan Oper 16(2):94–107
40. Cheung BSN, Chow KP, Hui LCK, Yong AMK (1999) Railway track possession assignment using constraint satisfaction. Eng Appl Artif Intell 12(5):599–611
41. Rodriguez J (2007) A constraint programming model for real-time train scheduling at junctions. Transp Res Part B: Methodol 41(2):231–245
42. Sparing Daniel, Goverde Rob MP (2017) A cycle time optimization model for generating stable periodic railway timetables. Transp Res Part B: Methodol 98:198–223
43. Högdahl J, Bohlin M, Fröidh O (2019) A combined simulation-optimization approach for minimizing travel time and delays in railway timetables. Transp Res Part B: Methodol 126:192–212
44. Burdett RL, Kozan E (2009) Techniques for inserting additional trains into existing timetables. Transp Res Part B: Methodol 43(8–9):821–836
45. Anoop KP, Reddy M, Bhatia MS, Jain AK, Gopalakrishnan R, Salekin M, Singh SP, Satwi RV, Pulapadi S, Bobade C, Sidhartha K, Shivasubramanian G, Appaiah K, Rangaraj N, Belur M, (2023) Rationalized timetabling using a simulation tool: A paradigm shift in Indian Railways. INFORMS J Appl Anal 53(4):295–306
46. Nishi T, Ohno A, Inuiguchi M, Takahashi S, Ueda K (2017) A combined column generation and heuristics for railway short-term rolling stock planning with regular inspection constraints. Comput Oper Res 81:14–25
47. Canca D, Barrena E (2018) The integrated rolling stock circulation and depot location problem in railway rapid transit systems. Transp Res Part E: Logistics Transp Rev 109:115–138
48. Subramanian S (2022) Utilization, maintenance and retrofit planning of passenger rakes on railway systems. Phd thesis, Indian Institute of Technology Bombay, Mumbai, India
49. Hansen IA, Pachl J (2014) Railway timetabling and operations: Analysis, modelling, optimisation, simulation, performance evaluation. Europress
50. Salsingikar S, Rangaraj N (2020) Reinforcement learning for train movement planning at railway stations. In: Adaptive and learning agent workshop. International conference on autonomous agents and multi-agent systems, Auckland, New Zealand, May 09–13, 2020. https://ala2020.vub.ac.be/papers/ALA2020_paper_11.pdf
51. Durazo-Cardenas I, Starr A, Turner CJ, Tiwari A, Kirkwood L, Bevilacqua M, Tsourdos A, Shehab E, Baguley P, Xu Y et al (2018) An autonomous system for maintenance scheduling data-rich complex infrastructure: Fusing the railways' condition, planning and cost. Transp Res Part C: Emerg Technol 89:234–253
52. De Martinis V, Corman F (2018) Data-driven perspectives for energy efficient operations in railway systems: Current practices and future opportunities. Transp Res Part C: Emerg Technol 95:679–697

53. Lusby R, Larsen J, Ehrgott M, Ryan D (2011) Railway track allocation: Models and methods. OR Spectr 33(4):843–883
54. Bussieck M, Winter T, Zimmermann U (1997) Discrete optimization in public rail transport. Math Program 79(1):415–444
55. Dong X, Li D, Yin Y, Ding S, Cao Z (2020) Integrated optimization of train stop planning and timetabling for commuter railways with an extended adaptive large neighborhood search meta-heuristic approach. Transp Res Part C: Emerg Technol 117:102681
56. Haahr J, Lusby R (2017) Integrating rolling stock scheduling with train unit shunting. Eur J Oper Res 259(2):452–468
57. Lusby R, Larsen J, Bull S (2018) A survey on robustness in railway planning. Eur J Oper Res 266(1):1–15
58. Zhang Y, Ng ST (2022) Robustness of urban railway networks against the cascading failures induced by the fluctuation of passenger flow. Reliab Eng Syst Saf 219:108227
59. Sedghi M, Kauppila O, Bergquist B, Vanhatalo E, Kulahci M (2021) A taxonomy of railway track maintenance planning and scheduling: A review and research trends. Reliab Eng Syst Saf 215:107827
60. Myneni MB, Dandamudi R (2020) Harvesting railway passenger opinions on multi themes by using social graph clustering. J Rail Transp Plan Manag 13:100151
61. Chen Z-S, Liu X-L, Chin K-S, Pedrycz W, Tsui K-L, Skibniewski MJ (2021) Online-review analysis based large-scale group decision-making for determining passenger demands and evaluating passenger satisfaction: Case study of high-speed rail system in China. Inf Fusion 69:22–39

Chapter 20
Railyard Optimization

Krishna C. Jha and Ravindra K. Ahuja

Abstract Railyards serve as the nerve center for freight railroads in moving goods from their origins to their destinations. Railcars are switched and new trains are formed at yards either by using humps to build outbound railcar blocks using gravity or by manually ordering railcars as per their destinations. Inefficiencies in railyard operations impact congestions in the yard resulting in excessive delays, higher operational cost per railcar switching, and lower throughput. We have developed simulation-driven optimization algorithms which in real time decide (i) where to place an inbound train in the yard, (ii) which blocks should be built where in the yard, and (iii) in which order outbound trains should be built. These algorithms use metaheuristics including very large-scale neighborhood search for the optimization. Our algorithms have been validated for a very large yard for a railroad in North America demonstrating significant improvements in yard key performance indicators.

20.1 Introduction

Freight railroad operations centered around carrying millions of goods railcars (also referred as wagons) from their origins to their destinations. For a large railroad, this requires an intricate set of handling railcars at terminals and operations of hundreds of trains every day. A typical Class I railroad (a large railroad in North America) has the regulatory requirements to provide trip options to shippers for every origin and destination pair in its network consisting of tens of thousands miles of rail tracks and hundreds of terminals. Sometimes, this leads to more than half a million possible origin and destination combinations. The origin–destination pair of movements is similar to sending packages in a package delivery network. The high cost of train operations and the network limitations prohibit carrying each shipment on direct

K. C. Jha (✉) · R. K. Ahuja
Optym, Gainesville, FL, USA
e-mail: krishna.jha@optym.com

R. K. Ahuja
e-mail: ravi.ahuja@optym.com

F. Hamid (ed.), *Optimization Essentials*, International Series in Operations Research & Management Science 353, https://doi.org/10.1007/978-981-99-5491-9_20

train service from its origin to destination. Shipments, customer orders consisting of moving one or more railcars, are instead collected together at local yards and then moved to large classification yards where railcars with destinations in close proximity to each other are grouped together. Each group (referred to as a block) of railcars is then moved in a train toward its destination region where, upon arrival, the group is disassembled and local trains deliver the railcars to each destination station.

Shipments within a block do not share a common final destination but do share the intermediate stops at which they are separated from each other and consolidated into different blocks. The blocking plan consists of designing a blocking network over the physical railroad network so that each individual shipment can be routed from its origin to its destination through a sequence of blocks. In a railroad network, yards are analogous to airline hubs, shipments are analogous to passengers, and reclassification of a shipment from one block to another corresponds to a passenger connecting from one flight to another. However, the main difference is that railcars need to be moved from one track to another using the tracks in the yard as against passengers moving from one gate to another by themselves. Railyards (also referred as yards) are the crucial facilities of the railroad network and are often referred to as the nerve centers of the railroad system. In North America, in a typical trip of a railcar from its origin and its destination, the railcar spends around 80% of time in yards.

Yards are physically capacitated with respect to the number of cars that can be classified and the number of block classifications that can be maintained in a given time period. Railroad service planners are often constrained by the trade-off of the number of blocks and trains that can be built in a yard in a day and the dwell time of railcars impacting network productivity. If a yard manager can squeeze more throughput out of a given facility, planners can take advantage of that extra capacity to decrease overall network costs. Yards must also classify cars and build originating trains in a timely manner to ensure that the overall transportation service product meets the shippers' expectations. Consequently, service designers are again faced with a tension between holding outbound trains long enough to accumulate enough traffic to justify running a train versus the cost of delay associated with the cars that arrived early waiting on the cars that arrived late. Yard managers are frequently criticized for delaying outbound trains and are also criticized when incoming shipments miss their scheduled outbound connections. These conflicting factors impacting not only the yard operations but also the rail network efficiency are motivations for the optimization of railyard operations in the real time.

Railcars are classified into two types of yards: hump yards and flat yards. As a rule of thumb, hump yards are considered more efficient if the number of cars handled per day exceeds one thousand. Otherwise, flat yards are considered more efficient. This contrast in efficiency based on number of cars handled can be explained by the fact that hump yards use an automated humping system to switch railcars while flat yards use yard engines. The installation and operational costs of an automated humping system exceed those of a yard engine, and therefore hump yards are justified for large volume only. Also, hump yards require more footprint than flat yards do to switch railcars. A hump yard must have separate receiving, classification, and departure

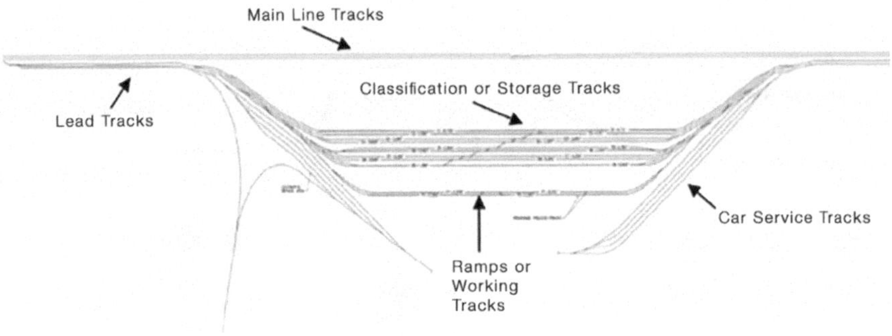

Fig. 20.1 Layout of a large flat yard handling a few hundred railcars per day

yards, whereas this distinction is not necessary for a flat yard. Fig. 20.1 presents the layout of a typical flat yard.

The subject of this chapter is the optimization of decisions made in a hump yard. Hump yards have been used for more than a hundred years to classify railcars into blocks (groups of railcars traveling together between two yards). As the name indicates, the fundamental feature of a hump yard is a mechanism which classifies cars by rolling them down from a small hump to their designated classification tracks. For a railcar, yard operations consist of the following steps: railcar is brought by a train and placed in the receiving yard; railcar undergoes inbound inspection; railcar is humped to its designated track in the classification yard; corresponding outbound train is built in the departure yard by pulling out blocks from the classification yards; outbound train undergoes outbound inspection; and train departs from the yard. A hump yard consists of sub-yards which place inbound trains (receiving yard), classify railcars into blocks (classification yard), and build outbound trains (departure yard). The operational practice and productivity of a hump yard greatly depend upon the geographical layout of these sub-yards and the configuration of lead-tracks facilitating humping and outbound train building processes. Figure 20.2 shows a logical representation of a hump yard. Figure 20.3 shows some variations in the layouts of hump yards in North America.

The dynamic of railcar movements in a yard and the coordination of activities among different resources make yard operations very difficult. This is connected to the fact that operations in different sub-yards are highly interdependent. For example, congestion in the departure yard leads to congestion in the classification yard. This in turn may cause humping to stop and ultimately create congestion in the receiving yard. Furthermore, since yards operate around the clock, disruptions like these take days to fix, further reducing efficiency. In this chapter, some of the optimization problems arising in yard operations are investigated in terms of problem definitions, solution approaches, and computational analysis. In the next section, findings from the literature review are discussed.

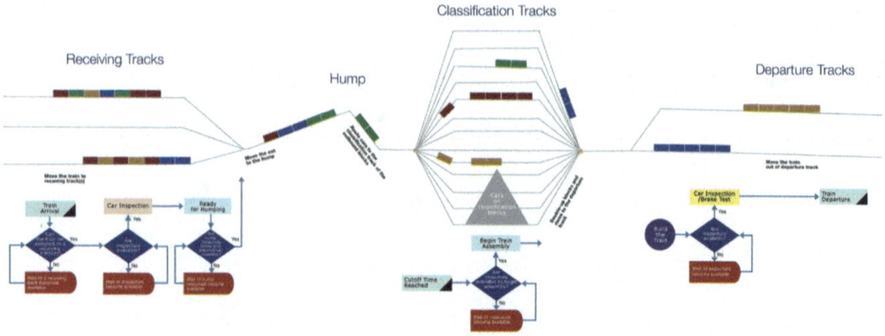

Fig. 20.2 Schematic illustration of activities performed in a hump yard

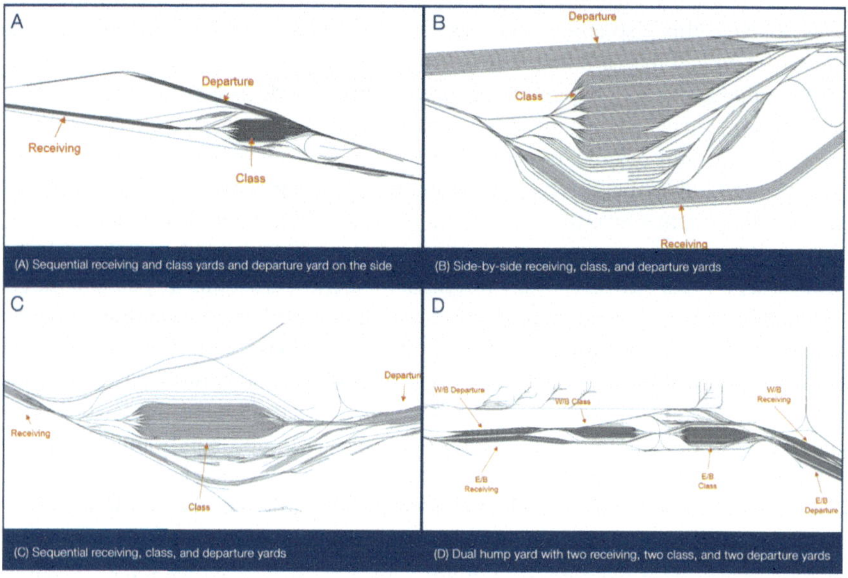

Fig. 20.3 Examples of variation in hump yard layouts

20.2 Literature Review

Railroads have been in operations for hundreds of years, however, the importance of yards has become prominent for a few decades in view of the change in the traffic mix. As railroads realized that yard operations are very vital for their operations, academicians, and practitioners have taken interest in studying and optimizing railyard operations. The past research in yard operations are of two types: (1) Capital expenditure focused in which yard layout, resources, and operational methodologies are analyzed and the improvement alternatives are identified. (2) Operational expenditure and service quality focused in which classification processes in yards have been studied and new methods have been proposed to improve the timeliness, classification operations, and utilization of existing yard resources.

In studies performed in the 1970's and the 1980's, queuing models and statistical analysis techniques have been used to predict the yard performance and to evaluate the priority of resources. Among these studies, the PMAKE analysis proposed by Martland [1] remained very popular for the long time in estimating yard capacity. The PMAKE analysis uses a probabilistic distribution model of time that a railcar spends in a yard for the intermediate switching. A queuing model is used in Petersen [2] to predict the yard throughput time. This queuing model is extended in Petersen [3] in relation to a yard's infrastructure to understand capacity limitations and capabilities of railroad yards. Turnquist and Daskin [4] also use queuing models in which the classification time of a railcar and train build time are considered as delay components in the lifecycle of railcar in the yard. These queuing models are used to predict the effectiveness of alternate strategies in performing classification and train building operations in a yard. These studies have helped in understanding whether or not a yard can handle a given traffic pattern under a set of predetermined decisions and resource levels. None of these studies have considered using tools to construct an operational plan and to make associated scheduling and resource allocation decisions.

The block-to-classification track assignment problem has always been considered a challenge and assumed to be the most prominent aspect in determining the usable blocking capacity of a yard. For decades, the railroads have been handling this issue with predetermined and historic assignment schemes. However, researchers have approached this problem algorithmically in order to find alternate assignment schemes that change the way classification tracks are allocated rather than plans that work in coordination with the current scheme. The proposed schemes do not necessarily subsume the one-to-one (many-to-one) track assignments. Instead, such schemes may distribute the railcars of a particular outbound block to several tracks in the classification yard. Therefore, these allocation schemes may not have been implemented due to the perceived risk of dramatic changes.

The earliest study for sorting techniques is by Siddiqee [5]. This study evaluates and compares different sorting and classification schemes that are implemented by railroads and discusses the advantages and disadvantages of such schemes. Daganzo et al. [6] proposed a multi-stage sorting study, which is a complicated classification scheme that separates car segments of a particular outbound block to several tracks initially and collects them at the time of assembly with other concurrent assemblies. Daganzo has carried out a series of studies on identifying effective classification schemes based on what is called geometric yard switching. In the earliest study, Daganzo [7] compares several classification schemes with respect to their track requirements and the total number of switches required for an average assembly process. In later studies, Daganzo [8, 9] proposed dynamic classification schemes for the first time. These schemes are based on the complex multistage sorting idea in their first study.

An early study by Yagar et al. [10] that focuses on the humping sequence decisions is among the first studies that have approached the problem from a cost minimization perspective. They develop a dynamic programming-based approach which simultaneously accounts for the compatibility between the inbound trains and outbound trains and takes into account service priorities for shipments. Shughart and Lawrence [11] developed a multi-variate econometric model that related the costs of operating a yard to the size of the yard, the type of traffic that was classified, the volume of cars, and the number of blocks that were maintained at that yard. This is the only known study that placed an economic cost on adding incremental blocking work, such as the cost of classifying a fixed number of railcars into 40 blocks versus 50 blocks.

In contrast to all these studies, works by Kraft are different in the sense that they aim to develop operational yard planning tools. Kraft [12] studied the hump sequencing problem with a real-time planning perspective. Kraft [13, 14] studied the railcar commitment and block-to-track assignment problems, respectively. The latter focuses on the assignment problem with guaranteed connections (i.e., no railcars are allowed to miss their connections even if it means delaying some trains) independent of the hump and assembly sequencing decisions.

A study by Sahin and Ahuja [15] develops integer programming formulations for the hump sequencing and assembly sequencing problems; they also attempt to introduce a dynamic block-to-track assignment scheme by making use of well-known optimization techniques. Railroads have been spending time and resources to understand how they can improve their efficiency and timeliness. An article by Gormick [16] established that railroads perceive the effectiveness of yard operations not only as an internal asset to a yard's performance by increasing a particular yard capacity but also as an essential element in reducing the travel times over the network. This article also points out that this achievement is not only possible by changing the ways things are done; it also depends on how railroads will make use of information technology in this transition phase.

In the last decade, the Rail Application Section of The Institute for Operations Research and the Management Sciences (INFORMS) and railroads have taken interest in optimizing railyard operations. In 2013, researchers from all over the world participated in a problem-solving competition[1] related to optimization of yard operations to maximize the capacity utilization of yard resources. The optimization goals were to minimize the total waiting time of railcars in the yard and maximize the total number of railcars processed during a certain period. In 2014, the problem-solving competition focused on one of the optimization problems in the yard, the block-to-track assignment problem. The goal of this optimization problem was to find assignments of blocks to tracks so that scarce tracks in the classification yard can be properly utilized and train building operations can be efficient. Submissions by participants in these competitions adopted modeling techniques ranging from simulation-guided heuristic approaches to Mixed Integer Programming approaches. Outside INFORMS, Bohlin et al. [17] have studied models for optimizing operations in marshaling yards in Ludwigshafen, Germany, and Hallsberg, Sweden. Another research by Aliakbari et al. [18] focuses on every specific problem on finding the shortest path for a set of railcars from its origin track to the destination track.

[1] https://connect.informs.org/railway-applications/new-item3/problem-repository16.

While attempts have been made in either improving the yard capacity or efficiency in the offline mode using simulation modeling or optimization modeling, no serious effort has been made in developing optimization approaches which can work on real-life hump yard problems in real time. The approaches discussed in this chapter were the first of their kind, which focused on the real-life validation of optimized outputs.

20.3 Yard Optimization Business Problem

Classification yards play an important role in the movement of railcars in a freight rail network. Within a yard, railcars from inbound trains are reclassified to their next destination and then placed to depart on the correct outbound train. For merchandise traffic, a railcar spends around 80% of its time within yards during its trip from origin to destination. In a yard, train masters and yardmasters are entrusted to make quick and complex decisions that have large implications on yard's efficiency. In the dynamic and impactful environment in a yard, decisions made at a point of time have a cascading impact on the yard for days to come. For example, an inefficient placement of inbound railcars and their classifications may lead to congestion in the yard for days to clear up. Similarly, the disruptions in the rail network may change the arrival and departure patterns in the yard requiring yardmasters to dynamically adjust classification plans impacting the whole rail network. These decisions are often made based on the previous experience of yardmasters, which are generally myopic in nature. These factors have led to research interests in optimizing business problems related to yard operations.

Yardmasters make many decisions that affect yard overall throughput and performance. These decisions can be categorized into four major parts (Fig. 20.4): (i) when to bring in a train and where to place in the yard, (ii) in what order the incoming trains should be humped, (iii) on which tracks and in which order cars should be classified, and (iv) when and how to build outbound trains. In these decision-makings, the primary objectives are to use track spaces in the yard efficiently, to minimize handlings of railcars, and to depart railcars on time. In addition, the secondary objective is to use yard resources efficiently so that more railcars can be processed with same set of yard resources—locomotives, conductors, inspectors, etc. A prime example of the optimization opportunity is classifying railcars in such a way that all safety/regulatory rules are honored in the outbound trains. In the absence of this consideration in the decision-making, several railcars need to be flat switched before being assembled in the output train.

Among the four decision-making parts, optimization of the second part (humping) and the third part (classification) are the most complex components and have the highest impact on the yard performance. The decision problem for sequencing of trains for humping in a hump yard is often referred as hump sequencing problem and that associated with the assignment of classification tracks to blocks as block-to-track assignment problem. In this chapter, these two problems are discussed in more

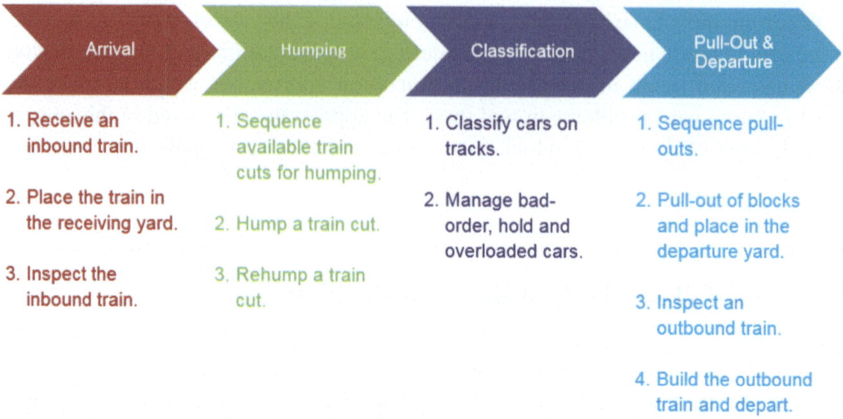

Fig. 20.4 Decision-making stages in a hump yard

details along with solution approaches. The solution approaches discussed below use sophisticated meta-heuristics and parallel computing techniques to solve these problems in real time for a large hump yard.

20.4 Hump Sequencing Problem

The hump sequencing problem (HSP) is to decide in which order train-cuts in the receiving yard will be humped. The sequencing problem also includes train-cuts on rehumping tracks in the classification yard. Rehumping implies reclassification of railcars because of misrouting during the first time classification or being classified to a mixed blocks track because of the lack of dedicated track for each block in the classification yard. In HSP, rehump tracks are treated in the same way as any receiving track while computing missed connections and train placement rule violations.

20.4.1 Hump Sequencing Problem Overview

The hump sequencing problem is a combinatorial optimization problem consisting of several objective function factors and constraints. While some factors are equally applicable in the planning or the real-time mode, there are several real-time specific requirements making the problem harder to solve.

Major objectives in HSP are centered around the utilization of resources in the yard and connections of railcars from the inbound train to the outbound train. These objectives are further explained below:

- **Maximize Car Connection**: Hump inbound railcars in such a way that railcars scheduled for an outbound train are ready for pullout at the time when the train is being built. This is the primary decision factor in the HSP module. The business impact of this objective is to reduce the missed connections and to improve the yard fluidity.
- **Receiving Yard Dwell**: While maximizing railcar connection is the primary objective, optimizing this objective only will lead to small train-cuts being left behind repeatedly as missing the connections of only a few railcars is not weighted as important. This causes two issues: tracks in the receiving yard are not freed up quickly for the next train placement, and the service quality of those left behind is poor. Minimizing receiving yard dwell addresses these situations. This is a non-linear factor in which the penalty for dwell increases non-linearly with the dwell time.
- **Receiving Yard Availability**: The yard fluidity greatly depends on having tracks available for incoming traffic in the receiving, class, and departure yards. To promote the availability of such open tracks, optimizing the receiving yard availability maximizes the availability of receiving tracks in the yard. This objective criteria is most impactful when the receiving yard is congested, with little space for inbound trains.

Any solution to the hump sequencing problem needs to conform to several business rules which are either logistical in nature or related to operating procedures of the yard. For example, a train-cut at the front end of a track must be humped before any train-cut on the same track at the rear end. Important business rules for the yard in the case study used in this chapter are described below. These business rules may not be applicable to all yards or there may be some additional business rules for different yards.

- **Pullout cutoff time**: To estimate whether a railcar can connect to an outbound train, this business rule dictates what should be the cutoff time before which a railcar must be humped to be included in its intended outbound train. If a railcar misses the cutoff time, it is considered to be a missed connection case and thus adversely impacting the objective function for HSP.
- **Inbound Inspection**. All inbound trains in a hump yard are visually inspected to determine whether railcars in the train are fit to be classified using humping process. This business rule ensures a train must be present in the receiving yard for a minimum time for the inspection, prior to the train-cut hump start time.
- **Enroute Traffic**: Enroute traffic is defined as railcars on trains arriving at the yard in the near future. Railcar-consists for these trains are generated from the estimated time of arrival (ETA) data when the trains leave the latest crew change point. Arriving trains are placed in the receiving yard without delay; however, it may take some additional time before an arriving train may be humped.
- **Time to hump a train-cut**: The humping time of a train-cut must include the following four components:

- Humping time: The duration between the first railcar humped and the last railcar humped from the train-cut. Typically, this is computed using a fixed hump speed parameter and the number of railcars in the train-cut.
- Trimming time: Trimming time is defined as the average time spent by the hump job trimming stalled railcars in the bowl in between two train-cuts humping. While the hump job may not trim in between each humped train-cut, the average time spent in trimming throughout the day is calculated and amortized in between each train-cut humped.
- Move to receiving end: This denotes the time taken by the hump job locomotive to move from the hump crest to the other end of the receiving and coupled with the next train-cut.
- Move train-cut to hump: This denotes the time taken in releasing hand brakes, cab transportation of personnel to the hump crest, switch alignment time, and moving the train-cut to hump.

The solution approach discussed in the next section has considered the above factors. It can be noted that the approach has been designed in such a way that yard-specific additional requirements can easily be incorporated.

20.4.2 Solution Approach

For the given use case, a heuristic optimization approach was adopted considering the need to solve HSP in real time and considering the complexity involved with objective function factors. The approach uses the neighborhood search algorithm (Fig. 20.5) with sophisticated modules for the evaluation of a solution. This approach requires construction of an initial feasible solution and improving it with a local neighborhood search algorithm. Both the construction and improvement algorithms use a common evaluation module which evaluates every candidate solution against the objective function defined in the previous section.

20.4.2.1 Solution Evaluation

The most important component in the heuristic approach above is the evaluation module which algorithmically estimates different components. In a real-time optimization system, it is important that all dynamic conditions be considered in evaluating the fitness of a humping sequence. Major factors impacting the objective function evaluation are described below.

Railcar Connections

All railcars (except railcars going to special-purpose bowl tracks) from inbound trains are assigned to an outbound train based on block cutoff time. If a railcar is humped after the cutoff time for a block, it misses its connection and is reassigned to

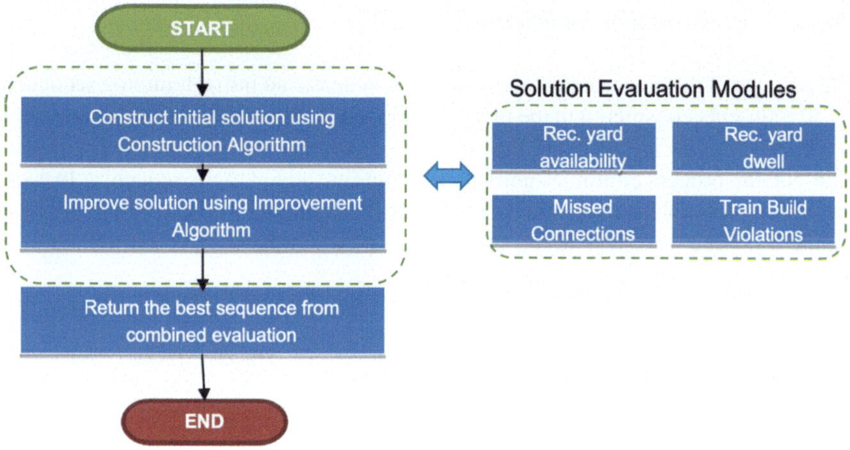

Fig. 20.5 Solution approach overview for HSP

the next available train. Therefore, for each candidate hump sequence, the algorithm computes how many railcars may miss their original connections.

Receiving Yard Dwell

Receiving yard dwell time is computed as a duration of time from train arrival to the receiving yard to the proposed humping time of the train-cut. Train-cuts with longer dwell time receive a higher priority based on this objective function term. This secondary objective is weighted with a non-linear function of dwell to prioritize the train-cuts for humping. This prevents small blocks of railcars from staying in the receiving yard for long periods of time, when newly arrived large train-cuts of railcars otherwise would have been given a higher priority.

Receiving Yard Availability

The receiving yard availability objective term accounts for the availability of tracks in the receiving yard. Depending on whether the receiving yard is congested, the algorithm gives a higher priority to the train-cuts that occupy longer tracks, as pulling out longer tracks will free up more receiving yard space. The algorithm considers two factors for computing receiving yard availability:

- Receiving yard congestion-level multiplier determining the importance of the receiving yard track availability based on the congestion level. This multiplier considers the following factors: the number of available tracks in the receiving yard, ETA of enroute trains, and humping sequence.
- Estimate the total length of tracks in the receiving yard for the inbound trains over the time. This requires algorithms to heuristically determine the tracks which will be used by inbound trains and the tracks which will be available as train-cuts are estimated to be humped. An algorithm similar to the maximum saving algorithm is used to determine the fitness of the inbound trains on the given set of tracks.

20.4.2.2 Construction Algorithm

The purpose of the construction algorithm is to generate an initial humping sequence for all train-cuts considered in the planning horizon. In this study, a greedy algorithm in which train-cuts are added to the solution sequence iteratively. Starting with one train-cut, the process continues until all train-cuts are added to the sequence. In each iteration, all eligible rehump train-cuts are also considered in the similar way as a regular train-cut. The steps in this algorithm can be summarized as follows:

```
Start
Create a list of all train cuts to be available to be sequenced (regular + rehump tracks)
Set current position to decide in sequence to one
Set train cuts as yet to be sequenced train cuts
Do
    For all yet to be sequenced train cuts
            Check whether the train cut is feasible to be sequenced at the current position
            If yes,
                    Compute the objective function value for the train cut at the current position
    End For
    Identify the train cut with highest favorable objective function value
    Assign the train cut to the current position
    Remove the train cut from yet to be sequenced train cut list
Repeat until all train cuts are sequenced
End
```

If there is no train-cut available for humping or rehumping, the algorithm will set the start time of the next humping at the earliest time a train-cut becomes available after inspection.

20.4.2.3 Improvement Algorithm (Neighborhood Search)

After the construction algorithm generates a humping sequence it is improved iteratively by swapping the positions of train-cuts within the sequence. In each iteration of the improvement algorithm, two train-cuts available for humping are swapped and the new solution is selected if it results in an improved objective function value. Note that two train-cuts can be swapped only if both of them are available for humping at the given time. The algorithm always maintains feasibility and does not sequence a train-cut at a position if it has not been in the yard for less than the inspection threshold value. The steps to find an improved solution are as follows:

Start
Get an initial humping sequence generated by the construction algorithm and set it as the current solution
Set current humping sequence equals to initial humping sequence
Do
    For each pair of train cuts in the current solution
        Swap positions of train cuts in the pair
        Check the feasibility of new sequence
        If feasible, compute the objective function value
    Repeat for the next pair
    Among all feasible options, identify the pair exchanges that resulted in objective values greater than the current solution
    Select the new sequence with the most improvement over the current solution; call this solution "best current sequence"
    Update the current solution by the "best current sequence"
Repeat until no improvements in the current solution be identified
End

The neighborhood search is a widely used improvement heuristics in which one solution leads to another solution with a small perturbation. Empirically, this iterative approach converges on a local optimal solution. Several approaches for further refinement of this algorithm are adopted. These approaches either make the algorithm faster or help the algorithm converge to a better solution. Some of the neighborhood search strategies adopted for HSP are first improvement, nearest neighbor swap, multiple starting solutions, and multi-exchange neighborhood search.

20.5 Block to Track Assignment Problem

The block-to-track assignment (BTA) problem is one of the most studied optimization problems to improve yard efficiency. A large railyard has a central sub-yard, called classification yard, which is used to sort railcars by their outbound blocks. After sorting, these blocks are combined together to form an outbound train to be departed from the yard after attaching locomotives. Decisions to build which block on which track greatly impact the capacity utilization of a yard and the efforts involved in building outbound trains. This problem is like deciding moves in a chess play in which current decision will have cascading impacts creating a significant impact on the game. It has been observed that poorly managed classification yard led to congestions requiring several days of efforts in bringing the yard operations to normal state. Followings are a couple of examples to demonstrate how the selection of classification tracks impacts the yard utilization and efficiency. In the first example (Fig. 20.6), Track 4 is underutilized while Block 1 needs to be classified on two tracks. In the second example (Fig. 20.7), building a train with Block 1 at the front and Block 2 at the rear requires occupying lead tracks connecting all classification tracks. This takes more time to build trains and obstructs any other train building operation in the yard. When Block 1 and Block 2 are classified on close-by tracks, only lead tracks in that part of the classification yard are used in the train building process.

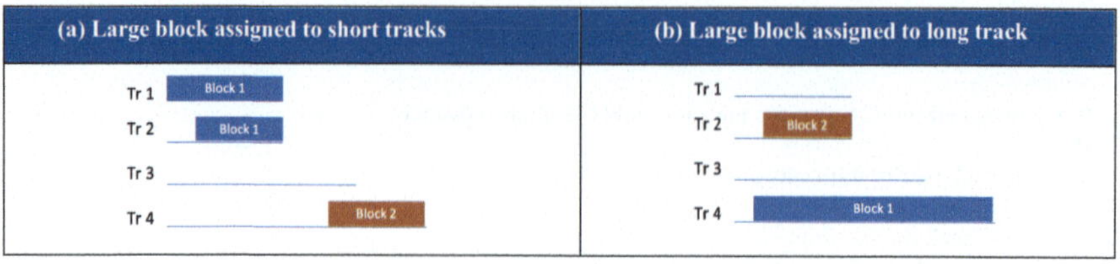

Fig. 20.6 Impact of BTA decision on track utilizations. **a** A large block is assigned to two short tracks and a small block to a long track, **b** small block assigned to the short track and large block to the long track. It can be noted that more tracks are needed to classify same number of railcars

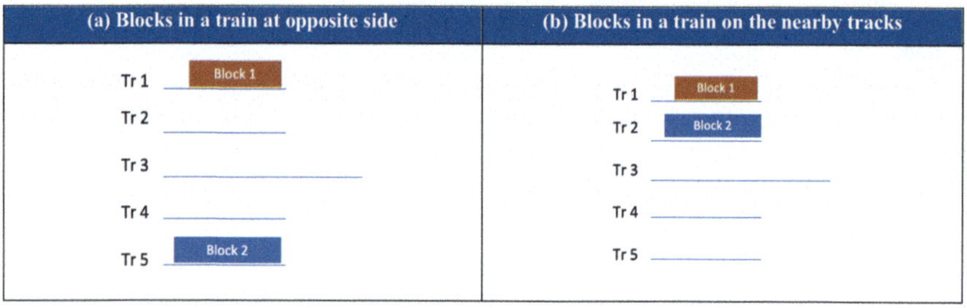

Fig. 20.7 Impact of BTA decision on efficiency of yard operations. **a** Blocks to be assembled in a train are on tracks at the opposite side of tracks, **b** blocks to be assembled in a train are on in the nearby tracks. The train building process in the first case takes more time and creates more interferences, whereas that in the second case requires less time and does not interfere with operations on Track 3–Track 5

Historically, static block-to-track assignments were in use in which classification tracks for blocks were pre-specified and stay the same. However, that had resulted in inefficient use of classification tracks and hence lower throughputs of yards if there are changes in the block volume and/or change in the outbound train configurations. Therefore, gradually, railroads are transitioning to the dynamic block-to-track assignment in which tracks for blocks are dynamically decided based on the current status of the yard and future traffic demand. In this section, this important decision problem is discussed and a solution approach is presented.

20.5.1 Block-To-Track Assignment Problem Overview

The block-to-track assignment (BTA) module decides the classification track on which a block code is sorted during humping. This is a dynamic decision which may change multiple times a day. The overall objective is to effectively use classification tracks to get maximum railcar processing throughput while maintaining the fluidity of the yard. The planning model considers the volume of all blocks in the classification yard (aka bowl) and enroute to the terminal for the next *k* hours (typically 10–12 h).

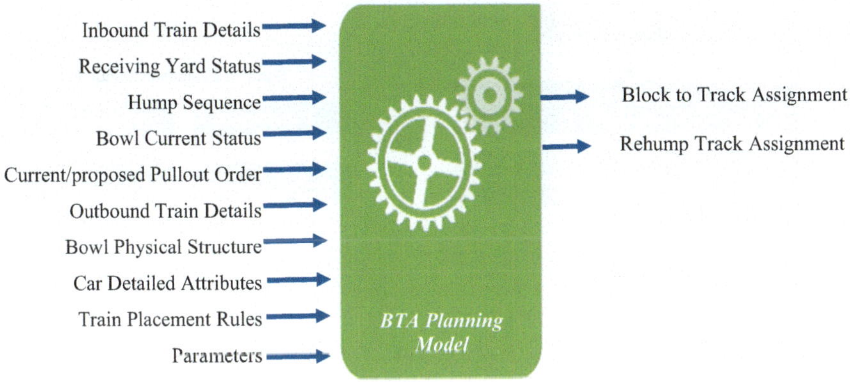

Fig. 20.8 Inputs and outputs in BTA model

Block-to-track assignment is done considering the pullout time of all the blocks in the bowl. Figure 20.8 outlines inputs and outputs to the BTA planning model.

The primary decisions of the BTA problem for a railcar are two sequential steps: whether the railcar needs to be classified on the regular or rehump track and to which track the railcar should be classified to. Making these decisions also indirectly requires deciding whether a new classification track be assigned to a block and if so deciding the track assignement. The BTA problem takes into consideration several operational factors and depends upon the correct estimations of the projected behavior of the yard. The important inputs are humping/rehumping sequences of train-cuts, the train building plan dictating in what order blocks will be pulled out, and the train placement rules. Among these, train placement rules are of special importance as a wrong placement of a railcar in the classification yard requires yard resources to switch out such cars before the train being built. This is an expensive operation in terms of time it takes and resources used.

20.5.2 BTA Solution Approach

Similar to the hump sequencing problem, a heuristic optimization approach has been adopted for the BTA problem. This approach uses a combination of construction algorithm and neighborhood search-based improvement algorithms. This approach has proven to be most suitable for solving BTA problems in real time and for the flexibility in adopting complex business requirements including non-linear objective function factors. Figure 20.9 provides an overview of the approach:

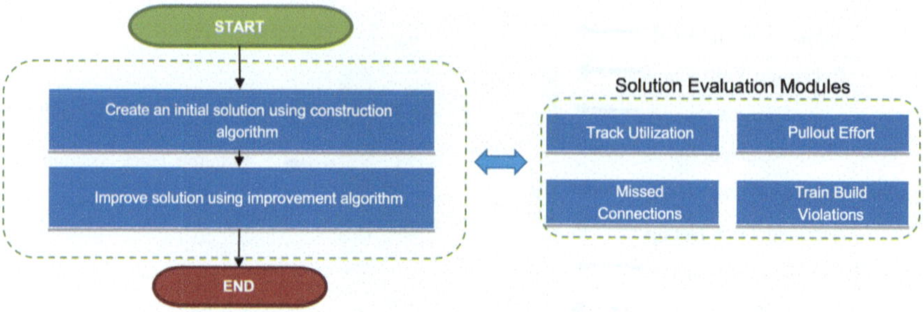

Fig. 20.9 Algorithm Overview for BTA Problem

20.5.2.1 Solution Evaluation

Considering that the heuristic optimization approach adopted as given above, the most crucial component is the solution evaluation module. The comprehensiveness and computational efficiency of this module determine the strategy for enumerating candidate solutions in the construction and improvement of algorithms.

Track Utilization

Track utilization is one of the most critical measures of the BTA model's impact on the fluidity and the ability of the yard to handle extra traffic. In general, track utilization can be measured by calculating the average number of empty tracks over time. The block-to-track assignment resulting in a higher number of empty tracks will receive a higher track utilization score. The track utilization evaluation uses weighted factors such as empty, not usable, and stackable track configurations.

Pullout Effort Estimation

Pullout effort defines how much time is required to build all trains from the blocks located on classification tracks based on their placement zone and length. The following two scenarios can reduce the pullout time and effort:

- Doubling blocks and pulling them out together. This action reduces pullout time and effort for a pair of blocks if they are conveniently placed in the classification yard for a pullout job to approach.
- Parallel pullout. The simultaneous pullout of two long blocks for the same train (or long blocks for different trains pulled out close to each other in time) if both blocks are in the same zone. Thus, it is beneficial to place long blocks, which will be pulled out at a similar time in different zones to avoid potential pullout conflicts.

Train Build Violations

There is a set of rules, every outbound train must follow based on the railcar type and the commodity, which are referred to as train build rules. In actual yard operations, crews spend a considerable amount of time during the pullout to fix train build violations. The majority of the train build violation rules apply to hazardous, heavy, or empty railcars that are in close proximity to the locomotive and/or sequence of railcars that are placed close to the end of an outbound train. One of the goals of the class track assignment is to reduce build violations through intelligent assignments and deciding the block-splitting strategy to minimize the effort needed in fixing build violations.

Missed Connection Estimation

The quality of a solution is proportional to the number of railcars that make their outbound train connections. A solution in which the yard is fluid, yet causes many railcars to miss their outlets from the yard, should not be considered a good solution. The algorithm to estimate the missed connections uses exploration techniques in identifying potential outlets of railcars and whether a railcar will miss the connection in each of these scenarios.

20.5.2.2 Construction Algorithms

The construction algorithm iteratively decides the placement of block-cuts (a set of railcars in a block which will be pulled together for an outbound train). This algorithm is executed in three steps: linking by block instance (combination of block code and outbound train), linking by block code, and the placement of remaining railcars.

The first step identifies what are block placements in the classification yard and decides whether block-cuts in the given humping sequence should be assigned to the same track. In the second step, a new track for a given block is placed behind another block, if they are conducive to be pulled together for the outbound trains. If none of these two steps work for the placement of a block-cut, a new track is being assigned for the block-cut in the classification yard.

Start
Sort blocks by hump start time
While there are block cuts remaining to be assigned
  Select the next block-cut to be placed assigned
  Determine which tracks have length availability for the time needed by the block
  For each feasible track for the block
    Estimate the objective function value
    Add to the feasible list of assignments
    If no feasible track is available for the block
      Perform priority swap to identify whether swapping track with another block is feasible
      Perform block splitting to identify whether placement on two tracks is feasible
      Estimate the objective function value for feasible cases
    Assign to the best feasible tarck(s) for the block
    If no feasible track, **assign** the block to the rehump track
End

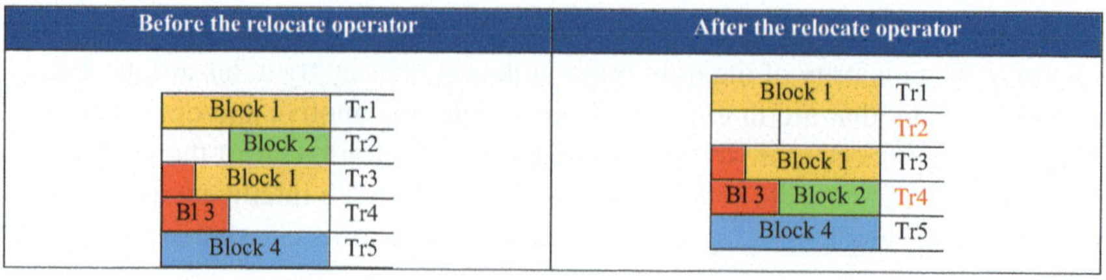

Fig. 20.10 Relocate Operator. Block 2 is reassigned to Tr4

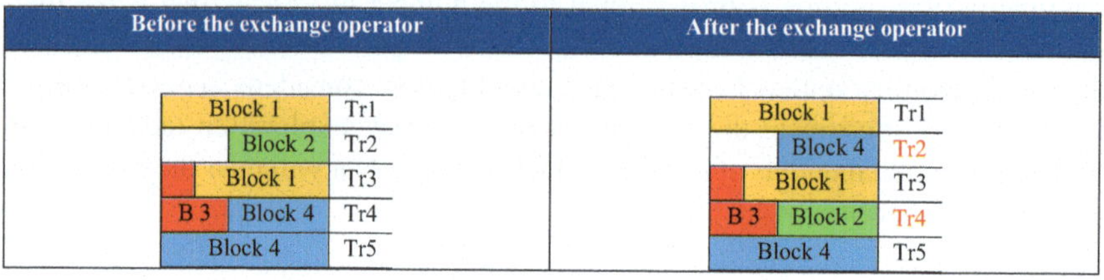

Fig. 20.11 Swap Operator. Track assignments for Block 2 and Block 4 are swapped

20.5.2.3 Improvement Algorithm (Neighborhood Search)

A neighborhood search begins with an initial solution and makes small changes in the current solution to enumerate and adopt better solutions compared to the current solution. The new solution generated by changing the current solution called a "neighbor" of the current solution. For BTA problem, neighboring solutions are generated using two operators:

1. Relocate operator: change the assignment of a block from one track to another track (Fig. 20.10)
2. Swap operator: swap the assigned tracks for two blocks (Fig. 20.11).

The neighborhood search is an iterative process in which the above two operators are used to create a set of neighboring solutions. For each neighboring solution, feasibility validations and solution evaluations are performed. The current solution is replaced with the best neighboring feasible solution. This process is repeated until either a certain time limit is reached, or the relevant improvement between two successive solutions drops below a certain predetermined threshold.

Start
While iteration count < max number of neighborhood steps
  Perform prescpecified number of relocate operations
  Perform prespecified number of swap operations
  Validate and evaluate each candidate solution
  Select the best feasible neigboring solution and set that as new current solution
end

20.5.3 Combined HSP and BTA Evaluation

The solution evaluation functions for HSP and BTA problem are interdependent. This implies the setup of the classification yard determines the quality of a humping sequence and similarly a given humping solution determines the quality of a solution of the BTA problem. To further improve the solution quality of real-time optimization, a hybrid approach is adopted in which multiple humping sequences are generated and used as the input to the BTA problem. Considering each of these humping sequences as the starting point, multiple BTA solutions are generated and the best among them is adopted as the final solution (Fig. 20.12).

20.6 Computational Analysis

The real-time yard optimization approach discussed in this chapter has been validated and analyzed using a realistic scenario for one of the major hump yards in the United States. This yard is one of the busiest in the country and processes more than 1,700 railcars per day. To analyze the algorithms, a simulated environment has been used to create different scenarios. For simulation, Optym's proprietary Hump Yard Simulation System (YardSYM) has been used. For the yard in the study, YardSYM has been validated to be more than 95% accurate. Figure 20.13 provides a high-level overview of YardSYM.

For the algorithmic performance analysis, a baseline scenario has been created and has been validated by the subject matter experts as the realistic digital representation of the yard operations. The baseline verification process was crucial because it was the system in which the yard optimization has been executed and evaluated. In order for the optimization system to be evaluated correctly, the simulation system must first be established as accurate. After the baseline was created, optimization algorithm analyses were performed to determine the improvement in yard operations for Key Performance Indicators (KPI) and for the algorithmic behavior.

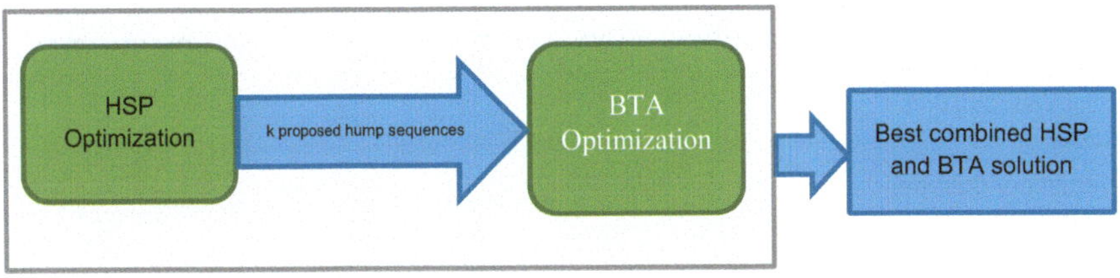

Fig. 20.12 Combined HSP and BTA optimization approach

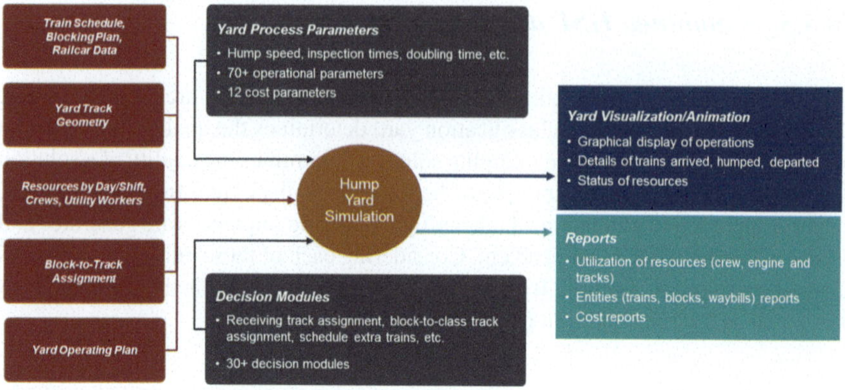

Fig. 20.13 Overview of YardSYM system

20.6.1 KPI Improvement

The yard performance is measured using multiple performance measures. These measures are important as they jointly reflect whether the yard operation is efficient in terms of resource utilization and whether the yard operation is conducive to the network operations. Table 20.1 provides the improvement by yard optimization algorithms. Optimization algorithms are measured in two cases: when HSP and BTA algorithms are applied independently and when HSP and BTA algorithms are applied jointly.

Table 20.1 The impact of the yard optimization modules on the key hump yard KPIs

KPI (per day)	Simulation	BTA		HCS + BTA	
		Value	Improvement (%)	Value	Improvement (%)
RCRT (%)	60.7	61.7	2	64.2	6
Cars processed(cars)	1,248	1,257	1	1,257	1
Average car dwell (hours)	47.8	47.8	0	47.8	0
Average number of clear classification tracks	5.15	7.58	47	7.31	42
Number of doubling opportunities	3.05	2.98	−2	3.09	1
Number of parallel pullout conflicts	50	47	7	47	6
Number of train build violations on regular tracks	15.6	2.5	84	3.0	81

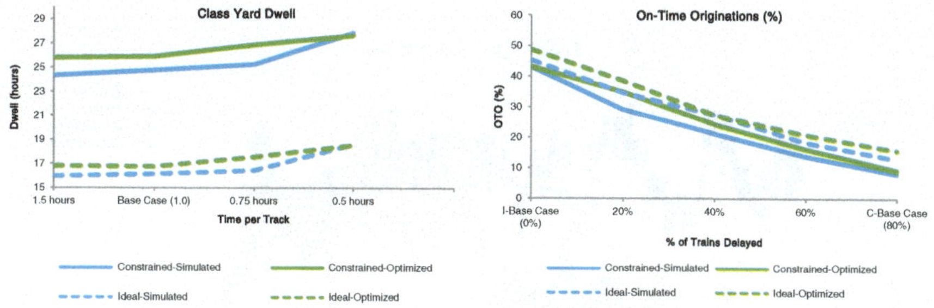

Fig. 20.14 Measurement of the robustness of the solution approach

20.6.2 Algorithm Analysis

One of the important criteria for the real-time adaptation of optimization algorithms was whether the algorithm behaves in the similar manner as in real life as the ground realities change. In that effect, it is also important to understand how the algorithms behave in the normal operations (aka Ideal Scenario) and in the congested operations (aka Constrained Scenario). Plots in Fig. 20.14 confirm that the algorithms discussed in this chapter are robust with respect to different scenarios. The first chart depicts the impact on the dwell time as the time to pull out track changes and the second chart depicts the impact on the on-time origination percentage of outbound trains when trains are delayed because of crew or locomotive unavailabilities.

Another important finding from the algorithmic analysis was to determine the threshold value when the yard operation changes from the normal mode to the congested mode. This finding is very important for yard personnel to avoid leading to the congested mode. Figure 20.15 displays the timestamps in which the traffic changes from minimal to normal and normal to dense. The trigger times are denoted by a vertical line. As the figure depicts, the yard traffic changes from a minimal traffic level to a normal traffic level at about 75 h into the simulation, and then changes from a normal to a heavy level after about 500 h into the simulation.

Another metric investigated is the Classtrack Train Build Violations. The resulting chart for this metric is displayed in Fig. 20.16. As the figure displays, the optimized results track well below the simulated results for class track build violations. Although the impact is less significant in the low-traffic region, the algorithms still benefit the system. This implies that the benefit of the algorithm exists at all traffic levels. This assertion is also supported when viewing the results of the Departure Yard Train Build Violations metric across the simulation run time as seen in the figure below.

Across all of the analyses discussed in this section, the results of the optimization algorithms show improvement across multiple metrics and did not appear to be significantly hindered by a change in track availability or a lack of some data, or any of the other what-if tests conducted. For this reason, it can be concluded that the yard optimization approach discussed in this chapter is robust and responsive in situations representing real-life situations faced in hump yards.

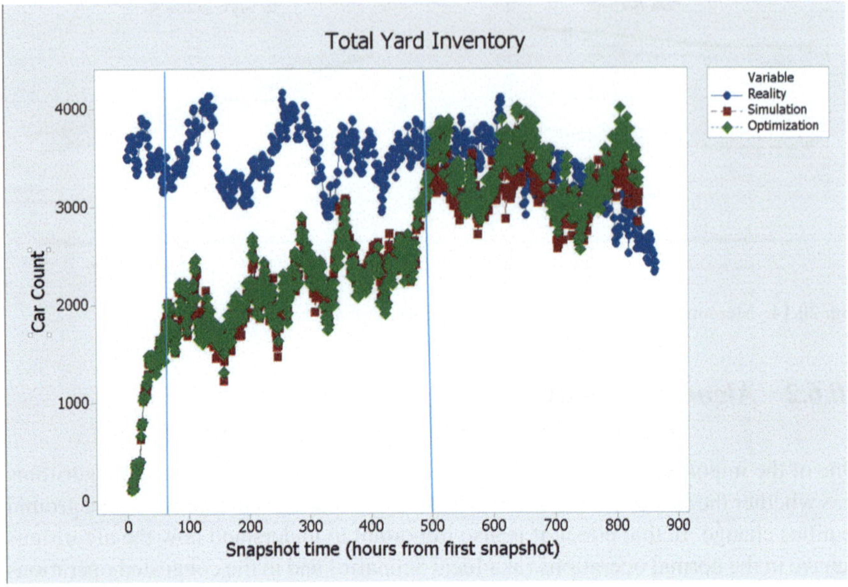

Fig. 20.15 Car count in the yard denoting congestion state

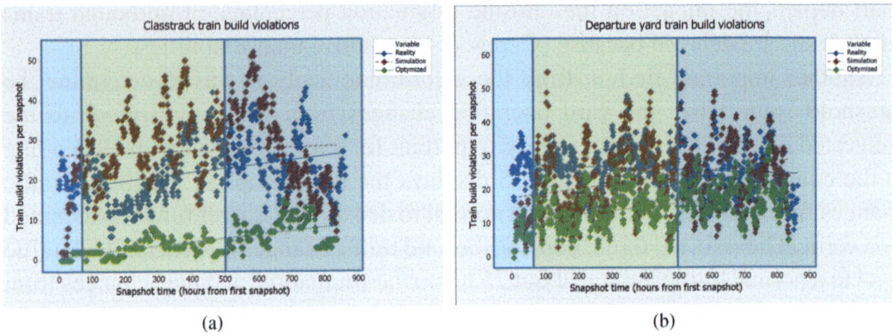

Fig. 20.16 **a** Class track train build violations across time with the differing traffic levels; **b** Departure yard train build violations across time with the differing traffic levels

20.7 Conclusions

Terminals and yards have long been recognized as a challenging aspect of railroad operations. Hump yards, the largest and most asset-intensive type of yard, are responsible for most of the railcar switching that is accomplished on a Class I railroad. Consequently, any improvements that can be made in the efficiency and effectiveness of hump yard operations will improve a railroad's return on assets and improve the railroad's service product. These improvements can be achieved through more efficient yard management, optimally scheduling yard resources, and understanding the symbiotic relationship between the yard operation and the overall railroad network.

The yard optimization framework discussed in this chapter is structured to enable various decision criteria and various optimization logic to be adapted in real time by decision-makers in yards. This research demonstrates that sophisticated problem-solving techniques can be developed, calibrated, tested, and adopted in real-time environment.

Future versions of the optimization algorithms will include the ability to model generic hump and flat yard operations. Also, additional yard resources such as crew taxis, locomotive servicing centers, railcar shops, and cleanout tracks can also be included in modeling the yard optimization. Finally, a stochastic overlay to the entire optimization will better represent the day-to-day variation inherent in any complex process.

References

1. Martland CD (1982) PMAKE analysis: Predicting rail yard time distributions using probabilistic train connection standards. Transp Sci 16:476–506
2. Petersen ER (1977) Railyard modeling: Part I. Predict put-through time. Transp Sci 11:37–49
3. Petersen ER (1977) Railyard modeling: Part II. The effect of yard facilities on congestion. Transp Sci 11:50–59
4. Turnquist MA, Daskin MS (1982) Queuing models of classification and delay in railyards. Transp Sci 16:207–230
5. Siddiqee MW (1971) Investigation of sorting and train formation schemes for a railroad hump yard. Traffic Flow Transp 377–388
6. Daganzo CF, Dowling RG, Hall RW (1983) Railroad classification yard throughput: The case of multistage triangular sorting. Transp Res Part A 17:95–106
7. Daganzo CF (1986) Static blocking at railyards: Sorting implications and tracks requirements. Transp Sci 20:186–199
8. Daganzo CF (1987) Dynamic blocking for railyards: Part I. Homogeneous traffic. Transp Res Part B 21:1–27
9. Daganzo CF (1987) Dynamic blocking for railyards: Part II. Heterogeneous traffic. Transp Res Part B 21:29–40
10. Yagar S, Saccomano FF, Shi Q (1983) An efficient sequencing model for humping in a rail yard. Transp Res Part A 17:251–262
11. Shughart LA, Lawrence MB (1993) A railroad terminal cost model to support network management strategy. J Transp Res Forum 33–1
12. Kraft ER (2000) A Hump sequencing algorithm for real time management of train connection reliability. J Transp Res Forum 30:95–115
13. Kraft ER (2002) Priority-based classification for improving connection reliability in railroad yards—part I: Integration with car scheduling. J Transp Res Forum 56:93–105
14. Kraft ER (2002) Priority-based classification for improving connection reliability in railroad yards—part II: Dynamic block to track assignment. J Transp Res Forum 56:107–119
15. Sahin G, Ahuja RK A mathematical modeling framework and solution approaches for efficient railroad yard operations planning. In: University of Florida, Department of Industrial and Systems Engineering, working paper
16. Gormick G (2005) How fluid is your yard? Railway age. https://connect.informs.org/railway-applications/new-item3/problem-repository16. Accessed 18–28 Aug 2005

17. Bohlin M, Hansmann R, Zimmermann UT (2018) Optimization of railway freight shunting. In: Borndörfer R, Klug T, Lamorgese L, Mannino C, Reuther M, Schlechte T (eds) Handbook of optimization in the railway industry. International series in operations research & management science, vol 268. Springer, Cham
18. Aliakbari M, Geunes J, Sullivan KM (2021) The single train shortest route problem in a railyard. Optim Lett 15:2577–2595

Chapter 21
A Neighborhood Search Algorithm for the Tool Indexing Problem without Tool Duplication

Deepti Mohan and Diptesh Ghosh

Abstract In automated machining centers, a job is processed by having multiple cutting tools work on it in a pre-determined sequence. The total time to process such jobs is the sum of the machining time required by the cutting tools, and the non-machining time, required for other activities. The machining time on the job is fixed, and so such machining centers can be made more effective by reducing non-machining times. A significant portion of the non-machining time in machining centers is due to indexing, which is the operation by which tools kept in slots in tool changers are brought to the location from which the tool arm can pick them up and replace them after use. Tool changers can be visualized as disks with slots along their circumference in which tools are kept. Optimal positioning of tools in the slots can significantly speed up non-machining time. This optimization problem, called the tool indexing problem, is studied in this chapter. Specifically, a version of the problem in which only one copy of a tool can be present in the tool changer is studied here. The tool indexing problem is computationally difficult, and the chapter presents local search and tabu search algorithms to obtain good quality solutions to the problem. Two innovative bookkeeping techniques that were used to reduce the search time by two orders of magnitude have been proposed here. Computational results on benchmark tool indexing instances show that the algorithms proposed here are competitive when compared with the state-of-the-art heuristic algorithms for this problem. It is shown that some of the algorithms proposed here are both faster and generate better quality solutions for many of the benchmark instances.

Keywords Tool indexing · Metaheuristics · Local search · Tabu search

D. Mohan · D. Ghosh (✉)
Operations & Decision Sciences, Indian Institute of Management, Ahmedabad, India
e-mail: diptesh@iima.ac.in

D. Mohan
e-mail: phd17deeptim@iima.ac.in

F. Hamid (ed.), *Optimization Essentials*, International Series in Operations Research & Management Science 353, https://doi.org/10.1007/978-981-99-5491-9_21

21.1 Introduction

Modern manufacturing setups often use automatic machining centers in manufacturing. In such automatic machining centers, several tools are available to perform various operations on a job sequentially. Automatic machining centers employ automatic tool changers to hold the tools and change them when required without manual intervention. Automatic tool changers are also called turret magazines, tool magazines, or simply, tool changers.

A tool changer can be visualized as a disk that has multiple pockets called slots located at equal distances along its circumference (see Fig. 21.1). It can rotate in either direction, i.e., both clockwise and counterclockwise.

The tools are arranged in the slots on the tool changer. Whenever a tool is required by the machining center, the tool changer rotates to bring the slot containing the tool to a fixed index position, from which the tool is picked, used, and returned after use by a tool change arm. The rotation is either clockwise or counterclockwise depending on the direction of rotation which makes it faster for the tool to reach the index position. Human intervention is not required in these operations. Hence such tool changers help reduce non-productive time by changing the tools very quickly. Tool changers can be either drum-type, used when the number of tools required is less than 30, or chain-type when the number of tools required is larger than 30. Chain-type tool changers can have up to 150 slots to store tools.

The processing time of a job at a machining center includes actual machining time as well as non-machining time. The tooling operation described above contributes to the non-machining time. Profits, however, are generated only during the machining time, so it is essential to perform the tooling operation as efficiently as possible. It is estimated that tooling accounts for 25–30% of fixed as well as variable costs in automated machining environments [17]. During the tooling operation, selecting the next tool requires the tool changer to rotate by an amount equal to the angular distance between the relative positions of the two tools. This time elapsed during the rotation of the tool changer from one tool to the next tool is known as indexing time. It is obvious that the indexing time can be minimized by arranging the tools

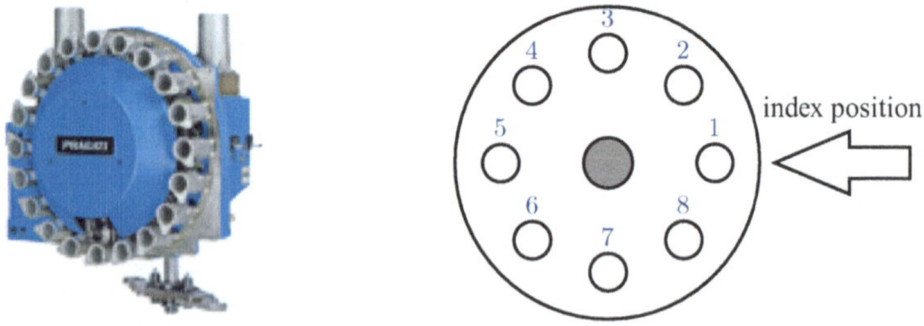

(a) An automatic tool changer (Source: Pragati Automation Pvt. Ltd)

(b) A schematic representation of a tool changer with 8 slots

Fig. 21.1 Automatic tool changer

in the slots in an efficient manner. In this work, the problem of minimizing the tool indexing time is addressed by finding an optimal allocation of tools to the slots. This problem is referred to as the tool indexing problem.

The number of slots in a tool changer and the number of tools required by a job may be different. Tool indexing problems are ones in which the number of slots in a tool changer is either equal to or more than the number of tools required for a job. All tool indexing problems can be converted into problems in which the number of tools equals the number of slots by using dummy tools, which have no use in processing the job at hand and are just placeholders in empty slots. Problems in which the number of tools exceeds the number of slots in a tool changer are dealt with as a related but different problem called the tool switching problem (see, e.g., [8]).

Tool indexing problems themselves can be of two types; one in which tool duplication is allowed and the other in which each tool has only one copy in the tool changer. In the first type, formally called the Tool Indexing Problem with Tool Duplication (TIPTD), there is usually an upper limit to the number of copies of a tool that can be present in the tool changer. In such situations, the copies of each tool used for individual operations may be different even when the sequence of tools used to perform a job may remain unchanged. The amount of rotation that a tool changer undergoes for performing a job depends on the copy of each tool used in the process. The ideal choice of the copies of particular tools to be used for individual operations reduces to a shortest path problem (see, e.g., [5]).

In the second type in which each tool has a single copy in the tool changer, simply called the Tool Indexing Problem (TIP), the amount of rotation that the tool changer undergoes for performing a job is unique given an assignment of tools to slots. Since there is only one copy of each tool, the exact sequence of operations is not required when computing the tool indexing time for a particular assignment of tools to slots; one simply needs information about the number of times that two tools are used in consecutive operations during the completion of the job. This type of tool indexing problem without tool duplication (TIP) is addressed in this paper.

The rest of the paper is organized as follows. Section 21.2 explains how the problem can be formulated as a Quadratic Assignment Problem and discusses the literature on it. Section 21.3 describes how the problem characteristics can be exploited to develop efficient swap and insert neighborhood search procedures for the problem that bring down the search complexity from $\mathcal{O}(N^4)$ time to $\mathcal{O}(N^2)$ time, which is quite possibly the best that can be achieved. A local search algorithm and a tabu search algorithm are designed for the problem based on our neighborhood search procedures and their implementation details are provided in Sect. 21.4. The results of the computational experiments are presented in Sect. 21.5 and the performance of our algorithms is briefly discussed. Finally, Sect. 21.6 summarizes the contributions of this paper and points to directions for further work in this area.

21.2 Problem Formulation

As mentioned earlier, the objective of TIP is to find an optimal allocation of tools to slots in the tool changer. Also, as discussed previously, the number of tools required for a machining center's operation is assumed to be equal to the number of slots in the tool changer; if it is less, then the shortfall is compensated with dummy tools that have no interchanges with the actual tools during the operations on the job being processed.

Consider an allocation of tools in a tool changer with N slots. Also, suppose that the tool changer rotates at a constant angular velocity, ω. The time taken to bring a tool in slot k to the index position immediately after the tool change arm has replaced a tool in slot l is given by the minimum angular distance $d(k, l)$ between slots k and l divided by ω. Assuming both ω and the angular distance between two consecutive slots in the tool changer as unity without loss of generality, $d(k, l)$ can be used to measure the time taken for this movement.

A tool in slot k can be brought to the index position that was occupied by the tool in slot l immediately before rotating the tool changer either clockwise or counterclockwise. In one direction, the angular distance will be $|k - l|$, and in the other direction, it will be $N - |k - l|$. Since $d(k, l)$ is the minimum angular distance between the two slots, the tool changer rotates in the direction which requires the least amount of rotation, $d(k, l) = \min\{|k - l|, N - |k - l|\}$. For example, in Fig. 21.1b, $d(1, 3) = 2$, which is the minimum of $|1 - 3|$, the distance when the tool changer rotates clockwise, and $8 - |1 - 3|$, the distance when the tool changer rotates counterclockwise. In the same figure, $d(7, 2) = 3$, which is the minimum of $|7 - 2|$ and $8 - |7 - 2|$, and is obtained when the tool changer rotates counterclockwise.

The total angular distance through which the tool changer rotates during the execution of a job depends on the number of times one slot follows another to the index position. The frequency with which tool i and tool j are used consecutively is denoted as $f(i, j)$, and the matrix of $f(i, j)$ values is denoted as $F = [f(i, j)]$. The terms $f(i, i)$ equal 0 for all i. Given an arrangement of tools to slots, the angular distance covered to change from one tool (say tool i) to the next (say tool j) is the same as the angular distance to change from tool j to tool i. So the F matrix is symmetric with $f(i, j) = f(j, i)$ for all i, j pairs. Also, since there is only one copy of each tool in the tool changer, the angular distance covered to change from one tool to another does not depend on when this exchange occurs in the tool sequence for the job. Hence the frequency information in F is sufficient to compute the complete tool indexing time for a job sequence. Given an allocation of tools to slots represented by $\Pi = (\pi_1, \pi_2, \ldots, \pi_k, \ldots, \pi_N)$ in which tool π_k occupies slot k, the total indexing time is given by $z(\Pi) = \sum_{k=1}^{N} \sum_{\substack{l=1 \\ l \neq k}}^{N} f(\pi_k, \pi_l) d(k, l)$ units. So, the tool indexing problem is one of obtaining a permutation $\Pi^* = \text{argmin}_{\Pi}\{z(\Pi\}$.

Let indices i and j be defined on the set of tools, and indices k and l on the set of slots in the tool changer. Let $x_{ik} \in \{0, 1\}$ take a value of 1 if tool i is in slot k and 0 otherwise. Then the tool indexing problem can be formulated as an integer programming problem as follows:

$$\text{Minimize} \quad \sum_{k=1}^{N}\sum_{\substack{l=1\\l\neq k}}^{N}\sum_{i=1}^{N}\sum_{\substack{j=1\\j\neq i}}^{N} f(i,j).d(k,l).x_{ik}.x_{jl} \qquad \text{// cost of the allocation} \tag{21.1a}$$

subject to

$$\sum_{k=1}^{N} x_{ik} = 1 \quad \forall i \qquad \text{// each tool occupies exactly one slot} \tag{21.1b}$$

$$\sum_{i=1}^{N} x_{ik} = 1 \quad \forall k \qquad \text{// each slot contains exactly one tool} \tag{21.1c}$$

$$x_{ik} \in \{0,1\} \quad \forall i,k \qquad \text{// binary variables} \tag{21.1d}$$

The formulation above is a quadratic assignment problem (QAP; please refer to [7, 21] for an introduction to this problem.) Since QAPs are NP-hard, metaheuristic algorithms are preferred over exact algorithms to find good solutions to tool indexing problems of practical sizes. References [9, 10] address the tool indexing problem using a genetic algorithm. The related tool indexing problem with tool duplication is considered in [4–6]. Atta et al. [3] developed a harmony search algorithm for solving the tool indexing problem. References [27, 28] considered tool indexing problem along with tool switching problem.

In this work, neighborhood search-based algorithms are proposed to solve the tool indexing problem. Methods to search solution neighborhoods efficiently that reduce the complexity of such searches by two orders of magnitude compared to naïve search are proposed. These methods are then used to present a tabu search algorithm to solve the tool indexing problem.

21.3 Searching Solution Neighborhoods

Neighborhood search algorithms work by searching neighborhoods of solutions, i.e., solutions that are in some sense, close to a solution. They do this by defining moves, i.e., operations that transform one solution into another. Given a set of moves that are allowed, all solutions that can be formed by applying those moves on a solution form the neighborhood of that solution. All solutions in the neighborhood are called neighbors of the solution.

The TIP is a permutation problem, in that all solutions to an instance of the problem are permutations of the tools (including dummy tools if the number of tools is less than the number of slots). There are common moves associated with creating neighbors of solutions to permutation problems. These are the swap and the insert moves, that result in swap and insert neighborhoods of solutions.

A swap move involves interchanging the positions of a pair of tools in a solution. All solutions generated by interchanging the positions of two tools in a solution form the swap neighborhood of the solution. Since there are $\binom{N}{2}$ pairs of tools in an instance with N tools, the swap neighborhood of a solution has $\binom{N}{2} = \mathcal{O}(N^2)$

solutions. Evaluating a solution using the expression in (21.1a) involves computing the angular rotation required for interchanging each pair of tools at the index position, and hence the effort required to evaluate each solution naïvely is $\mathcal{O}(N^2)$. Thus, the effort required to perform a naïve neighborhood search of a swap neighborhood is $\mathcal{O}(N^4)$.

An insert move involves picking a tool from its current position in the solution and inserting it in another position of the solution. All solutions generated by using such insert moves in a solution form the insert neighborhood of the solution. Note that in a solution with N tools, each tool that is picked up can be inserted in $N-1$ positions in the solution to obtain solutions that are distinct from the original solution. Since there are N tools in the solution, the insert neighborhood of a solution has $N(N-1) = \mathcal{O}(N^2)$ solutions. Once a tool is removed from its position and inserted in a different position, the slots occupied by the tools in the intermediate positions change. Hence evaluating each insert neighbor naïvely involves computing the angular rotation among $\mathcal{O}(N^2)$ pairs of tools, so that $\mathcal{O}(N^4)$ effort is required to naïvely search the insert neighborhood of a solution.

Clearly, an $\mathcal{O}(N^4)$ time search is prohibitively expensive, and hence a method is now described to reduce this search effort by two orders of magnitude. In the description, the presence of a frequency matrix $F = [f(i, j)]$ where $f(i, j)$ denotes the frequency with which tools i and j are used consecutively in the tool sequence, and a distance matrix $D = [d(k, l)]$ representing the distances between pairs of slots in the tool changer are assumed. Notice that both matrices F and D are symmetric.

21.3.1 Searching Swap Neighborhoods Efficiently

The number of neighbors in a swap neighborhood is $\mathcal{O}(N^2)$. Each of these neighbors must be searched while searching the swap neighborhood of a solution for a better solution, and the only way in which the search can be made more efficient is to compute the cost of a neighbor in an incremental manner from the cost of the current solution or the cost of another neighbor computed before it.

Consider a solution $\Pi_c = (\pi_1, \pi_2, \ldots, \pi_p, \ldots, \pi_q, \ldots \pi_N)$, which is a permutation of N tools in the tool changer. Swapping tools π_p and π_q at positions p $(1 \le p \le N)$ and q $(p+1 \le q \le N)$ in Π_c will create a swap neighbor $\Pi_{pq} = (\pi_1, \pi_2, \ldots, \pi_q, \ldots, \pi_p, \ldots \pi_N)$. The move of interchanging the tools at positions p and q is denoted as *swap*$[p, q]$. Consider the cost $z(\Pi_{pq})$ of the swap neighbor Π_{pq} obtained by applying the move *swap*$[p, q]$ on Π_c. Instead of computing $z(\Pi_{pq})$ *ab initio*, the cost of *swap*$[p, q]$ can be computed as $z(\Pi_{pq}) - z(\Pi_c) = \sum_{k=1}^{N} f(q, k)d(q, k) - \sum_{k=1}^{N} f(p, k)d(p, k)$. Once the cost $z(\Pi_c)$ is computed and stored, the cost of all its swap neighbors can be computed in linear time, making the total effort of computing the costs of all swap neighbors of Π_c to be $\mathcal{O}(N^3)$. This method has been proposed in [20].

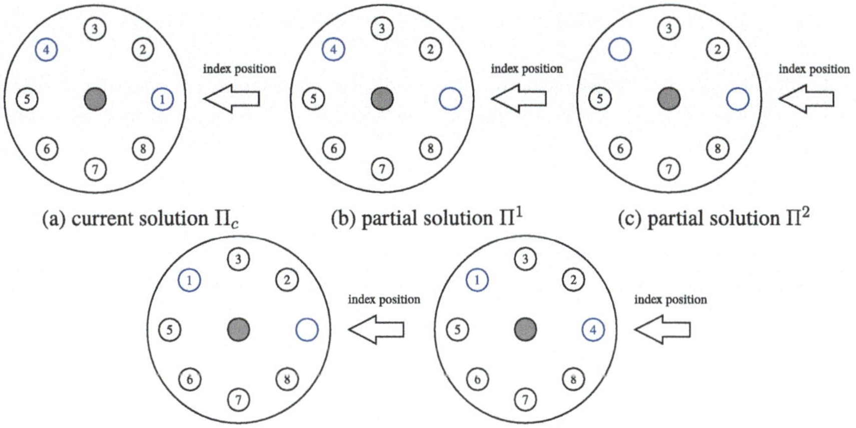

(a) current solution Π_c (b) partial solution Π^1 (c) partial solution Π^2

(d) partial solution Π^3 (e) swap neighbor Π_{14}

Fig. 21.2 Illustration of basic moves in generating swap neighbor Π_{14} of solution Π_c

To reduce the effort of searching the swap neighborhood of Π_c further, each swap move is broken down into a sequence of "basic" moves involving adding or removing a single tool in a solution. Different swap neighbors can be generated by combining these basic moves. Once the cost implications of these basic moves are pre-computed, the cost of evaluating neighbors takes constant time per neighbor, and the total cost of computing the costs of all solutions in the neighborhood requires $\mathcal{O}(N^2)$ time. This procedure is now explained in detail.

Let Π_{pq} be the swap neighbor of Π_c obtained by swapping the positions of π_p and π_q in Π_c. This swap can be obtained by performing the following steps:

1. Obtain a partial solution $\Pi^1 = (\pi_1, \pi_2, \ldots, \pi_{p-1}, *, \pi_{p+1}, \ldots, \pi_q, \ldots \pi_N)$ by removing tool π_p from Π_c. Let $Rem(p) = z(\Pi_c) - z(\Pi^1)$.
2. Obtain a partial solution $\Pi^2 = (\pi_1, \pi_2, \ldots, \pi_{p-1}, *, \pi_{p+1}, \ldots, \pi_{q-1}, *, \pi_{q+1}, \ldots \pi_N)$ by removing tool π_q from Π^1. Let $Rem(q) = z(\Pi^1) - z(\Pi^2)$.
3. Obtain a partial solution $\Pi^3 = (\pi_1, \pi_2, \ldots, \pi_{p-1}, *, \pi_{p+1}, \ldots, \pi_{q-1}, \pi_p, \pi_{q+1}, \ldots \pi_N)$ by adding back tool π_p in position q in Π^2. Let $Add(p, q) = z(\Pi^3) - z(\Pi^2)$.
4. Obtain $\Pi_{pq} = (\pi_1, \pi_2, \ldots, \pi_{p-1}, \pi_q, \pi_{p+1}, \ldots, \pi_{q-1}, \pi_p, \pi_{q+1}, \ldots \pi_N)$ by adding back tool π_q in position p in Π^3. Let $Add(q, p) = z(\Pi_{pq}) - z(\Pi^3)$.

In Steps 1, 2, and 3, "$*$" represents an empty position. The value $f(*, .) = 0$ for all tools in the solution.

For example, if $\Pi_c = \{1, 2, 3, 4, 5, 6, 7, 8\}$, then swap neighbor $\Pi_{14} = \{4, 2, 3, 1, 5, 6, 7, 8\}$ is obtained through partial solutions $\Pi^1 = \{*, 2, 3, 4, 5, 6, 7, 8\}$, $\Pi^2 = \{*, 2, 3, *, 5, 6, 7, 8\}$, and $\Pi^3 = \{*, 2, 3, 1, 5, 6, 7, 8\}$ (see Fig. 21.2).

The cost of $swap[p, q]$ can now be expressed as

$$z(\Pi_{pq}) - z(\Pi_c) = Add(p, q) + Add(q, p) - Rem(p) - Rem(q). \tag{21.2}$$

In Step 1, $Rem(p)$ is the reduction in cost after removing tool π_p from Π_c. Since tool π_p is removed, the cost of the partial solution Π^1 will not have terms involving the index p. Hence,

$$Rem(p) = \sum_{k=1}^{N} f(\pi_k, \pi_p)d(k, p). \tag{21.3}$$

In Step 2, tool π_q is removed from Π^1. If tool π_q was removed from Π_c, the reduction in cost would have been

$$R = \sum_{k=1}^{N} f(\pi_k, \pi_q)d(k, q). \tag{21.4}$$

However, since the removal is from Π^1, the supposed reduction due to the interaction between π_p and π_q in expression (21.4) should not be present. Hence,

$$Rem(q) = R - f(\pi_p, \pi_q)d(p, q) = \sum_{k=1}^{N} f(\pi_k, \pi_q)d(k, q) - f(\pi_p, \pi_q)d(p, q). \tag{21.5}$$

In Step 3, tool π_p is added in position q in partial solution Π^2. Note that both π_p and π_q are absent in Π^2. The increase in cost after this step is denoted as $Add(p, q)$. If a copy of tool π_p is placed at position q in Π_c, the sum of the costs of interaction between that copy and the remaining tools would be $\sum_{k=1}^{N} f(\pi_k, \pi_p)d(k, q)$. In this sum, the contribution of the term involving the second copy of π_p vanishes, since $f(\pi_p, \pi_p) = 0$, and the contribution of the term involving π_q vanishes, since $d(q, q) = 0$. Hence this sum is indeed the increase in cost when tool π_p is added to position q in Π^2. So,

$$Add(p, q) = \sum_{k=1}^{N} f(\pi_k, \pi_p)d(k, q). \tag{21.6}$$

The cost of adding tool π_q at position p in Π^3 is similar to the cost $Add(p, q)$. The only difference is that there is tool π_p in position q. Hence,

$$Add(q, p) = \sum_{k=1}^{N} f(\pi_k, \pi_q)d(k, p) + f(\pi_q, \pi_p)d(q, p). \tag{21.7}$$

Substituting Eqs. (21.3) to (21.7) in Eq. (21.2), cost of *swap*$[p, q]$ can be written as

$$z(\Pi_{pq}) - z(\Pi_c) = \sum_{k=1}^{N} f(\pi_k, \pi_p)d(k, q) + \sum_{k=1}^{N} f(\pi_k, \pi_q)d(k, p) + f(\pi_q, \pi_p)d(q, p)$$
$$- \sum_{k=1}^{N} f(\pi_k, \pi_p)d(k, p) - \sum_{k=1}^{N} f(\pi_k, \pi_q)d(k, q) + f(\pi_p, \pi_q)d(p, q). \quad (21.8)$$

Note that Eq. (21.8) requires the computation of terms $\sum_{k=1}^{N} f(\pi_k, \pi_p)d(k, q)$ and $\sum_{k=1}^{N} f(\pi_k, \pi_p)\, d(k, p)$ to be reused. Each of these terms is used multiple times when computing the cost of all swap neighbors of a solution. If these values are pre-computed and stored, then the costs of each swap neighbor of Π_c can be computed in constant time.

Let

$$a(i, j) = \sum_{k=1}^{N} f(\pi_k, \pi_i)d(k, j),\ i, j = \{1, 2, \ldots, N\}; \quad (21.9)$$

and let $A = [a(i, j)]$ be the matrix of all $a(i, j)$ values. In terms of entries in A, the cost of *swap*$[p, q]$ can be expressed as

$$z(\Pi_{pq}) - z(\Pi_c) = a(p, q) + a(q, p) - a(p, p) - a(q, q) + 2f(\pi_q, \pi_p)d(q, p) \quad (21.10)$$

and can be computed in constant time given the matrix A. This implies that if the matrix A is available, then the effort of searching the swap neighborhood of Π_c is $\mathcal{O}(N^2)$.

Computing entries of matrix A directly requires $\mathcal{O}(N^3)$ time, as it has $\mathcal{O}(N^2)$ elements and computation of each element involves summation of N frequency matrix elements weighted by the distance. However, through careful bookkeeping techniques, the elements of A can be computed $\mathcal{O}(N^2)$ time. To do this, two auxiliary matrices $B = [b(i, j)]$ and $C = [c(i, j)]$ are defined. Here,

$$b(i, j) = a(i, j + 1) - a(i, j) \quad 1 \le j \le N - 1 \quad (21.11)$$

and

$$c(i, j) = b(i, j + 1) - b(i, j) = a(i, j + 2) - 2a(i, j + 1) + a(i, j) \quad 1 \le j \le N - 2 \quad (21.12)$$

Without loss of generality, let $p = 1$. This assumption is valid since indices are being considered along a circular tool changer. The analysis for any other value of p can be reduced to this case simply by a shift in the numbering of the slots.

For even values of N, the expression for $b(1, j)$ obtained by expanding $a(1, j)$ and $a(1, j + 1)$ and subtracting the former from the latter is

$$b(1, j) = \sum_{k=1}^{N/2} f(1, (j + N/2 + k)|_{\text{mod } N}) - \sum_{k=1}^{N/2} f(1, (j + k)|_{\text{mod } N}). \quad (21.13)$$

Expanding $b(1, j)$ and $b(1, j + 1)$ and subtracting the former from the latter, the expression for $c(1, j)$ is

$$c(1, j) = 2f(1, j + 1) - 2f(1, (j + N/2 + 1)|_{\text{mod } N}). \quad (21.14)$$

For odd values of N, using very similar technique, the values of $c(1, j)$ are computed as

$$c(1, j) = 2f(1, j + 1) - f(1, (j + \lfloor N/2 \rfloor + 1)|_{\text{mod } N}) - f(1, (j + \lceil N/2 \rceil + 1)|_{\text{mod } N}). \quad (21.15)$$

Notice that $c(1, j)$, and therefore more generally $c(i, j)$, can be computed in constant time for both even and odd values of N. Therefore, populating the C matrix takes $\mathcal{O}(N^2)$ time. Once the C matrix is available, the B matrix can be populated in $\mathcal{O}(N^2)$ time. To see this, consider the first row of the B matrix. The values of $a(1, 1)$ and $a(1, 2)$ can be computed using expression (21.9) in $\mathcal{O}(N)$ time. These two values are used to compute $b(1, 1) = a(1, 2) - a(1, 1)$. The value of $b(1, 2)$ can be computed in constant time as $b(1, 2) = b(1, 1) + c(1, 2)$. In general, each of the other entries $b(1, j), 2 \leq j \leq N - 1$ in the first row of the B matrix can be computed in constant time as $b(1, j - 1) + c(1, j - 1)$, and the whole row can be populated in $\mathcal{O}(N)$ time. Since B has N rows, populating the B matrix requires $\mathcal{O}(N^2)$ time.

Once the B matrix is populated, the A matrix is populated as follows. For each row i, the value of $a(i, 1)$ is computed using expression (21.9) in $\mathcal{O}(N)$ time. Then each of the other elements $a(i, j)$ $2 \leq j \leq N$ of the row is populated in constant time using the formula $a(i, j) = a(i, j - 1) + b(i, j - 1)$. Hence populating each row of the A matrix requires $\mathcal{O}(N)$ time, and the A matrix can be populated in $\mathcal{O}(N^2)$ time. An illustration of the process of populating the A matrix using an example is given in Appendix A.1.

Once the A matrix is populated in $\mathcal{O}(N^2)$ time, as stated earlier, the evaluation of the complete swap neighborhood of a solution requires $\mathcal{O}(N^2)$ time. The following pseudocode formalizes the method for searching the swap neighborhood of a solution with $\mathcal{O}(N^2)$ effort:

Algorithm 1: pseudocode for searching swap neighborhood of a solution

Input: Frequency matrix $F = [f(i, j)], i, j \in \{1, \ldots, N\}$, distance matrix $D = [d(k, l)], k, l \in \{1, \ldots, N\}$, current solution $\Pi_c = [\pi_i]$, cost function $z(.)$.

Output: The best swap neighbor Π_{Best} of Π_c.

$c(i, j) \leftarrow c(i, j)$ computed using Eq. (21.14) (if N is even) or Eq. (21.15) (if N is odd) $\forall\ p \in \{1, 2, ..., N\}$ and $q \in \{1, 2, ..., N-2\}$;

foreach $p \in \{1, \ldots, N\}$ **do**

  $a(p, 1) \leftarrow \sum_{k=1}^{N} f(\pi_k, \pi_1).d(k, 1)$;

  $a(p, 2) \leftarrow \sum_{k-1}^{N} f(\pi_k, \pi_2).d(k, 2)$;

  $b(p, 1) \leftarrow a(p, 2) - a(p, 1)$;

  foreach $q \in \{2, \ldots, N-1\}$ **do**

    $b(p, q) \leftarrow b(p, q-1) + c(p, q-1)$;

    $a(p, q+1) \leftarrow a(p, q) + b(p, q)$;

  end

end

$z(\Pi_{pq}) \leftarrow z(\Pi_c) + a(q, p) - a(p, p) - a(q, q) + 2f(\pi_q, \pi_p)d(q, p)\ \forall\ p$ and q;

$\Pi_{\text{Best}} \leftarrow \Pi_{pq}$ with best $z(\Pi_{pq})$;

return Π_{Best};

21.3.2 Searching Insert Neighborhoods Efficiently

As with the swap neighborhood, the number of neighbors in the insert neighborhood of a solution is $\mathcal{O}(N^2)$ and searching it naïvely requires $\mathcal{O}(N^4)$ time. The remainder of this section will show a method of reducing the effort of searching the insert neighborhood of the solution $\Pi_c = (\pi_1, \pi_2, \ldots, \pi_p, \ldots, \pi_q, \ldots \pi_N)$ to $\mathcal{O}(N^2)$.

Consider an insert neighbor $\Pi^{pq} = (\pi_1, \pi_2, \ldots, \pi_{p-1}, \pi_{p+1} \ldots, \pi_q, \pi_p, \pi_{q+1}, \ldots \pi_N)$ obtained by removing the tool at position p in Π_c and inserting it between the tools at positions q and $q+1$ in Π_c. It is assumed here that $q = p + r > p$, i.e., position q is r slots to the right of position p. The treatment when $q < p$ can be done in a largely similar manner and is not elaborated here.

Since q is r slots to the right of p, the insert neighbor can be obtained by performing r swap moves between adjacent elements, i.e., first swap tools π_p and π_{p+1} to obtain one intermediate solution, Π^1, then swap tools π_{p+1} and π_{p+2} in Π^1 to obtain a solution Π^2 and so on until tools π_{q-1} and π_q in Π^{r-1} are swapped to obtain Π^{pq}. Note that tools in positions $p+1$ of Π^1, $p+2$ of Π^2, etc., are all the same, i.e., tool π_p in Π_c. An important observation here is that intermediate neighbors $\Pi^1, \Pi^2, \ldots, \Pi^{r-1}$ are the same as the insert neighbors $\Pi^{p,p+1}, \Pi^{p,p+2}, \ldots, \Pi^{p,q-1}$, respectively, of Π_c. For example, if $\Pi_c = (1, 2, 3, 4, 5, 6, 7, 8)$, then $\Pi^{15} = (2, 3, 4, 5, 1, 6, 7, 8)$

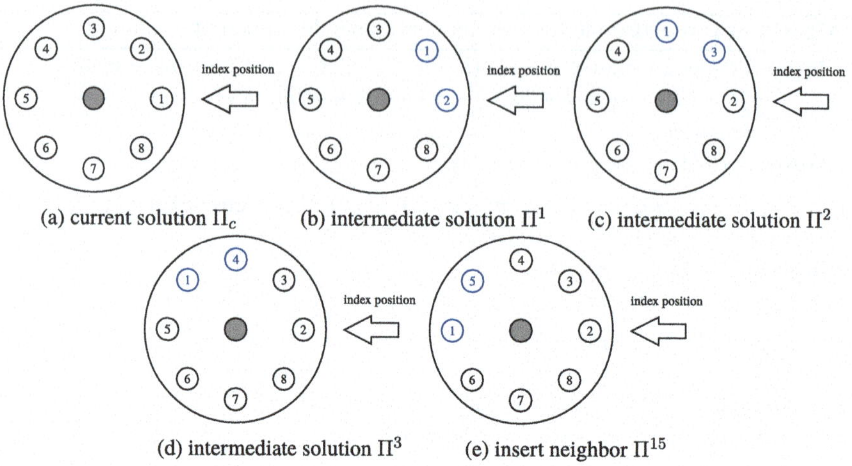

Fig. 21.3 Illustration of intermediate moves in generating insert neighbor Π^{15} of solution Π_c

is obtained through intermediate solutions $\Pi^1 = \Pi^{12} = (2, 1, 3, 4, 5, 6, 7, 8)$, $\Pi^2 = \Pi^{13} = (2, 3, 1, 4, 5, 6, 7, 8)$, and $\Pi^3 = \Pi^{14} = (2, 3, 4, 1, 5, 6, 7, 8)$ (see Fig. 21.3).

The difference $z(\Pi^{pq}) - z(\Pi_c)$ when $q = p + r$ can be computed as

$$z(\Pi^{pq}) - z(\Pi_c) = (z(\Pi^{pq}) - z(\Pi^{r-1})) + (z(\Pi^{r-1}) - z(\Pi^{r-2})) + \ldots + (z(\Pi^1) - z(\Pi_c)).$$

Each of the compound terms on the right-hand side computes the increase in cost if two adjacent tools are swapped; for example, the term $(z(\Pi^1) - z(\Pi_c))$ computes the increase in cost if the tool at position p is swapped with the tool at position $p + 1$.

The expressions for the costs of insert neighbor Π^{pq} of a solution Π_c depend on whether the value of N is even or odd, and on the relative positions of p and q. The expressions derived here are for even values of N, and for the case where $1 \leq p < q < K = N/2$. The expressions of other cases can be derived using a similar approach, but are not mentioned here explicitly.

Without loss of generality, assume $p = 1$, i.e., tool π_1 in position 1 is being removed from its position in Π_c and re-inserted between tools π_2 and π_3 to create the solution Π^{12}. This causes the distance between tool π_1 and tools π_3 through π_{K+1} to decrease by one unit each, and the distance between π_1 and tools π_{K+2} through π_N to increase by one unit each. It also causes the distance between tool π_2 and tools π_{K+2} through π_N to decrease by one unit each, and the distance between π_2 and tools π_3 through π_{K+1} to increases by one unit each. So,

$$z(\Pi^{12}) - z(\Pi_c) = \sum_{i=K+2}^{N} f(\pi_1, \pi_i) - \sum_{i=3}^{K+1} f(\pi_1, \pi_i) + \sum_{i=3}^{K+1} f(\pi_2, \pi_i) - \sum_{i=K+2}^{N} f(\pi_2, \pi_i). \tag{21.16}$$

Suppose Π^{12} and $z(\Pi^{12})$ are available. Consider the solution Π^{13}, obtained by removing tool π_1 in Π_c and inserting it between tools π_3 and π_4 in Π_c. Observe that Π^{13} can also be obtained by removing tool π_1 from its position in Π^{12} and re-inserting it after tool π_3. Following an argument similar to the one in the previous paragraph, the expression obtained for $z(\Pi^{13}) - z(\Pi^{12})$ is

$$z(\Pi^{13}) - z(\Pi^{12}) = \sum_{i=K+3}^{N} f(\pi_1, \pi_i) + f(\pi_1, \pi_2) - \sum_{i=4}^{K+2} f(\pi_1, \pi_i) + \sum_{i=4}^{K+2} f(\pi_3, \pi_i) - \sum_{i=K+3}^{N} f(\pi_3, \pi_i) - f(\pi_3, \pi_2). \tag{21.17}$$

Next, suppose Π^{13} and $z(\Pi^{13})$ are available. Consider the solution Π^{14}, obtained by removing tool π_1 in Π_c and inserting it between tools π_4 and π_5 in Π_c. Following an argument similar to the one in the previous paragraph, the expression obtained for $z(\Pi^{14}) - z(\Pi^{13})$ is

$$z(\Pi^{14}) - z(\Pi^{13}) = \sum_{i=K+4}^{N} f(\pi_1, \pi_i) + \sum_{i=2}^{3} f(\pi_1, \pi_i) - \sum_{i=5}^{K+3} f(\pi_1, \pi_i) + \sum_{i=5}^{K+3} f(\pi_4, \pi_i) - \sum_{i=K+4}^{N} f(\pi_4, \pi_i) - \sum_{i=2}^{3} f(\pi_4, \pi_i). \tag{21.18}$$

In general, for solution Π^{1j}, $2 < j < K$, obtained by removing tool π_1 in Π_c and inserting it between tools π_j and π_{j+1} in Π_c, the expression for $z(\Pi^{1j}) - z(\Pi^{1,j-1})$ is obtained as

$$z(\Pi^{1j}) - z(\Pi^{1,j-1}) = \sum_{i=K+j}^{N} f(\pi_1, \pi_i) + \sum_{i=2}^{j-1} f(\pi_1, \pi_i) - \sum_{i=j+1}^{K+j-1} f(\pi_1, \pi_i) + \sum_{i=j+1}^{K+j-1} f(\pi_j, \pi_i) - \sum_{i=K+j}^{N} f(\pi_j, \pi_i) - \sum_{i=2}^{j-1} f(\pi_j, \pi_i). \tag{21.19}$$

Note that Eq. 21.19 requires the computation of terms of the form $\sum_{k=k_L}^{k_U} f(\pi_1, \pi_k)$, $1 < k_L < k_U \le N$. Further, $\sum_{k=k_L}^{k_U} f(\pi_1, \pi_k) = \sum_{k=1}^{k_U} f(\pi_1, \pi_k) - \sum_{k=1}^{k_L-1} f(\pi_1, \pi_k)$. If the values of the terms $\sum_{k=1}^{j} f(\pi_i, \pi_k), i, j \in \{1, 2, \cdots, N\}$ are pre-computed and stored, then the costs of each insert neighbor of Π_c can be computed in constant time. Let

$$s(i, j) = \sum_{k=1}^{j} f(\pi_i, \pi_k); \tag{21.20}$$

and let $S = [s(i, j)]$ be the matrix of all the $s(i, j)$ values. Note that $s(i, j+1) = s(i, j) + f(\pi_i, \pi_{j+1})$, and so $s(i, j+1)$, can be computed in constant time once $s(i, j)$ is known. Starting with $s(i, 1) = f(\pi_i, \pi_1)$, each row of the S matrix can be computed in $\mathcal{O}(N)$ time, and hence all entries of S matrix can be computed in $\mathcal{O}(N^2)$ time. In terms of entries of S matrix, Eq. (21.16) can be written as

$$\begin{aligned} z(\Pi^{12}) - z(\Pi_c) = \{s(1, N) - s(1, K+1)\} - \{s(1, K+1) - s(1, 2)\} \\ + \{s(2, K+1) - s(2, 2)\} - \{s(2, N) - s(2, K+1)\} \end{aligned} \quad (21.21)$$

and Eq. (21.19) as

$$\begin{aligned} z(\Pi^{1j}) - z(\Pi^{1,j-1}) = \{s(1, N) - s(1, K+j-1)\} + \{s(1, j-1) - s(1, 1)\} \\ - \{s(1, K+j-1) - s(1, j)\} + \{s(j, k+j-1) - s(j, j)\} \\ - \{s(j, N) - s(j, K+j-1)\} - \{s(j, j-1) - s(j, 1)\}. \end{aligned} \quad (21.22)$$

Next, suppose that the S matrix is available. First, the cost of insert neighbor Π^{12} is evaluated using Eq. (21.21) . Then, the costs of insert neighbors Π^{1j} for $j = \{3, 4, \ldots, N\}$ are evaluated successively using Eq. (21.22). Each of these can be computed in constant time as the number of terms in Eqs. (21.21) and (21.22) are constant and not increasing in N. Similarly, each of the other insert neighbors of Π_c can be evaluated in constant time. Thus the cost of all insert neighbors of Π_c can be computed in $\mathcal{O}(N^2)$ time. The computation of the cost of insert neighbors using S matrix entries is further illustrated with the help of an example in Appendix A.2.

The following pseudocode formalizes the method for searching the insert neighborhood of a solution with $\mathcal{O}(N^2)$ effort.

Algorithm 2: pseudocode for searching insert neighborhood of a solution

Input: Frequency matrix $F = [f(i, j)]$, $i, j \in \{1, \ldots, N\}$, distance matrix $D = [d(k, l)]$, $k, l \in \{1, \ldots, N\}$, current solution $\Pi_c = [\pi_i]$, cost function $z(.)$.

Output: The best insert neighbor Π_{Best} of Π_c.

$s(i, j) \leftarrow \sum_{k=1}^{j} f(\pi_i, \pi_k)\ \forall\, i$ and j ; $S = [s(i, j)]$;
foreach $p \in \{1, \ldots, N\}$ **do**
  costdiff $\leftarrow$ computed using entries of matrix S (see Eq. 21.21);
  $z(\Pi^{p,p+1}) \leftarrow z(\Pi_c) +$ costdiff;
  foreach $q \in \{p+2, \ldots, N\}$ **do**
    costdiff $\leftarrow$ computed using entries of matrix S (see Eq. 21.22);
    $z(\Pi^{pq}) \leftarrow z(\Pi^{p,q-1}) +$ costdiff;
  end
end
$\Pi_{\text{Best}} \leftarrow \Pi^{pq}$ with best $z(\Pi^{pq})$;
return Π^{Best};

21.4 Neighborhood Search Algorithms

The TIP is a hard combinatorial optimization problem, and exact algorithms are not guaranteed to solve practical size instances of these problems in a reasonable time. Hence, as in the case of other hard optimization problems, heuristics are used to obtain good quality solutions to real-size instances of the TIP. Broadly, heuristics are of two types: construction heuristics, and improvement heuristics. Construction heuristics start with a trivial partial solution to a problem and develop a feasible solution, while improvement heuristics start with one or more feasible solutions, often generated at random or through construction heuristics, and improve upon them to obtain a good quality solution. The literature has mostly focused on improvement heuristics as a preferred approach for practical problem-solving. Improvement heuristics usually are of two types: neighborhood search heuristics and population heuristics.

Neighborhood search algorithms will be used in this paper to solve the TIP. They start with a solution to the problem as the current solution and search for its neighboring solutions to improve upon the current solution. The neighboring solutions are defined in terms of moves, which are operations that allow the algorithm to convert a solution to another solution. Once the predetermined termination criterion is met, the best solution found during the execution of the algorithm is output as an approximation to an optimal solution to the problem.

The most basic of neighborhood search algorithms is the local search algorithm. It is an algorithm that terminates when it encounters a local optimum, i.e., a solution that is not worse than any of its neighboring solutions. Local search is usually very fast, but it explores a very small fraction of the solution space. Hence it is often implemented in a multi-start mode, in which several starting solutions are chosen (usually in diverse regions of the solution space), and local search is executed from each of these starting solutions. The output of a multi-start local search is the best among the terminal solutions from all these starts. Other algorithms, such as tabu search and simulated annealing, are modifications of the local search algorithm that allow the search to move out of local optima. Tabu search is often the best performing algorithm among neighborhood search algorithms for most problems, and so is used in this work.

Other common improvement algorithms for combinatorial problems are population heuristics such as genetic algorithms, particle swarm algorithms, ant colony algorithms, etc. These algorithms do not make use of neighborhood structures and hence cannot take advantage of the speedups that have been proposed in Sect. 21.3. Hence they are not included in this computational setup.

The remainder of the section describes the local search and tabu search algorithms used in the computational experiments.

21.4.1 Local Search Algorithm

A local search algorithm starts with an initial solution which is set as the current solution. Construction heuristics are not available for the TIP, and hence in our implementations, a random permutation of the tools is taken as an initial solution. The algorithm proceeds by searching the neighborhood of the current solution and identifying the best solution among the neighbors. If the best solution is better than the current solution, then the local search marks that solution as the current solution and continues exploring the neighborhood of the new current solution. Otherwise, the search terminates and outputs the current solution as an approximation to an optimal solution. Notice that the quality of the current solution keeps improving as the local search progresses, and hence each move from a current solution to the next current solution is an improving move.

The pseudocode for the local search algorithm applied to the tool indexing problem is given in Algorithm 3. A local search algorithm is called swap-based or insert-based depending on the neighborhood that it explores. Our swap-based local search and insert-based local search algorithms are referred to as `SwapLS` and `InsertLS`, respectively.

Algorithm 3: pseudocode for the local search algorithms `SwapLS` and `InsertLS`

Input: Frequency matrix $F = [f(i, j)], i, j \in \{1, \dots, N\}$, distance matrix $D = [d(k, l)], k, l \in \{1, \dots, N\}$, starting solution $\Pi_c = [\pi_i]$, cost function $z(.)$.

Output: Π_{Best}, a good solution to the tool indexing problem

$\Pi_{\text{Best}} \leftarrow \Pi_c$;
$\Pi_{\text{Curr}} \leftarrow \Pi_c$;
Improving $\leftarrow$ TRUE;
while Improving **do**
    Improving $\leftarrow$ FALSE;
    foreach swap (or insert) neighbor $\Pi_{p,q}$ of Π_{Curr} **do**
        if $z(\Pi_{p,q}) < z(\Pi_{\text{Best}})$ **then**
            $\Pi_{\text{Best}} \leftarrow \Pi_{p,q}$;
            Improving $\leftarrow$ TRUE
        end
    end
    $\Pi_{\text{Curr}} \leftarrow \Pi_{\text{Best}}$;
end
return Π_{Best};

21.4.2 Tabu Search Algorithm

Since local search performs improving moves throughout its execution, it terminates at the first local optimum solution it encounters. A local optimal solution is one that is not worse than any of its neighbors. This local optimum may not be of good quality compared to an optimal solution to the problem, and hence the output of the local search may not be of high quality.

Tabu search, introduced and formalized in [13, 14], is a neighborhood search-based metaheuristic search technique equipped with features to extricate neighborhood search from local optima. It allows moves from a solution to a neighboring solution with a worse objective function value if none of the neighbors can improve on the current solution. These moves are called worsening moves and allow the search to move away from local optima. However, after a worsening move, the search can go back to the immediate past local optimum if it is not restricted from doing so. Tabu search restricts such a reversal by allowing the search to consider only a subset of all neighboring solutions in order to find a new current solution. This subset is constructed in a way that it excludes a few of the recently visited solutions, by maintaining a list of tabu, or restricted, moves that cannot be performed on the current solution to reach a neighbor.

The TIP is a specially structured quadratic assignment problem. So a tabu search implementation for the quadratic assignment problem is adapted for the indexing problem. To the best of our knowledge, the tabu search implementation in [26] is the only well-known tabu search implementation for the quadratic assignment problem, and hence it is this implementation that is modified to solve the TIP.

The remainder of this section describes memory structures used in the standard tabu search algorithm followed by their adaptation and implementation details for our swap and insert-based tabu search algorithms.

Tabu List As mentioned above, tabu search discourages the search from returning to solutions encountered immediately previously by restricting certain moves. In particular, tabu search restricts the reverse of the moves executed in recent moves. For example, in a swap neighborhood, if in a previous iteration, the search moved to a neighbor by swapping tools π_p and π_q at positions p and q, then allowing tool π_p to return to position p and tool π_q to return to position q negates that move. These moves are thus disallowed unless there is a pressing reason to allow them, and certain criteria that characterize these are met. These criteria, called aspiration criteria, are described later in this section. A move present in the tabu list is called tabu active. Note that there are many ways of defining tabu moves, and each implementation attempts to define tabu moves in a way that yields good results.

Tabu search thus maintains a list of disallowed moves, or tabu moves, in a list called the tabu list. The membership of a move in the tabu list is not permanent, and it is removed from the tabu list after a pre-specified number of iterations. This number of iterations for which a move remains in the tabu list is called the tabu tenure. Early tabu search implementations worked with a fixed tabu tenure, but later implementations, including the implementations described in this work, use a random tabu tenure drawn from within a pre-specified range.

The tabu search implementation in [26] defines moves in the tabu list as follows. A swap move *swap*$[p, q]$ is tabu if it assigns tool π_p as well as tool π_q to positions that they occupied in any of the last s iterations. To implement this definition of tabu list, move (π_i, j) is defined as an operation that places tool π_i in position j. Given a solution represented by Π, *swap*$[p, q]$ on Π constitutes moving tool π_p to position q, and moving tool π_q to position p and are indicated by move (π_p, q) and move (π_q, p), respectively. As per the definition of tabu list in [25], the reverse of *swap*$[p, q]$ is composed of move (π_p, p) and move (π_q, q). Hence, if the swap neighbor chosen at the end of an iteration is Π_{pq}, then moves (π_p, p) and (π_q, q) are added to the tabu list. A swap move *swap*$[p, q]$ is set as "tabu" if both the moves (π_p, q) and (π_q, p) are tabu active.

To the best of our knowledge, there is no tabu search implementation in the literature that uses the insert neighborhood. Hence the following definition of tabu moves using insert neighborhoods is new. In an insert move *insert*$[p, q]$, tool π_p moves to position q, and all the tools between tool π_p and π_{q+1} shifts one position to the left if $p < q$ and one position to the right if $p > q$. Hence the positions of $|q - p| + 1$ tools gets changed in *insert*$[p, q]$. Forbidding the reversal of all these moves might over-constrain the search. Hence only two of these moves are considered: move (π_{p+1}, p) and move (π_p, q). Then the moves that should be marked as tabu are the reverse moves: move $(\pi_p, p - 1)$ and move (π_q, p). However, forbidding one or both of these moves is not enough to prevent cycling back to earlier solutions. Hence, in addition to these moves, the cost of the solution is also added to the tabu list. Thus, if either of the moves $(\pi_p, p - 1)$ or (π_q, p) of $insert(p, q)$ are tabu active or if the cost of Π_{pq} is tabu active, then the move $insert(p, q)$ is marked tabu.

Tabu Tenure From the discussion on tabu list, recall that tabu tenure (denoted as s) is the duration (in terms of the number of iterations) for which a move remains in the tabu list. At the end of iteration i, moves that were added in iteration $(i - s)$ or earlier, are removed from the tabu list. A small value of tenure may cause frequent cycling over previously searched solutions while a large value may be too restrictive in exploring good quality solutions. Following [26], tabu tenure s is kept as a random number between S_{min} and S_{max} and is changed after every $2S_{max}$ iterations to ensure that at least a few iterations are performed with $s = S_{max}$. Based on computational experiments, S_{min} and S_{max} were set as $0.9N$ and as $1.1N$, respectively, for tabu search using swap neighborhoods and N and $3N$, respectively, for tabu search using insert neighborhoods.

Aspiration Criteria Recall from the discussion on tabu list that moves in the tabu list are not executed to generate neighbors in tabu search. However, such a policy may sometimes lead to the search ignoring high-quality solutions. Tabu search allows a mechanism called aspiration to let the search consider exceptional quality solutions even though moves to take the search from the current solution to that solution is tabu active. The criteria that are used to certify a move worthy of aspiration are called aspiration criteria. The classical aspiration criteria to override the tabu status on a move is to check if such a move leads to a solution better than the best solution found by tabu search thus far. This has been implemented in the tabu search. Additionally, in our implementations, the tabu status is also overridden for the move resulting in the best solution in the neighborhood when all the moves that constitute the neighborhood are tabu.

Termination criterion Unlike local search, which terminates on reaching a local optimum, tabu search can proceed indefinitely. Hence there is a need for pre-specified termination criteria, satisfying which, the search terminates and outputs the best solution it has found thus far. Commonly used termination criteria include a maximum time limit, a maximum number of search iterations or a maximum number of iterations without improvement in the best solution found so far. In our tabu search implementation, the number of iterations is specified and the search terminates when those many iterations have been completed. The number of possible solutions to an instance of the TIP with N slots in the tool changer is $\mathcal{O}(N!)$ which increases extremely quickly with increasing values of N. Hence larger size instances require more tabu search iterations to search the solution space effectively. In our implementations, the maximum number of iterations is used as the termination criteria is set to be N^2.

The pseudocodes for our implementation of swap-based tabu search, called `SwapTS`, and our insert-based tabu search, called `InsertTS`, developed based on the above description are presented in Algorithms 4 and 5, respectively.

Algorithm 4: pseudocode for `swapTS`

Input: Frequency matrix $F = [f(i, j)]$, $i, j \in \{1, \ldots, N\}$, distance matrix $D = [d(k, l)]$, $k, l \in \{1, \ldots, N\}$, starting solution $\Pi_c = [\pi(i)]$, cost function $z(.)$, Parameters: S_{min}, S_{max}, and MaxIter.

Output: Π_{Best}, a good solution to the tool indexing problem

```
Π_Curr ← Π_c;
Π_Best ← Π_Curr;
TabuList ← Ø;
for iter ← 1 to MaxIter do
    if iter is equal to 1 or a multiple of 2S_max then
        TabuTenure ← a random number between S_min and S_max;
    end
    Π_BestNbr ← Π_Curr;
    foreach swap neighbor Π_{p,q} of Π_Curr do
        MoveAllowed ← FALSE;
        MoveAspiring ← FALSE;
        if (π_p, q) ∉ TabuList AND (π_q, p) ∉ TabuList AND z(Π_{p,q}) < z(Π_BestNbr) then
            MoveAllowed ← TRUE;
        end
        if MoveAllowed = FALSE AND z(Π_{p,q}) < z(Π_Best) then
            MoveAspiring ← TRUE;
        end
        if MoveAllowed = TRUE OR MoveAspiring = TRUE then
            Π_BestNbr ← Π_{p,q};
            TabuMove1 ← (π_p, p);
            TabuMove2 ← (π_q, q);
        end
    end
    Π_Curr ← Π_BestNbr;
    if z(Π_Curr) < z(Π_Best) then
        Π_Best ← Π_Curr;
    end
    add TabuMove1 and TabuMove2 to TabuList for the next TabuTenure iterations;
    iter ← iter + 1;
end
return Π_Best;
```

Algorithm 5: pseudocode for `insertTS`

Input: Frequency matrix $F = [f(i, j)], i, j \in \{1, \dots, N\}$, distance matrix $D = [d(k, l)], k, l \in \{1, \dots, N\}$, starting solution $\Pi_c = [\pi(i)]$, cost function $z(.)$, Parameters: S_{min}, S_{max}, and MaxIter.

Output: Π_{Best}, a good solution to the tool indexing problem

$\Pi_{\text{Curr}} \leftarrow \Pi_c$;
$\Pi_{\text{Best}} \leftarrow \Pi_{\text{Curr}}$;
TabuList $\leftarrow \emptyset$;
for iter $\leftarrow$ 1 to MaxIter **do**
    if iter is equal to 1 or a multiple of $2S_{max}$ **then**
        TabuTenure $\leftarrow$ a random number between S_{min} and S_{max};
    end
    $\Pi_{\text{BestNbr}} \leftarrow \Pi_{\text{Curr}}$;
    foreach insert neighbor $\Pi_{p,q}$ of Π_{Curr} **do**
        MoveAllowed $\leftarrow$ *FALSE*;
        MoveAspiring $\leftarrow$ *FALSE*;
        if $(\pi_p, q) \notin$ TabuList AND $(\pi_q, p) \notin$ TabuList AND $z(\Pi_{p,q}) \notin$ TabuList AND $z(\Pi_{p,q}) < z(\Pi_{\text{BestNbr}})$ **then**
            MoveAllowed $\leftarrow$ *TRUE*;
        end
        if MoveAllowed = *FALSE* AND $z(\Pi_{p,q}) < z(\Pi_{\text{Best}})$ **then**
            MoveAspiring $\leftarrow$ *TRUE*;
        end
        if MoveAllowed = *TRUE* OR MoveAspiring = *TRUE* **then**
            $\Pi_{\text{BestNbr}} \leftarrow \Pi_{p,q}$;
            TabuMove1 $\leftarrow (\pi_p, p - 1)$;
            TabuMove2 $\leftarrow (\pi_p, p + 1)$;
            TabuCost $\leftarrow z(\Pi_{p,q})$;
        end
    end
    $\Pi_{\text{Curr}} \leftarrow \Pi_{\text{BestNbr}}$;
    if $z(\Pi_{\text{Curr}}) < z(\Pi_{\text{Best}})$ **then**
        $\Pi_{\text{Best}} \leftarrow \Pi_{\text{Curr}}$;
    end
    add TabuMove1, TabuMove2, TabuCost to TabuList for next TabuTenure iterations;
end
return Π_{Best};

21.5 Computational Experiments

The four swaps and insert-based local search and tabu search algorithms for the tool indexing problem described in Sect. 21.4: `SwapLS` , `InsertLS`, `SwapTS`, and `InsertTS`, were coded in C++ for our computational experiments. A machine with Intel i7-7700 CPU at 3.60 GHz was used to run the experiments. Instances used in the literature for the tool indexing problem (that are called the "Single" instances), and in [12] are used for our computational experiments. The instances in [12] include instances used in computational experiments for related problems adapted to the tool indexing problem (that are called the "Anjos" and "Sko" instances). Some new instances (that are called the "Large" instances") were also generated for our tests. Brief descriptions of these instances are given below.

Single instances These are three instances taken from the literature for the tool indexing problem. First is a small instance for the tool indexing problem from [9] 10 tools and 16 slots in the tool changer. The remaining two are instances for the tool indexing problem with tool duplication used in [4], of which one is a small instance with 8 tools and 12 slots, and the other is a larger instance with 50 tools and 70 slots.

Anjos instances These instances used in [12] were adapted from instances used in [2] for the single row facility layout problem (SRFLP). The matrix denoting the frequency of interaction between a pair of facilities in the SRFLP instance is used as the frequency matrix in the tool indexing problem instance. The number of slots is taken as 100 for the experiments.

Sko instances These instances used in [12] were adapted from instances used in [1] used for computational experiments on the SRFLP. The instances in [1] contain multiple instances with the same frequency matrix and different length matrices meaningful for the SRFLP. Discarding instances in which the frequency matrices were duplicates resulted in seven instances that were used for our experiments. To allow for comparison with the `Ghosh_SwapTS` algorithm, the tool magazine was assumed to have 60 slots when the number of tools was less than 60, and 100 slots when the number of tools exceeded 60.

Large instances Since a sufficient number of large-sized instances could not be found in the literature for comparing the performance of our algorithms, sets of large-sized problem instances were created by randomly generating frequency matrices for three different problem sizes; $N = 50, 75, \text{and} 100$. For each problem size, 100 frequency matrices numbered 1–100 were generated such that the entries of matrices 1–25 and 51–75 take integer values between and including 0 and 20, whereas those in matrices 26–50 and 76–100 take integer values between and including 0 and 10. Further, frequency matrices 51–100 were made sparse by keeping only less than half of the entries non-zero. The instances with frequency matrices numbered 1–25, 26–50, 51–75, and 76–100 were named as "Large(A)" , "Large(B)" , "Large(C)", and "Large(D)", respectively. Note that due to the way in which the frequency matrices were constructed, the sum of the entries in Large(A), Large(B), Large(C), and Large(D) were in decreasing order. In practice, this means that instances in Large(A)

Table 21.1 Performance comparison on single instances

Instance	Our swap algorithms				Our insert algorithms			
	SwapLS		SwapTS		InsertLS		InsertTS	
	Cost	Time (s)	Cost	Time (s)	Cost	Time (s)	Cost	Time (s)
Single-16-10-1	13	0.0	13	0.1	13	0.0	13	0.1
Single-12-8-1	53	0.0	53	0.1	53	0.0	53	0.1
Single-70-50-1	404	1.1	404	6.9	396	0.7	383	17.7

correspond to the longest tool sequences, and the instances in Large(D) correspond to the shortest tool sequences.

Each instance used in our experiments is denoted as `I-X-Y-Z`, where `I` is the name of the instance set (Single, Anjos, Sko, or Large), `X` is the number of slots, `Y` is the number of tools, and `Z` is the instance number. For example, the instance Anjos-100-75-2 refers to the Anjos instance with 100 slots, 75 tools, and frequency matrix labeled 2.

Notice that in many of the instances, the number of tools is fewer than the number of slots in the tool changer. As mentioned in Sect. 21.1, this is handled by adding an appropriate number of dummy tools as placeholders to convert the instance into one in which the number of tools equals the number of slots. All entries in the frequency matrix involving these dummy tools have zero values.

For each instance, the algorithm being tested is replicated 51 times, each replication starting with a different randomly generated solution. The number of replications is high in order to have a more reliable estimate of the algorithm's performance. It is chosen to be 51 following the experimental setup in [22].

The performance of our algorithms for the Single instances is given in Table 21.1. Against each instance given in the first column, the best cost reported (cost) and the time taken in CPU seconds (time) for `SwapLS`, `SwapTS`, `InsertLS`, and `InsertTS` are displayed in the subsequent columns. [9] used a genetic algorithm to solve Single-16-10-1 and obtained a solution with a cost of 13 in 80 s. Our neighborhood search algorithms `SwapLS` and `InsertLS` arrived at the same cost in under 0.005 CPU seconds, and our tabu search algorithms `SwapTS` and `InsertTS` obtain this value within 0.160 CPU seconds. Costs obtained by our algorithms for instances Single-12-8-1 and Single-70-50-1 cannot be compared with those in [4] since [4] have considered these instances for a tool indexing problem with tool duplication while in this study the instances have been used for a tool indexing problem without tool duplication.

A comparison of the performances of our algorithms with the one reported in [12], which is called the `Ghosh_SwapTS` algorithm, is presented in Tables 21.2 and 21.3. Against each instance listed in the first columns of Tables 21.2 and 21.3, the best cost reported (cost) and the time taken in CPU seconds (time) for `Ghosh_SwapTS`,

Table 21.2 Performance comparison on Anjos instances

Instance	Ghosh_SwapTS		Our swap algorithms				Our insert algorithms			
			SwapLS		SwapTS		InsertLS		InsertTS	
	Cost	Time (s)	Cost	Time (s)	Cost	Time (s)	Cost	Time (s)	Cost	Time (s)
Anjos-100-60-1	54053	179.5	54053	2.5	54053	15.8	54053	1.3	54054	33.2
Anjos-100-60-2	31278	180.1	31285	3.3	31285	16.0	**31274**	1.3	**31274**	31.6
Anjos-100-60-3	23510	180.1	23514	2.8	23514	13.9	23510	1.5	23511	32.6
Anjos-100-60-4	11592	178.2	11610	2.9	11608	15.3	11592	1.1	11592	31.8
Anjos-100-60-5	15182	178.2	15181	3.4	15181	14.4	**15168**	1.4	**15168**	31.1
Anjos-100-70-1	42297	178.5	42370	4.1	42370	25.7	42298	2.1	42301	56.8
Anjos-100-70-2	51723	178.0	51723	4.2	51723	23.4	51723	2.5	51727	56.0
Anjos-100-70-3	43795	178.5	43820	3.4	43819	26.0	**43794**	2.3	43795	56.3
Anjos-100-70-4	27705	178.8	27733	24.5	**27701**	25.4	**27701**	2.3	27703	56.9
Anjos-100-70-5	134269	181.5	134348	4.7	134348	25.3	**134238**	2.1	134243	57.3
Anjos-100-75-1	66656	182.2	66658	4.5	66658	33.5	**66630**	2.7	66637	74.7
Anjos-100-75-2	111806	181.5	111806	5.5	111806	33.6	111806	2.8	111810	73.5
Anjos-100-75-3	38158	181.9	38153	3.5	38153	28.7	**38151**	2.9	38152	74.4
Anjos-100-75-4	106341	179.5	106341	4.3	106341	29.1	106341	3.0	106343	75.1
Anjos-100-75-5	47031	179.1	47032	4.5	47032	28.5	47020	2.9	**47018**	77.0
Anjos-100-80-1	54463	178.9	54469	4.2	54469	37.1	54463	4.0	54464	98.7
Anjos-100-80-2	52853	179.0	52866	3.8	52866	40.2	52855	4.3	52853	96.2
Anjos-100-80-3	95091	182.6	95133	4.6	95133	36.5	95091	3.9	95093	94.5
Anjos-100-80-4	100950	184.2	100845	4.0	100845	37.0	**100828**	4.4	100829	94.6
Anjos-100-80-5	36227	185.5	36249	4.5	36249	36.1	36220	4.0	**36215**	95.1
Average	57249.0	180.3	57259.4	5.0	57257.7	27.1	57237.8	2.7	57239.1	64.9

Table 21.3 Performance comparison on Sko instances

Instance	Ghosh_SwapTS		Our swap algorithms				Our insert algorithms			
			SwapLS		SwapTS		InsertLS		InsertTS	
	Cost	Time (s)	Cost	Time (s)	Cost	Time (s)	Cost	Time (s)	Cost	Time (s)
Sko-60-42-1	24408	38.3	24435	0.5	24416	4.0	24475	0.3	24417	8.4
Sko-60-49-1	36582	38.1	36713	0.6	36713	5.9	36599	0.6	**36547**	28.0
Sko-60-56-1	52974	38.0	52889	0.7	52879	13.0	53042	1.0	**52794**	42.0
Sko-100-64-1	95337	182.1	95329	3.1	95329	19.7	95347	1.9	**95269**	43.1
Sko-100-72-1	133607	182.6	133074	3.8	133074	26.4	132897	2.8	**132636**	64.3
Sko-100-81-1	185852	183.1	185500	5.7	185476	40.6	185168	5.5	**184897**	97.9
Sko-100-100-1	290496	185.0	289990	6.2	289886	98.2	289017	9.9	**288855**	236.9
Average	117036.6	121.0	116847.1	2.9	116824.7	29.7	116653.6	3.1	116487.9	74.3

SwapLS, SwapTS, InsertLS, and InsertTS are displayed in the subsequent columns. For some of the instances, our algorithms were able to obtain better solutions than those obtained in [12]. These are displayed in bold font. Based on the comparison of the hardware used for our computational experiments with that used in [12], the time reported in [12] has to be scaled down by a factor of 1.66 (on the basis of CPU performance scores from [24]) to compensate for the lower processing power of the hardware. However, all experiments in [12] use 20 starts, while this study uses 51 starts. So for example, when Ghosh_SwapTS reports requiring 179.5 s for Anjos-100-60-1, it is understood that it would require $179.5/1.66 = 108.13$ s when run on our machine. Since that algorithm requires 20 starts, the average time it would require per start would be 5.4 s on our machine. If it was to use 51 starts, then it would have taken 275.7 s on our machine. This shows that the algorithms described here are much faster than Ghosh_SwapTS.

For Anjos instances (see Table 21.2), the InsertLS algorithm found better solutions than the Ghosh_ SwapTS algorithm in 10 out of the 20 instances and found worse solutions in only 2 of the 20 instances. It was much faster than the Ghosh_SwapTS algorithm. Interestingly, the InsertTS algorithm showed worse performance than InsertLS. This is possible because of the restrictions enforced by the tabu structure. SwapLS and SwapTS algorithms performed much worse. This superiority of the insert neighborhood over the swap neighborhood is consistent with observations in many related problems (see, for example, [19]). Our algorithms were faster than the Ghosh_SwapTS algorithm at least by 4.25 times on average across all Anjos instances.

For Sko instances (see Table 21.3), the InsertTS algorithm outperforms the Ghosh_SwapTS algorithm in six out of the seven instances. Here too the tabu search algorithm using the insert neighborhood output better solutions than the one using the swap neighborhood. Our local search algorithms and tabu search algorithms were faster than Ghosh_SwapTS algorithm by 62 times and 3.6 times on average, respectively.

Table 21.4 summarizes the performance of our algorithms on large instances. Tables 21.4(a) and 21.4(b) present the performance summary of our algorithms for Large(A), Large(B), Large(C), and Large(D) instances, while Tables 21.4(c) and 21.4(d) display the performance summary of our algorithms for the instances grouped by their problem size. Average costs and average times in CPU seconds are reported for each of our algorithms in Tables 21.4(a) and 21.4(c) for each instance type and problem size, respectively. For each instance, algorithms are ranked based on solution

Table 21.4 Performance summary on large instances

(a) Average cost and time for each instance type

Instance type	SwapLS		SwapTS		InsertLS		InsertTS	
	Cost	Time (s)	Cost	Time (s)	Cost	Time (s)	Cost	Time (s)
LargeA	430016.7	2.9	430003.1	28.0	429878.8	3.0	429377.8	75.7
LargeB	202642.8	2.9	202633.4	27.7	202576.1	3.0	202310.3	77.1
LargeC	101813.3	2.9	101799.4	29.7	101432.8	2.9	101016.1	81.8
LargeD	47852.8	2.8	47836.8	28.8	47671.8	2.9	47428.7	89.8

(b) Average rank for each instance type

Instance type	SwapLS	SwapTS	InsertLS	InsertTS
LargeA	4.6	4.4	4.5	2.6
LargeB	4.8	4.4	4.5	2.5
LargeC	4.8	4.6	4.2	2.1
LargeD	5.0	4.6	4.2	1.9

(c) Average cost and time for each problem size

Problem size	SwapLS		SwapTS		InsertLS		InsertTS	
	Cost	Time (s)	Cost	Time (s)	Cost	Time (s)	Cost	Time (s)
50	59531.9	0.6	59520.5	7.4	59489.6	0.8	59371.4	19.0
75	141220.4	2.1	141212.5	19.8	141053.6	2.1	140791.4	64.6
100	277171.9	4.2	2777154.5	41.5	276914.2	4.2	276415.0	112.8

(d) Average rank for each problem size

Problem size	SwapLS	SwapTS	InsertLS	InsertTS
50	4.9	4.4	4.7	1.4
75	4.9	4.6	4.4	1.8
100	4.9	4.5	4.3	2.1

cost. Algorithms with lower values of average costs were given higher (numerically lower) ranks. The summary of average rank grouped by instance type and problem size is presented in Tables 21.4(b) and 21.4(d), respectively.

Across instance types and problem sizes, `SwapTS` and `InsertTS` output lower cost solutions on average than `SwapLS` and `InsertLS`, respectively, but on average took 9.9 times and 27.5 times more execution times, respectively. Similarly, `InsertLS` and `InsertTS` output lower cost solutions on average than their corresponding counterparts `SwapLS` and `SwapTS`. The execution times of `SwapLS` and `InsertLS` are comparable while `InsertTS` on average took 2.9 times more execution time than `SwapTS`. The improvement in solution costs was confirmed by conducting paired t-test between the outputs of the algorithms. The differences are statistically significant as the corresponding p-values for the paired t-tests are all less than 0.0005.

The performance difference between our insert algorithms and their swap counterparts is more pronounced in problem sizes where the number of tools is 75 or more. The same is also true when frequency matrices are sparse (Large(C), and Large(D) instances). The best performing algorithm of all is `InsertTS` as it ranks better on average than the other algorithms across all categories of instances.

21.6 Conclusion

This chapter addresses a version of the tool indexing problem in which only a single copy of a tool is allowed in the tool changer. This indexing problem is an important layout problem in automated machining centers of production facilities. It can be modeled as a quadratic assignment problem with a special distance matrix structure. The problem is difficult for instances of practical size and hence local search and tabu search algorithms using the swap and insert neighborhood structures are developed to provide good quality solutions for the problem.

Searching the swap and insert neighborhoods for this problem naïvely requires $\mathcal{O}(N^4)$ time when there are N tools in the tool changer. This is computationally expensive; and one of the contributions in this chapter in Sect. 21.3 is to devise bookkeeping techniques that bring down this time complexity to $\mathcal{O}(N^2)$ and make the searches feasible for practical size instances.

Section 21.4 describes local search and tabu search algorithms using these neighborhood structures. In Sect. 21.5, results from computational experiments using these algorithms are described. To the best of our knowledge, very few instances are available in the literature for this problem, so benchmark instances from related problems are adapted to run our computational experiments. Our results are compared with those presented in [12] and it is evident that our algorithms often produce better solutions, i.e., solutions with lower costs, in less time. As with other layout problems, e.g., the single row facility layout problem, algorithms using the insert neighborhood are found to generate better quality solutions than those using the swap neighborhood, although they require more execution time.

There are several directions in which the work presented here can be extended. The tabu search algorithm proposed by us makes use of elementary memory structures. For example, the tabu list used in the chapter is a short-term memory structure. There are other medium and long-term memory structures that can be used to better guide tabu search (see, e.g., [16]). Other mechanisms like the ejection chains (see, e.g., [15]) and multi-start tabu search (see, e.g., [18]) are used in many implementations to improve the performance of tabu search. These approaches may be experimented to obtain better performing algorithms for the problem.

Apart from tabu search, one can use other neighborhood search algorithms. Simulated annealing (see, e.g., [4, 5]), for example, does not exhaustively search the neighborhood of a solution, but searches neighbors randomly. Thus, such algorithms may not be able to make use of the speeding up that is described in Sect. 21.4.1 but are fast algorithms nonetheless. Variable neighborhood search (see [23]) is another neighborhood search algorithm that is promising for these problems. Apart from neighborhood search, one can also look at population heuristics such as genetic algorithms to solve this problem (see, e.g., [10, 11]). However, as with simulated annealing, such methods cannot use the techniques suggested here to speed up the search for better solutions.

The variant of tool indexing that has been addressed in this paper is one of many related problems. The most obvious generalization is one in which multiple copies of a tool can be present in the tool changer. In this variant, the cost of an arrangement of tools can be obtained only by solving a shortest path problem. This variant is thus more challenging than the variant addressed in this paper.

Another closely related problem is the tool switching problem. In this problem, the number of tools required to complete a job is more than the number of slots in the tool changer. Hence, in the tool switching problem, one needs to find out how tools need to be batched and how each batch has to be fitted into the tool changer so that the total non-machining time is minimized. The number of copies of a tool in the tool changer can be one or more, depending on the variant. It is conjectured that this problem is strongly NP-hard and thus is a prime candidate for using metaheuristics.

A. Appendices

A.1 Illustration of Swap Neighborhood Search Procedure

Using a small example, first, the entries of matrices A, B, and C are shown and then the process of populating matrix A described in Sect. 21.3.1 is illustrated.

Consider a problem with $N = 8$, (i.e., even valued) and a current solution $\Pi_c = \{1, 2, 3, 4, 5, 6, 7, 8\}$. The entries of first row of matrix $A = a(i, j)$ computed using Eq. (21.9) are given below.

$a(1,1) = 1f(1,2) + 2f(1,3) + 3f(1,4) + 4f(1,5) + 3f(1,6) + 2f(1,7) + 1f(1,8);$
$a(1,2) = 0f(1,2) + 1f(1,3) + 2f(1,4) + 3f(1,5) + 4f(1,6) + 3f(1,7) + 2f(1,8);$
$a(1,3) = 1f(1,2) + 0f(1,3) + 1f(1,4) + 2f(1,5) + 3f(1,6) + 4f(1,7) + 3f(1,8);$
$a(1,4) = 2f(1,2) + 1f(1,3) + 0f(1,4) + 1f(1,5) + 2f(1,6) + 3f(1,7) + 4f(1,8);$
$a(1,5) = 3f(1,2) + 2f(1,3) + 1f(1,4) + 0f(1,5) + 1f(1,6) + 2f(1,7) + 3f(1,8);$
$a(1,6) = 4f(1,2) + 3f(1,3) + 2f(1,4) + 1f(1,5) + 0f(1,6) + 1f(1,7) + 2f(1,8);$
$a(1,7) = 3f(1,2) + 4f(1,3) + 3f(1,4) + 2f(1,5) + 1f(1,6) + 0f(1,7) + 1f(1,8);$ and
$a(1,8) = 2f(1,2) + 3f(1,3) + 4f(1,4) + 3f(1,5) + 2f(1,6) + 1f(1,7) + 0f(1,8).$

It is clear that the computation of each element of the matrix A requires $\mathcal{O}(N)$ time. Defining $b(i, j)$ as per Eq. (21.11), following expressions are obtained for first row of matrix B for the problem.

$b(1,1) = a(1,2) - a(1,1) = -f(1,2) - f(1,3) - f(1,4) - f(1,5) + f(1,6) + f(1,7) + f(1,8);$
$b(1,2) = a(1,3) - a(1,2) = f(1,2) - f(1,3) - f(1,4) - f(1,5) - f(1,6) + f(1,7) + f(1,8);$
$b(1,3) = a(1,4) - a(1,3) = f(1,2) + f(1,3) - f(1,4) - f(1,5) - f(1,6) - f(1,7) + f(1,8);$
$b(1,4) = a(1,5) - a(1,4) = f(1,2) + f(1,3) + f(1,4) - f(1,5) - f(1,6) - f(1,7) - f(1,8);$
$b(1,5) = a(1,6) - a(1,5) = f(1,2) + f(1,3) + f(1,4) + f(1,5) - f(1,6) - f(1,7) - f(1,8);$
$b(1,6) = a(1,7) - a(1,6) = -f(1,2) + f(1,3) + f(1,4) + f(1,5) + f(1,6) - f(1,7) - f(1,8);$ and
$b(1,7) = a(1,8) - a(1,7) = -f(1,2) - f(1,3) + f(1,4) + f(1,5) + f(1,6) + f(1,7) - f(1,8).$

The coefficient of terms is either $+1$ or -1 in the expressions for $b(i, j)$, but it still requires linear time to compute an element of matrix B. Further defining matrix C, where entries $c(i, j)$ are as defined in Eq. (21.12), first row of matrix C for the problem are given below.

$c(1,1) = b(1,2) - b(1,1) = 2f(1,2) - 2f(1,6);$
$c(1,2) = b(1,3) - b(1,2) = 2f(1,3) - 2f(1,7);$
$c(1,3) = b(1,4) - b(1,3) = 2f(1,4) - 2f(1,8);$
$c(1,4) = b(1,5) - b(1,4) = 2f(1,5) - 2f(1,1);$
$c(1,5) = b(1,6) - b(1,5) = 2f(1,6) - 2f(1,2);$ and
$c(1,6) = b(1,7) - b(1,6) = 2f(1,7) - 2f(1,3).$

Note that the expression for each entry of matrix C has only two frequency terms and hence can be computed in constant time.

Now taking a problem with $N = 7$ (i.e., odd valued), and a current solution $\Pi_c = \{1, 2, \ldots, 7\}$, the entries of first rows of matrix A, B, and C computed similarly are given below.

$a(1,1) = 1f(1,2) + 2f(1,3) + 3f(1,4) + 3f(1,5) + 2f(1,6) + 1f(1,7);$
$a(1,2) = 0f(1,2) + 1f(1,3) + 2f(1,4) + 3f(1,5) + 3f(1,6) + 2f(1,7);$
$a(1,3) = 1f(1,2) + 0f(1,3) + 1f(1,4) + 2f(1,5) + 3f(1,6) + 3f(1,7);$
$a(1,4) = 2f(1,2) + 1f(1,3) + 0f(1,4) + 1f(1,5) + 2f(1,6) + 3f(1,7);$

$a(1,5) = 3f(1,2) + 2f(1,3) + 1f(1,4) + 0f(1,5) + 1f(1,6) + 2f(1,7);$
$a(1,6) = 3f(1,2) + 3f(1,3) + 2f(1,4) + 1f(1,5) + 0f(1,6) + 1f(1,7);$ and
$a(1,7) = 2f(1,2) + 3f(1,3) + 3f(1,4) + 2f(1,5) + 1f(1,6) + 0f(1,7).$

$b(1,1) = a(1,2) - a(1,1) = -f(1,2) - f(1,3) - f(1,4) + f(1,6) + f(1,7);$
$b(1,2) = a(1,3) - a(1,2) = f(1,2) - f(1,3) - f(1,4) - f(1,5) + f(1,7);$
$b(1,3) = a(1,4) - a(1,3) = f(1,2) + f(1,3) - f(1,4) - f(1,5) - f(1,6);$
$b(1,4) = a(1,5) - a(1,4) = f(1,2) + f(1,3) + f(1,4) - f(1,5) - f(1,6) - f(1,7);$
$b(1,5) = a(1,6) - a(1,5) = f(1,3) + f(1,4) + f(1,5) - f(1,6) - f(1,7);$ and
$b(1,6) = a(1,7) - a(1,6) = -f(1,2) + f(1,4) + f(1,5) + f(1,6) - f(1,7).$

$c(1,1) = b(1,2) - b(1,1) = 2f(1,2) - f(1,5) - f(1,6);$
$c(1,2) = b(1,3) - b(1,2) = 2f(1,3) - f(1,6) - f(1,7);$
$c(1,3) = b(1,4) - b(1,3) = 2f(1,4) - f(1,7) - f(1,1);$
$c(1,4) = b(1,5) - b(1,4) = 2f(1,5) - f(1,1) - f(1,2);$ and
$c(1,5) = b(1,6) - b(1,5) = 2f(1,6) - f(1,2) - f(1,3).$

The expression for each entry of matrix C has only three frequency terms which again makes it computable in constant time. To illustrate the procedure for computing entries of the matrix A using entries of matrix C, a concrete tool sequence for the example problem with $N = 8$ is considered; the F matrix for the example problem is taken as

$$F = \begin{bmatrix} f(1,1) & f(1,2) & f(1,3) & \dots & f(1,8) \\ f(2,1) & f(2,2) & f(2,3) & \dots & f(2,8) \\ \vdots & & \ddots & & \vdots \\ f(8,1) & f(8,2) & f(8,3) & \dots & f(8,8) \end{bmatrix} = \begin{bmatrix} 0&1&3&4&1&0&1&2 \\ 1&0&2&5&2&1&0&4 \\ 3&2&0&1&2&4&1&0 \\ 4&5&1&0&2&1&1&3 \\ 1&2&2&2&0&1&5&1 \\ 0&1&4&1&1&0&2&3 \\ 1&0&1&1&5&2&0&4 \\ 2&4&0&3&1&3&4&0 \end{bmatrix} \tag{21.23}$$

Using this example problem, each step of the algorithm is explained below. First, all entries of matrix C are computed using Eq. (21.14) as

$c(1,1) = 2f(1,2) - 2f(1,6) = 2 \times 1 - 2 \times 0 = 2;$
$c(1,2) = 2f(1,3) - 2f(1,7) = 2 \times 3 - 2 \times 1 = 4;$
$\vdots$
$c(1,5) = 2f(1,6) - 2f(1,2) = 2 \times 0 - 2 \times 1 = -2;$
$c(1,6) = 2f(1,7) - 2f(1,3) = 2 \times 1 - 2 \times 3 = -4;$
$\vdots$

Evaluating entries of the first row of matrix A is illustrated below (same procedure is applied to compute the elements of matrix A row by row). The first two elements in row 1 of matrix A are computed as

$$\begin{aligned} a(1,1) &= 1f(1,2) + 2f(1,3) + 3f(1,4) + 4f(1,5) + 3f(1,6) + 2f(1,7) + 1f(1,8) \\ &= 1 + 6 + 12 + 4 + 0 + 2 + 2 \\ &= 27; \text{ and} \\ a(1,2) &= 0f(1,2) + 1f(1,3) + 2f(1,4) + 3f(1,5) + 4f(1,6) + 3f(1,7) + 2f(1,8) \\ &= 0 + 3 + 8 + 3 + 04 + 3 + 4 \\ &= 21. \end{aligned}$$

Now $b(1,1)$ can be calculated as $b(1,1) = a(1,2) - a(1,1) - 21 - 27 = -6$. Once $c(1,1)$ and $b(1,1)$ are known, $b(1,2)$ can be obtained as $b(1,2) = b(1,1) + c(1,1) = -6 + 2 = -4$. The third element in the first row of matrix A can now be calculated using $b(1,2)$ as $a(1,3) = a(1,2) + b(1,2) = 21 + -4 = 17$. Again, $b(1,3) = b(1,2) + c(1,2) = -4 + 4 = 0$ and $a(1,4) = a(1,3) + b(1,3) = 17 + 0 = 17$. This way, all entries in the first row of matrix A can be computed. Matrix A obtained by following this procedure for each row is given below.

$$A = \begin{bmatrix} a(1,1) & a(1,2) & a(1,3) & \ldots & a(1,8) \\ a(2,1) & a(2,2) & a(2,3) & \ldots & a(2,8) \\ \vdots & & \ddots & & \vdots \\ a(8,1) & a(8,2) & a(8,3) & \ldots & a(8,8) \end{bmatrix} = \begin{bmatrix} 27 & 21 & 17 & 17 & 21 & 27 & 31 & 31 \\ 34 & 31 & 26 & 25 & 26 & 29 & 34 & 35 \\ 27 & 30 & 29 & 26 & 25 & 22 & 23 & 26 \\ 23 & 24 & 33 & 42 & 45 & 44 & 35 & 26 \\ 26 & 28 & 32 & 30 & 30 & 28 & 24 & 26 \\ 23 & 21 & 21 & 25 & 25 & 27 & 27 & 23 \\ 35 & 35 & 31 & 29 & 21 & 21 & 25 & 27 \\ 34 & 35 & 38 & 33 & 34 & 33 & 30 & 35 \end{bmatrix}$$

The cost of any swap neighbor of Π_c can now be computed using Eq. 21.10. For example, cost of the swap neighbor Π_{12} is

$$\begin{aligned} z(\Pi_{12}) &= z(\Pi_c) + a(1,2) + a(2,1) - a(1,1) - a(2,2) + 2f(1,2)d(1,2) \\ &= z(\Pi_c) + 21 + 34 - 27 - 31 + 2 \\ &= z(\Pi_c) - 1. \end{aligned}$$

The solution cost of Π_c evaluated using the cost function $z(.)$ is 123. So, $z(\Pi_{12}) = 123 - 1 = 122$.

21.6.1 A.2 Illustration of Insert Neighborhood Search Procedure

The insert neighborhood search procedure is illustrated by taking the same example problem as used in Sect. 21.6 i.e., $N = 8$, current solution $\Pi_c = \{1, 2, \ldots, 8\}$ and frequency matrix F as given in Eq. 21.23. The cost of solution Π_c evaluated using the cost function $z(.)$ is, $z(\Pi_c) = 123$. The entries of matrix S are first computed using Eq. (21.20) as shown below.

$$s(1,1) = f(1,1) = 0;$$
$$s(1,2) = f(1,1) + f(1,2) = 1;$$
$$s(1,3) = f(1,1) + f(1,2) + f(1,3) = 4;$$
$$s(1,4) = f(1,1) + f(1,2) + f(1,3) + f(1,4) = 8;$$
$$\vdots$$
$$s(1,8) = f(1,1) + f(1,2) + \cdots + f(1,8) = 12;$$
$$\vdots$$

In matrix form,

$$S = \begin{bmatrix} s(1,1) & s(1,2) & s(1,3) & \ldots & s(1,8) \\ s(2,1) & s(2,2) & s(2,3) & \ldots & s(2,8) \\ \vdots & & & \ddots & \vdots \\ s(8,1) & s(8,2) & s(8,3) & \ldots & s(8,8) \end{bmatrix} = \begin{bmatrix} 0 & 1 & 4 & 8 & 9 & 9 & 10 & 12 \\ 1 & 1 & 3 & 8 & 10 & 11 & 11 & 15 \\ 3 & 5 & 5 & 6 & 8 & 12 & 13 & 13 \\ 4 & 9 & 10 & 10 & 12 & 13 & 14 & 17 \\ 1 & 3 & 5 & 7 & 7 & 8 & 13 & 14 \\ 0 & 1 & 5 & 6 & 7 & 7 & 9 & 12 \\ 1 & 1 & 2 & 3 & 8 & 10 & 10 & 14 \\ 2 & 6 & 6 & 9 & 10 & 13 & 17 & 17 \end{bmatrix}. \quad (21.24)$$

Consider insert move Π^{12} where tool 2 is inserted between tool 2 and tool 3 in Π_c. The incremental cost of this move, say $\Delta(1, 2)$, can be computed using the entries of matrix S as follows. Since the positions of only tool 1 and tool 2 change while the rest of the tools are in the same positions before and after the move, say $\Delta(1, 2)$, is the sum of incremental cost due to the change in position of tool 1 and that of tool 2. Before the insert operation, the distance of tool 1 to each of the tools from 1 to 8 is 0, 1, 2, 3, 4, 3, 2, 1 respectively, while that of tool 2 to each of these tools is 1, 0, 1, 2, 3, 4, 3, 2. After the insert operation, the distance of tool 1 to these tools changes to 0, 1, 1, 2, 3, 4, 3, 2, respectively, while that of tool 2 to these tools changes to 1, 0, 2, 3, 4, 3, 2, 1, respectively. Thus,

$$\begin{aligned}\Delta(1,2) = &(f_{12} + f_{13} + 2f_{14} + 3f_{15} + 4f_{16} + 3f_{17} + 2f_{18}) - \\ &(f_{12} + 2f_{13} + 3f_{14} + 4f_{15} + 3f_{16} + 2f_{17} + f_{18}) + \\ &(f_{21} + 2f_{23} + 3f_{24} + 4f_{25} + 3f_{26} + 2f_{27} + f_{28}) - \\ &(f_{21} + f_{23} + 2f_{24} + +3f_{25} + 4f_{56} + 3f_{27} + 2f_{28}) \\ = &(f_{16} + f_{17} + f_{18}) - (f_{13} + f_{14} + f_{15}) + (f_{23} + f_{24} + f_{25}) - (f_{26} + f_{27} + f_{28}) \\ = &(s(1,8) - s(1,5)) - (s(1,5) - s(1,2)) - (s(2,8) - s(2,5)) - (s(2,5) - s(2,2)).\end{aligned} \tag{21.25}$$

Note that the same expression is obtained using Eq. (21.21) by substituting $N = 8$ and $K = N/2 = 4$. Thus,

$$\begin{aligned}z(\Pi^{12}) - z(\Pi_c) &= \{s(1,8) - s(1,5)\} - \{s(1,5) - s(1,2)\} \\ &\quad + \{s(2,5) - s(2,2)\} - \{s(2,8) - s(2,5)\} \\ &= (12 - 9) - (9 - 1) + (10 - 1) - (15 - 10) \\ &= -1.\end{aligned} \tag{21.26}$$

So, $z(\Pi^{12}) = z(\Pi_c) + \Delta(1,2) = 123 - 1 = 122$. Now using Eq. (21.22), the difference of $z(\Pi^{1,2})$ from $z(\Pi^{13})$ can be computed as

$$\begin{aligned}z(\Pi^{13}) - z(\Pi^{12}) = &\{s(1,8) - s(1,6)\} + \{s(1,2) - s(1,1)\} \\ &- \{s(1,6) - s(1,3)\} + \{s(3,6) - s(3,3)\} \\ &- \{s(3,8) - s(3,6)\} - \{s(3,2) - s(3,1)\} \\ = &(12 - 9) + (1 - 0) - (9 - 4) + (12 - 5) \\ &- (15 - 12) - (5 - 3) \\ = &1.\end{aligned} \tag{21.27}$$

So, $z(\Pi^{13}) = 122 + 1 = 123$. Similarly,

$$\begin{aligned}z(\Pi^{14}) - z(\Pi^{13}) = &\{s(1,8) - s(1,7)\} + \{s(1,3) - s(1,1)\} \\ &- \{s(1,7) - s(1,4)\} + \{s(4,7) - s(4,4)\} \\ &- \{s(4,8) - s(4,7)\} - \{s(4,3) - s(4,1)\} \\ = &(12 - 10) + (3 - 1) - (10 - 8) + (12 - 5) \\ &- (17 - 14) - (10 - 4) \\ = &0.\end{aligned} \tag{21.28}$$

Hence, $z(\Pi^{14}) = 123 + 0 = 123$. By proceeding in this manner, all insert neighbors of the solution Π_c can be computed, and the computation of the solution cost of each neighbor takes only constant time.

References

1. Anjos MF, Yen G (2009) Provably near-optimal solutions for very large single-row facility layout problems. Optim Methods Softw 24(4–5):805–817. https://doi.org/10.1080/10556780902917735
2. Anjos MF, Kennings A, Vannelli A (2005) A semidefinite optimization approach for the single-row layout problem with unequal dimensions. Discret Optim 2(2):113–122. https://doi.org/10.1016/j.disopt.2005.03.001
3. Atta S, Sinha Mahapatra PR, Mukhopadhyay A (2019) Solving tool indexing problem using harmony search algorithm with harmony refinement. Soft Comput 23(16):7407–7423. https://doi.org/10.1007/s00500-018-3385-5
4. Baykasoğlu A, Dereli T (2004) Heuristic optimization system for the determination of index positions on CNC magazines with the consideration of cutting tool duplications. Int J Prod Res 42(7):1281–1303. https://doi.org/10.1080/00207540310001622557
5. Baykasoğlu A, Ozsoydan FB (2016) An improved approach for determination of index positions on CNC magazines with cutting tool duplications by integrating shortest path algorithm. Int J Prod Res 54(3):742–760. https://doi.org/10.1080/00207543.2015.1055351
6. Baykasoğlu A, Ozsoydan FB (2017) Minimizing tool switching and indexing times with tool duplications in automatic machines. Int J Adv Manuf Technol 89(5–8):1775–1789. https://doi.org/10.1007/s00170-016-9194-z
7. Burkard RE, Çela E, Pardalos PM et al (1998) The quadratic assignment problem. In: Handbook of combinatorial optimization. Springer US, Boston, MA, pp 1713–1809. https://doi.org/10.1007/978-1-4613-0303-9_27
8. Calmels D (2019) The job sequencing and tool switching problem: State-of-the-art literature review, classification, and trends. https://doi.org/10.1080/00207543.2018.1505057
9. Dereli T, Filiz İ (2000) Allocating optimal index positions on tool magazines using genetic algorithms. Robot Auton Syst 33(2–3):155–167. https://doi.org/10.1016/S0921-8890(00)00086-5
10. Dereli T, Baykasoğlu A, Gindy NN et al (1998) Determination of optimal turret-index positions of cutting tools by using genetic algorithms. In: Proceedings of the second international symposium on intelligent manufacturing systems, pp 743–750
11. Ghosh D (2016a) A new genetic algorithm for the tool indexing problem. IIMA Working Papers WP2016-03-17, Indian Institute of Management Ahmedabad, Research and Publication Department
12. Ghosh D (2016b) Allocating tools to index positions in tool magazines using tabu search. IIMA Working Papers WP2016-02-06, Indian Institute of Management Ahmedabad, Research and Publication Department
13. Glover F (1989) Tabu search-Part I. ORSA J Comput 1(3):190–206. https://doi.org/10.1287/ijoc.1.3.190
14. Glover F (1990) Tabu search-Part II. ORSA J Comput 2(1):4–32. https://doi.org/10.1287/ijoc.2.1.4
15. Glover F (1992) New ejection chain and alternating path methods for traveling salesman problems. In: Computer science and operations research. Pergamon, Amsterdam, pp 491–509. https://doi.org/10.1016/B978-0-08-040806-4.50037-X
16. Glover F, Laguna M (1993) Tabu search. In: Modern heuristic techniques for combinatorial problems. Wiley, USA, pp 70–150. https://doi.org/10.5555/166648.166655
17. Gray AE, Seidmann A, Stecke KE (1993) A synthesis of decision models for tool management in automated manufacturing. Manag Sci 39(5):549–567. https://doi.org/10.1287/mnsc.39.5.549
18. James T, Rego C, Glover F (2009) Multistart tabu search and diversification strategies for the quadratic assignment problem. IEEE Trans Syst Man Cybern—Part A: Syst Humans 39(3):579–596. https://doi.org/10.1109/TSMCA.2009.2014556

19. Kothari R, Ghosh D (2013) Tabu search for the single row facility layout problem using exhaustive 2-opt and insertion neighborhoods. Eur J Oper Res 224(1):93–100. https://doi.org/10.1016/j.ejor.2012.07.037
20. Levitin G, Rubinovitz J (1993) Genetic algorithm for linear and cyclic assignment problem. Comput Oper Res 20(6):575–586. https://doi.org/10.1016/0305-0548(93)90113-W
21. Loiola EM, de Abreu NMM, Boaventura-Netto PO et al (2007) A survey for the quadratic assignment problem. Eur J Oper Res 176(2):657–690. https://doi.org/10.1016/j.ejor.2005.09.032
22. Martı R, Reinelt G (2011) The linear ordering problem: Exact and heuristic methods in combinatorial optimization. Appl Math Sci 175
23. Mladenović N, Hansen P (1997) Variable neighborhood search. Comput Oper Res 24(11):1097–1100. https://doi.org/10.1016/S0305-0548(97)00031-2
24. PassMark CPU Benchmarks (2022) High mid range CPUs. https://www.cpubenchmark.net/mid_range_cpus.html. Accessed 15 Jan 2022
25. Taillard ÉD (1990) Some efficient heuristic methods for the flow shop sequencing problem. Eur J Oper Res 47(1):65–74. https://doi.org/10.1016/0377-2217(90)90090-X
26. Taillard ÉD (1991) Robust taboo search for the quadratic assignment problem. Parallel Comput 17(4–5):443–455. https://doi.org/10.1016/S0167-8191(05)80147-4
27. Wang J, Xi XC, Qin L et al (2021) Non-productive time optimization for 5-axis EDM drilling using HVNTS algorithm. Int J Prod Res 59(16):5068–5082. https://doi.org/10.1080/00207543.2020.1779961
28. Wang S, Zou H, Wang S (2022) A Tabu-GA-based parallel machine scheduling with restrained tool resources. Proc Inst Mech Eng Part B: J Eng Manuf 236(1–2):39–50. https://doi.org/10.1177/0954405420928691

Chapter 22
Solving a Large-Scale Multi-Depot Vehicle Routing Problem Heuristically

Buşra Baytur, Eren Özceylan, Çağrı Koç, and Güneş Erdoğan

Abstract This chapter focuses on the distribution plan of a large-scale distributor of care and cleaning products to its customers located in the eastern and south eastern regions of Turkey. The distribution network consists of three depots and 502 customers. The vehicle fleet consists of homogeneous vehicles. The problem is to determine which depot should serve which customers including the routing decisions, which is an instance of the well-known Multi-Depot Vehicle Routing Problem (MDVRP). The authors use a cluster-first, route-second approach to solve the model. To do so, we first use the capacitated *p*-median formulation for clustering and assignment of customers to each depot. Next, we use a single-depot VRP to solve the routing problem for each depot and its cluster of customers. For this, a Guided Local Search metaheuristic is implemented and Google-OR-Tool is utilized as a solver. Real data of the company including demands, vehicle capacities, exact coordinates of depots and customers is utilized. Detailed computational experiments and their results are presented.

Keywords Capacitated *p*-median; vehicle routing · Multi-depot · Clustering · Meta-heuristic

---

Source Code: https://github.com/BusraBaytur/OR_CVRP.git.

B. Baytur · E. Özceylan
Department of Industrial Engineering, Gaziantep University, Gaziantep, Türkiye
e-mail: b.baytur@hotmail.com

E. Özceylan
e-mail: eozceylan@gantep.edu.tr

Ç. Koç
Department of Business Administration, Social Sciences University of Ankara, Ankara, Türkiye
e-mail: cagri.koc@asbu.edu.tr

G. Erdoğan (✉)
School of Management, University of Bath, Bath, United Kingdom
e-mail: ge277@bath.ac.uk

F. Hamid (ed.), *Optimization Essentials*, International Series in Operations Research & Management Science 353, https://doi.org/10.1007/978-981-99-5491-9_22

22.1 Introduction

The Fast-Moving Consumer Goods (FMCG) sector is one of the critical sectors where logistics activities are of great importance. The main aim of FMCG companies is to carry out sales and distribution activities together to deliver the product groups in their portfolio to the distribution network on a point basis. FMCG companies try to speed up their distribution processes and reduce related costs while improving their customer service [32].

Distribution planning requires the evaluation of multiple alternative routes known as the Vehicle Routing Problem (VRP), which is a complex NP-hard problem [9]. The classical VRP aims to determine an optimal routing plan for a fleet of homogeneous vehicles to serve a set of customers, such that each vehicle route starts and ends at the depot, each customer is visited once by one vehicle, and some side constraints are satisfied. Many variants and extensions of the VRP have been intensively studied in the scientific literature [3, 19, 35]. The Multi-Depot Vehicle Routing Problem (MDVRP), one of the variants of the VRP, aims to satisfy customer demands from more than one depot located in different places in a scattered area [23]. The FMCG industry inherently involves the MDVRP. The delivery teams often operate out of multiple depots to distribute orders to the customers.

We study the distribution problem of a Turkish FMCG company, specializing in personal care and cleaning products that consist of 502 customer locations and three depots. To solve this complex optimization problem effectively, we divide the problem into two parts and create single-depot sub-problems. We first use Capacitated *p*-Median Problem (CPMP) to assign customers to the depots. We then create distribution plan of each depot independently by solving the standard single-depot VRP.

The rest of this chapter is structured as follows. Section 22.2 provides a brief literature review on the MDVRP. Section 22.3 details the CPMP and Google OR-Tools methodologies and gives the mathematical model for the Capacitated VRP (CVRP). Section 22.4 presents the problem data and the computational results of the case study. Finally, Sect. 22.5 summarizes the results obtained.

22.2 A Brief Literature Review

The MDVRP was introduced as the single and multiple terminal delivery problem by authors in [34], who used a modified Clarke and Wright savings algorithm to find solutions. Several papers in the literature solved the MDVRP by using the partitioning method [38]. Raft [28] solved the MDVRP with a 3-step heuristic algorithm. In the first step, routes to meet all demands are estimated and centers are assigned to these routes. In the second step, the assigned points are placed in the nearest depot. In the last step, routes for each depot are generated using M-TOUR heuristic of [31] and refined with 3-optimal exchange method. Authors in [2] studied MDVRP for the case of randomly generated customer locations in the plane and unlimited vehicle capacity. In their paper, they showed that a heuristic that first clusters customers around the nearest depot and then performs the Traveling Salesman Problem (TSP) routing finds the most suitable tour with near optimality.

Li and Simchi-Levi [21] performed a worst-case analysis of tour partitioning heuristics for the MDVRP. Cordeau et al. [6] proposed a tabu search heuristic to solve the Periodic VRP, the Periodic TSP and the MDVRP. The authors obtained promising results for all three problems. Pisinger and Ropke [26] proposed an Adaptive Large Neighborhood Search heuristic that can be used for large-scale optimization problems of various types of VRPs, including MDVRP. Giosa et al. [11] addressed a MDVRP with a time windows variant. The authors designed six different heuristics to assign customers to depots and compared their performance. Crevier et al. [7] studied the MDVRP with Inter-Depot Routes (MDVRPI) and proposed a three-step methodology based on integer programming. Mirabi et al. [22] presented three hybrid heuristics for the MDVRP that include both constructive heuristics and improvement techniques to minimize delivery time. Vidal et al. [39] developed an effective and very flexible hybrid genetic algorithm with local search procedures to solve three VRP variants: MDVRP, Periodic VRP, and Multi-Depot Periodic VRP. Reiners [29] proposed a constraint programming-based genetic algorithm to solve the MDVRP with time windows. Rabbani et al. [27] considered a MDVRP with time windows and fuzzy requests and proposed several algorithms.

There are also several studies that used the Ant Colony Optimization (ACO) algorithm, which is a probability-based metaheuristic to solve the MDVRP. Zhen and Zhang [41] proposed a hybrid meta-heuristic algorithm to solve the MDVRP with time windows. To improve the performance of the hybrid ant colony algorithm, they developed a local search improvement algorithm that explores a large neighborhood of the current solution and proposed a two-step solution methodology. Stodola [33] developed a hybrid ACO algorithm which combines both probabilistic and exact techniques. The results revealed that the algorithm is suitable for the MDVRP and outperforms state-of-the-art methods.

Increasing environmental sensitivity has also been reflected in research in the field of MDVRP. Fan et al. [10] introduced the Multi-Depot Green Vehicle Routing Problem and proposed a hybrid genetic algorithm to reduce the costs of companies in the FMCG sector caused by long-distance distribution and to provide sustainable transportation by addressing their fuel consumption.

We summarize the literature on the MDVRP in Table 22.1. The acronyms used and their meanings are provided again within the table to simplify the presentation.

Table 22.1 Summary of the MDVRP literature

Author	Problem type	Method	Application area
Tillman and Cain [34]	MDVRP	Clarke and wright savings algorithm	
Raft [28]	MDVRP	Russell' M-TOUR and 3-optimal exchange method	
Li and Simchi-Levi [21]	MDVRP	Tour-partitioning heuristics	Designing telecommunication networks
Cordeau et al. [6]	MDVRP, PVRP and PTSP	Tabu search heuristic	
Giosa et al. [11]	MDVRPTW	Clustering and assignment heuristics	Dairy industry
Baltz et al. [2]	MDVRP	Probabilistic analysis	
Crevier et al. [7]	MDVRPI	Adaptive memory, tabu search and integer programming	
Pisinger and Ropke [26]	VRPTW, CVRP, MDVRP, SDVRP, OVRP	An adaptive large neighborhood search	
Zhen and Zhang [41]	MDVRPTW	Hybrid ant colony algorithm	
Mirabi et al. [22]	MDVRP	Deterministic, stochastic and simulated annealing	
Vidal et al. [39]	MDVRP, PVRP and MDPVRP	Hybrid genetic algorithm	
Reiners [29]	MDVRPTW	Constraint programming and genetic algorithm	
Rabbani et al. [27]	MDVRPTW	Genetic algorithm and particle swarm optimization	Home health care services
Stodola [33]	MDVRP	Hybrid ant colony optimization	
Fan et al. [10]	MDGVRP	Hybrid genetic algorithm	FMCG and cargo transportation
This study	MDVRP	CPMP and guided local search	FMCG industry

22.3 Methodology

The MDVRP is known to be difficult to solve to optimality [30]. Only very small-scale instances with a few customers can be solved exactly by commercial solvers such as CPLEX and GUROBI. To tackle this problem effectively, we use a cluster-first, route-second method. In our problem, we have three depots and firstly CPMP is solved to perform customer assignments to the depots. In this way, three sub-problems are obtained. We then acquire routes for each depot using the CVRP by applying the Guided Local Search algorithm. In this section, we first provide the mathematical model of the CPMP. We then present the mathematical model for the CVRP, before we provide details about Google OR-Tools.

22.3.1 Capacitated P-Median Problem

CPMP is the problem of locating p facilities on the network consisting of n nodes, considering the facility capacities and customer demands, and determining the demand points that will receive service from these placed facilities with minimum total distance. The problem has various applications in the field of location-allocation problems [1, 5]. Since the p-median problem is computationally difficult to solve with exact methods, similar complexity applies to its capacity-constrained version, CPMP [16]. Let $G = (E, A)$ be a graph in which a direct edge exists between all nodes. E is set of the nodes and A is set of the edges defined between these nodes $(A = \{(i, j)|i \neq j \ and \ i, j \in E\})$. Set E consists of the combination of N (set of the customers) and K (set of the potential facility points) clusters $(E = N \cup K)$. The notation and the mathematical model of [24] for the CPMP are as follows:

Parameters

w_i :	Demand of customer i
d_{ik} :	Distance between locations i and k
p :	Number of facilities to be located
c_k :	Capacity of facility (potential depot) k

Decision Variables

$$x_{ik} = \begin{cases} 1, \text{ if customer } i \text{ is assigned to facility } k \\ 0, \text{ otherwise} \end{cases}$$

$$y_k = \begin{cases} 1, \text{ if a facility is opened at location } k \\ 0, \text{ otherwise} \end{cases}$$

Model

$$\text{Min} \sum_{i \in I} \sum_{k \in K} w_i d_{ik} x_{ik} \tag{22.1}$$

$$\sum_{k \in K} x_{ik} = 1 \qquad \forall i \in N \tag{22.2}$$

$$x_{ik} \leq y_k \qquad \forall i \in N,\ k \in K \tag{22.3}$$

$$\sum_{k \in K} y_k = p \tag{22.4}$$

$$\sum_{i \in I} w_i x_{ik} \leq c_k y_k \qquad \forall k \in K \tag{22.5}$$

$$x_{ik}, y_k \in \{0, 1\} \qquad \forall i \in N, \forall k \in K \tag{22.6}$$

The objective function (22.1) aims to minimize the total demand-weighted distance between the facilities and the customers. Constraints (22.2) ensure that all customer demands are met from a single facility. Constraints (22.3) guarantee that no demand point is assigned to the facility that is not opened. Constraints (22.4) allow p facilities to be opened. Constraints (22.5) ensure that the sum of the demands cannot exceed the capacity of the facility. Constraints (22.6) define the domain of the decision variables.

22.3.2 The Capacitated VRP Mathematical Model

In this section, we present the mathematical model of CVRP [20]. Let $G = (V, A)$ be a graph that delimits the solution region with its edges. V is the set of all nodes including the depot v_0. $V_C = \{v_1, v_2, ..., v_n\}$ is the customer set, which consists of n customers. A is the set of arcs between each pair of nodes denoted by $A = \{(v_i, v_j) : v_i, v_j \in V, i \neq j\}$. Distance between nodes i and j is indicated by d_{ij}. Each customer has a nonnegative demand w_i. There is a homogeneous fleet with total of L vehicles, each one with a capacity Q.

In the CVRP, each vehicle route starts and ends at the depot, each customer is serviced exactly once by a vehicle and the total demand of each route does not exceed the vehicle capacity. CVRP aims to minimize the total distribution cost.

The decision variable x_{ij} used in the model is equal to 1 if the pair of nodes i and j are visited consecutively in the route of the vehicle, and 0 otherwise. The temporary variable u_i is used for sub-tour elimination and capacity constraints.

The mathematical model for the CVRP is then:

$$\text{Minimize} \sum_{i \in V} \sum_{j \in V} x_{ij} d_{ij}\ ; i \neq j \tag{22.7}$$

subject to:

$$\sum_{i \in V} x_{ij} = 1 \qquad \forall j \in V_c; i \neq j \tag{22.8}$$

$$\sum_{j \in V} x_{ij} = 1 \qquad \forall i \in V_c; i \neq j \tag{22.9}$$

$$\sum_{j \in V_c} x_{0j} \leq L \tag{22.10}$$

$$\sum_{i \in V_c} x_{i0} = \sum_{j \in V_c} x_{0j} \tag{22.11}$$

$$u_i \geq u_j + w_i - Q(1 - x_{ij}) \qquad \forall i, j \in V_c, i \neq j \tag{22.12}$$

$$w_i \leq u_i \leq Q \qquad \forall i \in V_c \tag{22.13}$$

$$x_{ij} \in \{0, 1\} \qquad \forall i, j \in V \tag{22.14}$$

Objective function (22.7) minimizes the total distribution costs. Constraints (22.8) and (22.9) ensure that a customer is visited exactly once. Constraint (22.10) limit the number of vehicles leaving the depot to L. Constraint (22.11) equalizes the number of vehicles (number of routes) entering and exiting the depot. Constraints (22.12) are the flow constraints to make certain that the demand of each customer is met by taking into account the vehicle capacities. Constraints (22.13) guarantee that load on the vehicle at node is between the demand of node i and the vehicle capacity (Q). Constraints (22.14) indicate that x_{ij} are binary variables.

22.3.3 Google-OR Tools

Google OR-Tools [13] is an open-source software package that is developed for solving difficult combinatorial optimization problems. Originally created by [25], this package includes solvers for:

- Linear programming,
- Mixed integer programming,
- Constraint programming,
- Bin packing and knapsack algorithms,
- TSP,
- VRP and variants,
- Graph algorithms (shortest paths, min cost flows, max flows, linear sum assignment),
- Scheduling.

Google OR-Tools can be used with C++, Python, Java or C# programming languages according to users' preference and it is distributed under the Apache License 2.0. In this study, we have used the library named VrpCapacity developed for the CVRP and written in C# [12, 14]. Although various solution methods have been proposed for VRPs until today, we use heuristic algorithms in this study since problems with a maximum of 135 demand points can be solved with algorithms that give optimal results [40]. Standard VrpCapacity library of Google OR-Tools uses the Path Cheapest Arc method to create an initial solution and the Guided Local Search (GLS) metaheuristic algorithm to produce better results [37]. Path Cheapest Arc method (also known as Nearest Neighbor Heuristic) aims to continually travel to the nearest unvisited point, beginning from the starting point. If the demand of the nearest point exceeds the vehicle capacity, that point is bypassed and the next point is targeted. Tours are created until the vehicle capacity is full or all demands are met [36]. After the initial solution is generated, the GLS algorithm is activated. GLS is a meta-heuristic algorithm that works by penalizing searches that revisit previously visited local minimums, by adding the penalty factor to the objective function [17].

The flow chart of the general methodology used in the solution of this study is as shown in Fig. 22.1.

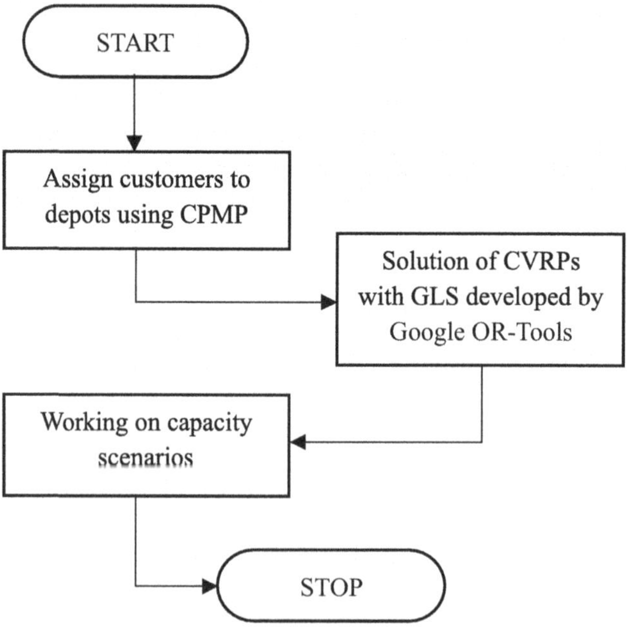

Fig. 22.1 Flow chart of the methodology

22.4 Case Study

This section provides the details of the application and presents the computational results.

22.4.1 Problem Description

We now attempt to solve the MDVRP for a distributor of self-care and cleaning products to its customers located in the eastern and southeastern regions of Turkey. The company operates sales and distribution activities of products in the categories of home and personal care in 18 provinces. It has more than 2,500 product items in its portfolio. It sells and distributes to wholesalers, sub-distributors, supermarkets, and local stores from its three depots located in Erzurum, Van and Mardin (Headquarter).

A shipment plan is needed to ensure customer satisfaction and to reduce costs. In this context, the real coordinates of 502 local store customers, which the company serves every month and distributes to ensure the availability of products, were collected by the sales team during customer visits. Customers with large turnover, whose orders are filled by one full truck or vehicle, are excluded. We have used Google Maps to find the distances between the customers and depots. As a result, a 505 $\times$ 505 distance matrix was obtained. The distance information between the nodes (d_{ij}) is in kilometers. Figure 22.2 shows the map of the region of interest, where markers in large red pins correspond to the depot locations, whereas small yellow pins depict the customer locations.

The average pallet demands were calculated based on 6-month sales data of the customers. The total demand of the customers addressed in the problem is 1,415 pallets. Demands of customers are given in Table A22.1 of the Appendix. Depot capacities are given in Table 22.2 in terms of pallets. All vehicles have a capacity Q of 10 pallets. Depending on the delivery demand, extra vehicles may be outsourced when necessary.

22.4.2 Solving the CPMP and CVRP

Experiments were conducted on a server with Intel Core i5 processor at 3.6 GHz processor. The CPMP model (Sect. 22.3.1) was coded in the GAMS Studio Version 34 commercial package program and solved using branch-and-bound. The optimal solution was realized in 16.703 s, showing that 237 customers were assigned to the Mardin depot, 138 customers to the Erzurum depot and 127 customers to the Van depot. The total demands assigned are 643 pallets to Mardin depot, 441 pallets to Erzurum depot and 331 pallets to Van depot. These assignments are presented in Table A22.2 of the Appendix. The graph of customer depot assignments is given in Fig. 22.3.

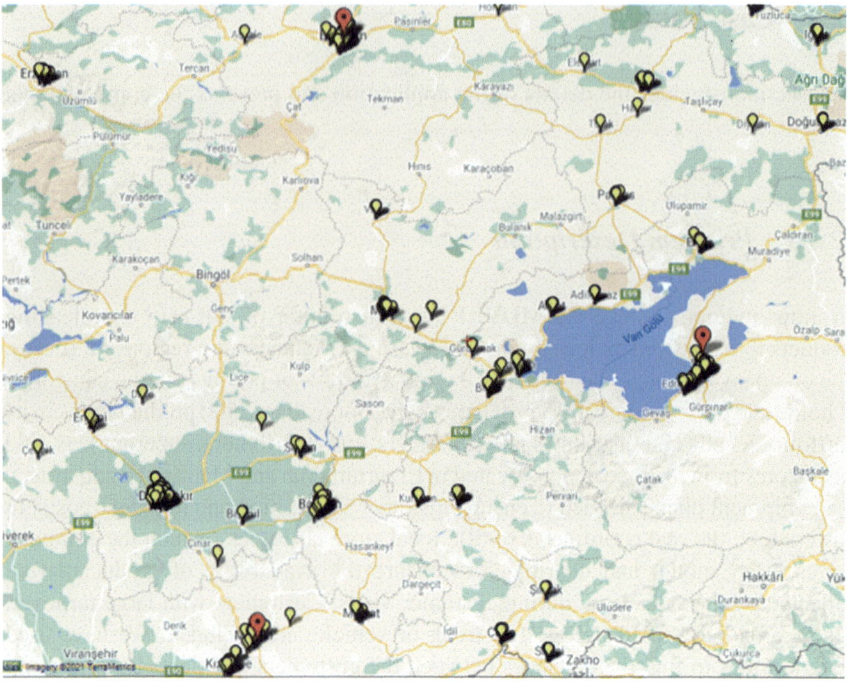

Fig. 22.2 Distribution of nodes in the area

Table 22.2 Depot capacities

Depot	Capacity (pallet)
Mardin	3,750
Erzurum	1,900
Van	1,500

In the next step, we have solved the CVRP for each sub-problem. While solving the CVRP for each depot, data entry was made in the code block of the distance information between the nodes, the vehicle capacity and the number of vehicles. An initial solution is created with Path Cheapest Arc and better solutions are explored with GLS. The problem is a large-scale real-life case. During the interviews with the managers, it was decided that it required a real-time solution and the working time for the developed software was determined as 240 s. The program was run 30 times for each depot. It was observed that there was no significant improvement in the objective function when run after these times. The graph of the observed values is shown in Fig. 22.4. The lowest objective function value calculated for Mardin is 21,686 km, for Erzurum is 9,964 km and for Van is 7,349 km, only those that differ from these values are also shown in the graph.

The shortest and longest total route distances found are given in Table 22.3. The best solution is obtained as 38,999 km which consists of a total of 144 routes (65 for

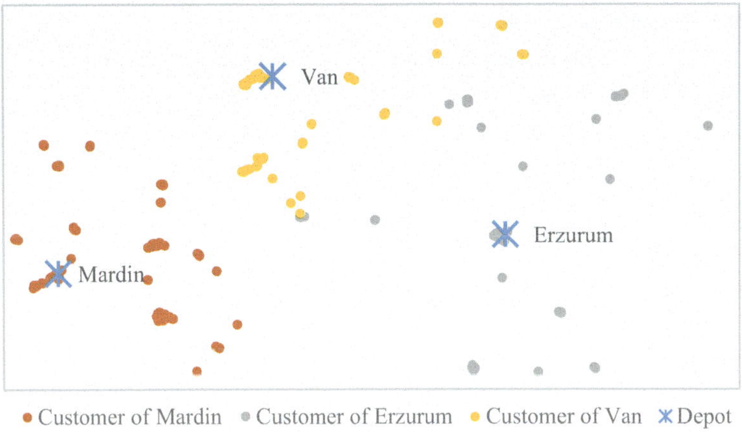

Fig. 22.3 Customer assignments scatterplot

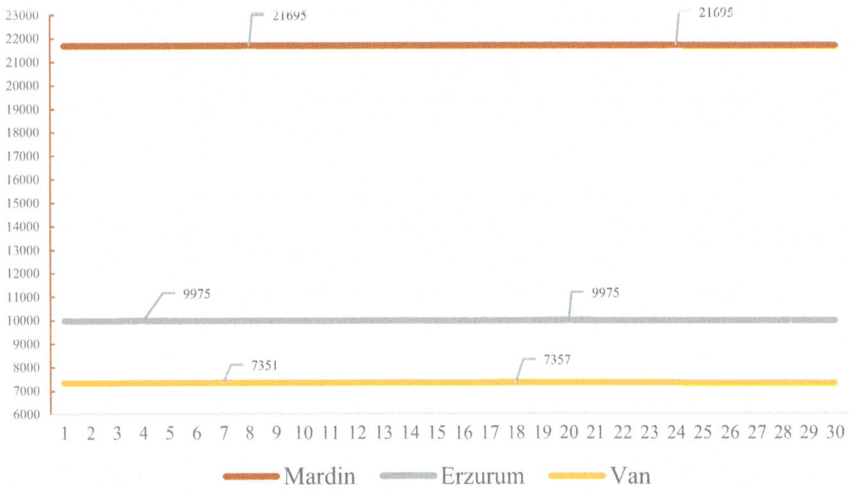

Fig. 22.4 The variation of objective values in 30 runs

Mardin depot, 45 for Erzurum depot and 34 for Van depot). Tables A22.3, A22.4 and A22.5 of the Appendix show the results obtained by the GLS algorithm. The average distance of 65 tours obtained for Mardin depot is 334 km, the average distance of the 45 tours obtained for Erzurum depot is 221 km, and the average distance of the 34 tours obtained for Van depot is 216 km.

In addition to the above results, we now generate a scenario looking at changes in vehicle capacities and their effect on the performance measures. For this purpose, we gradually increase the current capacity of the vehicles ($Q = 10$ pallets) in increments of 10% up to 50%, keeping other parameters constant. Each scenario is run for 30

Table 22.3 Objective values of depots

Depot	Shortest distance (km)	Longest distance (km)
Mardin	21,686	21,695
Erzurum	9,964	9,975
Van	7,349	7,357
Total	38,999	39,027

Table 22.4 Number of routes according to the changing capacity

Depot	Pallet capacity (Q)					
	10	11	12	13	14	15
Mardin	65	59	54	50	46	43
Erzurum	45	41	37	34	32	30
Van	34	31	28	26	24	23
Total	144	131	119	110	102	96

times with 240 s as the time limit. The best overall route distances and the route numbers of the depots are recorded. Table 22.4 and Fig. 22.5 illustrate the results of these experiments. Reduction between 9.02% and 33.3% was observed in the number of routes.

The computational results indicate that the increase in vehicle capacity creates large differences in the total route length. Furthermore, there is a significant decrease in the number of routes obtained. It can be said that increasing the vehicle capacity will create observable reductions in logistics expenses which is one of the largest cost items of the company. Furthermore, this situation will also create a change in the

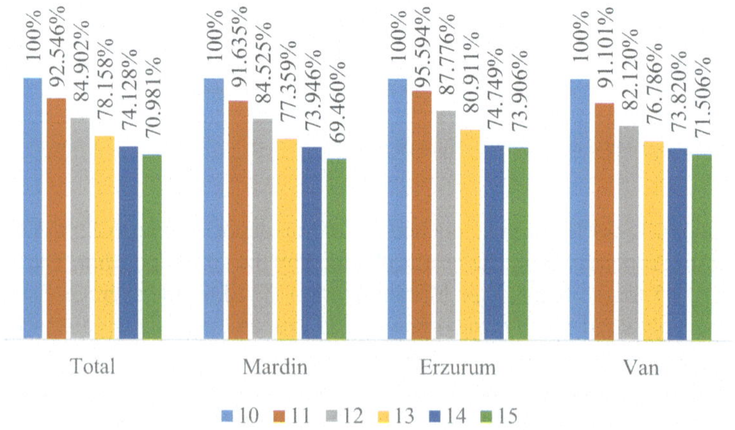

Fig. 22.5 Percentage change of the objective values of the depots (in km) according to the vehicle capacity

amount of carbon footprint of the company depending on the number of vehicles used and the travelled distances. As the number of routes decreases, a smaller number of vehicles will be used, which will result in lower fuel consumption and a consequent decrease in carbon emissions [4]. The company should make a cost analysis on the increase in vehicle capacity considering the growth targets in their strategic plan. From this point of view, it is possible to develop a vision that considers the environmental concerns of society by reducing the carbon footprint as well as the economic concerns of the company. An alternative is to optimize the packing of the orders to send higher quantities with the same fleet.

22.5 Conclusions

We have studied the distribution problem of a Turkish FMCG company, specializing in personal care and cleaning products that consist of 502 customer locations and three depots. We divided this MDVRP into two parts.

First, CPMP is used to assign customers to depots. We have then used GLS to create distribution plans of three different single-depot VRPs. Currently, we have determined the shipping routes of 502 customers from 3 depots for vehicles with a capacity of 10 pallets.

We have applied scenario analysis to the data by changing the vehicle capacity. We have seen that improvement can occur as the objective function value decreased at the rates of 7.5, 15.1, 21.8, 25.9 and 29.0%, respectively, for five different capacities that we have addressed by increasing the vehicle capacity. Furthermore, the decrease in the number of routes with the increase in capacity is remarkable. We determined that we could reduce the total number of routes between 9.02% and 33.3% with the increase in capacity. This may lead to a decrease in the number of vehicles needed for shipments and thus in the amount of CO_2 to be released into the environment. As a result, we conclude that by investing in higher-capacity vehicles, the company could achieve higher efficiency and shorter distances. We also observe that solution for a large scale VRP can be found efficiently using the Google OR-Tools library.

One of the limitations of this study is that the distance matrix between the points is created once and the changing conditions are not taken into account instantly. The creation of route plans that can adapt to changing conditions (traffic conditions, possible road errors and alternative routes as a result of temporary closure of the roads used in the routes, etc.) can be targeted in further studies. In addition, customers' demand uncertainties were not addressed. Stochastic handling of some parameters in the problem, such as demand and distance, may yield more realistic results for real-life problems.

In future studies, different optimization algorithms may be evaluated for the company to survive against difficult market conditions and to reduce its costs. For this, comparisons could be made by considering other metaheuristic methods, and time window constraints can be included by targeting the completion of shipments within certain deadlines.

Appendix

See Tables A22.1, A22.2, A22.3, A22.4 and A22.5.

Table A22.1 Customer demands

N	D	N	D	N	D	N	D	N	D	N	D	N	D	N	D	N	D	N	D
C1	2	C52	5	C103	1	C154	9	C205	1	C256	5	C307	1	C358	1	C409	6	C460	4
C2	3	C53	1	C104	1	C155	2	C206	2	C257	1	C308	8	C359	1	C410	3	C461	1
C3	7	C54	1	C105	1	C156	4	C207	1	C258	9	C309	1	C360	1	C411	3	C462	6
C4	4	C55	1	C106	1	C157	4	C208	2	C259	10	C310	4	C361	1	C412	1	C463	3
C5	4	C56	9	C107	1	C158	1	C209	2	C260	3	C311	4	C362	1	C413	2	C464	2
C6	10	C57	1	C108	10	C159	1	C210	3	C261	5	C312	6	C363	8	C414	1	C465	2
C7	2	C58	1	C109	1	C160	1	C211	3	C262	3	C313	3	C364	1	C415	1	C466	2
C8	1	C59	5	C110	1	C161	2	C212	1	C263	2	C314	2	C365	1	C416	1	C467	1
C9	1	C60	3	C111	1	C162	7	C213	4	C264	1	C315	2	C366	2	C417	4	C468	1
C10	2	C61	2	C112	2	C163	1	C214	8	C265	2	C316	1	C367	1	C418	1	C469	1
C11	1	C62	1	C113	1	C164	1	C215	5	C266	1	C317	3	C368	2	C419	1	C470	2
C12	4	C63	1	C114	1	C165	9	C216	5	C267	4	C318	2	C369	1	C420	2	C471	1
C13	1	C64	1	C115	1	C166	2	C217	3	C268	3	C319	2	C370	4	C421	2	C472	1
C14	1	C65	6	C116	1	C167	10	C218	2	C269	3	C320	1	C371	5	C422	1	C473	1
C15	4	C66	1	C117	1	C168	3	C219	3	C270	1	C321	2	C372	4	C423	1	C474	1
C16	1	C67	1	C118	2	C169	4	C220	4	C271	1	C322	1	C373	1	C424	1	C475	1
C17	1	C68	1	C119	10	C170	1	C221	3	C272	5	C323	1	C374	1	C425	1	C476	1
C18	2	C69	6	C120	1	C171	2	C222	2	C273	8	C324	3	C375	1	C426	2	C477	1
C19	2	C70	3	C121	1	C172	5	C223	8	C274	4	C325	10	C376	2	C427	2	C478	5
C20	1	C71	2	C122	1	C173	1	C224	5	C275	3	C326	1	C377	3	C428	6	C479	1

(continued)

Table A22.1 (continued)

N	D	N	D	N	D	N	D	N	D	N	D	N	D	N	D	N	D	N	D
C21	3	C72	1	C123	4	C174	6	C225	6	C276	1	C327	3	C378	8	C429	1	C480	1
C22	1	C73	3	C124	1	C175	2	C226	5	C277	7	C328	3	C379	1	C430	2	C481	9
C23	3	C74	1	C125	1	C176	1	C227	1	C278	6	C329	4	C380	1	C431	4	C482	1
C24	1	C75	1	C126	1	C177	2	C228	1	C279	5	C330	1	C381	2	C432	1	C483	1
C25	1	C76	1	C127	1	C178	1	C229	1	C280	3	C331	9	C382	10	C433	1	C484	7
C26	1	C77	2	C128	1	C179	5	C230	1	C281	3	C332	1	C383	2	C434	1	C485	1
C27	2	C78	1	C129	1	C180	3	C231	1	C282	5	C333	1	C384	10	C435	1	C486	1
C28	1	C79	3	C130	7	C181	5	C232	1	C283	5	C334	1	C385	1	C436	5	C487	4
C29	3	C80	5	C131	1	C182	3	C233	2	C284	3	C335	1	C386	1	C437	4	C488	3
C30	3	C81	1	C132	5	C183	1	C234	1	C285	2	C336	1	C387	1	C438	1	C489	1
C31	3	C82	4	C133	1	C184	5	C235	3	C286	1	C337	6	C388	1	C439	6	C490	3
C32	1	C83	2	C134	3	C185	7	C236	3	C287	6	C338	5	C389	1	C440	1	C491	1
C33	1	C84	1	C135	7	C186	5	C237	2	C288	6	C339	1	C390	2	C441	4	C492	6
C34	1	C85	2	C136	3	C187	2	C238	1	C289	5	C340	1	C391	3	C442	1	C493	3
C35	1	C86	3	C137	1	C188	2	C239	1	C290	3	C341	3	C392	1	C443	4	C494	6
C36	1	C87	2	C138	1	C189	2	C240	3	C291	1	C342	1	C393	7	C444	3	C495	4
C37	1	C88	1	C139	3	C190	1	C241	10	C292	6	C343	1	C394	2	C445	9	C496	6
C38	3	C89	2	C140	5	C191	3	C242	9	C293	3	C344	8	C395	2	C446	6	C497	2
C39	3	C90	9	C141	4	C192	1	C243	2	C294	1	C345	1	C396	6	C447	4	C498	5
C40	4	C91	1	C142	1	C193	4	C244	1	C295	7	C346	3	C397	3	C448	3	C499	3
C41	7	C92	1	C143	1	C194	1	C245	1	C296	5	C347	2	C398	2	C449	7	C500	1

(continued)

Table A22.1 (continued)

N	D	N	D	N	D	N	D	N	D	N	D	N	D	N	D	N	D	N	D
C42	1	C93	1	C144	1	C195	2	C246	2	C297	6	C348	1	C399	1	C450	1	C501	2
C43	1	C94	1	C145	1	C196	2	C247	1	C298	6	C349	7	C400	2	C451	3	C502	1
C44	5	C95	3	C146	2	C197	5	C248	4	C299	5	C350	10	C401	1	C452	8		
C45	1	C96	6	C147	1	C198	3	C249	1	C300	2	C351	1	C402	6	C453	2		
C46	1	C97	2	C148	1	C199	3	C250	3	C301	4	C352	7	C403	4	C454	2		
C47	1	C98	2	C149	2	C200	1	C251	9	C302	3	C353	9	C404	1	C455	4		
C48	1	C99	1	C150	9	C201	10	C252	4	C303	4	C354	1	C405	4	C456	3		
C49	1	C100	1	C151	1	C202	1	C253	5	C304	1	C355	1	C406	1	C457	1		
C50	10	C101	1	C152	5	C203	4	C254	4	C305	2	C356	9	C407	2	C458	2		
C51	2	C102	1	C153	1	C204	1	C255	3	C306	1	C357	1	C408	1	C459	1		

N: Node, D: Demand

Table A22.2 Results of customers assignments to depots

Mardin						Erzurum				Van			
C32	C72	C157	C197	C337	C377	C1	C243	C283	C384	C5	C114	C456	C496
C33	C73	C158	C198	C338	C378	C2	C244	C284	C385	C7	C115	C457	C497
C34	C74	C159	C199	C339	C379	C3	C245	C285	C386	C9	C116	C458	C498
C35	C75	C160	C200	C340	C380	C4	C246	C286	C389	C12	C117	C459	C499
C36	C76	C161	C201	C341	C381	C6	C247	C287	C390	C13	C118	C460	C500
C37	C122	C162	C202	C342	C407	C8	C248	C288	C391	C14	C119	C461	C501
C38	C123	C163	C203	C343	C408	C10	C249	C289	C392	C16	C120	C462	C502
C39	C124	C164	C204	C344	C409	C11	C250	C290	C393	C18	C121	C463	
C40	C125	C165	C205	C345	C410	C15	C251	C291	C394	C21	C308	C464	
C41	C126	C166	C206	C346	C411	C17	C252	C292	C395	C23	C309	C465	
C42	C127	C167	C207	C347	C412	C19	C253	C293	C396	C24	C310	C466	
C43	C128	C168	C208	C348	C413	C20	C254	C294	C399	C25	C311	C467	
C44	C129	C169	C209	C349	C414	C22	C255	C295	C400	C86	C312	C468	
C45	C130	C170	C210	C350	C415	C26	C256	C296	C401	C87	C313	C469	
C46	C131	C171	C211	C351	C416	C27	C257	C297	C402	C88	C314	C470	
C47	C132	C172	C212	C352	C417	C28	C258	C298	C404	C89	C315	C471	
C48	C133	C173	C213	C353	C418	C29	C259	C299	C405	C90	C316	C472	
C49	C134	C174	C214	C354	C419	C30	C260	C300	C406	C91	C317	C473	
C50	C135	C175	C215	C355	C420	C31	C261	C301		C92	C387	C474	
C51	C136	C176	C216	C356	C421	C77	C262	C302		C93	C388	C475	
C52	C137	C177	C217	C357	C422	C78	C263	C303		C94	C397	C476	
C53	C138	C178	C218	C358	C423	C79	C264	C304		C95	C398	C477	
C54	C139	C179	C219	C359	C424	C80	C265	C305		C96	C403	C478	
C55	C140	C180	C220	C360	C425	C81	C266	C306		C97	C439	C479	
C56	C141	C181	C221	C361	C426	C82	C267	C307		C98	C440	C480	
C57	C142	C182	C222	C362	C427	C83	C268	C318		C99	C441	C481	
C58	C143	C183	C223	C363	C428	C84	C269	C319		C100	C442	C482	
C59	C144	C184	C224	C364	C429	C85	C270	C320		C101	C443	C483	
C60	C145	C185	C225	C365	C430	C231	C271	C321		C102	C444	C484	
C61	C146	C186	C226	C366	C431	C232	C272	C322		C103	C445	C485	
C62	C147	C187	C227	C367	C432	C233	C273	C323		C104	C446	C486	
C63	C148	C188	C228	C368	C433	C234	C274	C324		C105	C447	C487	
C64	C149	C189	C229	C369	C434	C235	C275	C325		C106	C448	C488	
C65	C150	C190	C230	C370	C435	C236	C276	C326		C107	C449	C489	
C66	C151	C191	C331	C371	C436	C237	C277	C327		C108	C450	C490	
C67	C152	C192	C332	C372	C437	C238	C278	C328		C109	C451	C491	
C68	C153	C193	C333	C373	C438	C239	C279	C329		C110	C452	C492	
C69	C154	C194	C334	C374		C240	C280	C330		C111	C453	C493	
C70	C155	C195	C335	C375		C241	C281	C382		C112	C454	C494	
C71	C156	C196	C336	C376		C242	C282	C383		C113	C455	C495	

Table A22.3 Routes for Mardin depot

Route ID	Customers served	Cost (in Km)	Load (in Pallet)	Route ID	Customers served	Cost (in Km)	Load (in Pallet)
1	Depot-175-177-224-194-Depot	436	10	19	Depot-202-228-141-169-Depot	562	10
2	Depot-144-150-Depot	446	10	20	Depot-211-132-230-170-Depot	611	10
3	Depot-157-180-221-Depot	410	10	21	Depot-181-220-128-Depot	420	10
4	Depot-210-152-212-142-Depot	447	10	22	Depot-203-159-215-Depot	439	10
5	Depot-206-198-179-Depot	441	10	23	Depot-135-155-204-Depot	439	10
6	Depot-160-214-133-Depot	408	10	24	Depot-213-225-Depot	440	10
7	Depot-143-136-139-199-Depot	447	10	25	Depot-205-165-Depot	421	10
8	Depot-356-Depot	194	9	26	Depot-191-188-149-168-Depot	433	10
9	Depot-146-223-Depot	441	10	27	Depot-217-184-209-Depot	452	10
10	Depot-186-163-227-229-173-151-Depot	433	10	28	Depot-145-156-197-Depot	410	10
11	Depot-208-218-131-166-219-Depot	420	10	29	Depot-216-182-176-200-Depot	440	10
12	Depot-190-158-222-174-Depot	446	10	30	Depot-148-178-130-183-Depot	428	10
13	Depot-358-171-195-172-Depot	426	10	31	Depot-207-187-140-189-Depot	444	10
14	Depot-137-226-196-161-Depot	409	10	32	Depot-127-123-193-36-Depot	393	10
15	Depot-154-164-Depot	436	10	33	Depot-331-357-Depot	264	10
16	Depot-185-134-Depot	441	10	34	Depot-378-347-Depot	261	10
17	Depot-147-153-162-138-Depot	419	10	35	Depot-360-335-338-374-361-348-Depot	264	10
18	Depot-50-Depot	206	10	36	Depot-341-337-339-Depot	260	10

(continued)

Table A22.3 (continued)

Route ID	Customers served	Cost (in Km)	Load (in Pallet)	Route ID	Customers served	Cost (in Km)	Load (in Pallet)
37	Depot-350-Depot	215	10	52	Depot-58-38-59-64-Depot	205	10
38	Depot-368-362-377-372-Depot	258	10	53	Depot-44-39-61-Depot	206	10
39	Depot-334-340-355-352-Depot	228	10	54	Depot-366-381-371-354-Depot	214	10
40	Depot-349-379-373-359-Depot	228	10	55	Depot-46-74-69-72-62-Depot	209	10
41	Depot-367-353-Depot	255	10	56	Depot-345-363-Depot	91	9
42	Depot-201-Depot	439	10	57	Depot-409-411-Depot	240	9
43	Depot-428-432-422-426-Depot	304	10	58	Depot-333-364-375-380-376-336-342-365-Depot	265	9
44	Depot-63-41-129-192-Depot	401	10	59	Depot-408-410-418-419-416-412-76-47-Depot	286	10
45	Depot-167-Depot	434	10	60	Depot-415-413-417-414-407-Depot	188	10
46	Depot-40-49-54-125-126-122-124-Depot	320	10	61	Depot-423-435-421-437-434-Depot	229	9
47	Depot-42-37-52-75-51-Depot	218	10	62	Depot-429-436-431-Depot	175	10
48	Depot-35-56-Depot	204	10	63	Depot-433-438-424-425-430-420-427-Depot	177	10
49	Depot-32-33-71-70-73-Depot	211	10	64	Depot-344-351-Depot	187	9
50	Depot-65-45-66-57-67-Depot	209	10	65	Depot-370-343-369-332-346-Depot	188	10
51	Depot-48-55-60-68-34-43-53-Depot	215	9				

Table A22.4 Routes for Erzurum depot

Route ID	Customers served	Cost (in Km)	Load (in Pallet)	Route ID	Customers served	Cost (in Km)	Load (in Pallet)
1	Depot-270-275-256-276-Depot	19	10	19	Depot-244-288-285-Depot	16	9
2	Depot-301-287-Depot	24	10	20	Depot-81-82-79-84-247-Depot	276	10
3	Depot-246-324-326-318-323-Depot	474	9	21	Depot-78-83-85-80-Depot	271	10
4	Depot-77-292-306-307-Depot	486	10	22	Depot-15-11-1-2-Depot	370	10
5	Depot-19-20-3-Depot	377	10	23	Depot-259-Depot	8	10
6	Depot-30-330-327-328-Depot	526	10	24	Depot-258-Depot	18	9
7	Depot-253-261-Depot	23	10	25	Depot-384-Depot	427	10
8	Depot-271-303-291-255-257-Depot	22	10	26	Depot-251-286-Depot	30	10
9	Depot-8-4-22-10-17-Depot	402	9	27	Depot-264-267-283-Depot	26	10
10	Depot-302-278-249-Depot	25	10	28	Depot-300-262-280-263-Depot	27	10
11	Depot-234-243-236-235-239-240-Depot	407	10	29	Depot-296-279-Depot	18	10
12	Depot-238-237-231-232-233-240-Depot	403	10	30	Depot-297-254-Depot	17	10
13	Depot-6-Depot	375	10	31	Depot-393-284-Depot	516	10
14	Depot-294-242-Depot	392	10	32	Depot-269-250-248-Depot	16	10
15	Depot-245-405-394-406-399-401-Depot	441	10	33	Depot-382-Depot	431	10
16	Depot-29-28-27-31-26-Depot	462	10	34	Depot-325-Depot	475	10
17	Depot-304-396-390-389-Depot	431	10	35	Depot-277-260-Depot	9	10
18	Depot-265-281-272-Depot	29	10	36	Depot-322-320-319-321-329-Depot	465	10

(continued)

Table A22.4 (continued)

Route ID	Customers served	Cost (in Km)	Load (in Pallet)	Route ID	Customers served	Cost (in Km)	Load (in Pallet)
37	Depot-241-Depot	392	10	42	Depot-383-402-386-Depot	316	9
38	Depot-282-293-Depot	14	8	43	Depot-268-295-Depot	9	10
39	Depot-289-299-Depot	14	10	44	Depot-274-252-Depot	12	8
40	Depot-385-392-400-391-404-395-Depot	434	10	45	Depot-298-266-290-Depot	20	10
41	Depot-273-305-Depot	19	10				

Table A22.5 Routes for Van depot

Route ID	Customers served	Cost (in Km)	Load (in Pallet)	Route ID	Customers served	Cost (in Km)	Load (in Pallet)
1	Depot-479-502-495-441-Depot	36	10	9	Depot-460-439-Depot	36	10
2	Depot-469-491-501-493-490-Depot	71	10	10	Depot-119-Depot	307	10
3	Depot-463-484-Depot	51	10	11	Depot-114-113-103-96-116-Depot	355	10
4	Depot-99-118-89-498-Depot	305	10	12	Depot-448-500-446-Depot	34	10
5	Depot-107-97-100-110-112-94-121-104-Depot	359	10	13	Depot-480-445-Depot	35	10
6	Depot-499-468-466-457-451-Depot	42	10	14	Depot-485-486-492-482-471-Depot	46	10
7	Depot-115-388-387-398-397-111-105-Depot	481	10	15	Depot-494-497-Depot	27	8
17	Depot-489-91-102-109-92-403-101-Depot	389	10	26	Depot-478-450-458-454-Depot	203	10
18	Depot-481-467-Depot	32	10	27	Depot-95-87-86-98-Depot	346	10
19	Depot-449-470-Depot	35	9	28	Depot-12-25-93-120-106-117-477-Depot	394	10
20	Depot-461-473-483-442-462-Depot	44	10	29	Depot-456-443-453-Depot	199	9
21	Depot-496-444-440-Depot	34	10	30	Depot-88-90-Depot	308	10
22	Depot-464-452-Depot	33	10	31	Depot-13-317-315-311-Depot	616	10
23	Depot-465-487-488-Depot	29	9	32	Depot-108-Depot	302	10
24	Depot-308-Depot	444	8	33	Depot-475-474-14-5-7-Depot	304	9
25	Depot -9-24-18-21-23-Depot	431	10	34	Depot-455-472-459-447-Depot	36	10

References

1. Abareshi M, Zaferanieh M (2019) A bi-level capacitated *P*-median facility location problem with the most likely allocation solution. Transp Res Part B: Methodol 123:1–20
2. Baltz A, Dubhashi D, Tansini L, Srivastav A, Werth S (2005) Probabilistic analysis for a multiple depot vehicle routing problem. In: Foundations of software technology and theoretical computer science. Lecture notes in computer science, vol 3821, pp 360–371
3. Battarra M, Erdoğan G, Vigo D (2014) Exact algorithms for the clustered vehicle routing problem. Oper Res 62(1):58–71
4. Bektaş T, Laporte G (2011) The pollution-routing problem. Transp Res Part B: Methodol 45(8):1232–1250
5. Caccetta L, Dzator M (2005) Heuristic methods for locating emergency facilities. In: Proceedings Modsim, vol 2005
6. Cordeau J-F, Gendreau M, Laporte G (1997) A tabu search heuristic for periodic and multi-depot vehicle routing problems. Networks 30(2):105–119
7. Crevier B, Cordeau J-F, Laporte G (2007) The multi-depot vehicle routing problem with inter-depot routes. Eur J Oper Res 176(2):756–773
8. Du J, Li X, Yu L, Dan R, Zhou J (2017) Multi-depot vehicle routing problem for hazardous materials transportation: A fuzzy bilevel programming. Inf Sci 399:201–218
9. Erdoğan G (2017) An open source spreadsheet solver for vehicle routing problems. Comput Oper Res 84:62–72
10. Fan H, Zhang Y, Tian P, Lv Y, Fan H (2021) Time-dependent multi-depot green vehicle routing problem with time windows considering temporal-spatial distance. Comput Oper Res 129:105211
11. Giosa ID, Tansini IL, Viera IO (2002) New assignment algorithms for the multi-depot vehicle routing problem. J Oper Res Soc 53:977–984
12. Github OR-Tools (2021). https://github.com/google/or-tools. Accessed 22 Oct 2021
13. Google OR-Tools (2021). https://developers.google.com/optimization. Accessed 22 Oct 2021
14. Google OR-Tools CVRP (2021). https://developers.google.com/optimization/routing/cvrp. Accessed 22 Oct 2021
15. Google Maps (2021). https://www.google.com/maps
16. Kariv O, Hakimi SL (1979) An algorithmic approach to network location problems II: The P-medians. SIAM J Appl Math 37:539–560
17. Kilby P, Prosser P, Shaw P (1999) Guided local search for the vehicle routing problem with time windows. In: Meta-heuristics: advances and trends in local search paradigms for optimization, pp 473–486
18. Kliewer N, Mellouli T, Suhl L (2006) A time-space network based exact optimization model for multi-depot bus scheduling. Eur J Oper Res 175(3):1616–1627
19. Koç Ç, Laporte G, Tükenmez İ (2020) A review of vehicle routing with simultaneous pickup and delivery. Comput Oper Res 122:104987
20. Laporte G (1992) The vehicle routing problem: An overview of exact and approximate algorithms. Eur J Oper Res 59:345–358
21. Li C-L, Simchi-Levi D (1990) Worst-case analysis of heuristics for multidepot capacitated vehicle routing problems. ORSA J Comput 2(1):64–73
22. Mirabi M, Ghomi SF, Jolai F (2010) Efficient stochastic hybrid heuristics for the multi-depot vehicle routing problem. Robot Comput-Integr Manuf 26(6):564–569
23. Montoya-Torres JR, Franco JL, Isaza SN, Jiménez HF, Herazo-Padilla N (2015) A literature review on the vehicle routing problem with multiple depots. Comput Ind Eng 79:115–129
24. Mulvey JM, Beck MP (1984) Solving capacitated clustering problems. Eur J Oper Res 18(3):339–348
25. Perron L (2011). Operations research and constraint programming at Google. In: Lecture notes in computer science, pp 2–2
26. Pisinger D, Ropke S (2007) A general heuristic for vehicle routing problems. Comput Oper Res, 34(8):2403–2435

27. Rabbani M, Tohidi Fard M, Partovi M, Farrokhi-Asl H (2018) Solving a multi-depot vehicle routing problem with time windows and fuzzy demands using metaheuristic algorithms in home health care services. J Decis Oper Res 3(2):114–127
28. Raft OM (1982) A modular algorithm for an extended vehicle scheduling problem. Eur J Oper Res II:67–76
29. Reiners C (2016) Constraint programming-based heuristics for the multi-depot vehicle routing problem with a rolling planning horizon (PhD Thesis, Universitätsbibliothek Duisburg-Essen)
30. Renaud J, Laporte G, Boctor FF (1996) A tabu search heuristic for the multi-depot vehicle routing problem. Comput Oper Res 23(3):229 235
31. Russell RA (1977) An effective heuristic for the M-tour traveling salesman problem with some side conditions. Oper Res 25(3):517–524
32. Simchi-Levi D, Kaminsky P, Simchi-Levi E, Shankar R (2008) Designing and managing the supply chain: concepts, strategies and case studies. Tata McGraw-Hill Education
33. Stodola P (2020) Hybrid ant colony optimization algorithm applied to the multi-depot vehicle routing problem. Nat Comput 19(2):463–475
34. Tillman FA, Cain TM (1972) An upperbound algorithm for the single and multiple terminal delivery problem. Manage Sci 18(11):664–682
35. Toth P, Vigo D (eds) (2014) Vehicle routing: problems, methods, and applications. MOS-SIAM Series on Optimization, Philadelphia
36. Van Breedam A (2002) A parametric analysis of heuristics for the vehicle routing problem with side-constraints. Eur J Oper Res 137(2):348–370
37. Van Omme N, Perron L, Furnon V (2014) OR-Tools user's manual. Rapp. tech. Google
38. Venkata Narasimha K, Kivelevitch E, Sharma B, Kumar M (2013) An ant colony optimization technique for solving min-max multi-depot vehicle routing problem. Swarm Evol Comput 13:63–73
39. Vidal T, Crainic TG, Gendreau M, Lahrichi N, Rei W (2012) A hybrid genetic algorithm for multi-depot and periodic vehicle routing problems. Oper Res 60(3):611–624
40. Zarandi MHF, Hemmati A, Davari S, Turksen IB (2013) Capacitated location-routing problem with time windows under uncertainty. Knowl-Based Syst 37:480–489
41. Zhen T, Zhang Q (2009) A hybrid metaheuristic algorithm for the multi-depot vehicle routing problem with time windows. In: 2009 international conference on networks security, wireless communications and trusted computing, vol 2. IEEE, pp 798–801

Chapter 23
Effective Metaheuristic Algorithms for Platelet Collection Routing Problem

Milad Elyasi, Ramin Talebi Khameneh, Ali Ekici, and Okan Örsan Özener

Abstract In medical treatments, patients may need a particular blood product to continue their treatment process. The only source for all blood products is blood donation. In order to extract any product from donated blood, there is a specific time limit after donation occurs. One of the blood products is platelet, which is a vital element in cancer therapies, organ transplant procedures, and different surgical operations. Platelet production is the most time-sensitive process among all other blood products. Donated blood should be processed in blood processing centers within six hours after donation; otherwise, the donated blood will not be appropriate for platelet production. Therefore, blood collection vehicles should pick up and carry the donated blood into the blood processing center in a proper period to save blood products from perishing. Because of the perishability of donated blood and the accumulative nature of blood donation, the scheduling of blood collection vehicles to collect the maximum number of blood units for platelet production is a complicated routing problem. In this study, two metaheuristic methods, namely the *hybrid flower pollination algorithm* (HFPA) and *simulated annealing* (SA), are implemented to solve a variant of the *maximum blood collection problem* (MBCP). The effectiveness of the proposed algorithms through exhaustive computational studies is investigated. The proposed algorithms show an average improvement of around 4% and 6% in the objective function value over the benchmark algorithm.

M. Elyasi · R. T. Khameneh · A. Ekici · O. Ö. Özener (✉)
Industrial Engineering Department, Özyeğin University, Istanbul, Turkey
e-mail: orsan.ozener@ozyegin.edu.tr

M. Elyasi
e-mail: milad.elyasi@ozyegin.edu.tr

R. T. Khameneh
e-mail: ramin.khameneh@ozu.edu.tr

A. Ekici
e-mail: ali.ekici@ozyegin.edu.tr

F. Hamid (ed.), *Optimization Essentials*, International Series in Operations Research & Management Science 353, https://doi.org/10.1007/978-981-99-5491-9_23

23.1 Introduction

Blood and its products are needed by patients in surgical operations, emergencies, or even regular medical treatments. Moreover, one out of seven patients in hospitals needs at least one of blood products [4]. Furthermore, one unit of whole blood donation, which is the only source of whole blood and its components, can save up to three lives [29]. Therefore, blood is a vital commodity that requires a well-managed supply chain.

After blood donation, the blood products should be extracted from whole blood in less than six hours to avoid perishing. Furthermore, this perishability is not the same for different blood products. Blood components, such as plasma, red cells, and platelets have different perishability from hours to months. Therefore, all blood products are considered perishable, while some blood products such as platelets are highly perishable, and their life span is in hours. In other words, the platelet is the most perishable blood product [36]. The perishable nature of blood and its components makes *blood supply chain* (BSC) management complicated.

In addition, the platelet is not only the most time-dependent blood product, but it also has no synthetic substitute while it is needed with great demand as a significant factor in cancer treatments, organ transplants, and other surgical procedures. It is noteworthy to emphasize that according to [30] every 15 s, someone in the U.S. needs platelets. Thus, an effective BSC, especially for platelet, seems essential. The BSC distributes blood products, including platelets, plasma, and red blood cells. Each product has a particular structure and characteristics that requires designing a different supply chain from the other blood products. Among the blood products, platelets are one of the essential products in healthcare, but the high perishability makes the extraction procedure time-dependent, causing hardship in designing an effective supply chain for the product. To cope with these problems, many studies propose different modeling and solution approaches. Some studies have focused on approaches that survey on time-sensitivity of perishable products. At the same time, most of the literature explores the cost of designing a BSC. *Blood collection problem* (BCP) literature presents solutions to minimize the cost of routing for blood collection vehicles or minimize the tours and maximize collected blood units. In addition, since the BSC has a nature of uncertainty in all aspects, many papers consider robust modeling defining disruption scenarios over each part of the supply chain. In order to have an effective platelet supply chain, it is needed to focus on time wastage of the supply chain because of its nature of high perishability. The literature gap in the routing aspect of the platelet supply chain motivated us in this study to focus on a routing problem that addresses maximizing the number of collected blood considering platelet process limitations.

Moreover, the donated whole blood from donors in collection sites must be separated into the components in less than 18 hours; but for platelet extraction, it is recommended to be done within six hours after donation. Considering this limitation [23] define a problem named MBCP as a variant of BCP. In MBCP, it is considered that donors visit the collection sites in a predefined scheduling pattern in order to have the minimum waste of time for donors. The objective of MBCP is to maximize platelet production by optimizing the routing of blood collection vehicles. This problem assumes that collected blood units from the blood donation center should be collected and processed at the processing center within six hours of donation; otherwise, the collected blood units cannot be used in platelet production. Özener and Ekici [23] show that MBCP is an NP-hard problem, and it cannot be solved in polynomial time. Consequently, they solve MBCP through several types of heuristics. In all of the methods, they do a clustering for the donation sites and assign a vehicle for each cluster to carry the donated blood. Needless to say, in case there is no initial clustering, more donated blood could be collected because of the freedom of the movements of the vehicles. This study proposes the idea of solving a new variant of the MBCP without clustering to have a greater feasible region for the solution. Elimination of clustering causes a tremendous increase in the size of the feasible region. Because of the number and dispersion of blood donation sites and time window limits for collecting donations for platelet production, metaheuristic methods have been proposed to optimize blood collection operations. The variant of MBCP is inherently a VRP problem, and VRP is proven to be NP-hard [14]. Therefore, this problem cannot be solved in polynomial time. Consequently, metaheuristic approaches are inevitable to solve the variant of MBCP. Therefore, SA and HFPA algorithms are implemented as solution approaches for this problem due to several reasons, such as comprehensibility, ease of implementation, and their capability in handling combinatorial problems.

The contribution of this study are as follows: (i) proposing a new variant of maximal blood collection problem which avoids the suboptimal solutions because of the initial clustering, (ii) proposing effective HFPA and SA algorithms to maximize the collected blood units suitable for platelet extraction without clustering blood donation sites that outperform the clustered solutions. Moreover, in order to evaluate the performances of the proposed heuristic algorithms and the effect of elimination of clustering the donation sites in the problem, a comparison between the results of [23] and this study is done. The results are better in most cases than their ones; emphasizing a larger feasible region improves the quality of the solution. In addition, a comparison between HFPA and SA is made. HFPA provides better results than OEH (Özener & Ekici Heuristic proposed by [23]), which is the benchmark, and SA has better performance in both runtime and quality of results compared to OEH.

The rest of this study is organized into five sections: Sect. 23.2 provides a review of the relevant literature on blood donation problems. Section 23.3 presents the definition and mathematical modeling of the problem. The solution approaches are discussed in Sect. 23.4. Section 23.5 is dedicated to implementing the proposed HFPA and SA heuristics and computationally demonstrates how the methods perform on several randomly generated instances providing some numerical analyses and managerial insights. Finally, concluding remarks are provided in Sect. 23.6.

23.2 Literature Review

BSC management was introduced for the first time in a dissertation on inventory control for perishable commodities by [31]. Pirabán et al. [24] investigate the BSC considering different decision-making environments, modeling and solution methods, different variants of the problem, and various criteria for optimizing decisions in it. The reader is referred to [32] for the most recent review of the literature on BSC.

A broad range of papers concentrate on minimizing the costs related to the BSC. Nagurney et al. [20] consider a generalized supply chain network optimization model for a complex BSC, including collection sites, testing and processing facilities, storage facilities, distribution centers, and hospitals. They design a network to determine the optimal allocation of blood to the demand points to reduce the costs associated with blood wastage. Cheraghi et al. [7] present a mixed-integer linear programming model in a network of BSCs to lessen the total cost of the platelet supply chain. Hamdan and Diabat [11] propose a bi-objective robust optimization model for the design of a disaster-resilient BSC when disruption occurs in blood facilities and transportation routes to minimize the time and cost of delivering blood to hospitals. Zahiri and Pishvaee [36] propose a bi-objective mathematical programming model to minimize the total cost as well as the maximum unsatisfied demand of hospitals. They design a BSC network considering blood group compatibility. Ensafian and Yaghoubi [9] presents an integrated age-differentiated platelet supply chain problem using two mixed-integer linear programming models based on first-in-first-out and last-in-first-out policies. Also, they develop a bi-objective model to maximize the platelet's freshness and minimize the total cost. Yaghoubi et al. [33] present a bi-objective model that makes a trade-off between the network costs and platelets' freshness considering the uncertainty of network facilities' performance. The result of their robust model obtains a considerable reduction in the level of relative regret while disrupting the network.

The BSC consists of four different stages: collection, production, inventory, and delivery [22], out of which distribution is of high significance due to the perishable nature of the blood. Although this stage of the BSC has been studied in depth by researchers [5, 21], MBCP has not been heeded so profoundly. Some of the more relevant works associated with the study include minimizing the shortage and wastage of blood products in BSCs. Blake et al. [3] focus on a model for ordering platelet considering shortage and waste acceptable range. Stanger et al. [28] make recommendations to minimize losses of red blood cells due to time expiry by managing inventory in hospital transfusion laboratories. Gunpinar and Centeno [10] present integer programming models to minimize the total cost, shortage, and wastage levels of blood products at a hospital within a planning horizon. Rajendran and Ravindran [25] develop a mixed-integer stochastic programming model to minimize wastage of platelets while maintaining a specified service level. To achieve this objective, they propose three heuristic methods for determining the platelet ordering policy at the hospitals. Dehghani et al. [8] address a two-stage stochastic programming model to decrease blood wastage and shortage costs considering the uncertain demand of hospitals from a central blood bank.

In addition, some publications regarding blood donor appointment scheduling are more relevant to this study since MBCP deals with the pattern of donor visits to donation sites. Alfonso et al. [1] address modeling and simulation of blood collection systems in France for blood collection, emphasizing the scheduling of plasma and platelet donors. Mobasher et al. [19] present a mixed-integer linear programming model for coordinating collection and appointment scheduling operations at blood donation sites to maximize platelet production considering processing time requirements for platelet. Baş et al. [2] propose a mixed-integer linear programming model for preallocating time slots to blood types and an online prioritization policy for the appointment reservation that accounts for both booked donors and donors arriving without a reservation. Li et al. [16] propose deterministic and stochastic demand models of blood donors according to the number of donors arriving at donation sites to schedule blood collection staff based on predefined scenarios. The results show that the proposed models need few staff when facing adequate staff scenarios and reduce the number of donors when facing inadequate staff scenarios.

The most notorious VRP combined by time is the *VRP with time windows* (VRPTW), where a customer's service can begin within the time window defined by the earliest and the latest times when the customer will permit the start of service. Lot of heuristic and metaheuristic methods are proposed to solve the VRPTW [15, 27]. The study by [35] is the first one concentrating on the routing aspect of the BSC. The author develops a model for a VRPTW to collect blood from donation sites considering a lower bound for extractable platelet amount from the collection and solve it using a column-generation method. The study reduces the cost of the collection up to 60% in a real-life case study in the US. Gunpinar and Centeno [10] present an integer programming model that simultaneously determines the number of bloodmobiles and minimizes the total distance traveled by vehicles. In order to solve the problem, the authors use a commercial solver software and a branch-and-price algorithm. Şahinyazan et al. [26] address a new model for a selective VRP to determine the best tours of the bloodmobiles and the shuttle and their lengths of stay at each stop that helps to collect blood from bloodmobiles considering logistic costs. Besides, they propose another model to maximize the total amount of collected blood in a time horizon. The proposed model is diffident from the MBCP because they do not consider platelet production and perishability of blood products. Lodree et al. [17] address a host of common sense heuristic policies to generate routes for the donation collections problem, which is a network routing problem motivated by specific challenges that arise during the response phase immediately following large-scale disaster events. Chao et al. [6] present a multiple tour maximum collection problem called team orienteering problem. The goal of their problem is to maximize the total collected profit by satisfying a time limit for each vehicle. The main difference between team orienteering problem and MBCP is the time limit. In MBCP, the donated blood in each node has a limited time to be collected and processed in the blood processing center. However, in team orienteering problem, each vehicle has a limited time to travel on edges and to visit nodes. In other words, in team orienteering problem time limit is on collection vehicles, while in MBCP, the time limit is on nodes.

There is scarce literature considering the different points of view on the BCP. Katsaliaki and Brailsford [12] analyze policies for managing the blood inventory system using discrete-event simulation to determine ordering policies leading to reductions in shortages and wastage of blood in hospitals.

As the literature reveals, a small proportion of the literature is dedicated to the perishability of the platelet in designing a BSC. Although considering the number of collected blood donations for platelet extraction has been only yet considered by [23], the methodology of this work is based on a preassigned clustering method that limits the solution space.

Furthermore, existing literature has a slight focus on dealing with the routing aspect of blood collection, while, in almost all real-life platelet supply chains, the time parameter and the amount collected are essential factors. Therefore, ignoring time in the total number of collected/processed blood may impose high risks on decision-makers. A novel mathematical programming model for a non-clustered MBCP is elaborated in this paper to address the literature gap. Also, for the first time, HFPA and SA are developed for MBCP.

23.3 Problem Definition[1]

Blood donations commonly come about via static or temporary donation sites. The uninterrupted changes in locations of donation sites lead to having continuous changes in blood collector vehicle scheduling. In the case of extracting platelets from the blood, the platelet's perishability causes an increase in the importance of the time of shipment of blood units to the blood center. In order to have an efficient platelets supply chain, it is necessary to optimize the BSC in two main areas as follows:

- Assigning an optimal number of blood collection vehicles for blood shipment between donation sites and blood processing center.
- Finding optimal schedule for vehicles shipping the blood from donation sites to the blood processing center.

This section provides a formal definition of the non-clustered variant of MBCP as well as an upper bound on the number of tours that a vehicle can perform. Our definition of a route is a sequence of blood donation sites visited by a vehicle starting at the blood processing center and ending there. Accordingly, a tour is defined as a series of routes driven by a single vehicle in order. An example of a tour and route is presented in Fig. 23.1. The shape rectangle is a tour containing three routes, namely A, B, and C. It means that in the tour, the blood collection vehicle visit routes A, B,

[1] The problem definition follows a previous study [23], nevertheless we present here for the sake of completeness.

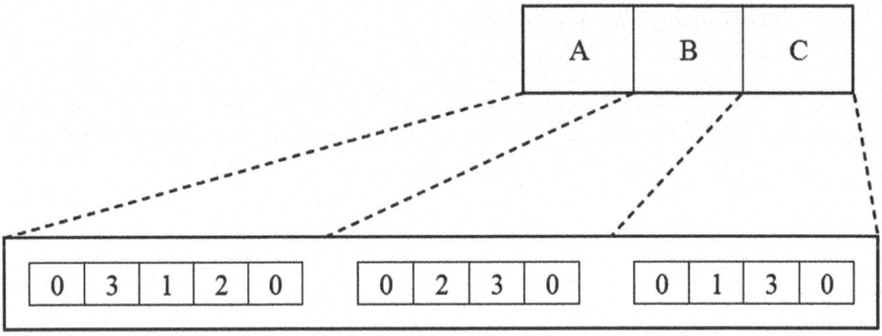

Fig. 23.1 An example of tour (solution) and route

and C in order in the time horizon. Each route contains a set of donation sites that the vehicle visits. For instance, in this example, the vehicle starts traveling on route B as the second route of its tour from node 0 (processing center) and then visits nodes 2 and 3 (donation sites), and finally, it returns to node 0.

The assumptions regarding the variant of MBCP are as follows:

- The daily donations at blood donation sites are known, and the centers have enough capacity to accommodate the daily collected donations.
- The donation sites would have a fixed location. The location of donation sites would not change during the planning horizon, even for the mobile ones.

Given a graph $G = (V_0, E)$ where V_0 is the union of the set of blood processing center (0) and V which is the set of donation sites, and E is the set of edges attaching the nodes in V_0. A tour begins from the blood processing center by a blood collection vehicle. The vehicle travels on a combination of edges in E in t_{ij} units of time for each edge (i, j) and collects blood units from a subset of V with a given order. At the end, the vehicle returns to the processing center. In this model, t_{ij} satisfies the triangle inequality. Let T be the *processing time limit* (PTL) considered in the routing of any of the m blood collection vehicles. The collected blood units will be wasted and cannot be used in platelet production if the summation of associated t_{ij}'s is more than T. Hence, the objective of the variant of MBCP is to collect the maximum donated blood units within T time units of donation time for every vehicle, without any restraint on the routing of the vehicles in visiting the blood donation sites. That is to say, a primary characteristic of the non-clustered variant of MBCP compared to MBCP is that blood collection vehicles have a wider degree of freedom in visiting blood donation sites in the variant of MBCP. Besides, an upper bound is needed on the number of tours that a vehicle can take during the planning horizon to provide the mathematical model. The upper bound is a parameter denoted by L.

23.3.1 Mathematical Modeling

The proposed model for non-clustered variant of MBCP is formulated as a mixed-integer programming problem. Accordingly, indices, parameters, sets, variables and the model are presented in the following sections.

23.3.1.1 Parameters

- t_{ij} travel time on edge (i, j)
- a_i opening time associated with donation site $i\,(a_i \in Z^+)$
- b_i closing time associated with donation site $i\,(b_i \in Z^+)$
- T_s start time of planning horizon zone $(T_s = \min_{i \in V_0}\{a_i\})$
- T_e end time of planning horizon zone $(T_e = \max_{i \in V_0}\{b_i\})$
- d_t^i number of blood donations at time t at donation site i
- T processing time limit
- M a large enough number
- L upper bound on the number of tours by a vehicle in a single day $\left(L = \dfrac{T_e - T_s + 2T}{\min_{i \in V}\{2t_{0i}\}}\right)$

The upper bound L is obtained as follows. A vehicle would never leave the processing center earlier than $T_s - T$ since the earliest time a donated unit becomes available is T_s. The latest time a donation is available for pickup is T_e, and the donation must be delivered to the processing center no later than $T_e + T$. Hence, a vehicle would travel for $T_e + T - (T_s - T)$ time units at most.

23.3.1.2 Sets

- V set of donation sites $\{1, 2, \ldots, n\}$ indexed by i and j
- V_0 set of processing center and donation sites $\{0, 1, 2, \ldots, n\}$ indexed by i and j
- E set of edges connecting the nodes in V_0
- F set of the vehicles $\{1, 2, \ldots, m\}$ indexed by k
- $\mathcal{T}$ set of tours performed by a vehicle in a single day $\{1, 2, \ldots, L\}$
- T_i set of t (discrete time units) in site i

23.3.1.3 Decision Variables

y_{ijlk} $\begin{cases} 1 & \text{if location } j \text{ is visited after location } i \text{ in tour } l \text{ of vehicle } k \\ 0 & \text{otherwise} \end{cases}$

x_{lk}^{it} $\begin{cases} 1 & \text{if donations completed at time } t \text{ at donation site } i \text{ are picked up in tour } l \text{ of vehicle } k \\ 0 & \text{otherwise} \end{cases}$

z_{lk} $\begin{cases} 1 & \text{if vehicle } k \text{ performs tour } l \\ 0 & \text{otherwise} \end{cases}$

e_{lk} arrival time of vehicle k at the processing center after completing tour l

v_{ilk} arrival time of vehicle k at donation site i in tour l

The problem is formulated as follows:

$$\max \quad \sum_{m \in \mathcal{T}} \sum_{k \in F} \sum_{i \in V} \sum_{t \in T_i} d_i^t x_{mk}^{it} \tag{23.1}$$

$$\text{s.t.} \quad \sum_{j \in V_0 \setminus \{i\}} y_{ijmk} = \sum_{j \in V_0 \setminus \{i\}} y_{jimk} \qquad \forall i, m, k \tag{23.2}$$

$$\sum_{j \in V_0 \setminus \{i\}} y_{jimk} \leq 1 \qquad \forall i, m, k \tag{23.3}$$

$$\sum_{j \in V} y_{0jmk} = z_{mk} \qquad \forall m, k \tag{23.4}$$

$$\sum_{j \in V} y_{j0mk} = z_{mk} \qquad \forall m, k \tag{23.5}$$

$$z_{mk} \leq z_{m-1,k} \qquad \forall m, k \tag{23.6}$$

$$\sum_{i \in V_0} \sum_{j \in V_0} y_{ijmk} \leq M z_{mk} \qquad \forall m, k \tag{23.7}$$

$$v_{imk} + t_{ij} \leq v_{jmk} + M(1 - y_{ijmk}) \qquad \forall i, j, m, k \tag{23.8}$$

$$e_{m-1,k} + t_{0i} \leq v_{imk} + M(1 - y_{0imk}) \qquad \forall i, m, k \tag{23.9}$$

$$v_{i1k} \geq a_i \qquad \forall i, k \tag{23.10}$$

$$a_i + M \sum_{j \in V_0} y_{ji1k} \geq v_{i1k} \qquad \forall i, m, k \tag{23.11}$$

$$v_{i,m-1,k} + M \sum_{j \in V_0} y_{jimk} \geq v_{imk} \qquad \forall i, m, k \tag{23.12}$$

$$v_{imk} \geq v_{i,m-1,k} \qquad \forall i, m, k \tag{23.13}$$

$$v_{imk} + t_{i0} \leq e_{mk} + M(1 - y_{i0mk}) \qquad \forall i, m, k \tag{23.14}$$

$$x_{mk}^{it} \le \sum_{j \in V_0} y_{jimk} \qquad \forall i, m, k, t \qquad (23.15)$$

$$\sum_{m \in \mathcal{T}} \sum_{k \in F} x_{mk}^{it} \le 1 \qquad \forall i, t \qquad (23.16)$$

$$e_{mk} - K \le t + M(1 - x_{mk}^{it}) \qquad \forall i, m, k, t \qquad (23.17)$$

$$v_{imk} \ge t - M(1 - x_{mk}^{it}) \qquad \forall i, m, k, t \qquad (23.18)$$

$$x_{mk}^{it} \in \{0, 1\} \qquad \forall i, m, k, t \qquad (23.19)$$

$$y_{ijmk} \in \{0, 1\} \qquad \forall i, j, m, k \qquad (23.20)$$

$$z_{mk} \in \{0, 1\} \qquad \forall m, k \qquad (23.21)$$

$$v_{imk}, e_{mk} \ge 0 \qquad \forall i, m, k \qquad (23.22)$$

The objective function (23.1) seeks to maximize the total number of donations collected. Constraints (23.2) ensure that if a vehicle visits a donation site, it must leave the donation site. Constraints (23.3) guarantee that any donation site is visited at most once in a single tour by a vehicle. Constraints (23.4) and (23.5) state that if a vehicle is not taking a tour, it will not leave the center. It also ensures that a vehicle can visit only one site from the center. Constraints (23.6) ensure that a vehicle cannot begin a new tour if it has not completed the previous one. Constraints (23.7) refer to a logical constraint emphasizing that if a tour is not performed by a vehicle, the tour can be visited by the vehicle in that tour. In a tour, the arrival time of a vehicle is determined by constraints (23.8) and (23.9). Constraints (23.10) guarantee that it is not possible to visit a donation site before it opens. By constraints (23.11), if a donation site is not visited during the first tour of a vehicle, the time of its first visit is set to a_i. Constraints (23.12) and (23.13) ensure that v_{imk} is equal to $v_{i,m-1,k}$ if donation site i is not visited by vehicle k in tour m. Constraints (23.14) determine the arrival time of a vehicle to the processing center after performing a tour. Constraints (23.15) and (23.16) assure that a site's donation can only be collected once and that it can be collected by a vehicle visiting the donation site. Constraints (23.17) and (23.18) impose the visiting time of a donation site in a tour as well as the completion time of this tour if a donation is picked up from this donation site on this tour at a specific time. Finally, constraints (23.19)–(23.22) impose binary, integer, and non-negativity restrictions on decision variables.

23.4 Solution Approaches

Elimination of the clustering phase causes the growth of feasible region of MBCP and increases the complexity of the problem. Besides, the visit time of blood donors to the donation sites is not uniform. The arbitrary pattern of donation makes the problem even more complex. Since the donation pattern is arbitrarily distributed, it becomes difficult to determine the impact of such changes (skipping a donation site or altering the tour start/end time) on a solution, while the literature assumes a uniform donation pattern. Furthermore, using several blood collection vehicles simultaneously results in some partial overlaps in tours. When blood collection vehicles start tours from the blood processing center to collect blood units from donation sites, some vehicles may meet at a given donation site in their tours. Assume two vehicles visit the donation site at different times. The amount of blood units available to collect for the second vehicle is dependent on the meeting time of the first vehicle. In other words, vehicles' performance affects the performance of each other. This donation site has an arbitrary blood donation pattern. Hence, when it is allowed to have overlap on tours, the complexity of the problem expands tremendously.

To solve the problem, a framework is proposed to solve subproblems for each vehicle and combine the subproblems to reach the solution to solve the variant of MBCP. Each subproblem is the blood collection of a single blood collection vehicle. Solving the subproblem for each vehicle separately is an excellent way to eliminate the overlap of tours and reduce the size of the feasible region. However, this approach ignores the overlap of tours for several vehicles. It is proposed to solve subproblems in order and update the amount of blood available for the next subproblem according to the amount of collected blood by the current subproblem to cope with the overlap issue.

23.4.1 *Hybrid Flower Pollination Algorithm*

Flower pollination algorithm is a numerical stochastic optimization metaheuristic method inspired by the pollination mechanism of flowers in nature introduced by [34]. Plants have male and female parts. Male parts of plants have an element called pollen used to fertilize the female part of the plants. Pollination is defined as the transmission of pollen from male parts to female parts of flowers during the fertilization process, which needs an external agent. The most common ways of pollen transmission are transferring by animals and wind, which are called pollinators. Pollination, which is relied on wind and other natural pollinators such as water, is called *abiotic* pollination. On the other hand, some animals such as bees and hummingbirds help the flowers to make the pollination when these animals are going to feed themselves.

Pollination, which is relied on animal pollinators, is called *biotic* pollination. Furthermore, plants have two common types of pollination, namely *cross-pollination*, and *self-pollination*. When pollen transmits from a flower in a given plant to a flower in a different plant, the transmission is called cross-pollination, while for transmission of pollen between flowers of the same plant is called self-pollination. The aim of using different types of pollination by flowering plants arises from the survival endeavors of plants. Flower pollination is the only way that most plants can reproduce and survive. Hence, different pollination methods help plants have a chance to reproduce a more adaptive plant. Inspiring from the behavior of plants' pollination, the flower pollination algorithm can reproduce fitter solutions over time by selecting the fittest solution to survive and reproduce. Despite the well-known genetic algorithm where reproduction occurs only for a couple of parents, in flower pollination algorithm, a set of solutions reproduce in each iteration. In other words, flower pollination algorithm uses a pool of solutions to find a better solution by reproducing lots of new generation solutions.

In this study, flower pollination algorithm is used as a solution approach. The superiority of this algorithm to the well-known powerful algorithms such as genetic algorithm and particle swarm optimization algorithm is shown by [34]. The flower pollination algorithm is a population-based algorithm, which starts with a random initial population. After creating the initial population, the flower pollination algorithm finds the fitness of each flower in the initial population. The fitness value is determined by counting the units of blood collected and processed within PTL. Then, using some techniques, the algorithm pollinates the flowers and creates new solutions. If the new solution is fitter than solutions in the initial population, they will be replaced in the population. This procedure continues for a given number of iterations. The best solution in the latest iteration is the solution derived from the flower pollination algorithm.

In order to fit the flower pollination algorithm to the variant of MBCP, an HFPA presented in Algorithm 1, is proposed. The problem has a two-dimensional solution, namely tours and departure times. The first dimension is the tour composed of routes, and the second dimension of the solution is associated with the departure times of each route. In other words, a solution for the problem includes a tour with associated departure times for the tour.

The flower pollination algorithm is used to find a high-quality set of tours for vehicles, and an algorithm that is inspired by an *invasive weed optimization* (IWO) algorithm is used to find high-quality departure times for these tours.

Algorithm 1 *Hybrid Flower Pollination Algorithm*

```
N := min(Number of sufficient routes,Number of feasible routes)
R := Set of routes with one donation site
while Size(R) ≤ N do
    R := R+Set of routes with one more site than R
end while
σ_initial := 1, σ_final := 0.01, iter_max := 20, P_max := 200
for vehicle ∈ Vehicles do
    ρ := Random tours/flowers from R
    t_ρ := Random departure times for ρ
    F_ρ,t := Fitness value of ρ&t
    for iter ∈ iter_max do
        F_max := Max(F_ρ,t)
        F_min := Min(F_ρ,t)
        BestSol := flower_F_max
        σ_iter = ⌈σ_final + (σ_initial − σ_final) × ((iter_max − iter)/iter_max)^n⌉
        for flower ∈ Flowers do
            p :=random (0, 1)
            if p ≤ α then
                swap two random pollens between two flowers of the same plant
            else if p ≤ α then
                swap a random pollen of flower with pollen of BestSol.
            else
                swap a random route in flower with a random route of a random flower ∈ Flowers
            end if
            NewFlower := flower
            t_NewFlower := Random departure times for NewFlower
            for i ∈ ⌈10 × σ_iter⌉ do
                t_temp := Random departure times for NewFlower
                if F_NewFlower,t_temp ≥ F_NewFlower,t_NewFlower then
                    t_NewFlower := t_temp
                end if
                add ρ_NewFlower,t_NewFlower to ρ
            end for
        end for
        while size(ρ) ≥ P_max do
            Remove the weakest weeds according to the fitness from ρ
        end while
    end for
    ρ* := ρ_maxFitness
    t_ρ* := t_maxFitness
    MaximumCollected_vehicle := F_ρ*,t_ρ*
    Update blood collection schedule regarding ρ* & t_ρ*
end for
MaximumCollected := Σ_{vehicle∈Vehicles} MaximumCollected_vehicle
```

23.4.1.1 Initial Population

The non-clustered variant of MBCP is a problem with large amount of possible solutions. In HFPA, the proposed solution representation is a set of tours and a set of departure times associated with the tours. Each tour consists of several routes that a vehicle can complete blood collection on them. In order to create an initial

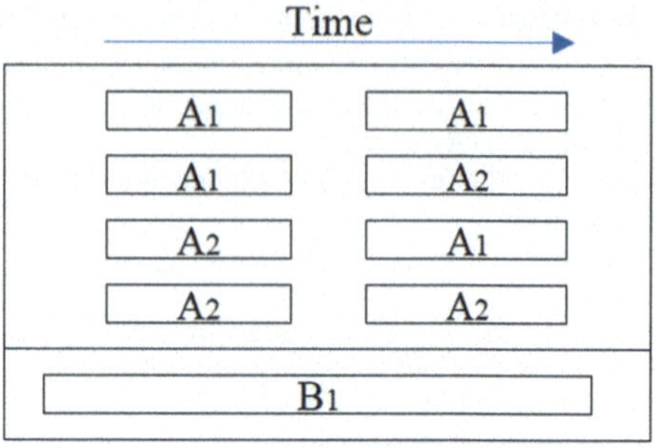

Fig. 23.2 An example of short and long route effect on tour creation

population for HFPA, a sufficiently large number of possible routes is needed. Then, using the HFPA set of the best routes as a tour for a given vehicle will be selected. In other words, as it is illustrated in Algorithm 1, HFPA will find a good combination of routes for each vehicle to collect a large number of blood units suitable for platelet production. Therefore, it is necessary to have a sufficiently large number of possible routes for the problem as input data for HFPA. The proposed HFPA starts the initial population by creating a set of possible routes. Then HFPA uses the set to create a combination of routes for each vehicle as a tour. Then set of these tours is considered as the solution for the problem.

It might be impractical to create all the possible routes. Hence, the proposed HFPA limits the number of possible routes using a simple rule. Due to the fact that shorter routes can create more combinations of routes than more extensive routes, the HFPA starts creating routes using the shortest one and then add new, more extensive routes to the set of possible routes until the set has a large number of routes. This process loses some larger routes in some instances, but it saves lots of time in creating possible routes while the created possible routes set has a good enough diversity to create a high-quality combination among them as a solution for a vehicle. When the set of possible routes is created, HFPA randomly selects several routes from the set and creates a tour for each vehicle. A simple example of this issue is shown in Fig. 23.2. In this example, two short routes, namely A_1 and A_2, are able to create four tours, while for the setting, an extensive tour denoted by B_1 only can produce one tour. However, different departure times for these routes can produce several different solutions using any of the tours in the Fig. 23.2. This example shows that short routes can create a larger combination of routes (tours). As a result, using shorter routes to create tours will give a larger number of solutions.

The randomly selected tours need associated departure times to be a set of solutions. In order to create an associated solution, HFPA uses a Dirichlet distribution. At first, the HFPA calculates the idle time of blood collection vehicles, and then it divides the idle time among the routes of a tour randomly using the Dirichlet distribution. Therefore, the initial population is formed by sets of randomly created tours with associated departure times. In this solution approach, each set of the tours is a flower in HFPA.

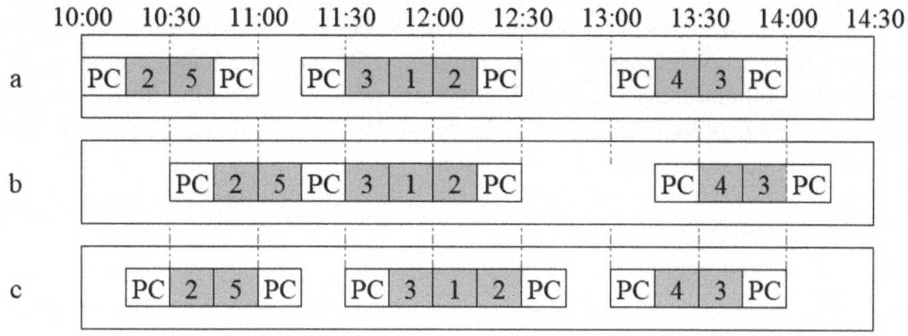

Fig. 23.3 An example of creating random departure times for a tour

As a simple example of associating departure times, a tour with three randomly created associated departure times is presented in Fig. 23.3. A vehicle will perform a tour on three determined routes as a tour. For instance, the first route of the tour "a" starts from the processing center at 10:00 toward site 2. After visiting the site 2 at 10:15, the vehicle goes on to visit the site 5 at 10:30. Next, the vehicle returns to the processing center with collected blood units. In the second route, the vehicle leaves the processing center at 11:15 to visit the site 3 at 11:30, and after visiting the site 2 at 12:00, it returns to the processing center. Accordingly, the vehicle visits the third route and returns to the processing center at 14:00. In this example, the vehicle is idle for 120 min out of 270 min available for the tour. Using the Dirichlet's distribution, the 120 min of idle time distributed between the routes is as follows:

- Tour a: 0, 30, 45, 45
- Tour b: 30, 0, 69, 30
- Tour c: 15, 30, 30, 45

It is crystal clear that the tour in the example for each departure time set will have different collected blood units that can be used for platelet production (fitness value). Therefore, it is essential to find good associated departure times for a tour.

In the next step, the HFPA determines each solution's fitness level by calculating the number of blood units collected by the solution. HFPA finds the best solution from the initial population based on the fitness value. The best solution in iteration should be determined because the HFPA will use the best flower to reproduce new plants.

23.4.1.2 Pollination and Reproduction

A given number of flowers forms the initial population. The flowering plants are distributed in different places in a field, while each one has several flowers. These flowers are pollinated by the global pollination process or the local pollination. Each flower represents a solution. In HFPA, it is considered that biotic pollinators are used for cross-pollination. In other words, it is assumed that each flower has one pollen, and the pollens are carried by animals such as bees between the different plant's flowers which is considered as cross-pollination. On the other hand, it is assumed that abiotic pollinators such as wind only carry pollen between the flowers of the same plant, a self-pollination type of pollination. The local pollination and the global pollination occur randomly in each iteration.

In the proposed HFPA shown in Algorithm 1, three pollination techniques are used. In each iteration of the algorithm, only one of three methods can occur randomly. The pollination techniques of HFPA are as follows:

- Local pollination method that exchanges the pollens of a given flower by a random flower of the same plant by an abiotic pollinator. In order to fit the shape of the problem to the solution approach, in this method, a random route from a given tour will be swapped with a random route from the population.
- Global pollination method that swaps the pollen of a given plant's flower by a random flower from a different plant. Then, a route from a given tour will be swapped by another route from different possible routes set.
- Swapping method that swaps the pollen of a given flower by the highest fertility flower. This method increases the fitter pollens' chance of reproduction.

When all flowers in the population are pollinated, the next iteration will iterate, and a set of new plants and flowers will grow accordingly.

23.4.1.3 Spatial Dispersal

Each tour needs departure times, and the pollination only has created the tours while tours are not considered a solution without departure times. Therefore, HFPA should be able to handle fitting intense departure times for the tours to increase the number of collected blood units by the vehicles. In HFPA, it is proposed to use an operator of IWO algorithm called spatial dispersal to find high-quality departure times for blood collection vehicles. IWO algorithm is a metaheuristic optimization algorithm inspired by weed colonization introduced by [18]. In IWO, weeds represent solutions like flowers in flower pollination algorithm. Weeds in the fields tend to grow invasively and occupy all the cropping fields, while there are several limitations on the resources of field and free spaces of field. Weeds with higher fitness to the environment reproduce rapidly, while weak weeds are extincted. Spatial dispersal is a powerful tool in IWO used to avoid being stuck in a local optimum.

Spatial dispersal is a random solution generation method that uses a normal distribution with a zero mean and a decreasing variance over time. The zero mean causes flowers to pollinate randomly in the field while the decreasing variance by iteration converges the associated departure times for a tour to a higher quality. As iterations go further, using an equation, the variance of the distribution goes lower. When this method is used in associated departure times generation, it helps to have a great diversity for tours in initial iterations. Also, it concentrates on departure times that show a high quality during the iterations, and in the final iterations, a small number of good enough associated departure times are introduced for the tours from the previous step of HFPA. The spatial dispersal saves lots of time by proposing fewer random departure times as the iterations go further. Besides, when a departure time is proposed for a tour, the same fitness function as in the initial population is used to determine the fitness level of each solution. The fitness function calculates the quality of each solution containing the tour and associated departure times.

In HFPA, initial amount for the standard deviation which is represented by σ is given. The update rule of σ in each iteration is according to the following equation:

$$\sigma_{iter} = \sigma_{final} + (\sigma_{initial} - \sigma_{final}) \times (\frac{iter_{max} - iter}{iter_{max}})^n$$

where σ_{iter}, $\sigma_{initial}$, and σ_{final} are the standard deviation of current, initial, and final iterations respectively. $\sigma_{initial}$ is set to one and σ_{final} has a value less than $\sigma_{initial}$.

In order to fit spatial dispersal with associated departure times, the departure times are considered as seeds of weeds. Spatial dispersal is going to distribute the seeds in the cropping field. The method to create departure times for a tour is the same method used in initial population iterations. In other words, spatial dispersion is used to select fit departure times within a proper timeframe by creating multiple randomly generated departure times for a single tour. In each iteration, a number ($\lceil 10 \times \sigma_{iter} \rceil$) of randomly generated departure times associated with the tours are created using a Dirichlet distribution. IWO proposes several departure times associated with each tour. The availability of different departure times for a single tour increases the chance of finding higher quality solutions, but the random generation of several departure times takes much time to be tested by each tour. However, it is a good idea to use lots of departure times for a tour, but in order to save time as well as the quality of weeds, IWO is utilized to decrease the number of randomly distributed seeds iteration by iteration. When new seeds mature, the departure times will be fitter than the previous ones; hence IWO will decrease the number of random departure times after a while. The on-hand departure times, in combination with the associated tour, probably will be strong enough to generate a high-quality solution. Consequently, considering the behavior of associated departure times generation, which is iteration-based, the HFPA uses the σ_{iter} value in each iteration to disperse the seeds reproduced by weeds to the solution space. The σ_{iter} derived from the above-mentioned equation helps the HFPA to keep a large amount of diversity in the initial iteration for the seeds, leading to the avoidance of HFPA from getting stuck in local optimum solutions. In addition, in the later iterations, the σ_{iter} saves time and causes the reproduction of the solution with higher quality.

23.4.1.4 Competitive Deprivation

In the last step of HFPA iterations, the quality of flowers in the population is checked, and with a greedy selection, the weak flowers are eliminated. The weak flowers are replaced by newly born flowers that are more adapted to the field and have a better chance to reproduce high-quality seeds. This prevents the time consumption by pollination of weak flowers that do not have a high chance of reproducing high-quality solutions. The population size remains the same in every iteration.

Removing the flowers with the lowest fitness value will not reduce the fitness of the population. The fitness of all the solutions in the population will increase or remain the same in every iteration so that the algorithm will find the solution with the highest fitness value in its last iteration. In other words, The solution with the largest fitness value will be selected as the final solution when the iteration finishes.

23.4.2 A Simulated Annealing Algorithm

The SA proposed by [13] is an iterative metaheuristic approach to solve non-convex optimization problems. SA is inspired by metallurgical annealing, which is a technique to increase metal quality by controlling heat and cold. In SA, the procedure starts with an initial sufficient temperature to have a wide range of neighbor solutions. Temperature decreases slowly by iterations leading to a reduction in the neighbor solutions range, and the algorithm updates the current solution with a better neighbor solution. A fitness function evaluates the quality of each solution. After a while, the algorithm terminates at a near-optimal solution. Additionally, SA accepts some transitions to a worse neighbor solution with a probability determined by Boltzmann in the equation ($P = e^{-\theta/\mathcal{T}}$) to avoid getting stuck in local optima [13]. The difference between the best solution fitness value so far and the newly generated solution fitness value is denoted by θ, and $\mathcal{T}$ is the temperature that is decreasing during the algorithm iterations till the termination criteria are activated. The approach is presented in Algorithm 2.

In order to create new solutions by the SA algorithm, initializing from a random route, a tour (solution) is being created. Then, using the route in the tour, a new route will be created through three local search methods. The newly created route will be added up to the tour. This process will be terminated either when the tour gets too large or when a random probability meets a given amount. Then, using the Dirichlet distribution, departure times for the routes inside the tour will be randomly determined. Finding associated departure time is the same method used in the HFPA algorithm.

Algorithm 2 uses three local search methods to create a new tour, namely *swap*, *2-opt*, and *skip* operators as follows:

- Swap operator creates a new solution by selecting two random routes from a tour and swapping the order of two selected routes. As shown in Fig. 23.4, a simple example of this operator for a tour includes routes represented by numbers in an order from 1 to 6. If two randomly selected routes in the tour are 2 and 5, then the order of the new solution is 1, 5, 3, 4, 2, 6.
- 2-opt operator works like a swap operator and selects two random routes among the tour. Then it reverses all the order between these two selected routes. For the example mentioned above, the results will be 1, 5, 4, 3, 2, 6.

Algorithm 2 Simulated Annealing Algorithm

δ := a small enough number
S := random route
t_s := random start time for route
add $S\&t_s$ to ψ
F_ψ :=, Fitness value of ψ
for $v \in$ Vehicles **do**
  while $\mathcal{T} \geq \delta$ **do**
    while $Rand(0, 1) \geq 0.3\&$ Duration (ψ+ShortestRoute) $\leq$ SitesWorkingHours **do**
      $p := Random(0, 1)$
      if $p \leq 0.3$ **then**
        NewSol := 2-opt (S)
      else if $0.3 < p \leq 0.6$ **then**
        NewSol := Skip (S)
      else if $0.6 < p \leq 0.9$ **then**
        NewSol := Swap (S)
      else
        NewSol := a random solution
      end if
      S := NewSol
      t_s := random start time for route
      add $S\&t_s$to ψ
    end while
    $\theta := F_\psi - F_{\psi_*}$
    if $\theta \geq 0$ **then**
      $\psi_* := \psi$
    else if $e^{\frac{-\theta}{\mathcal{T}}} > 0.8$ **then**
      $\psi_* := \psi$
    end if
    $\mathcal{T} := \mathcal{T} - \delta$
  end while
  $MaximumCollected_{vehicle} := F_{\psi_*}$
  Update blood collection schedule regarding ψ_*
end for
$MaximumCollected := \sum_{vehicle \in Vehicles} MaximumCollected_{vehicle}$

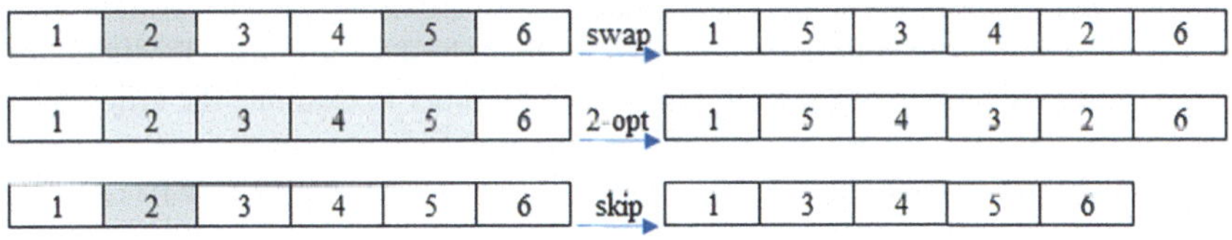

Fig. 23.4 A chromosome representation example of three operators for creating a new solution

- Skip operator is used to skip a route in a tour to create a neighbor solution. This operator selects a random route and removes it from the tour to create the new solution. For the example mentioned for the swap operator, if the operator randomly selects route 2, the new solution will be 1, 3, 4, 5, 6.

The proposed SA keeps diversity by using an entirely random solution creation method in $\alpha\%$ of iterations. There is $\alpha\%$ chance to create a new solution randomly without any dependency on previous solutions in each iteration. This feature helps SA to avoid getting stuck at local optima. Each iteration (temperature) randomly selects one of the three operators or the random method and generates a new route to create a new solution. Also, using the Dirichlet distribution, 10 possible visit times are randomly generated for any new solution, and the one with the highest fitness value, as the visit time of the solution is selected. In other words, the setting up of departure times for vehicles from the processing center is based on a random process.

The proposed SA, as shown in Algorithm 2 for the non-clustered variant of MBCP, has a deterministic initial temperature.

In Algorithm 2, the initial temperature is set to $\mathcal{T}$ degrees, and it decreases in each iteration by δ degrees. Based on a sensitivity analysis, the initial temperature and cooling are determined. The solution representation is a set of chromosomes in which each gene of a chromosome represents a donation site and its order to be visited, and the whole chromosome is the solution for a single vehicle (subproblem). Then, solutions for each vehicle will affect the collected blood units from the blood donation sites, and the number of accumulated blood units will be updated according to the amount collected by a given vehicle. This updating will be done for each vehicle in sequential order. When the collection process finishes, the algorithm combines the number of blood units collected by all the vehicles to find the total collected blood units during the day. In order to find the amount, the algorithm checks all the blood donation sites visited by more than one vehicle. If there is an overlap between the collections routes, the algorithm will assign the blood units to the vehicles according to the time sequence of the visit. The values of $\mathcal{T}$, δ, and α are set to 200, 0.01, and 10 based on a Taguchi method.

23.5 Computational Study

This section presents some experimental analyses of HFPA and SA performance for the non-clustered variant of MBCP compared with the best algorithms in the literature for the clustered version of the problem. Hence, in order to examine the efficacy of the proposed HFPA and SA algorithms, a comprehensive computational study is used. The problem is solved using proposed algorithms for different PTLs. The performance of the proposed algorithms is studied compared to the best benchmark solution in the literature.

The results of the computation study using the proposed algorithms are presented in Table 23.1. The computational experiments are carried out on a computer running 64-bit Windows with two 1.8 GHz i7-8550 CPUs and 8 GB of RAM. The algorithms are implemented using Python 3.8 programming language.

Table 23.1 Performance of two proposed algorithms compared to OEH

PTL	# BDS	# BCV	SA				HFPA				OEH		
			ARI[1]	CP[2]	PFP[3]	Runtime (s)	ARI	CP	PPP	Runtime (s)	CP	PPP	Runtime (s)
300	**28**	**6**	1,94%	108	17,73%	59	9,71%	116	19,05%	80	106	17,41%	16
	32	**6**	5,39%	181	24,83%	180	13,77%	195	26,75%	191	172	23,59%	27
	36	**7**	4,50%	212	28,04%	163	13,00%	229	30,29%	184	203	26,85%	26
	38	**8**	−2,66%	378	44,11%	676	1,60%	394	45,97%	450	388	45,27%	45
	43	**9**	6,43%	303	29,36%	413	13,21%	322	31,20%	404	287	27,81%	45
	46	**10**	14,55%	254	23,09%	250	17,73%	261	23,73%	273	222	20,18%	46
	51	**10**	14,98%	317	25,84%	490	19,48%	329	26,81%	457	276	22,49%	57
	Average		**6,45%**	**250**	**27,57%**	**319**	**12,64%**	**264**	**29,11%**	**292**	**236**	**26,23%**	**37**
500	**28**	**6**	1,81%	225	36,95%	110	2,26%	226	37,11%	196	221	36,29%	70
	32	**6**	2,57%	359	49,25%	470	1,14%	254	34,84%	362	350	48,01%	79
	36	**7**	7,34%	380	50,26%	318	10,73%	392	51,85%	342	354	46,83%	90
	38	**8**	−5,77%	526	61,38%	824	−3,06%	541	63,13%	575	558	65,11%	88
	43	**9**	4,83%	578	56,01%	771	5,58%	582	56,40%	617	551	53,39%	108
	46	**10**	8,85%	547	49,73%	950	7,24%	539	49,00%	712	502	45,64%	109
	51	**10**	−2,36%	629	51,26%	756	−2,36%	629	51,26%	590	644	52,49%	131
	Average		**2,47%**	**463**	**50,59%**	**600**	**3,08%**	**452**	**49,08%**	**485**	**454**	**49,68%**	**96**
700	**28**	**6**	14,94%	400	55,58%	492	13,79%	396	65,02%	319	348	57,14%	91
	32	**6**	1,07%	473	54,38%	612	−1,50%	461	63,24%	376	470	64,47%	104
	36	**7**	7,93%	517	58,39%	464	5,01%	503	66,53%	413	480	63,49%	111
	38	**8**	−1,89%	674	78,65%	729	−0,15%	686	80,05%	484	687	80,16%	113
	43	**9**	−2,53%	764	74,03%	759	−3,20%	759	73,55%	565	783	75,87%	144
	46	**10**	1,94%	787	71,55%	797	0,14%	773	70,27%	623	772	70,18%	142
	51	**10**	−7,51%	835	68,05%	1191	−6,32%	846	68,95%	729	903	73,59%	169
	Average		**1,99%**	**636**	**70,18%**	**721**	**1,11%**	**632**	**69,66%**	**501**	**634**	**69,27%**	**70**
Average			**3,64%**	**450**	**49,48%**	**546**	**5,61%**	**449**	**49,29%**	**426**	**442**	**48,39%**	**110**

[1] Average Relative Improvement w.r.t. OEH (ARI) = $\frac{\sum_{1}^{R} \frac{alg_i - OEH}{OEH}}{R} \times 100$, where alg_i is the result of run i for a given instance and R is the number of runs by the algorithm

[2] Collected Platelet Units

[3] Percentage of Produced Platelet out of Donated Blood Units

23.5.1 Dataset Configuration

A dataset for MBCP generated by [23] is used to assess the performance of the proposed HFPA and SA. There are seven different types of blood donation sites maps, and each map has a randomly generated blood donation pattern. Each map is solved using a given number of blood collection vehicles for different PTLs. The reported results are the average values for 15 times solving any instance. The dataset consists of several different numbers of blood collection vehicles for each instance as follows: $\{6, 7, 8, 9, 10\}$. Besides, in order to avoid wasting the platelet extraction possibility, the blood's PTL is considered 300, 500, and 700 min for any instance. Seven instances are generated for each PTL, and each instance has a specific number of blood collection vehicles. In addition, different numbers of blood donation sites are designed for each instance. The smallest instance for PTL having 300 min is using six vehicles to collect donated blood units from 28 blood donation sites distributed randomly, while the largest instance includes ten blood collection vehicles to collect blood units from 51 blood donation sites in 700 min PTL. Besides, travel distances are Euclidean distances.

Moreover, the same setup parameters and conditions for both proposed algorithms are carried out to obtain reliable experimental results. The measure of merit to evaluate the performance of the proposed algorithms for the variant of MBCP is the comparison with OEH, which is the best solution in the literature for the MBCP. It is assumed that all blood collection vehicles are available for all donation sites throughout the working hours.

23.5.2 Results and Discussion

Avoiding clustering of blood donation sites makes MBCP a more challenging problem with a larger feasible region than MBCP. In other words, the growth of the feasible region in MBCP increases the complexity of the problem, making it hard to solve compared to solving MBCP. The results in Table 23.1 present the average relative percentage improvement of the proposed SA and HFPA over OEH for 21 different randomly generated instances. These instances are solved for three different PTL values for seven different blood donation sites maps and a randomly generated blood donation pattern for each map.

According to the results of 15 runs, the proposed HFPA and SA outperform OEH on average with 5.61% and 3.64%, respectively. On average, HFPA provides better solutions than SA whenever the PTL values are 300 and 500. However, for the PTL value of 700, SA provides slightly better solutions. Out of 21 instances, SA outperforms HFPA and OEH in five instances, while HFPA outperforms SA and OEH in ten instances. Table 23.1 also represents the runtime of the proposed algorithms. On average, HFPA is faster than SA. However, for the instance with 300 PTL, 28 blood donation sites, and six blood collection vehicles, SA can find a solution in 59 s, while it takes HFPA 80 s to find the solution. For this instance, the improvement in solution by HFPA is 9.71% compared to merely 1.94% by SA. Furthermore, OEH's superior run time compared to HFPA and SA is due to the parallel processing of operations in OEH. Since OEH benefits from clustering blood donation sites, it solves a significantly reduced-size MBCP. In spite of the significantly higher complexity of non-clustered MBCP, HFPA and SA perform well in run time.

23.6 Conclusion

Healthcare systems need never-ending blood products reservoirs to keep their medical treatments. Despite all technological developments, blood donation is the only source of blood products that are extracted from donated blood. This study focuses on platelets as a blood product that plays a vital role in the survival of patients in lots of medical areas. An essential attribute of platelet production is the perishable nature of blood after donation, making it complicated to collect the donated units and extract platelets from the blood. The donated blood should be extracted from its components in less than six hours to avoid perishability. In addition, blood donors always visit donation sites with an arbitrary pattern to donate blood. Therefore, scheduling the blood collection vehicles to collect donated blood units from the donation site considering the time constraint for each vehicle as well as the donation pattern is a complex blood collection problem. This study proposes a non-clustered maximal blood collection problem that focuses on the maximization of produced platelet. Ignoring the initial clustering of the donation sites in this problem increases the problem's size, but it leads to a larger feasible region and higher quality solutions. In order to solve the problem, two metaheuristic algorithms, namely simulated annealing and hybrid flower pollination algorithms, are implemented for this problem for the first time to maximize the number of collected blood suitable for platelet production. The hybrid flower pollination algorithm has the advantage of being able to deal with two-dimensional solutions by using a powerful invasive weed optimization technique in setting departure times of blood collection vehicles in blood collection routing. This algorithm shows superior efficacy in solving the problem. Besides, the proposed simulated annealing shows a good quality to solve problems with a larger processing time limit. The obtained numerical results demonstrate that the proposed model and algorithms, notably the hybrid flower pollination algorithm, are significantly more cost-effective and practical than the currently available literature solutions for clustered problems by 5.61%.

Future directions may be based on a more realistic view of MBCP, taking drivers' behavior into account. Drivers who are familiar with a route might perform better than those who are new to it. Moreover, considering the uncertainties of disruptions in routes such as floods and accidents can lead to more resilient solutions to this problem. Furthermore, sharing collected blood units between vehicles can create a new problem with possible better solutions than current ones.

References

1. Alfonso E, Xie X, Augusto V, Garraud O (2012) Modeling and simulation of blood collection systems. Health Care Manag Sci 15(1):63–78
2. Baş S, Carello G, Lanzarone E, Yalçındağ S (2018) An appointment scheduling framework to balance the production of blood units from donation. Eur J Oper Res 265(3):1124–1143
3. Blake J, Heddle N, Hardy M, Barty R (2010) Simplified platelet ordering using shortage and outdate targets. Int J Health Plann Manag 1:144–156
4. Blood Centers of the Pacific (2016) http://www.bloodcenters.org
5. Do Carmo BBT, Gurgel JLM, Junior AFS, De Sena DC (2013) Demand forecast and inventory management: Sizing inventory of blood products in a blood bank in Brazil. In: Schoenherr T (ed) Proceedings of the 24th meeting of the production and operations management society
6. Chao IM, Golden BL, Wasil EA (1996) The team orienteering problem. Eur J Oper Res 88(3):464–474
7. Cheraghi S, Hosseini-Motlagh SM, Ghatreh Samani M (2017) Integrated planning for blood platelet production: A robust optimization approach. J Ind Syst Eng 10:55–80
8. Dehghani M, Abbasi B, Oliveira F (2021) Proactive transshipment in the blood supply chain: A stochastic programming approach. Omega 98:102112
9. Ensafian H, Yaghoubi S (2017) Robust optimization model for integrated procurement, production and distribution in platelet supply chain. Transp Res Part E: Logistics Transp Rev 103:32–55
10. Gunpinar S, Centeno G (2016) An integer programming approach to the bloodmobile routing problem. Transp Res Part E: Logistics Transp Rev 86:94–115
11. Hamdan B, Diabat A (2020) Robust design of blood supply chains under risk of disruptions using lagrangian relaxation. Transp Res Part E: Logistics Transp Rev 134:101764
12. Katsaliaki K, Brailsford SC (2007) Using simulation to improve the blood supply chain. J Oper Res Soc 58(2):219–227
13. Kirkpatrick S, Gelatt CD, Vecchi MP (1983) Optimization by simulated annealing. Science 220(4598):671–680
14. Lenstra JK, Kan AR (1981) Complexity of vehicle routing and scheduling problems. Networks 11(2): 221–227
15. Li X, Tian P, Leung SCH (2010) Vehicle routing problems with time windows and stochastic travel and service times: Models and algorithm. Int J Prod Econ 125(1):137–145
16. Li X, Fan H, Liu J, Xun Q (2020) Staff scheduling in blood collection problems. Ann Oper Res 1–36
17. Lodree EJ, Carter D, Barbee E (2016) The donation collections routing problem. In: International conference on dynamics of disasters, Springer, Berlin, pp 159–189
18. Mehrabian AR, Lucas C (2006) A novel numerical optimization algorithm inspired from weed colonization. Ecol Inf 1(4):355–366
19. Mobasher A, Ekici A, Özener OÖ (2015) Coordinating collection and appointment scheduling operations at the blood donation sites. Comput Ind Eng 87:260–266
20. Nagurney A, Masoumi AH, Yu M (2012) Supply chain network operations management of a blood banking system with cost and risk minimization. Comput Manag Sci 9(2):205–231
21. Nahmias S (2011) Perishable inventory systems. Springer, Berlin
22. Osorio AF, Brailsford SC, Smith HK (2015) A structured review of quantitative models in the blood supply chain: A taxonomic framework for decision-making. Int J Prod Res 53(24):7191–7212
23. Özener OÖ, Ekici A (2018) Managing platelet supply through improved routing of blood collection vehicles. Comput Oper Res 98:113–126
24. Pirabán A, Guerrero WJ, Labadie N (2019) Survey on blood supply chain management: Models and methods. Comput Oper Res 112:104756
25. Rajendran S, Ravindran AR (2017) Platelet ordering policies at hospitals using stochastic integer programming model and heuristic approaches to reduce wastage. Comput Ind Eng 110:151–164

26. Şahinyazan FG, Kara BY, Taner MR (2015) Selective vehicle routing for a mobile blood donation system. Eur J Oper Res 245(1):22–34
27. Savelsbergh MWP (1985) Local search in routing problems with time windows. Ann Oper Res 4(1):285–305
28. Stanger SH, Yates N, Wilding R, Cotton S (2012) Blood inventory management: Hospital best practice. Trans Med Rev 26(2):153–163
29. The American Red Cross Blood Services (2021a) https://www.redcrossblood.org/donate-blood/how-to-donate/how-blood-donations-help/blood-needs-blood-supply.html
30. The American Red Cross Blood Services (2021b) https://www.redcrossblood.org/donate-blood/how-to-donate/types-of-blood-donations/platelet-donation.html
31. Van ZGJJ (1963) Inventory control for perishable commodities. Technical report, North Carolina State University. Department of Statistics
32. Williams EP, Harper PR, Gartner D (2020) Modelling of the collections process in the blood supply chain: A literature review. IISE Trans Healthcare Syst Eng (Just-accepted):1–24
33. Yaghoubi S, Hosseini-Motlagh SM, Cheraghi S, Larimi NG (2020) Designing a robust demand-differentiated platelet supply chain network under disruption and uncertainty. J Ambient Intell Humanized Comput 11(8):3231–3258
34. Yang XS (2012) Flower pollination algorithm for global optimization. In: International conference on unconventional computing and natural computation, Springer, Berlin, pp 240–249
35. Yi J (2003) Vehicle routing with time windows and time-dependent rewards: A problem from the American Red Cross. Manuf Serv Oper Manag 5(1):74–77
36. Zahiri B, Pishvaee MS (2017) Blood supply chain network design considering blood group compatibility under uncertainty. Int J Prod Res 55(7):2013–2033

Chapter 24
NSGA-II and TOPSIS for a Multi-objective Vehicle Routing Problem with Ecological Considerations

Sanja Petrovic, Kingshuk Jubaer Islam, and Alexander Trautrims

Abstract Reducing CO_2 emissions in transport and logistics is currently a goal of utmost importance in vehicle routing. Environmental awareness has grown in recent years, and organizations are more willing to consider sustainability in their decision-making at all levels, from strategic to operational. This research is concerned with the vehicle routing problem (VRP) with simultaneous pickup and delivery (VRPSPD), which considers not only economic objectives but also environmental ones. Although environmental issues have been included in VRP models, traditionally multiple objectives have been combined in a single-objective function. In these approaches, a decision maker (e.g. the logistic manager) has to determine the importance of objectives a priori, which is often a difficult task. In our research, the VRPSPD is modeled as a multi-objective optimization problem where the objectives are minimization of travelled distance and fuel consumption, the latter being equivalent to CO_2 emissions. The development of the model is informed by a real-word VRPSPD from our industrial collaborator. A method for multi-objective optimization based on NSGA-II is developed, with a novel constructive heuristic that uses a multi-criteria decision-making method TOPSIS to generate initial solutions. Statistical analysis proves that initial populations that combine different ratios of solutions generated by TOPSIS to randomly generated solutions produce significantly different final Pareto fronts. Our algorithm produces solutions in which the travelled distance is comparable or smaller than the distance in the solutions generated by the company, but in a majority of instances reduces the fuel consumption. In addition, the performance of the developed algorithm is evaluated on problem instances of larger size taken from the literature. The obtained Pareto fronts with multiple solutions give an opportunity

S. Petrovic (✉) · K. J. Islam · A. Trautrims
Division of Operations Management and Information Systems, Nottingham University Business School, Jubilee Campus, Wollaton Road, Nottingham NG8 1BB, UK
e-mail: sanja.petrovic@nottingham.ac.uk

K. J. Islam
e-mail: kingshuk.islam@gmail.com

A. Trautrims
e-mail: alexander.trautrims@nottingham.ac.uk

F. Hamid (ed.), *Optimization Essentials*, International Series in Operations Research & Management Science 353, https://doi.org/10.1007/978-981-99-5491-9_24

to the logistics manager to express a posteriori preferences toward the importance of objectives and choose a solution correspondingly.

Keywords Multi-objective vehicle routing problem · Simultaneous pickup and delivery · Multi-objective evolutionary optimization · NSGA-II · Initialization of GA · TOPSIS · CO_2 emission

24.1 Introduction

In light of the continuously growing environmental awareness, companies are facing challenges to reduce or even eliminate the environmental impact of their operations. One of the key concepts proposed is "low energy consumption, low pollution and low emissions" [45], which leads to the concept of back flow of products or reverse logistics. Reverse Logistics (RL) is the field of supply chain management that aims to integrate the forward flow of products from producers to customers with a backward flow of used products or waste. Re-filling of containers and re-using and recycling of materials has become one of the essential initiatives in RL. The Vehicle Routing Problem (VRP) denotes a class of combinatorial optimization problems where customers are served by a number of vehicles, each customer being characterized by a certain demand [15]. In RL, this demand could be expressed in terms of the number of products to be picked up and delivered at the same time. Such types of problems are referred to as Vehicle Routing Problem with Simultaneous Pickup and Delivery (VRPSPD). The VRPSPD naturally arises in the planning of RL operations, for example, in the planning of the delivery of full bottles and collection of empty ones [32]. In the planning of routes for vehicles, distances between the depot and customers and the capacity of each vehicle have to be considered. Single objectives commonly used in the VRP are to minimize the travel distance, travel time, number of vehicles, or transportation cost.

The reported statistics show that fuel cost represents a significant proportion of total transportation cost. Xiao et al. [43] pointed out that for a road transportation company in Shanghai, China, the fuel cost (amounting to about US$1,550,000 per year) accounted for 67.41% of the total transportation cost. Following the increase of awareness of environmental issues in transport and logistics, new extensions of the classical VRP have emerged, which have been grouped into three groups: VRP in RL, Green VRP, and Pollution Routing Problem [28]. In these problems, additional objectives related to environmental issues were considered. Different models to calculate the CO_2 emissions were developed that could be classified as macroscopic and microscopic models [29]. The macroscopic models use simple expressions for calculating the CO_2 emission based on curb weight, vehicle load, road angle, gravitational constant, etc. [10]. The microscopic models incorporate more details such as the characteristics of vehicles, speed of vehicles, and characteristic of roads. Barth et al. [3] proposed a microscopic model that considered characteristics of heavy-goods vehicles such as their load and average speed. This model was often in use in the VRP due to its easy application, for example in [2, 12, 24, 37]. The CO_2 emission is directly proportional to the amount of fuel consumed by a vehicle. Tiwari and Chang [39] calculated the total emitted CO_2 on a rather crude level based on the total load of the vehicle on the whole route, the average distance travelled by the vehicle, and average value of CO_2 emission factor. The CO_2 emission is often calculated taking into account the load of the vehicle and travel distance [22, 24, 26]. In the last 10–15 years, there has been an increased interest in the VRP community to consider a trade-off between the speed of vehicles and the incurred cost, i.e. lower speed incurs longer driving time and consequently increases in the cost and vice versa [24, 36]. However, the current practise in the majority of industries that deal with the road freight transportation does not impose any control (optimization) of the speed of vehicles, simply because the drivers would not adhere to it.

In the majority of approaches to model VRP, multiple objectives that addressed economic and environmental costs were combined in a single-objective function (for example, [31]). In these approaches, a decision maker (e.g. the logistic manager) has to determine the importance of the objectives a priori, which is often a difficult task. Single-objective optimization methods produce only one solution, which may not address adequately the preferences of the decision maker. The decision maker has only an option to change the importance of the objectives in a multi-objective scenario and investigate the effect this produces. Additionally, weights usually have to be expressed by numerical values, which often creates an added burden for the decision maker.

Multi-objective optimization algorithms generate Pareto fronts with multiple solutions that present a trade-off between objectives' values. No objective's value of a solution on the Pareto front can be improved without worsening the value of another objective. Multi-objective optimization approaches enable the decision maker to select one solution depending on his/her preferences regarding the relative importance of the objectives. Changing the weights of objectives combined in a single-objective function can generate the whole Pareto front only if the Pareto front is convex (see, for example, [20]). If the Pareto front is non-convex, it is possible that parts of the front cannot be obtained by the weighted sum of objectives. Therefore, its use is problematic in cases where no a priori knowledge of the shape of the Pareto front is given.

Non-dominated Sorting Genetic Algorithm (NSGA-II) proposed by [9] has often been used in solving a variety of multi-objective optimization problems. It has been applied to the VRP with environmental considerations but to a very limited extent. To the best of our knowledge, one of the rare research in which a standard NSGA-II algorithm was developed to minimize the travel distance and CO_2 emission cost in a typical VRP problem was conducted by [21], but considered only delivery of products. We have included both pickup and delivery which makes it a more complex problem. Also, we introduced a new initialization of the NSGA-II to reflect better the presence of two objectives in the problem.

The research into development and application of population-based metaheuristics to a variety of optimization problems has resulted in a large volume of publications. However, the research into the impact of the quality of solutions in the initial population on the final result has been very scarce and only conducted for single-objective metaheuristics [6]. In the population-based single-objective metaheuristics developed for VRP, constructive heuristics have been often used in the generation of the initial population. The solutions they create are often referred to as "structured solutions", and they are expected to be of good quality. For example, [7] used Savings heuristics, [30] developed a Parallel Route Building Algorithm, and [33] used Nearest Neighbor, Savings Algorithm, and Random Insertion construction heuristics. Often, a part of the initial population (e.g. 80%) consists of structured solutions, while the remaining part contains randomly generated solutions to avoid a premature convergence of the algorithm. However, there has been no analysis of what the impact of the structured solution on the final Pareto front is, and consequently what an appropriate ratio of the number of solutions constructed by heuristics to the number of randomly generated solutions should be.

This research is concerned with a road freight transportation/reverse logistic problem present with our industrial collaborator. The company is facing a daily problem of generating routes for a large fleet of homogenous capacitated vehicles to serve all customers with pickup and delivery demands for gas cylinders. We model this problem as multi-objective VRPSPD with two objectives, including travel distance and fuel consumption. These objectives are incommensurable, i.e. expressed in different measurement units, kilometres, and litres, respectively. The developed algorithm is based on NSGA-II. In order to improve the quality of the final Pareto front, we employ technique for order preference by similarity to ideal solution (TOPSIS), a well-known multi-criteria decision-making method [19], to construct initial solutions of NSGA-II iteratively. In each iteration, the finite set of unserved customers is ranked by considering both the criteria simultaneously, the distance from the previous customer in the route (or depot in the first iteration), and the corresponding fuel consumption. The first ranked customer is inserted into the route. A statistical analysis of the effect that initial solutions generated by TOPSIS have on the quality of the final result, i.e. Pareto front, is performed. The contributions of our research are multiple. First, contribution is the advancement of knowledge of Multi-objective Optimization of VRP, which is still rather limited. Second, Pareto fronts generated by NSGA-II provide logistic managers with managerial insights and enable them to better understand the trade-off between the typical VRP objectives and environmental ones. Third, contribution is to the field of population-based metaheuristics for multi-objective optimization by employing a multi-criteria decision-making method in the initial population generation.

We evaluate the performance of our algorithm on the VRP instances provided by the collaborating company. Our algorithm produces solutions in which the travelled distance is comparable or even smaller than the distance in the solutions generated by the company, and in majority of instances, reduces the fuel consumption. Further, the performance of the algorithm is evaluated on problem instances of larger size taken from the literature.

The chapter is organized as follows. Section 24.2 presents a mathematical model for the VRPSPD. A detailed description of the developed NSGA-II algorithm is provided in Sect. 24.3. Section 24.4 presents the computational results and evaluates the performance of the developed algorithm, followed by conclusions and discussion about the future research work, which are presented in Sect. 24.5.

24.2 Problem Formulation

The VRPSPD with a homogeneous vehicle fleet, which is investigated in this research, is defined as follows. Let $G = (V, A)$ be a complete directed graph, with vertex set $V = \{0, 1, 2, \ldots., n\}$ and an arc set $A = \{(i, j) : i, j \in V, i \neq j\}$. A distance matrix $\left[c_{ij}\right]_{n+1\times n+1}$ is associated with A to denote distance between nodes i and j, $i \neq j$, $i, j \in V$.

The notations used in this model are defined below.

Sets and Parameters	
V	Set of nodes which denote customers, $V = \{0, 1, \ldots, n\}$, 0 denotes a depot, $V\backslash\{0\}$ denotes customers
K	Set of vehicles indexed by k, $k \in K$
L	Set of capacity measures used to measure the load of vehicle and customer demand (this set can contain multiple measures)
C^l	The capacity of vehicle given in units of capacity measure $l \in L$
c_{ij}	Distance between nodes i and j,$i, j \in V$
p_i^l	Demand for pickup from customer i, $i \in V\backslash 0$, l denotes capacity measure,$l \in L$
d_i^l	Demand for delivery to customer i, $i \in V\backslash 0$, l denotes capacity measure,$l \in L$
T	Maximum driving time for each vehicle
v	Speed of vehicle (constant)
M	Sufficiently large number

Decision Variables	
$Z_j^{k,l}$	Load of vehicle k at customer j, $k \in K$, $j \in V$ expressed in the capacity measure l, $l \in L$
S_j	Variable used to avoid sub-tours, which can be interpreted as position of node $j \in V$ in the route
x_{ij}^k	Binary decision variable that indicates whether vehicle k travels from i to $j, k \in K, i, j \in V, i \neq j$

The task is to create a set of routes, each starting and ending at the depot, with simultaneous pickup and delivery for n customers. Each customer is visited only once by exactly one vehicle. When a customer is visited, their complete demand for pickup and delivery is met by the same vehicle. A fleet of homogeneous vehicles K is available to serve the customers. Each vehicle $k \in K$ has a fixed capacity, C^l, which can be expressed by different capacity measures $l, l \in L$. For example, capacity can be measured by the number of containers that can be placed in the vehicle, weight that can be loaded, and number of pallets. The total load of a vehicle must not exceed its capacity, $C^l, l \in L$, at any time in the allocated route. The total time that a vehicle spends on the route must not exceed a pre-defined maximum driving time, T. It is also assumed that no transhipments are made at the locations of any of the customers.

In our model, we adopted the calculation of CO_2 emissions proposed by [3] and later used by [25]. The CO_2 emission is equivalent to fuel consumption. In the remaining part of the paper, the term "fuel consumption" will be used instead of "CO_2 emissions". The parameters used in the objective function to calculate the fuel consumption are presented in Table 24.1.

In our study, the distances c_{ij} between nodes i and j, $i, j \in V$, are the actual driving distances calculated using Google Maps API. The distances are symmetric, i.e.$c_{ij} = c_{ji}$ and satisfy metric triangular inequality property $c_{ij} \le c_{ik} + c_{kj}$. The formula used for distance measurements is given in the following URL: https://maps.googleapis.com/maps/api/distancematrix/json?origins=&destinations=&key.

Decision variables, objective functions, and constraints of the developed model are as follows.

The solution to the problem is a route for each vehicle determined by decision variables $x_{ij}^k, i, j \in V, k \in K$, which take value 1 if vehicle k travels from i to j, and 0 otherwise. Variables $Z_j^{k,l}$ determine the load of vehicle k at customer j, $k \in K$, $j \in V$, expressed in the capacity measure $l, l \in L$. In addition, decision variables S_j, $j \in V$, are used to avoid sub-tours.

The objectives in this model are as follows:

Total distance travelled by all the vehicles to be minimized:

$$F_1 = \sum_{i,j \in V} \sum_{k \in K} c_{ij} x_{ij}^k \tag{24.1}$$

Total fuel consumption to be minimized:

$$F_2 = \sum_{i,j \in V} \sum_{k \in K} \lambda k N V_e \frac{c_{ij}}{v} x_{ij}^k + \sum_{i,j \in V} \sum_{k \in K} \lambda \gamma \beta v^2 x_{ij}^k + \sum_{i,j \in V} \sum_{k \in K} \lambda \gamma \alpha \left(w + Z_j^{k,l} \right) c_{ij} x_{ij}^k \tag{24.2}$$

where c_{ij} is distance between customers i and $j, i, j \in V$, $Z_j^{k,l}$ is load of vehicle $k \in K$, at customer $j \in V$, $Z_j^{k,l} = Z_{j-1}^{k,l} + p_j^l - d_j^l$, v is speed of the vehicle.

Objective function given in expression (24.2) (adaptation of [3]) includes the components relevant to the calculation of the fuel consumption. The first, second, and third terms consider the engine of the vehicle, the speed of the vehicle, and the load of the vehicle, respectively. Using the typical values of the parameters given in Table 24.1 transforms (24.2) into

$$\sum_{i,j\in V}\sum_{k\in K} 0.00145 \times \frac{c_{ij}}{v} x_{ij}^k + \sum_{i,j\in V}\sum_{k\in K} 7.35^{-7} \times v^2 x_{ij}^k + \sum_{i,j\in V}\sum_{k\in K} 1.33^{-8} \times \left(w + Z_j^{k,l}\right) c_{ij} x_{ij}^k \quad (24.3)$$

The following constraints are defined:

$$\sum_{i\in V}\sum_{k\in K} x_{ij}^k = 1, \forall j \in V\backslash\{0\} \quad (24.4)$$

$$\sum_{i\in V} x_{ih}^k - \sum_{j\in V} x_{hj}^k = 0, \forall h \in V\backslash\{0\}, k \in K \quad (24.5)$$

$$Z_0{}^{k,l} = \sum_{i\in V}\sum_{j\in V\backslash\{0\}} d_j x_{ij}^k, \forall k \in K, l \in L \quad (24.6)$$

$$Z_j^{k,l} \geq Z_i^{k,l} - d_j + p_j - M\left(1 - x_{ij}^k\right), \forall i, j \in V\backslash\{0\}, i \neq j, k \in K, l \in L \quad (24.7)$$

$$Z_j^{k,l} \leq C^l, \forall j \in V, k \in K, l \in L \quad (24.8)$$

$$\sum_{i\in V}\sum_{j\in V} \frac{c_{ij}}{v} x_{ij}^k \leq T, \forall k \in K \quad (24.9)$$

Table 24.1 Parameters used in the calculation of fuel consumption for heavy vehicles; source: [3, 25]

Notation	Description	Typical values
ξ	Fuel to air mass ratio	1
η	Efficiency parameter for diesel engine	0.45
ρ	Air density (kg/m^2)	1.2041
κ	Heating value of a typical diesel fuel (kJ/g)	44
Ψ	Conversion factor (g/s to l/s)	737
w	Curb weight (kg)	14,000
k	Engine fraction factor (kJ/rev/l)	0.15
N	Engine speed (rev/s)	30.0
V_e	Engine displacement (l)	10.5
C_d	Coefficient of aerodynamics drag	0.9
C_r	Coefficient of rolling resistance	0.01
A	Frontal surface area (m^2)	10
n_{tf}	Vehicle drive train efficiency	0.45
θ	Road angle	0
τ	Acceleration (m/s^2)	0

$$S_j \geq S_i + 1 - M\left(1 - \sum_{k \in K} x_{ij}^k\right) \forall i, j \in V \backslash \{0\}, j \neq i \tag{24.10}$$

$$S_j \geq 0, Z_j^{k,l} \geq 0, \forall j \in V \backslash \{0\}, k \in K, l \in L \tag{24.11}$$

$$x_{ij}^k \in \{1, 0\}, \forall i, j \in V, k \in K \tag{24.12}$$

Constraints (24.4) ensure that each customer is served exactly once. Constraints (24.5) guarantee that for each customer, the same vehicle arrives at and leaves from the customer. In constraints (24.6), the initial load of each vehicle is calculated as the sum of demands of all customers assigned to that vehicle. The load of each vehicle after serving other customers is calculated by using expression (24.7). Constraints (24.8) impose limits on the vehicle capacity. Drivers are not allowed to drive for more than a specific period of time, which is defined in constraints (24.9). Constraints (24.10) ensures the sub-tour elimination [38]. Restrictions (24.11) maintain the non-negativity of decision variables S_j and $Z_j^{k,l}, j \in V, k \in K, l \in L$, while the binary status of decision variables $x_{ij}^k, i, j \in V, k \in K$, is ensured by (24.12).

It would be difficult to combine two incommensurable objectives, the distance and fuel consumption, whose values are expressed in kilometres and litres, respectively, into a single-objective function. The mapping of the objectives' values to one measurement unit, for example monetary units, is not straightforward. Also, the fuel consumption depends not only on the travelled distance but also on the load of the vehicle. Two routes with different sequences of customers to visit can have the same distance but different fuel consumption due to different loads of the vehicle between the customers. For these reasons, two objective functions are required, which can be in conflict, and they are treated separately instead of being combined in a single sum.

24.3 An NSGA-II for VRPSPD

In this research, we adopt a standard architecture of NSGA-II [9]. A brief description of the algorithm is provided with the focus of components of the algorithm that are specially designed for the VRPSPD. A block diagram that presents one iteration in the NSGA-II is given in Fig. 24.1. In each iteration t, NSGA-II starts with a solution set P_t of size Pop_{size}. Crossover and mutation operations are applied to solutions in P_t to generate new solutions Q_t. These two sets are combined together to undergo non-dominated sorting, which groups them into different levels of Pareto front starting from PF^1, PF^2, and so on. Solutions which are non-dominated are placed in PF^1. Then solutions from PF^1 are disregarded. The solutions in the remaining set are compared and the non-dominated ones are placed in PF^2, and so on. As the number of solutions (the population size) in each iteration is limited to Pop_{size}, the worst solutions have to be removed from the solution set. Solutions to be kept for the next

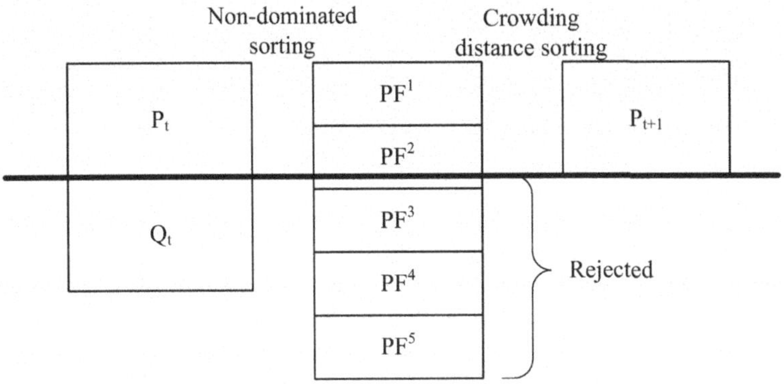

Fig. 24.1 A block diagram of NSGA-II. *Source* [9]

iteration are taken starting from the Pareto front PF^1. Each Pareto front is checked to see whether all of its solutions can be included. If there are more solutions in the Pareto front than needed, the Crowding distance metric is used to estimate the density of each solution, keeping solutions from less crowded areas. Density refers to the number of solutions in a certain area. The new created population P_{t+1} is passed to the next iteration. For more details about the Non-dominated sorting and the Crowding distance, the reader is referred to [9].

The components of the developed NSGA-II tailored particularly for VRPSPD will be described, followed by a description of the novel constructive heuristic, which uses TOPSIS in the initialization phase of NSGA-II.

Representation of solutions. Each chromosome is represented as an integer string that denotes the sequence of customers to be visited in each route. The routes are split by delimiter "0", which represents the depot.

Crossover operator: In general, GA searches the solution space for better solutions by using crossover and mutation operators. The crossover operator simulates reproduction between two chromosomes, where the offspring inherit some characteristics of each chromosome. We employ Alternating Edge Crossover (AEX), which has been successfully implemented in VRP since it was first introduced by [16]. In AEX, two parent chromosomes are selected, one is chosen randomly from the Pareto front PF^1, while the second parent is chosen randomly from the whole population. Edges from two parents are exchanged subsequently to produce an offspring. An example of AEX crossover is presented in Fig. 24.2. The two parents with removed delimiters chosen for the crossover operation are denoted by P_1 and P_2, and are given in part (a). In the first step, the first two genes (5 and 1) from P_1 are chosen to initialize the offspring shown in row (b). The arc in P_2 emanating from 1 and going to 9 is chosen for the offspring. The arc from 9 going to 6 is now chosen from P_1, and 6 is added to the offspring. The process continues. If the chosen customer creates infeasibility (already appears in the sequence), any unvisited node is chosen randomly, and the insertion of customers continues. Instead of inserting 6 in the

(a)	P_1	5	1	7	8	4	9	6	2	3
	P_2	3	6	2	5	1	9	8	4	7
(b)	offspring	5	1	9	6	2	3	7	8	4

Fig. 24.2 An example of alternating edge crossover (AEX)

offspring, let us assume that the next randomly chosen customer is 7 and the process continues until all of the customers are inserted.

Mutation operators: In our NSGA-II, we use three Swap Mutations proposed by [27] in their study of the identical parallel machine scheduling problem. They are illustrated in Fig. 24.3. Each of them is performed with a given probability. Two positions in a chromosome, i and j, are chosen randomly. The first mutation circularly shifts customers to the left, within the i and j positions, the second mutation circularly shifts customers to the left, this time outside i and j, and the third mutation swaps the positions of customers in i and j.

Both crossover and mutation satisfy the constraints that all customers are visited only once (as elements in the permutations are not repeated), but they cannot guarantee the satisfaction of other constraints, including the vehicle capacity and driving time constraints. To address this problem, each parent chromosome is transformed into a corresponding giant tour by removing the route delimiters, i.e. assuming there is one vehicle with a sufficiently large capacity. After application of crossover and mutation operators, Split Algorithm, which was introduced by [35], is used to divide the obtained solution into optimal routes so that the constraints on vehicle capacity and driving time are not violated.

Split Algorithm: This algorithm divides a giant tour into smaller feasible routes respecting the sequence of the customers, and the constraints imposed on the VRP with the distance as the objective function to be minimized. In our NSGA-II, Split Algorithm for distance objective will be used, but we need to adapt it for the fuel consumption objective.

Split Algorithm creates an auxiliary directed acyclic graph, $H = (V, A, R, F)$. Set V contains $n + 1$ nodes indexed from 0 to n, and A contains arc $(i, j), i < j$, if a route visiting customers i and j are feasible in terms of load and driving time (customer j does not need to be visited immediately after i).

Set A contains arcs (i, j) such that

$$\sum_{k=i+1}^{j} d_k \leq C \tag{24.13}$$

$$c_{0,i+1} + \sum_{k=i+1}^{j-1} c_{k,k+1} + c_{j0} \leq D \tag{24.14}$$

where

C is the capacity of the vehicle

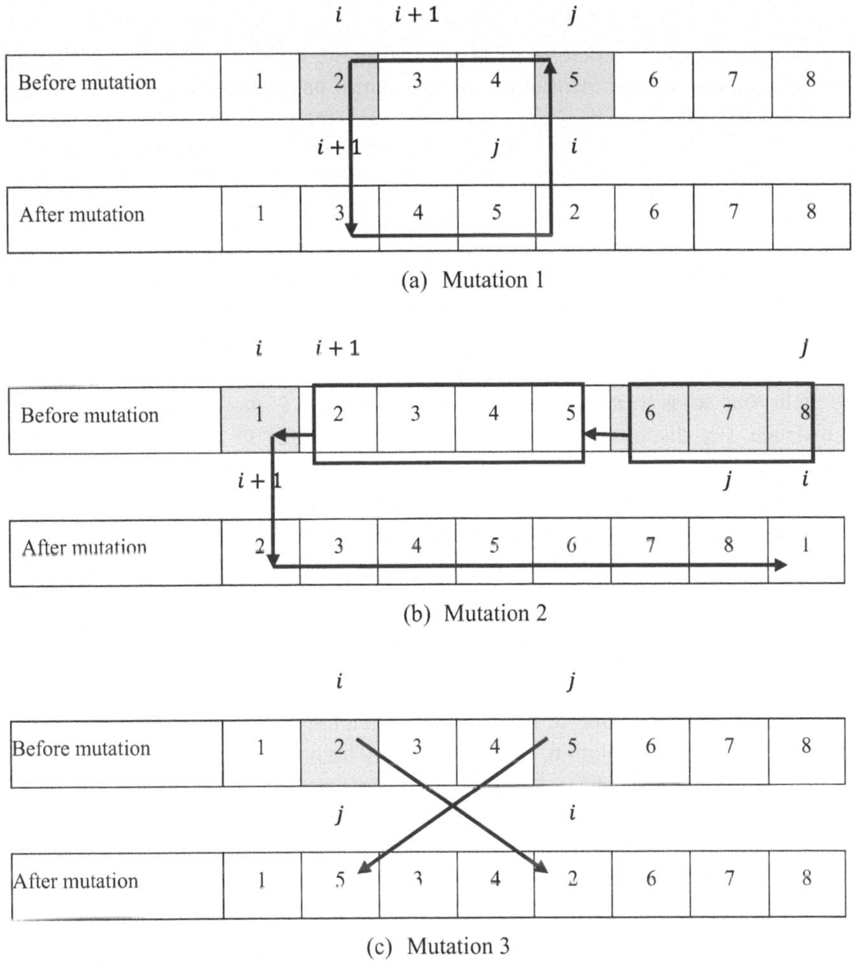

Fig. 24.3 An example of Swap Mutations

c_{ij} is the distance to travel between customers i and j, $i, j \in V$

d_i is the demand of customer i (note: for the sake of simplicity it is assumed that customers only need delivery and not pickup services)

D is the maximum allowed distance to travel for each vehicle.

Constraint (24.13) ensures that the load of the vehicle does not exceed its capacity, while constraint (24.14) imposes the limit on the driving time (equivalent to the distance to travel). Arcs represent feasible routes between customers.

A simple example to demonstrate Split Algorithm is provided and shown in Fig. 24.4. Let us suppose that there are five customers labeled by $1, \ldots, 5$, with giant tour $S = (1, 2, 3, 4, 5)$. The demands of the customers are given in the brackets in the corresponding nodes. Triangular inequalities between customers' distances hold. The capacity of each vehicle is 10000 kg, while the limit on driving time is 10 h; therefore, a vehicle can potentially travel 800 km if its speed is 80 km/h.

In this example V = {0, 1, 2, ..., 5}. R is the total distance of the routes, F is the total fuel consumption, and both are to be minimized. For example, $(1, 2) \in A$, because the corresponding route $(0, 1, 2, 0)$ has distance 55. The task is to select arcs which will form a route with a minimal total distance. The labels *a*:*b* assigned to arcs denote the route *a* and the incurred distance *b*. Split Algorithm generates three routes $(0, 1, 2, 0, 3, 0, 4, 5, 0)$ with minimum total distance 205, which are shown in Fig. 24.4b. This solution has fuel consumption of 39.63 litres. For more details about Splitting Algorithm for minimal distance, see [35].

Next, a route with minimal fuel consumption has to be determined. Splitting the giant tour so as to minimize the fuel consumption is more complex compared to distance. For distance, an addition of a new customer to a route requires only an update of the last leg of the route. However, the fuel consumption depends on the load and when a new customer is added to the route, the calculation of fuel consumption needs to be updated considering the load starting from the depot, i.e. the delivery load of the new customer needs to be added to the calculation. Using the same example as for distance, the calculations for generating the routes with minimal fuel consumption are shown in Fig. 24.5. The labels *a*:*b* assigned to arcs denote the route *a* and the incurred fuel consumption *b*. The solution $(0, 1, 2, 0, 3, 4, 0, 5, 0)$ incurs minimum fuel consumption, 39.42 litres, with distance 220 km.

Initialization phase: One of the research problems in the design of a GA is the quality of the initial population. The presence of a higher quality initial population in a GA is likely to lead to better solutions within a reasonable amount of time. However, randomly generated solutions, which may be of lower quality, can still contribute to the evolution of solutions and enable search of the wider solution space. In the majority of population-based metaheuristics for VRPs, initial populations are generated using a sequential insertion heuristic in which customers are inserted in a random order at randomly chosen positions within routes [4, 13, 17]. Constructive heuristics for VRP, which can be used to generate initial solutions of good quality, include the well-known Savings Algorithm and Sweep Algorithm. The Savings Algorithm, introduced by [8], is one of the best-known algorithms for generating VRP solutions. The main idea behind this algorithm is to merge two routes of two different vehicles which contribute the most to the cost saving. The idea of the Sweep Algorithm, proposed by [14], is to group and serve together customers that are geographically close to each other where "closeness" is measured by their polar angles defined with respect to the depot.

New constructive heuristic based on TOPSIS: We propose a new iterative constructive heuristic, which considers simultaneously the identified objectives to build an initial population. In each iteration, a customer with the best trade-off between the defined criteria, namely the distance to be travelled from the depot or the previous customer in the route, and incurred fuel consumption is selected. A multi-criteria decision-making method TOPSIS is used to rank the customers not assigned to a route yet [19]. This method calculates the distance of each alternative (customer)

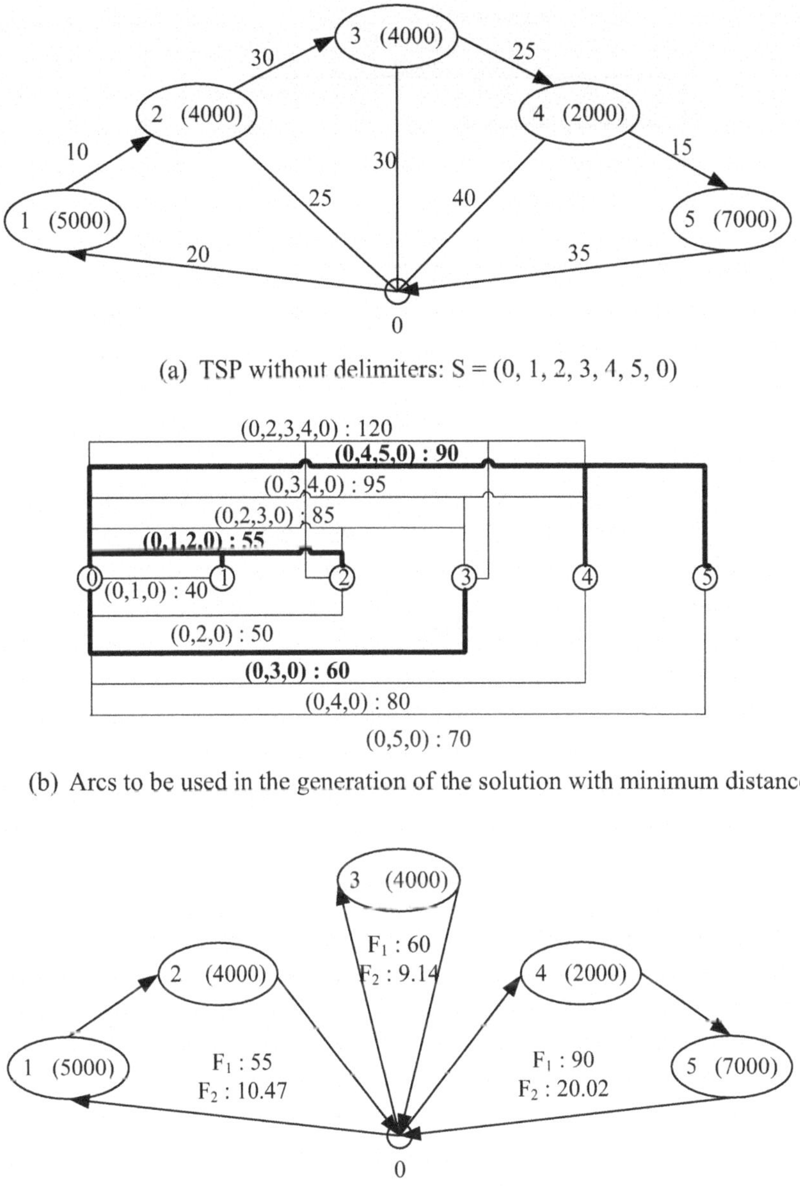

(a) TSP without delimiters: S = (0, 1, 2, 3, 4, 5, 0)

(b) Arcs to be used in the generation of the solution with minimum distance

(c) TSP broken into smaller routes with minimum total distance 205 km and fuel consumption 39.63 litres

Fig. 24.4 An example of the Split Algorithm for minimum distance

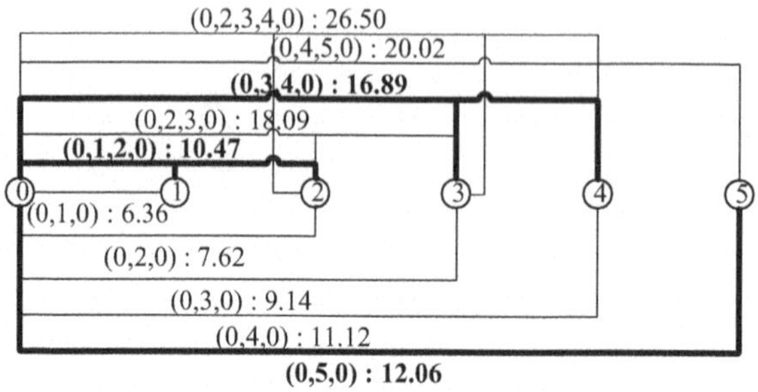

(a) Arcs to be used in the generation of the solution with minimum fuel consumption

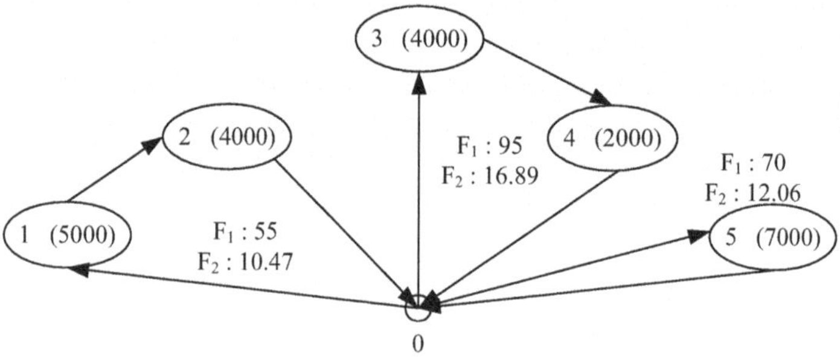

(b) TSP broken into smaller routes with minimum fuel consumption 39.42 litres and distance 220 km

Fig. 24.5 An example of Split Algorithm for minimum fuel consumption

from the ideal and nadir points. The ideal (nadir) point is composed of all best (worst) criteria values attainable (the assumption is that the utility is monotonically increasing or decreasing). TOPSIS is based on the concept that the best ranked customer should have the shortest distance from the ideal point and the largest from the nadir. Typically, the Euclidean distance is used, while the criteria values of alternatives are normalized by using the vector normalization. Let us denote by S_{i+}, and S_{i-} the distance from the ideal and nadir points, respectively. The relative closeness of customer i to the ideal point is calculated as

$$C_i = \frac{S_{i-}}{S_{i+} + S_{i-}}, i \in \{1, \ldots, n\}. \tag{24.15}$$

The customers are sorted by the C_i values, where the "best" customer has the largest C_i. The ranked customers form a list which we will refer to as the candidate list.

The steps of the TOPSIS algorithm are presented below using a simple example with only 5 customers. The decision matrix with the distances between customers

Table 24.2 Distances and fuel consumption

	Distance (km)	Fuel consumption (l)
Customer 1	80	5
Customer 2	75	7
Customer 3	85	10
Customer 4	88	6
Customer 5	79	8

and the depot and the fuel consumption incurred are given in Table 24.2. Let us also suppose that both criteria are of the same importance, i.e. $w = [0.5, 0.5]$.

A normalized decision matrix is created by vector normalization. In this example, the normalized values r_{ij} are calculated as follows:

$$r_{ij} = \frac{n_{ij}}{\sqrt{\sum_{i=1}^{5} n_{ij}^2}}, \forall i = 1, \ldots, 5, j = 1, 2 \tag{24.16}$$

where n_{ij}, $i = 1,\ldots,5, j = 1,2$, are the corresponding values in the decision matrix. The weighted normalised decision matrix is generated by multiplying each column of the matrix by the weight of the corresponding criterion. The values of the weighted normalized matrix, b_{ij}, $i = 1,\ldots,5, j = 1,2$ are given in Table 24.3.

The ideal and nadir points are $A^{+} = \{0.1931, 0.3625\}$ and $A^{-} = \{0.1646, 0.1812\}$, respectively. The Euclidean distances between the customers and the ideal and nadir points are calculated as follows:

$$S_{i+} = \sqrt{\sum_{j=1}^{2} \left(b_{ij} - a_j^{+}\right)^2}, \forall i = 1, \ldots., 5 \tag{24.17}$$

$$S_{i-} = \sqrt{\sum_{j=1}^{2} \left(b_{ij} - a_j^{-}\right)^2}, \forall i = 1, \ldots.., 5 \tag{24.18}$$

The values are given in Table 24.4.

The relative closeness of customers to the ideal point are given in Table 24.5.

Table 24.3 Weighted normalized decision matrix

	Distance (km)	Fuel consumption (l)
Customer 1	0.1755	0.1812
Customer 2	0.1646	0.2537
Customer 3	0.1865	0.3625
Customer 4	0.1931	0.2175
Customer 5	0.1733	0.2900

Table 24.4 Distances between the customers and ideal and nadir points

	S_{i+}	S_{i-}
Customer 1	0.1821	0.0110
Customer 2	0.1124	0.0725
Customer 3	0.0066	0.1826
Customer 4	0.1450	0.0461
Customer 5	0.0751	0.1091

Table 24.5 Relative closeness of customers to the ideal point

i	1	2	3	4	5
C_i	0.0568	0.392	0.965	0.241	0.5922

Therefore, TOPSIS ranks the customers as follows: Customer 3, Customer 5, Customer 2, Customer 4, and Customer 1.

The proposed constructive heuristic is deterministic. In order to generate multiple elements in the initial population, a greedy randomization is employed to randomly select one customer from the α percentage of top solutions in the candidate list. A larger α value represents greater randomness, while α value of 0 denotes pure greediness. The selected customer is inserted into the route and becomes the new starting point in the next iteration of the constructive heuristic. The process continues until either all customers are allocated to a route, or the time and/or vehicle capacity constraints are violated. If either constraint is violated, a new route starting from the depot is created.

24.4 Computational Experiments

For evaluation of the developed NSGA-II algorithm, two sets of problem instances are used: real-world instances and artificially generated ones taken from the literature.

Our collaborator provided us with 20 real-world instances, each describing a daily VRPSPD with the number of customers between 8 and 20. In each problem instance, the locations of the central depot and the customers are given together with the delivery and pickup demand requirements of each customer.

The following assumptions are made:

(1) No customer has a demand larger than the capacity of the vehicle; therefore no customer will be served by more than one vehicle.
(2) Drivers are not allowed to drive more than 10 h ($T = 10$ h).
(3) The customers are located in different cities. Consequently, after the consultation with the company and following the motor vehicle guide, the motorway speed limit is set to be 80 km/h (equivalent to 50 miles/h).

The company outsources delivery, and has no limit on the number of vehicles at its disposal. The cost is determined based on the distance traveled by all the vehicles, and not on the number of vehicles used. The load of vehicles is evaluated in three different measures: containers, weight, and pallets. The relationship between these measures is as follows. Each pallet can hold up to 12 cylinders. The capacity of each vehicle is 24 pallets, or 288 cylinders, which is equivalent to 24,000 kg. However, cylinders that are destined for different customers cannot share the same pallet. For example, if one customer requires only one cylinder and another customer requires 11 cylinders, then all 12 cylinders cannot be put on one pallet, but must instead be put on two different pallets. The number of cylinders to be delivered from the depot to customer $i, i = 1, \ldots, n$, and the number of cylinders to be picked up from the customer i are denoted by d_i^c and p_i^c, respectively. Different varieties of cylinders are assumed to be compatible with each other, and can be loaded on the same vehicle. The pickup and delivery demand expressed as the number of cylinders is converted to a corresponding number of pallets, p_i^p and d_i^p, and weight, p_i^w and d_i^w, respectively. Due to different measures of the load, the cylinders, pallets, and weight have to be considered separately as capacity constraints. The fuel consumption is calculated based on the load of vehicle given in kilograms. The characteristics of the problem instances are given in Table 24.6.

Hypervolume is often used as a metric for evaluating the quality of the generated Pareto fronts, because it encapsulates in a single value the spread of the solutions along the Pareto front as well as their closeness to the Pareto front. Hypervolume measures the region defined by an n-dimensional set of points (where n is the number of objectives) that is dominated by the generated Pareto front [46]. The higher the value of the Hypervolume, the better the Pareto front. Many different algorithms have been developed to calculate the Hypervolume metric, each with its own strengths and weaknesses. Algorithms reported in the literature include Hypervolume via Inclusion Exclusion [42], the LebMeasure Hypervolume Algorithm [11], Hypervolume using Overmars, Yap Algorithm and Hypervolume Approximation [18], and the Optimal 3D Hypervolume Algorithm [5]. In this research, the Exact Exclusive Hypervolume method developed by [41] is utilized. In the Hypervolume calculation, the points which are weakly dominated by others are replaced by the ones that dominate those points.

Figure 24.6 illustrates the Exact Exclusive Hypervolume metric of two Pareto fronts for a minimization problem. Let us assume that the circles in Fig. 24.6 represent the first Pareto front (P_1), while the squares represent a second Pareto front (P_2). The second Pareto front is obtained when points from P_1 are excluded. Point B in P_1 weakly dominates point A, which has the same f1 value. As a result, for the purpose of Exact Exclusive Hypervolume calculation, the bounded region for the Hypervolume calculation will now ignore point A. To compare the two Pareto fronts (P_1 and P_2), the regions that comprise the points on the Pareto fronts and the reference point must be bounded. The region bounded by P_1, indicated by solid lines, is larger than the region bounded by P_2, indicated by dotted lines, which implies that Pareto front P_1 has a higher quality of solutions than Pareto front P_2. The choice of reference point is a disadvantage of the Hypervolume calculation, and there is no consensus in

Table 24.6 Characteristics of real-world problem instances

Problem instance	No. of customers	Maximum distance from depot (km)	Minimum distance from depot	Average distance from depot	Maximum distance between customers	Minimum distance between customers	Average distance between customers	Required collection/ delivery (in containers)	Required collection/ delivery (in kg)	Required collection/ delivery (in palettes)
1	7	100.26	23.85	68.77	80.57	1.11	44.82	130/174	9891/15,344	11/16
2	8	92.68	10.00	58.93	98.60	2.36	57.62	278/256	20,931/ 21,515	23/22
3	11	116.23	11.65	74.02	120.35	8.58	59.41	315/471	22,902/ 37,406	25/40
4	11	127.36	10.00	71.33	127.12	10.99	58.70	511/512	29,648/ 32,955	37/33
5	12	100.26	24.25	58.56	102.85	1.35	41.94	359/301	26,048/ 25,165	29/27
6	12	127.36	10.00	64.87	127.12	4.99	45.14	518/535	32,254/ 41,229	40/45
7	13	101.12	40.86	70.71	103.36	3.25	45.32	324/239	20,935/ 17,703	28/24
8	13	107.81	37.86	70.34	105.45	0.67	49.62	282/300	18,967/ 21,348	24/25
9	15	96.84	10.00	57.21	101.47	4.38	44.89	406/461	25,526/ 30,122	31/38
10	15	97.61	10.00	58.75	128.90	1.40	51.86	561/451	45,761/ 38,335	47/40

(continued)

Table 24.6 (continued)

Problem instance	No. of customers	Maximum distance from depot (km)	Minimum distance from depot	Average distance from depot	Maximum distance between customers	Minimum distance between customers	Average distance between customers	Required collection/delivery (in containers)	Required collection/delivery (in kg)	Required collection/delivery (in palettes)
11	15	114.03	8.81	62.96	119.34	0.28	52.16	628/663	47,709/53,266	52/58
12	15	116.23	10.39	73.33	149.05	0.68	61.76	428/294	30,829/25,594	38/27
13	16	101.13	11.65	70.09	106.10	0.68	48.05	326/318	24,006/29,084	27/29
14	16	116.23	26.68	68.57	129.75	1.24	45.87	278/370	21,520/32,725	22/33
15	16	116.23	26.68	79.00	119.12	1.38	47.03	207/254	15,165/20,927	17/26
16	16	232.75	35.01	119.54	332.85	4.93	137.95	469/310	19,651/22,108	32/27
17	17	101.13	16.51	53.99	108.22	1.06	43.24	189/256	12,980/20,835	17/28
18	18	101.13	23.35	62.82	123.55	1.11	49.49	233/310	14,777/27,450	23/33
19	18	127.92	10.00	69.10	132.79	0.22	51.23	328/338	24,638/29,286	27/28
20	19	116.23	23.35	61.05	117.61	1.25	50.85	282/411	19,230/38,816	28/38

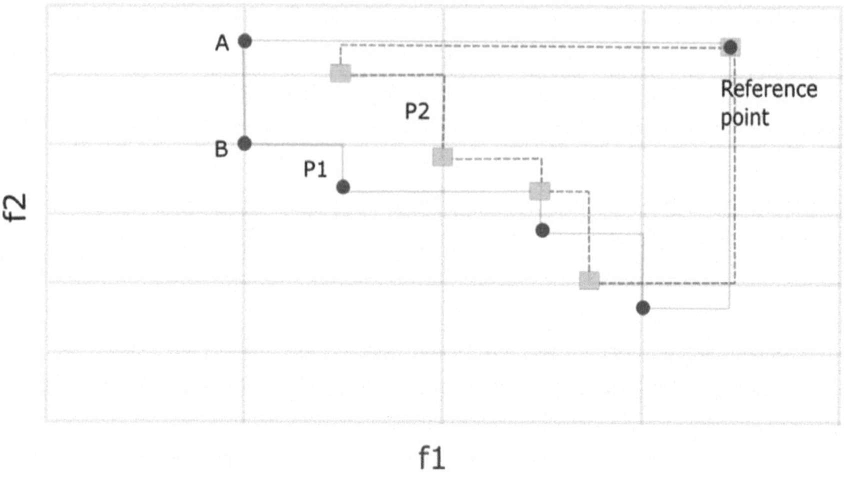

Fig. 24.6 A graphical illustration of Exact Exclusive Hypervolume of two different Pareto fronts

the literature how to properly define it. The approach that is adopted here takes the corresponding coordinates of the one corner point of the Pareto front to define the reference point.

The experiments have been performed on a laptop running Windows 8 with an Intel® Core (TM) i3-4000 M CPU @ 2.40 GHz processor, and the NSGA-II was implemented using Matlab R2015b. The values of the parameters used in the developed NSGA-II are tuned experimentally or taken from the literature. They are presented in Table 24.7.

The initial population generation involves both TOPSIS and random construction of solutions. An investigation is conducted to establish what would an adequate combination of solutions generated by TOPSIS and greedy randomization (such solutions will be referred to as TOPSIS in the remaining part of the section) and randomly generated solutions be.

The aim of the first set of experiments is to investigate if TOPSIS can provide a good initial population for NSGA-II that would lead to a better final Pareto front than would a completely randomly generated initial population. Experiments are run with different percent of TOPSIS solutions and the randomly generated solutions

Table 24.7 Parameters of NSGA-II and TOPSIS

Population size (Popsize)	50
Crossover rate	0.9
Mutation rate	0.1
Termination condition (number of iterations)	400
Discretization step	100
α in the greedy randomization	10%

in the initial population. A total of 11 combinations are considered, starting with T0, which denotes that all solutions are generated randomly, up to T100, which denotes that the initial population consists solely of TOPSIS solutions. For example, T80 (80% TOPSIS) implies that out of 50 solutions in the initial population, 40 are generated using TOPSIS, while the remaining 10 are randomly generated. For each combination, NSGA-II is run 10 times. There are in total 110 Pareto fronts (11 combinations run 10 times). The Pareto fronts generated by the NSGA-II are ranked by their Hypervolume values, where rank 1 denotes the largest Hypervolume with the best Pareto front. The average of ranks of Pareto fronts produced by different ratios of TOPSIS solutions to randomly generated solutions are presented in Table 24.8. The minimum possible value is 10 (maximum possible value is 110) which implies that in all runs the given combination of TOPSIS and randomly generated solutions have the best (worst) Pareto front in terms of Hypervolume. The smallest values of the average of ranks obtained for each instance are shown in bold. The combination T80 provides the best result in 17 out of 20 problem instances.

Table 24.8 Average of ranks of Pareto fronts obtained using different percentages of TOPSIS solutions in the initial population

Problem instance	T0	T10	T20	T30	T40	T50	T60	T70	T80	T90	T100
1	52.4	46.4	54.4	64.0	67.1	52.3	55.1	72.7	**44.8**	51.8	49.5
2	60.9	63.2	66.3	62.3	56.4	70.0	54.4	50.6	41.7	**38.6**	46.1
3	66.7	50.9	66.8	45.9	46.9	55.9	44.6	70.0	**43.7**	50.7	68.4
4	50.9	66.4	46.4	51.3	58.0	77.5	44.7	61.7	**42.7**	57.3	53.6
5	75.3	56.2	56.2	50.0	46.0	53.6	51.4	60.9	**41.3**	74.5	45.1
6	60.7	59.4	47.7	48.7	49.9	62.0	60.3	60.0	**47.6**	54.1	60.1
7	65.6	56.5	49.5	39.0	65.3	64.6	64.7	51.9	**38.2**	59.2	56.0
8	60.7	61.4	71.6	40.4	55.4	61.2	65.2	48.9	**32.8**	57.0	55.9
9	61.7	68.7	58.8	44.7	46.9	42.7	74.2	60.6	**43.9**	47.2	61.1
10	67.3	40.9	59.3	46.8	58.7	49.4	63.6	70.8	**44.7**	63.6	45.4
11	68.2	55.2	51.7	57.4	65.0	68.1	39.7	66.8	**40.0**	49.2	49.2
12	55.6	57.6	57.9	51.8	49.2	60.4	52.3	57.9	**48.5**	57.4	61.9
13	43.1	60.2	50.6	58.9	72.8	67.4	57.2	**42.8**	53.0	56.3	48.2
14	52.5	46.9	51.9	53.6	81.3	55.4	53.0	48.9	**41.8**	70.7	54.5
15	56.6	55.5	51.9	57.3	66.9	60.7	53.6	62.5	**34.5**	45.1	65.9
16	55.1	50.4	56.7	40.6	58.6	61.3	72.5	69.4	**39.9**	52.8	53.2
17	69.8	44.7	69.2	38.8	50.3	63.0	52.3	52.3	**43.8**	63.8	62.5
18	47.8	60.6	52.2	55.6	73.3	59.0	58.7	47.7	**41.0**	60.5	54.1
19	70.9	59.9	50.6	58.8	65.7	36.8	71.5	53.3	**42.5**	49.7	50.8
20	48.4	49.8	50.2	52.1	52.9	**47.2**	65.9	69.4	51.6	69.3	53.7

A statistical test is performed to prove the hypothesis that the solutions on the Pareto fronts are statistically different using One Way ANOVA. The results of the Kolmogorov–Smirnov test also confirmed that the population of data follows the normal distribution, which is a requirement for the application of the One Way ANOVA test. The ANOVA results show that the generated Pareto fronts are indeed statistically different with significance level $p = 0.05$. Consequently, T80 will be used in the initial population in the remaining experiments.

In the next set of experiments, different initial population generation methods and TOPSIS are analysed in terms of the quality of the final Pareto front the NSGA-II generates. This approach is compared to the two well-known heuristics for VRP, namely Savings Algorithm and Sweep Algorithm, and also with randomly generated initial populations. One of the major drawbacks of using the Savings Algorithm with a GA is that it produces a single solution, whereas the GA is a population-based metaheuristic. However, we investigate whether the single solution generated by the Savings Algorithm appears in the Pareto front in the initial population, and as such has a role in the final solution produced by the NSGA-II. Experimental results show that solutions generated by the Savings Algorithm do not appear in the Pareto front level 1, PF^1, in the initial population and are discarded before the end of the search in most experiments. Therefore, the Savings Algorithm is not used as the initial population generation method.

The Pareto fronts generated by the developed NSGA-II with different initialization methods are evaluated by Hypervolume. The Hypervolume values of the produced Pareto fronts are given in Table 24.9. In the experiments, the size of population with randomly generated initial population and T80 is 50, while with Sweep Algorithm the population size equals the number of customers (to avoid repeated solutions in the population).

The largest Hypervolume for each problem instance is given in bold. The proposed T80 method for initialization always produces the Pareto front that has the largest Hypervolume compared to the other two methods. ANOVA test proved that the average Hypervolume values are also different with significance level $p = 0.05$.

In the next experiment, the quality of results produced by NSGA-II are compared to the results which are optimal with respect to a single-objective function, distance, and fuel consumption, separately. However, CPLEX could solve only the instances in which the number of customers were smaller or equal to 11 (4 out of 20). The values of the minimum distance and fuel consumption produced by CPLEX and NSGA-II for these instances together with the CPU time required are presented in Table 24.10. CPLEX required less time for the distance minimization problem than for the fuel minimization problem due to the larger number of variables in the objective function of the latter. Note that often NSGA-II does not generate a single solution with minimum distance and minimum fuel consumption, but one corner solution on the Pareto front has minimum distance and other minimum fuel consumption. The column difference shows the difference between the value of the objective functions produced by CPLEX and NSGA-II given as %. It is difficult to set a strict threshold for the difference values which will support the claim that NSGA-II is of good quality. Nevertheless, the % values between 4.33 and 8.47 indicate that

Table 24.9 Hypervolumes of Pareto fronts generated using three initial population generation methods

Problem instances	Random generation	Sweep algorithm	T80
1	2.67E+04	2.68E+04	**2.76E+04**
2	4.92E+04	4.87E+04	**4.94E+04**
3	7.60E+03	6.57E+03	**7.85E+03**
4	8.37E+03	8.16E+03	**9.41E+03**
5	3.13E+03	3.15E+03	**3.15E+03**
6	5.18E+03	5.18E+03	**5.19E+03**
7	4.98E+03	2.66E+03	**6.54E+03**
8	4.70E+03	7.09E+03	**7.07E+03**
9	1.19E+04	1.20E+04	**1.35E+04**
10	1.33E+04	1.33E+04	**1.40E+04**
11	2.14E+04	2.87E+04	**2.12E+04**
12	1.33E+04	8.81E+03	**1.57E+04**
13	2.94E+03	2.98E+03	**3.56E+03**
14	1.03E+04	5.86E+03	**1.14E+04**
15	1.01E+04	7.50E+03	**1.05E+04**
16	3.61E+05	3.94E+05	**6.10E+05**
17	7.53E+03	6.41E+03	**7.54E+03**
18	9.38E+03	9.54E+03	**1.03E+04**
19	1.07E+04	9.64E+03	**1.20E+04**
20	1.99E+04	1.98E+04	**2.28E+04**

the results obtained by NSGA-II are comparable to single-objective optimization solutions.

Getting a single solution that minimizes both objectives is not always possible. The generated solutions in the Pareto front provide the logistics manager with insights into the trade-off between the distance and the fuel consumption that can be achieved, i.e. enable them to choose a solution on the Pareto front that reflects the relative importance given to the objectives. The approaches in which multiple objectives are combined into a single-objective function require the preferences to be made at the beginning of the decision-making process. Contrary to that, approaches which generate a Pareto front enable the manager to express their preferences a posteriori.

Next, we compare the solutions generated by NSGA-II with those provided by the collaborator company. The company considered distance as the only objective to be minimized. They calculated the Euclidian distance between customers, contrary to the actual driving distances employed in our model, which are calculated using Google Maps API. In 11 instances, the NSGA-II produces better solutions with respect to both objectives. In the other nine instances there are mixed

Table 24.10 Comparison of results by CPLEX and NSGA-II

Problem instances	No. of customers	Distance (km)			Time CPLEX (sec)	Fuel consumption (l)			Time CPLEX (s)	Time NSGA-II (s)
		CPLEX	T80	Difference		CPLEX	T80	Difference		
1	7	210.23	227.87	−7.74%	03	110.5	117.45	−5.92%	15	52
2	8	280.43	306.71	−8.57%	01	151.03	162.12	−6.84%	17	63
3	11	450.43	486.25	−7.37%	350	230.08	251.36	−8.47%	7952	87
4	11	401.32	428.98	−6.45%	603	215.23	227.91	−5.56%	12,398	98

results, i.e. the NSGA-II improves one objective compared to the company's solution at the expense of deterioration of another objective. We illustrate the interpretation of results using instance 16. The company produced a solution with distance 1295.00 km and fuel consumption 531.23l. There are 8 solutions generated by NSGA-II, which provide a trade-off between the distance (in the range 1095.15–1298.5 km) and fuel consumption (in the range 530.69–777.65l). Solutions 1 and 8 have the shortest distance and smallest fuel consumption, respectively. The logistic manager can calculate the marginal rate of substitution (MRS), which is a term generally used in economics. It shows the rate at which a manager can give up some amount of one objective in exchange for improvement of another objective. MRS values are shown in Table 24.11 with respect to distance and fuel consumption together with the corresponding percentage of deterioration (positive %) and improvement (negative %) of the objectives. For example, the MRS values for distance are calculated as the differences between the distances of each solution from 1 to 7 relative to the distance in solution 8, which has the minimum fuel consumption. As an illustration, for solution in row 7, the difference is 1296.02–1298.53 = −2.51, which implies (1296.02–1298.53)/1298.53 × 100 = −0.19%.

The majority of benchmark VRP instances available in the literature require optimization of a single objective [15, 23, 32, 38], or multiple objectives combined into a single objective [1, 24, 34, 40]. It is observed that no specific data set is available that deals with multi-objective VRPSPD optimization problems. Nevertheless, in order to evaluate the developed NSGA-II on larger instances we use the VRPSPD instances by [44], which are modifications of the instances given by [32]. The distances between customers are Euclidean. The characteristics of these instances are presented in Table 24.12.

Again, our experiments prove that in all instances NSGA-II with T80 outperforms the other two initialization approaches, random generation and Sweep algorithm, in terms of Hypervolumes. Next, results produced by the NSGA-II with T80 are compared with results reported in [44]. The latter results were generated by Arc

Table 24.11 MRS with respect to distance and fuel consumption for instance 16

Solution	Distance (km)	Fuel consumption (l)	Difference between distance	%	Difference between fuel consumption	%
1	1095.15	777.64	−203.37	−15.66	246.96	46.54
2	1272.48	664.23	−26.05	−2.01	133.54	25.16
3	1272.48	596.03	−26.04	−2.01	65.35	12.31
4	1273.10	592.59	−25.42	−1.96	61.90	11.67
5	1275.94	546.17	−22.59	−1.74	15.48	2.92
6	1296.02	541.16	−2.50	−0.19	10.47	1.97
7	1296.02	534.47	−2.50	−0.19	3.78	0.71
8	1298.53	530.68				

Table 24.12 Characteristics of benchmark problem instances in [44]

Problem instance	Number of customers	Vehicle capacity	Total delivery demand	Total pickup demand
r101	100	200	1458	2339
r201	100	1000	1458	2262
c101	100	200	1810	3070
c201	100	700	1810	2910
R1_2_1	200	200	3513	4406
R2_2_1	200	1000	3513	4358
C1_2_1	200	200	3530	5370
C2_2_1	200	700	3770	6010
R1_4_1	400	200	7109	10,433
R2_4_1	400	1000	7109	9571
C1_4_1	400	200	7190	12,470
C2_4_1	400	700	7560	10,050

Promise Algorithm (APA) with distance being used as a single objective to be minimized. We calculated fuel consumption in their solutions. In the benchmark instances, we interpret distance as kilometres. The main aim of this comparison is to investigate what price has to be paid if the fuel consumption is considered simultaneously with the distance in the optimization. The value of parameter α to be used by TOPSIS and the population size are tuned to better fit the instances of larger size and are set to be $\alpha = 2\%$ and the population size is 200 (Table 24.13).

It does not come as a surprise that NSGA-II with T80 is outperformed by APA with respect to the travelled distance in 11 out of 12 instances, because the latter considered only distance as objective function. However, in 10 out of 12 instances the fuel consumption is improved (in corner solutions on Pareto front), the largest improvement being 50% for instance c201.

In order to investigate further the behaviour of the developed NSGA-II with T80, analysis of customer locations in these problem instances is required. Namely, the location of customers is uniformly generated around the depot, which is not the case in the real-world problem of our collaborator. Also, the coordinates of the customers are not given in real measurement units such as in meters or miles. As the calculation of fuel consumption is based on realistic meter or miles measures, having no measurable units here jeopardizes the calculation of fuel consumption.

Table 24.13 Comparison of results by APA [44] and by NSGA-II with T80 on benchmark problem instances

Problem instance		APA	T80	% difference(%)
r101	Distance (km)	1009.95	1009.95	**0.00**
	Fuel cons. (l)	0.53	0.51	**−3.92**
r201	Distance (km)	666.20	671.70	0.82
	Fuel cons. (l)	0.48	0.41	**−17.07**
c101	Distance (km)	1220.18	1261.44	3.27
	Fuel cons. (l)	0.57	0.50	**−14.00**
c201	Distance (km)	662.06	693.05	4.47
	Fuel cons. (l)	0.54	0.36	**−50.00**
R1_2_1	Distance (km)	3375.19	3378.76	0.11
	Fuel cons. (l)	1.95	1.57	**−24.20**
R2_2_1	Distance (km)	1665.58	1668.81	0.19
	Fuel cons. (l)	1.04	1.00	**−4.00**
C1_2_1	Distance (km)	3641.88	3837.75	5.10
	Fuel cons. (l)	1.8	1.56	**−15.38**
C2_2_1	Distance (km)	1726.73	1790.26	3.55
	Fuel cons. (l)	0.99	0.99	**0.00**
R1_4_1	Distance (km)	9668.18	10,074.55	4.03
	Fuel cons. (l)	5.81	4.25	**−36.71**
R2_4_1	Distance (km)	3560.72	3882.17	8.28
	Fuel cons. (l)	2.18	2.23	2.24
C1_4_1	Distance (km)	11,125.14	12,354.73	9.95
	Fuel cons. (l)	5.12	4.85	**−5.57**
C2_4_1	Distance (km)	3549.20	3845.45	7.70
	Fuel cons. (l)	2.06	2.09	1.44

24.5 Conclusions and Future Work

This paper presents a multi-objective vehicle routing problem with simultaneous pickup and delivery, which is particularly useful in the reverse logistics. The travel distance and fuel consumption are objectives to be minimized. A multi-objective evolutionary algorithm NSGA-II was developed with Alternating Edge Crossover and Swap Mutation operators tailored for the VRP and modified Split Algorithm, which was used to preserve the feasibility of the evolved solutions.

A particular novelty in the design of the NSGA-II was the initial population generation, which employed TOPSIS as a constructive heuristic and combined it with greedy randomization. The selected combination of 80% TOPSIS and 20%

randomly generated solutions was evaluated against two common initial population generation methods including the Sweep Algorithm and random generation. The proposed algorithm performed better than these two alternatives with respect to Hypervolume values of the generated Pareto fronts on both problem instances provided by the collaborating company and on the problem instances taken from the literature.

The output of the NSGA-II is a Pareto front, which consists of solutions that represent trade-offs between distance and incurred fuel consumption. The Pareto front with multiple solutions provides an opportunity to the logistics manager to express a posteriori preferences toward the importance of objectives, and consequently choose a preferred solution.

In the next stage of the research, the proposed algorithm may be evaluated further on instances with a larger number of customers. Also, problem instances with "special" distribution of locations should be analyzed. In the studied problem, a homogenous fleet of vehicles was used. The effect of a heterogeneous fleet of vehicles on the distance and fuel consumption cost need to be evaluated. A thorough analysis of the managerial implications of the considerations of the environmental issues should be investigated.

An important factor in environment protection is also the type of vehicles. Electric vehicles can be the future for road transportation with zero emissions, but they bring a different dimension to the problem. New constraints will be required as recharging stations are still not that much available as petrol stations.

References

1. Ai TJ, Kachitvichyanukul V (2009) A particle swarm optimization for the vehicle routing problem with simultaneous pickup and delivery. Comput & Oper Res 36(5):1693–1702
2. Akçelik R, Biggs DC (1985) A discussion on the paper on fuel consumption modeling by post et al. Transp Res Part B: Methodol 19(6):529–533
3. Barth M, Younglove T, Scora G (2005) Development of a heavy-duty diesel modal emissions and fuel consumption model (Technical Report. University of California, Berkeley, California Partners for Advanced Transportation Technology)
4. Berger J, Barkaoui M (2003) A hybrid genetic algorithm for the capacitated vehicle routing problem. In: Cantú-Paz E, et al (eds) The Proceeding of Genetic and Evolutionary Computation-GECCO 2003. Springer, pp 646–656
5. Beume N, Fonseca CM, Lopez-Ibanez M, Paquete L, Vahrenhold J (2009) On the complexity of computing the hypervolume indicator. IEEE Trans Evol Comput 13(5):1075–1082
6. Cinar D, Ervural BC, Gakis K, Pardalos MP (2017) Constructive algorithms for the cumulative vehicle routing problem with limited duration. In: Cinar D, Gakis K, Pardalos PM (eds) Sustainable logistics and transportation: Optimization models and algorithms, Springer optimization and its applications, vol 129. Springer, pp 57–86
7. Chiang T-C, Hsu W-H (2014) A knowledge-based evolutionary algorithm for the multiobjective vehicle routing problem with time windows. Comput Oper Res (COR) 45:25–37
8. Clarke G, Wright JW (1964) Scheduling of vehicles from a central depot to a number of delivery points. Oper Res (OR) 12(4):568–581
9. Deb K, Pratap A, Agarwal S, Meyarivan T (2002) A fast and elitist multiobjective genetic algorithm: NSGA-II. IEEE Trans Evol Comput 6(2):182–197

10. Demir E, Bektaş T, Laporte G (2014) The bi-objective pollution-routing problem. Eur J Oper Res 232(3):464–478
11. Fleischer M (2003) The measure of Pareto optima applications to multi-objective metaheuristics. In: Fonesca CM, Fleming PJ, Zitzler E, Thiele L, Deb K (eds) Lecture notes in computer science (LNCS). Springer, Berlin, pp 519–533
12. Franceschetti A, Honhon D, Van Woensel T, Bektaş T, Laporte G (2013) The time-dependent pollution-routing problem. Transp Res Part B: Methodological 56:265–293
13. García-Villoria A, Pastor R (2010) Solving the response time variability problem by means of a genetic algorithm. Eur J Oper Res 202(2):320–327
14. Gillett BE, Miller LR (1974) A heuristic algorithm for the vehicle-dispatch problem. Oper Res 22(2):340–349
15. Golden B, Assad A, Levy L, Gheysens F (1984) The fleet size and mix vehicle routing problem. Comput Oper Res 11(1):49–66
16. Grefenstette JJ, Gopal R, Rosmaita BJ, Gucht DV (1985) Genetic algorithms for the traveling salesman problem. In: The proceedings of the 1st international conference on genetic algorithms. L. Erlbaum Associates Inc., pp 160–168
17. Hicks C (2006) A Genetic Algorithm tool for optimising cellular or functional layouts in the capital goods industry. Int J Prod Econ (IJPE) 104(2):598–614
18. Hochstrate B, Rudolph G (2006) Faster s-metric calculation by considering dominated hypervolume as Klee's measure problem. Evol Comput 17(4):477–492
19. Hwang C-L, Yoon K (1981) Multiple attribute decision making: Methods and applications: a state-of-the-art survey. In: Fandel G, Trockel W (eds) Lecture notes in economics and mathematical systems. Springer Science & Business Media, pp 1–278
20. Jakob W, Blume C (2014) Pareto optimization or cascaded weighted sum: A comparison of concepts. Algorithms 7(1):166–185
21. Jemai J, Zekri M, Mellouli K (2012) An NSGA-II algorithm for the green vehicle routing problem. In: Hao J-K, Middendorf M (eds) In the proceeding of European conference on evolutionary computation in combinatorial optimization. Springer, Berlin, pp 37–48
22. Kara, İ., Kara, B.Y., Yetis, M.K. (2007). Energy minimizing vehicle routing problem. In Lecture Notes in Computer Science (LNCS), A. Dress, Y. Xu, and B. Zhu, eds. pp. 62–71.
23. Karaoglan I, Altiparmak F, Kara İ, Dengiz B (2011) A branch and cut algorithm for the location-routing problem with simultaneous pickup and delivery. European Journal of Operational Research (EJOR). 211(2):318–332
24. Koç Ç, Bektaş T, Jabali O, Laporte G (2014) The fleet size and mix pollution-routing problem. Transportation Research Part B: Methodological. 70:239–254
25. Koç Ç, Bektaş T, Jabali O, Laporte G (2016) The fleet size and mix location-routing problem with time windows: Formulations and a heuristic algorithm. European Journal of Operational Research (EJOR). 248(1):33–51
26. Kopfer, H.W., Kopfer, H. (2013). Emissions minimization vehicle routing problem in dependence of different vehicle classes. In Dynamics in Logistics, H.J. Kreowski, B. Scholz-Reiter, and K.D. Thoben, eds. (Berlin: Springer, Berlin, Heidelberg), pp. 49–58.
27. Lee W-C, Wang J-Y, Lee L-Y (2015) A hybrid genetic algorithm for an identical parallel-machine problem with maintenance activity. Journal of the Operational Research Society 66(11):1906–1918
28. Lin C, Choy KL, Ho GTS, Chung SH, Lam HY (2014) Survey of green vehicle routing problem: Past and future trends. Expert Syst Appl 41(4):1118–1138
29. Marrekchi, E., Besbes, W., Dhouib, D. (2016). A new taxonomy for the green vehicle routing problem. In The Proceeding of the 6th International Conference on Metaheuristics and Nature Inspired Computing, META'2016., p.
30. Melián-Batista B, De Santiago A, AngelBello F, Alvarez A (2014) A bi-objective vehicle routing problem with time windows: A real case in Tenerife. Appl Soft Comput 17:140–152
31. Molina JC, Eguia I, Racero J, Guerrero F (2014) Multi-objective vehicle routing problem with cost and emission functions. Procedia Soc Behav Sci 160:254–263

32. Montané FAT, Galvão RD (2006) A Tabu Search algorithm for the vehicle routing problem with simultaneous pick-up and delivery service. Computers & Operations Research 33(3):595–619
33. Oliveira, P.R.D.C., Mauceri, S., Carroll, P., Pallonetto, F. (2017). A genetic algorithm for a green vehicle routing problem. In The Proceeding of the 8th International Network Optimization Conference (INOC), Lisbon, Portugal, pp. 65–74.
34. Ombuki B, Ross BJ, Hanshar F (2006) Multi-objective genetic algorithms for vehicle routing problem with time windows. Appl Intell 24(1):17–30
35. Prins C (2004) A simple and effective evolutionary algorithm for the vehicle routing problem. Computers & Operations Research 31(12):1985–2002
36. Qian J, Eglese R (2016) Fuel emissions optimization in vehicle routing problems with time-varying speeds. European Journal of Operational Research 248(3):840–848
37. Soysal M, Bloemhof-Ruwaard JM, Bektaş T (2015) The time-dependent two-echelon capacitated vehicle routing problem with environmental considerations. International Journal of Production Economics 164:366–378
38. Tasan AS, Gen M (2012) A Genetic Algorithm based approach to vehicle routing problem with simultaneous pick-up and deliveries. Computers & Industrial Engineering 62(3):755–761
39. Tiwari A, Chang P-C (2015) A block recombination approach to solve green vehicle routing problem. International Journal of Production Economics 164:379–387
40. Wang H-F, Chen Y-Y (2012) A Genetic Algorithm for the simultaneous delivery and pickup problems with time window. Computers & Industrial Engineering 62(1):84–95
41. While L, Bradstreet L, Barone L (2012) A fast way of calculating exact Hypervolumes. IEEE Trans Evol Comput 16(1):86–95
42. Wu J, Azarm S (2000) Metrics for quality assessment of a multiobjective design optimization solution set. J Mech Des 123(1):18–25
43. Xiao Y, Zhao Q, Kaku I, Xu Y (2012) Development of a fuel consumption optimization model for the capacitated vehicle routing problem. Computers & Operations Research 39(7):1419–1431
44. Zachariadis EE, Kiranoudis CT (2011) A local search metaheuristic algorithm for the vehicle routing problem with simultaneous pick-ups and deliveries. Expert Syst Appl 38(3):2717–2726
45. Zhang J, Zhao Y, Xue W, Li J (2015) Vehicle routing problem with fuel consumption and carbon emission. International Journal of Production Economics 170:234–242
46. Zitzler, E., Thiele, L. (1998). Multiobjective optimization using evolutionary algorithms—A comparative case study. In The Proceeding of the Parallel Problem Solving from Nature—PPSN V, Lecture Notes in Computer Science (LNCS), Vol 1498, A.E. Eiben, T. Bäck, M. Schoenauer, and H.-P. Schwefel, eds. (Springer, Berlin, Heidelberg.), pp. 292–301.

Chapter 25
Modeling and Solving Difficult-to-Represent Optimization Problems

Akhil Yadav and Partha Chakroborty

Abstract Most optimization problems one encounters in reference books such as this can be, and are generally, formulated as mathematical programming (MP) problems. For some problems, the modeling requires deep insights into mathematical programming techniques or the way the variables of the problems interact with one another. Such exercises, however, involved at the end, yield mathematical programming formulations that have well defined methods for their solution. These solution methods may themselves be complex and require a good understanding of calculus as well as numerical techniques. Often these methods are computationally expensive and in itself requires go-around tools. However, there are some optimization problems that arise in engineering that cannot be formulated as mathematical programming problems. Often these require external procedure-based declarations to evaluate the system performances or define interactions between various subparts of the problem. Such problems are often solved using optimization metaheuristics like genetic algorithms that are structurally quite versatile and typically require information on only system performance to drive the optimization process. In this chapter the simple everyday problem of efficient urban mobility is introduced as an example that poses unique difficulties when modeled as an optimization problem. The question that is sought to be answered is whether a combination of one-way and two-way roads can be obtained such that an urban area provides the least travel time to its users. This chapter presents a formulation as well as a genetic algorithm-based solution methodology for the problem.

Keywords Transportation · Urban network design · Genetic algorithm · Metaheuristic

A. Yadav · P. Chakroborty (✉)
Department of Civil Engineering, Indian Institute of Technology, Kanpur, India
e-mail: partha@iitk.ac.in

A. Yadav
e-mail: akhily93@gmail.com

F. Hamid (ed.), *Optimization Essentials*, International Series in Operations Research & Management Science 353, https://doi.org/10.1007/978-981-99-5491-9_25

25.1 Introduction

There are many engineering problems that involve optimization. Some problems can be formulated as linear programing problems, others as integer programing problems, and yet others as non-linear mixed integer programing problems. There are some problems, however, that cannot be formulated as traditional mathematical programing problems. The reasons can be many. For example, there can be problems where how good an alternative solution is cannot be evaluated using an algebraic function that relates the various decision variable values of an alternative solution to the goodness of the solution. Such problems are generally solved using optimization metaheuristics like genetic algorithms and others. In this chapter one such problem from the area of mobility on urban networks is introduced. It may be mentioned that a better appreciation of the contents of this chapter can be had if the reader familiarizes himself/herself with some basic concepts of transportation engineering (for example, see [7]).

An urban road transport system comprises of a network of roads with many at-grade and some grade-separated intersections. Generally, the roads are two-way roads. On two-way undivided roads, during heavy traffic, there is considerable friction between the opposing flows. This in turn reduces the speeds of vehicles as well and increases safety hazards. The situation is worse at intersections where two-way approaches carrying large flows necessitate a greater number of phases and longer cycle lengths. Consequently, delays are high and the quality of service is bad. With most cities facing a steep increase in vehicular traffic, congestion has become a perennial problem. Further, the most congested regions of the city are those where roads cannot be expanded due to space restrictions. Some cities have resorted to building a second tier of roads on certain corridors to improve traffic flow. These capacity augmentation schemes are expensive and often not feasible from financial and engineering standpoints.

One of the low-cost, easy-to-implement ways of partially alleviating the congestion on city roads is to reduce friction on the roads and simplify intersections through the conversion of certain two-way roads to one-way roads. One-way roads have numerous advantages. Such roads can carry larger traffic volumes for a given level-of-service as compared to two-way undivided roads. This is because, due to the absence of opposing flows, speeds on one-way undivided roads at a given volume are greater than those on undivided two-way roads for the same total volume. This happens as drivers view on-coming traffic (on two-way undivided roads) as greater threat than those moving in the same direction. Further, overtaking opportunities on two-way undivided roads are often less than those on one-way undivided roads because of changes in overtaking behaviour. Sheffi [33] too discusses the impact of two-way roads on travel time. Also, the chances of incidents are less; this improves the reliability of travel time. Intersections with one-way approaches are simpler and offer lesser average delay to vehicles crossing the intersection as compared to intersections with two-way approaches. This is because, with one-way approaches, the number of turning movements at intersections reduces; leading to simpler phasing schemes and lesser lost times. Also, the saturation flow value for the lanes increases due to a lesser number of shared movements on the lanes. Overall, with one-way roads, the average speeds of vehicles in the network increase.

On the other hand, the presence of one-way roads causes some vehicles to travel longer distances to reach their destinations since the number of possible paths between node pairs is reduced. Therefore, the introduction of one-way roads creates a trade-off between longer paths and higher average travel speeds. The efficiency of mobility in the network depends on achieving a balance between higher speeds and longer distances through a meticulous choice of one-way roads. The problem of choosing the right mix of one-way and two-way roads without any infrastructural augmentation is referred to here as the network topology design problem or simply the network design problem. The network design problem is a discrete optimization problem with the direction of roads as decision variables and the efficiency of mobility as the objective.

While developing an efficient (or optimum) network design, care needs to be taken to ensure that certain service-related restrictions are satisfied. For example, it is reasonable to expect that there is traffic demand between every pair of nodes in the network. In order to satisfy or serve this demand, the mix of one-way and two-way roads should be so chosen that there exists at least one path between every pair of nodes. That is, the network is strongly connected. Further, for traffic control purposes and to reduce significant turning volumes at intersections, traffic engineers may demand that certain contiguous road sections that form a part of an arterial road (or some other corridor) cannot have frequently opposing directions. Many other such service-related constraints may exist. Urban roads have frequent intersections. An optimum network design problem must therefore incorporate the impact of change in the mix of one-way and two-way roads on the delay at intersections in a realistic manner. Finally, any network design problem must represent the impact of network design on link flows realistically. That is, at the very least, assignment algorithms used to calculate the flows on road sections or links must consider the impact of flow on travel time. Thus the problem here is to develop an optimum network design (i.e., a specific distribution of one-way and two-way roads) that provides efficient mobility while considering service-related constraints, intersection delay and realistic link flow calculations.

The chapter is divided into six sections of which this is the first. The next section presents a review of relevant literature along with the motivation behind this work. Section 25.3 describes the formulation of the problem as a constrained optimization problem. Section 25.4 discusses the proposed solution methodology. In Sect. 25.5, results are presented for various networks on which the proposed algorithm is tested. This section also discusses various aspects of the proposed solution methodology. Section 25.6 concludes the chapter.

25.2 Literature Review and Motivation

Farahani et al. [14] provides, among other things, a comprehensive review of the work done on the network design problem. From [14]'s review, it can be seen that while many have worked on this problem, none have considered all the service restrictions or constraints (discussed here) in their formulations. Further, some of the authors have not included intersection delay in their objective function. Since creation of one-way roads impact intersection delay significantly, approaches that ignore intersection delay are not realistic. In the following, a comprehensive review of specific relevant works are discussed.

Drezner and his team [10–12] have worked extensively on the network design problem. However, in none of their work, service-related restrictions or other constraints have been taken into account. Objective functions do not consider intersection delay. Network efficiency is evaluated in terms of the total travel time on roads. Traffic assignment (a procedure that assigns flows on different roads) is carried out using a simplistic flow-independent strategy like all-or-nothing. Various optimization algorithms are used to solve the network design problem. Drezner and Wesolowsky [12] have used branch and bound algorithms, two heuristic algorithms and simulated annealing. Drezner and Salhi [10] have used tabu search method. In a later work, Drezner and Salhi [11] have used modified simulated annealing and problem-specific heuristic augmented genetic algorithm (GA) based methods. They conclude that the heuristic augmented GA-based method provides the best results. The use of heuristics in their GA-based method makes the solution technique specific to the unconstrained problems they have tried. It is difficult to ascertain how the method will work when say the objective function (other than that used by the authors) changes, or traffic assignment becomes flow-dependent, etc. Even though, it is assumed that Drezner and his team have ways of identifying and weeding out infeasible solutions, their articles are silent on this issue.

Lee and Yang [22] have considered total travel time (that includes intersection delay) as their objective function and traffic assignment is done using the more realistic, flow-dependent, deterministic user equilibrium assignment. Two solution algorithms have been proposed, one is a heuristic and the other is based on simulated annealing. The authors mention the inclusion of a procedure used to eliminate infeasible solutions but did not provide any details. Although [22] made an attempt to calculate intersection delay, the signalization design employed did not use flow and intersection geometry-dependent phasing scheme. Saturation flows were also independent of intersection geometry. Service-related restrictions were not considered.

Cantarella et al. [4] have evaluated the network efficiency in terms of total travel time. Traffic assignment is carried out using the flow-dependent deterministic user equilibrium assignment. A number of solution algorithms such as hill climbing, tabu search, simulated annealing and genetic algorithms were tried. Cantarella et al. [4] do not mention how they identify and weed out infeasible solutions. As mentioned earlier, none of the service-related restrictions have been considered.

Gallo et al. [17] have also evaluated network efficiency in terms of total travel time and traffic assignment is done using flow-dependent stochastic user equilibrium assignment. The solution algorithm employed is scatter search. Regarding the infeasible solutions, the authors have simply discarded them and replaced them with randomly generated feasible solutions.

References [5, 32, 36, 40] have recast the problem as that of allocating different number of lanes in either direction of the road. Zhang and Gao [40] employed particle swarm optimization to minimize the total travel time. Cantarella and Vitetta [5] evaluated the efficiency of the network by multi-objective criteria that include travel times of vehicles, travel time of pedestrians (after car parking) and pollutant emission. The authors propose reapplying crossover and mutation if an infeasible solution is generated by the GA process.

Among the many previous work in this area, intersection delay, an important component of urban travel time, has been incorporated in the travel time calculations by [4, 5, 17, 22] only. However, most have made simplifying assumptions such as flow-independent phasing schemes that do not change with the number of shared movements on a lane, etc. All except [22] also assume flow-independent cycle length. The over-simplification of these important factors adversely impacts the delay calculations at intersections and ultimately the quality of the solutions.

Even though the network design problem does not involve infrastructural augmentation, some earlier works have included infrastructural interventions along with road orientations as decision variables. For example, [24] considers lane additions and road orientation as the decision variables. The efficiency of a network is evaluated in terms of reserve capacity[1] of the network. Certain budgetary constraints have been imposed for lane additions along with the maximum possible number of lanes that can be added. The authors have not considered any of the service-related restrictions described here. Intersection delay is also not included.

Several authors considered only infrastructural interventions [8, 18, 21, 30, 31, 37, 39] as decision variables (arc direction is not considered). They have proposed a network design problem involving simultaneous choice of arc (road segment) additions and capacity improvements.

Drezner and Wesolowsky [13] have looked at the academic problem of creating an optimal network connecting points (nodes) that have travel demand between them, but initially, only the nodes exist in space with no roads connecting them. The objective is to minimize total construction and transportation costs. This work has limited application to any real network as all of them have a road network in place.

Marciano et al. [23] have considered signal setting optimization of a network (whose topology is given). The objective function considered is the total travel time, which includes link travel time and delay at intersection.

References [19, 25, 26] have assessed the trip serving capacities and vehicle operations of various one-way and two-way networks. The authors have assumed a homogeneous grid network (all links are the same). Such assumptions of idealized

---

[1] Reserve capacity is the largest multiplier applied to a demand matrix, that can be allocated to a network without violating the road segment arc/link capacities.

grid network and lane discipline may be applicable in highly planned cities but most cities do not have grid networks and lane discipline is only weakly followed in many cities in developing economies.

In summary, the review shows that only a few methods refer to the issue of connectivity while the rest are silent. One therefore does not know whether these methods ensure strong connectivity in the network. Further, methods that refer to the issue of connectivity seem to handle it in an ad hoc manner by relying on randomized techniques to chance upon networks that are strongly connected. Secondly, the review indicates some methods like those that use flow-independent assignment techniques ignore the relationship that exists between flow and travel time and yet use the latter as the objective. Such assumptions raise questions about the applicability and utility of these methods. Third, many methods ignore delays at intersections in their analysis. Those that include delay make simplifying assumptions, like movement and flow-independent phasing scheme and saturation flow, thereby severely compromising the usefulness of such methods. Finally, even though most methods use metaheuristics (traditional mathematical programming is unsuitable for this problem), many include problem-specific heuristic-based pre- or post-processing. This limits the applicability of such methods.

In this chapter, a metaheuristic-based connectivity conserving optimization algorithm that can efficiently provide optimal network designs while incorporating various service-related restrictions, realistically computed delays at intersections and link flows is developed.

25.3 Problem Formulation

In the first of the two subsections of this section, the network design problem is stated more formally. Next, the problem is formulated as a constrained optimization problem.

25.3.1 Problem Statement

A road network is represented as a weighted digraph (see [15] or any other textbook on graph theory for a formal definition of digraphs). The nodes are generally intersections or points (often referred to as centroids) representing an area that is a demand generator. The directed arcs are generally road segments or centroid connectors (virtual roads representing the roads of the area depicted using the centroid) with the allowed direction of movement. Sometimes, however, the arcs represent movements at intersections, and nodes, locations on approaches just upstream of the intersection. For example, Fig. 25.1 shows two types of intersection representation. In the first (Fig. 25.1a), a four-way intersection (say Intersection i) with all two-way approach roads is represented as a part of a digraph. In this case, the weights on an

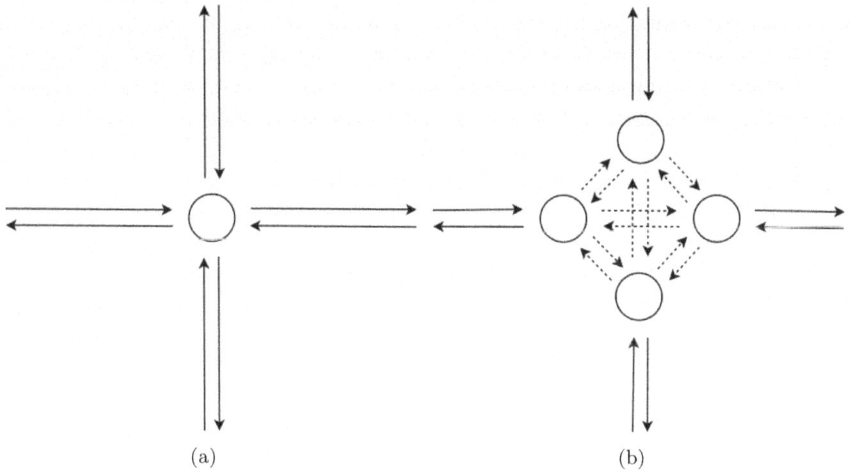

Fig. 25.1 Network representations of an intersection; **a** single node representation and **b** expanded representation

arc are the sum of the travel time on the approach road it represents and the average delay at the downstream intersection. The assumption here is that all the movements (left-turn, through and right-turn) that go through an intersection face more or less the same average delay. If this assumption is not true then in the most general case (where each movement at the intersection has a different delay) the intersection can be represented as shown in Fig. 25.1b. Not all arcs in the figure represent physical roads, nor do the nodes represent the intersection. In this representation, the nodes represent points on the approach roads immediately upstream of the intersection and the arcs drawn using broken lines represent various movements at the intersection. The arcs drawn using solid lines represent physical road sections. The weight on a solid line arc is the travel time on the road section it represents. The weight on the broken line arc is the average delay for the movement it represents.

In general, the network design problem can have three types of arcs. The first type (or Type 1) of arcs are those like the ones shown in Fig. 25.1a and their weights include both travel time on the road sections they represent and the delay at the downstream intersection. Type 2 arcs are like the solid line arcs of Fig. 25.1b and Type 3 arcs are the broken line arcs of the same figure. The reader should note that although Fig. 25.1b type of representation of intersections are versatile (in that they can represent delays at the intersection most realistically), they should be used sparingly since they increase the number of nodes and arcs of the corresponding diagraph. Centroids can be treated as nodes like the one shown in Fig. 25.1a. A centroid connector arc can be treated as an arc in Fig. 25.1a or as a solid arc in Fig. 25.1b depending on the representation of the downstream intersection.

As will be seen later, the present problem is a discrete optimization problem that is computationally intensive and notoriously difficult to solve [29]. Hence, indiscriminately increasing the network size to incorporate different delays at an intersection may be counterproductive. Further, since studies such as this are done at a planning level (or strategic stage) and not at an operational level, sometimes, loss of accuracy in delay representation due to Fig. 25.1a type description can be considered acceptable.

From the earlier description, it is clear that the weight on an Arc(i, j) that goes from Node i to Node j, w_{ij}, is:

$$w_{ij} = \begin{cases} TT_{ij} + d_j, & \text{if Arc}(i, j) \text{ is of Type I} \\ TT_{ij}, & \text{if Arc}(i, j) \text{ is of Type II} \\ d_i^{m(i,j)}, & \text{if Arc}(i, j) \text{ is of Type III} \end{cases} \tag{25.1}$$

where, TT_{ij} is the travel time on the road segment from Intersection i (or Centroid i) to Intersection j (or Centroid j); d_j is the average delay at Intersection j; $d_i^{m(i,j)}$ is the delay at the intersection for the movement $m(i, j)$. TT_{ij} is a function of the per lane flow on the road segment from i to j and also on whether Arc j to i exists and the road is physically undivided. In this chapter, the well studied BPR equation [3] is used as the base expression for TT_{ij}. Here,

$$TT_{ij} = K \times TT_{ij}^{o} \left(1 + \alpha \left(\frac{v_{ij}}{b}\right)^{\beta}\right) \tag{25.2}$$

where $K > 1$ (and user-defined) if also $(j, i) \in A$ (where A is the set of all the arcs in the digraph) and the road between i and j is undivided, else $K = 1$; TT_{ij}^{o} is the free flow travel time along the road from i to j, b is the practical capacity of a lane, α and β are generally taken as 0.15 and 4, respectively; v_{ij} is the per lane flow on the road from i to j. Note, the factor K ensures that if i to j link is part of a two-lane undirected road, then for a given per lane volume, it will have a higher travel time than if i to j was a part of a one-way undivided road at the same per lane volume. The reader should note that irrespective of whether a road is divided, the number of lanes (hence v_{ij}) moving traffic from i to j changes based on whether $(j, i) \in A$. For example, when a road joining nodes i and j has four lanes then it may have two lanes in each direction if $(i, j) \in A$ and $(j, i) \in A$, but will have all four lanes in a given direction when either (i, j) or (j, i) is in A.

The average delay at a signalized intersection depends on a number of factors such as the traffic flow per lane, the saturation flow, the phasing scheme, the cycle length, and the effective green time. The phasing scheme is typically chosen from various discrete strategies on how to signalize an intersection; the choice is made by engineers based primarily on turning movement volumes and perceived threat to safety. The choice of phasing scheme impacts cycle length and delay. In the proposed method, an algorithmic procedure is prescribed to determine delay based on the various movements that come to an intersection, their flow values and the chosen phasing scheme. The algorithm implements the standard process of signal timing design (see any textbook on transportation such as [7] or [28]) and the delay

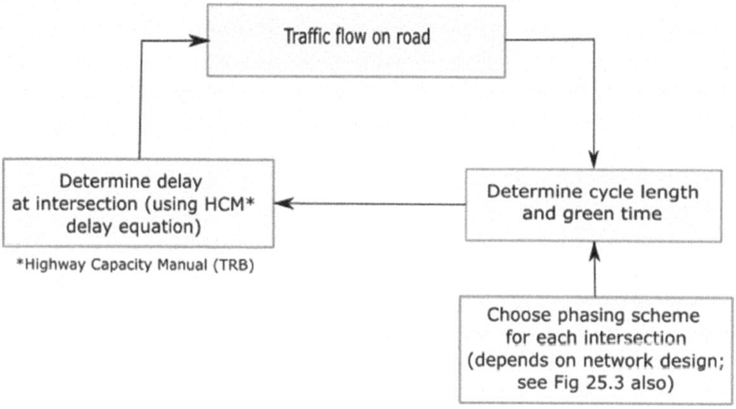

Fig. 25.2 The algorithm to find delay at intersections

is calculated using the equations in Highway Capacity Manual [35]. The details are skipped here for the sake of brevity; Fig. 25.2 shows the algorithm to determine the delay that is implemented here; the cycle in the figure is continued till the delay estimates (and flows) stabilize.

Thus far, the discussion concentrates on how the weights of the digraph representing the road network are calculated. It can be seen that the weights, among other things, depend on (i) flow on the arcs, (ii) whether the arc represents a direction of travel on an undivided two-way road or one-way road or a divided two-way road, (iii) the number of lanes and (iv) the combination of movements at intersections. All of these depend on the design of the network; or alternatively the direction of the arcs - the decision variables. Another point that needs to be reiterated is that the inclusion of K in Eq. 25.2 (see point (ii) also) implies that all link travel time functions are not necessarily independent of one another and link interactions exist.

Finally, it needs to be noted that the relationship between flow and network design is an outcome of the route choice process of drivers and is implemented in practice through traffic assignment algorithms. One of the widely accepted flow dependent traffic assignment algorithms is the user equilibrium method. It assumes that each traveler chooses a path so as to minimize his/her travel time which in turn depends on how many people use that path. For a detailed description of the user equilibrium process one may refer to [33]. The user equilibrium assignment process is an optimization problem and can become computationally onerous when used repeatedly. It is proposed that the incremental assignment technique, which when implemented with appropriate increments can approximate the user equilibrium assignment results be used here. The reader may refer to [7] or [27] to get a better understanding of the incremental assignment method. Use of incremental assignment method also allows one to easily incorporate a strategy to concurrently determine the signal timings and the flows on links (note that signal timings are dependent on flows and impact delays at intersections; the latter influences travel time and through it the link flows).

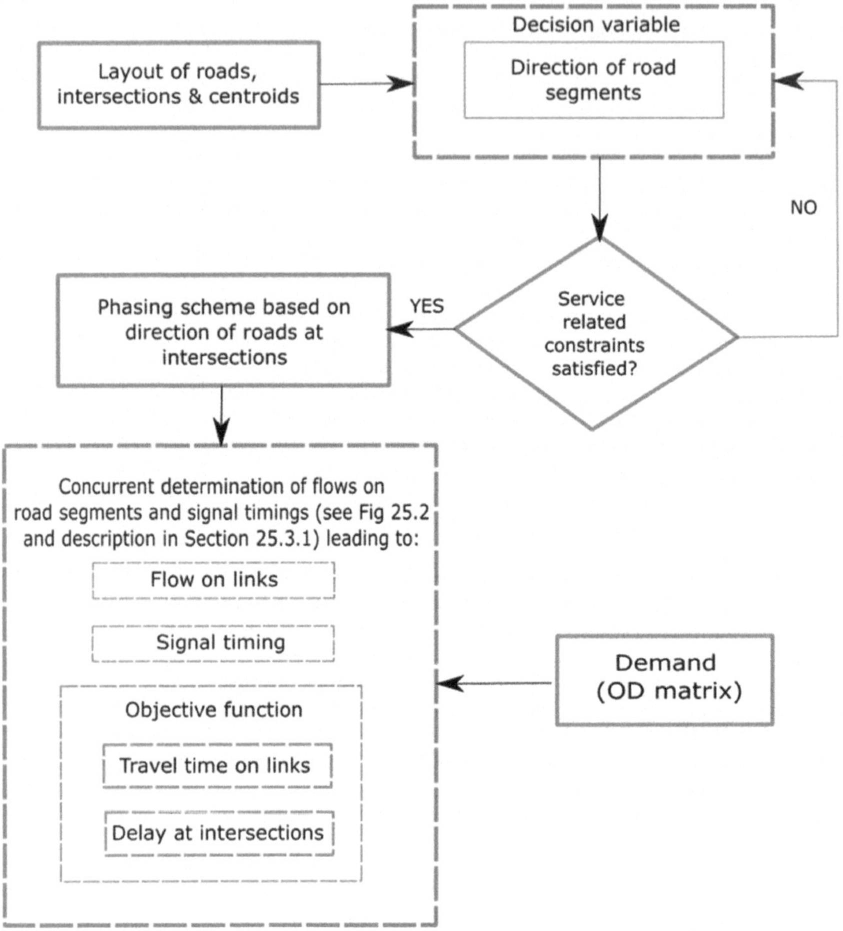

Fig. 25.3 Various features of the network design problem and their interdependence

There are aspects of the problem that are not dependent on the weight of the arcs but dependent primarily or wholly on the direction of the arcs. Two of them are (i) strong connectivity,[2] and (ii) direction of allowed movements on arterials/corridors made up of many contiguous arcs (road segments). Figure 25.3 shows a schematic of the problem delineating its various features and their interdependence. For the present, the reader may ignore the boxes drawn with broken lines.

As indicated in Fig. 25.3, the layout of physical roads, intersections, centroids and the origin-destination (OD) matrix—a table of travel demands from the various origins to destinations, are taken as input to the problem. The direction of road

[2] According to [15], a digraph is said to be *strongly connected*, if every two of its distinct nodes, u and v say, are such that u is reachable from v and v is reachable from u.

segments are chosen in such a way that the resultant network satisfies the service-related constraints while minimizing the sum total of travel time on links and delays at intersection. Figure 25.3 also attempts to express how the decision variable (direction of road segments) affects the objective function. Unlike in previous studies, here, for a given choice of direction of road segments, phasing schemes at every intersections are obtained. Next, the demand and the phasing scheme are used as inputs to concurrently determine the signal time and flows on the road segments. The classical incremental assignment is slightly modified to allow such a determination. In the first stage of the incremental assignment, flows are loaded on the network based on the free flow travel times and zero delays at intersections. These flows are next used not only to update the travel time on the road links but also the delays at intersections. In order to find delays at intersections, the flows are first used to determine the signal time. This is done using Webster's equation to calculate the cycle length and flow ratios to determine green time (the reader may refer to [7] or [28] for more details). With the updated travel time and delays, the next iteration of the incremental assignment process is done. This process of incrementally assigning the flows and updating the travel time, signal timing and delays at intersections are carried on till the entire demand has been loaded. The impact of the particular network design (i.e., the choice of road segment directions) is measured from the total travel time the network offers to the users. Broadly speaking, the network design is changed till one obtains the design that offers the minimum travel time to all its users.

As described earlier, an optimal network design problem looks at the task of improving mobility of the network users. In the next section this problem is formulated as an optimization problem.

25.3.2 Optimization Formulation

The efficiency of mobility afforded by the network to its users is generally measured through either total travel time of all the users or the average travel time per user. Thus, the objective function representing efficiency of mobility must include both the travel time on road segment and the delay at intersection (see Fig. 25.3). If M users are in the network, the total travel time, TTT or the average travel time, ATT is given as:

$$TTT = \sum_{(i,j)\in A} w_{ij} \times f_{ij} \tag{25.3}$$

$$ATT = \frac{TTT}{M} \tag{25.4}$$

where A is the set of all arcs in the digraph and f_{ij} represents the flow from i to j and w_{ij} is the weight of Arc(i, j) (defined in Eq. 25.1). The objective function, say Z, is either TTT or ATT.

Physically, the decision variables, are the direction of the road segments (see Fig. 25.3). Mathematically, the decision variable δ_{ij} is a 0–1 variable indicating whether the Arc(i, j) is a part of the digraph representing the network; δ_{ij} is zero when Arc(i, j) is not a part and it is unity otherwise. Those (i, j) pairs that represent road segments (i.e., Type 1 and Type 2 arcs) are decision variables. If set A_{12} represents all (i, j) pairs of Type 1 and Type 2 arcs, then the following constraint must hold:

$$\delta_{ij} + \delta_{ji} \geq 1 \qquad \forall (i, j) \in A_{12} \tag{25.5}$$

The existence of arcs that are introduced to represent delay more accurately (Type 3 arcs) are dependent on the configuration of intersections and hence on the approach directions (or the values of δ_{ij}'s for Type 1 and Type 2 arcs). These (i, j) pairs are chosen through external procedures that represent an intersection (if the need arises) with known Type 1 and Type 2 arcs in the form shown in Fig. 25.1b.

As discussed before, while developing an efficient network design, care needs to be taken to ensure that certain service-related restrictions are maintained. The two service-related constraints used in the proposed formulation are described here.

Connectivity constraint—A network must have at least one path between every pair of nodes. Mathematically, the network should be strongly connected. For more details on connected graphs, one can refer to any book on Graph theory (e.g., [15]). Mathematical representation is avoided as checks for connectivity of the digraph is done using external procedures, namely Tarjan's algorithm [34]

Directional corridor constraint—Urban road networks typically have arterial roads that run through significant portions of the city and a large proportion of arterial traffic have long trip lengths. From traffic control purposes, traffic engineers may require that a certain set of the connected roads (forming the arterial or otherwise) be designated as a directional corridor. A directional corridor, as defined here, is a set of connected roads where all the constituent road sections are either all two-way or all one-way in a given direction. Let Δ_{ij} represent the direction of the road link between i and j. Further let

$$\Delta_{ij} = \begin{cases} 0, & \text{if } \delta_{ij} = 1 \text{ and } \delta_{ji} = 0 \\ 1, & \text{if } \delta_{ij} = 0 \text{ and } \delta_{ji} = 1 \\ 2, & \text{if } \delta_{ij} = 1 \text{ and } \delta_{ji} = 1 \end{cases} \tag{25.6}$$

that is, $\Delta_{ij} = (\delta_{ij} + \delta_{ji}) \times \delta_{ji}$. Then if road links connecting j_1 and j_2, j_2 and j_3 and so on to j_{n-1} and j_n are part of the corridor connecting j_1 and j_n that is to be designed as a directional corridor, then the digraph must satisfy the following constraint:

$$\Delta_{j_1 j_2} = \Delta_{j_2 j_3} \ldots = \Delta_{j_{n-1} j_n} \tag{25.7}$$

Although the problem is set up as an optimization problem and some constraints are defined mathematically to give the reader a flair of the problem, a complete mathematical programming formulation is avoided. This is because (as shown in

Fig. 25.3) many of the relations between the decision variables and the problem features such as ATT and connectivity are procedure-based. For example, for a given set of δ_{ij}, the values of f_{ij} and w_{ij} can be determined only after deciding the appropriate phasing scheme and signal timings. The values cannot be obtained using algebraic relations and must be obtained through external procedures.

25.4 Solution Methodology

The problem of determining the optimum network design is a discrete combinatorial optimization problem. Classical methods such as branch and bound work efficiently for only small sized networks [5, 12, 14] with simplifying assumptions that make them unrealistic. Additionally, for the more realistic problem described here many features cannot be easily related to the decision variables through algebraic expressions. For example, the choice of phasing scheme, saturation flow, etc. can only be decided through external procedures. Further, ensuring that a network is strongly connected through the use of constraints in the form of algebraic expressions is at best difficult, if not impossible. Hence for large, real-world networks, one has to rely on heuristics and metaheuristics to obtain good solutions in reasonable time [5]. However, it is preferable not to use heuristics as they are problem specific and limit the applicability of the algorithm.

Drezner and Salhi [11] have worked on many metaheuristic and heuristic augmented metaheuristic solution algorithms such as tabu search, simulated annealing and GA. They conclude that GA gives the best result. In similar combinatorial problems with external procedure-based declarations within the optimization process, past experience has shown that genetic algorithms work effectively [1, 6]. Given the unsuitability of integer programming methods and past experiences with the effectiveness of genetic algorithms, in this chapter a GA (metaheuristic) based solution methodology is used. The reader may refer to [20] for an introduction to GA. The choice of GA in the current work does not imply that other metaheuristic-based techniques that have similar performance properties cannot be used.

In the following subsections, the GA based procedure is described in detail. For the purpose of this description, intersections are assumed to be represented as shown in Fig. 25.1a.

25.4.1 String Representation

In order to solve the optimization problem using GA, the network needs to be represented in the form of a string of bits. The string representation is somewhat different from the digraph representation of the network. The direction of each road segment on the network is represented by a bit in the string. The position of the bit in the string stands for the road segment it represents. Since the road segments do not change (only

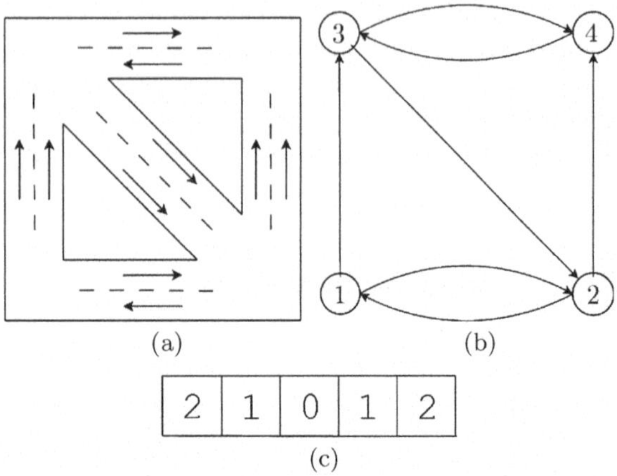

Fig. 25.4 **a** A simple illustrative road network (India follows keep-left policy), **b** digraph representation of (**a**) and (**c**) string representation of (**a**) or (**b**)

their directions change), a map between bit positions and roads exists and needs to be created only once.

The string proposed here is a ternary string with each bit representing the value of Δ_{ij}, where i and j are nodes present in the network and i is always assumed to be less than j (note that in the Δ_{ij} representation, unlike in the δ_{ij} representation, (i, j) does not imply i to j. It only implies i and j). If a bit has a value of two, then the road segment that the bit represents is considered to be two-way. If the value is one, then the road segment is considered to be a one-way road directed from the lower numbered node to the higher numbered node. If the value is zero, then the road segment is considered to be a one-way road directed from the higher numbered node to the lower numbered node. Figure 25.4 illustrates a simple road network, its digraph and string representations. Note, that the string representation is based on the following map: the first bit represents the road segment between Nodes 1 and 2, the second bit between Nodes 1 and 3, the third bit between Nodes 2 and 3, the fourth bit between Nodes 2 and 4, and the fifth bit between Nodes 3 and 4. Note the implication of second bit being 1 is that the Arc (1, 3) exists in the digraph but not Arc (3, 1); if this bit was 0 it would imply the opposite; while if it was 2 it would imply that both Arcs (1, 3) and (3, 1) exist in the digraph.

All road segments that are part of a directional corridor must have the same direction (if they are one-way) or must be two-way roads. In the present representation, only one bit is assigned for all the road segments of a directional corridor. Operationally, this implies that the map between bit positions and road segments is one-to-many and one bit may represent the direction of more than one road segment. Through this representation, the constraint on each road segment of a directional corridor (i.e., $\Delta_{j_1 j_2} = \Delta_{j_2 j_3} \ldots = \Delta_{j_{n-1} j_n}$ constraint) is automatically satisfied thereby increasing the efficiency of the search process.

25.4.2 Initial Population Generation

GA works with a population of strings. The initial population is generated randomly. Each bit of a string is randomly assigned one of the three possible values, 0, 1 or 2. Generating the initial population in this manner may lead to initial solutions (or strings) that do not represent strongly connected digraphs. Since the problem requires each solution to be strongly connected, strings in the initial solutions may have to be modified to become strongly connected. The procedure to identify whether a digraph (network) is strongly connected and if not, the procedure to make it strongly connected are explained next.

25.4.3 Connectivity Function

Before proceeding further, the reader may to refer to [15] for an introduction to graph theory. Figure 25.5 shows the flowchart of the connectivity function. The connectivity function takes a string as an input and returns a string as an output. First, the string is converted into an adjacency matrix. The adjacency matrix of a string or its corresponding Digraph D, with n nodes is the matrix $B = (b_{ij})_{n\times n}$, in which $b_{ij} = 1$, if Arc(i, j) is in D, and 0 otherwise. The conversion process utilizes the values of the bits, the bit position and the bit position to node-pair map to generate matrix B.

Given the adjacency matrix representing the digraph, the next step is to identify all the strongly connected components in the digraph. A *strongly connected component* (or a strong component) in a digraph D is a maximal strong subdigraph[3] of D. For every strong component in a digraph, there exists a path between every ordered pair of vertices belonging to that strong component. All the strongly connected components (say S) in a digraph are identified using Tarjan's algorithm [34]. If there is only one strong component (i.e., $S = 1$), then the digraph on the whole (or the entire network) is strongly connected. If $S > 1$, then the network on the whole is not strongly connected; it has S sub-networks that are strongly connected. In this case the connectivity function makes certain changes in the network design (i.e. in the digraph or adjacency matrix) in order to make the entire network strongly connected (or make $S = 1$).

An iterative procedure is employed in order to make $S = 1$. As shown in Fig. 25.5, the procedure first randomly selects any two strong components (from S) in the digraph. It then proceeds to determine the number of arcs between the two strong components (i.e., arcs that start from nodes belonging to one of the two components and ends at nodes belonging to the other component). If there are no arcs between the two strong components, the connectivity function again randomly selects two strong components until there is at least one arc between the two selected components. If

[3] A digraph $D' = (V', A')$, is termed a subdigraph of digraph $D = (V, A)$, if $V' \subseteq V$ and $A' \subseteq A$ and D' is a digraph.

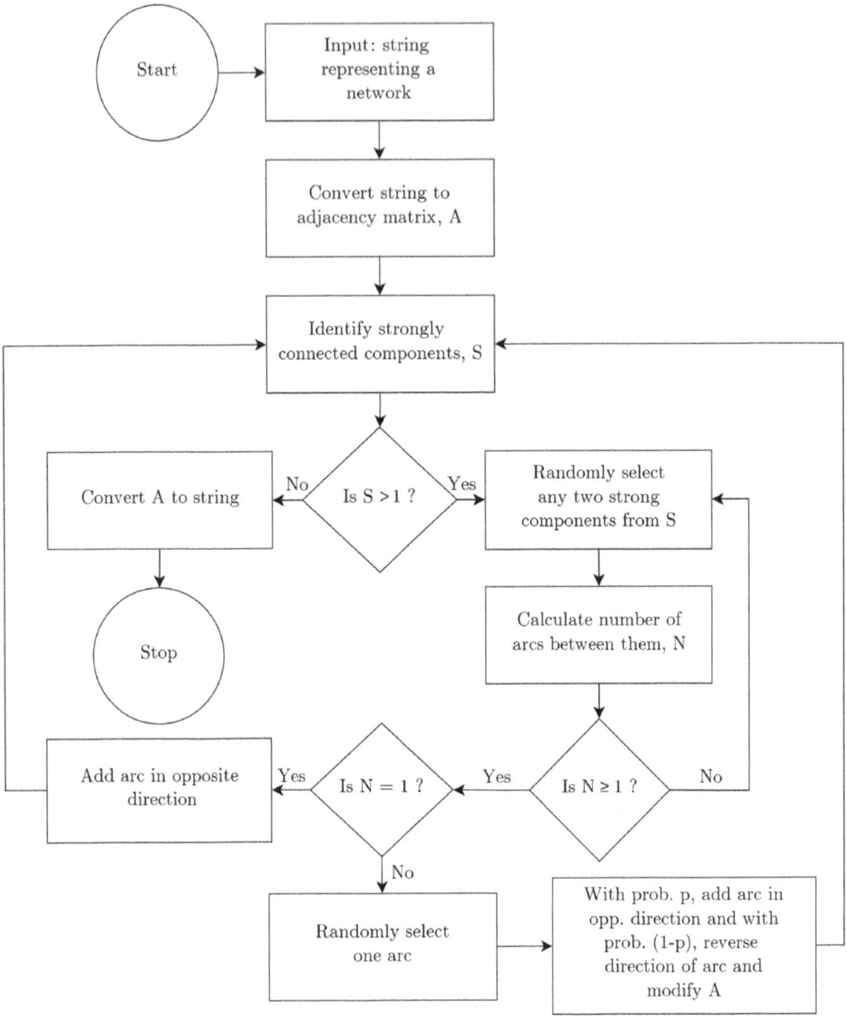

Fig. 25.5 Connectivity function flowchart

there is only one arc between the two strong components, then the connectivity function converts the two strong components into a single strong component by adding another arc in the opposite direction between the same pair of nodes. Physically this means that a one-way road between the two nodes is converted into a two-way road. If there are more than one arc between the two strong components, then it can be shown that all of them necessarily start from nodes of one strong component and end at the nodes of the other strong component. The connectivity function randomly selects one arc and either has its direction reversed (i.e., turns the corresponding road into a one-way road in the opposite direction) or adds an arc in the opposite direction (i.e., turns the corresponding road into a two-way road). In this manner, the two selected strong subdigraphs (or sub-networks) become one strongly connected network (or bigger sub network). The Tarjan's algorithm again identifies the strong components in this digraph. This process is repeated until there is only one strong component in the digraph; i.e., the entire network is strongly connected. Note the above process assumes that the given network is strongly connected when all the road segments are two-way. This is always true for real-world road networks.

The string representation and the connectivity function ensures that the strings at the beginning of every generation of the GA is feasible from the stand point of connectivity and directional corridor constraints. The quality of a string therefore depends on what kind of travel times and travel paths the network it represents provides to its users. This quality is measured through the fitness function described next.

25.4.4 Fitness Function

In general the fitness function of a string defines how good that string is when evaluated using the objective function and functions measuring the violation of feasibility constraints. The total travel time, TTT (see Eq. 25.3) is used as the objective function for the strings (or networks) in the present problem.

Once the fitness for each string is obtained, three GA operators are applied in sequence to obtain the population of strings for the next generation. These operators are briefly described next. Elitism is used in the proposed GA based method. In this, the best string (network) found till a given generation is always passed on to the next generation. The reason for using elitism is to preserve the best solution since the network design problem is not only a discrete combinatorial problem but is also ill-behaved. That is, a small change in the string may cause a large change in the performance of the solution.

25.4.5 GA Operators

The three GA operators used are reproduction, crossover and mutation. Tournament selection of size two is used as the reproduction or selection operator. Single point crossover with probability p_c is chosen and bit-wise mutation with probability p_m is applied. Elitism is implemented by identifying the best string in a given generation, say Generation Z, and comparing it with the best string in the Generation $Z-1$. If the best string in Generation $Z-1$ is better than the best string in Generation Z, then the worst string of given Generation Z is replaced with the best string of Generation $Z-1$. If the best string in the Generation Z is better, then no replacements are made.

Figure 25.6 is a flow chart of the optimization process described here. In the next section, results are presented for three synthetic networks and two real-world-based networks.

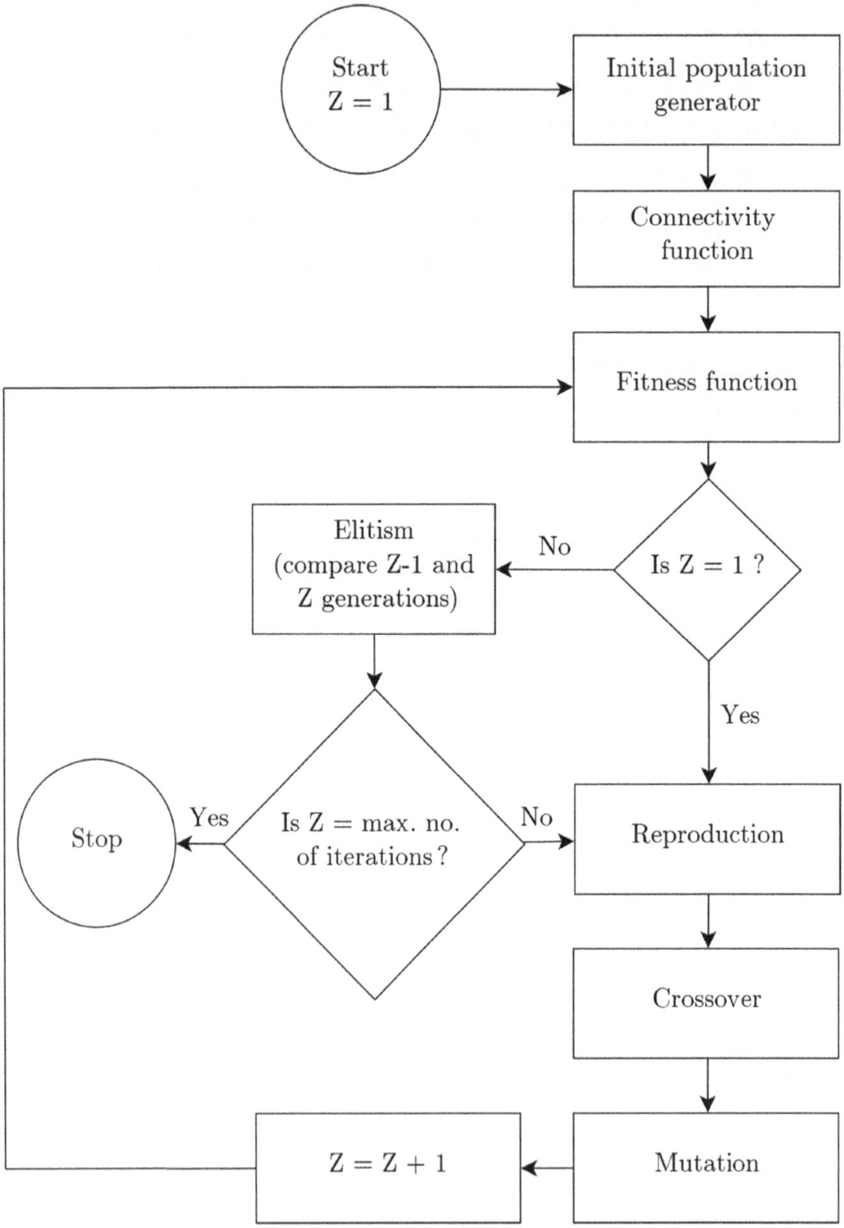

Fig. 25.6 The GA based solution methodology proposed for the network design problem

25.5 Results and Discussion

In this section the results obtained using the proposed method are presented. The proposed method has been applied to nine different problems with five different networks. All these results may be found in [38]. Here some of the results from five networks are described. Three of these are from past studies (even though many authors have worked on the problem, network details and demand matrix data was only found for three synthetic networks used by [11]). In this work, two networks are **created** based on real-world road networks of Indian cities. The results from both these networks are included here. The details of these two networks and the demand matrices are available in [38].

In this section, while describing the size of the networks, the word *intersection* denotes actual intersections and centroids, and the phrase *road segment* denotes actual road segments and centroid connectors. The synthetic networks from [11]'s study are referred to as the DS-Sparse, DS-Medium and DS-Dense networks; they have 65, 99 and 164 road segments, respectively. All of them have 40 intersections. The networks created from the real-world Indian cities are referred to as RW-1 and RW-2. RW-1 has 32 intersections and 45 road segments while RW-2 network has 223 intersections and 361 road segments. Note that the RW-2 network is comparable with the largest networks in the literature (for example, largest network used by [4, 5, 17], respectively have 176 intersections and 510 road segments, 150 intersections and 141 road segments and 254 intersections and 449 road segments). Before presenting the results, values for the various parameters used here are stated.

25.5.1 Parameter Values

The various parameter values used here are those that are generally adopted in the literature or are applicable for the given network.

- Parameters related to TT_{ij} equation (see Eq. 25.2)
  - b = 450 vph
  - $\alpha = 0.15$
  - $\beta = 4$
  - $K = 1.2$
  - TT_{ij}^{o} = Length of Arc(i, j) / Free flow speed
  - Free flow speed = 40 kmph
- Parameters related to d_j or $d_i^{m(i,j)}$
  - Default parameter values suggested by [35] are used for delay calculations
  - Saturation flows for various movements from a lane that only allows

    Through movement = 1500 vphg (veh. per hour of green)
    Left turn = 1200 vphg (India follows a keep-left policy)

Left turn + through movement = 1200 vphg
Right turn = 1100 vphg
Right turn + through movement = 1100 vphg
Left turn + through movement + right turn = 1000 vphg

- Parameters related to GA operators (selected after studying the performance)
  - Crossover probability, $p_c = 0.95$
  - Mutation probability, $p_m = 0.01$
  - Population size = 10
- Parameter related to Connectivity function
  p (probability with which one-way road is converted to two-way road) = 0.75

25.5.2 Results for Drezner and Salhi's Networks

The purpose of this section is to compare the performance of the proposed metaheuristic-based method with the performance of [11]'s heuristic assisted algorithm. It should be noted that while comparing, the following aspects are ignored—(i) delay at intersections, (ii) flow-dependent travel time, and (iii) directional corridor restrictions, etc. The reason for ignoring these important considerations is that Drezner and Salhi's heuristic-based method does not have the ability to handle these conditions. The results obtained using the proposed procedure and the results obtained by Drezner and Salhi are provided in Table 25.1 (note that these are only for Drezner's $\alpha = 0.7$; also [11] have not provided any unit for travel time). In the table, the columns titled "Best" report the travel time of the best solution obtained by a method from its various runs. The columns marked "Average" report the average of the best of each of the various runs of a method. For the purposes of this comparison, the proposed method was run 10 times (note: [11] conducted 10 runs for their GA based method).

The table shows that the best results obtained using the GA based procedure developed here are similar to the best results obtained by Drezner and Salhi; only for the DS-Dense, the best solution for the proposed method is 0.45% higher than

Table 25.1 Comparison between results obtained by the proposed method and that obtained by [11]

Network	Optimal total travel time			
	Proposed method		Drezner and Salhi [11]	
	Best	Average	Best	Average
DS-Sparse	181356	182070	181356	181356
DS-Medium	146431	146855	146431	146484
DS-Dense	129515	129791	128935	129172

Drezner and Salhi's best result. The averages are also only marginally higher than Drezner and Salhi's. Given that the proposed method, unlike [11], does not use a problem-specific heuristic-based algorithm and is capable of solving problems that are more realistic and general than those solved by [11], the performance is considered acceptable. In the next section, results from the real-world-based networks RW-1 and RW-2 are presented.

25.5.3 *Results for Real-World-Based RW-1 and RW-2 Networks*

In this section four problems using the RW-1 and RW-2 networks are solved. In the first two problems, only the connectivity constraint is considered for the two networks. In the last two problems all constraints introduced in the problem description are considered. The results are presented in the following two subsections.

25.5.3.1 Minimizing Total Travel Time with Only Connectivity Constraint

In this section, results are presented for the cases when only connectivity constraint is considered. The results are presented in two parts. In the first part, the optimal designs obtained for the RW-1 and RW-2 networks using the proposed method are presented and their performance is analyzed. The optimal design (of network) is the best solution, S^* obtained from 30 different runs of the GA based procedure. That is, if S_i represents the solution with the best objective function value for the ith run of the proposed procedure, then S^* is the best among all the S_i's.

In the second part, results are presented to show that the GA based procedure is stable and the results do not vary much from one run to the other. In this part the mean performance of the proposed procedure and its variability are presented.

The optimal design for the RW 1 network is shown in Fig. 25.7. Similarly the optimal design for the RW-2 network is shown in Fig. 25.8a. Figure 25.8b shows the magnified version of the boxed part of Fig. 25.8a for greater clarity. In these figures, the arrows indicate the direction of the road segment; absence of arrows indicate that the road segment is two-way.

Table 25.2 shows the performance of the optimal networks obtained using the proposed method and compares the results with the performance of the same road networks when all roads are two-way (referred to as *all two-way network*). The all two-way network is considered as the base since a road network in its natural state consists of only two-way roads. Table 25.2 is divided into two halves. The first half pertains to the RW-1 network while the second half is for RW-2. The table has three columns. The last two columns respectively list the values of the various performance measures for the all two-way network and the optimal network.

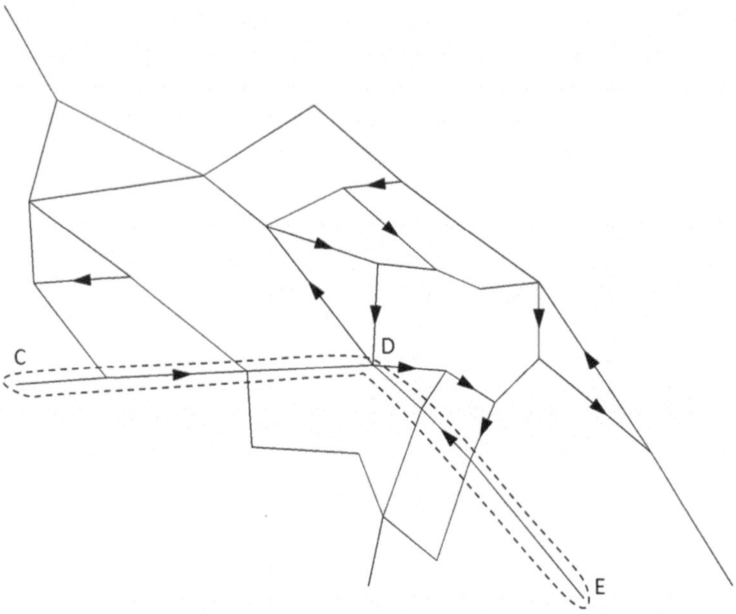

Fig. 25.7 Optimal network orientation with only connectivity constraint for the RW-1 network (32 intersections and 45 road segments)

Table 25.2 Comparison of the performance measures of the optimal network with only connectivity constraint and the all two-way networks for the RW-1 and RW-2 networks

Network	Performance measure (PM_j)	All two-way network	Optimal network
RW-1	PM_1: ATT (min)	27.36	20.68 (+24.4%)
	PM_2: AD (km)	5.25	5.51 (–5.0%)
	PM_3: AS (km/hr)	11.52	15.98 (+38.7%)
	PM_4: AID (min)	14.07	2.95 (+79.0%)
RW-2	PM_1: ATT (min)	33.95	23.69 (+30.2%)
	PM_2: AD (km)	10.50	11.34 (–8.0%)
	PM_3: AS (km/hr)	18.55	28.73 (+54.9%)
	PM_4: AID (min)	11.90	1.75 (+85.3%)

The second column lists all the performance measures. The following performance measures (PM_j) are used: (i) PM_1: average travel time of a vehicle, ATT, (ii) PM_2: average distance travelled by a vehicle, AD, (iii) PM_3: average speed of the vehicles, AS (note, speed of a vehicle is the ratio of the travel distance to the travel time and the latter includes delays at intersections), and (iv) PM_4: average time spent (delay) at intersections, AID. The percentage improvement in a performance measure in the optimal network with respect to the all two-way network is given in parentheses;

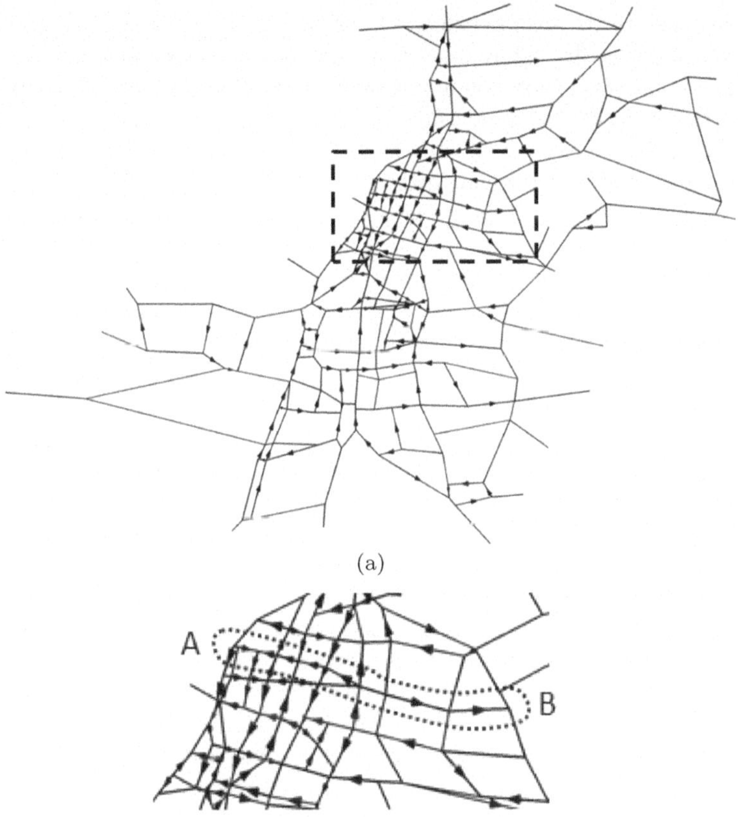

Fig. 25.8 Optimal network orientation with only connectivity constraint for the RW-2 network (223 intersections and 361 road segments); (**a**) entire optimal network and (**b**) enlarged view of the boxed section in (**a**)

when there is a worsening then a negative sign precedes the percentage. Note that improvement may sometimes imply an increment in the value while at other times it may imply a decrement.

Given that the ATT is proportional to the objective function, it is no surprise that it has decreased in the optimal designs for both the RW-1 and the RW-2 networks. The extent of decrease (24.4% for the RW-1 network and 30.2% for the RW-2 network) indicates that large gains in the efficiency of the network can be achieved by optimizing the design. The results also show that intersection delay is a significant part of the travel times of users. For example, when all roads are two-way, on an average, intersection delay, AID is 35–50% of the travel times of users (ATT). The largest gain also comes from reduction in intersection delay (AID). This is expected as in any urban network the bottlenecks are the intersections and simplifying them by reducing the types of turning movements they have to handle (which happen when some of the approach roads become one-way) improves their efficiency. The efficiency improvement comes through reduced loss times and higher saturation flows. Loss times reduce since fewer types of turning movements allow the optimization process to select fewer phases. Saturation flow increases because there are fewer types of movements at the intersection which leads to fewer shared lanes. The importance of intersections toward the part it plays in determining travel time highlights the need for optimization formulations of the network design problem to represent intersection delay carefully.

It can be seen that in each of the optimal design the average distance travelled per vehicle (AD) is higher compared to the corresponding all two-way network. This is expected as the presence of one-way roads causes vehicles to travel longer distances in order to reach their destinations.

Finally, the results show that the optimal network may have arterial roads (corridor) like CDE in Fig. 25.7 and AB in Fig. 25.8b with road segments that are often in opposing directions. As discussed earlier, such frequent changes in directions are not acceptable from a traffic engineering standpoint as they are uncomfortable for the road users. This is precisely the reason why it is felt that the directional corridor constraints (see description earlier) are necessary.

Since GA based methods use probabilistic search schemes, one needs to establish that such methods consistently provide good results over different runs. In this study the proposed algorithm was run 30 times for each of the cases.

Let PM_j^i represent the value of the performance measure, j (see Table 25.2 and its preceding discussion) for the ith run. Let APM_j, SPM_j, MPM_j and IPM_j respectively represent the average, standard deviation, median and inter-quartile range of the PM_j^i values. Table 25.3 shows the values of PM_j for the best result obtained from the thirty runs (same as the "optimal" network of Table 25.2) as well as APM_j, SPM_j, MPM_j and IPM_j values. As with Table 25.2, the top half of Table 25.3 pertains to results from the RW-1 network while the bottom half contains results from the RW-2 network. The results show that the proposed method performs well in repeated runs and generates networks that are comparable to the "optimal" network. The average and the median values are close to the best. The standard deviation and the inter-quartile range are small indicating less variability from one run to another.

25.5.3.2 Minimizing Total Travel Time with Connectivity and Directional Corridor Constraint

In this section, results for the problem that includes both connectivity and directional corridor constraints are presented. The road segments between C and D and E in Fig. 25.7 and between A and B in Fig. 25.8b are considered to be part of a directional corridor for the RW-1 network and the RW-2 network, respectively. The network performance measures for the optimal solution are listed in Table 25.4. The optimal design for the RW-1 network is shown in Fig. 25.9 and that of RW-2 network is shown in Fig. 25.10.

Table 25.3 Performance of the proposed method in repeated applications for RW-1 and RW-2 networks with only connectivity constraint

Network	Performance measure, (PM_j)	PM_j values in optimal network	APM_j	SPM_j	MPM_j	IPM_j
RW-1	PM_1: ATT (min)	20.68	22.25	1.20	22.02	1.51
	PM_2: AD (km)	5.51	5.54	0.14	5.52	0.15
	PM_3: AS (km/hr)	15.98	15.00	0.89	15.19	1.25
	PM_4: AID (min)	2.95	4.38	1.33	4.07	1.37
RW-2	PM_1: ATT (min)	23.69	25.03	0.74	24.84	1.11
	PM_2: AD (km)	11.34	11.54	0.21	11.51	0.31
	PM_3: AS (km/hr)	28.73	27.66	0.59	27.60	0.85
	PM_4: AID (min)	1.75	1.97	0.33	1.97	0.41

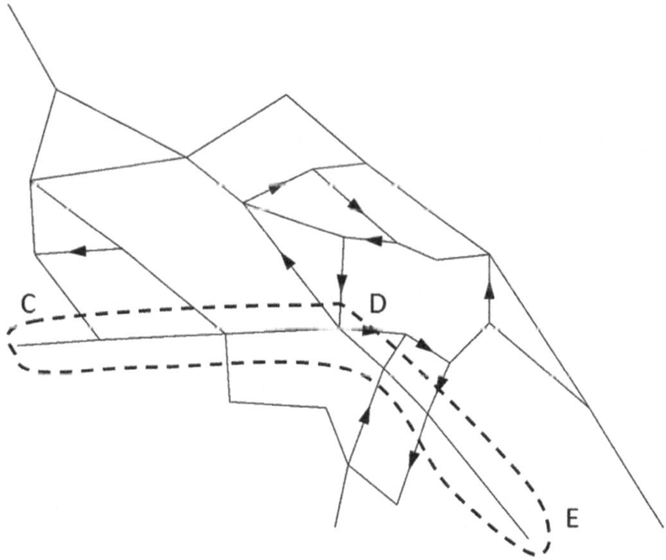

Fig. 25.9 Optimal network orientation with connectivity and directional corridor constraint for the RW-1 network (32 intersections and 45 road segments)

Table 25.4 Comparison of performance measures of the optimal network and the all two-way network of RW-1 and RW-2 networks when connectivity constraint and directional corridor constraint are considered

Network	Performance measure (PM_j)	All two-way network	Optimal network
RW-1	PM_1: ATT (min)	27.36	21.40 (+21.8%)
	PM_2: AD (km)	5.25	5.49 (–4.6%)
	PM_3: AS (km/hr)	11.52	15.39 (+33.6%)
	PM_4: AID (min)	14.07	2.01 (+85.7%)
RW-2	PM_1: ATT (min)	33.95	24.21 (+28.7%)
	PM_2: AD (km)	10.50	11.53 (–9.8%)
	PM_3: AS (km/hr)	18.55	28.58 (+54.1%)
	PM_4: AID (min)	11.90	1.57 (+86.8%)

Table 25.5 Performance of the proposed method in repeated applications for RW-1 and RW-2 networks when connectivity constraint and directional corridor constraint are considered

Network	Performance measure, (PM_j)	PM_j values in optimal network	APM_j	SPM_j	MPM_j	IPM_j
RW-1	PM_1: ATT (min)	21.40	23.53	2.10	22.62	3.40
	PM_2: AD (km)	5.49	5.39	0.17	5.40	0.31
	PM_3: AS (km/hr)	15.39	13.84	1.01	14.19	1.42
	PM_4: AID (min)	2.01	6.25	2.75	4.75	4.48
RW-2	PM_1: ATT (min)	24.21	24.90	0.43	24.86	0.55
	PM_2: AD (km)	11.53	11.50	0.16	11.48	0.23
	PM_3: AS (km/hr)	28.58	27.73	0.38	27.75	0.41
	PM_4: AID (min)	1.57	1.95	0.23	1.97	0.30

The results show similar trends as in the previous case (see Table 25.2). Also Fig. 25.9 and Fig. 25.10b show that unlike in Figs. 25.7 and 25.8b, respectively, all the road sections are in the same direction (or two way) as demanded by the directional corridor constraint.

Results on the performance of the algorithm for these problems in 30 repeated runs are shown in Table 25.5. As in Table 25.3, this table also includes the best result

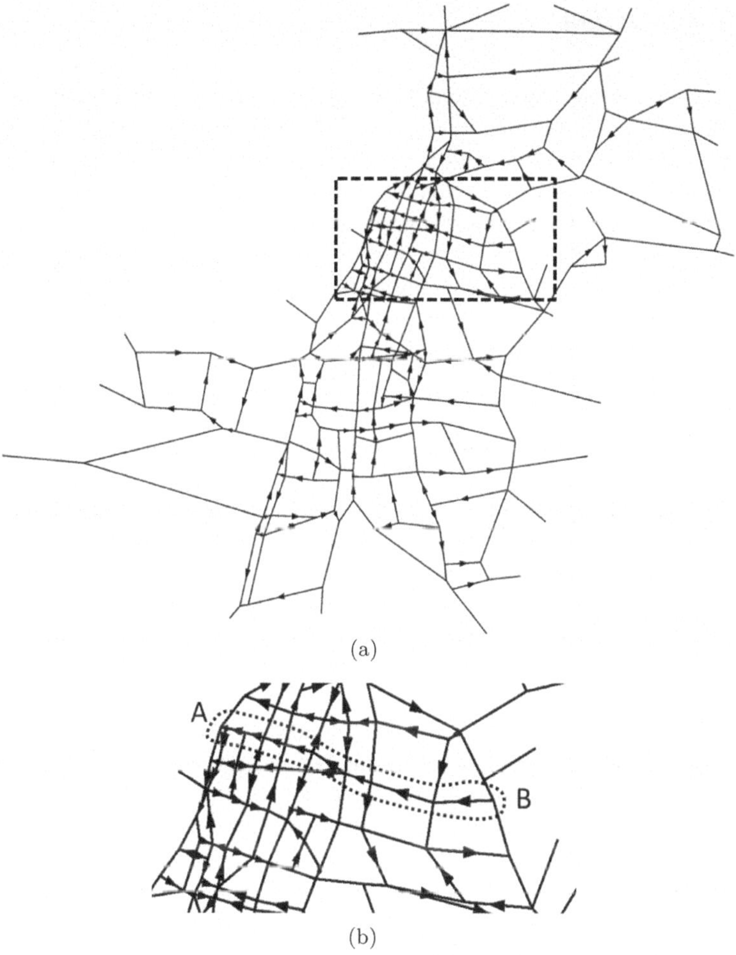

Fig. 25.10 Optimal network orientation with connectivity and directional corridor constraint for the RW-2 network (223 intersections and 361 road segments); (**a**) entire optimal network and (**b**) enlarged view of the boxed section in (**a**)

from the thirty different runs under the column "PM_j values in optimal network". These results show trends similar to Table 25.3.

A t-test reveals that the mean of the optimal travel time (APM_1) obtained in the 30 runs in Tables 25.5 and 25.3 are not statistically different.

25.5.4 *Discussion*

So far this section presented results from different networks that highlighted the ability of the proposed formulation to repeatedly find network designs that provide efficient mobility to its users while incorporating various engineering and service-related constraints. In this subsection discussions that relate to various other facets of the performance of the proposed algorithm are provided.

First the effect of the connectivity function on the GA based search process is studied. Second, the impact of population size on the efficiency of the optimization process is discussed. Finally, a discussion on the computation time requirements of the proposed methodology is provided.

25.5.4.1 Connectivity Function and Its Impact on the Search Process

Recall, the connectivity function converts networks that are not strongly connected, into strongly connected networks by modifying the value of certain bits in the GA strings. Since the connectivity function is not part of the standard GA and yet it makes changes to the strings during the optimization process, it is important to study the effect of these changes. The percentage of bits (in the population) whose values are changed by the connectivity function is taken as a measure of its interference in the GA search process.

As can be seen from Fig. 25.6, the first application of the connectivity function happens before the GA based search (optimization) process starts. In a way, the first application can be thought of as a part of the initial population generation. Subsequent applications of this function happens after the search process is underway. For the RW-1 and the RW-2 networks the connectivity function changes respectively about 20% and 10% of the bits only in its first application. Each of the subsequent applications affect changes to less than 1% of the bits for both the networks.

Given that mutation probabilities are generally around 1% (of the same order as the proportion of changes made by the connectivity function) and the connectivity function is a feasibility preserving operation, the authors feel that the disruption created by these changes do not interfere with the search process significantly. If anything, it helps channelize the search more efficiently.

25.5.4.2 Impact of Population Size on the Search Process

For a given population size, the GA based optimization process is run three times (with different random number seeds). For each of these runs, at every generation, the objective function value (TTT_i) for the best string is noted. For a given number of generations, the average (over the three runs) of these best objective function values are plotted against the generation number in order to see the impact of population size on the efficiency of the GA based process. Such a plot for the RW-2 network

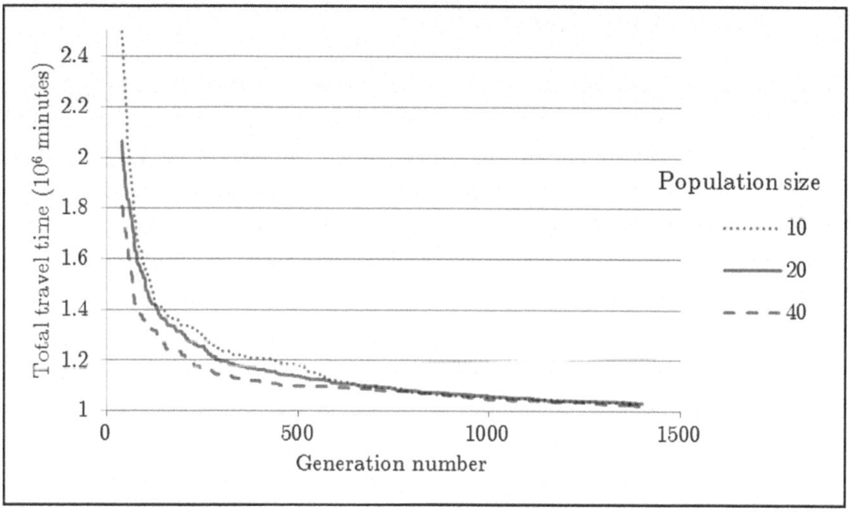

Fig. 25.11 Comparison between different population sizes for RW-2 network

is shown in Fig. 25.11. As can be seen from the figure, a larger population leads to better solutions for the initial generations. However, by the 800th generation, the objective function values for the different population sizes are almost same. Similar trend is observed for the RW-1 network in which by the 300th generation, the objective function values for the different population sizes are same. For the RW-1 network, a figure similar to Fig. 25.11 is not shown in order to avoid repetition. Therefore, considering the reduced computational effort for a smaller population, it is recommended that a population size of 10 can be utilized, even for large networks.

25.5.4.3 Computational Time

The algorithm was run on a shared High Performance Computing (HPC) system with 20 parallel cores. The HPC is based on Intel Xeon E5-2670V (2.5 GHz) on HP-Proliant-SL-230s-Gen8 servers with 128 GB of RAM per node. For the smaller RW-1 network with 32 intersections and 45 road segments [38], the computation time is about 0.36 hrs. For the larger RW-2 network (with 223 intersections and 361 road segments), the computation time is 32.9 hrs. In both cases a population size of 10 is used. In the case of RW-1 network, the process was stopped at 1000 generations (the best solution had been found by then) while for the RW-2 network the process was stopped at 1400 generations. In any case, the results show that the proposed method can yield near-optimum network designs within reasonable time even for large networks.

25.6 Summary and Conclusions

In this chapter a practical problem was presented. The nature of the optimal urban network design problem presented here was such that traditional mathematical programing techniques/formulations could not be used. A metaheuristic (Genetic Algorithm) formulation with external procedure-based declarations is proposed to solve the optimal network design problem. A review of past literature suggests that while the network design problem has been solved before, there are various shortcomings that need to be addressed. Some of the solution methods in the literature employ heuristics; this makes their solution technique specific to the problem. Service-related restrictions such as ensuring a series of road segments (that, for example may constitute an arterial) be either two-way or all in the same direction are not considered in any of the earlier attempts. While some of the articles are silent about network feasibility issues (i.e., ensuring that the network connects all node pairs), others who did consider it, tried to maintain feasibility in an ad hoc fashion—either by discarding the infeasible networks or by redoing their search process. While many choose to ignore intersection delay, others who do include it make simplifying assumptions such as flow-independent cycle lengths, fixed phasing schemes, and fixed saturation flows, etc.

Here the metaheuristic-based feasibility conserving optimization algorithm that is proposed incorporates the following service-related restrictions: (i) strong connectivity (i.e., the network connects all node pairs), and (ii) same direction or directions for specified road segments (like roads that constitute an arterial). Although only two service-related restrictions are considered, the proposed method is generic and other restrictions can also be incorporated with ease. For example, restrictions such as limiting the total traffic volume at certain intersections, restricting certain movements at intersections, barring certain roads from converting to one-way, etc. can also be considered. Feasibility conservation is achieved with minimal interference with the GA based search process using a novel method based on Tarjan's algorithm [34].

Finally the intersections have been represented in a realistic manner. Various intricacies of an intersection design and operation such as flow-depended phasing scheme and cycle length, and movement dependent saturation flows are incorporated. Although in the current implementation the "keep-left" driving policy is used, the algorithm can be used for the "keep-right" driving scenarios, and also with minor modification to the signalized intersection representation.

The proposed method is tested on a number of networks. The first three networks are synthetic networks that are described in existing literature [11]. Comparison of the results show that the GA based procedure developed in this article produces efficient network designs. The proposed method is also tested on two real-world networks (RW-1 and RW-2) based on two different cities of India. The size of the network for RW-2 is comparable with the largest network used in existing literature. Good results are obtained within reasonable computation time in repeated runs of the proposed algorithm.

The proposed method can be improved upon by using user equilibrium assignment techniques (currently it uses incremental assignment). Certain other service-related restrictions (some of which are identified here) can also be included. The authors believe the network design problem must also include performance design parameters either in the objective function or in the constraint that defines the average travel complexity.

References

1. Amiripour SMM, Mohaymany AS, Ceder A (2015) Optimal modification of urban bus network routes using a genetic algorithm. ASCE J Transp Eng 141(3):04014081
2. Ben-Ayed O, Boyce DE (1988) A general bilevel linear programming formulation of the network design problem. Transp Res Part B 22(4):311–318
3. Bureau of Public Roads (1964) Traffic assignment manual. U.S, Department of Commerce
4. Cantarella GE, Pavone G, Vitetta A (2006) Heuristics for urban road network design: Lane layout and signal setting. Eur J Oper Res 175(3):1682–1695
5. Cantarella GE, Vitetta A (2006) The multi-criteria road network design problem in an urban area. Transportation 33(6):567–588
6. Chakroborty P, Dwivedi T (2002) Optimal route network design for transit systems using genetic algorithms. Eng Optim 34(1):83–100
7. Chakroborty P, Das A (2017) Principles of transportation engineering. PHI Learning Private Limited, New Delhi
8. Chen M, Alfa AS (1991) A network design algorithm using a stochastic incremental traffic assignment approach. Transp Sci 25(3):215–224
9. Deb K (2000) An efficient constraint handling method for genetic algorithms. Comput Methods Appl Mech Eng 186(2–4):311–338
10. Drezner Z, Salhi S (2000) Selecting a good configuration of one-way and two-way routes using tabu search. Control Cybern 29(3):725–740
11. Drezner Z, Salhi S, Using hybrid metaheuristics for the one-way and two-way network design problem. Naval Res Logistics (NRL) 49(5):449-463
12. Drezner Z, Wesolowsky GO (1997) Selecting an optimum configuration of one-way and two-way routes. Transp Sci 31(4):386–394
13. Drezner Z, Wesolowsky GO (2002) Network design: Selection and design of links and facility location. Transp Res Part A 37(3):241–256
14. Farahani RZ, Miandoabchi E, Szeto WY, Rashidi H (2013) A review of urban transportation network design problems. Eur J Oper Res 229(2):281–302
15. Foulds LR (1992) Graph theory applications. Springer Science and Business Media
16. Friesz TL, Anandalingam G, Mehta NJ, Nam K, Shah SJ, Tobin RL (1993) The multiobjective equilibrium network design problem revisited: A simulated annealing approach. Eur J Oper Res 65(1):44–57
17. Gallo M, D'Acierno L, Montella B (2010) A meta-heuristic approach for solving the urban network design problem. Eur J Oper Res 201(1):144–157
18. Gao Z, Wu J, Sun H (2005) Solution algorithm for bi-level discrete network design problem. Transp Res Part B 39(6):479–495
19. Gayah VV, Daganzo CF (2012) Analytical capacity comparison of one-way and two-way signalized street networks. J Transp Res Board 2301:76–85
20. Goldberg DE (1989) Genetic algorithm in search, optimization and machine Learning. Addison Wesley, Reading, Mass
21. LeBlanc LJ (1975) An algorithm for the discrete network design problem. Transp Sci 9(3):183–199

22. Lee C, Yang K (1994) Network design of one-way streets with simulated annealing. Pap Reg Sci 73(2):119–134
23. Marciano FA, Musolino G, Vitetta A (2015) Signal setting optimization on urban road transport networks: The case of emergency evacuation. Saf Sci 72:209–220
24. Miandoabchi E, Farahani RZ (2011) Optimizing reserve capacity of urban road networks in a discrete network design problem. Adv Eng Softw 42(12):1041–1050
25. Ortigosa J, Menendez M, Gayah VV (2015) Anaysis of network exit functions for various urban grid network configurations. J Transp Res Board 2491:12–21
26. Ortigosa J, Gayah VV, Menendez M (2017) Analysis of one-way and two-way street configurations on urban grid networks. Transport Dynamics, Transportmetrica B
27. Ortuzar JD, Willumsen LG (1996) Modelling transport. Wiley, New York
28. Papacostas CS, Prevedouros P (2005) Transportation engineering and planning. Pearson/Prentice Hall
29. Parker RG, Rardin RL (1988) Discrete optimization. Academic
30. Poorzahedy H, Rouhani OM (2007) Hybrid meta-heuristic algorithms for solving network design problem. Eur J Oper Res 182(2):578–596
31. Possel B, Wismans LJJ, Berkum ECV, Bliemer MCJ (2018) The multi-objective network design problem using minimizing externalities as objectives: Comparison of a genetic algorithm and simulated annealing framework. Transportation 45(2):545–572
32. Russo F, Vitetta A (2006) A topological method to choose optimal solutions after solving the multi-criteria urban road network design problem. Transportation 33:347–370
33. Sheffi Y (1985) Urban transportation networks: Equilibrium analysis and mathematical programming methods. Prentice-Hall Inc, Englewood Cliffs, N.J., p 07632
34. Tarjan R (1972) Depth-first search and linear graph algorithms. SIAM J Comput 1(2):146–160
35. Transportation Research Board (TRB) (2000) Highway capacity manual. National Research Council, Washington DC, USA
36. Wu JJ, Sun HJ, Gao ZY, Zhang HZ (2009) Reversible lane-based traffic network optimization with an advanced traveller information system. Eng Optim 41(1):87–97
37. Xiong Y, Schneider JB (1992) Transportation network design using a cumulative algorithm and neural network. Transp Res Board 1364:37–44
38. Yadav A (2015) Optimal road network configuration with service related constraints. M.Tech, Thesis, Indian Institute of Technology Kanpur, India
39. Yang H, Bell MGH (1998) Models and algorithms for road network design: A review and some new developments. Transp Rev 18(3):257–278
40. Zhang H, Gao Z (2007) Two-way road network design problem with variable lanes. J Syst Sci Syst Eng 16(1):50–61

Chapter 26
Coordinating a Behavioral Supply Chain Under Revenue-Sharing Contract

Mohd Mujahid Khan and B. Vipin

Abstract The chapter introduces supply chain contracts involving nonlinear optimization in the stochastic demand setting. A literature review on prospect theory in supply chain context and theoretical results on the traditional supply chain are provided. This chapter analyzes the performance of a supply chain that includes a behavioral retailer in the context of a revenue-sharing contract using a decision-theoretic framework. We investigate the framework of prospect theory with a nonlinear stochastic reference point to model the retailer's behavior. We derive the optimal behavioral retailer option under the revenue-sharing contract. The analysis demonstrates that the behavioral supply chain's revenue-sharing contract can coordinate the supply chain. The chapter concludes with managerial implications and future research initiatives.

Keywords Behavioral operations management · Prospect theory · Stochastic optimization · Revenue-sharing contract

26.1 Introduction

Chester I. Barnard's idea of business strategy [4] separates business policy as an independent discipline of study. From the perspective of supply chains, Barnard proposes that supply chain participants' individual goals should align with the general objectives of the entire supply chain. This alignment of objectives is usually achieved by designing proper incentives to the members through supply chain contracts (thereby reducing double marginalization[1]). A supply chain reaches the coordination state

[1] Where agents of the same supply chain mark up prices, lowering the whole supply chain's performance for their own advantage.

Mohd Mujahid Khan · B. Vipin (✉)
Department of Industrial and Management Engineering, Indian Institute of Technology, Kanpur 208 016, India
e-mail: vipin@iitk.ac.in

Mohd Mujahid Khan
e-mail: mujahid@iitk.ac.in

F. Hamid (ed.), *Optimization Essentials*, International Series in Operations Research & Management Science 353, https://doi.org/10.1007/978-981-99-5491-9_26

when the total profit of the segregated firms or decentralized supply chain equals that of the integrated firm or centralized supply chain, wherein the order quantities of individual firms or supply chain agents should be equal to the integrated firm's order quantity. A variety of contracts documented in the literature, such as buyback [34] (publishing industry), advance-purchase discount [12] (for instance, airline ticketing), and revenue-sharing [13] (as prominent in the video-rental industry) can coordinate a supply chain.

Most traditional contract models rely on the premise that human decision-makers are rational and maximize their expected profits. This assumption is challenged in many recent research studies such as [5, 6, 8, 9, 27, 36, 47], wherein subjects are found to be prone to decision biases while making inventory decisions.[2] In the experimental newsvendor problem settings, which is the underlying model in analyzing contracts under stochastic demand settings, it is common to observe the behavioral biases of human decision-makers, for instance, the pull-to-center (PTC) bias or effect.

In this chapter, we model the retailer's behavior through the prospect theory framework [25]. Prospect theory is one of the most popular behavioral theories which can describe the behavior of agents in different decision-making contexts under risk and ambiguity. People compare prospects with a certain expectation or aspiration level termed as a reference point in decision-making (prospect theory, [25]) (for example, experiments on negotiations on salary increase as in [24]). Decision makers value losses more than the same amount of gains and are risk-seeking in the loss frame and risk-averse in gains. In addition, people overweight low and underweight high probabilities, which are modeled through the probability weighting function for human decision-makers [42]. In this research, we use the reference point aspect of the prospect theory in modeling the behavior of the retailer facing stochastic demand under revenue-sharing contracts.

In real-world scenarios, decision-makers drift from the order quantities, as predicted by conventional models, which are expected to maximize profits. For instance, [14] find that retailers are responsible for 72% of stock-outs arising due to erroneous inventory management practices. Further, [2] analyze 40 days of sales data for 1.5 million transactions involving 228 different products for a pharmacy firm in India. They observe a significant gap between optimal inventory management policies and actual inventory management decisions [3]. Hence, there is a need for models incorporating the biases of human decision-makers for better management of the supply chains, as by capturing the actual behavior, models can help design the optimal contracts, thus maximizing the system's performance.

Revenue-sharing contracts, as discussed, are popular coordinating contracts in the real world. Recently, Bharat Sanchar Nigam Limited switched their FTTH fiber network, MobiCall, and other services to third-party service providers on a revenue-sharing basis [11]. In 2016, the Indian Cabinet approved revenue-sharing contracts

[2] Despite systematic *biases*, there could also be the presence of random *noise* in individual decision-makers as suggested by [26]. Human behavior is modeled as rational but biased and noisy. In this chapter, we concentrate on biases only and not on the noise part, as we primarily aim to model our revenue-sharing contract settings analytically.

for all hydrocarbon explorations [18]. Moreover, the revenue-sharing contract has emerged as the front-line coordinating mechanism among developers and launching platforms in the software industry. Google shares revenue with app publishers and content creators using the revenue-sharing model. For instance, through *AdSense*, Google guarantees revenue sharing to the publishers [20]. With the income from ticket sales and merchandise, several movie projects and sports leagues implement revenue-sharing contracts. Revenue-sharing contracts must have limits; otherwise, we would see them in every industry. In industries where the retailer's effort influences the demand and/or where the cost to administer the supply chain under a revenue-sharing contract is considerably high, the benefits of revenue-sharing may not outweigh a simple wholesale price contract [13].

This chapter adds to the existing literature by analyzing a dyadic supply chain consisting of a risk-neutral (RN) supplier and a behavioral retailer and demonstrating that a revenue-sharing contract may be used to coordinate the supply chain. We derive the conditions for such coordination and illustrate our results using numerical analysis. The rest of the chapter is divided into the following sections: Sect. 26.2 provides a brief literature review. In Sect. 26.3, the mathematical model is presented. Section 26.4 provides the numerical illustration, and Sect. 26.5 brings the chapter to a close with managerial inferences.

26.2 Literature Review

In this part of our study, we review the behavioral contract literature referring to the context of inventory decisions modeled using the prospect theory framework. We have classified this section into two—prospect theory in the newsvendor model and in supply chain contracts.

26.2.1 Prospect Theory in the Newsvendor Model

We aim to examine the prospect theory framework's behavioral newsvendor problem. Furthermore, we restrict our review to single-agent system (or monopoly market) settings and thus exclude the multi-agent competitive newsvendor settings. The classical newsvendor model is the most popular mathematical model for inventory management under stochastic demand [1]. Furthermore, as [46] mentions "if there is one common lens through which the field of operations management views the firm, it is the newsvendor model."

Schweitzer and Cachon [36] contribute the pioneering work to trigger human behavioral research in inventory decision-making. Under newsvendor settings, [36] observe a "too low/too high" "aggregated" pattern of ordering, where the average order quantity lies "in between the mean demand and optimal order quantities"; and this phenomenon is first coined as the "pull-to-center" (PTC) effect by [10]. Authors prove (based on their "natural" reference point equals current wealth, that was status-quo) that the famous prospect theory [25] "cannot explain this pattern in the high-demand range condition" (p. 417) for low-profit margin products. Two explanations fit their observed data:: (1) the preference to minimize the ex-post inventory error and (2) the anchoring and insufficient adjustment heuristic.

Bolton and Katok [8] observe that the newsvendors' experience and feedback can improve their ordering behavior. Surprisingly, [31] argue that, in practical environments, more frequent feedback may lead to decline in performance. Both [8, 31] confirm PTC. Bolton et al. [9] discover a similar behavioral bias in managers, i.e., the PTC effect, as students do. While [9] focus on education, experience, and managerial position, [32] use cognitive reflection [17] to predict an individual's performance. Benzion et al. [7] observe the PTC effect under various experimental designs, including high and low-margin products under uniform and normal demand distributions. Su [39] investigates the classical newsvendor models by applying bounded rationality.[3] Mathematically, [39] shows that "the average behavioral order quantity is pulled toward the mean demand."

The capability of the prospect theoretic approach to describe the behaviors in the newsvendor problem substantially depends upon defining the reference point in explaining the observed behavior. Ho et al. [22] analyze the "ex-post inventory error-based reference model" in centralized and decentralized settings. Long and Nasiry [30] use the reference point, which is "the weighted sum of maximum and minimum pay-offs."

Ockenfels and Selten [33] propose impulse balance[4] and model the newsvendor problem, which describes the PTC, nonlinearity, and asymmetry in ordering. Vipin and Amit [44] describe the aforementioned observations and heterogeneity in the ordering at the individual level using the prospect theory framework with a reference-dependent model relying on the behavioral parameters of the agent. Although it seems tedious to segregate the models of individual heterogeneity among human decision-makers, studies such as [30, 43, 44] prove that individual heterogeneity in order quantities can be accommodated by the framework of prospect theory. Table 26.1 summarizes the notable papers on the newsvendor problem in the behavioral domain.

26.2.2 Prospect Theory in Supply Chain Contracts

Katok and Wu [27] contribute the pioneering work investigating wholesale price, buyback, and revenue-sharing contracts in a two-echelon decentralized chain in controlled laboratory settings. The PTC effect is found to be consistent in this study, and in the actual world, the wholesale price contract outperforms what is predicted

[3] *Bounded rationality* refers that human rationality is limited (or bounded) when individuals make decisions [38]. It is because of individuals' information, cognitive, and time and cost constraints. The most general model applied to the newsvendor problem based on bounded rationality is the quantal choice model [39].

[4] Ockenfels and Selten [33] propose that "order quantities different than demand realizations, generate impulses which are proportional to the cost associated to the mismatch."

Table 26.1 Brief summary of prospect theory in the newsvendor problem

Paper	Data	Supporting framework (Decision biases, Utility functions, Heuristics etc.)
Hoskin [23]	EX	Learning, Mental accounting
Schweitzer and Cachon [36]	EX	Anchoring and demand chasing
Bostian et al. [10]	EX	Learning, Anchoring and demand chasing
Benzion et al. [7]	EX	Learning, Anchoring and demand chasing
Bolton and Katok [8]	EX	Learning, Anchoring and demand chasing
Su [39]	TH	Bounded rationality
Lurie and Swaminathan [31]	EX	Demand chasing, Learning
Ho et al. [22]	EX	Reference dependence
Kremer et al. [28]	EX	Quantal choice
Bolton et al. [9]	EX	Learning, Individual heterogeneity
Moritz et al. [32]	EX	Demand chasing, Individual heterogeneity
Feiler et al. [16]	EX	Censorship bias, Risk preferences
Ockenfels and Selten [33]	EX	Impulse balance
Rudi and Drake [35]	EX	Observation bias, Demand chasing, Learning
Gurnani et al. [21]	EX	Quantal choice
Käki et al. [29]	EX	Risk preferences
Long and Nasiry [30]	TH	PT
Vipin and Amit [44]	TH	PT, Individual heterogeneity
Surti et al. [40]	EX	PT, Non-monetary preferences
Goldschmidt et al. [19]	EX	Mental accounting, Risk preferences, Learning
Shinde et al. [37]	EX	Ambiguity aversion

EX = Experimental, TH = Theoretical, PT = prospect theory

by theory. Moreover, the authors note that whereas buyback contracts and revenue-sharing contracts are mathematically equivalent, they do not result in identical supply chain performance.

Becker-Peth et al. [6] analyze buyback contracts extensively. Their model is based on prospect theory- anchoring on mean demand [25] and mental accounting, i.e., decision-makers value different income sources differently [41]. Becker-Peth and Thonemann [5] analyze the contracts through a behavioral model and prove the role of reference points in describing the behavior experimentally.

Table 26.2 Prospect theory in contracts for the supply chain

Paper	Wholesale price	Buyback	Revenue sharing	Data	Supporting behavioral framework
Cachon and Lariviere [13]			Yes	TH	
Becker-Peth et al. [6]		Yes		EX	Mental accounting
Becker-Peth and Thonemann [5]			Yes	EX	Reference dependence
Katok and Wu [27]	Yes	Yes	Yes	EX	Anchoring and adjustment bias, Minimizing ex-post inventory errors
Zhang et al. [47]		Yes	Yes	EX	Loss aversion

(EX = Experimental, TH = Theoretical/Mathematical, PT = Prospect theory)

Zhang et al. [47] collate the performance of buyback and revenue-sharing contracts for a loss-averse supplier who sets the contract parameters for his retailer. Using behavioral bias of loss aversion, in a setting with a high critical ratio, the supplier's revenue-sharing agreement is more profitable than its corresponding buyback deal.

Davis et al. [15] study in experimental settings the inventory risk effects of the push contract (where retailer incurs risks), the pull contract (where supplier incurs risks), and the advanced purchase discount contract (where both share inventory risks). The authors conclude that "a pull contract achieves higher channel efficiency." Moreover, for long-term collaborative relationships, the advanced purchase discount contract is recommended over the push contract. Table 26.2 presents a brief summary of the behavioral studies in supply chain contracts.

26.3 Mathematical Model

First, we review the results of the centralized decision-making environment, the wholesale price, and the revenue-sharing contract in the risk-neutral supply chain setting and then introduce our model on behavioral supply chain under revenue-sharing contract.

26.3.1 Centralized Supply Chain

In a centralized supply chain setting, a single agent is responsible for both procurement and sales and hence, is free from any double-marginalization effect. Considering the classical newsvendor model[5] for RN centralized decision maker, let $X \in [0, b]$ be the stochastic demand having cumulative distribution function (cdf) as $F(x)$ and probability density function (pdf) as $f(x)$. Let Q represent the quantity of orders placed long before the beginning of the selling season, and p, c, and v are the selling price, procurement cost, and salvage value per unit. Let the stock-out cost be neglected for the sake of simplicity in analysis and assume $p > c > v \geq 0$.

The pay-off function of the RN centralized supply chain can be written as:

$$\pi_c(x, Q) = \begin{cases} px - cQ + v(Q - x) & \text{if } x \leq Q \\ (p - c)Q & \text{if } x > Q \end{cases} \tag{26.1}$$

And the expected profit of the RN centralized supply chain is:

$$E(\pi_c(x, Q)) = \int_0^Q (px - cQ + v(Q - x)) f(x)dx + \int_Q^b ((p - c)Q) f(x)dx, \tag{26.2}$$

The optimal order quantity[6] is

$$F(Q_c^*) = \frac{p - c}{p - v} \tag{26.3}$$

26.3.2 Supply Chain—RN Supplier-RN Retailer

Consider a decentralized supply chain with one retailer who orders quantity Q from his supplier well in advance of the selling season. The stochastic demand $X \in [0, b]$ is characterized by pdf $f(x)$ and cdf $F(x)$. Table 26.3 provides notations for the variables and parameters used in this chapter.

As we discussed earlier in Sect. 26.1, an RN supply chain achieves coordination when the RN decentralized supply network's overall profit equals that of the RN centralized supply chain, and individual RN agents' order quantities equal the RN centralized supply chain's order quantity. Mathematically, coordination in the RN supply chain is achieved when,

---

[5] Also termed as single-period model, the classical newsvendor problem is deciding on the order quantity wherein the trade-off is between overstocking and understocking [1]. Newsvendor model serves as the basis to model inventory problems with uncertain demand.

[6] The ratio representing $F(Q^*)$ is known as critical fractile or service level 1 or cycle service level. By definition, for low-cost (high-cost) products, the critical fractile is more (less) than 0.5".

Table 26.3 Notations

p	Per unit selling price
w	Per unit wholesale price
w_R	Wholesale price per unit under the revenue-sharing contract
c	Per unit procurement cost
v	Per unit salvage value
$Q^*_{rB_R}$	Behavioral retailer's optimal order quantity under revenue-sharing contract
$r_R(Q)$	Retailer's reference point under revenue-sharing contract
ϕ	Pessimism level of behavioral retailer
η	Perception level of behavioral retailer
$q_R(Q)$	Break-even order quantity under revenue-sharing contract
$V(\cdot)$	Value function
$\pi(\cdot)$	Pay-off function
$1-\rho$	Fraction of the retailer's revenue shared with the supplier

1. $Q_r^* = Q_s^* = Q_c^*$
2. $E(\pi_r) + E(\pi_s) = E(\pi_c)$
3. All agents should earn a positive profit, i.e., $E(\pi_r) > 0$, $E(\pi_s) > 0$

The thing to note here is that the above three prerequisites for coordination are valid for the supply chain with RN supplier and RN retailer. For the case of behavioral agents involved in the supply chain (for instance, as we discuss in Sect. 26.3.3, the dyadic supply chain settings with a behavioral retailer and RN supplier), the expected value function comes to the role instead of the expected profit function, with the above mentioned third coordination condition changed, as now the behavioral agents should gain the expected value as positive, i.e., $E(V(\pi_r)) > 0$.

26.3.2.1 Wholesale Price Contract

Keeping the general setting the same as discussed earlier, in this contract, the supplier fulfills the order Q put in by the retailer before the start of the selling season at a wholesale price w such that $p > w > c > v \geq 0$.

Retailer's perspective:
The pay-off function of the RN retailer is as follows:

$$\pi_r(x, Q, w) = \begin{cases} px - wQ + v(Q - x) & \text{if } x \le Q \\ (p - w)Q & \text{if } x > Q \end{cases} \tag{26.4}$$

Under the terms of the wholesale price contract, the RN retailer's expected profit function is expressed as

$$E(\pi_r(x, Q, w)) = \int_0^Q (px - wQ + v(Q - x))f(x)dx + \int_Q^b (p - w)Qf(x)dx, \tag{26.5}$$

Under the wholesale price contract, the optimal order quantity of RN retailer is determined by

$$F(Q_r^*) = \frac{p - w}{p - v} \tag{26.6}$$

Supplier's perspective: Under the wholesale price contract, RN supplier's pay-off function is:

$$\pi_s(Q, w) = (w - c)Q \tag{26.7}$$

With exogenous and endogenous wholesale price settings, one can verify that the risk-neutral supply chain is not possible to coordinate. For an elaborate discussion on this, refer [45].

26.3.2.2 Revenue-Sharing Contract

In the revenue-sharing type contract, a fraction of revenue of the retailer is shared with the supplier. Consider the part of the retailer's revenue that goes to the supplier to be $(1 - \rho)$, with the remaining ρ staying in the retailer's hands. The analysis of the contract is as follows:

Retailer's perspective:
Under the revenue-sharing type contract, the RN retailer's pay-off function is given as:

$$\pi_{r_r}(x, Q, w_R, \rho) = \begin{cases} \rho px - w_R Q + \rho v(Q - x) & \text{if } x \le Q \\ (\rho p - w_R)Q & \text{if } x > Q \end{cases} \tag{26.8}$$

Under the revenue-sharing contract, the expected pay-off function of the RN retailer is:

$$E(\pi_{r_R}(x, Q, w_R, \rho)) = \int_0^Q (\rho px - w_R Q + \rho v(Q - x))f(x)dx + \int_Q^b (\rho p - w_R)Qf(x)dx, \tag{26.9}$$

Under the revenue-sharing contract, the optimal order quantity of the RN retailer is as follows,

$$F(Q^*_{r_R}) = \frac{\rho p - w_R}{\rho p - \rho v} \tag{26.10}$$

Supplier's perspective:
Under the revenue-sharing contract, the expected pay-off function of the RN supplier is:

$$E(\pi_{s_R}(x, Q, w_R, \rho)) = (w_R - c)Q + (1-\rho)p\left(\int_0^Q xf(x)dx + \int_Q^b Qf(x)dx\right)$$
$$+(1-\rho)v\int_0^Q (Q-x)f(x)dx \tag{26.11}$$

For coordination, $F(Q^*_{r_R})$ should be equal to $F(Q^*_c)$. For $0 \le \rho \le 1$, from Eqs. 26.10 and 26.3,

$$\frac{\rho p - w_R}{\rho p - \rho v} = \frac{p-c}{p-v} \implies w_R = \rho c \tag{26.12}$$

Hence, a revenue-sharing contract can coordinate a supply chain between RN suppliers and RN retailers at a wholesale price lower than the procurement cost. The supplier incurs a loss in the initial transaction with the retailer; however, the initial loss is compensated with the additional revenue shared by the retailer with the supplier.

26.3.3 Behavioral Supply Chain—Revenue-Sharing Contract

In this part, we examine the coordination issues for a behavioral supply chain comprising an RN supplier and a behavioral retailer using a conventionally coordinating revenue-sharing contract. Under the framework of prospect theory, the retailer's behavior is modeled using a stochastic reference point that is affected by subjective parameters. The retailer considers the pay-off as gain (loss) if the pay-off is greater (less) than the subjective reference level set by the behavioral retailer.

Under revenue-sharing contract, the behavioral retailer's expected value function is:

$$E(V(\pi_{r_R}(x, Q, w_R, \rho))) = \int_{q_R(Q)}^{Q} (\rho px - w_R Q + \rho v(Q-x) - r_R(Q))f(x)dx$$
$$+\int_Q^b ((\rho p - w_R)Q - r_R(Q))f(x)dx$$
$$-\int_0^{q_R(Q)} (r_R(Q) - (\rho px - w_R Q + \rho v(Q-x)))f(x)dx \tag{26.13}$$

The first two terms represent the projected gains, and the third term represents the predicted loss relative to the reference point, as specified below. The stochastic reference point $r_R(Q)$ is given as:

$$r_R(Q) = \phi \times \left(\frac{w_R - \rho v}{\rho(p-v)}\right)^{\eta} (\rho p - w_R)Q + \left(1 - \phi \times \left(\frac{w_R - \rho v}{\rho(p-v)}\right)^{\eta}\right)(\rho v - w_R)Q \quad (26.14)$$

The reference point is "the convex combination of the pay-offs corresponding to maximum and minimum demands, wherein the weights capture the behavioral parameters of the retailer." $0 \leq \phi \leq 1$ and $\eta \geq 0$ represent the "behavioral parameters" of the retailer. ϕ=0 implies that the retailer is highly optimistic and $\phi = 1$ corresponds to a pessimistic retailer. η captures the perception of the retailer about the understocking cost. $\eta = 0$ implies that the retailer does not worry about the understocking cost, and as η increases, the concern regarding understocking increases. The aforementioned reference point has successfully captured agent's behavior in the newsvendor [44] and the supply chain contract [45] settings.

RN supplier's expected pay-off function remains the same as given by Eq. 26.11. Under the revenue-sharing contract, the behavioral retailer's optimal order quantity is determined by Proposition 26.1.

Proposition 26.1 *Under the revenue-sharing contract, the behavioral retailer's unique optimal order quantity is given by* $F(Q^*_{B_R}) = 1 - \phi \times \left(\frac{w_R - \rho v}{\rho(p-v)}\right)^{\eta}$

Proof Under the revenue-sharing contract, the behavioral retailer's expected value function is:

$$E(V(\pi_{r_R}(x, Q, w_R, \rho))) =$$
$$\int_{q_R(Q)}^{Q} (\rho p x - w_R Q + \rho v(Q - x) - r_R(Q)) f(x)dx + \int_{Q}^{b} ((\rho p - w)Q - r_R(Q)) f(x)dx$$
$$- \int_{0}^{q_R(Q)} (r_R(Q) - (\rho p x - w_R Q + \rho v(Q - x))) f(x)dx$$

Under the revenue-sharing contract, the reference point $r_R(Q)$ and break-even quantity $q_R(Q)$ are:

$$r_R(Q) = \phi \times \left(\frac{w_R - \rho v}{\rho(p-v)}\right)^{\eta} (\rho p - w_R)Q + \left(1 - \phi \times \left(\frac{w_R - \rho v}{\rho(p-v)}\right)^{\eta}\right)(\rho v - w_R)Q$$
$$q_R(Q) = \frac{r_R(Q) + (w_R - \rho v)Q}{\rho(p-v)}$$

Also, $r'_R(Q) = \phi \times \left(\frac{w_R - \rho v}{\rho(p-v)}\right)^{\eta} (p-v)\rho - (w_R - \rho v)$

$$E'(V(\pi_{r_R}(x, Q, w_R, \rho))) = -\left((w_R - \rho v) + r'_R(Q)\right)\left(\int_0^Q f(x)dx\right)$$
$$+\left((\rho p - w_R) - r'_R(Q)\right)\left(\int_Q^b f(x)dx\right)$$

$E''(V(\pi_{r_R}(x, Q, w_R, \rho))) = -(\rho(p-v))f(Q) < 0 \Rightarrow E'(V(\pi_{r_R}(x, Q, w_R, \rho))) = 0$ yields unique optimal order quantity.

$$E'(V(\pi_{r_R}(x, Q, w_R, \rho))) = 0 \Rightarrow$$
$$F(Q^*_{r_{B_R}})(-\rho(p - v)) + \rho(p - v) - \phi \times \left(\frac{w_R - \rho v}{\rho(p - v)}\right)^{\eta} \rho(p - v) = 0 \Rightarrow$$
$$F(Q^*_{r_{B_R}}) = 1 - \phi \times \left(\frac{w_R - \rho v}{\rho(p - v)}\right)^{\eta}$$

□

Proposition 26.2 examines the possibility of coordination in the behavioral supply chain under revenue-sharing type contracts.

Proposition 26.2 *Under the following conditions, the revenue-sharing contract can set up the coordination of a behavioral supply chain:*

$$\phi \times \left(\frac{w_R - \rho v}{\rho(p - v)}\right)^{\eta} = \frac{c - v}{p - v} \tag{26.15}$$

According to Proposition 26.2, a revenue-sharing contract with $\rho \in [0, 1]$ can coordinate a behavioral supply chain so long as the wholesale price is lower than the selling price. It is important to note that in a behavioral supply chain, coordination under a revenue-sharing type contract can also be established by keeping the wholesale price higher than the procurement cost.

Proof Under coordination, $Q^*_{r_{B_R}} = Q^*_c \Rightarrow$

$$1 - \phi \times \left(\frac{w_R - \rho v}{\rho(p - v)}\right)^{\eta} = \frac{p - c}{p - v}$$
$$\Rightarrow \frac{c - v}{p - v} = \phi \times \left(\frac{w_R - \rho v}{\rho(p - v)}\right)^{\eta}$$

If the preceding requirement is met, a revenue-sharing contract may be utilized to coordinate a behavioral supply chain. □

Corollary 26.1 *When a pessimistic retailer has a perceived value of understocking cost equal to the actual understocking cost, the revenue-sharing contract for a behavioral supply chain reduces to that of a risk-neutral supply chain.*

Proof We have, from Proposition 1, the coordinating revenue-sharing fraction given by Eq. 26.15. When $\eta = \phi = 1$, Eq. 26.15 reduces to $w_R = \rho c$, which is the same coordinating revenue-sharing as that of a supply chain with RN agents. □

Corollary 26.1 shows that the revenue-sharing type contract in the behavioral supply chain reduces to that of an RN supply chain. As discussed in Sect. 26.3.2.2, in an RN supply chain, a revenue-sharing type contract can achieve coordination in the chain even if the wholesale price is less than the procurement cost.

26.4 Numerical Illustration

In this section, we illustrate the aforementioned supply chain coordination using a revenue-sharing contract with the following context. For the analysis, we consider the demand to be uniformly distributed, i.e., $X \in U[0, 200]$, with selling price per unit $p = 40$, purchase cost per unit $c = 25$, and per unit salvage value $v = 1$. We evaluate the coordination possibility of the contract in the behavioral retailer setting as given by Table 26.4.

In the centralized setting, optimal order quantity $Q_c^* = 76.92$, and the expected profit for the entire supply chain is equal to 576.92. In the aforementioned cases, one can observe that the decentralized or segregated firms' supply chain under the revenue-sharing achieves the centralized system's profit level wherein the profits of individual agents and the value function for the retailer are positive. Hence, we conclude that the decentralized supply chain achieves coordination under the revenue-sharing contract. A critical observation from the numerical analysis is that, when interacting with certain behavioral retailers, the supply chain is coordinated under a revenue-sharing contract even with a wholesale price greater than the procurement cost, which is not the case in the RN supply chain setting.

From Table 26.5, it can be inferred that the supplier can be better off when interacting with a behavioral retailer, in the instance documented for the same revenue-sharing fraction, when compared to the scenario of interacting with an RN retailer as the supplier earns more profit in the former case. Furthermore, it is interesting to

Table 26.4 Numerical analysis establishing coordination under revenue-sharing contract

ϕ	η	w	ρ	Q^*_{BR}	$E(\pi_r)$	$E(\pi_s)$	$E(\pi_r) + E(\pi_s)$	$E(V(\pi_r))$
0.8	1.2	18	0.5567	76.92	6.81	570.11	576.92	321.10
0.85	1.1	19	0.6317	76.92	117.76	459.16	576.92	364.45
0.9	1.05	20	0.7103	76.92	237.50	339.43	576.92	409.84
0.9	1.05	26	0.9235	76.92	308.75	268.18	576.92	532.79
0.85	1.1	27	0.8977	76.92	167.34	409.58	576.92	517.91
0.8	1.2	28	0.8658	76.92	10.59	566.32	576.92	499.49

Table 26.5 RN and behavioral retailer supply chain performance comparison

Retailer type	ρ	w	$E(\pi_r)$	$E(\pi_s)$	$E(\pi_r) + E(\pi_s)$	$E(V(\pi_r))$
Behavioral ($\phi = 0.8, \eta = 1.2$)	0.8658	28	10.59	566.32	576.92	499.49
RN	0.8658	21.64	499.49	77.44	576.92	499.49

note that the behavioral retailer's expected value function is equal to the expected profit function of the RN retailer at the same revenue-sharing fraction (ρ) and the respective corresponding coordinating wholesale price (w_R).

26.5 Conclusions and Managerial Implications

The chapter explores the effects of behavioral factors in a supply chain with a risk-neutral supplier and a behavioral retailer under a revenue-sharing contract. The chapter provides a thorough review focusing on the behavioral supply chain contract literature. The study analyzes the supply chain coordination with the reference-dependent behavioral retailer and proves that the revenue-sharing contract can coordinate such a supply chain, although not always guaranteed. The basic introduction analyzing the centralized supply chain and a dyadic decentralized system in the RN settings under a wholesale price and a revenue-sharing contract is provided. Furthermore, the condition for the coordination under the revenue-sharing type contract for a behavioral retailer is mathematically established and is illustrated under uniform demand settings. It is well known that a revenue-sharing type contract has the coordination ability for an RN supply chain so long as the wholesale price is lower than the procurement cost; however, in the case of an RN supplier and a behavioral retailer, the supply chain can be coordinated even at a wholesale price that is more than the procurement cost, an exciting insight obtained from the numerical analysis.

The managerial implications include insights to the supplier to explore the behavioral retailer's bias in decision-making to attain coordination in the supply chain under a revenue-sharing type contract. The supplier could exploit the key learning from the model to achieve higher profits considering the behavioral aspects of the retailer while designing the contracts. If the cost and effort required to administer the contract are pretty high, practically, the contract designers could consider relatively simple contracts, such as wholesale price contracts, even though it is not coordinating. An interesting future research is to extend the study to multiple behavioral retailers and design their customized contracts based on individual retailers' behavioral biases. As an extension of this work, we can also question the RN assumption of the supplier to include the supplier's behavioral biases in our model.

References

1. Arrow KJ, Harris T, Marschak J (1951) Optimal inventory policy. Econometrica 19(3):250–272
2. Bandi C, Han E, Nohadani O (2019) Sustainable inventory with robust periodic-affine policies and application to medical supply chains. Manag Sci 65(10):4636–4655
3. Bandi C, Han E, Nohadani O (2019b) There's a better way to manage your inventory. Kellogg Insight. https://insight.kellogg.northwestern.edu/article/smarter-pharmacy-inventory-management. Accessed 30th June 2021
4. Barnard CI (1938) The functions of the executive. Harvard University Press, Cambridge, MA
5. Becker-Peth M, Thonemann UW (2016) Reference points in revenue sharing contracts-How to design optimal supply chain contracts. Eur J Oper Res 249(3):1033–1049
6. Becker-Peth M, Katok E, Thonemann UW (2013) Designing buyback contracts for irrational but predictable newsvendors. Manag Sci 59(8):1800–1816
7. Benzion U, Cohen Y, Peled R, Shavit T (2008) Decision-making and the newsvendor problem: An experimental study. J Oper Res Soc 59(9):1281–1287
8. Bolton GE, Katok E (2008) Learning by doing in the newsvendor problem: A laboratory investigation of the role of experience and feedback. Manuf Serv Oper Manag 10(3):519–538
9. Bolton GE, Ockenfels A, Thonemann UW (2012) Managers and students as newsvendors. Manag Sci 58(12):2225–2233
10. Bostian AA, Holt CA, Smith AM (2008) Newsvendor "pull-to-center" effect: Adaptive learning in a laboratory experiment. Manuf Serv Oper Manag 10(4):590–608
11. BSNL (2018) Tender of expression of interest. http://tender.bsnl.co.in/bsnltenders/pdf/Mobicall_Open_EOI.pdf. Accessed 10 Jan 2022
12. Cachon GP (2004) The allocation of inventory risk in a supply chain: Push, pull, and advance-purchase discount contracts. Manag Sci 50(2):222–238
13. Cachon GP, Lariviere MA (2005) Supply chain coordination with revenue-sharing contracts: Strengths and limitations. Manag Sci 51(1):30–44
14. Corsten D, Gruen T (2004) Stock-outs cause walkouts. Harvard Business Review, issue month: May. https://hbr.org/2004/05/stock-outs-cause-walkouts. Accessed 30 June 2021
15. Davis AM, Katok E, Santamaría N (2014) Push, pull, or both? A behavioral study of how the allocation of inventory risk affects channel efficiency. Manag Sci 60(11):2666–2683
16. Feiler DC, Tong JD, Larrick RP (2013) Biased judgment in censored environments. Manag Sci 59(3):573–591
17. Frederick S (2005) Cognitive reflection and decision making. J Econ Perspect 19(4):25–42
18. GOI (2021) Model revenue sharing contract. https://mopng.gov.in/files/Whatsnew/Draft_MRSC_OALP.pdf. Accessed 06 Dec 2022
19. Goldschmidt K, Kremer M, Thomas DJ, Craighead CW (2020) Strategic sourcing under severe disruption risk: Learning failures and under-diversification bias. In: Manufacturing and service operations management, articles in advance, pp 1–20
20. Google (2022) Adsense revenue sharing. https://support.google.com/adsense/answer/180195?hl=. Accessed 06 Dec 2022
21. Gurnani H, Ramachandran K, Ray S, Xia Y (2014) Ordering behavior under supply risk: An experimental investigation. Manuf Serv Oper Manag 16(1):61–75
22. Ho TH, Lim N, Cui TH (2010) Reference dependence in multilocation newsvendor models: A structural analysis. Manag Sci 56(11):1891–1910
23. Hoskin RE (1983) Opportunity cost and behavior. J Account Res 21(1):78–95
24. Kahneman D (2011) Thinking, fast and slow. Farrar, Straus and Giroux, New York
25. Kahneman D, Tversky A (1979) Prospect theory: An analysis of decision under risk. Econometrica 47:263–291
26. Kahneman D, Rosenfield AM, Gandhi L, Blaser T (2016) Noise: How to overcome the high, hidden cost of inconsistent decision making. Harvard Bus Rev October issue:36-43
27. Katok E, Wu DY (2009) Contracting in supply chains: A laboratory investigation. Manag Sci 55(12):1953–1968

28. Kremer M, Minner S, Wassenhove LNV (2010) Do random errors explain newsvendor behavior? Manuf Serv Oper Manag 12(4):673–681
29. Käki A, Liesiö J, Salo A, Talluri S (2015) Newsvendor decisions under supply uncertainty. Int J Prod Res 53(5):1544–1560
30. Long X, Nasiry J (2015) Prospect theory explains newsvendor behavior: The role of reference points. Manag Sci 61(12):3009–3012
31. Lurie NH, Swaminathan JM (2009) Is timely information always better? The effect of feedback frequency on decision making. Organ Behav Hum Decis Process 108:315–329
32. Moritz BB, Hill AV, Donohue KL (2013) Individual differences in the newsvendor problem: Behavior and cognitive reflection. J Oper Manag 31:72–85
33. Ockenfels A, Selten R (2014) Impulse balance in the newsvendor game. Games Econ Behav 86:237–247
34. Pasternack BA (1985) Optimal pricing and return policies for perishable commodities. Mark Sci 4(2):166–176
35. Rudi N, Drake D (2014) Observation bias: The impact of demand censoring on newsvendor level and adjustment behavior. Manag Sci 60(5):1334–1345
36. Schweitzer ME, Cachon GP (2000) Decision bias in the newsvendor problem with a known demand distribution: Experimental evidence. Manag Sci 46(3):404–420
37. Shinde A, Mehta P, Amit RK (2020) An experimental investigation of newsvendor decisions under ambiguity. Int J Prod Res 1–12
38. Simon HA (1955) A behavioral model of rational choice. Q J Econ 69(1):99–118
39. Su X (2008) Bounded rationality in newsvendor models. Manuf Serv Oper Manag 10(4):566–589
40. Surti C, Celani A, Gajpal Y (2020) The newsvendor problem: The role of prospect theory and feedback. Eur J Oper Res 287(1):251–261
41. Thaler RH (1985) Mental accounting and consumer choice. Mark Sci 4:199–214
42. Tversky A, Kahneman D (1992) Advances in prospect theory: Cumulative representation of uncertainity. J Risk Uncertainity 5(4):297–323
43. Uppari BS, Hasija S (2019) Modeling newsvendor behavior: A prospect theory approach. Manuf Serv Oper Manag 21(3):481–500
44. Vipin B, Amit RK (2019) Describing decision bias in the newsvendor problem: A prospect theory model. Omega 82:132–141
45. Vipin B, Amit RK (2021) Wholesale price versus buyback: A comparison of contracts in a supply chain with a behavioral retailer. Comput Ind Eng 162:107689
46. Zhang Y, Siemsen E (2019) A meta-analysis of newsvendor experiments: Revisiting the pull-to-center asymmetry. Prod Oper Manag 28(1):140–156
47. Zhang Y, Donohue K, Cui TH (2016) Contract preferences and performance for the loss-averse supplier: Buyback versus revenue sharing. Manag Sci 62(6):1734–1754

The manufacturer's authorised representative in the EU is Springer Nature Customer Service Centre GmbH, Europaplatz 3, 69115 Heidelberg, Germany. If you have any concerns regarding our products, please contact ProductSafety@springernature.com

Printed and bound by CPI Group (UK) Ltd, Croydon, CR0 4YY
01/10/2026
02225934-0018